Eberhard Breitmaier, Wolfgang Voelter

Carbon-13 NMR Spectroscopy

High-Resolution Methods
and Applications in Organic Chemistry
and Biochemistry

Third, completely revised edition

Prof. Dr. Eberhard Breitmaier
Institut für Organische Chemie und Biochemie
der Universität
Gerhard-Domagk-Straße 1
D-5300 Bonn 1
Federal Republic of Germany

Prof. Dr. Wolfgang Voelter
Abteilung für Physikalische Biochemie
Physiologisch-Chemisches Institut
der Universität
Hoppe-Seyler-Straße 1
D-7400 Tübingen
Federal Republic of Germany

First Edition: 1974
Second Edition: 1978
Third, completely revised edition: 1987

Editorial Director: Dr. Hans F. Ebel

Library of Congress Card No. 86-28098

Deutsche Bibliothek Cataloguing-in-Publication Data

Breitmaier, Eberhard:
Carbon-13 NMR spectroscopy : high resolution methods and applications in organ. chemistry and biochemistry / Eberhard Breitmaier ; Wolfgang Voelter. – 3., completely rev. ed. – Weinheim ; New York, NY : VCH, 1987.
  Bis 2. Aufl. u.d.T.: Breitmaier, Eberhard: 13 C NMR spectroscopy
  ISBN 3-527-26466-3 (Weinheim)
  ISBN 0-89573-493-1 (New York)
NE: Voelter, Wolfgang

© VCH Verlagsgesellschaft mbH, D-6940 Weinheim (Federal Republic of Germany), 1974, 1978, 1987.
All rights reserved (including those of translation into other languages). No part of this book may be reproduced in any form – by photoprint, microfilm, or any other means – nor transmitted or translated into a machine language without written permission from the publishers. Registered names, trademarks, etc. used in this book, even when not specifically marked as such, are not to be considered unprotected by law.

Composition, printing, and bookbinding: Graphischer Betrieb Konrad Triltsch, D-8700 Würzburg
Printed in the Federal Republic of Germany

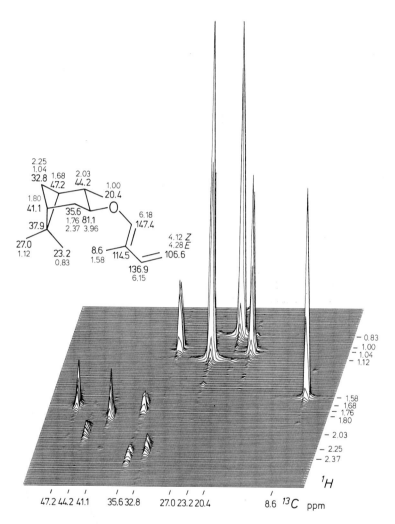

The cover displays a stacked plot of a two-dimensional carbon-proton shift correlation. The plot is reprinted here with its numerical data. Isopinocampheoxyisoprene in deuteriochloroform is the sample. The aliphatic carbon and proton shift ranges are shown. All carbon-proton connectivities of the bicyclic substructure can be clearly derived. For example, the carbon atom whose signal occurs at 35.6 ppm is attached to the protons with signals at 1.76 and 2.37 ppm. These signals make up an AB system in the proton NMR spectrum and overlap with other proton signals in the one-dimensional spectrum.

Preface to the Third Edition

The development of carbon-13 NMR during the last eight years has been characterized by a continual increase in the sensitivity and quality of spectra. A reduction in measuring time – equivalent to an enhancement in sensitivity – has been achieved mainly by cryomagnet technology. The efficiency with which NMR information can be obtained has been substantially improved by new computer-controllable pulse sequences for one- and two-dimensional NMR experiments. A selection of these new methods, in particular, those used for multiplicity analysis and homo- or heteronuclear shift correlations, is presented in chapter 3 of this edition.

The large number of carbon-13 NMR applications published since 1978 made a careful selection necessary. We have chosen to give a representative and systematic rather than an exhaustive survey of the subject. Older values have been updated whenever possible. New sections about coupling constants, organophosphorus and organometallic compounds as well as synthetic polymers have been added. The scope remains limited to high-resolution methods and molecular systems. For solid state systems, references are given to other monographs.

We wish to thank Dr. C. Dyllick-Brenzinger for proofreading our manuscript and for helpful comments.

Bonn and Tübingen,
November 1986

Eberhard Breitmaier
Wolfgang Voelter

Contents

1	*Introduction to NMR Spectroscopy*	1
1.1	Nuclear Magnetism	1
1.2	Nuclear Precession	2
1.3	Nuclear Magnetic Energy Levels	3
1.4	Nuclear Magnetic Resonance	4
1.5	Relaxation	5
1.5.1	Equilibrium of Nuclear Spins in the $\boldsymbol{B}_0$ Field	5
1.5.2	Spin-Lattice Relaxation	5
1.5.3	Spin-Spin Relaxation	6
1.5.4	Saturation	6
1.6	Magnetization Vectors	7
1.7	Bloch Equations	8
1.7.1	Motion of the Magnetization Vector in a Fixed Coordinate System	8
1.7.2	Motion of the Magnetization Vector in the Rotating Coordinate System	9
1.7.3	NMR in the Rotating Frame of Reference	10
1.7.4	Relaxation in the Rotating Frame of Reference	12
1.8	NMR Spectra	13
1.8.1	Nuclear Induction	13
1.8.2	Absorption and Dispersion Spectra	13
1.8.3	Magnitude Spectra	14
1.9	Chemical Shift	15
1.9.1	Shielding of Nuclei in Atoms and Molecules	15
1.9.2	Calibration of NMR Spectra	16
1.9.3	Reference Standard	17
1.10	Spin-Spin Coupling	17
1.10.1	Multiplicity of Signals	17
1.10.2	Coupling Constants	18

| 1.10.3 | Comparison between Chemical Shifts and Coupling Constants | 18 |
| 1.10.4 | Origin of Spin-Spin Coupling | 18 |

2 Instrumental Methods of ^{13}C NMR Spectroscopy ... 21

2.1	Sensitivity of ^{13}C NMR Spectroscopy	21
2.2	Methods of Sensitivity Enhancement in ^{13}C NMR Spectroscopy	21
2.3	Continuous Wave NMR Spectroscopy	22
2.4	Pulsed NMR Spectroscopy	22
2.4.1	Magnetization	22
2.4.2	$90°_x$ Pulse	22
2.4.3	Transverse Magnetization	23
2.4.4	Free Induction Decay	24
2.4.5	Pulse Interferograms	25
2.5	Pulse Fourier Transform (PFT) NMR Spectroscopy	28
2.5.1	FID Signal and NMR Spectrum as Fourier Transforms	28
2.5.2	Position, Width and Phase of FT NMR Signals	29
2.5.3	Acquisition of Pulse Interferograms for Fourier Transformation	30
2.5.3.1	Digitization	30
2.5.3.2	Dwell Time and Pulse Interval	30
2.5.3.3	Filtering of Frequencies Outside of the Spectral Width	30
2.5.4	Optimization of Pulse Interferograms for Fourier Transformation	31
2.5.4.1	Adjustment of Pulse Frequency	31
2.5.4.2	Adjustment of Pulse Width	32
2.5.5	Data Transformation and Subsequent Manipulations	33
2.5.5.1	Fourier Transformation	33
2.5.5.2	Phase Correction	33
2.5.5.3	Computation of Magnitude Spectra	36
2.5.6	Controlling Signal to Noise and Resolution in PFT NMR Spectroscopy	36
2.5.6.1	Digital Filtering	36
2.5.6.2	Number of FID Data Points and Digital Resolution	36
2.5.7	Spin-Lattice Relaxation and Signal to Noise in PFT NMR Spectroscopy	39
2.5.8	Comparison between CW and PFT	41
2.6	Double Resonance Techniques used in ^{13}C NMR Spectroscopy as Assignment Aids	43
2.6.1	Basic Concept of Spin Decoupling	43
2.6.2	Proton Broad Band Decoupling in ^{13}C NMR Spectroscopy	44
2.6.3	Nuclear Overhauser Effect in $^{13}C\{^1H\}$ NMR Experiments	46
2.6.4	Quenching Nuclear Overhauser Effects in $^{13}C\{^1H\}$ NMR Experiments	47
2.6.5	Proton Off-Resonance Decoupling	47
2.6.6	Pulsed Proton Broadband Decoupling	50
2.6.6.1	Measurement of NOE Enhanced Coupled ^{13}C NMR Spectra	50

2.6.6.2	Measurement of Proton-Decoupled ^{13}C NMR Spectra with Suppressed NOE	50
2.6.6.3	Measurement of Nuclear Overhauser Enhancements	51
2.6.7	Selective Proton Decoupling	53
2.7	Measurement of ^{13}C Relaxation Times	55
2.7.1	Spin-Lattice Relaxation Times	55
2.7.1.1	Inversion-Recovery or 180°, τ, 90° Method	55
2.7.1.2	Saturation-Recovery Method	59
2.7.1.3	Progressive Saturation or 90°, τ,... Method	60
2.7.2	Spin-Spin Relaxation Times	63
2.7.2.1	CPMGSE Experiments	63
2.7.2.2	Spin-Locking Fourier Transform Experiments	66
2.8	Instrumentation	67
2.8.1	Magnet	69
2.8.2	Stabilization Channel (Lock)	70
2.8.3	Observation Channel	71
2.8.4	Decoupling Channel	71
2.8.5	Sample	71
2.9	From the First to the Second Dimension of ^{13}C NMR Spectroscopy	73
2.9.1	The Spin-Echo	73
2.9.2	J-Modulated Spin-Echo	75
2.9.3	Polarization Transfer	78
2.9.3.1	Selective Polarization Transfer	79
2.9.3.2	Non-Selective Polarization Transfer	80
2.9.4	Measurement of Carbon-Carbon Coupling Constants Without ^{13}C Enrichment: INADEQUATE	84
2.10	The Second Dimension	87
2.10.1	Basic Concept	87
2.10.2	J-Resolved Two-Dimensional ^{13}C NMR Spectra	89
2.10.3	Two-Dimensional Carbon-Proton Shift Correlation	92
2.10.4	Two-Dimensional Proton-Proton Shift Correlation: The COSY Experiment	96
2.10.5	Two-Dimensional $C_A H_A - C_H H_H$ Correlation: The RELAY Experiment	100
2.10.6	Two-Dimensional Carbon-Carbon Shift Correlation: The Second Dimension of the INADEQUATE Experiment	102
2.10.7	Carbon-13 NMR Spectroscopy: Strategy for Structure Elucidation	104
3	^{13}C NMR Spectral Parameters and Structural Properties	107
3.1	Chemical Shifts	107
3.1.1	Comparison of ^{13}C and ^{1}H Shifts	107
3.1.2	Referencing ^{13}C Chemical Shifts	108

3.1.3	Survey of ^{13}C Chemical Shifts	110
3.1.3.1	Carbon Hybridization	111
3.1.3.2	Electronegativity	111
3.1.3.3	Crowding of Alkyl Groups and Substituents	112
3.1.3.4	Unshared Electron Pairs at Carbon	112
3.1.3.5	Electron Deficiency at Carbon	113
3.1.3.6	Mesomeric Effects	113
3.1.3.7	Conjugation	114
3.1.3.8	Steric Interactions	115
3.1.3.9	Electric Fields of Charged Substituents	116
3.1.3.10	Anisotropic Intramolecular Magnetic Fields	116
3.1.3.11	Heavy Atoms	117
3.1.3.12	Isotope Effect	117
3.1.3.13	Intramolecular Hydrogen Bonding	117
3.1.3.14	Substituent Increments and Functional Group Shifts	118
3.1.4	Medium Shifts	120
3.1.4.1	Dilution Shifts	120
3.1.4.2	Solvent Shifts	120
3.1.4.3	pH Shifts	121
3.1.5	Isotropic Shifts	123
3.1.6	Intramolecular Mobility and Temperature Dependence of ^{13}C Chemical Shifts and Line Widths	127
3.1.6.1	Introduction	127
3.1.6.2	Temperature Dependence of ^{13}C NMR Spectra	131
3.2	^{13}C Coupling Constants	133
3.2.1	Basic Theoretical Considerations	133
3.2.2	Carbon-Proton Coupling	134
3.2.2.1	One-Bond Coupling (J_{CH})	134
3.2.2.2	Longer-Range Carbon-Proton Couplings: The J_{CH}/J_{HH} Ratio	140
3.2.2.3	Two-Bond Coupling ($^2J_{CH}$)	141
3.2.2.4	Three-Bond Coupling ($^3J_{CH}$)	143
3.2.3	Carbon-Deuterium Coupling	147
3.2.4	Carbon-Carbon Coupling	147
3.2.4.1	One-Bond Coupling (J_{CC})	147
3.2.4.2	Longer-Range Carbon-Carbon Couplings ($^2J_{CC}$, $^3J_{CC}$)	152
3.2.5	^{13}C – ^{15}N Coupling Constants	155
3.2.5.1	One-Bond Couplings (J_{CN})	155
3.2.5.2	Longer-Range Couplings ($^2J_{CN}$, $^3J_{CN}$)	157
3.2.6	Coupling between Carbon and Other Heteronuclei X (X $\neq$ C, H, D)	160
3.3	Spin-Lattice Relaxation Times	163
3.3.1	Mechanisms of ^{13}C Spin-Lattice Relaxation	163
3.3.1.1	Relaxation Resulting from Chemical Shift Anisotropy (CSA Mechanism)	163
3.3.1.2	Relaxation by Scalar Coupling (SC Mechanism)	163
3.3.1.3	Relaxation by Spin Rotation (SR Mechanism)	163

3.3.1.4	Relaxation by Internuclear Dipole-Dipole Interaction (DD Mechanism)	164
3.3.1.5	Electron Spin-Nucleus Interactions and Consequences	165
3.3.2	Influence of Molecular Motion on Dipole-Dipole Relaxation	166
3.3.3	Information Content of ^{13}C Spin-Lattice Relaxation Times	168
3.3.3.1	Degree of Alkylation and Substitution of C Atoms	168
3.3.3.2	Molecular Size and Relaxation Mechanisms	168
3.3.3.3	Anisotropy of Molecular Motion	169
3.3.3.4	Internal Molecular Motion	172
3.3.3.5	Association and Solvation of Molecules and Ions	178
3.3.3.6	Determination of Quadrupole Relaxation Times and Coupling Constants from ^{13}C Spin-Lattice Relaxation Times	180
3.3.4	Medium and Temperature Effects	181

4 ^{13}C NMR Spectroscopy of Organic Compounds ... 183

4.1	Saturated Hydrocarbons	183
4.1.1	Alkanes	183
4.1.2	Cycloalkanes	186
4.1.3	Polycycloalkanes	189
4.2	Alkenes	192
4.2.1	Open-Chain Alkenes and Dienes	192
4.2.2	Cycloalkenes	194
4.3	Alkynes and Allenes	196
4.4	Halides	198
4.4.1	Alkyl Halides	198
4.4.2	Cycloalkyl Halides	203
4.4.3	Alkenyl Halides	205
4.4.4	Carbon-Fluorine Coupling Constants in Alkyl and Cyloalkyl Fluorides	205
4.5	Alcohols	206
4.5.1	Alkanols	206
4.5.2	Cycloalkanols	209
4.6	Ethers	213
4.6.1	Dialkyl Ethers	213
4.6.2	Enol Ethers and Alkynyl Ethers	215
4.7	Carbonyl Compounds	215
4.7.1	General	215
4.7.2	Aldehydes and Ketones	216
4.7.3	Quinones and Annulenones	222
4.7.4	Carboxylic Acids and Derivatives	226

| 4.8 | Aliphatic Organosulfur Compounds | 233 |

4.9	Aliphatic Organonitrogen Compounds	236
4.9.1	Amines	236
4.9.2	Enamines, Enaminoaldehydes, Cyanines	238
4.9.3	Imines	240
4.9.4	Nitriles and Isonitriles	242
4.9.5	Azacumulenes	244
4.9.6	Nitroso and Nitro Compounds	246

4.10	Organophosphorus Compounds	247
4.10.1	Survey of Carbon-13 Shifts	247
4.10.2	Carbon-Phosphorus Couplings	250

4.11	Aromatic Compounds	254
4.11.1	General	254
4.11.2	Substituted Benzenes	255
4.11.3	Substituted Naphthalenes	263
4.11.4	Benzocycloalkenes and Hydroaromatic Compounds	264
4.11.5	Fused and Bridged Aromatic Rings	265
4.11.6	Typical Coupling Constants	266
4.11.7	^{13}C- and ^{12}C-Enriched Aromatic Compounds for Structural Assignments and Mechanistic Studies	270

4.12	Heterocyclic Compounds	272
4.12.1	Heterocycloalkanes	272
4.12.2	Non-Aromatic Heterocycles with sp^2 Ring Carbons	276
4.12.3	Heteroaromatic Compounds	281
4.12.4	Typical Coupling Constants	286

4.13	Organometallic Compounds	293
4.13.1	General	293
4.13.2	Group I Organometallic Compounds	295
4.13.3	Group II Organometallic Compounds	295
4.13.4	Group III Organometallic Compounds	296
4.13.5	Group IV Organometallic Compounds	297
4.13.6	Organo-Transition-Metal Compounds	300

4.14	Ions	302
4.14.1	Carbocations	302
4.14.2	Carbanions	305

4.15	Synthetic Polymers	308
4.15.1	Tacticity	308
4.15.2	Configurational Isomerism	311
4.15.3	Segmental Mobility	313

4.16	Substituent Increments: Summary and Application	313
4.16.1	Substituted Alkanes	314
4.16.2	Substituted Cyclohexanes and Bicyclo[2.2.2]heptanes	316

4.16.3	Substituted Alkenes	318
4.16.4	Substituted Benzenes	319
4.16.5	Substituted Pyridines	322
4.16.6	Nitrogen Increments in Fused Heterocycles	322

5 ^{13}C NMR Spectra of Natural Products 327

5.1	Terpenes	327
5.2	Steroids	337
5.2.1	Androstanes, Pregnanes and Estranes	338
5.2.2	Cholestanes and Cholanes	340
5.2.3	Cardenolides and Sapogenins	358
5.3	Alkaloids	360
5.4	Carbohydrates	379
5.4.1	Monomeric Aldoses and Ketoses	380
5.4.2	Di- and Polysaccharides	394
5.4.3	Polyols	397
5.4.4	Aldonic Acids	397
5.4.5	Inositols	400
5.5	Nucleosides and Nucleotides	401
5.5.1	Assignment of the Purine Resonances	402
5.5.2	Assignment of the Pyrimidine Resonances	409
5.5.3	Assignment of the Isoalloxazine Resonances	409
5.5.4	Assignment of the Sugar and Polyol Carbon Atoms	409
5.5.5	Correlations of ^{13}C Chemical Shifts with Other Physicochemical Parameters	410
5.5.6	Nucleic Acids	412
5.6	Amino Acids	414
5.6.1	^{13}C Chemical Shifts of Amino Acids	414
5.6.2	pH-Dependence of the ^{13}C Chemical Shift Values of Amino Acids	420
5.6.3	Prediction of Carbon Shifts and their Correlation with Other Physicochemical Parameters	421
5.7	Peptides	422
5.7.1	Oligopeptides	422
5.7.2	Homopolymeric Polypeptides	436
5.7.3	Proteins	437
5.8	Porphyrins	441
5.9	Coumarins	41
5.10	Flavonoids	450
5.11	Elucidation of Biosynthetic Pathways	457
5.11.1	Radicinin	457

5.11.2	Asperlin	457
5.11.3	Cycloheximide	457
5.11.4	Averufin, Versicolorin A and their Relation to Aflatoxin B_1	459
5.11.5	Virescenosides	459
5.11.6	Methyl Palmitoleate	459
5.11.7	Sepedonin	462
5.11.8	Antibiotic X-537 A	463
5.11.9	Cephalosporin	463
5.11.10	Prodigiosin	465
5.11.11	Myxovirescin A_1	465
5.12	Appendix	467

6 References 469

Subject Index 499

1 Introduction to NMR-Spectroscopy

1.1 Nuclear Magnetism

When placed in a static magnetic field of flux density B_0, a nucleus may undergo nuclear magnetic resonance (NMR) [1–5] if it possesses an angular momentum p. This angular momentum is referred to as nuclear spin. The component of p in the direction of B_0 (Fig. 1.1), denoted as p_0, can only take on values which are half-integral or integral multiples m of $h/2\pi$:

$$p_0 = m \frac{h}{2\pi}; \quad m = \pm n \frac{1}{2}; \quad n = 0, 1, 2, \ldots \tag{1.1}$$

The values of m are further limited by the total spin quantum number I:

$$m = I, I-1, \ldots, -I. \tag{1.1a}$$

I is a constant characteristic of the ground state of every nucleus.

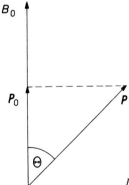

Fig. 1.1. Component p_0 of the torque p in the direction of the B_0 field.

According to an empirical rule, the magnitude of I depends on whether the atomic number z and the mass number a of an atom ^a_zX are both even, both odd or mixed (Table 1.1).

Table 1.1. Atomic Number z, Mass Number a, and Spin Quantum Number I of Selected Nuclei.

z	a	I	^a_zX
even	even	0	$^{12}_6C$, $^{16}_8O$
odd	odd	$\frac{1}{2}, \frac{3}{2}, \frac{5}{2}, \ldots$	^1_1H, $^{15}_7N$, $^{19}_9F$, $^{31}_{15}P$
even	odd	$\frac{1}{2}, \frac{3}{2}, \frac{5}{2}, \ldots$	$^{13}_6C$, $^{17}_8O$
odd	even	$1, 2, 3, \ldots$	^2_1H, $^{14}_7N$

Nuclei with total spin quantum number $I \neq 0$ interact with magnetic fields due to their magnetic moment $\boldsymbol{\mu}$. Its magnitude μ is related to the spin p by eq. (1.2).

$$\mu = \gamma p. \tag{1.2}$$

A constant γ, called the gyromagnetic ratio, characterizes each nucleus. In keeping with eqs. (1.1) and (1.2), the component of $\boldsymbol{\mu}$ in the direction of $\boldsymbol{B}_0$, μ_0, is quantized.

$$\mu \cos \Theta = \mu_0 = \gamma I \frac{h}{2\pi}. \tag{1.3}$$

Table 1.1 shows that the nucleus of major interest in organic chemistry, ^{12}C, does not have a nuclear spin and cannot be used as an NMR nucleus. In contrast, its isotope ^{13}C, whose natural abundance is only 1.1%, has a nuclear spin of $\frac{1}{2}$.

1.2 Nuclear Precession

When exposed to a static magnetic field $\boldsymbol{B}_0$, a spinning nucleus behaves like a gyroscope in a gravitational field. As illustrated by Fig. 1.2, the spin axis – which coincides with the magnetic moment vector $\boldsymbol{\mu}$ – precesses about $\boldsymbol{B}_0$. The frequency of precession, ν_0, is known as the Larmor frequency of the observed nucleus.

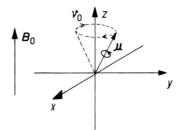

Fig. 1.2. Larmor precession of a spinning nucleus in a static magnetic field.

In contrast to bar magnets, the magnetic moments of spinning nuclei do not align in the direction of $\boldsymbol{B}_0$, no matter how strong this field is. Instead, the Larmor precession is accelerated by increasing the strength B_0 of the field vector $\boldsymbol{B}_0$:

$$v_0 \propto B_0. \tag{1.4}$$

1.3 Nuclear Magnetic Energy Levels

The energy of a magnetic moment $\boldsymbol{\mu}$ in a field $\boldsymbol{B}_0$ is given by the product of B_0 and μ_0. Thus from eq. (1.3),

$$E = -\mu_0 B_0 = -\gamma \frac{h}{2\pi} I B_0. \tag{1.5}$$

Following quantum mechanical rules, a nucleus with total spin quantum number I may occupy $(2I+1)$ different energy levels when placed in a magnetic field. For nuclei with $I = \frac{1}{2}$, e.g. ^{1}H, ^{13}C, ^{15}N, ^{19}F, ^{31}P, two spin alignments relative to B_0 arise (Fig. 1.3); these are symbolized by $+\frac{1}{2}$ and $-\frac{1}{2}$.

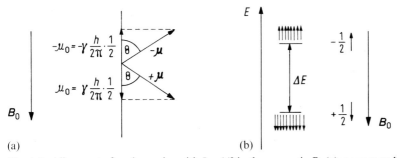

Fig. 1.3. Alignment of nuclear spins with $I = 1/2$ in the magnetic B_0 (a); corresponding energy levels and spin populations (b).

Precession of the nuclear spins about $\boldsymbol{B}_0$ is energetically favored (Fig. 1.3(b)) since the component of the nuclear magnetic moment vector $\boldsymbol{\mu}$ in the direction of $\boldsymbol{B}_0$ reinforces the magnetic field (Fig. 1.3(a)).

The energies $E_{+1/2}$ and $E_{-1/2}$ and the energy difference ΔE between the two levels are derived from eq. (1.5):

$$E_{+1/2} = -\mu_0 B_0 = -\gamma \frac{h}{4\pi} B_0.$$

$$E_{-1/2} = \mu_0 B_0 = \gamma \frac{h}{4\pi} B_0. \tag{1.6}$$

$$\Delta E = E_{-1/2} - E_{+1/2} = 2\mu_0 B_0 = \gamma \frac{h}{2\pi} B_0. \tag{1.7}$$

1.4 Nuclear Magnetic Resonance

ΔE in eq. (1.7) and Fig. 1.3(b) is the difference between the energies of precession along and opposite to B_0. Using eq. (1.7), the Larmor precession frequency v_0 of nuclei with $I = \frac{1}{2}$ can be calculated, recalling that ΔE also equals $h v_0$:

$$v_0 = \frac{\gamma}{2\pi} B_0. \tag{1.8}$$

For a field strength of 2.13 Tesla, the Larmor frequency of ^{13}C is in the order of 22 to 23 MHz, much lower than that of ^1H (90 MHz). This is in the radio-frequency (rf) range.

An alternating magnetic field B_1 with frequency v_1 irradiating an ensemble of nuclear spins precessing in the static field B_0 may overcome the energy difference ΔE if it meets two conditions: The vector of the alternating field B_1 must rotate in the plane of precession with the Larmor frequency v_0 of the nuclei to be observed (Fig. 1.4(a)).

$$v_1 = v_0. \tag{1.9}$$

As a result, the spins originally precessing with B_0 flip over, and now precess against B_0. Absorption of energy ΔE from B_1 takes place (*nuclear magnetic resonance*).

In order to observe NMR, a sample containing nuclear spins (*e.g.* ^1H, ^{13}C) is placed in a static magnetic field B_0. An alternating field B_1 with radio frequency v_1 is applied perpendicularly to B_0. Usually, v_1 is increased or decreased slowly and continuously

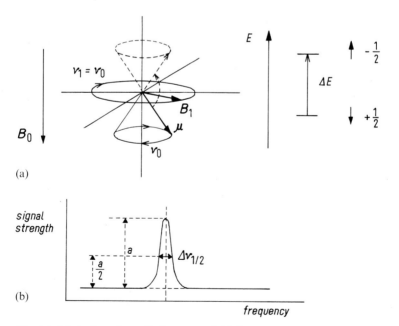

Fig. 1.4. (a) Action of a radio-frequency field B_1 on a nucleus precessing about direction B_0.
(b) NMR signal ($\Delta v_{1/2}$ is the half-maximum intensity line width).

during observation (frequency sweep). When v_1 matches the Larmor frequency of the nucleus to be observed (eq. (1.9)), an absorption signal is recorded in the receiver of the NMR spectrometer (Fig. 1.4(b)).

Continuous irradiation by the radio frequency $\boldsymbol{B}_1$ would soon cause all nuclei to precess against $\boldsymbol{B}_0$, and no further absorption of energy would occur, if there were no processes at work to restore the energetically favored orientation of the spins. In fact, energy absorption from radio frequency fields due to NMR is observed for long periods if the rf power is not too high. The processes responsible are referred to as relaxation.

1.5 Relaxation

1.5.1 Equilibrium of Nuclear Spins in the $\boldsymbol{B}_0$ Field

At equilibrium, the nuclear magnetic energy levels are populated according to a Boltzmann distribution which favors the lower state. For the two orientations relative to $\boldsymbol{B}_0$ of nuclei with $I = \frac{1}{2}$, the spin populations may be symbolized by N_+ and N_- (Fig. 1.3(b)). The distribution N_+/N_- can be expressed by the Boltzmann factor, recalling that $\Delta E = 2\mu_0 B_0$:

$$\frac{N_+}{N_-} = e^{\frac{\Delta E}{kT}} \approx 1 + \frac{\Delta E}{kT} = 1 + \frac{2\mu_0 B_0}{kT}. \tag{1.10}$$

At room temperature, $\Delta E = 2\mu_0 B_0$ is much less than 4.2 Joule, even at the highest field strengths now achievable. As a result, the term $\dfrac{2\mu_0 B_0}{kT}$ is very small, and $\dfrac{N_+}{N_-}$ is not much greater than 1.

1.5.2 Spin-Lattice Relaxation

At resonance, the rf field $\boldsymbol{B}_1$ causes a spin transfer from the lower to the upper level. The equilibrium distribution of the spins in the static field $\boldsymbol{B}_0$ is disturbed. Following any disruption, the nuclear spins relax to be in equilibrium with their surroundings (the "lattice"), which, of course, includes $\boldsymbol{B}_0$. This relaxation is assumed to be a first-order rate process with a rate constant $1/T_1$ characterizing each kind of nucleus. T_1 is called the *spin-lattice relaxation time*. It covers a range of about 10^{-4} to 10^4 s.

For liquids with rapid intermolecular motion, T_1 is a measure of the life-time of a nucleus in a particular spin state. According to the Heisenberg uncertainty relation,

$$\Delta E\, \Delta t = h\, \Delta v_{1/2}\, T_1 \gtrsim h, \tag{1.11}$$

the minimum width $\Delta v_{1/2}$ at half-maximum intensity of an NMR signal (Fig. 1.4(b)) can be estimated by eq. (1.12):

$$\Delta v_{1/2} \gtrsim \frac{1}{T_1}. \tag{1.12}$$

1.5.3 Spin-Spin Relaxation

In solids and liquids with slowly tumbling molecules internuclear dipole-dipole interactions may become important. Furthermore, energy quanta $\Delta E = 2\mu_0 B_0$ are exchanged between nuclei to a certain degree. Both factors tend to shorten the life-times of all spin states, again leading to line broadening. This second type of first-order relaxation process, called spin-spin relaxation and characterized by the time constant $1/T_2$, competes with spin-lattice relaxation. T_2 is referred to as the *spin-spin relaxation time*.

Since spin-spin relaxation reduces the life-time of a nucleus in a particular spin state to $T_2 \leqq T_1$, the half-maximum intensity line width is expressed more precisely by eq. (1.13), which accounts for spin-spin relaxation.

$$\Delta v_{1/2} = \text{const.} \frac{1}{T_2}. \tag{1.13}$$

The observed line width of an NMR signal depends additionally on the field inhomogeneity ΔB_0, whose contribution to $\Delta v_{1/2}$ arises from eq. (1.8):

$$\Delta v_{1/2 \,(\text{inhom.})} = \frac{\gamma \Delta B_0}{2\pi} = \text{const.} \frac{\gamma \Delta B_0}{2}. \tag{1.14}$$

By adding eqs. (1.13) and (1.14) one obtains the observed half-maximum intensity line width $\Delta v_{1/2 \,(\text{obs.})}$:

$$\Delta v_{1/2 \,(\text{obs.})} = \Delta v_{1/2} + \Delta v_{1/2 \,(\text{inhom.})} = \text{const.} \left(\frac{1}{T_2} + \frac{\gamma \Delta B_0}{2} \right). \tag{1.15}$$

Spin-lattice relaxation of nuclei (*e.g.* ^{1}H, ^{13}C) in a molecule may be accelerated

a) by interaction with adjacent nuclei having spin 1 or greater (*e.g.* ^{2}H, ^{14}N): the electric quadrupole moments of such nuclei result in additional magnetic fields in the tumbling molecule;

b) by interaction with unpaired electrons in paramagnetic compounds (radicals, some metal chelates).

Spin-spin relaxation of nuclei is accelerated when the nuclei participate in a dipolar bond (O–^{1}H, N–^{1}H, ^{13}C–^{1}H). Spin-spin relaxation involving dipole-dipole interaction is very effective in solids and viscous liquids with slow molecular motion, since the magnetic fields caused by slowly tumbling dipoles change very slowly.

All these interactions cause considerable line broadening, making the observation of NMR signals rather difficult in some cases (*e.g.* in paramagnetic compounds). Briefly summarized, T_2 affects NMR signal line widths, but not the energy level population as does T_1.

1.5.4 Saturation

If the radio-frequency power is too high, relaxation cannot compete with the disruption of the equilibrium of spins. The population difference between the nuclear magnetic energy levels decreases to zero, and so does the intensity of the absorption signal (saturation).

1.6 Magnetization Vectors

The Boltzmann factor $\dfrac{2\mu_0 B_0}{kT}$ only slightly favors the lower spin state. It is more realistic, therefore, to study the influence of magnetic fields on an assembly of a large number of identical nuclei than on one spinning nucleus.

For an assembly of identical nuclei with $I = \tfrac{1}{2}$, two orientations with respect to $\boldsymbol{B}_0$ are possible for each nucleus. Due to the favoring of the lower state with spin alignment more nearly parallel to $\boldsymbol{B}_0$, more nuclei precess about the direction of $\boldsymbol{B}_0$, defined as the $+z$ direction (Fig. 1.5). A net macroscopic magnetization $\boldsymbol{M}_0$ along the $+z$ axis arises as shown in Fig. 1.5(a). This orientation of the magnetization vector $\boldsymbol{M}_0$ parallel to $\boldsymbol{B}_0$ is characteristic of equilibrium of the spin ensemble. Upon disturbance of equilibrium by an rf field $\boldsymbol{B}_1$ with the proper frequency v_1, the magnetic moment vectors $\boldsymbol{\mu}$ are forced to precess in phase (Fig. 1.5(b)). The resultant magnetization vector $\boldsymbol{M}$ is no longer parallel to $\boldsymbol{B}_0$ (Fig. 1.5(b)), and is now composed of three components along the axes x, y, and z. These components are related to $\boldsymbol{M}$ by eq. (1.16), using the unit vectors $\boldsymbol{i}, \boldsymbol{j}$, and $\boldsymbol{k}$ along x, y, and z, as illustrated in Fig. 1.5(b).

$$\boldsymbol{M} = M_x \boldsymbol{i} + M_y \boldsymbol{j} + M_z \boldsymbol{k}. \tag{1.16}$$

$M_z \boldsymbol{k}$ is the longitudinal magnetization along the z axis; $M_x \boldsymbol{i}$ and $M_y \boldsymbol{j}$ make up the transverse magnetization in the xy plane.

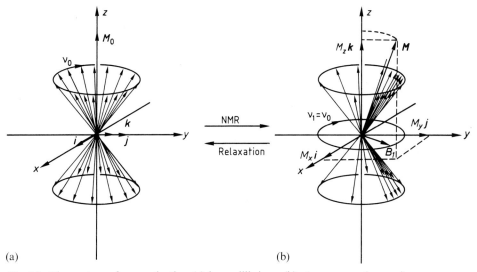

Fig. 1.5. The vectors of magnetization (a) in equilibrium, (b) at resonance ($v_1 = v_0$).

Relaxation can be described in terms of the magnetization vector components. At resonance, the equilibrium magnetization M_0 parallel to $\boldsymbol{B}_0$ decreases to M_z, due to the transitions between the nuclear magnetic energy levels caused by the alternating field $\boldsymbol{B}_1$. Following resonance, the equilibrium of the nuclear spins with their lattice and with each other is restored by relaxation.

By spin-lattice relaxation, the longitudinal magnetization M_z increases to its equilibrium value M_0 (longitudinal relaxation). T_1 is therefore often called the longitudinal relaxation time.

By spin-spin relaxation, the nuclei relax to equilibrium among themselves (*i.e.* precession occurs without phase coherence). The vectors dephase (Fig. 1.5(b) → Fig. 1.5(a)), and the components of transverse magnetization, M_x and M_y, decay to zero as a result (transverse relaxation). The spin-spin relaxation time T_2 is thus also referred to as the *phase memory time* or the *transverse relaxation time*.

Longitudinal and transverse relaxations have been assumed by Bloch et al. [6] to be first-order rate processes. Following this assumption, the increase of M_z to M_0 and the decay of M_x and M_y to zero may be expressed in terms of spin-lattice and spin-spin relaxation times, T_1 and T_2:

$$\frac{dM_z}{dt} = -\frac{M_z - M_0}{T_1}, \qquad (1.17\,\mathrm{a})$$

$$\frac{dM_x}{dt} = -\frac{M_x}{T_2}, \qquad (1.17\,\mathrm{b})$$

$$\frac{dM_y}{dt} = -\frac{M_y}{T_2}. \qquad (1.17\,\mathrm{c})$$

1.7 Bloch Equations

1.7.1 Motion of the Magnetization Vector in a Fixed Coordinate System

Upon irradiation with the rf field $\boldsymbol{B}_1$, at resonance,

$$\nu_1 = \nu_0$$

transitions between the nuclear magnetic energy levels occur. The nuclear spins change their directions relative to $\boldsymbol{B}_0$: The direction of the nuclear spin $\boldsymbol{p}$ is now time dependent. Following the equation of motion of a spin $\boldsymbol{p}$ in a magnetic field $\boldsymbol{B}$, this time dependence is given by the vector product of the magnetic moment $\boldsymbol{\mu}$ due to the spin $\boldsymbol{p}$ and the total field $\boldsymbol{B}$ resulting from the static field $\boldsymbol{B}_0$ and the rf field $\boldsymbol{B}_1$.

$$\frac{d\boldsymbol{p}}{dt} = \boldsymbol{\mu} \times \boldsymbol{B}. \qquad (1.18)$$

By multiplying eq. (1.18) with the gyromagnetic ratio γ one obtains the time dependence of the magnetic moment $\boldsymbol{\mu}$, remembering that $\boldsymbol{\mu} = \gamma \boldsymbol{p}$:

$$\frac{d\boldsymbol{\mu}}{dt} = \gamma \boldsymbol{\mu} \times \boldsymbol{B}. \tag{1.19}$$

For an assembly of nuclei with magnetic moment $\boldsymbol{\mu}$ the resultant vector sum is the magnetization vector $\boldsymbol{M}$. Its time dependence is also given by eq. (1.19).

$$\frac{d\boldsymbol{M}}{dt} = \gamma \boldsymbol{M} \times \boldsymbol{B}. \tag{1.20}$$

1.7.2 Motion of the Magnetization Vector in the Rotating Coordinate System

In the eyes of a distant observer using a fixed coordinate system, a meteorite falling in the gravitational field of the earth describes a parabolic path. An observer "standing" on earth uses the rotating frame of reference of the earth. For him, the complicated path of the falling meteorite simplifies to a straight vertical line.

Correspondingly, the path of the magnetization vector $\boldsymbol{M}$ subjected to magnetic fields is simplified in a coordinate system rotating at the angular velocity $\omega_1 = 2\pi\nu_1$ of the alternating field $\boldsymbol{B}_1$. In this case, the directions of the unit vectors $\boldsymbol{i}, \boldsymbol{j}$, and $\boldsymbol{k}$ change (they rotate), whereas their magnitudes remain constant. Due to the time dependence of the unit vectors, the derivative $\dfrac{d\boldsymbol{M}}{dt}$ of the magnetization vector,

$$\boldsymbol{M} = M_x \boldsymbol{i} + M_y \boldsymbol{j} + M_z \boldsymbol{k},$$

with respect to time is obtained using the product rule of differentiation:

$$\begin{aligned}\frac{d\boldsymbol{M}}{dt} &= \frac{\partial M_x}{\partial t}\boldsymbol{i} + M_x \frac{\partial \boldsymbol{i}}{\partial t} + \frac{\partial M_y}{\partial t}\boldsymbol{j} + M_y \frac{\partial \boldsymbol{j}}{\partial t} + \frac{\partial M_z}{\partial t}\boldsymbol{k} + M_z \frac{\partial \boldsymbol{k}}{\partial t} \\ &= \frac{\partial M_x}{\partial t}\boldsymbol{i} + \frac{\partial M_y}{\partial t}\boldsymbol{j} + \frac{\partial M_z}{\partial t}\boldsymbol{k} + M_x \frac{\partial \boldsymbol{i}}{\partial t} + M_y \frac{\partial \boldsymbol{j}}{\partial t} + M_z \frac{\partial \boldsymbol{k}}{\partial t}.\end{aligned} \tag{1.21}$$

Looking at eq. (1.21), it is seen that the time dependence of the magnetization vector $\boldsymbol{M}$ results from two contributions. One is the partial time derivate of $\boldsymbol{M}$ in the rotating coordinate system:

$$\left(\frac{\partial \boldsymbol{M}}{\partial t}\right)_{\text{rot.}} = \frac{\partial M_x}{\partial t}\boldsymbol{i} + \frac{\partial M_y}{\partial t}\boldsymbol{j} + \frac{\partial M_z}{\partial t}\boldsymbol{k}. \tag{1.22}$$

The other arises from the rotation of the unit vectors $\boldsymbol{i}, \boldsymbol{j}$, and $\boldsymbol{k}$ with angular velocity ω. Since the time derivatives of the unit vectors are the vector products of ω and $\boldsymbol{i}, \boldsymbol{j}$, and $\boldsymbol{k}$, respectively,

$$\frac{\partial \boldsymbol{i}}{\partial t} = \omega \times \boldsymbol{i}; \quad \frac{\partial \boldsymbol{j}}{\partial t} = \omega \times \boldsymbol{j}; \quad \frac{\partial \boldsymbol{k}}{\partial t} = \omega \times \boldsymbol{k}, \tag{1.23}$$

the second term simplifies as follows:

$$M_x \frac{\partial i}{\partial t} + M_y \frac{\partial j}{\partial t} + M_z \frac{\partial k}{\partial t} = M_x \omega \times i + M_y \omega \times j + M_z \omega \times k$$
$$= \omega \times (M_x i + M_y j + M_z k) \quad (1.24)$$
$$= \omega \times M.$$

Using eqs. (1.22; 1.24), and recalling eq. (1.20), the time derivative of M is

$$\frac{dM}{dt} = \left(\frac{\partial M}{\partial t}\right)_{\text{rot.}} + \omega \times M = \gamma M \times B. \quad (1.25)$$

From this relation, the time derivative of M in the rotating frame of reference can be calculated:

$$\left(\frac{\partial M}{\partial t}\right)_{\text{rot.}} = \gamma M \times B - \omega \times M = \gamma \left(M \times B - \frac{\omega \times M}{\gamma}\right). \quad (1.26)$$

With $\omega \times M = -M \times \omega$, eq. (1.26) changes to

$$\left(\frac{\partial M}{\partial t}\right)_{\text{rot.}} = \gamma M \times \left(B + \frac{\omega}{\gamma}\right). \quad (1.27)$$

Since only terms with identical dimensions are allowed to be added, ω/γ must have the dimension of a magnetic field. Thus, in the rotating frame of reference, the effective field $B_{\text{eff.}}$ experienced by M differs from B by a term ω/γ arising from rotation,

$$B_{\text{eff.}} = B + \frac{\omega}{\gamma}, \quad (1.28)$$

and eq. (1.29) replaces eq. (1.27):

$$\left(\frac{\partial M}{\partial t}\right)_{\text{rot.}} = \gamma M \times B_{\text{eff.}}. \quad (1.29)$$

Eq. (1.19) described the precession of M about the total magnetic field B using a coordinate system with fixed axes x, y, and z. Correspondingly, eq. (1.29) describes the magnetization vector M as it precesses about the effective field $B_{\text{eff.}}$ [7] in a coordinate system rotating with frequency $\omega = 2\pi\nu$ about the z axis and symbolized as the x', y', z frame of reference with the rotating unit vectors i', j', and k.

1.7.3 NMR in the Rotating Frame of Reference

In the absence of B_1, the vector M keeps its equilibrium value and position M_0 in the z direction. M is thus time-invariant in the rotating frame of reference, so that

$$\left(\frac{\partial M}{\partial t}\right)_{\text{rot.}} = \gamma M \times B_{\text{eff.}} = 0 \quad (1.30)$$

1.7 Bloch Equations 11

and consequently

$$B_{\text{eff.}} = 0, \tag{1.31}$$

since

$$M = M_0 \neq 0.$$

If the frame rotates at the Larmor frequency $\omega_0 = 2\pi v_0$, the rotational field term of eq. (1.28) reaches ω_0/γ. Since the effective field is zero, the Larmor equation (1.8a) is obtained:

$$B_{\text{eff.}} = \left(B_0 + \frac{\omega_0}{\gamma}\right)k = 0, \quad \text{so that} \quad \omega_0 k = -\gamma B_0 k. \tag{1.8a}$$

This means further that the rotational field ω/γ opposes $B_0 k$ in the rotating frame of reference (Fig. 1.6(a)), finally cancelling $B_0 k$ when the coordinate system rotates at Larmor frequency ω_0.

If an rf field B_1 with $\omega_1 = 2\pi v_1$ is applied perpendicularly to B_0 along the x' axis, the effective field $B_{\text{eff.}}$ in the frame of reference rotating at ω_1 is obtained from Fig. 1.6(a):

$$B_{\text{eff.}} = \left(B_0 + \frac{\omega_1}{\gamma}\right)k + B_1 i'. \tag{1.32}$$

If the frame of reference rotates at the radio frequency v_1 matching the Larmor frequency v_0 (NMR), the term $\left(B_0 + \frac{\omega_1}{\gamma}\right)k$ becomes zero according to eq. (1.8a). The remaining effective field at resonance is now

$$B_{\text{eff.(res.)}} = B_1 i', \tag{1.33}$$

so that eq. (1.29) changes to (1.34):

$$\left(\frac{\partial M}{\partial t}\right)_{\text{rot.}} = \gamma M \times B_1 i'. \tag{1.34}$$

This relation tells us that, at resonance, the magnetization vector M precesses about the field vector $B_1 i'$ of the radio-frequency field (Fig. 1.6(b)).

Since the coordinate system and B_1 are chosen to rotate at the same frequency, B_1 can be assigned along the rotating x' axis. According to eq. (1.34), the magnetization vector

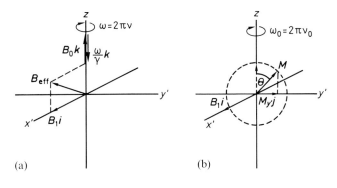

Fig. 1.6. The effective field in the rotating frame (a) off resonance ($\omega \neq \omega_0$), (b) on resonance ($\omega = \omega_0$).

M precesses about the x' axis (Fig. 1.6(b)). The precession frequency ω_1 of M about x' also follows the Larmor equation:

$$\omega_1 = \gamma B_1. \qquad (1.8\,b)$$

Thus, an rf field B_1, applied at resonance for t_p seconds, causes the vector M to precess about the x' axis by an angle Θ (Fig. 1.6(b)):

$$\Theta = \omega_1 t_p = \gamma B_1 t_p \text{ (radians)}. \qquad (1.35)$$

1.7.4 Relaxation in the Rotating Frame of Reference

If only B_0 is applied, the nuclear moments precess without any phase coherence. No resultant component of the magnetization in the $x'y'$ plane is observed and M_z equals M_0 (Fig. 1.7(a)).

A radio-frequency field B_1 applied perpendicularly to B_1 forces the nuclei to precess in phase, tipping the vector M_0 toward the y' axis by an angle Θ (Fig. 1.7(b)). A transverse magnetization $M_{y'}\boldsymbol{j}$ in the $x'y'$ plane arises, and the magnitude of the longitudinal magnetization $M_{z'}$ decreases (Fig. 1.7(b)). When restoring equilibrium, the nuclei exchange energy with each other (spin-spin relaxation). They dephase, causing $M_{y'}\boldsymbol{j}$ to spread out in components in the $x'y'$ plane and, finally, to decay to zero with time constant $1/T_2$ (Fig. 1.7(c, d)). Dephasing may be accelerated due to field inhomogeneities, so that

$$(1/T_2^*)_{(\text{inhom.})} > 1/T_2.$$

Moreover, the nuclear moments lose energy to their surroundings (spin-lattice relaxation), causing $M_z \boldsymbol{k}$ to increase to M_0 (Fig. 1.7(d) → Fig. 1.7(a)).

The decay of the magnitude of transverse magnetization, $M_{y'}$, due to spin-spin relaxation (T_2) or due to field inhomogeneities (T_2^*) may be faster but cannot be slower than the spin-lattice relaxation (T_1):

$$T_2^* \leq T_2 \leq T_1. \qquad (1.36)$$

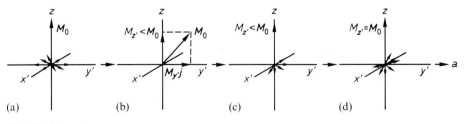

Fig. 1.7. Relaxation.

1.8 NMR Spectra

1.8.1 Nuclear Induction

At resonance, the magnetization vector M precesses about the vector $B_1 i'$ of the alternating field according to eq. (1.34). As a result, a component of transverse magnetization $M_{y'} j$ rotates in the $x' y'$ plane at the Larmor frequency v_0. If a receiver coil is placed in the $x' y'$ plane, the rotating magnetic vector $M_{y'} j$ induces an electromotive force measurable as an inductance current. This process is called nuclear induction [5, 8]. The orientation of the coil axis will affect the phase relative to $B_1 i'$ but not the magnitude of the induction current.

In the rotating frame of reference, the field vector $B_1 i'$ of the rf field, rotating at angular velocity ω_1 in phase with the rotating x axis (x'), is used as the reference. The transverse magnetization $M_x i + M_y j$ in the fixed coordinate system is then resolved into two components u and v in the rotating frame of reference: u rotates in phase, v rotates 90° or $\pi/2$ out of phase with $B_1 i'$ (Fig. 1.8).

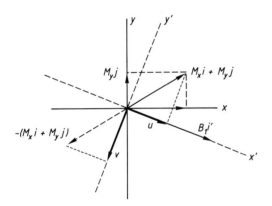

Fig. 1.8. The components u and v of transverse magnetization in the rotating frame.

Following eq. (1.34), the magnetization vector M is rotated toward the y' axis by the oscillation of $B_1 i'$ in phase with the x' axis of the rotating frame of reference. In agreement with the Lenz rule of induction, the current $I_{ind.}$ due to the induced EMF opposes the inducing magnetization. At resonance, the magnetization vector rotates $90° = \pi/2$ behind $B_1 i'$ (Fig. 1.6). The maximum induction current, however, is observed $90° = \pi/2$ ahead of phase relative to $B_1 i'$ in the v direction.

1.8.2 Absorption and Dispersion Spectra

Maximum induction current due to resonance between rf field and Larmor precession corresponds to maximum absorption of energy. Thus, the plot of the induction current in the v direction ($\pi/2$ ahead of the vector $B_1 i'$) as a function of frequency is the NMR

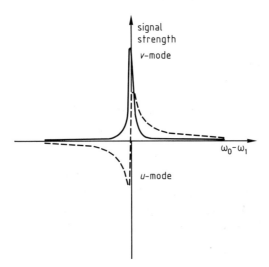

Fig. 1.9. Absorption (v, ———) and dispersion (u, – – –) spectrum.

absorption spectrum, called the v mode. The absorption curve has a Lorentzian shape, as shown in Fig. 1.9.

Before and after resonance, there are magnetization components of opposite sign 180° out of phase and in phase ($+\boldsymbol{u}$ and $-\boldsymbol{u}$ direction) with the rf field $B_1\,\boldsymbol{i}'$ (Fig. 1.8). At resonance, there is no magnetization in the $\boldsymbol{u}$ direction. If the receiver coil obtains the inductance current in phase with $B_1\,\boldsymbol{i}'$ (the $\boldsymbol{u}$ direction), a dispersion curve (Fig. 1.9) results, called the u mode. When the absorption or out of phase spectrum (v) reaches its maximum ($I_{\text{ind.}}(\omega) = \text{max.}$), the dispersion or in phase spectrum (u) goes through zero and changes its sign, as illustrated in Fig. 1.9.

1.8.3 Magnitude Spectra

If NMR spectra are computed by Fourier transformation of pulse interferograms (Chapter 2), complex quantities are used during computation. Real and imaginary components $v(\omega)$ and $i\,u(\omega)$ of the NMR spectrum are obtained as a result. Magnitude or power spectra $P(\omega)$ can be computed from the real and imaginary parts as follows:

$$P(\omega) = \sqrt{[v(\omega)]^2 + [i\,u(\omega)]^2}. \tag{1.37}$$

If the real part $v(\omega)$ of the NMR spectrum is computed in the absorption (v) mode, the imaginary part is usually displayed in the dispersion (u) mode. The magnitude spectrum is therefore related to the v and u modes as indicated in eq. (1.37).

1.9 Chemical Shift

1.9.1 Shielding of Nuclei in Atoms and Molecules

The Larmor frequency of a "free" nuclear spin is given by the Larmor equation:

$$v_0 = \frac{\gamma}{2\pi} B_0 \quad \left(I = \frac{1}{2}\right). \tag{1.8}$$

In atoms and molecules, a nucleus i is shielded by electrons. It does not experience the static field B_0 applied but an individual field B_i, arising from superposition of the B_0 field and an additional field $B_{\text{ind.}i}$ induced by the shielding electrons:

$$B_i = B_0 - B_{\text{ind.}i}. \tag{1.38}$$

The strength of $B_{\text{ind.}i}$ induced by the electrons is proportional to the strength of the applied field B_0.

$$B_{\text{ind.}i} = \sigma_i B_0. \tag{1.39}$$

The factor σ_i [9] is called the *magnetic shielding constant* for the nucleus i, and characterizes the chemical environment of that nucleus. The effective field experienced by the nucleus i follows from eqs. (1.38, 1.39):

$$B_i = B_0 (1 - \sigma_i). \tag{1.40}$$

Thus, the nucleus i precesses at the Larmor frequency

$$v_{0_i} = \frac{\gamma}{2\pi} B_0 (1 - \sigma_i) \tag{1.8 c}$$

when exposed to the static magnetic field B_0.

Eq. (1.8 c) tells us that nuclei having different chemical environments precess at different Larmor frequencies. The shift of Larmor frequencies due to chemical nonequivalence of nuclei in molecules is called the *chemical shift*.

For n chemically nonequivalent nuclei in a molecule, n absorption signals are observed in the NMR spectrum due to n different Larmor frequencies. For example, acetone has two kinds of chemically nonequivalent ^{13}C nuclei, one carbonyl nucleus and two methyl nuclei. Two signals are observed in the ^{13}C NMR spectrum as shown in Fig. 1.10(a). The signal intensity is proportional to the number of equivalent nuclei of a particular type giving rise to the considered signal. In acetone, for instance (Fig. 1.10(a)), the ratio of signal intensities is equal to the ratio of the numbers of chemically nonequivalent ^{13}C nuclei:

$$n_{CO} : n_{CH_3} = 1 : 2.$$

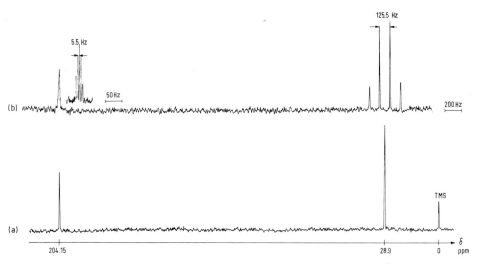

Fig. 1.10. Natural abundance ^{13}C NMR spectrum of acetone, neat liquid, (a) decoupled, 512 scans, (b) coupled, 4096 scans; swept radio frequency 22.628 MHz corresponding to field strength of 2.13 Tesla; the frequency differences relative to tetramethylsilane (TMS) are $\Delta v_{CH_3} = 655$ Hz and $\Delta v_{CO} = 4632$ Hz; the values according to eq. (1.43) are therefore $\delta_{CH_3} = \dfrac{655}{22.628} \cdot 10^{-6} = 28.9$ ppm and $\delta_{CO} = \dfrac{4632}{22.628} \cdot 10^{-6} = 204.15$ ppm.

1.9.2 Calibration of NMR Spectra

In order to measure chemical shifts, the absorption signal of a reference compound R appearing at frequency v_{OR} is assigned the shift zero. The chemical shift (or Larmor frequency) of chemically equivalent nuclei of a sample having their signal at frequency v_{OS} may then be measured as the frequency difference Δv_S in Hertz (Fig. 1.10(b), legend).

$$\Delta v_S = v_{OS} - v_{OR} \text{ (in Hz)}. \tag{1.41}$$

According to the Larmor equation (1.8), chemical shifts can be related to field differences ΔB_S, measurable in millitesla.

$$\Delta B_S = B_S - B_R \text{ (in mT)}. \tag{1.42}$$

Also because of the Larmor equation (1.8), the frequency or field differences Δv_S or ΔB_S are proportional to the swept radio frequency v_1 (in MHz) or the field strength of $\boldsymbol{B}_0$ (in T). Therefore, chemical shifts Δv_S (or ΔB_S) obtained at different radio frequencies v_1 (or field strengths B_0) have to be adjusted to the same radio frequency (or field) before comparison. In order to get chemical shift values which are independent of the frequency or field strength used, the δ scale of chemical shifts is introduced. δ values are obtained by dividing the frequency differences Δv_S (in Hz) by the frequency v_1 used (in MHz = 10^6 Hz).

$$\delta_S = \frac{\nu_{OS} - \nu_{OR}}{\nu_1} 10^{-6}. \tag{1.43}$$

Because $\Delta \nu_S$ between nuclei of the same type is very small (several Hz) compared to ν_1 (several MHz), the shifts on the δ scale are given in ppm (parts per million = units of 10^{-6}). Examples for the calculation of δ values in ppm are given for the ^{13}C chemical shifts of acetone in the legend of Fig. 1.10(b).

1.9.3 Reference Standard

A common reference compound used for calibrating ^1H and ^{13}C NMR spectra is tetramethylsilane (TMS), $Si(CH_3)_4$. Signals with small δ_S ($\Delta \nu_S$, ΔB_S) relative to TMS are said to appear at high shielding field $\sigma_i B_0$. The corresponding nuclei are strongly shielded; the amount of σ_i in eq. (1.40) is large. Signals with large δ_S are said to be at low field. Their nuclei are weakly shielded ("deshielded"); the amount of σ_i in eq. (1.40) is small.

The reference compound can be added to the sample solution (internal reference) or kept separate from the sample in a sealed capillary (external reference, Fig. 2.36). If an external reference is necessary, a correction term accounting for the difference between the bulk susceptibilities of reference (χ_R) and sample solution (χ_S) must be added to the observed shift, $\delta_{obs.}$:

$$\delta_{corrected} = \delta_{obs.} + \frac{2\pi}{3}(\chi_R - \chi_S). \tag{1.44}$$

1.10 Spin-Spin Coupling

1.10.1 Multiplicity of Signals

The ^{13}C NMR spectrum of acetone shown in Fig. 1.10(a) was obtained by proton decoupling. For the two nonequivalent nuclei two sharp singlets are observed. If proton decoupling is not applied, a proton-coupled ^{13}C spectrum is obtained, and both ^{13}C signals of acetone split into multiplets as shown in Fig. 1.10(b). A large quartet is found for the methyl carbons, which are each directly bonded to three protons with $I_H = \frac{1}{2}$. The signal of the carbonyl carbon atom, which is separated from six hydrogen atoms by two bonds, splits into a narrow septet.

A system containing n equivalent nuclei X with total spin quantum number I_X and m equivalent nuclei A with I_A is said to be of type $A_m X_n$ if the chemical shift difference between X and A is large. The number of lines in the NMR spectrum for each of these nuclei follows the multiplicity rule (1.45).

$$\left.\begin{array}{l} 2nI_X + 1 \quad \text{for nucleus A} \\ 2mI_A + 1 \quad \text{for nucleus X} \end{array}\right\} \tag{1.45}$$

The ^{13}C NMR spectrum of a methyl group, representing an AX$_3$ system, is a quartet according to eq. (1.45), since $I_X = \frac{1}{2}$ and $n = 3$. The ^{1}H NMR spectrum of a methyl group, however, is a doublet because one ^{13}C nucleus is adjacent to three equivalent protons, so that $I_A = \frac{1}{2}$ for ^{13}C and $m = 1$. Due to the low natural abundance of ^{13}C, the doublets (^{13}C satellites) arising from coupling of carbon-13 with protons are usually lost in the noise of ^{1}H NMR spectra.

1.10.2 Coupling Constants

The multiplet lines in the ^{13}C spectrum of acetone are equidistant. The distance (quoted in Hz) between any two neighboring lines is termed the coupling constant. The coupling constant of an A$_m$X$_n$ system is given the symbol J_{AX}.

In the ^{13}C NMR spectrum of acetone (Fig. 1.10(b)), the coupling constant is 125.5 Hz for the methyl signal and 5.5 Hz for the carbonyl signal. This illustrates that the magnitude of the coupling constant J_{AX} decreases with increasing number of bonds separating the nuclei A and X. If A and X are more than 4 to 5 bounds apart, the multiplet structure of the A and X spectrum cannot be resolved.

1.10.3 Comparison between Chemical Shifts and Coupling Constants

Chemical shifts depend on the strength of the applied magnetic field B_0 according to eq. (1.8 c) when measured on the frequency scale. Coupling constants remain constant when the strength of the magnetic field B_0 changes.

Because of intramolecular mobility (rotations, inversions) and intermolecular interactions, chemicals shifts depend on temperature, solvent, and concentration. Coupling constants, however, for the most part do not depend on these conditions.

1.10.4 Origin of Spin-Spin Coupling [10]

The magnitude of ^{13}C–^{1}H coupling constants depends on the hybridization of the carbon atom (sp^3, sp^2, sp) coupled to the proton. This is indicative of a mechanism of spin-spin coupling involving the bonding electrons, as is illustrated in Fig. 1.11 for a group AX with nuclei A and X, each with a total spin quantum number of $I = \frac{1}{2}$.

The multiplets arise from splitting of nuclear energy levels: Those of nucleus A split due to the additional magnetic field caused by the nucleus X precessing with or against B_0 when $I_X = \frac{1}{2}$. Precession of X with B_0 is more stable (Fig. 1.11 (a)). – For nucleus A, only transitions with $I_A = \pm 1$ and $I_X = 0$ are allowed (Fig. 1.11 (a)) due to quantum mechanical rules. Therefore, only two transitions are possible for A, both of them requiring the same amount of energy. No splitting arises yet ($J_{AX} = 0$, Fig. 1.11 (a)), and the influence of other factors must be taken into account.

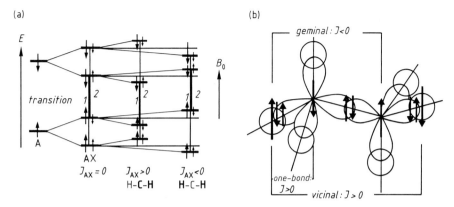

Fig. 1.11. Energy levels of an AX spin system (a) and interaction of the nuclear spins A and X with $I = 1/2$, involving the bonding electrons (ethane molecular orbital model).

Additionally, spin-correlation by the way of bonding electrons plays a role as can be rationalized in Fig. 1.11 (b), recalling the Pauli principle. In the more stable state, any spin, including the spins of both bonding electrons occupying the bonding molecular orbital, is antiparallel to an adjacent one (Fig. 1.11 (b)). Thus, for nuclei separated by one and three bonds (*one-bond* and *vicinal* coupling) a stabilization occurs for antiparallel precession, and J_{AX} is defined to be positive ($J_{AX} > 0$, Fig. 1.11 (a)). The stabilization energy corresponds to a frequency of $\frac{1}{4}hJ_{AX}$, as can be seen from Fig. 1.11 (a). The parallel precession of A and X, on the other side, is destabilized by the same amount of energy. Two transitions of different frequencies take place, and the A signal splits into a doublet as a result.

For *geminal* nuclei A and X, the Pauli principle favors a parallel precession (Fig. 1.11 (b)). A stabilization-destabilization pattern opposite to the directly bonded and *vicinal* nuclei arises (Fig. 1.11 (a)), and the coupling constant is defined to be negative ($J_{AX} < 0$). There are experimental techniques to determine the relative sign of coupling constants [5].

Often, more than one nucleus X with $I_X = \frac{1}{2}$, each with spin aligned with or opposed to B_0, couples with nucleus A. This is the case in CH_2 and CH_3 groups, representing AX_2 and AX_3 systems. In this case, all possible spin configurations of the X nuclei relative to B_0 must be considered, as illustrated in Fig. 1.12. Thus, a total spin characterizes each spin configuration relative to B_0. Each total spin has a particular weight arising from the number of spin configurations producing that total spin. The multiplicity of the signal A is given by the number of unequal total spins of the coupling nuclei X, and the multiplicity rule (1.45) is obtained.

The intensities of the multiplet lines arise from the number of spin configurations belonging to each total spin (Fig. 1.12). For n coupling nuclei X the intensity ratios are equal to the nth binominal coefficients (Table 1.3).

The analysis of multiplets by means of the multiplicity rules given above applies to systems where chemical shift differences between nuclei A and X in the frequency scale are large compared to the coupling constant J_{AX}. This is expected for carbon-13-proton

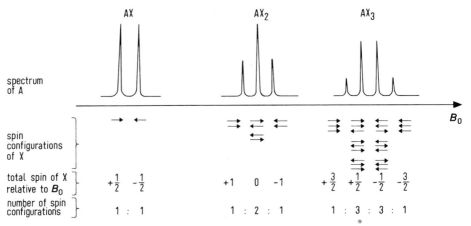

Fig. 1.12. Multiplicity and multiplet line intensities of the A signals in AX, AX$_2$, and AX$_3$ systems.

Table 1.2. Multiplicity of Signal A due to Coupling with Nuclei X ($I_X = \frac{1}{2}$).

System	Multiplicity
AX (CH)	$2 = 2 \cdot 1 \cdot \frac{1}{2} + 1$
AX$_2$ (CH$_2$)	$3 = 2 \cdot 2 \cdot \frac{1}{2} + 1$
AX$_3$ (CH$_3$)	$4 = 2 \cdot 3 \cdot \frac{1}{2} + 1$
AX$_n$	$2n I_X + 1$

Table 1.3. Multiplet Line Intensity Ratios of Signal A due to Coupling with n Equivalent Nuclei X ($I_X = \frac{1}{2}$).

$n = 0$			1			
1			1	1		
2			1	2	1	
3		1	3	3	1	
4		1	4	6	4	1
5	1	5	10	10	5	1

multiplets, as the difference in Larmor frequencies between protons (about 90 MHz at 2.13 Tesla) and carbon-13 nuclei (about 22.6 MHz at 2.13 Tesla) is of the order of several MHz; the coupling constants, however, are between 0 and 300 Hz. Natural abundance ^{13}C NMR spectra should therefore be easily analyzed following the multiplicity rules (1.45). But, as large one-bond ^{13}C-^{1}H couplings are involved, some energy levels may approach each other so that they mix. As a result, spectra of higher than first order are frequently observed for longer range couplings. The analysis of these spectra may require quantum mechanical calculations [5], computer simulation of spectra [11, 12] or proton decoupling and two-dimensional experiments as outlined in section 2.6.

2 Instrumental Methods of ^{13}C NMR Spectroscopy

2.1 Sensitivity of ^{13}C NMR-Spectroscopy

The main difficulty in ^{13}C NMR is the low natural abundance of the carbon-13 nucleus (1.108%) and its low gyromagnetic ratio γ, which yields a much smaller Boltzmann exponent $2\gamma p_0 B_0/kT$ than that of protons. Low natural abundance and small gyromagnetic ratio are the reasons why ^{13}C NMR is much less sensitive (1.59%) than ^{1}H NMR (100%). A common measure of sensitivity in NMR is the signal to noise of a reference sample, *e.g.* 1% ethyl benzene in deuteriochloroform. Several methods are available for improving the signal: noise in ^{13}C NMR.

2.2 Methods of Sensitivity Enhancement in ^{13}C NMR-Spectroscopy

The number of ^{13}C nuclei in the homogeneity range of the magnet can be increased by ^{13}C *enrichment* of the compound to be measured. Further, the *sample concentration* at a given volume or the *sample volume* at a given concentration can be increased; the former is limited by limited solubility, the latter by the air gap of the magnet.

In keeping with the Boltzmann equation, the sensitivity of ^{13}C NMR can be slightly increased by *lowering the sample temperature*. Care must be taken with temperature dependent spectra. The Boltzmann exponent can be also increased by *increasing the magnetic field strength*. Fields of up to 12 Tesla having the necessarily high homogeneity are now attainable in superconducting solenoids cooled with liquid helium.

The signal strength increases with the *rf power* as long as relaxation effects are adequate to restore equilibrium. The method is therefore limited by saturation.

Accumulation of spectra in a digital computer involving averaging of noise is known as the CAT method (computer averaged transients). The signal: noise ratio, $S:N$, increases with the number of accumulated scans n according to eq. (2.1).

$$(S:N)_n = (S:N)_1 \sqrt{n}. \tag{2.1}$$

The most economical and efficient method of sensitivity enhancement in ^{13}C NMR of organic molecules is the *pulse Fourier transform technique* (PFT) in combination with *decoupling methods* such as proton broad band decoupling and *polarization transfer*. These methods will be described in the following sections.

2.3 Continuous Wave NMR Spectroscopy

In the conventional NMR experiment, a radio-frequency field is applied continuously to a sample in a magnetic field. The radio-frequency power must be kept low to avoid saturation. An NMR spectrum is obtained by sweeping the rf field through the range of Larmor frequencies of the observed nucleus. The nuclear induction current (Section 1.8.1) is amplified and recorded as a function of frequency. This method, which yields the frequency domain spectrum f(ω), is known as the *steady-state absorption* or *continuous wave* (CW) NMR spectroscopy [1–3].

2.4 Pulsed NMR Spectroscopy

2.4.1 Magnetization

Remember the origin of any NMR experiment: When placed in a static magnetic field, the magnetic moments of atomic nuclei precess with Larmor frequency v_0 about the direction of this field (Fig. 1.2). Due to directional quantization, nuclei with spin quantum number of $I = 1/2$ (e.g. ^{1}H and ^{13}C) are distributed among two precession states: Some nuclei will precess about the direction of the magnetic field vector $\boldsymbol{B}_0$, some will precess in the opposite direction (Fig. 2.1. (a)). Precession about $\boldsymbol{B}_0$ is the state with lower potential energy and will be more populated, according to the Boltzmann distribution. Thus, an equilibrium magnetization $\boldsymbol{M}_0$ is built up along the direction of the magnetic field defined as the z axis (Fig. 2.1 (a)).

2.4.2 $90°_x$ Pulse

A radio-frequency (rf) alternating field initiates any NMR experiment. At resonance, the field vector $\boldsymbol{B}_1$ rotates with Larmor frequency ($v_1 = v_0$) perpendicularly to the static field vector $\boldsymbol{B}_0$, as shown in Fig. 2.1. In this situation, the nuclear magnetic moments will precess about both fields $\boldsymbol{B}_0$ and $\boldsymbol{B}_1$. Provided $\boldsymbol{B}_1$ extends along the x axis at a certain instant, the double cone of precession will rotate about the x axis in the yz plane (Fig. 2.1 (a) → (b)). The *flip angle* Θ relative to the z axis at a given field strength $\boldsymbol{B}_1$ depends on the *pulse width* t_p of the rf field, in the range of some μs, so that

$$\Theta = 2\pi v_1 t_p = \omega_1 t_p.$$

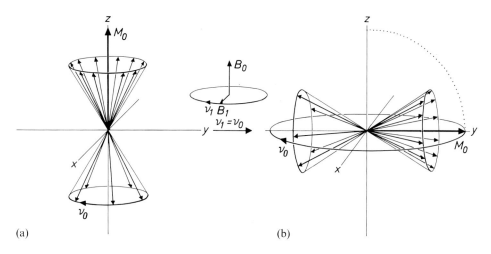

Fig. 2.1. Preferred nuclear precession about the direction of the magnetic field (a) and the effect of a 90°_x radio frequency pulse (b).

A pulse width exactly adjusted to turn the double cone of precession by 90° about the x axis is called a 90°_x or $\pi/2$ pulse.

2.4.3 Transverse Magnetization

The process depicted in Fig. 2.1. simplifies when considering only the motion of the magnetization vector M_0 in the x, y, z coordinate system (Fig. 2.2): The 90°_x pulse rotates M_0 by 90° from z to y. The longitudinal magnetization along z is converted into *transverse magnetization* along y, as given by eq. (2.2.):

$$M_y = M_0 \sin \omega_1 t_p$$
$$= M_0 \qquad \text{for } \omega_1 t_p = \pi/2 \ (90°). \qquad (2.2)$$

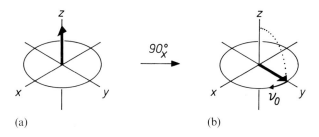

Fig. 2.2. Magnetization before (a) and after (b) the 90°_x pulse.

As shown in Fig. 2.1 (b), the nuclear moments still precess with Larmor frequency v_0 about the z axis in the xy plane, as does the resultant transverse magnetization (Figs. 2.1 (b) and 2.2 (b)). In the rotating frame (Section 1.7.3), the transverse magnetization with reference frequency v_0 stands while faster or slower components with $v_i > v_0$ or $v_i < v_0$ will rotate clockwise or counterclockwise, respectively, as shown in Fig. 2.3.

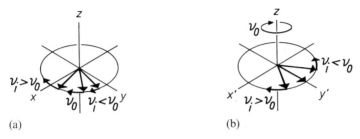

Fig. 2.3. Components of transverse magnetization with different chemical shifts in a fixed coordinate system x, y, z (a) and in the rotating frame x', y', z (b).

2.4.4 Free Induction Decay

The rotating transverse magnetization as portrayed in Figs. 2.2. and 2.3. induces a nuclear induction current in the receiver coil of the spectrometer, known as the NMR signal. When the rf pulse has ended, this signal decays because of field inhomogeneities and relaxation processes with time constant T_2^*, referred to as *effective transverse relaxation time*. Correspondingly, the receiver detects a steadily decreasing nuclear induction current called the *free induction decay* (*FID*) or the transverse relaxation function [7]. At resonance ($v_1 = v_0$), the FID will be a simple exponential (Fig. 2.4 (d)), while in the off-resonance situation ($v_1 \neq v_0$) the FID signal will be modulated by the frequency difference $1/(v_1 - v_0)$ (Fig. 2.4 (e)).

Fig. 2.4 outlines the concept of pulsed NMR, including the formation of transverse magnetization $M_{y'}$ by the rf pulse (b), followed by the free induction decay (c) and the corresponding time-dependent signal detectable in the resonance and off-resonance situation (d, e).

Due to chemical shielding, each nucleus may resonate within a range of Larmor frequencies, $2\pi\Delta = \omega_0 - \omega$, depending on its chemical environment. In order to rotate all nuclear spins within that range by the same angle Θ, the strength of the rf pulse must meet the requirement

$$\gamma B_1 \gg 2\pi\Delta. \tag{2.3}$$

Furthermore, the pulse width must be much shorter than the relaxation times,

$$t_p \ll T_1, T_2, \tag{2.4}$$

so that relaxation is negligible during the pulse.

According to eq. (2.2), the FID signal initiated by an rf pulse increases with the pulse width t_p, reaching a maximum for $\omega_1 t_p = \pi/2$, a 90° pulse, and decreasing to zero for

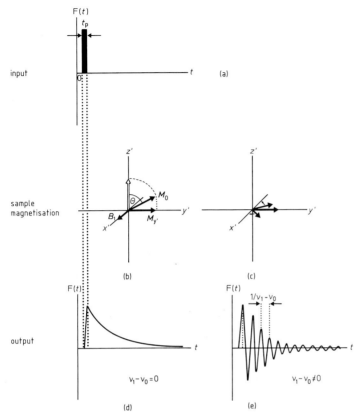

Fig. 2.4. Free induction decay.
(a) rf pulse as input signal; (b) sample magnetization during the rf pulse; (c) free induction decay following the rf pulse; (d) output signal for rf at resonance; (e) output signal for rf off resonance.

$\omega_1 t_p = \pi$, a 180° pulse. Pulse widths can thus be adjusted for 90° pulses by maximizing the FID signals.

2.4.5 Pulse Interferograms

For a liquid sample containing identical nuclei, the transverse magnetization M'_y arises from one Larmor frequency, which is actually the maximum of a very small frequency distribution caused by spin-spin relaxation and slight field inhomogeneities. The FID signal F(t) of this sample decays exponentially according to eq. (2.5), in keeping with the differential equations (1.17) describing transverse relaxation. F(0) is the amplitude of the FID signal at the time the pulse has stopped. For short pulses ($t_p \ll T_2$), the pulse width is negligible and F(0) approximates the FID amplitude at time zero.

$$F(t) = F(0) e^{-t/T_2}. \tag{2.5}$$

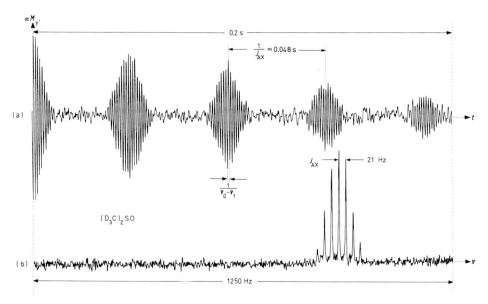

Fig. 2.5. (a) FID signal of hexadeuteriodimethyl sulfoxide; neat liquid; natural ^{13}C abundance; 22.63 MHz; 30 °C; pulse width: 10 μs for about 30° as pulse angle; observation time: 0.8 s; 0.2 s of the decay are shown; 2048 accumulated scans;
(b) Fourier transform of (a); recorded width: 1250 Hz; the distance between two beat maxima in (a) is about 0.048 s; this corresponds to a carbon-deuterium coupling constant of about 21 Hz.

If a sample contains equivalent nuclei A (^{13}C) subject to spin-spin coupling with nuclei X (^{1}H), the transverse magnetization arises from two or more Larmor frequencies, depending on the multiplicity. The corresponding magnetization vectors periodically rephase and dephase with the field vector $\boldsymbol{B}_1$, as in the off-resonance case with one Larmor frequency (Section 2.4.1). The FID signal is thus modulated by the frequency of the coupling constant J_{AX} [7, 13] as illustrated in Fig. 2.5 (a) for hexadeuteriodimethyl sulfoxide.

Similarly, in a sample containing two nonequivalent nuclei A_1 and A_2, the transverse magnetization results from two components due to two Larmor frequencies. In this case, the FID signal is modulated by the chemical shift difference of Larmor frequencies, $\Delta v = v_1 - v_2$. This modulation is illustrated in Fig. 2.6 (a) by the FID signal of a sample of 1-^{13}C-D-glucose after mutarotation. The product mixture of mutarotation contains α- and β-glucopyranose with differently shielded glycosidic carbons separated in the ^{13}C NMR spectrum by 87.5 Hz.

FID signals caused by rf pulses and modulated due to spin-spin coupling and chemically shifted Larmor frequencies are referred to as pulse interferograms. As illustrated in Figs. 2.5 (a) and 2.6 (a), the structural parameters of a CW NMR spectrum such as chemical shifts and coupling constants can be obtained by analysis of pulse interferograms. Most organic molecules, however, contain more than two nonequivalent nuclei; additionally, spin-spin coupling may complicate the patterns. As an example, the

2.4 Pulsed NMR Spectroscopy

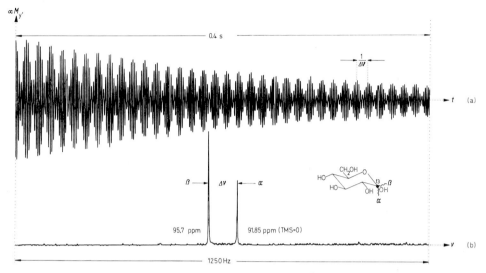

Fig. 2.6. (a) FID signal of mutarotated 1-^{13}C-D-glucose (60% ^{13}C); 22.63 MHz; 50 mg/mL deuterium oxide; proton decoupled; 30 °C; pulse width adjusted for maximum signal (12 µs); observation time: 0.4 s; 32 accumulated scans;
(b) Fourier transform of (a); observed width: 1250 Hz; the distance between two beat maxima in (a) is about 0.0115 s; this corresponds to a chemical shift difference of about 87.5 Hz or 3.85 ppm at 22.63 MHz.

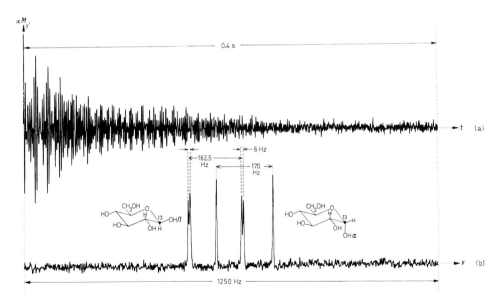

Fig. 2.7. (a) FID signal of mutarotated 1-^{13}C-D-glucose (60% ^{13}C); same sample and conditions as in Fig. 2.6, however without proton decoupling; (b) Fourier transform of (a).

pulse interferogram in Fig. 2.6(a) changes to the pattern shown in Fig. 2.7(a) due to ^{13}C-^{1}H coupling if proton decoupling is not applied while pulsing 1-^{13}C-D-glucose. In these cases, visual analysis of pulse interferograms becomes difficult. Fourier transformation is used to obtain chemical shifts and coupling constants. Examples of Fourier transformed pulse interferograms are given in Figs. 2.5(b), 2.6(b), and 2.7(b).

2.5 Pulse Fourier Transform (PFT) NMR Spectroscopy

2.5.1 FID Signal and NMR Spectrum as Fourier Transforms

In most pulsed NMR experiments, the rf field is applied off-resonance. Modulated pulse interferograms (Fig. 2.4(e), 2.5(a), 2.6(a), and 2.7(a)) arise because the vectors of transverse magnetization do not precess with a constant phase shift of $\pi/2$ relative to the vector $\boldsymbol{B}_1$. This is demonstrated in Fig. 2.8. The transverse magnetization is then a resultant of two components, $v(t)$ with a phase shift of $\pi/2$ relative to $\boldsymbol{B}_1$ and $u(t)$ in phase with $\boldsymbol{B}_1$:

$$v(t) = M_0 \sin \Theta \cos \omega t$$
$$u(t) = M_0 \sin \Theta \sin \omega t . \tag{2.6}$$

In mathematical treatments of FID and NMR signals, F(t) and f(ω), it is convenient to use complex quantities. The time domain signal is then defined by eq. (2.7).

$$F(t) = v(t) + i u(t) \quad \text{where} \quad i = \pm \sqrt{-1} . \tag{2.7}$$

Combining eqs. (2.6) and (2.7) yields

$$F(t) = M_0 \sin \Theta (\cos \omega t + i \sin \omega t) \tag{2.8}$$

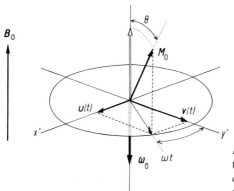

Fig. 2.8. The components $v(t)$ and $u(t)$ of the transverse magnetization in the rotating frame of reference for the off-resonance case ($\omega_1 - \omega_0 = \omega \neq 0$).

or, recalling that

$$e^{i\omega t} = \cos \omega t + i \sin \omega t, \qquad (2.9)$$

$$F(t) = M_0 \sin \Theta \, e^{i\omega t}. \qquad (2.10)$$

Since NMR spectra are not sequences of lines representing discrete Larmor frequencies but sequences of Lorentzian frequency distributions $f(\omega)$ (Fig. 1.9), eq. (2.10) must be replaced by eq. (2.11): $M_0 \sin \Theta \, e^{i\omega t}$ is multiplied by the frequency function $f(\omega)$, where ω represents the difference between the frequency ω_1 and the Larmor frequency distribution $\omega_0 + \Delta\omega$, $\omega = \omega_1 - (\omega_0 + \Delta\omega)$. Further, $M_0 \sin \Theta f(\omega) e^{i\omega t}$ must be integrated over the Larmor frequency distribution. Given a Lorentzian line shape as in Fig. 1.9, the limits of integration are $\pm \infty$:

$$F(t) = M_0 \sin \Theta \int_{-\infty}^{+\infty} f(\omega) e^{i\omega t} \, d\omega. \qquad (2.11)$$

Eq. (2.11) can be solved by developing it as a complex series of sines and cosines according to relation (2.9). This is a Fourier series [14]. Thus, an exponential in the time domain, $F(t)$, and a Lorentzian in the frequency domain, $f(\omega)$, are Fourier transforms of each other [15–17].

2.5.2 Position, Width and Phase of FT NMR Signals

Each component of transverse magnetization, *i.e.* every kind of nucleus with individual chemical shift, will be characterized by its *Larmor frequency*, its *phase shift* relative to the rotating frame (Fig. 2.9 (a–d)), and its *lifetime* defined by the effective transverse relax-

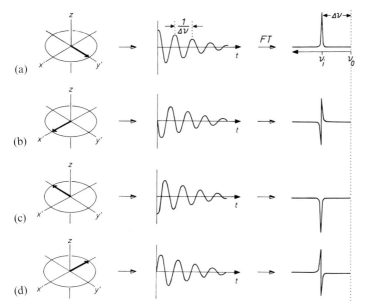

Fig. 2.9. (a–d) Phase shifts of transverse magnetization relative to the rotating frame (left), corresponding free induction decays (center), and phase errors of the FT NMR signals (right).

ation time T_2^*. The position of the signal in the FT NMR spectrum corresponds to the individual Larmor frequency (Fig. 2.9(a)). The lifetime of the transverse magnetization determines the signal width: A slow free induction decay will cause a sharp signal after Fourier transformation, and a broad signal corresponds to a fast FID. The phase shift of the transverse magnetization relative to the rotating frame at the instant of FID acquisition defines the phase error of the FT NMR signal (Fig. 2.9(a–d)) which has to be corrected (Section 2.5.5.2). Phase sensitive detectors or controlling the detector phase by a program permit the selection of certain components of transverse magnetization (phase selection), as is necessary for many experiments involving pulse sequences.

2.5.3 Acquisition of Pulse Interferograms for Fourier Transformation

2.5.3.1 Digitization

Fourier transformations of pulse interferograms are normally performed in digital computers. Consequently, the FID analog signal must pass an analog to digital converter. The pulse interferogram is then recorded digitally as a series of several thousand (N) data points. Adjusted to the memory structure of digital computers, N is usually a power of 2, e.g. $2^{12} = 4096$ or 4 "K".

2.5.3.2 Dwell Time and Pulse Interval

The sampling time during which FID data points must be collected in order to obtain the true NMR spectrum after Fourier transformation depends on the spectral width Δ. According to information theory [18], the sweep time per data point, called the *dwell time* t_{dw}, must satisfy the minimum given by the Nyquist equation (2.12).

$$t_{dw} \leq \frac{1}{2\Delta} \text{ s/point}. \tag{2.12}$$

Following this relation, the spectral width of $\Delta = 5$ kHz often used in ^{13}C NMR requires a dwell time of 100 μs. Multiplying the dwell time with the number of data points N to be collected during the FID yields the *acquisition time* τ required for recording the interferogram digitally.

$$\tau = N t_{dw} \leq \frac{N}{2\Delta}. \tag{2.13}$$

When several pulse interferograms must be accumulated in order to improve the signal: noise ratio, τ is the minimum repetition time between two pulses or the minimum pulse interval.

2.5.3.3 Filtering of Frequencies Outside of the Spectral Width

Sometimes, Larmor frequencies not included in Δ contribute to the FID signal. This is the case particularly when only partial spectra are desired. All frequencies outside of the

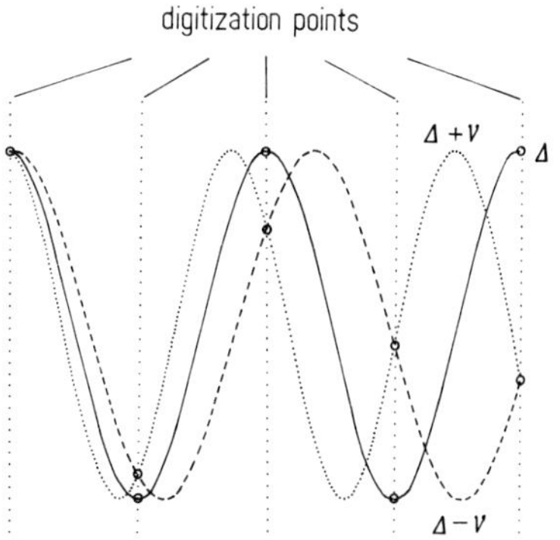

Fig. 2.10. "Folding back": At a sampling time given by the spectral width Δ (Hz) and the number of data points N, $\tau = N/2\Delta$, two frequencies higher and lower than Δ, $\Delta + v$ and $\Delta - v$, have identical digital values. The higher frequency is folded back on the lower one on Fourier transformation.

spectral width given by eq. (2.12) at a certain dwell time are "folded back" within the range Δ of the Fourier transformed FID. This is also true for high and low frequency noise.

Folding back of frequencies is illustrated in Fig. 2.10 for three sine waves arising from the frequencies Δ, $\Delta + v$, and $\Delta - v$. During the dwell time t_{dw}, the frequency Δ is sampled twice per cycle according to eq. (2.12). The lower frequency, $\Delta - v$, is sampled more, the higher one, $\Delta + v$, less than twice per cycle. The digital values of lower and higher frequency, however, are exactly equal (Fig. 2.10), so that both frequencies, $\Delta + v$, and $\Delta - v$, are indistinguishable digitally. After Fourier transformation, the higher frequency $\Delta + v$ behaves as if it had been folded back onto the frequency $\Delta - v$.

In order to avoid folding back in PFT NMR spectra, frequency components higher than Δ must be filtered before digitization.

2.5.4 Optimization of Pulse Interferograms for Fourier Transformation

2.5.4.1 Adjustment of Pulse Frequency

The frequency of the rf pulse must be outside of the Larmor frequency range to be observed as indicated in Fig. 2.11. This requirement is due to the experimental arrangement, which measures frequency differences relative to the rf field using phase sensitive detectors. Positive and negative frequencies relative to the irradiating frequency cannot be distinguished in the FID. Fourier transformation of an interferogram obtained by an rf pulse of frequency v'_1, which is within the spectral width Δ, yields a distorted NMR spectrum in which the frequencies on both sides of the irradiating pulse overlap.

For routine PFT NMR measurements, the pulse frequency is adjusted by changing the frequency of the transmitter until the CW spectrum of a reference sample with signals at both ends of the spectral width (*e.g.* acetone, Fig. 1.10) is reproduced exactly by Fourier transformation of the FID signal.

If the pulse frequency offset is chosen in the center of Larmor frequency range Δ, both halves $\Delta/2$ of the spectrum are sampled simultaneously and folded into each other. Folding can be corrected by a modification of the Fourier transformation software and

32 2 Instrumental Methods of ^{13}C NMR Spectroscopy

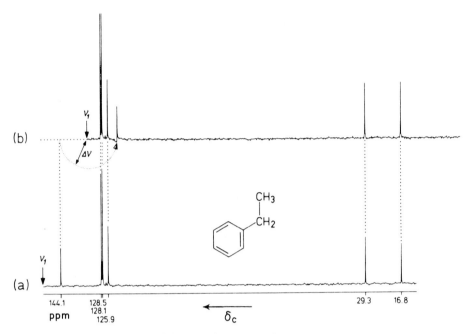

Fig. 2.11. Correct adjustment of the pulse frequency offset v_1 (a) and maladjustment (b). In (b) the signal at 144.1 ppm is folded to $\delta < 125$ ppm. Sample: 90% ethylbenzene in hexadeuteriobenzene at 20 MHz; single scan experiments with 90° pulses.

hardware. Simultaneous sampling and folding of both halves, however, gives a signal: noise enhancement of up to $\sqrt{2}$ (about 40%), as illustrated in Fig. 2.18. This is the concept of *digital quadrature detection* (DQD or QD).

2.5.4.2 Adjustment of Pulse Width

For a maximum FID signal the pulse width t_p must be adjusted for a 90° pulse according to eq. (2.2), so that $\gamma B_1 t_p = \pi/2$. This is usually achieved by displaying the pulse interferogram of a reference sample on an oscilloscope and changing the pulse width t_p until the maximum FID signal is observed. For further examination, the 90° pulse width found in this manner is doubled (180° pulse), and minimum signal should be observed if the adjustment is correct. More conveniently, the 90° pulse can be determined by a computer-controlled series of PFT spectra of varying pulse width, as is illustrated in Fig. 2.12 for the ^{13}C-^{1}H quartet of methanol.

In order to rotate the magnetization vectors of all nuclear spins within the range of Larmor frequencies to be observed, the pulse must not only be adjusted for 90°, so that $\gamma B_1 t_p = \pi/2$ (eq. 2.2)), but must also be very strong, so that $\gamma B_1 \gg 2\pi\Delta$ (eq. (2.3)). These requirements give the relation between pulse width and spectral width:

$$t_p \ll \frac{1}{4\Delta}. \qquad (2.14)$$

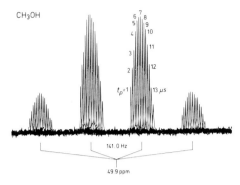

Fig. 2.12. Adjustment of the 90° pulse width; sample: methanol, 80% by vol. in deuterium oxide at 30 °C and 20 MHz; computer-controlled experiment with variable pulse width. The 90° pulse width is found to be 7 μs.

Maximum FID signal for a large spectral width (e.g. $\Delta = 5$ kHz for ^{13}C) thus requires very short rf pulses (e.g. $t_p \ll 1/4 \cdot 5 \cdot 10^3 = 50$ μs for ^{13}C). Usually, square wave bursts of several μs width are used in PFT ^{13}C NMR.

2.5.5 Data Transformation and Subsequent Manipulations

2.5.5.1 Fourier Transformation

Routine Fourier transformations of pulse interferograms are achieved in digital computers with a memory size of 4–256 K. Using eq. (2.9), a Fourier series of sines and cosines from eq. (2.11) is developed, and the transformation is computed for all data points of the FID signal. The transformation program makes use of the Cooley-Tukey algorithm [19, 20], which requires a minimum of time-consuming multiplications and efficiently uses the computer memory. This computation is called the fast Fourier transformation (FFT) and, depending on the speed of the computer, requires less than one sec for transforming an 8 K interferogram.

The solution of eq. (2.11) is a complex function. FFT computation therefore yields both real and imaginary PFT NMR spectra, $v(\omega)$ and $i\,u(\omega)$, which are related to the absorption and dispersion modes of CW spectra. The two parts of the complex spectrum are usually stored in different blocks of the memory and can be displayed on an oscilloscope to aid in further data manipulations.

2.5.5.2 Phase Correction

The real and imaginary spectra obtained by Fourier transformation of FID signals are usually mixtures of the absorption and dispersion modes as shown in Fig. 2.13(a). These phase errors mainly arise from frequency-independent maladjustments of the phase sensitive detector and from frequency-dependent factors such as the finite length of rf pulses, delays in the start of data acquisition, and phase shifts induced by filtering frequencies outside the spectral width Δ.

One method of phase correction assumes a linear dependence of the phase φ on the frequency, as in eq. (2.15); φ_A is the phase at frequency difference zero, φ_B is the phase shift across the total spectral width from zero to Δ Hz.

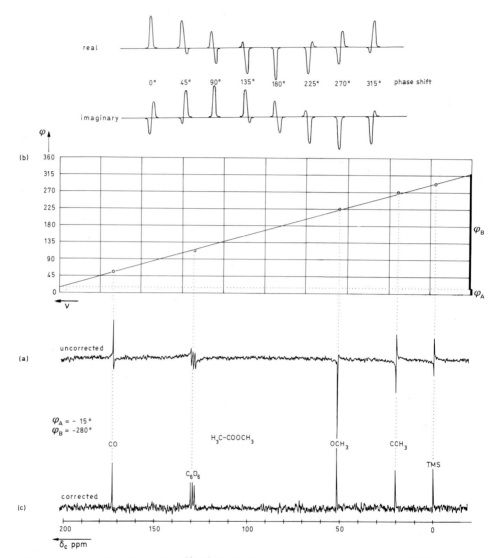

Fig. 2.13. (a, c) 22.63 MHz PFT ^{13}C{^{1}H} NMR spectrum of methyl acetate (20%) in hexa-deuteriobenzene (75%) and tetramethylsilane (5%); 256 accumulated pulse interferograms;
(a) real part before phase correction;
(b) phase correction according to eq. (2.15), achieved by using the phase shifts indicated above; extrapolation of the linear plot $\varphi(v)$ yields $\varphi_A = +15°$ and $\varphi_B = +280°$; for correction, the signs have to be changed, thus $\varphi_A = -15°$ and $\varphi_B = -280°$;
(c) real part after phase correction according to (b).

2.5 Pulse Fourier Transform (PFT) NMR Spectroscopy 35

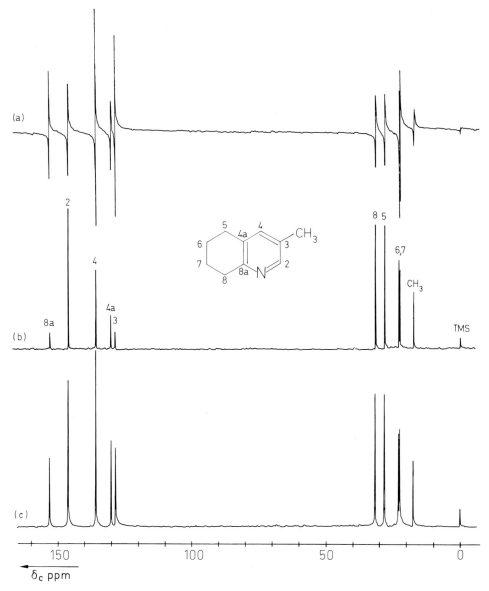

Fig. 2.14. 22.63 MHz PFT ^{13}C NMR spectra of 3-methyl-5,6,7,8-tetrahydroquinoline; 200 mg/mL deuteriochloroform; proton decoupled; 512 accumulated pulse interferograms; pulse width: 10 µs; pulse interval: 0,4 s;
(a) dispersion (i · u) mode, phase corrected;   (b) absorption (v) mode, phase corrected;
(c) magnitude ($\sqrt{v^2 + u^2}$) spectrum.

$$\varphi = \varphi_A + \varphi_B v. \tag{2.15}$$

ig. 2.13 (b, c) illustrates a phase correction. For correcting the phase, either the real or the imaginary part of the spectrum can be used. Correction of the real part for the absorption mode yields the dispersion mode in the imaginary part and *vice versa* (Fig. 2.13).

2.5.5.3 Computation of Magnitude Spectra

Magnitude or power spectra can be computed from the real and imaginary parts of the spectrum according to eq. (1.37).

$$P(v) = \sqrt{[v(v)]^2 + [u(v)]^2}. \tag{1.43}$$

Due to the quadratic form of eq. (1.37), the magnitude spectrum is phase independent, and a manipulated signal: noise improvement relative to the pure u and v modes is attained (Fig. 2.14). If only magnitude information is desired, phase correction is not necessary, and $P(v)$ can be computed immediately after Fourier transformation.

Weak lines are sometimes more easily localized in dispersion and magnitude spectra, as illustrated in Fig. 2.14 (b, c). However, the signals of dispersion and magnitude spectra suffer from "tailing" (Fig. 2.14 (b, c)), and a weak line closely spaced to a strong one may be lost or distorted in the tail of the strong signal.

2.5.6 Controlling Signal to Noise and Resolution in PFT NMR Spectroscopy

2.5.6.1 Digital Filtering

The signal: noise ratio or, alternatively, the resolution of the interferogram and its Fourier transform can be improved by digital filtering. This involves multiplication of the FID with an exponential $e^{\pm at}$ [21 a]. Negative values of at enhance the signal: noise ratio while causing some artificial line broadening. This is illustrated in Fig. 2.15.

Positive values of at improve the resolution at the expense of sensitivity. Another method of resolution enhancement without significant reduction of signal: noise is referred to as Gauss multiplication [21 b]. This involves multiplication of the FID signal with an exponential of second order, e^{-at-bt^2} where $a < 0$ and $b > 0$. The best value of a is the negative digital resolution (Section 2.5.6.2), while the optimum of b is related to the time after which the FID is practically zero.

2.5.6.2 Number of FID Data Points and Digital Resolution

For a dwell time t_{dw} limited by the spectral width according to eq. (2.12), the resolution dv of a PFT NMR spectrum depends on the number N of data points constituting the

2.5 Pulse Fourier Transform (PFT) NMR Spectroscopy

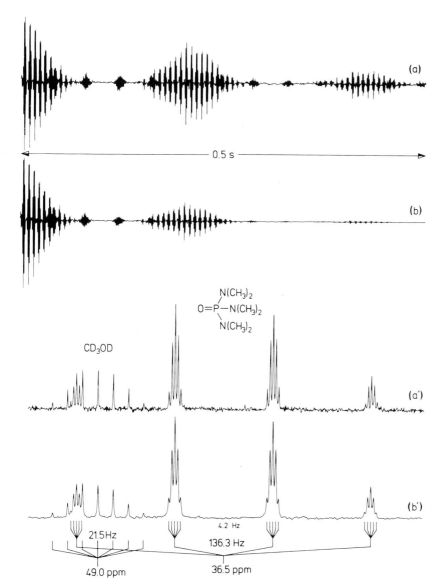

Fig. 2.15. Signal: noise enhancement by exponential multiplication at the cost of resolution; sample: hexamethylphosphoramide (70%) in tetradeuteriomethanol (30%) at 30 °C and 15.08 MHz; coupled;
(a) 5000 accumulated interferograms without exponential multiplication;
(a') Fourier transform of (a);
(b) 5000 accumulated interferograms with exponential multiplication for a line broadening of 1 Hz;
(b') Fourier transform of (b).

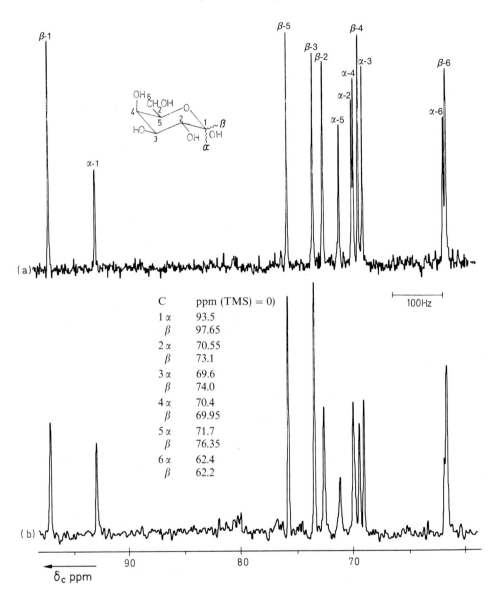

Fig. 2.16. Phase corrected 22.63 MHz PFT ^{13}C NMR spectrum of mutarotated D-galactose; 100 mg/mL deuterium oxide; 30 °C; proton decoupled; 512 accumulated pulse interferograms;
(a) obtained from an 8 K interferogram (pulse interval: 0.8 s);
(b) obtained from a 2 K interferogram (pulse interval: 0.2 s).

digitized FID signal to be transformed:

$$dv = \frac{1}{N\, t_{dw}} = \frac{2\Delta}{N} \text{ Hz}. \tag{2.16}$$

If an 8 K interferogram (8192 points) is transformed, the digital resolution is $dv = 1.22$ Hz/point for a spectral width of 5 kHz. In the Fourier transform of a 2 K interferogram (2048 points) resolution is reduced to $dv = 4.88$ Hz/point. This is illustrated by Fig. 2.16, which shows the proton broad-band decoupled PFT ^{13}C NMR spectrum of D-galactose. This carbohydrate with six nonequivalent carbons undergoes mutarotation in water. The equilibrium mixture contains two anomers, so that twelve ^{13}C signals are expected. All 12 signals are observed in the Fourier transform of the 8 K interferogram; the spectrum arising from the 2 K interferogram, however, shows only 10 well resolved signals.

2.5.7 Spin-Lattice Relaxation and Signal to Noise in PFT NMR Spectroscopy

For signal: noise improvement of weak samples, pulse interferograms are coherently accumulated before Fourier transformation. According to eq. (2.13), the minimum repetition time of an 8 K ^{13}C interferogram at a spectral width of 5 kHz is $\tau = 0.82$ s. Spin-lattice relaxation times of some ^{13}C nuclei, such as quaternary ^{13}C atoms, are as large as 100 s, so that $T_1 \gg \tau$. These nuclei cannot relax within the pulse interval τ. Using the Bloch equations, it can be shown that a stationary state is reached in this case [22]; the equilibrium magnetization is attenuated and the signal strength depends on the ratio $\tau/2\, T_1$. The signal: noise ratio of the FID and its Fourier transform decrease as a result.

An obvious way to avoid signal: noise attenuation for slowly relaxing carbons is to add a delay time to the pulse interval. This relaxation delay should be at least in the order of $3\, T_1$. However, accumulation of pulse interferograms becomes time consuming by this method.

A more practicable way [17] is to decrease the flip angle of M_0 to a value of $\Theta < 90°$ by reducing the pulse width. Restoration of the equilibrium magnetization then requires shorter periods. However, the transverse magnetization is also decreased by reducing Θ according to eq. (2.2). The signal: noise of the total FID and its Fourier transform again decrease. The best way to use this method is to optimize the flip angle Θ by adjusting the pulse width for optimum signal: noise of all signals in the PFT NMR spectrum.

The third method involves a three pulse sequence, $90° - \tau - 180° - \tau - 90°$, with a repetition time of t_r s. This pulse sequence refocuses the magnetization vector M_0 into its equilibrium position within the repetition time, thus representing a pulse driven relaxation acceleration. This technique, known as DEFT NMR [23, 24] (driven equilibrium Fourier transform NMR) can be understood by following the behavior of the magnetization vector M_0 under the influence of the pulse sequence in the rotating frame of reference (Fig. 2.17 (a–e)).

Starting with equilibrium of nuclear spins (a), a $90°$ pulse rotates the magnetization vector M_0 by $90°$ (b). Due to spin-spin relaxation, the spins dephase in (c), causing an FID

40 2 Instrumental Methods of ^{13}C NMR Spectroscopy

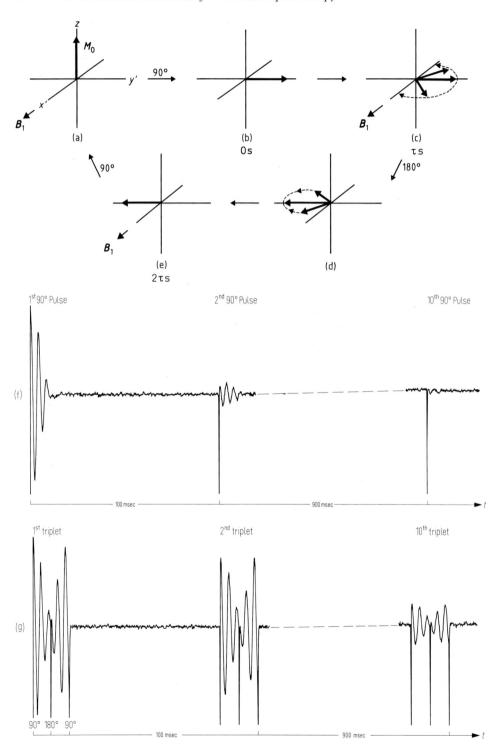

signal. The 180° pulse turns the spin system towards the negative y' axis (d). The spins then refocus, because the direction of dephasing in (c) remains the same. The inverted FID signal is produced as seen in (g), and a spin echo arises in (e) which may dephase again. Following a subsequent 90° pulse, the spin system returns to equilibrium (a). The pulse sequence can then be repeated for accumulation of FID signals. Fig. 2.17 (g) illustrates a driven equilibrium interferogram in comparison to the FID signal following a 90° pulse (Fig. 2.17(f)); reduction in signal: noise is shown to be less for the sequenced FID than for the single 90° pulse FID during accumulation.

The repetition time t_r of the pulse sequence is independent of T_1, which may be different for nonequivalent nuclei. The optimum repetition time has been found to be $t_r = 4\tau$ [22]. DEFT NMR requires careful adjustment of pulse widths for 90° and 180° pulses and (computer-controlled) pulse programming for accurate timing between pulses and pulse sequences. Other methods for improving signal: noise using other pulse sequences and spin echo trains have been described [22, 25]. DEFT NMR, however, appears to be the most efficient method so far, as long as T_1 and T_2 are of the same order of magnitude.

2.5.8 Comparison between CW and PFT

As any time domain function $F(t)$, a square wave rf pulse of width t_p can be approximated by a Fourier series of sines and cosines with frequencies $n/2 t_p (n = 1, 2, 3, 4, 5, ...)$ [14, 7]. An rf pulse of width t_p thus simulates a multifrequency transmitter of frequency range $\Delta = 1/4 t_p$ (eq. (2.14)). Accordingly, an rf pulse of 250 µs simultaneously rotates the M_0 vectors of all Larmor frequencies within a range of at least $\Delta = 1$ kHz. It simulates at least 1000 simultaneously stimulating transmitters, the resolution in the Fourier transform depending on the number of FID data points (eq. (2.16)), not the stimulation time t_p.

During the 250 µs needed to stimulate the Larmor frequencies within a range of 1 kHz by a single rf pulse, only $0.5 \cdot 10^{-6}$ of a 1 kHz scan is stimulated in a CW experiment using a 500 s/1 kHz sweep. A more realistic comparison accounts for the time required for Fourier transformation: NMR information at a spectral width of 1 kHz can be obtained

(1) in 10–25 s by Fourier transformation of a 8 K interferogram,
(2) in 500 s by a CW experiment, using a sweep of 1 kHz/500 s.

Furthermore, in the 500 s required for sweeping a 1 kHz spectrum, at least 1000 2 K interferograms can be accumulated before being Fourier transformed. The signal: noise of the PFT NMR spectrum is thus increased by a factor of $10\sqrt{10}$, according to eq. (2.1).

In summary, PFT NMR is much more sensitive for equivalent measuring times and much less time consuming for equivalent signal: noise ratios in comparison to CW NMR.

Fig. 2.17. Driven equilibrium FID; (a)–(b)–(c)–(d)–(e)–(a): the sample magnetization following a 90°, τ, 180°, τ, 90° pulse sequence; (f) and (g): comparison between the FID signals following repetitive 90° pulses and those following repetitive 90°, τ, 180°, τ, 90° pulse sequences; example: ^{13}C FID of enriched carbon tetrachloride, ^{13}CCl$_4$.
(f) and (g) recorded by B. Knüttel, Bruker Physik AG, Karlsruhe-Forchheim, Germany.

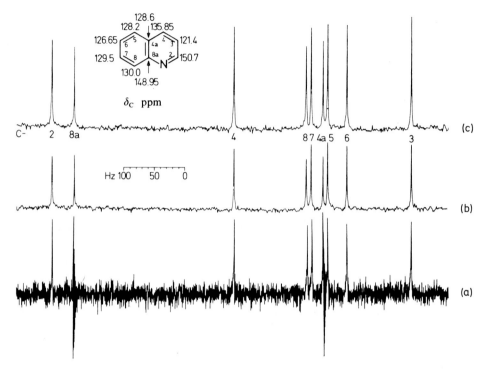

Fig. 2.18. 22.63 MHz ^{13}C{^{1}H} NMR spectrum of quinoline (70%) in hexadeuterioacetone (30%) at 25 °C; single scan experiments;
(a) CW spectrum; sweep: 1 Hz/s; spectral width: 1000 Hz; measuring time: 1000 s;
(b) PFT spectrum, obtained from one 90° pulse interferogram after Fourier transformation and phase correction;
(c) PFT spectrum, obtained as in (b), however using the quadrature detection technique.
The measuring and calculation time (Fourier transformation and phase correction for the 8 K interferogram/4 K spectrum) was less than 100 s in (b) and (c).

For illustration, the proton decoupled ^{13}C NMR spectra of quinoline, obtained from one CW scan and one pulse, respectively, are compared in Fig. 2.18. A further signal: noise enhancement of up to 40% arises from application of the QD technique outlined in Section 2.5.4.1 (Fig. 2.18(c)).

In requiring less measuring time and producing higher sensitivity in comparison to CW, PFT NMR follows the Fellgett principle [26, 27]: The signal: noise of any spectroscopic experiment increases if simultaneous multichannel excitation is applied. In the PFT technique, rf pulses simulate multichannel transmitters. If m transmitters stimulate simultaneously, the enhancement factor relative to one channel excitation ($m = 1$) is the square root of m (eq. (2.17), [26, 27]).

$$(S:N)_m = (S:N)_1 \sqrt{m} . \tag{2.17}$$

The minimum number of simultaneously exciting channels m required for resolving a spectrum of width Δ at a resolution dv, limited by natural line width and field inhomogeneity, is

$$m = \frac{\Delta}{\mathrm{d}\nu}. \qquad (2.18)$$

In addition, the resolution of a PFT NMR spectrum is governed by eq. (2.16), i.e.

$$\mathrm{d}\nu = \frac{2\Delta}{N}. \qquad (2.16)$$

The number of simultaneously stimulating channels simulated by an rf pulse is therefore

$$m = \frac{N}{2}, \qquad (2.19)$$

which depends only on the computer memory size and yields an enhancement factor of $\sqrt{N/2}$ according to eq. (2.17). Thus, according to the Fellgett principle, a PFT NMR spectrum obtained from an 8 K interferogram is expected to show a signal: noise enhancement factor of $\sqrt{4096} = 64$ relative to a CW spectrum of equal width, resolution, and measuring time.

Due to its greatly enhanced sensitivity in comparison to CW NMR, the PFT method has made ^{13}C NMR into a routine method of structure analysis for all molecules having the natural ^{13}C abundance of 1.1%. Additionally, phase-corrected PFT NMR spectra contain all spectral details without the lineskewing and ringing observed in CW spectra. Finally, short-lived molecules can be measured by PFT NMR, and sensitivity enhancement by accumulation of interferograms before Fourier transformation requires much less time than the accumulation of CW NMR spectra, due to the short time required for acquisition of FID signals.

2.6 Double Resonance Techniques used in ^{13}C NMR Spectroscopy as Assignment Aids

2.6.1 Basic Concept of Spin Decoupling

As was illustrated for the methyl and carbonyl signals in the ^{13}C NMR spectrum of acetone (Fig. 1.10), ^{13}C−^{1}H spin-spin coupling vanishes when proton broadband decoupling is applied. Proton broadband decoupling is the most important decoupling technique used in routine ^{13}C NMR, simplifying ^{13}C−^{1}H multiplet systems to spectra of up to z singlet lines for z nonequivalent ^{13}C nuclei of a sample.

Spin decoupling or nuclear magnetic double resonance (NMDR) is achieved by irradiating an ensemble of nuclei not only with a radio frequency B_1 at resonance with the nuclei to be observed, but additionally with a second alternating field B_2 at resonance with the nuclei to be decoupled (e.g. ^{1}H). Decoupling experiments can be carried out to convert homonuclear (^{1}H−^{1}H, ^{19}F−^{19}F) or heteronuclear multiplets (^{19}F−^{1}H,

$^{13}C-^{1}H$) into singlets. NMDR spectra are often symbolized by putting the nuclei to be decoupled between brackets besides the nuclei to be observed: A {X}. Proton decoupled ^{13}C NMR experiments are thus referred to as $^{13}C\{^{1}H\}$ NMR spectra.

Collapsing of spin multiplets AX_n (Fig. 1.10(b) → Fig. 1.10(a)) by NMDR can be explained using a frame of reference rotating at the angular velocity $\omega_2 = 2\pi\nu_2$ of the irradiating radio-frequency field $\boldsymbol{B}_2$. The effective field, which causes precession of the magnetization vectors, is then given by eq. (1.32).

$$\boldsymbol{B}_{\text{eff.}} = \boldsymbol{B}_0 + \frac{\omega_2}{\gamma} + \boldsymbol{B}_2. \tag{1.32}$$

If $\boldsymbol{B}_2$ is applied at resonance with the nuclei X to be decoupled, ω_2/γ cancels $\boldsymbol{B}_0$ so that $\boldsymbol{B}_{\text{eff.}} = \boldsymbol{B}_2$, as in eq. (1.33). As a result, the magnetization vectors of the irradiated nuclei X process perpendicularly to $\boldsymbol{B}_0$. Now, the irradiated nuclei X have their spins quantized perpendicularly to $\boldsymbol{B}_0$, while the spins of the observed nuclei A are still quantized along $\boldsymbol{B}_0$. Since the observed coupling between nuclei A and X is the scalar product of the spin quantizations I_A and I_X [28–30], the observed coupling is related to the angle α enclosed by the spin quantizations I_A and I_X.

$$J_{AX(\text{obs.})} \propto \cos\alpha. \tag{2.20}$$

The observed splitting $J_{AX(\text{obs.})}$ is equal to the coupling constant J_{AX} when both spins A and X are quantized in the same direction ($\alpha = 0°$, $\cos\alpha = 1$, NMR coupled). Coupling between A and X collapses ($J_{AX(\text{obs.})} = 0$), however, when the spins A and X are quantized perpendicularly to each other ($\alpha = 90°$, $\cos\alpha = 0$, NMDR, decoupled).

Another explanation of spin decoupling assumes that the perturbing field $\boldsymbol{B}_2$ induces rapid transitions between the spin levels of the irradiated nuclei X. The life-time t_x of X in its spin states is then short compared to the reciprocal coupling constant, so that the condition for spin-spin coupling.

$$t_x \geq \frac{1}{J_{AX}} \tag{2.21}$$

is not met anymore, and no coupling is observed [28, 29]. Decoupling of A_mX_n systems cannot be considered as to arise from saturation of the transitions of one nucleus, since decoupling fields are much stronger than saturating fields (see eq. (2.22)).

2.6.2 Proton Broad Band Decoupling in ^{13}C NMR Spectroscopy

In a $^{13}C\{^{1}H\}$ experiment, a $^{13}C-^{1}H$ multiplet can be selectively decoupled by an $\boldsymbol{B}_2$ field whose frequency matches the Larmor frequency of the coupling protons. In routine ^{13}C NMR, however, usually all $^{13}C-^{1}H$ multiplets are decoupled for reasons of sensitivity and simplicity. This is achieved when the decoupling field $\boldsymbol{B}_2$ covers the range of all proton Larmor frequencies. For the protons in organic compounds in a magnetic field $\boldsymbol{B}_0$ of 2.3 Tesla, this is at least 1 kHz at about 90 MHz. Decoupling fields with large

2.6 Double Resonance Techniques used in ^{13}C NMR Spectroscopy

frequency ranges can be realized either by a very large rf power B_2, so that

$$\frac{\gamma_X B_2}{2\pi} \geq \Delta_X = 1 \text{ kHz for } ^1H, \qquad (2.22)$$

or by modulation of the decoupling frequency by an audio-frequency [31] or white noise [32]. Modulation of the decoupling frequency distributes the rf power among the center band (e.g. 90 MHz) and numerous sidebands (e.g. 90 MHz ± 10, ± 20, ± 30, ± 40, ± 50 Hz etc.), with the spacing given by the modulation frequency (e.g. 10 Hz). Decoupling frequency bands of several kHz width are attainable by noise modulation. This is much greater than the range of the normal frequency distribution for protons according to eq. (2.22).

Double resonance techniques using large, noise-modulated frequency bands are referred to as broadband or noise decoupling. Decoupling frequency bands covering the range of all proton Larmor frequencies cause all $^{13}C-^1H$ multiplets to collapse. Examples of PFT $^{13}C\{^1H\}$ spectra obtained by *high power broadband decoupling* are shown in Figs. 1.10(a), 2.6(b), 2.11, 2.13, 2.14, 2.16, and 2.18. *Low power proton noise decoupling* broadens the signals of proton-coupled carbons, leaving only non-coupled ones with significant signal: noise. This is an aid in assigning quaternary carbons, as shown in Fig. 2.19 for colchicine.

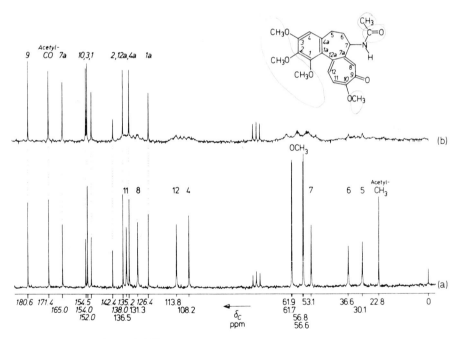

Fig. 2.19. Proton-decoupled ^{13}C NMR spectra of colchicine, 150 mg in 1 mL deuteriochloroform, 20 MHz, 25 °C, 200 scans;
(a) high power noise decoupled, ^1H decoupling frequency at 4 ppm;
(b) low power noise decoupled, ^1H decoupling frequency at 14 ppm, for localization of quaternary carbons (*italic type*) in the spectrum.

2.6.3 Nuclear Overhauser Effect in $^{13}C\{^1H\}$ NMR Experiments

Decoupling increases the sensitivity of NMR measurements because the intensities of all multiplet lines in a coupled spectrum are accumulated in one singlet signal in the decoupled spectrum. Thus the ^{13}C signal intensity in a $^{13}C\{^1H\}$ experiment on formic acid should be twice the intensity of one doublet line in the coupled ^{13}C spectrum. However, as demonstrated in Fig. 2.20, the intensity of the ^{13}C signal of formic acid increases much more than twice upon proton decoupling [33]. This additional sensitivity enhancement achieved during decoupling is attributed to the nuclear Overhauser effect (NOE) [34].

The nuclear Overhauser effect (NOE) observed upon proton decoupling of carbon-13 resonances results from *dipolar relaxation* induced by molecular motion in liquids. Rotation of a C−H bond in a tumbling molecule, for example, induces a fluctuating magnetic field. This arises from the magnetic moments of both ^{13}C and ^1H so that a rotating C−H bond is a rotating dipole. Provided molecular motion includes frequencies in the order of the Larmor frequencies of carbon-13 and proton nuclei, transitions between the energy levels of ^{13}C and ^1H are induced by these motions. Spin-lattice relaxation of nuclei in liquids essentially arises from such processes. − When the protons of C−H bonds are decoupled, rotations of these bonds due to molecular motion predominantly contribute to proton relaxation. Due to *dipolar proton-carbon coupling*, each carbon-13 nucleus attached to a relaxing proton is also relaxed. To conclude, proton decoupling induces additional carbon-13 relaxation within $^{13}C-{^1H}$ bonds. An increased population of the carbon-13 ground state arises, and the intensities of the signals of protonated carbons are enhanced.

The maximum NOE enhancement factor $\eta_A(X)$ observable for the A signal in an A{X} experiment mainly depends on the gyromagnetic ratios of A and X, if the A nuclei are

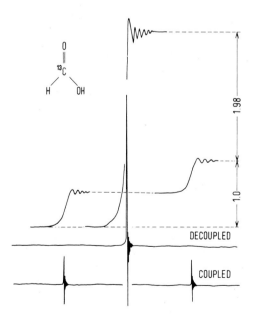

Fig. 2.20. Nuclear Overhauser effect observed on proton broad-band decoupling of ^{13}C-enriched formic acid [33]. (Reproduced by permission of the copyright owner from K.F. Kuhlmann and D.M. Grant, J. Am. Chem. Soc. *90*, 7355 (1968).)

predominantly relaxed by an intramolecular dipole-dipole mechanism during irradiation of the X nuclei [34].

$$\eta_A(X) = \frac{\gamma_X}{2\gamma_A}. \tag{2.23}$$

Using the gyromagnetic ratios of ^{13}C and ^{1}H, $\gamma_{13C} = 6.726$ and $\gamma_{1H} = 26.752$ rad s^{-1} gauss^{-1}, the maximum NOE enhancement factor observable in ^{13}C{^{1}H} spectra is almost 2, in good agreement with measurements (Fig. 2.20) [33].

$$\eta_{13C}(^{1}H) = 1.988 \quad [34].$$

In summary, NOE sensitivity enhancements are attained in heteronuclear NMDR when the nucleus with low γ is observed while the nucleus with high γ is decoupled. On the other hand, negligible enhancement of ^{1}H NMR sensitivity can be expected from ^{1}H{^{13}C} experiments, the enhancement factor being only $\eta_{1H}(^{13}C) = 0.216$ [34].

2.6.4 Quenching Nuclear Overhauser Effects in ^{13}C{^{1}H} NMR Experiments

^{13}C signal intensities in ^{13}C{^{1}H} NMR spectra are affected to varying extents by the NOE. The signals of quaternary carbons are weak, those of coupled carbons (CH, CH$_2$, CH$_3$) are very strong. The ratios of nonequivalent carbons cannot be determined by integration of ^{13}C{^{1}H} NMR spectra. This can be achieved, however, by adding small amounts of paramagnetic compounds such as *tert*-butyl nitroxide, cobalt(II) acetate, copper(II) and iron(III) acetylacetonate to the sample solution [37, 38].

Paramagnetic additives accelerate relaxation by way of dipole-dipole interaction between unpaired electrons and magnetic nuclei. The internuclear ^{13}C–^{1}H dipole-dipole mechanism does not predominate anymore in ^{13}C and ^{1}H relaxation during ^{13}C{^{1}H} experiments in the presence of paramagnetic additives, and nuclear Overhauser effects are reduced. Differences in signal intensities become smaller. Additionally, signal: noise enhancements for slowly relaxing nuclei are achieved in PFT ^{13}C measurements due to acceleration of relaxation. Paramagnetic additives in sample solutions are in this sense experimental alternatives to the signal: noise improvement of slowly relaxing carbons by DEFT (Section 2.5.7) [38]. Pseudo-contact and contact shifts as well as line broadenings are negligible as long as the concentrations of the paramagnetic substances are low [38].

2.6.5 Proton Off-Resonance Decoupling

Proton broadband decoupling simplifies and sensitizes CW and PFT ^{13}C NMR spectra. Further, the signals due to quaternary carbons can often be recognized in decoupled spectra, since they are weaker because of their lack of Overhauser effect, and in PFT spectra because of their long relaxation time T_1. However, information concerning the multiplicity of ^{13}C signals is lost. Primary, secondary and tertiary carbons are indistinguishable. In these cases, proton off-resonance decoupling may be applied.

Off-resonance decoupling is based on eq. (2.20): splitting of magnitude smaller than J_{AX} but larger than 0 Hz is observed for an AX system when the perturbing field B_2 is not at resonance with the nuclei X to be decoupled. In this case the effective field in the rotating frame of reference is given by eq. (1.32). With increasing offset $\Delta f_2 = \Delta v_2 - \Delta v_{0X}$ of the decoupling frequency band Δv_2 from the Larmor frequency range Δv_{0X} of the nuclei X to be decoupled, the angle between the spin quantizations of A and X decreases from 90° to 0°. According to eq. (2.20), the observed coupling increases under these conditions. This can be seen in Fig. 2.21 (a) for 1,1,3,3-tetraethoxypropane. The residual $^{13}C-^{1}H$ splittings J_R asymptotically approach the coupling constant J_{CH} with increasing offset Δf_2 of the decoupling frequency. On the other hand, the residual couplings decrease as the decoupling frequency approaches the proton resonances, smaller couplings vanishing first. Additionally, signal: noise increases gradually due to NOE, as complete decoupling approaches.

For off-resonance decoupling of ^{13}C NMR spectra, frequency offsets of about 0.5–1 kHz are used. In order to avoid complete collapsing of multiplets by large, noise-modulated frequency bands, non-modulated decoupling fields are usually applied. The decoupling frequency offset can be adjusted until the multiplets are so narrow that no or only slight overlapping occurs.

The practical use and the advantage of proton off-resonance decoupling – less multiplet overcrowding and more signal: noise relative to coupled spectra – is illustrated in Fig. 2.47. for a triterpene derivative in comparison to modern and more accurate methods for determination of CH multiplicity. An unequivocal assignment of the number of directly attached hydrogens may be possible for all carbons.

However, off-resonance multiplets are distorted, in that the multiplet lines on the side of the decoupling frequency are weaker than those on the opposite side. Further, off-resonance splittings of higher than first order may be observed for substructures such as $-CH=CH-$, $>CH-CH_2-$ or $-CH_2-CH_2-$. In the latter case, for example, the X part of an AA′BB′X system with X = ^{13}C and AA′ = BB′ = ^{1}H is observed, and the condition $J_{AX} \gg J_{AA'} \sim J_{BB'}$ is no longer met due to off-resonance decoupling. This can be seen for the methylene triplet of 1,2-dimethoxyethane in Fig. 2.21 (c). For off-resonance offsets smaller than 1 KHz, the intensities of higher order splittings successively increase.

The residual coupling J_R of an individual carbon with the carbon-proton coupling constant J_{CH} is a linear function of the decoupling frequency offset Δf_2 from the protons attached to that carbon, provided the constant decoupling power $\gamma B_2/2\pi$ is sufficiently

---

Fig. 2.21. Series of stacked proton off-resonance decoupled ^{13}C NMR spectra with varying frequency offset Δf_2.
(a) The residual C–H splittings of 1,1,3,3-tetraethoxypropane (50% by vol. in deuteriochloroform, 25 °C, 20 MHz, 500 scans/experiment) asymptotically approach the coupling constants J_{CH} with increasing offset from the TMS proton resonance.
(b) A J_R versus Δf_2 plot in the ppm or Hz scale for offsets between − and +1 KHz from TMS gives the chemical shifts of directly attached ^{1}H and ^{13}C nuclei.
(c) Residual splittings of higher order are observed for the triplet at 72.6 ppm of 1,2-dimethoxyethane (50% by vol. in hexadeuterioacetone, 25 °C, 15.08 MHz, 500 scans/experiment).

2.6 Double Resonance Techniques used in ^{13}C NMR Spectroscopy

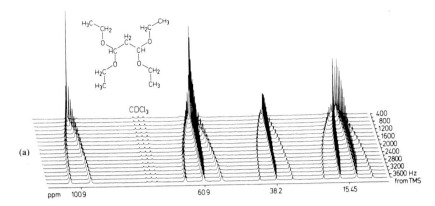

(a)

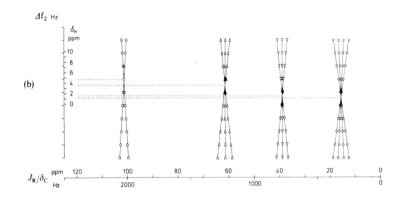

(b)

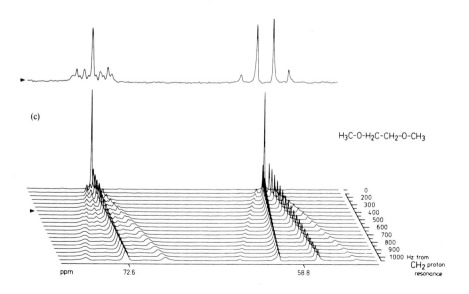

(c)

strong and the offset is not too large:

$$J_R = \frac{2\pi J_{CH}}{\gamma B_2} \Delta f_2 = \text{const.} \cdot \Delta f_2. \qquad (2.24)$$

This linear relation is an additional aid in assigning ^{13}C and ^{1}H shifts. After recording a series of off-resonance decoupled spectra with varying frequency offsets Δf_2 but with the same decoupling power, the residual splittings can be plotted *versus* the decoupling frequencies. For an $n + 1$ multiplet due to n protons, $n + 1$ straight lines will result. They intersect at one point, giving as coordinates the chemical shifts of both coupling nuclei (C and H), as shown in Fig. 2.21 (b) for another off-resonance spectral series of 1,1,3,3-tetraethoxypropane. This was an early method of *carbon-proton shift correlation*.

2.6.6 Pulsed Proton Broadband Decoupling

Proton decoupling begins immediately after the decoupler is switched on. The changes of spin populations responsible for NOE enhancements, however, are spin-lattice relaxation processes requiring considerably more time (up to some minutes). Correspondingly, proton decoupling stops instantly if the decoupler is switched off, but the NOE decreases slowly *via* spin-lattice relaxation. This is the basic concept underlying the pulsed decoupling techniques discussed in the following sections.

2.6.6.1 Measurement of NOE Enhanced Coupled ^{13}C NMR Spectra

The signal: noise of coupled ^{13}C NMR spectra is much lower than that of proton-decoupled ones because of signal splitting and lack of nuclear Overhauser enhancement. NOE enhanced coupled ^{13}C NMR spectra can be measured, however, by switching on the proton decoupler between successive ^{13}C transmitter pulses and switching it off during pulsing and FID aquisition. The timing is outlined in Fig. 2.22 (a). By this technique, referred to as alternately pulsed ^{13}C{^{1}H} double resonance or *gated decoupling* [35, 36], signal: noise enhancements of up 2 (eq. 2.23!) may be achieved, as can be clearly seen by comparing Figs. 2.22 (b) and (c).

2.6.6.2 Measurement of Proton-Decoupled ^{13}C NMR spectra with Suppressed NOE

Nuclear Overhauser enhancements and spin-lattice relaxation times are individual for each carbon. As a result, signal intensities cannot be evaluated from PFT ^{13}C NMR spectra obtained with continuous proton broadband decoupling.

If the decoupler is switched on only during FID acquisition as shown in the timing of Fig. 2.23 (a), proton-decoupled ^{13}C NMR spectra with suppressed NOE will be obtained. This technique is known as *inverse gated decoupling*. Additionally, the effects of relaxation on the signal intensity are reduced by waiting long enough between subsequent transmitter pulses (3 $T_{1\,max.}$), in order to permit relaxation to the "slowest" carbon (with $T_{1\,max.}$) of the sample molecule. Quantitative evaluation of proton-decoupled ^{13}C NMR spectra is then possible, as is illustrated in Fig. 2.23 (c) for the tautomers of acetylacetone.

2.6 Double Resonance Techniques used in ^{13}C NMR Spectroscopy

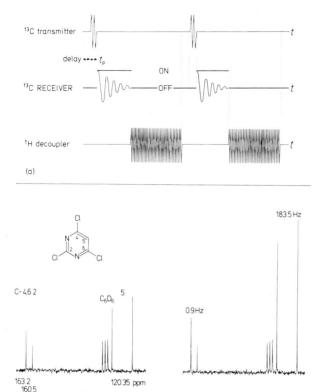

Fig. 2.22. Measurement of NOE enhanced coupled ^{13}C NMR spectra; sample: 2,4,6-trichloropyrimidine, 75% by vol. in hexadeuteriobenzene, 25 °C, 20 MHz;
(a) timing of the alternating transmitter and decoupler pulses for gated decoupling;
(b) single resonance spectrum (decoupler off, delay between successive transmitter pulses: 10 s, 50 scans);
(c) gated decoupled spectrum, obtained with the pulse timing of (a) (delay between successive transmitter pulses: 10 s, 50 scans, noise normalized to the level obtained in (b)).

2.6.6.3 Measurement of Nuclear Overhauser Enhancements

The measurement of nuclear Overhauser enhancement factors η_C is necessary sometimes, as these data provide information about relaxation mechanisms and the distance of protons to quaternary carbons. However, η_C cannot always be obtained as shown in Fig. 2.20 for the very simple case of formic acid. For larger molecules, integration of coupled spectra suffers from bad signal: noise and multiplet overcrowding, as is illustrated for menthol in Fig. 2.24 (b). In such cases, a modification of the inverse gated decoupling technique can be applied: The signal intensity is measured once with long decoupling time, *i.e.* $3 T_{1\,max.}$, accounting for the "slowest" carbon. In the second experiment, the decoupler is switched on only during FID acquisition as outlined in the timing of

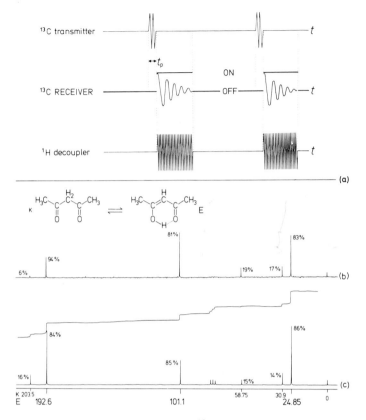

Fig. 2.23. Quantitative evaluation of ^{13}C NMR spectra with suppressed NOE; sample: acetylacetone, 50% by vol. in deuteriochloroform, 25 °C, 20 MHz;
(a) timing of the decoupler and transmitter pulses for inverse gated decoupling;
(b) ^{13}C NMR spectrum obtained with normal proton noise decoupling and without delay between successive 90° transmitter pulses, 200 scans;
(c) ^{13}C NMR spectrum obtained by inverse gated decoupling as outlined in (a); delay time between successive 90° transmitter pulses: 60 s; 200 scans.
Note that by integration of its ^{1}H NMR spectrum acetylacetone was found to be 85% enolized in deuteriochloroform.

Fig. 2.24 (a). For long decoupling times, the signal intensity I_1 benefits from the full NOE enhancement. In the second experiment with short decoupling time, practically no NOE enhancement builds up and a weaker signal intensity $I_2 < I_1$ is recorded. The intensity ratio I_1/I_2 is then the NOE factor η_C. The measurement can be averaged from a series of experiments as is shown in Fig. 2.24 (b) for the carbons of menthol with an η_C of less than 2.

2.6 Double Resonance Techniques used in ^{13}C NMR Spectroscopy

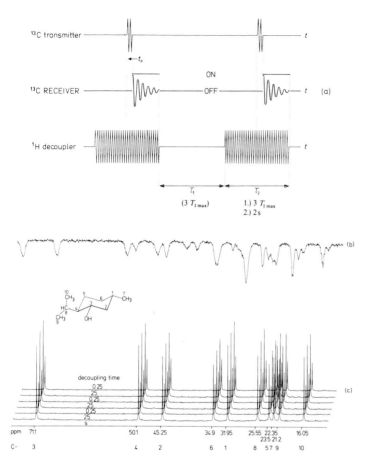

Fig. 2.24. Measurement of NOE enhancement factors by pulsed proton decoupling; sample: menthol, 200 mg in 1 mL of deuteriochloroform, 25 °C, 20 MHz;
(a) timing of decoupler and transmitter pulses for repetitive experiments with long and short decoupling times (3 $T_{1\,max.}$ and some ms, respectively); note that decoupling also takes place during FID acquisition; thus, τ_2 in (a) is the total decoupling time;
(b) coupled ^{13}C NMR spectrum of menthol, 500 scans;
(c) computer-controlled series of six ^{13}C experiments according to the timing (a) with alternating long (25 s) and short decoupling times (250 ms), 100 scans/experiment.

2.6.7 Selective Proton Decoupling

The ^{13}C signal of a particular carbon can be assigned unequivocally by selective decoupling of the protons attached to it, provided the proton resonance is unequivocally assigned and well separated in the ^1H NMR spectrum of the same compound in the same solvent. For selective proton decoupling, the ^1H NMR spectrum of the compound is

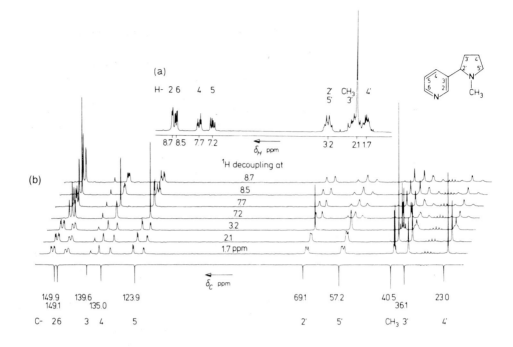

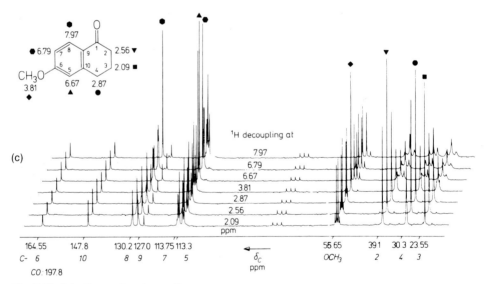

Fig. 2.25. Selective proton decoupling;
(a) ^{1}H PFT NMR spectrum of nicotine, 50% by vol. in hexadeuterioacetone, 25 °C, 80 MHz, indicating the assigned ^{1}H resonances to be decoupled;
(b) automatically recorded series of selectively proton-decoupled ^{13}C NMR spectra of nicotine, sample (a), 20 MHz, 500 scans/experiment; a proton broadband decoupled spectrum (bottom) is recorded for comparison with inverted phase;
(c) automatically recorded series of selectively proton-decoupled 20 MHz ^{13}C NMR spectra of 6-methoxy-α-tetralone, 200 mg in 1 mL of deuteriochloroform, 500 scans/experiment; the decoupling frequency offsets are calculated from the known proton chemical shifts relative to the offset determined for TMS.

recorded in order to determine the Larmor frequencies of the protons to be decoupled, as illustrated for nicotine in Fig. 2.25(a). These values are chosen as the decoupling frequencies. Noise modulation is switched off, and the decoupling power is reduced relative to that used for wide band decoupling. Typical values are adjusted so that $\gamma B_2/2\pi \approx 400 \pm 100$ Hz, ensuring that the ^{13}C satellites are also touched by the decoupling frequency distribution.

The stacked series of selectively proton-decoupled ^{13}C NMR spectra with varying decoupling frequency is conveniently recorded fully computer-controlled, as can be seen for nicotine in Fig. 2.25(b). The series clearly shows that ^{13}C signals, e.g. those of C-2, C-6, C-4, C-5 and C-4', are unequivocally assigned if the protons attached to these carbons do not overlap with others in the ^{1}H NMR spectrum. For overlapping proton resonances (e.g. the pairs 2'-H, 5'-H and 3'-H, CH$_3$) more than one carbon is affected by decoupling, of course, and other assignment aids such as two-dimensional CH correlation (Section 2.10) have to be taken into account (e.g. for the pairs C-2', C-5' and C-3', CH$_3$ of nicotine in Fig. 2.25(b)).

The offsets necessary for selective proton decoupling do not have to be measured, provided the decoupling frequency of TMS protons is known and the proton shifts of the compound are available from the literature. In this case, the decoupling frequency offsets are calculated from the proton shifts. This is performed for the protons of 6-methoxy-α-tetralone. The result of decoupling experiments with these values is shown in Fig. 2.25(c). A complete assignment of all protonated carbons is achieved.

2.7 Measurement of ^{13}C Relaxation Times

In addition to chemical shifts and coupling constants, relaxation times – particularly T_1 – are gaining increasing significance as molecular mobility parameters and aids in assigning ^{13}C NMR spectra. Several methods for measuring spin-lattice and spin-spin relaxation times T_1 and T_2 have been described [7, 13]. Many of these only permit selective determination of T_1 or T_2 for one kind of nucleus in simple molecules. For organic structural analysis, simultaneous determination and comparison of T_1 and T_2 for all nonequivalent nuclei present in a molecule is desirable. PFT methods meeting these requirements exist. Some of these will be described below.

2.7.1 Spin-Lattice Relaxation Times

2.7.1.1 Inversion-Recovery or 180°, τ, 90° Method

Spin-lattice relaxation times T_1 of individual nuclei (^{13}C, ^{1}H) present in a molecule can be obtained by Fourier transformation of the FID signal following a 180°, τ, 90° pulse sequence. The technique is referred to as inversion-recovery method [39, 40, 41] or

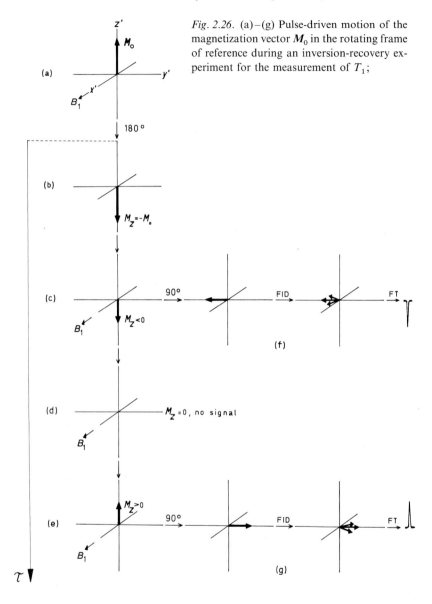

Fig. 2.26. (a)–(g) Pulse-driven motion of the magnetization vector M_0 in the rotating frame of reference during an inversion-recovery experiment for the measurement of T_1; partially relaxed NMR spectroscopy [42]; it can be explained by following the pulse-driven motion of the magnetization vector M_0 in the rotating frame of reference with the axes x', y', z (Fig. 2.26).

Starting at equilibrium (a), a 180° pulse along the x' axis inverts the magnetization vector ($M_0 \rightarrow -M_0$) (b). By subsequent longitudinal relaxation, the longitudinal magnetization M_z reequilibrates, going from $-M_0$ through zero to $+M_0$, as given by the Bloch equation (1.17a).

2.7 Measurement of ^{13}C Relaxation Times

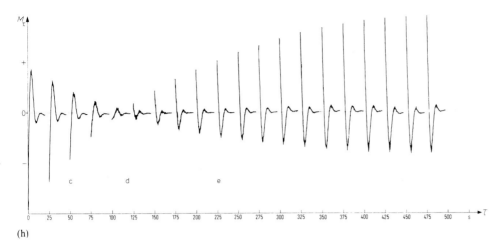

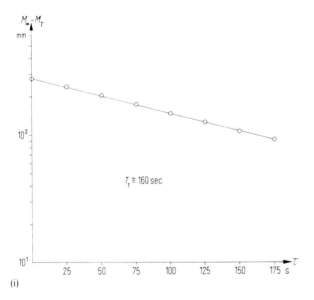

Fig. 2.26. (h–i)
(h) Initial amplitude of the FID after the 90° pulse, responding to M_τ, as a function of τ; the FID signals arise from 20 separate 180°, τ, 90° sequences with τ between 0 and 500 s; sample: 90% ^{13}C-enriched carbon tetrachloride, neat liquid, 25 °C, (i) $\lg(M_\infty - M_\tau)$ versus τ plot for the determination of T_1, using the FID amplitudes of (h).
(h) recorded by B. Knüttel, Bruker Physik AG, Karlsruhe-Forchheim, Germany.

$$\frac{dM_z}{dt} = -\frac{M_z - M_0}{T_1}. \tag{1.17a}$$

Depending on time, the magnetization has partially relaxed to an amount $M_z < 0$ (c), $M_z = 0$ (d), or $M_z > 0$ (e). τ seconds after the 180° pulse, a 90° pulse rotates the partially relaxed magnetization towards the negative ((f), $M_z < 0$) or the positive y' axis ((g), $M_z > 0$). The resulting transverse magnetization ($-M_{y'}$ for $M_z < 0$; $+M_{y'}$ for $M_z > 0$) precesses freely after the 90° pulse, giving rise to an FID signal which is Fourier transformed to the NMR spectrum. If the PFT NMR spectrum obtained by 90° pulses is phase corrected for the absorption mode before the experiment, Fourier transformation of the FID arising from the 180°, τ, 90° sequence gives an inverted absorption spectrum for $M_z < 0$ ($\tau < T_1 \ln 2$), no signal at all for $M_z = 0$ ($\tau_0 = T_1 \ln 2$), and an absorption spectrum for $M_z > 0$ ($\tau > T_1 \ln 2$).

The signal intensities A correspond to the transverse magnetization after the 180°, τ, 90° sequence. The transverse magnetization, in turn, arises from the partially relaxed longitudinal magnetization, given by integration of the Bloch equation (1.17a) between $M_z = -M_0$ at time $t = 0$ and $M_z = M_\tau$ at $t = \tau$:

$$\int_{-M_0}^{M_\tau} \frac{dM_z}{M_z - M_0} = -\frac{1}{T_1} \int_0^\tau dt$$

or

$$\ln \frac{(M_0 - M_\tau)}{2 M_0} = -\frac{\tau}{T_1}$$

or

$$\ln \frac{(A_\infty - A_\tau)}{2 A_\infty} = -\frac{\tau}{T_1}. \tag{2.25}$$

Here, A_τ and A_∞ are the signal intensities at pulse interval τ and ∞, respectively. The magnitude of A_∞, which equals the signal intensity at $\tau = 0$, can be determined by extrapolating the plot $A_\tau(\tau)$ either to $\tau = 0$ or to $\tau = \infty$.

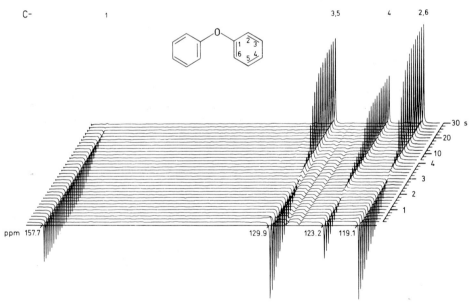

Fig. 2.27. Determination of ^{13}C spin-lattice relaxation times by the inversion-recovery technique; sample: diphenyl ether, 75% in hexadeuteriobenzene, 25 °C, degassed, 15.08 MHz, 10 scans/experiment. T_1, conveniently but not accurately obtained from the zero transition times τ_0 according to eq. (2.26), is 40.5 s for C-1, 4.6 s for C-2,6 and C-3,5, and 2.9 s for C-4, respectively.

2.7 Measurement of ^{13}C Relaxation Times

The spin-lattice relaxation time T_1 can be determined according to eq. (2.25), either from the slope $1/T_1$ of the semilogarithmic plot of $(A_\infty - A_\tau)$ versus τ (Figs. 2.26(i), 2.30(c)), or simply from the zero transition times τ_0 of the signal intensities following the solution (2.26) of eq. (2.25) for $A_\tau = 0$ (Fig. 2.27).

$$T_1 = \frac{\tau_0}{\ln 2} \approx \frac{\tau_0}{0.69}. \tag{2.26}$$

The first method averages deviations arising from maladjustments of pulse angles due to changing B_1 strength during the experiment. These errors may influence the $\tau_0/\ln 2$ method, which requires a good signal: noise ratio and small pulse interval steps for an accurate determination of the zero transitions.

The measurement of T_1 in $^{13}C\{^1H\}$ NMR is illustrated for the ^{13}C nuclei of diphenyl ether in Fig. 2.27. If accumulation of the FID signals is required for signal: noise improvement, the waiting period between two successive sequences should be at least $3\,T_1$ (180°, τ, 90°, $-3\,T_1$, $-180°$, τ, 90°, $-3\,T_1$, $-\cdots$), T_1 accounting for the slowest relaxing nucleus in the molecule. In this case, the experiment may require several hours, and it is advantageous when pulse sequencing, waiting periods, accumulations, Fourier transformation, phase correction and stacked recording of the spectra $A_\tau(v,\tau)$ are completely computer-controlled.

2.7.1.2 Saturation-Recovery Method [43]

Spin-lattice relaxation of nuclear spins may also be initiated by a perturbing 90° pulse along x' which rotates the magnetization to the $x'y'$ plane (Fig. 2.28(a) → (b)). The resultant transverse magnetization is then dispersed by a field gradient pulse (the *homospoil pulse*) along the z axis or by an *aperiodic pulse train*. This ensures that a second 90° pulse applied along x' after τ s monitors only the partially relaxed magnetization M_τ (Fig. 2.28(d) → (e)) without the contribution of the residual transverse magnetization built up by the first 90° pulse. The subsequent FID is Fourier transformed to the NMR signal with amplitude A_τ, which is related to the partially relaxed magnetization M_τ. – As known from the inversion-recovery techniques, waiting periods of $3\,T_{1\,\text{max}}$, permitting relaxation to the slowest carbon of the molecule, have to be programmed between successive pulse sequences when several FID's must be accumulated to obtain an adequate signal: noise ratio.

A stacked set of saturation recovery spectra is given in Fig. 2.29, which shows that the signal intensity approaches an equilibrium value A_∞ related to the fully relaxed magnetization M_0. The intensity A_τ as a function of τ is the integral of the Bloch equation (1.17a) between $M_{z'} = M_\tau$ and M_0, or in terms of intensities, from A_τ at time τ to A_∞ at time ∞:

$$\ln \frac{A_\infty - A_\tau}{A_\infty} = -\frac{\tau}{T_1}. \tag{2.27}$$

Thus, T_1 is obtained as the reciprocal slope of a semilogarithmic $(A_\infty - A_\tau)$ versus τ plot. Fig. 2.29(b) illustrates an example of a convenient graphical evaluation, also including the lg → ln conversion necessary when decadic semilogarithmic paper is used.

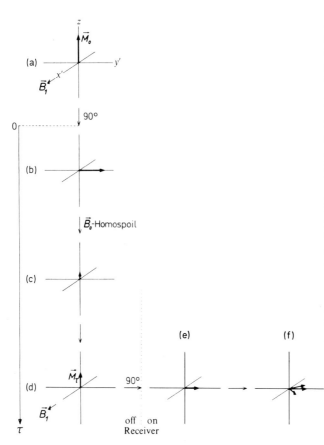

Fig. 2.28. Motion of the magnetization vector during a saturation-recovery experiment; the first 90° pulse rotates the magnetization vector M_0 to the $x' y'$ plane (a → b), where the resultant transverse magnetization is dispersed by a field gradient pulse (*homospoil*); after τ s (d), a second 90° pulse monitors the partially relaxed magnetization M_τ (d → f), and the resultant FID signal is Fourier transformed to the NMR signal with amplitude A_τ. (Reproduced by permission of the copyright owner from E. Breitmaier and G. Bauer: ^{13}C-NMR-Spektroskopie, eine Arbeitsanleitung mit Übungen. Georg Thieme Verlag, Stuttgart, 1977.)

2.7.1.3 Progressive Saturation or 90°, τ, ... Method

^{13}C spin-lattice relaxation times of individual nuclei can also be measured by PFT ^{13}C{^{1}H} experiments using a 90°, τ, 90°, τ, ... pulse train and noise modulation of the proton decoupling frequency. This method is known as progressive saturation [43] and is based on the following concept.

Recalling Section 2.5.7, the pulse interval τ of a 90°, τ, 90°, τ, ... sequence used for accumulation of FID signals in PFT ^{13}C NMR is often short compared to ^{13}C spin-lattice relaxation. After several 90° pulses, a steady state magnetization $M'_0 < M_0$ is established [22]; the signal intensity in the Fourier transform of the FID is attenuated. Using successive 90° pulse trains with increasing intervals τ, the steady state magnetization M'_0 approaches the relaxed magnetization M_0. The 90° pulse train successively rotates the M'_0 vectors towards the y' axis, giving rise to transverse magnetizations which yield a train of FID signals to be accumulated and Fourier transformed. In the PFT spectrum obtained for each τ, the signal intensity A_τ corresponds to the steady state magnetization M'_0. The intensity A_τ as a function of τ is the integral of the Bloch equation

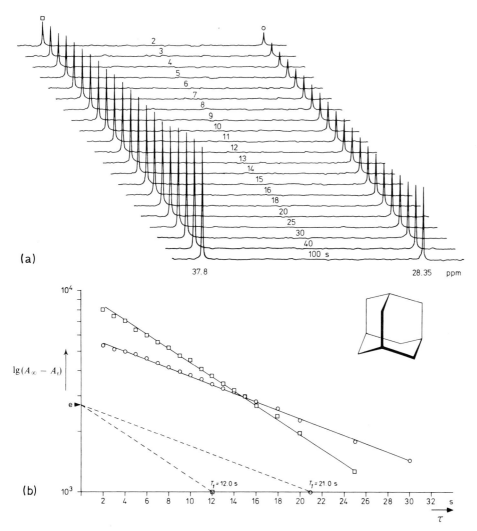

Fig. 2.29. Measurement of ^{13}C spin-lattice relaxation times by saturation-recovery;
(a) stacked set of saturation-recovery spectra; sample: adamantane, saturated solution in deuteriochloroform, 25 °C, 25 MHz, 50 scans/experiment (*i.e.* $(90°, \text{homospoil}, \tau, 90°)_{25}$);
(b) graphic evaluation of the intensities taken from spectra set (a); T_1 is 12.0 s for CH_2 and 21.0 s for CH.
(Reproduced by permission of the copyright owner from E. Breitmaier and G. Bauer: ^{13}C-NMR-Spektroskopie, eine Arbeitsanleitung mit Übungen. Georg Thieme Verlag, Stuttgart 1977.)

(1.17a) between $M_{z'} = M_0'$ and M_0, or, in terms of intensities, from A_τ at $t = \tau$ to A_∞ at $t = \infty$.

According to eq. (2.27), T_1 is obtained as the reciprocal slope of the semilogarithmic plot of $(A_\infty - A_\tau)$ versus τ (Fig. 2.30(c)). The equilibrium intensity A_∞ corresponding to the relaxed magnetization M_0 is approximated by recording a PFT spectrum arising from

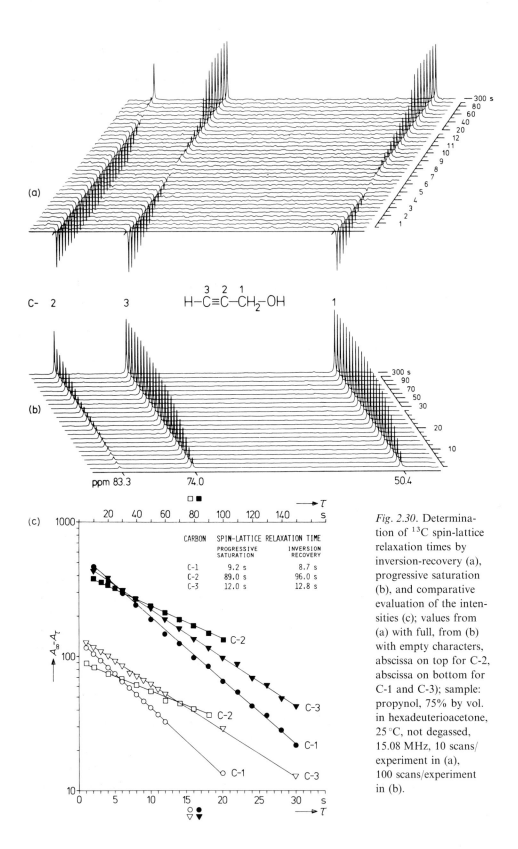

Fig. 2.30. Determination of ^{13}C spin-lattice relaxation times by inversion-recovery (a), progressive saturation (b), and comparative evaluation of the intensities (c); values from (a) with full, from (b) with empty characters, abscissa on top for C-2, abscissa on bottom for C-1 and C-3; sample: propynol, 75% by vol. in hexadeuterioacetone, 25 °C, not degassed, 15.08 MHz, 10 scans/experiment in (a), 100 scans/experiment in (b).

a 90° pulse train with an interval of at least $\tau = 3\,T_1$, T_1 accounting for the slowest relaxing nucleus.

Progressive saturation measurement of T_1 is restricted to nuclei with $T_1 > T_2$. Otherwise, the transverse magnetization remaining along the y' axis after the interval τ is rotated to the negative z' axis by the following 90° pulse. The resulting signal attenuation would invalidate eq. (2.27). The requirement $T_1 > T_2$ is met in ^{13}C{^{1}H} NMR when noise modulation of the decoupling field is applied: ^{13}C{^{1}H} spin-echo experiments on ^{13}C-enriched methyl iodide involving a 90°, τ, 180° pulse sequence show that proton noise decoupling accelerates the free induction decay, T_2 being decreased to $T_2^* \ll T_1, T_2$ [44].

T_1 measurements by progressive saturation can be completely computer-controlled. First, the sequence 90°, τ, ... is repeated several times for each τ without acquiring the FID signals, in order to establish the steady state. Then, the steady state FID signals are accumulated until a certain signal: noise ratio is reached. Following Fourier transformation and phase correction, the PFT NMR spectrum for each τ is recorded. Automatic stacking of the spectra can be programmed so that the panorama representation of $A_\tau(v, \tau)$ is obtained as shown in Fig. 2.30(b) for propynol [43]. Fig. 2.30(c) compares the results obtained from inversion-recovery and progressive saturation measurements. The two methods yield almost identical slopes and T_1 values [43]. Thus, the progressive saturation technique does not suffer from serious systematic errors provided $T_1 > T_2$.

The 180°, τ, 90° method is more general, not being restricted to nuclei with $T_1 > T_2$ or $T_1 > T_2^*$. However, long accumulation times due to long waiting periods between the pulse sequences are required for nuclei with small signal: noise ratio and long T_1, such as ^{13}C. These waiting periods are not necessary in the 90°, τ, ... method. Therefore, progressive saturation is more advantageous for measuring ^{13}C spin-lattice relaxation times of slowly relaxing carbons when proton noise decoupling can be applied.

2.7.2 Spin-Spin Relaxation Times

2.7.2.1 CPMGSE Experiments

Spin-spin relaxation times are obtained from spin-echo experiments [7, 44–46] using 90°, τ, 180° pulse sequences, as known from the DEFT method discussed in Section 2.5.7. T_2 values without serious errors due to maladjustment or changes of pulse angles are obtained by a pulse train of sequence 90°, τ, 180°, 3 τ, 180°, 5 τ, 180°, 7 τ, ... The initial 90° pulse is applied along the x' axis of the rotating frame. All subsequent 180° pulses are phase shifted by 90° relative to the 90° pulse, thus being applied along the y' axis of the rotating frame of reference. The method was developed by Carr, Purcell, Meiboom, and Gill [45, 46] and is referred to as the CPMG spin-echo or CPMGSE technique.

The concept underlying CPMGSE may be understood by following the pulse-driven motion of the magnetization vector M_0 in a frame of reference which, for the sake of clarity, is rotating slower than the smallest Larmor frequency of the nuclei to be observed (Fig. 2.31). An initial 90° pulse applied along x' rotates M_0 to y' (a), giving rise to an FID signal in the receiver due to transverse relaxation: The nuclear spins dephase with different angular velocities, some slower (↶), some faster (↷), due to different local fields

64   2 Instrumental Methods of ^{13}C NMR Spectroscopy

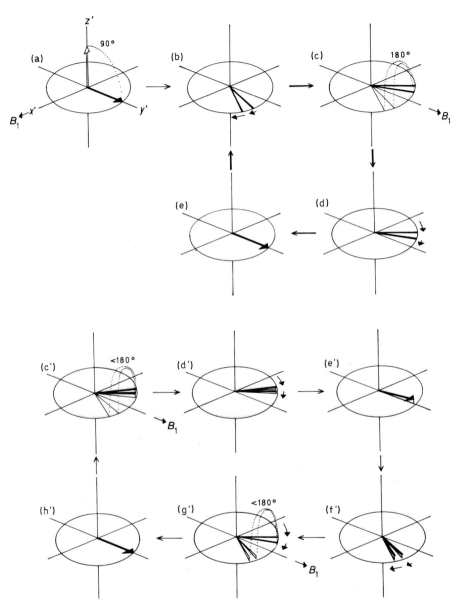

Fig. 2.31. Motion of the magnetization vector M_0 driven by a 90°, τ, 180°, 3 τ, 180°, 5 τ... pulse train during a CPMGSE experiment.

In (a)–(e), the 180° pulses are supposed to be exactly adjusted. In practice, this is usually not the case. Figs. (c')–(h') account for such maladjustments: If the pulse angles are slightly smaller than 180°, the first 180° pulse causes the spins to rephase above the x', y' plane (e'). As a result, the measurable component of transverse magnetization in the x', y' plane is slightly reduced, and so is the amplitude of the first echo and all subsequent odd echoes. The second 180° pulse, maladjusted like the first one, is just large enough to refocus the spins exactly in the x', y' plane (h'). Therefore, the second echo and all subsequent even echoes have the correct amplitude.

2.7 Measurement of ^{13}C Relaxation Times

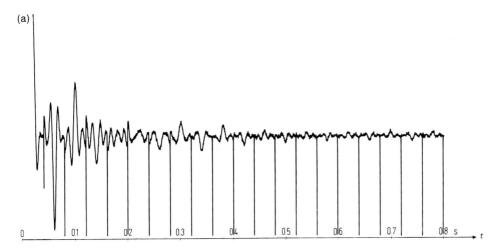

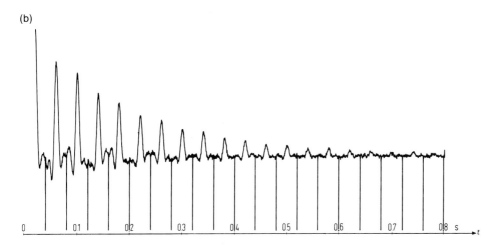

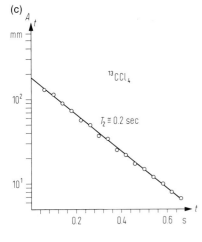

Fig. 2.32. Spin-echo trains for the determination of the ^{13}C spin-spin relaxation time T_2 of carbon tetrachloride; neat liquid sample, 90% ^{13}C enriched; 25 °C;
(a) Carr-Purcell sequence: the 180° pulses are applied in phase with the initial 90° pulse; the phase of subsequent echo signals alternates;
(b) Carr-Purchell-Meiboom-Gill sequence: the 180° pulses are applied perpendicularly to the initial 90° pulse; subsequent echoes have the same phase;
(c) Semilogarithmic plot A_t versus t (data from Fig. (b)) for the determination of T_2. (a) and (b) by B. Knüttel, Bruker Physik AG, Karlsruhe-Forchheim, Germany.

in the sample mainly caused by the inhomogeneity of B_0. In this situation (b), reached after τ s, a 180° pulse along y' rotates the spins 180° about y' (c). Since the direction of dephasing does not change after the 180° pulse, the spins rephase to a spin-echo in (d) after 2τ s. An echo FID is obtained which culminates in (e), followed by renewed dephasing (b). Another 180° pulse at time 3τ restricts dephasing, rotating the spins about y' (c) and causing another spin-echo (d). Thus, a 90°, τ, 180°, 3τ, 180°, 5τ, 180°, 7τ,... pulse train generates a train of echo FID signals spaced at 2τ, 4τ, 6τ,... as illustrated in Fig. 2.32 for the ^{13}C resonance of ^{13}C-enriched carbon tetrachloride, ^{13}CCl$_4$. The phase of the echo signals does not change, as all echos are along the positive y' axis. If the 180° pulses are applied in phase with the initial 90° pulse, the phase of subsequent echo signals alternates [7, 45], since rephasing alternately occurs in the positive and negative y' axis.

The echo amplitude decays with time. This decay is faster than transverse relaxation, since dephasing of nuclei is accelerated by varying local fields at different places in the sample due to inhomogeneity of B_0, and since diffusion of nuclei within the sample from one homogeneity range to another may take place. The echo amplitude $A_{t(\text{echo})}$ therefore does not decay as a simple exponential. Rather, the decay follows eq. (2.28), the term $f(t)$ accounting for inhomogeneity and diffusion.

$$A_{t(\text{echo})} = \text{const.} \, [e^{-t/T_2} - f(t)]. \tag{2.28}$$

T_2 of the observed nucleus is obtained as the reciprocal slope of the semilogarithmic plot $A_{t(\text{echo})}$ versus t for small values of t, since $f(t)$ can be neglected when t approaches zero.

2.7.2.2 Spin-Locking Fourier Transform Experiments

CPMGSE experiments are restricted to compounds containing one ^{13}C nucleus, as is the case in carbon disulfide or compounds containing one ^{13}C carbon at high concentration (e.g. ^{12}CH$_3$–^{13}COOH), so that only one resonance is detectable with significant signal: noise. If more chemically shifted or coupled ^{13}C nuclei are present in a molecule, the FID echo signals are modulated. In this case, the Fourier transformation of an FID signal somewhere in the echo train must be computed in order to obtain the T_2 of individual nuclei. Spin-locking Fourier transform experiments are a modification of the CPMGSE-FT concept.

If a CPMG pulse train is applied at very small pulse intervals τ, dephasing of the magnetization vectors away from y' is restricted. The spins are forced to precess almost in phase with the y' axis of the rotating frame ("spin-locking") [47]. This situation is approached when the 90° pulse along x' is followed by an rf field B_1 applied along y' continuously for t seconds. During this period t, the magnetization is aligned along y', decaying at a rate determined by the spin-lattice relaxation time in the rotating frame, $T_{1(\text{rot.})}$ [47]. For liquids, $T_{1(\text{rot.})}$ practically equals the spin-spin relaxation time T_2. When the continuous rf field is switched off after t s, the freely developing transverse magnetization gives rise to an FID signal, wich is Fourier transformed. Each individual line in the resulting PFT NMR spectrum has decayed to a fraction $e^{-t/T_{1(\text{rot.})}}$ of the equilibrium intensity A_0; A_0 is measurable by a PFT NMR experiment applying 90° pulses along x' to the totally relaxed spins. Since spin-locking largely reduces field inhomogeneity and the influence of diffusion on transverse relaxation, the signal intensity decay $A_t(t)$ can be

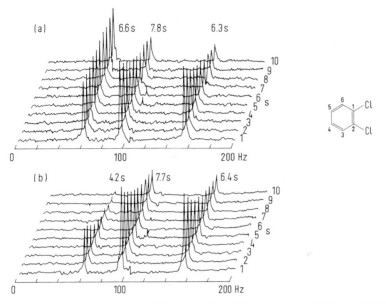

Fig. 2.33. Relaxation times of ^{13}C nuclei in *o*-dichlorobenzene [47] by ^{13}C{^{1}H} NMR.
(a) Spin-lattice relaxation times by an inversion-recovery experiment (the differences of signal amplitudes, $(A_\infty - A_\tau)$, are plotted *versus* δ and τ);
(b) Spin-spin relaxation times by a spin-locking FT experiment. (Reproduced by permission of the copyright owner from R. Freeman and H. D. W. Hill, J. Chem. Phys. 55, 1985 (1971).)

approximated by an exponential such as eq. (2.29).

$$A_t = A_0 e^{-t/T_{1(\text{rot.})}} = A_0 e^{-t/T_2}. \tag{2.29}$$

In analogy to the methods described for measuring T_1, T_2 is the reciprocal slope of a semilogarithmic plot of $(A_t - A_0)$ *versus* t, when the experiment is repeated for a series of different t.

The spin-locking FT method is illustrated in Fig. 2.33 for *o*-dichlorobenzene in a ^{13}C{^{1}H} experiment avoiding noise modulation of the decoupling frequency during application of the CW field [47]. The results are compared in Fig. 2.33 with T_1 measurements in order to demonstrate the relation between T_1 and T_2 in liquids: $T_1 \geqq T_2$.

2.8 Instrumentation

An NMR spectrometer capable of PFT measurements is a combination of a CW circuit such as found in conventional NMR spectrometers [1–5], a computer controllable pulse generator, and a small digital computer of core memory size 8–256 K. Fig. 2.34 outlines this experimental arrangement.

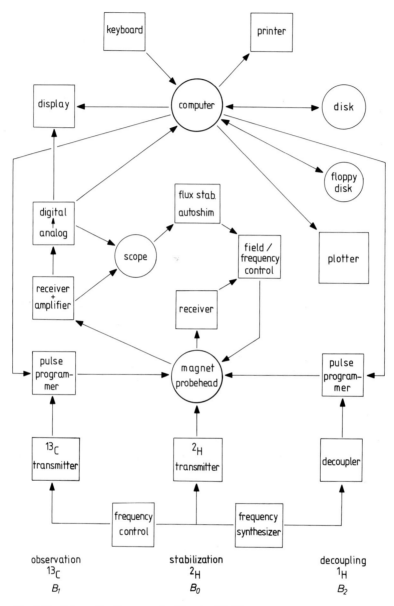

Fig. 2.34. Schematic diagram of a Fourier NMR spectrometer.

2.8 Instrumentation

Besides a strong magnet, modern NMR spectrometers consist of at least two rf channels: one for field/frequency stabilization and one for each nucleus to be observed. One additional channel is required for each nucleus to be decoupled. $^{13}C\{^1H\}$ experiments thus require three channels: one for observing ^{13}C NMR (B_1); one for proton decoupling (B_2); one for field/frequency stabilization $(B_0$, mostly operating at deuterium Larmor frequency). The observing and decoupling frequencies, v_1 and v_2, must be synchronized and phase-locked. This is achieved by deriving both from one master oscillator. The stabilization frequency v_0 is usually derived from the same master oscillator, in order to compensate for variations in the observing and decoupling frequencies.

2.8.1 Magnet

According to the Larmor equation (1.8), the strength of the magnetic field, B_0, determines the Larmor frequency of any nucleus, as shown in Fig. 2.35. Magnetic fields strengths of between 1 and 3 Tesla having the required homogeneity and stability are attainable by electromagnets.

Commercially available instruments include electromagnets of 2.114 and 2.35 Tesla (1 Tesla = 10 kilogauss). For these fields, the range of Larmor frequencies for nuclei of interest in organic structure analysis can be obtained from Fig. 2.35, *e.g.*:

2.114 Tesla: ^{13}C: 22.63 MHz; ^1H: 90 MHz; ^2H: 13.82 MHz.
2.35 Tesla: ^{13}C: 25.2 MHz; ^1H: 100 MHz; ^2H: 15.3 MHz.

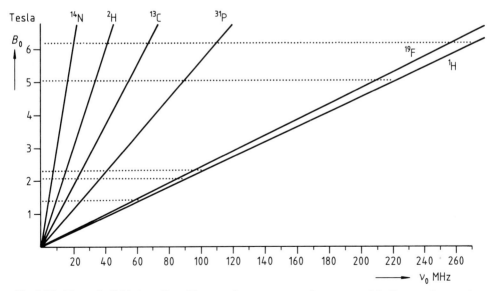

Fig. 2.35. Magnetic field strength and Larmor frequency range for some nuclei of interest in organic structure analysis.

The stonger the magnetic field B_0, the better the line separation of chemically shifted nuclei in the frequency scale, according to eq. (1.8 c). Since coupling constants remain unaffected by the magnetic field strength, multiplet overlapping decreases with increasing field strength, and homonuclear couplings (^{1}H$-^1$H, ^{19}F$-^{19}$F, ^{13}C$-^{13}$C) become small compared to chemical shift differences in Hz. Thus, the homonuclear multiplets also approach A_mX_n systems (Section 1.104), assignable by following the multiplicity rule (1.45). Moreover, the population of the lower spin level increases with increasing field according to eq. (1.10), leading to a corresponding increase in the sensitivity of the NMR experiment. In conclusion, analysis of homo- and heteronuclear coupled NMR spectra becomes much easier and sensitivity is increased by high-field NMR. High-field NMR is realized in superconducting solenoids giving rise to magnetic fields of 5–12 Tesla. The range of Larmor frequencies for common nuclei at these fields also follows from Fig. 2.35:

5.06 Tesla: ^{13}C: 55.45 MHz; ^{1}H: 220.00 MHz; ^{2}H: 33.65 MHz.
6.18 Tesla: ^{13}C: 67.88 MHz; ^{1}H: 270.00 MHz; ^{2}H: 41.46 MHz.
9.39 Tesla: ^{13}C: 100.58 MHz; ^{1}H: 400.13 MHz; ^{2}H: 61.40 MHz.
11.74 Tesla: ^{13}C: 125.72 MHz; ^{1}H: 500.00 MHz; ^{2}H: 76.75 MHz.

2.8.2 Stabilization Channel (Lock)

"CATing" of CW spectra, coherent accumulation of FID signals, and multipulse experiments necessary for measuring relaxation times require field/frequency stabilization, *i.e.* the ratio of magnetic field strength to transmitter frequency must be held constant. This is usually accomplished by electronically "locking on" to a strong, narrow signal and compensating any tendency for the signal to drift – caused by small field or frequency variations – with corresponding changes in the field strength.

Following from eq. (1.8) and Fig. 2.35, the field/frequency ratio must not necessarily be stabilized for the nuclei to be observed (homonuclear lock). It can also be set using any other nucleus (heteronuclear lock).

The homogeneity of the static field B_0 is optimized by changing the currents in shim coils placed in the air gap of the magnet until maximum signal and ringing is reached (optimum signal: noise and resolution). The current in the y gradient shim coils can be automatically controlled for maximum signal and homogeneity (autoshim) during the long measuring periods required for CAT experiments or FID signal accumulations of weak samples. Computer-controlled autoshim of other gradients is also possible.

Homonuclear stabilization causes a strong interference beat in the NMR spectrum stemming from the lock signal. This interference may lead to computer memory overflow during spectrum or FID accumulation before the sample signals have reached a significant signal: noise ratio. In view of the inherent insensitivity of the ^{13}C NMR experiment, a heteronuclear lock is usually chosen for such measurements. An intense deuterium signal is often used for heteronuclear stabilzation in ^{13}C NMR, as deuterated solvents are readily available and cause no signal at all (D$_2$O) or only weak narrow multiplets due to ^{13}C$-^2$H coupling in ^{13}C{^{1}H} experiments. Alternatively, the fluorine resonance of a fluorinated sample additive or solvent (*e.g.* hexafluorobenzene) can be used for locking, fluorine being a more sensitive nucleus than deuterium. Proton stabilization, however, obviously cannot be used during ^{13}C{^{1}H} experiments.

2.8.3 Observation Channel

The operating concept of PFT NMR can be recognized by following the arrows in Fig. 2.34, going from pulse programmer *via* probehead and computer to plotter.

The computer controllable pulse programmer generates DC pulses. These pulses gate the rf input signal of the B_1 transmitter, adjusted to the spectral width Δ according to Fig. 2.11 by the frequency control. The widths of the resulting rf pulses are adjustable for 90° or 180° by maximizing or minimizing FID signals shown on the display unit. In addition, various pulse sequences can be programmed. The rf pulses are amplified and fed to the tuned circuit of transmitter coil and receiver. The FID output signal is filtered, amplified, and digitized. Before performing the Fourier transformation, the accumulated FID signal can be observed on display in order to estimate if the signal: noise ratio is satisfactory.

After Fourier transformation, the effects on the spectrum of data manipulations, such as phase adjustments, can be controlled on a display before giving final calculating commands. Communication with the computer is generally *via* keyboard and graphic display. Light pen control *via* oscilloscope is also possible.

The PFT NMR spectrum can then be recorded in analog or digital form by a plotter controlled by the computer. If the computer core memory allows, additional subroutines may be stored which automatically compute signal intensities and positions (in ppm values relative to some standard or in Hz) and outputs them to the printer.

Programs and data are stored on disks or diskettes. Using a bank of chemical shift data and coupling constants of several thousand compounds, stored on magnetic tapes or disks, and the appropriate software, partial structural assignments by the computer are possible [48].

2.8.4 Decoupling Channel

According to eq. (2.22), the rf power of decoupling fields B_2 is much stronger than that of observing fields B_1. Therefore, the rf transmitter power of the decoupling channel must be amplified before being fed to the transmitter coil. For decoupling of large Larmor frequency ranges, wide frequency bands must be generated by modulation with white noise of adjustable band width ($\lesssim$ 30 kHz). For broad band decoupling, B_2 is amplified and noise modulated; off-resonance decoupling in $^{13}C\{^1H\}$ experiments is mostly achieved without noise modulation. In order to obtain coupling constants without sacrificing NOE signal: noise enhancements, B_1 and B_2 can be gated alternately by the pulse programmers, as outlined in Section 2.6.6.

2.8.5 Sample

The sample solution is usually placed in a glass tube of diameter 5, 8, 10, or 15 mm, which is rotated perpendiculary to the magnetic field B_0 in electromagnets and along the field in superconducting solenoids. This trick averages the magnetic field for each sample molecule, so that homogeneity is improved within the sample and sharper signals are obtained [49].

For ^{13}C experiments, the sample is usually prepared by dissolving the compound to be investigated in a deuterated solvent, which is usually required for field/frequency stabilization. For calibration, a small amount of a reference compound (usually tetramethylsilane) is added to the sample.

If the reference compound does not dissolve in the sample solution, it can be used externally in a capillary coaxially centered in the sample tube by teflon anti vortex plugs, as illustrated in Fig. 2.36. This is also advantageous if the reference signal causes or suffers from solvent shifts. The use of external references requires susceptibility corrections of chemical shifts according to eq. (1.44), with reference to volume susceptibility tables (*e.g.* in ref. [2]). The deuterated locking compound can also be used externally, for instance because of solubility reasons, if deuterium/hydrogen exchange is to be expected, or if the pH dependences of chemical shifts are to be measured. In order to obtain relaxation times accurately by multipulse experiments, the samples should be degassed, particularly to remove dissolved oxygen. The sample insert temperature can be registered thermoelectrically and adjusted for low or high temperatures by a stream of cooled or heated nitrogen gas. High rf powers used for decoupling considerably increase the insert temperature, so that cooling is required for sensitivity reasons (eq. (1.10)) and in order to protect sample and insert.

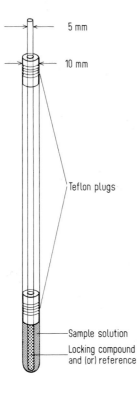

Fig. 2.36. Coaxially-centered sample tubes for external lock or reference.

2.9 From the First to the Second Dimension of ^{13}C NMR Spectroscopy

All experiments described previously deal with one-dimensional ^{13}C NMR spectra: Spectral parameters such as chemical shifts, coupling constants and relaxation times are evaluated from spectra with one frequency axis as one dimension. A second dimension comes into the NMR experiment when two spectral parameters, such as chemical shifts and coupling constants or chemical shifts of different nuclei, are represented in an area limited by two frequency axes. These experiments are based on pulse sequences facilitating multiplicity analysis and partly involving further sensitivity enhancement of the ^{13}C NMR experiment.

2.9.1 The Spin-Echo

The spin-echo pulse sequence $90°_x$, τ, $180°_x$ [23, 24, 44–46] is an essential part of many one- and two-dimensional NMR experiments. Consider a magnetization resulting from two components of frequencies, $v_0 \pm \Delta v$. Following a $90°_x$ pulse, the transverse magnetization dephases into a slower and a faster component, $v_0 - \Delta v$ and $v_0 + \Delta v$, respectively, so that the situation shown in Fig. 2.37(a) is attained after τ seconds. At this moment, a $180°_x$ pulse rotates both vectors about the x' axis (Fig. 2.37(b)). Retaining their frequencies and their direction of rotation, both components refocus after 2τ seconds to a *spin-echo* along the negative y'-axis (Fig. 2.37(c)). Relaxation processes will attenuate the echo amplitude relative to the initial transverse magnetization.

Fig. 2.37 illustrates the significance of a spin-echo sequence: Components of transverse magnetization with different frequencies due to chemical shift or spin-spin coupling can

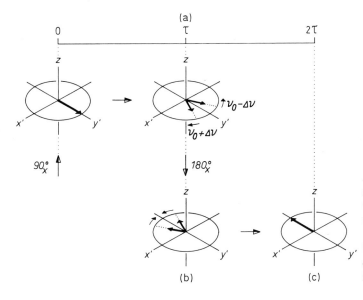

Fig. 2.37. Spin-echo pulse sequence: (a) components of transverse magnetization at time τ after the $90°_x$ pulse; (b) $180°_x$ rotation of the vectors by the $180°_x$ pulse; (c) spin-echo at time 2τ after the initial $90°_x$ pulse.

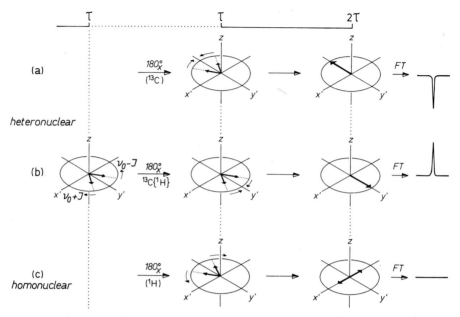

Fig. 2.38. Spin-echo responses of heteronuclear (a, b) and homonuclear (c) AX systems. (a) A $180°_x$ pulse in the ^{13}C shift range refocusses the CH doublet to an echo signal; (b) a $180°_x$ pulse in the proton shift range refocusses the doublet to a positive signal; (c) a $180°_x$ pulse in the proton shift range quenches the proton-proton doublet.

be refocussed. Many experiments involving the spin-echo sequence make use of spin-spin coupling effects. Therefore, the influence of this sequence on a two-spin system AX (Fig. 2.38) will be explained next.

In the case of a *heteronuclear AX system*, e.g. a CH bond with $A = {}^1$H and $X = {}^{13}$C, the transverse magnetization dephases to the doublet components with $v_0 + J_{CH}$ and $v_0 - J_{CH}$ because of CH coupling (Fig. 2.38(a)). A ^{13}C–^{1}H pair thus responds to the echo sequence shown in (Fig. 2.37(a–c)), provided the $180°_x$ pulse covers the range of carbon-13 shifts: At time 2τ after the initial pulse, a spin-echo builds up along the negative y' axis. Subsequent Fourier transformation computes a negative signal (Fig. 2.38(a)).

In the *homonuclear AX case*, e.g. a proton-proton two-spin system, the $180°_x$ pulse affects both nuclei. The doublet vectors are reflected at the $x'z$ plane. Inverting the precession states of all other coupling nuclei, the $180°_x$ pulse also inverts rotation of both AX components (Fig. 2.38(c)). At time 2τ, the vectors will be aligned antiparallel along $\pm x'$ (Fig. 2.38(c)); the resultant is zero and no signal will be detected.

A $180°_x$ pulse irradiating the protons just inverts the rotation of the doublet vectors (Fig. 2.38(b)). At time 2τ, the components refocus along the positive y'-axis (Fig. 2.38(b)), so that a positive signal arises from Fourier transformation.

Magnetic field inhomogeneities and chemical shifts additionally dephase each doublet component in Fig. 2.38. These effects are also refocussed by a $180°_x$ pulse, provided it rotates the components about the x'-axis, as is the case in experiments (a) and (c) of Fig. 2.38.

2.9.2 J-Modulated Spin-Echo

A combination of the spin-echo sequence (Fig. 2.37) and gated proton decoupling (Section 2.6.6) permits straightforward recognition of methyl, methylene, methine and quaternary carbon nuclei in ^{13}C NMR spectra, without the known problems of multiplet overlapping and artefacts induced by proton off-resonance decoupling (Section 2.6.5.). This gated spin-echo experiment is referred to as *J-modulated spin-echo* [50a], *spin-echo-FT* ("*SEFT*" [50b, c] and *attached proton test* ("*APT*" [50d]). The pulse sequence of a J-modulated spin-echo follows the pattern given in Fig. 2.39 and includes the three time periods characteristic of all modern pulse sequences – *preparation, evolution* and *detection*. The preparation period is usually a relaxation delay. It ends with the initial $90°_x$ pulse, generating transverse magnetization. Thereafter, the proton decoupler is switched off, so that the transverse magnetization evolves under the influence of proton-carbon coupling (*J-modulation*) during the first half of the evolution period. At the beginning of the second half (Fig. 2.39), a $180°_x$ pulse refocusses all effects other than J-modulation, such as field inhomogeneities. Synchronously, the decoupler is switched on again, in order to observe a proton-decoupled FID signal with nuclear Overhauser enhancement during the final detection period.

J-modulation is explained by Fig. 2.40: After switching off the proton decoupler, the multiplet components of transverse magnetization dephase with different velocities due to proton-carbon coupling. For CH doublets, two components rotate with frequencies $\pm J/2$ relative to the rotating frame. In the CH_2 triplet case, the central component remains aligned along y', while the peripheral components dephase with frequencies $\pm J$. For CH_3 quartets, two slower but stronger vectors rotate with $\pm J/2$, while two faster but weaker ones dephase with $\pm 3J/2$. The resultant magnetization of each multiplet depends on how long the proton decoupler is switched off, as shown in Fig. 2.40: At time $\tau = (2J)^{-1}$, the resultant is zero for all CH_n multiplets; at time $\tau = (J)^{-1}$, the vector sum is negative for CH_3 and CH but positive for CH_2. The transverse magnetization of non-protonated carbons, however, remains constant. Neglecting attenuations due to relaxation, the resultant transverse magnetizations and the responding signal intensities follow cosine functions, as drawn in Fig. 2.41 and as verified in Fig. 2.42(a) for the aliphatic carbon nuclei of D-camphor. Briefly summarized, J_{CH}-modulation converts J_{CH}

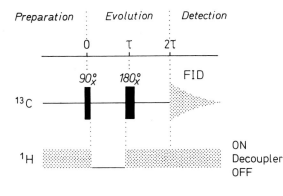

Fig. 2.39. Pulse sequence of the J-modulated spin-echo experiment [50a]. In "SEFT" experiments [50b], the proton decoupler is switched off during the second half of the evolution period.

76 2 Instrumental Methods of ^{13}C NMR Spectroscopy

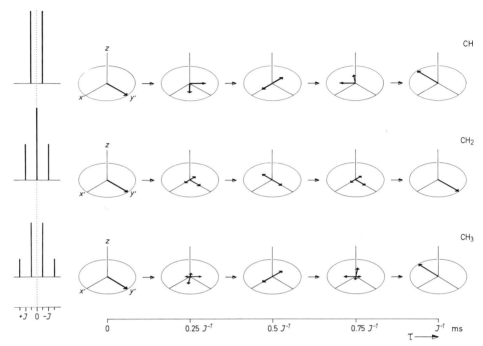

Fig. 2.40. J-Modulation: Components of transverse CH, CH$_2$, and CH$_3$ magnetizations, depending on the coupling delay τ.

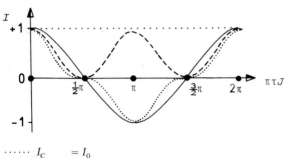

$\cdots\cdots\ I_C\ =\ I_0$
$\text{———}\ I_{CH}\ =\ I_0 \cos(\pi\tau J)$
$\text{----}\ I_{CH_2}\ =\ I_0 [1 + \cos(2\pi\tau J)]$
$\cdots\cdots\ I_{CH_3}\ =\ I_0 [3\cos(\pi\tau J) + \cos(3\pi\tau J)]$

Fig. 2.41. J-Modulation: Signal intensities of C, CH, CH$_2$, and CH$_3$ carbon nuclei, depending on the coupling delay τ.

coupling and *multiplicity* to *intensity* and *phase* information which is observed in the decoupled spectrum.

Figs. 2.41 and 2.42(a) make clear how the switch-off delay τ of the proton decoupler has to be adjusted in order to sort CH$_n$ multiplets: For $\tau = (2J)^{-1}$, only quaternary carbon atoms give rise to intense positive signals (Fig. 2.42(c)). For $\tau = (J)^{-1}$, non-protonated and CH$_2$ carbon nuclei appear positive in contrast to negative CH and CH$_3$

2.9 From the First to the Second Dimension of ^{13}C NMR Spectroscopy

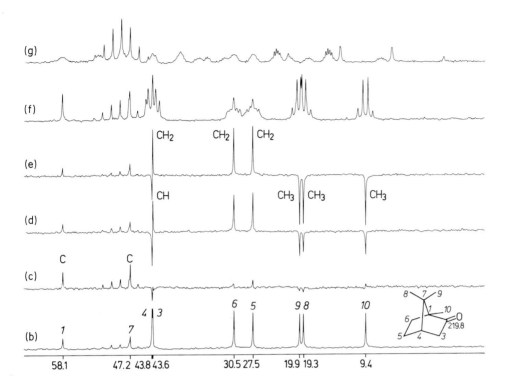

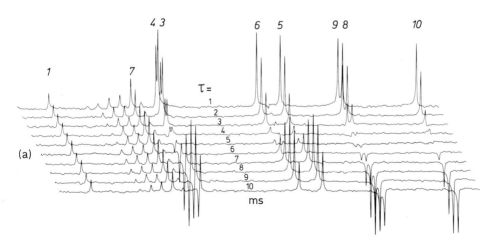

Fig. 2.42. ^{13}C NMR spectra of D-camphor in tetradeuteriomethanol at 15.08 MHz; (a) J-modulation of aliphatic carbon signals depending on the decoupling delay τ, a verification of Fig. 2.41; (b) proton broadband decoupled spectrum; (c–e) J-modulated spin-echo experiments with $\tau = 4, 6$, and 8 ms for CH multiplicity analysis; (f–g) spectra with off-resonance (f) and gated decoupling of protons (g) for comparison.

signals (Fig. 2.42(e)); an additional experiment with $\tau = 3(J)^{-1}/4$ may help to distinguish weaker methyl from stronger methine signals (Fig. 2.42 d)).

When compared with proton off-resonance decoupling (Section 2.6.5.), J-modulated spin-echo experiments permit more accurate CH_n multiplicity analysis. Fig. 2.42 provides a good example: The "*JMSE*" experiment (d) clearly identifies the carbon signals at 43.8 and 43.6 ppm as a CH with negative and a CH_2 group with positive amplitude, while the corresponding multiplets in the off-resonance and gated decoupled spectra cannot be analyzed unequivocally due to overlapping.

The optimum switch-off delays τ (Fig. 2.39) have to be adapted to the individual CH coupling constants, mostly spanning from 125 Hz for alkyl groups ($\tau = (J)^{-1} = 8$ ms) to 175 Hz for alkene and aromatic CH bonds ($\tau = (J)^{-1} = 6$ ms). But values of up to 250 Hz for alkynyl groups ($\tau = (J)^{-1} = 4$ ms) are also possible. To conclude, a JMSE experiment may be used to sort out CH_n multiplets with similar CH coupling constants, such as the aliphatic carbon nuclei of terpenes (Fig. 2.42) or steroids. However, the differentiation between CH and CH_3 groups, as shown in Fig. 2.42 (d), remains unsatisfactory, and the delays in the preparation period between successive experiments must be adjusted to permit carbon-13 relaxation ($\sim 3\, T_{1_{max}}$). Thus, FID accumulation for signal:noise improvement, which is necessary for less concentrated samples, becomes time consuming. This problem has been solved by application of initial pulse angles much smaller than $90°_x$ and an additional refocussing sequence Δ, $180°_x$, Δ, where $\Delta < \tau$, as is verified in the attached proton test modificaton [50 c] of the J-modulated spin-echo.

2.9.3 Polarization Transfer

The sensitivity of an atomic nucleus in the NMR experiment is related to its gyromagnetic ratio γ. It determines the energy difference ΔE between the precession states in a magnetic field of flux density B_0 (Figs. 2.1 and 2.35):

$$\Delta E = \frac{\gamma h B_0}{2\pi}.$$

The populations N_e and N_g of the excited (e) and the ground state (g) follow a Boltzmann distribution:

$$\frac{N_e}{N_g} = e^{-\Delta E/kT} = 1 - \frac{\gamma h B_0}{2\pi k T}.$$

As nuclear magnet, carbon-13 is about four times weaker than the proton ($\gamma_H \cong 4\gamma_C$). The population difference between the two ^{13}C precession states will be correspondingly smaller, so that carbon-13 nuclei are much *less polarized* by a magnetic field when compared with protons. Thus, the resultant carbon-13 magnetization and the observed ^{13}C signal will be weaker. A transfer of strong proton magnetization to weakly polarized nuclei, such as carbon-13, is the concept behind various methods of ^{13}C signal enhancement known as *polarization* or *population transfer*.

2.9 From the First to the Second Dimension of ^{13}C NMR Spectroscopy

2.9.3.1 Selective Polarization Transfer

Consider the $^{13}C-^{1}H$ bond as a two-spin system. CH coupling occurs between one nucleus with small population difference (^{13}C) and another one with large polarization (^{1}H). Fig. 2.43(a) illustrates this situation by the number of dots on the energy levels. Population inversion of the proton levels 1 and 3 connected by the transition $^{1}H_2$ is achieved by an appropriate $180°_x$ pulse, which turns the double cone of precession shown in Fig. 2.1 upside down. Thereafter, the inverted proton population difference controls both carbon-13 transitions (Fig. 2.43(b)). This is the *polarization* or *population transfer* making up an enhanced absorption signal for one transition (e.g. $^{13}C_1$ in Fig. 2.43(b)) and an enhanced emission on the other (e.g. $^{13}C_2$ in Fig. 2.43(b)).

Selective population or polarization transfer (abbreviation: *SPT*) was first achieved for chloroform [51] by brief irradiation of one of both ^{13}C satellites in the proton NMR spectrum before recording the $^{13}C-^{1}H$ doublet as FID signal. In this experiment, the decoupling pulse inverts the population of the proton precession states connected by the irradiated satellite transition (Fig. 2.43(b, c)). The two SPT experiments shown for the $^{13}C-H$ doublet of chloroform in Fig. 2.43(b, c) give an impression of the signal enhancement achievable by population transfer in comparison to a normal ^{13}C NMR spectrum

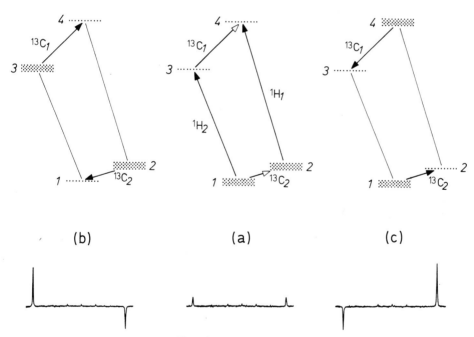

Fig. 2.43. *Top:* Energy levels of a $^{13}C-^{1}H$ pair and their populations according to a Boltzmann distribution (a) and after population inversion of levels *1* and *3* (b) and *2* and *4* (c), respectively. Differences of ^{13}C populations are neglected. *Bottom:* $^{13}C-^{1}H$ doublet of chloroform with Boltzmann distribution of nuclei (a) and after selective population inversion according to (b) and (c) [51].

Table 2.1. Intensities of CH_n multiplets in Coupled ^{13}C NMR Spectra with Boltzmann Distribution of Nuclei (a) and after Selective Polarization Transfer (b).

		(a)				(b)			
C	(singlet)	1				1			
CH	(doublet)	1	1			−3	5		
CH_2	(triplet)	1	2	1		−7	2	9	
CH_3	(quartet)	1	3	3	1	−11	−9	15	13

without proton decoupling (Fig. 2.43(a)). Enhancements of the usual CH_n multiplets are summarized in Table 2.1.

SPT experiments are valuable aids for signal assignments [52], for the determination of the signs of coupling constants [53] and, particularly, for the observation of insensitive nuclei. Their application to larger molecules is limited, however, as the ^{13}C satellites of all proton signals should be known, and because only one satellite transition is inverted in a single experiment. A non-selective population transfer enhancing the signals of all CH_n carbon atoms by one experiment would be more desirable.

2.9.3.2 Non-Selective Polarization Transfer

Pulse sequences for non-selective polarization transfer, not only useful for signal enhancement but also for multiplicity selection, are referred to as "*INEPT*" [54], abbreviated from "Insensitive Nuclei Enhanced by Polarization Transfer". An improved method denoted as "Distortionless Enhancement by Polarization Transfer" or "*DEPT*" [55] permits the cleanest multiplicity selection known so far, with full enhancement and low sensitivity to individual CH coupling constants. In addition, fully enhanced and undistorted coupled spectra can be recorded. Finally, subspectra for CH, CH_2 and CH_3 groups can be generated.

Fig. 2.44 illustrates the influence of the DEPT sequence on a $^{13}C - {}^1H$ two-spin system. A 90_x° pulse in the range of proton shifts initiates the experiment (a). Thereupon, the resulting transverse proton magnetization is modulated due to CH coupling until, at time $\tau = 1/2 J_{CH}$, the CH doublet vectors have attained a phase difference of 180° (b). At this moment, a 90_x° pulse in the carbon shift range rotates the vectors of carbon magnetization into the xy plane (c), while a 180_x° pulse in the proton channel causes refocussing of inhomogeneity effects (b). Now, z magnetization is zero for both protons and carbons; in fact the two nuclei are decoupled. Therefore, the vectors of ^1H and ^{13}C magnetization remain standing in the subsequent time period $\tau = 1/2 J_{CH}$ (d and e). At the end of this second τ period, a $\Theta_y = 90_y^\circ$ pulse in the proton shift range rotates both doublet components of proton magnetization to the $\pm z$ axis (f). This is a polarization transfer because the polarization of protons induced by the Θ_y pulse also controls the ^{13}C population due to $^{13}C - {}^1H$ coupling. Subsequently, the enhanced $^{13}C - {}^1H$ doublet vectors rotate about the $\pm z$ axis due to the coupling protons, refocussing along the x' axis after the third period of $\tau = 1/2 J_{CH}$ (h). A 180_x° pulse in the carbon-13 channel at the beginning of the

2.9 From the First to the Second Dimension of ^{13}C NMR Spectroscopy

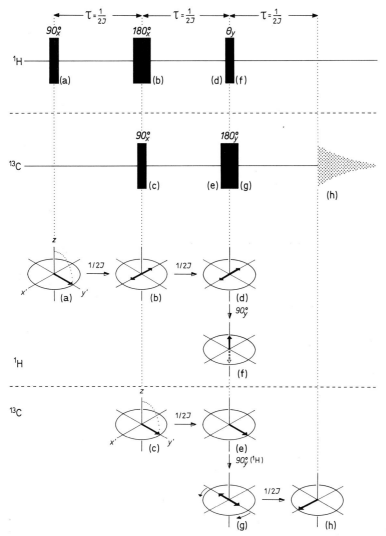

Fig. 2.44. The DEPT experiment for a ^{13}C-^{1}H doublet; pulse sequences in the proton and carbon-13 channel (a–g) and the motion of proton and carbon-13 magnetization, controlled by the pulses and by J-modulation (a–h).

third τ period (g) cancels inhomogeneity effects on ^{13}C magnetization, after which the FID signal is finally recorded and Fourier transformed.

The extent of polarization transfer is determined by the pulse angle Θ_y, giving a maximum with $\Theta_y = 90°$ for CH doublets, as is illustrated in Fig. 2.44. Under the same conditions, the resultant x' magnetization is zero for CH_2 and CH_3 carbon signals (Fig. 2.45) [56]. Further, the dependence of the intensities on the polarization transfer angle Θ_y varies for each CH_n signal. Fig. 2.45 gives an experimental verification of the different sine modulations of CH, CH_2 and CH_3 carbon signals induced upon variation

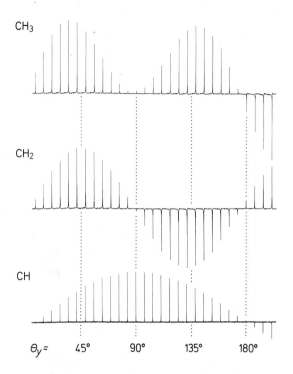

Fig. 2.45. Signal intensities of CH, CH$_2$ and CH$_3$ groups as functions of the polarization transfer angle Θ_y. The proton-decoupled signals of C-9 (CH), C-1 (CH$_2$) and C-18 (CH$_3$) in cholesteryl acetate (100 mg/mL in deuteriochloroform) are observed at 100.6 MHz [56].

of the polarization transfer angle Θ_y. To conclude, the DEPT sequence provides a clean method of CH$_n$ *multiplet analysis with the sensitivity enhanced by polarization transfer*: One experiment with $\Theta_y = 90°$ generates the subspectrum of CH carbon atoms. An additional one with $\Theta_y = 135°$ gives negative CH$_2$ but positive CH and CH$_3$ carbon signals (Fig. 2.45). Differentiation between CH and CH$_3$ is straightforward, as the CH subspectrum is known from the first experiment. Signals observed in the normal proton broadband-decoupled ^{13}C NMR spectrum but not in the DEPT spectra belong to quaternary carbon centers. Calibration of the decoupler pulse Θ_y for 90° is achieved by best possible nulling of all CH$_2$ (and CH$_3$) signals in a reference sample.

DEPT spectra can be detected with and without proton decoupling. Correspondingly, decoupled and coupled spectra with multiplicity selection are observable. Fig. 2.46 illustrates such experiments with (−)-menthol, also providing clear analysis of all CH coupling constants of the molecule.

The insensitivity of the DEPT sequence to different J_{CH} coupling constants, as illustrated in Fig. 2.46, makes it useful for editing ^{13}C NMR spectra. To edit a carbon-13 spectrum, three DEPT experiments for the polarization transfer angles

$$\Theta_1 = 45°, \quad \Theta_2 = 90° \quad \text{and} \quad \Theta_3 = 135°$$

are recorded. For carefully adjusted rf pulses, the Θ_2 experiment provides the CH carbon subspectrum. Methylene and methyl subspectra arise from linear combination of the three experiments; Fig. 2.45 explains why this works.

2.9 From the First to the Second Dimension of ^{13}C NMR Spectroscopy

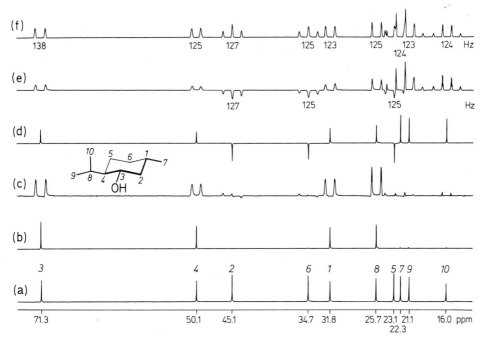

Fig. 2.46. ^{13}C NMR spectra of (−)-menthol (100 mg/mL deuteriochloroform; 100.6 MHz); (a) proton broadband-decoupled spectrum; (b, c) DEPT spectra with $\Theta_y = 90°$ for CH selection with and without proton decoupling; (d, e) DEPT experiments with $\Theta_y = 135°$ for positive CH and CH$_3$ but negative CH$_2$ signals with and without proton decoupling; (f) gated-decoupled spectrum for reference; (a–e) 16 scans; (f) 256 scans.

CH: Θ_2
CH$_2$: $\Theta_1 - \Theta_3$
CH$_3$: $\Theta_1 + \Theta_3 - 0.707\,\Theta_2$

Edited CH$_n$ subspectra generated in this manner are presented in Fig. 2.47 (c–e) for the hexacyclic triterpene anhydropanaxadiol [57]. To conclude, sorting of CH$_n$ multiplets by editing of DEPT spectra is achievable without problems, even in the case of complicated compounds or mixtures with signal overcrowding. A J-modulated spin-echo experiment (JMSE) for selection of quarternary carbon atoms may complete the ananlysis (Fig. 2.47 (b)). When compared with JMSE, DEPT offers at least three advantages: It is insensitive to J_{CH}; the recycling delay depends on faster proton (not on carbon-13) relaxation; this fact, as well as signal enhancements induced by polarization transfer, gives rise to considerable savings of measuring time.

84   2 Instrumental Methods of ^{13}C NMR Spectroscopy

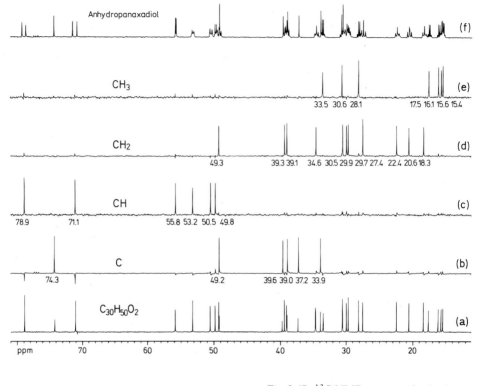

Fig. 2.47. ^{13}C NMR spectra of anhydropanaxadiol (15 mg/mL; deuteriochloroform; 100.6 MHz; (a–e) 400 scans; (f) 1600 scans): (a) proton broadband decoupled; (f) proton off-resonance decoupled; (b) subspectrum of quaternary carbon atoms resulting from a *J*-modulated spin-echo experiment; (c–e) edited CH$_n$ subspectra obtained from DEPT experiments with $\Theta_y = 45°$, 90° and 135° [57].

2.9.4 Measurement of Carbon-Carbon Coupling Constants Without ^{13}C Enrichment: INADEQUATE

If all carbon-carbon bonds of an organic compound are known, the carbon skeleton is defined, and an essential part of structure elucidation is achieved. One method of determining carbon-carbon bonds involves the measurement of ^{13}C–^{13}C coupling constants, since identical couplings of two carbon atoms indicate the presence of bonding between these carbons.

The measurement of a $^{13}C-^{13}C$ coupling constant implies the connection of two ^{13}C nuclei. The natural abundance of carbon-13, however, is 1.1% or about 10^{-2}. Thus, the probability of the $^{13}C-^{13}C$ linkage is 0,011% or about 10^{-4}, and splittings due to $^{13}C-^{13}C$ coupling will only appear as very weak satellites (0.5%) on both sides of the $^{13}C-^{12}C$ central signal. Usually, these satellites are lost in the noise. Even in the case of strong signals, localization of the satellites is difficult, due to overlapping spinning side bands, impurities and the linewidth of the strong central line. Thus, reliable measurements of $^{13}C-^{13}C$ coupling constants in natural abundance are only achievable by effective suppression of the strong $^{13}C-^{12}C$ central signal. This is the idea behind a pulse sequence denoted as *INADEQUATE* [58], abbreviated from „Incredible Natural Abundance Double Quantum Transfer Experiment".

Fig. 2.48 illustrates one variant of the INADEQUATE sequence [58]. Carbon-13 transverse magnetization generated by the initial $90°_x$ pulse (a) results from a strong component S_0 for the $^{13}C-^{12}C$ pairs ($\sim 99\%$) and a very weak one S_2 for the less abundant $^{13}C-^{13}C$ pairs ($\sim 1\%$). When proton decoupling is applied, the strong component S_0 remains standing in the rotating frame, while the weak magnetization S_2 is *J*-modulated due to $^{13}C-^{13}C$ coupling: Both $^{13}C-^{13}C$ doublet components dephase with the frequency of the carbon-carbon coupling constant J_{CC} until, at time $\tau = 1/4 J_{CC}$, they have attained a phase difference of 90° (b). At this moment, a $180°_y$ pulse (c) serves to refocus inhomogeneity effects during the second period of $t = 1/4 J_{CC}$. Thereafter, the $^{13}C-^{13}C$ doublet vectors align along the $\pm x$ axis (d). The $90°_x$ pulse irradiated at this moment rotates the strong magnetization S_0 to the $-z$ axis (e). Thereby, the $^{13}C-^{12}C$ magne-

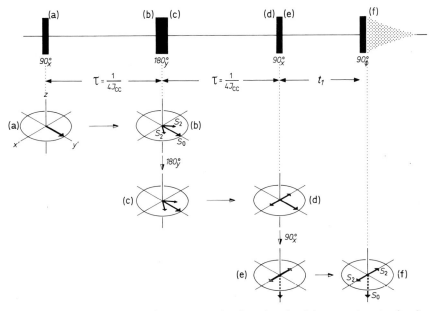

Fig. 2.48. INADEQUATE pulse sequence (a-f) and pulse-driven motion (a-f) of carbon-13 magnetizations S_0 for $^{13}C-^{12}C$ singlet and S_2 for $^{13}C-^{13}C$ doublet signals in the rotating frame. Proton broadband decoupling is applied throughout the experiment.

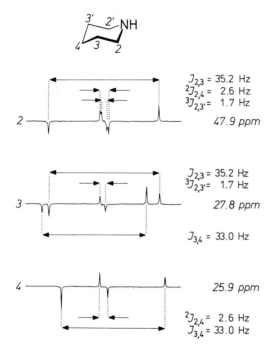

Fig. 2.49. INADEQUATE spectra of the $^{13}C-^{13}C$ doublets of piperidine (neat liquid, 50 MHz), including an analysis of one-, two- and three-bond carbon-carbon coupling constants [58].

tization is transferred to the $^{13}C-^{13}C$ doublet components, as known from polarization transfer experiments. After a short switching time t_1, the S_2 magnetization is monitored by a $90°_\Phi$ pulse (f). The phase Φ of this monitor pulse ($\Phi = \pm x, \pm y$) is varied in several cycles (a–f, at least 32 times), in order to permit coherent addition of the weak $^{13}C-^{13}C$ multiplets and optimum cancellation of the strong central $^{13}C-^{12}C$ signal.

As shown in Fig. 2.48(e), the vectors of $C^{13}-C^{13}$ doublet magnetizations are aligned in opposite directions when irradiated by the $90°_x$ pulse. Thus, in the INADEQUATE spectrum the $C^{13}-C^{13}$ doublet signals will appear with the corresponding antiphase relationship, as shown in Fig. 2.49, which also demonstrates the effective suppression of the strong $^{13}C-^{12}C$ signals of piperidine. Analysis of carbon-carbon coupling constants can be performed easily in this simple case.

The antiphase relationship of the $C^{13}-C^{13}$ doublet signals in the INADEQUATE spectrum can be eliminated by an additional spin-echo sequence ($-1/4\, J_{CC} - 180°_x - 1/4\, J_{CC} -$) before the $90°_\Phi$ monitor pulse [58]. The sensitivity of the experiment may be improved by the application of stronger magnetic fields or by using proton polarization transfer techniques [59].

INADEQUATE involves effective suppression of the strong central signal. The process (d) → (e) in Fig. 2.48, denoted as *double quantum transfer*, does not at all enhance the $^{13}C-^{13}C$ satellites. Their signal:noise ratio remains at 0.5% of the normal carbon-13 NMR sensitivity.

The INADEQUATE spectrum of a molecule collects all AB or AX systems due to carbon-carbon coupling. Identical $^{13}C-^{13}C$ couplings localize carbon-carbon bonds, so that the carbon skeleton of the sample molecule can be derived. Complete analysis of an

INADEQUATE spectrum, however, may become difficult, even in the case of smaller molecules: At a tertiary carbon for example, three doublets will appear, usually overlapping due to similar coupling constants. Further, isotope shifts and AB effects will remove the $^{13}C-^{13}C$ satellites from the shift position of the known $^{13}C-^{12}C$ signal. Thus, it was not until a second dimension was introduced to the experiment – that INADEQUATE became a practical method of structure elucidation.

2.10 The Second Dimension

2.10.1 Basic Concept

Preparation, *evolution* and *detection* are the time periods of pulse sequences described for J-modulated spin-echo and polarization transfer experiments. The basic FT NMR technique operates without any evolution period t_1: immediately after generation of transverse magnetization (preparation) its free induction decay $S(t_2)$ is detected. Subsequent Fourier transformation provides the FT NMR spectrum $S(\delta)$.

A *constant evolution period* t_1 is the new feature of J-modulated spin-echo, DEPT, and INADEQUATE sequences. During this time period, a J-modulation or a polarization transfer may evolve. Such pulse sequences provide FID signals $S(t_2)$ which are still functions of one variable time t_2. The Fourier transforms, however, are NMR spectra with specific information, depending on the constant evolution period t_1. One simple example is the generation of the quaternary carbon-13 subspectrum by means of a J-modulated spin-echo experiment with an evolution time of $t_1/2 = \tau = 1/2\,J_{CH}$, as in Fig. 2.42(c) or Fig. 2.47(b).

It is a *varying evolution time* t_1 which brings the second dimension into the NMR experiment, provided this second variable time t_1 induces *periodic changes of the signal amplitude or the signal phase*. In the J-modulated spin-echo experiment (Fig. 2.42(a)), for example, the switch-off delay $\tau = \frac{1}{2}t_1$ of the proton decoupler can be increased stepwise by a constant increment Δt_1 (e.g. 1 s in Fig. 2.42(a)) in a series of k subsequent single experiments. Thereby, the FID signals will become functions of two variable times, t_1 and t_2:

    FID signal $S = f(t_1, t_2)$.
    Parameters included J_{CH}, δ.

As usual, t_2 is related to the chemical shift δ; t_1 permits evolution of J-modulation, in this case acting as an information carrier of the coupling constant J_{CH}. Fourier transformations of these FID signals provide a series of NMR spectra in which the signal amplitudes periodically change due to J-modulation, following the cosine functions as drawn in Fig. 2.41. This kind of amplitude modulation was illustrated for the CH_n carbon atoms of camphor in Fig. 2.42(a): As usual, the carbon-13 shift defines the position of the signal, but the signal amplitude periodically changes with the evolution time t_1.

    First Fourier transformation $S' = f(t_1, \delta)$.

2 Instrumental Methods of ^{13}C NMR Spectroscopy

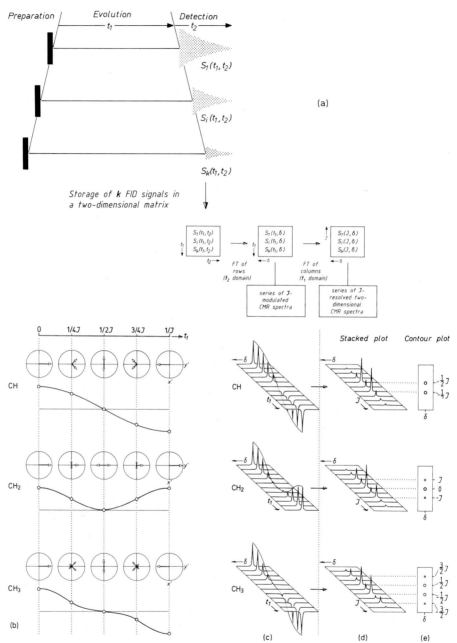

Fig. 2.50. Generation of J-resolved two-dimensional ^{13}C NMR spectra: (a) flux diagram; (b) J-modulation of CH doublets, CH$_2$ triplets and CH$_3$ quartets during evolution, vector diagrams in the $x'y'$ plane and cosine curves described by the signal maxima; (c) series of ^{13}C NMR spectra of CH$_n$ groups with t_1 dependent J-modulation of the signal amplitudes; (d) series of J-resolved two-dimensional ^{13}C NMR spectra for methine, methylene and methyl carbon atoms, stacked plots and contour plots (e).

It is the coupling constant J_{CH} which modulates the signal amplitude (Fig. 2.41). Thus, in order to recover the carbon-proton multiplets as the spectra of modulation frequencies J_{CH}, a second Fourier transformation in the time domain t_1 has to be performed.

Second Fourier transformation $S'' = f(J_{CH}, \delta)$.

The concept behind this two-dimensional ^{13}C NMR experiment is summarized in Fig. 2.50(a): A series of k FID signals with successively increasing evolution time t_1 is detected. The signals are digitized and stored on disk in a two-dimensional matrix limited by the time domain axes t_2 and t_1. If 256 FID signals, each with 1024 ($=$ 1 K) data points, are acquired, the matrix includes 256 rows and 1024 columns. The rows extend along t_2 (detection), the columns correspond to the t_1 domain (evolution). Fourier transformation in the t_2 domain (along the rows) provides a series of spectra $S'(t_1, \delta)$ with J-modulated signal amplitudes (Fig. 2.50(c)) due to J-modulation during the evolution period (Fig. 2.50(b)). The signal maxima describe the cosine curves characteristic of CH, CH_2 and CH_3, as drawn in Fig. 2.41. A second Fourier transformation in the t_1 domain (along the columns) generates a series of spectra $S''(J_{CH}, \delta)$ with maxima at the positions of the carbon-proton multiplets (Fig. 2.50(d)). Together, both Fourier transformations produce a new matrix limited by two frequency axes, J_{CH} and δ: Signals with different carbon-13 shifts appear on the rows (δ axis); multiplets belonging to the individual shifts are found on the corresponding columns (J_{CH} axis).

The two-dimensional matrix can be recorded by an analog plotter, or much faster by a digital plotter or a graphic display. *Stacked plots* as drawn in Fig. 2.50(d) can be obtained, providing a panoramic picture of the carbon-13 signals (δ) and their individual multiplets (J_{CH}). Useful for practical evaluation, the *contour plot* is like a contour map of the signal mountains, giving a type of aerial view of the CH_n multiplets (Fig. 2.50(e)).

2.10.2 J-Resolved Two-Dimensional ^{13}C NMR Spectra

The series of spectra drawn in Fig. 2.50(d, e) for CH, CH_2 and CH_3 groups are referred to as *J-resolved two-dimensional ^{13}C NMR spectra*. In this experiment, carbon-proton coupling is separated from carbon-13 shift information by a second dimension. The most convenient pulse sequence for obtaining J-resolved two-dimensional ^{13}C NMR spectra is similiar to that known from the J-modulated spin-echo or SEFT experiment (Section 2.9.2, Fig. 2.39). However, a *variable evolution time t_1* is used for J-modulation [60]. Fig. 2.51 outlines the effect of this sequence on a ^{13}C–^{1}H doublet pair. During the first half-period $t_1/2$ of the evolution time (Fig. 2.51(a–b)), proton decoupling is switched on. Usually, the vector of transverse CH carbon magnetization to be observed rotates faster or slower relative to the rotating frame due to its chemical shift. Thus, within the first half-period a faster vector will dephase clockwise by a certain angle φ from the y' axis (Fig. 2.51(b)). Thereafter, a $180°_x$ pulse rotates the magnetization about x', so that the effects of inhomogeneity are refocussed at the end of the evolution period t_1. When proton decoupling is switched off during the second half-period of the evolution, different CH multiplet components will precess away from the vector localized by the carbon-13 shift. These diverging components will not refocus after the second half-time $t_1/2$, when proton decoupling is switched on again in order to observe decoupled ^{13}C signals

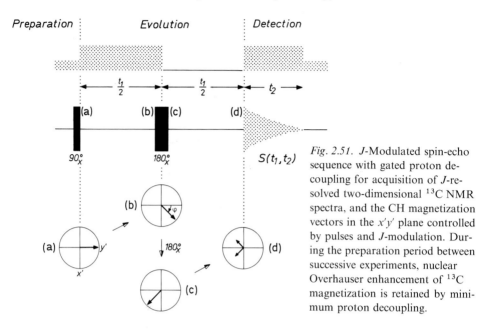

Fig. 2.51. J-Modulated spin-echo sequence with gated proton decoupling for acquisition of J-resolved two-dimensional ^{13}C NMR spectra, and the CH magnetization vectors in the $x'y'$ plane controlled by pulses and J-modulation. During the preparation period between successive experiments, nuclear Overhauser enhancement of ^{13}C magnetization is retained by minimum proton decoupling.

(Fig. 2.51 (d)). As a result, only the *vector sum of all multiplet components will be detected*, and the signal amplitudes will be J-modulated by carbon-proton coupling as t_1 is varied.

Since J_{CH} coupling is only in operation during one $t_1/2$ period in the J-modulated spin-echo sequence with gated proton decoupling, as drawn in Fig. 2.51, the second Fourier transformation provides only one half of the actual J_{CH} magnitudes. Therefore, resolution of smaller couplings will be poor.

J-resolved two-dimensional carbon-13 NMR spectra [60, 61] separate chemical shifts and coupling constants in the two-dimensional J_{CH}, δ_C plane. Their practical value can be estimated on looking at Fig. 2.52. In the one-dimensional ^{13}C NMR spectrum of

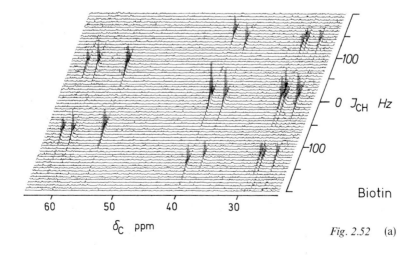

Fig. 2.52 (a)

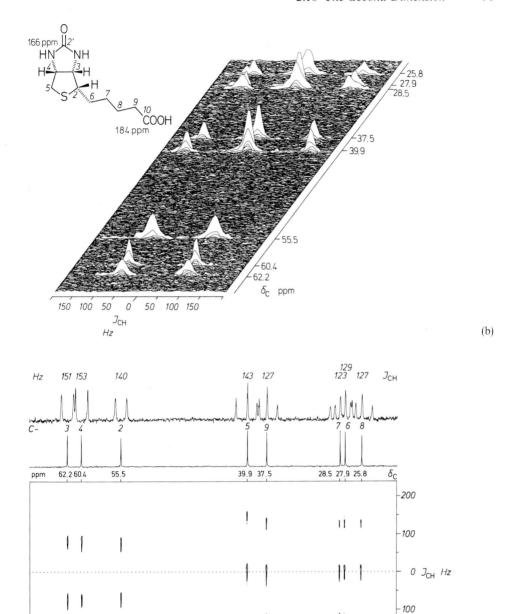

Fig. 2.52. *J*-Resolved two-dimensional ^{13}C NMR spectra of biotin (100.6 MHz for ^{13}C, 400.1 MHz for ^{1}H; 40 mg in 0.1 mol/L aqueous sodium hydroxide; measuring time: 25 min; transform time: 12.5 min); stacked plot before (a) and after rotation (b) of the matrix by 90° [62]; (c) contour plot with one-dimensional ^{13}C NMR spectra recorded with and without proton decoupling.

biotin crowding of signals between 25 and 30 ppm does not permit straightforward analysis of CH multiplets at first glance. The contour plot of the two-dimensional J_{CH}, δ_C matrix (Fig. 2.52(c)), however, clearly identifies three CH_2 triplets. A stacked plot is recorded for illustration in Fig. 2.52(a). After rotation of the matrix by 90° another relief plot is obtained, presenting the CH_n multiplet subspectra along the horizontal axis (Fig. 2.52(b)).

2.10.3 Two-Dimensional Carbon-Proton Shift Correlation

The ^{13}C signal of a particular carbon atom can be assigned by selective decoupling of the proton(s) attached to it (Section 2.6.7). This kind of one-dimensional carbon-proton shift correlation does not give unequivocal results in the case of crowded or overlapping proton signals. Moreover, differently shielded AB protons of a methylene group cannot be decoupled simultaneously, so that residual off-resonance splittings will remain for the methylene carbon signal. A *two-dimensional carbon-proton correlation with carbon-13 and proton shifts as frequency axes* would be the method of choice: Overlapping proton signals δ_H could be separated by the larger dispersion of carbon-13 shifts δ_C in the second dimension.

J-resolved two-dimensional ^{13}C NMR spectra arise from modulation of ^{13}C signal amplitudes by the J_{CH} coupling constant as modulation frequency (Section 2.10.2). Coupling information is recovered in a second dimension by Fourier transformation in the second time domain. Thus, a two-dimensional carbon-proton shift correlation will be based on modulation of carbon-13 signals by the Larmor frequencies of the attached protons. Fig. 2.53 outlines one pulse sequence suitable for data acquisition in order to obtain a two-dimensional heteronuclear shift correlation [63], using the simplest case of $^{13}C-^1H$ doublet magnetization for illustration.

A $90°_x$ pulse in the range of proton shifts initiates the experiment. It rotates the magnetization of both ^{13}C satellites to the y' axis (Fig. 2.53(b)). Thereafter, both components 1 and 2 will dephase in the $x'y'$ plane, due to coupling with the attached ^{13}C nuclei (Fig. 2.53(c)). The J_{CH} coupling constant will define the *dephasing angle* Φ. Additionally, both components will precess relative to the rotating frame with a frequency determined by the proton shielding δ_H (Fig. 2.53(c)), giving rise to the *phase shift* φ. A $180°_x$ pulse in the carbon-13 channel irradiated after the first half-period of the evolution time t_1 inverts the population of carbon-13 energy levels. As a result, the satellite components 1 and 2 interchange their coupling frequencies (Fig. 2.53(d)), in order to refocus at the end of the evolution period t_1 (Fig. 2.53(e)). Thus, the variable t_1 permits evolution of proton shift information with carbon-proton coupling removed.

A constant *mixing period* Δ follows evolution (Fig. 2.53). Its purpose is to transfer proton chemical shift information, as given by the phase angle φ, to carbon-13 magnetization. A polarization transfer takes place. Therefore, the mixing time Δ is adjusted to an average value of all possible J_{CH} coupling constants (usually 145 Hz), so that the half-periods will be

$$\Delta_1 = \Delta_2 = 1/2\, J_{CH}.$$

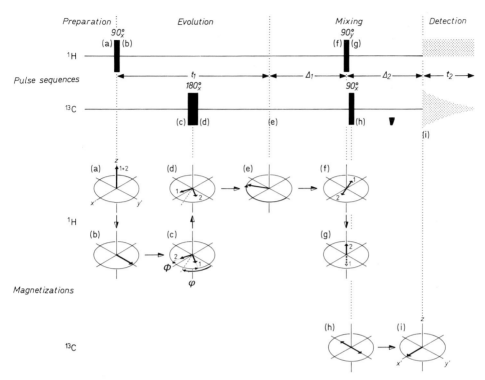

Fig. 2.53. Pulse sequences in the proton and carbon-13 channels and $^{13}C-^{1}H$ doublet vectors controlled by pulses and J-modulation for acquisition of FID signals in a two-dimensional carbon-proton shift correlation [63].

After the first half-period Δ_1 of the mixing time (Fig. 2.53(f)), the two doublet components 1 and 2 have precessed away from each other by 180°, due to coupling with the attached carbon-13 nuclei. In this situation, they are rotated to the $\pm z$ axis by a 90_y° pulse in the proton channel (Fig. 2.53(g)). This is the transfer of proton magnetization with shift information to the coupling carbon-13 nuclei. Subsequently, the enhanced ^{13}C doublet components are rotated to the $\pm y'$ axis by a 90_x° pulse in the carbon-13 shift range (Fig. 2.53(h)). Due to their coupling with the attached protons, they will refocus along x' at the end of the mixing period (Fig. 2.53(i)). After switching on the proton decoupler, an FID signal enhanced by polarization transfer is detected for a CH singlet.

It is an essential feature of the sequence (Fig. 2.53) that only the x' components of the doublet vectors 1 and 2 will be rotated by the 90_y° pulse in (f). The magnitudes of these components depend on the phase angle φ, which is related to the proton shift δ_H. To conclude, the extent of polarization transfer (Fig. 2.53(f-g)) is a function of proton chemical shift. After the first Fourier transformation in the t_2 domain, carbon-13 signals with modulated amplitudes will be obtained when t_1 is varied. Chemical shifts of the attached protons are the modulation frequencies. Therefore, a *second Fourier transformation* in the t_1 domain provides maximum signals located at the chemical shifts δ_H and δ_C of the coupling proton and carbon-13 nuclei.

The data flux of a two-dimensional carbon-proton shift correlation is similiar to that described in Fig. 2.50(a) for a J-resolved 2D CMR experiment, with one difference: Instead of carbon-proton couplings J_{CH}, proton chemical shifts δ_H are stored in the evolution time t_1. Fourier transformation in the t_2 domain thus yields a series of NMR spectra with carbon-13 signals modulated by the attached proton Larmor frequencies. A second Fourier transformation in the t_1 domain generates the δ_H, δ_C matrix of a two-dimensional carbon-proton correlation.

$$S(t_1, t_2) \xrightarrow{\text{1st FT}(t_2)} S'(t_1, \delta_C) \xrightarrow{\text{2nd FT}(t_1)} S''(\delta_H, \delta_C).$$

A series of ^{13}C NMR spectra with selective decoupling of individual protons is the known alternative to a two-dimensional carbon-proton shift correlation. The number of single proton decoupling experiments increases, however, with the number of CH bonds to be identified. As a result, full analysis of larger molecules by selective proton decoupling becomes time consuming. Further, doubtful results will arise for crowded or AB-type proton signals. In comparison, a two-dimensional carbon-proton shift correlation is unequivocal, more efficient and less time consuming. The clarity stems from resolution of overlapping proton resonances by the larger carbon-13 shifts in the second dimension. The efficiency results from the fact that one two-dimensional experiment identifies all carbon-proton connectivities. Finally, the measuring time (or the sensitivity) benefits from polarization transfer.

Fig. 2.54 presents a two-dimensional carbon-proton shift correlation of D-lactose after mutarotational equilibration (40% α-, 60% β-D-lactose in deuterium oxide), demonstrating the good resolution of overlapping proton resonances between 3.6 and 4 ppm by means of the larger frequency dispersion of carbon-13 shifts in the second dimension. The assignment known for one nucleus – carbon-13 in this case – can be used to analyze the crowded resonances of the other nucleus. This is the significance of the two-dimensional CH shift correlation, in addition to the identification of CH bonds. For practical evaluation, the contour plot shown in Fig. 2.54(b) proves to be more useful than the stacked representation (Fig. 2.54(a)). In the case of D-lactose, selective proton decoupling between 3.6 and 4 ppm would not afford results of similiar quality.

A two-dimensional CH shift correlation also reveals the fine structure of proton signals, provided that resolution in the proton shift dimension is good enough. In the contour plot of an experiment recorded for 5β-androstan-3,17-dione (Fig. 2.55), an AB system with $\delta_A = 2.02$ and $\delta_B = 2.69$ ppm is found for the protons of the methylene group in position 4 with carbon-13 resonance at $\delta_C = 42.4$ ppm. The equatorial proton A exhibits a doublet due to *geminal* coupling ($^2J_{HH}$) with the axial proton B which splits into a doublet of doublets (almost a triplet) due to an additional large *vicinal* coupling ($^3J_{HH}$) to the axial proton in position 5. Thus, the relative configuration of protons within the C-4-C-5 substructure of the molecule is exposed by the experiment. Complete correlation of carbon-13 and proton shifts, as achieved from the contour plot (Fig. 2.55), is summarized in Table 2.2. Proton shifts were correlated to the known assignments of carbon-13 shieldings [65].

Useful modifications of the basic experiment include CH correlations with proton-proton couplings removed in the t_1 (δ_H) domain [66] and long-range carbon-proton shift correlations with the mixing period in Fig. 2.53 adjusted to an average value of $^2J_{CH}$ and

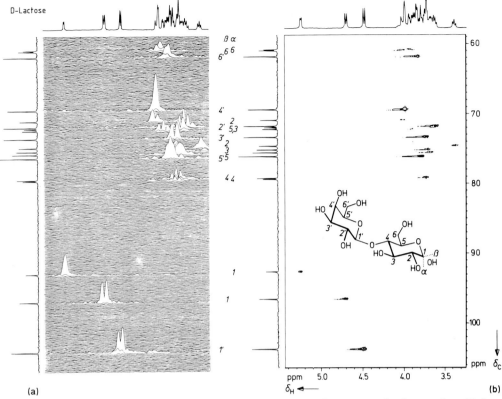

Fig. 2.54. Two-dimensional carbon-proton shift correlation of mutarotated D-lactose (1 mol/L in deuterium oxide; ^{13}C: 100.6 MHz; ^{1}H: 400.1 MHz; measuring time 90 min; transform time: 25 min); (a) stacked plot [64]; (b) contour plot.

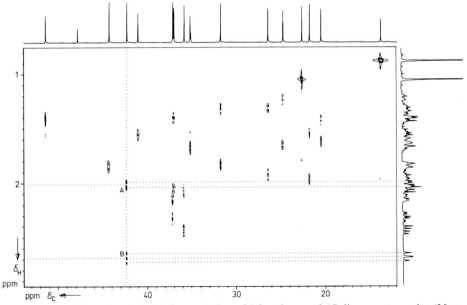

Fig. 2.55. Two-dimensional CH shift correlation of 5β-androstan-3,17-dione contour plot (25 mg in 0.4 mL of deuteriochloroform at 20 °C; measuring conditions as in Fig. 2.54).

96 2 Instrumental Methods of ^{13}C NMR Spectroscopy

Table 2.2. Correlation Table of Carbon-13 and Proton Shifts in 5β-Androstan-3,17-dione, derived from Fig. 2.55.

δ_C ppm			δ_H ppm		
C-1	CH$_2$	37.3	2.15; 2.32	(AB)	
2	CH$_2$	37.1	1.39; 2.05	(AB)	
3	C=O	209.5			
4	CH$_2$	42.4	2.02; 2.69	(AB)	
5	CH	44.2	1.84		
6	CH$_2$	26.5	1.30; 1.83	(AB)	
7	CH$_2$	31.9	1.30; 1.82	(AB)	
8	CH	35.2	1.67		
9	CH	51.3	1.41		
10	C	35.1			
11	CH$_2$	20.5	1.42; 1.60	(AB)	
12	CH$_2$	24.8	1.23; 1.65	(AB)	
13	C	47.8			
14	CH	41.0	1.55		
15	CH$_2$	21.8	1.54; 1.95	(AB)	
16	CH$_2$	35.9	2.08; 2.44	(AB)	
17	C=O	219.6			
18	CH$_3$	13.8	0.86		
19	CH$_3$	22.6	1.05		

$^3J_{CH}$ (5–10 Hz) [67]. The latter also localizes protons two or three bonds away from quarternary carbon centers, so that the investigation of connectivities *via* longer-range carbon-proton interactions becomes feasible. An improved sequence denoted as "*COLOC*" for *correlation via long-range coupling* inserts the evolution time into the mixing period in order to reduce relaxational attenuation of signal intensity [67b]. An example is shown in Fig. 5.17.

2.10.4 Two-Dimensional Proton-Proton Shift Correlation: The COSY Experiment

Key experiments useful for substructure determination by NMR include the DEPT sequence (*e.g.* Figs. 2.44–2.46) for analysis of CH$_n$ multiplicities, as well as the two-dimensional CH correlation for identification of all CH bonds (*e.g.* Fig. 2.55 and Table 2.2) and localization of individual proton shifts. If, in addition, vicinal and longer-range proton-proton coupling relationships are known, all CH substructures of the sample molecule can be derived. Classical identification of homonuclear proton coupling relationships involves homonuclear proton decoupling. A *two-dimensional proton-proton shift correlation* would be an alternative and the complementary experiment to carbon-proton shift correlation. Several methods exist [68]. Of those, the *COSY* sequence abbreviated from *Correlation spectroscopy* [69] is illustrated in Fig. 2.56.

2.10 The Second Dimension

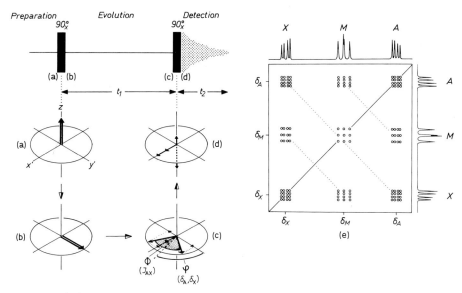

Fig. 2.56. (a–d) Pulse sequence of the COSY experiment for two-dimensional homonuclear proton shift correlation and doublet magnetization of nucleus A in an AX two-spin system [69]. The mixing pulse can also be 45°_x or less (COSY-45) in order to discriminate diagonal signals. (e) Schematic representation of a contour plot obtained from a COSY experiment. Fine structures of AMX multiplets can be recognized provided digital resolution (Hz/data point) is good enough.

Consider the A doublet of an AX system in the proton NMR spectrum. The first 90°_x pulse rotates both doublet magnetization vectors to the y' axis (Fig. 2.56(a–b)). Because of AX coupling, the doublet components of transverse magnetization precess away from each other in the $x'y'$ plane. The *spreading angle* Φ depends on the coupling constant J_{AX}. In addition, both vectors dephase relative to the rotating frame due to the chemical shift δ_A of the observed nucleus A. The *phase angle* φ thus depends on δ_A (Fig. 2.56(c)). The second 90°_x pulse – referred to as the *mixing pulse* – rotates the δ_A-dependent y' components of the doublet magnetizations to the $\pm z$ axis (Fig. 2.56(d)). Doublet magnetization components of the X protons are treated correspondingly. In conclusion, the mixing pulse transfers δ_A-dependent magnetization to nucleus X and δ_X-dependent magnetization to nucleus A. The extent of this *magnetization exchange* depends on the evolution time t_1, which is varied in a series of k single experiments. The FID stored for each single experiment thus includes A and X signals which are both modulated by the proton shieldings δ_X and δ_A of the coupling nucleus. Therefore, two Fourier transformations in the t_2 and the t_1 domains generate four signal maxima (due to $2^2 = 4$) for an AX two-spin system with chemical shifts δ_A and δ_X, neglecting multiplet fine structure at this point. The coordinates of these maxima correspond to the

diagonal signals $(\delta_A, \delta_A); (\delta_X, \delta_X)$ and the
cross signals $\quad (\delta_A, \delta_X); (\delta_X, \delta_A)$

As both frequency domains are proton chemical shifts, the two-dimensional COSY matrix is quadratic (Fig. 2.56(e)). Within the square, the one-dimensional proton NMR

98   2 Instrumental Methods of ^{13}C NMR Spectroscopy

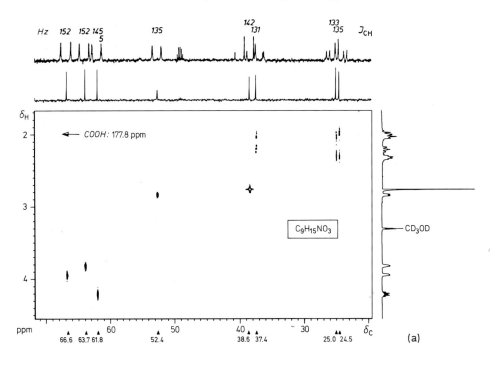

(a)

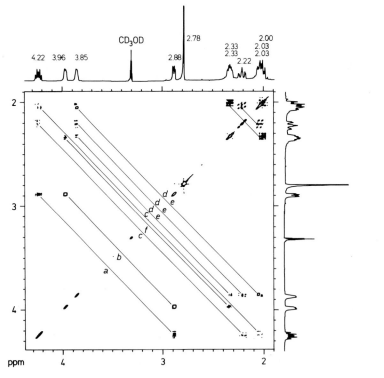

(b)

spectrum is projected on the diagonal, giving rise to the diagonal signals. In addition, homonuclear couplings generate cross signals. Diagonal and cross signals of a coupling proton pair are located at the corners of a square with the coordinates given above (Fig. 2.56(e)). Protons without couplings, such as those of isolated methyl groups ($-CO-CH_3$ or $-OCH_3$), will cause diagonal signals only. For closely spaced and coupling proton resonances, strong diagonal signals usually overlap with cross signals, and correlations are masked. In such cases, diagonal signals can be suppressed by using smaller mixing pulses ($45°_x$ or $30°_x$) [69] in order to favor the cross signals. Smaller couplings are emphasized in the COSY experiment by adding a delay to the evolution time (*COSY with delay*).

Fig. 2.57(b) displays a COSY experiment of sample compound $C_9H_{15}NO_3$, in addition to a two-dimensional carbon-proton shift correlation (Fig. 2.57(a)), demonstrating one strategy (among others) to deduce the CH backbone of an organic molecule from two-dimensional NMR. The COSY correlation (b) reveals six *vicinal* proton relationships **1** besides three *geminal* AB couplings of methylene protons, using the cross signals connected by a line for clarity in Fig. 2.57(b). No correlation is found for the proton signal at $\delta_H = 2.78$ ppm. Its protons are attached to the carbon atom with $\delta_C = 38.6$ ppm (Fig. 2.57(a)), which belongs to an N-methyl group because of its carbon shielding and quartet splitting of $J_{CH} = 142$ Hz. Full evaluation of carbon-proton and proton-proton shift correlations from Fig. 2.57 is summarized in Table 2.3, and yields the carbon-carbon network **2** (page 100).

Using carbon-proton multiplicities as obtained from DEPT experiments, or a *J*-resolved two-dimensional or a one-dimensional carbon-13 NMR spectrum, with gated proton decoupling in less complicated cases (top of Fig. 2.57(a)), the CH skeleton can be derived. A seven-membered ring is identified as partial structure **3**. Summation of all the information on carbon-proton multiplets, including N-methyl nitrogen and a carboxy function (see Table 2.3), yields $C_9H_{14}NO_2$. On comparison with the actual molecular formula, $C_9H_{15}NO_3$, an OH group turns out to be the additional substituent. Finally,

⬅

Fig. 2.57. Two-dimensional carbon-proton and proton-proton shift correlation for substructure determination of sample compound $C_9H_{15}NO_3$: (a) Carbon-proton shift correlation with proton broadband and gated decoupled ^{13}C NMR spectra on top; (b) COSY-45 proton-proton shift correlation with ^{1}H-NMR spectra for reference (400.1 MHz; 20 mg in 0.4 mL of tetradeuteriomethanol at 30 °C; measuring times: (a) 50 min; (b) 25 min; transform times for (a) and (b): 25 min).

Table 2.3. Carbon-Proton and Proton-Proton Correlations of Sample Compound $C_9H_{15}NO_3$, Taken From Fig. 2.57.

δ_C ppm	J_{CH} Hz		δ_H ppm
66.6	CH	152 ...	3.96
63.7	CH	152 ...	3.85
61.8	CH	145 ...	4.22
52.4	CH	135 ...	2.88
38.6	CH_3	142 ...	2.78
37.4	CH_2	131 ...	2.03 AB 2.22
25.0	CH_2	133 ...	2.03 AB 2.33
24.5	CH_2	135 ...	2.00 AB 2.33
	$C_9H_{15}NO_3$		

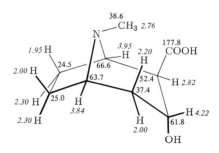

these functional groups can be connected with the seven-membered ring, taking into account that carbon atoms attached to the N-methyl group should display similar shieldings (63.7 and 66.6 ppm) and coupling constants (152 Hz), and that electron withdrawal of a hydroxy function is larger (61.8/4.22 ppm) compared with a carboxy group (52.4/2.88 ppm). Thus, the sample compound is ecgonine. All shifts can be assigned, sterically interacting methylene protons being less shielded within the AB systems.

2.10.5 Two-Dimensional $C_AH_A - C_XH_X$ Correlation: The RELAY Experiment

The COSY sequence for proton-proton shift correlation ($H_A - H_X$ relationships, Fig. 2.56) can be combined with a two-dimensional carbon-proton shift correlation (C_AH_A and C_XH_X bonds, Fig. 2.53). This experiment, referred to as *two-dimensional relayed coherence transfer* or *RELAY* [70], permits direct identification of $C_AH_A - H_XC_X$

substructures in one experiment. The pulse sequence first involves an exchange of magnetization (coherence transfer) between the coupling protons ($H_A - H_X$ correlation), as known from the COSY sequence. Thereupon, the mixed magnetization M (δ_A, δ_X) is transferred from proton H_A to the attached carbon C_A. The mixing time (Fig. 2.53) thus depends on both carbon-proton ($J_{C_A H_A}$) and proton-proton couplings ($^{2,3}J_{H_A H_X}$). If proton-proton couplings differ too much from each other, ranging, for example, from 2 to 15 Hz within the sample molecule, the experiment sometimes has to be repeated for appropriate magnitudes of $^{2,3}J_{H_A H_X}$ in order to trace out all $C_A H_A - H_X C_X$ substructures. In this case, a COSY experiment and a separately acquired two-dimensional CH correlation may be less time consuming.

Fig. 2.58 displays the result of a RELAY experiment acquired for 5-piperidinopentadienal. $C_A H_A - H_X C_X$ bonding relationships are indicated within the two-

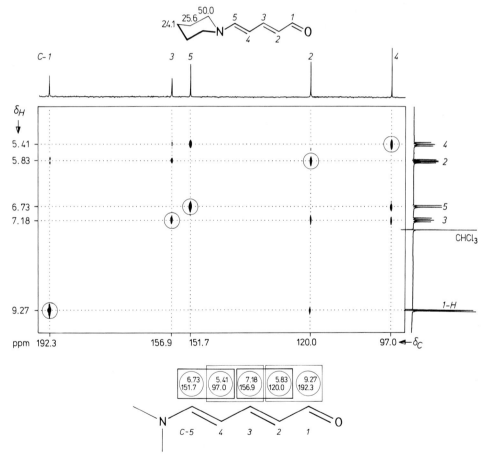

Fig. 2.58. RELAY experiment of piperidinopentadienal (300 MHz; 0.5 mol/L in deuteriochloroform; 30 °C; measuring and transform time: 3 h; submatrix for the pentadienal substructure; encircled correlation signals for CH bonds).

dimensional matrix limited by carbon-13 and proton shift range by the rectangle given by correlation maxima of the directly bonded carbon-proton pairs with δ_{C_A}, δ_{H_A} and δ_{C_X}, δ_{H_X} and the mixed correlation signals with δ_{C_A}, δ_{H_X} and δ_{C_X}, δ_{H_A}. The pentadienal substructure of the sample molecule thus arises from the rectangles formed by the encircled carbon-proton correlations and the mixed signals, as can easily be verified in Fig. 2.58.

2.10.6 Two-Dimensional Carbon-Carbon Shift Correlation: The Second Dimension of the INADEQUATE Experiment

The AB and AX systems of all $^{13}C-^{13}C$ bonds appear in one spectrum when the INADEQUATE pulse sequence (Fig. 2.48) is applied. Complete interpretation usually becomes difficult in practice due to signal overlapping, isotope shifts and AB effects (Section 2.9.4). A separation of the individual $^{13}C-^{13}C$ two-spin systems by means of a second dimension would be desirable. It is the frequency of the double quantum transfer (d–e) in Fig. 2.48 which introduces a second dimension to the INADEQUATE experiment. This *double quantum frequency* v_{DQ} characterizes each $^{13}C_A - ^{13}C_X$ bond, as it depends on the sum of the individual carbon shieldings v_A and v_X in addition to the frequency v_0 of the transmitter pulse located in the center of the spectrum if quadrature detection is applied [69c, 71]:

$$v_{DQ} = v_A + v_X - 2v_0.$$

To conclude, the second dimension is introduced if the *switching time* t_1 (Fig. 2.48) is incremented in a series of single experiments so as to reach all possible double quantum frequencies v_{DQ} within a sample molecule by the reciprocals $1/t_1$. Again, the acquired FID signals will depend on two variable times t_1 and t_2, respectively. A first Fourier transformation in the t_2 domain generates $^{13}C-^{13}C$ satellite spectra. The corresponding AB or AX type doublet pairs, however, are modulated by the individual double quantum frequencies which characterize each AB or AX pair. The second Fourier transformation in the t_1 domain liberates the double quantum frequency as the second dimension: Maximum AB or AX $^{13}C-^{13}C$ subspectra are observed at the corresponding double quantum frequencies, so that each doublet appears with unique coordinates,

$$\delta_A; \quad \delta_A + \delta_B \qquad \delta_B; \quad \delta_A + \delta_B.$$

Each pair of connected carbon atoms, $C_A - C_B$, is thus segregated according to its sum of chemical shifts, so that the actual magnitude of the carbon-carbon coupling constant is no longer required for identification of the bond. A straightforward method of tracing out the carbon skeleton of a compound is thus established.

The two-dimensional INADEQUATE experiment also suffers very much from low sensitivity given by the low natural abundance of carbon-13 (about 10^{-2}), so that only 0.01% or 10^{-4} of all carbon-carbon bonds contribute to the satellite signals. In fact, the basic experiment can be modified in order to reduce the data matrix and to save measuring time [72], giving COSY-like square correlations as shown in Fig. 2.60. Nevertheless, the two-dimensional INADEQUATE experiment requires several hours of measuring

2.10 The Second Dimension 103

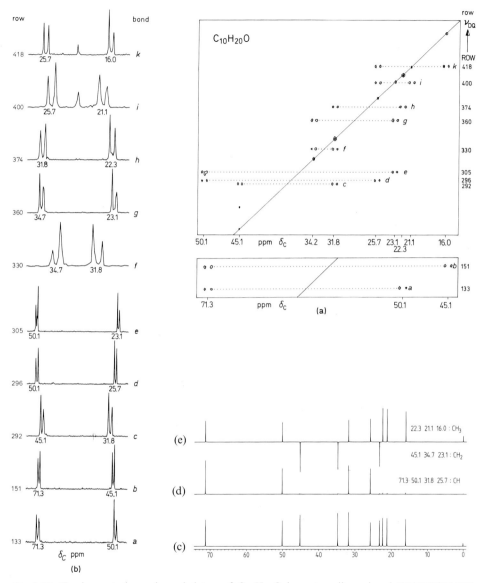

Fig. 2.59. Tracing out the carbon skeleton of $C_{10}H_{20}O$ by a two-dimensional INADEQUATE experiment in combination with two DEPT experiments (100.6 MHz; (a–b) 400 mg in 1 mL, (c–e) 50 mg in 0.4 mL of deuteriochloroform at 30 °C; measuring time for (a): 14 h; transform time: 50 min); (a) contour plot of the INADEQUATE matrix; (b) subspectra of all $^{13}C - ^{13}C$ bonds $a-k$; (c) proton-decoupled ^{13}C NMR spectrum; (d–e) DEPT experiments for generation of a CH carbon subspectrum (d) and for separation of CH_2 carbon atoms (negative) from CH and CH_3 groups (positive).

time, even in the case of very concentrated sample solutions (more than 100 mg/mL) and medium-sized molecules ($C_{10}-C_{20}$).

If enough sample is available, however, a two-dimensional INADEQUATE experiment with sufficient signal: noise permits straightforward and unequivocal determination of the carbon skeleton, without the need to search for repeating J_{CC} couplings. Fig. 2.59 provides an instructive example. Evaluation of the contour plot (Fig. 2.59(a)) identifies all ten C-C bonds a – k of the sample compound $C_{10}H_{20}O$. Without any additional information, drawing bonds between all correlating carbon-13 shifts produces the carbon skeleton of a cyclohexane which is at least disubstituted:

$$\begin{array}{ccc} & 34.7 \overset{g}{-} 23.1 & 16.0 \\ \text{f}/ & \backslash \text{e} & /\text{k} \\ 22.3 - 31.8 & 50.1 - 25.7 \\ \text{h} & \text{c}\backslash \quad /\text{a} \quad \text{d} \quad \backslash \text{i} \\ & 45.1 \underset{b}{-} 71.3 & 21.1 \end{array}$$

Using the CH_n multiplicities as obtained from two DEPT experiments (Fig. 2.59(d, e)), the CH skeleton can be derived in the next step, so that the compound turns out to be a 1,2,4-trisubstituted cyclohexane of the menthane type:

$$\begin{array}{l} 4 \text{ CH} : C_4H_4 \\ 3 \text{ CH}_2: C_3H_6 \\ 3 \text{ CH}_3: C_3H_9 \\ \hline C_{10}H_{19} \\ C_{10}H_{20}O \end{array} \qquad \begin{array}{c} H_2C-CH_2 \quad CH_3 \\ / \quad \backslash \quad / \\ H_3C-CH \quad CH-CH \\ \backslash \quad / \quad \backslash \\ H_2C-CH \quad CH_3 \\ \quad \text{(OH)} \end{array}$$

(OH)

Summation of information on all CH_n multiplets yields $C_{10}H_{19}$ as partial molecular formula. When compared with the actual molecular formula ($C_{10}H_{20}O - C_{10}H_{19} = OH$) a hydroxy group is identified as the third substituent. Thus, the compound is found to be menthol, after comparing the data with those already given in Fig. 2.46. It could also have been another stereoisomer, such as isomenthol, neomenthol or neoisomenthol, with different relative configurations of the substituents. In order to solve such configurational problems, a knowledge of proton and carbon-13 shieldings and coupling constants is still necessary.

Moreover, INADEQUATE experiments for analysis of carbon-carbon connectivities require careful interpretation when the magnitudes of two-bond and one-bond carbon-carbon coupling constants approach each other. Large two-bond CC couplings, for example, are observed for alkynyl, cyclobutyl, cyclobutenyl substructures and for carbonyl compounds substituted by electronegative groups in an α position.

2.10.7 Carbon-13 NMR Strategy for Structure Elucidation

In short, some of the classical and modern techniques of carbon-13 NMR can be summarized by a strategy proposed for the practice of structure elucidation in liquids:

2.10 The Second Dimension 105

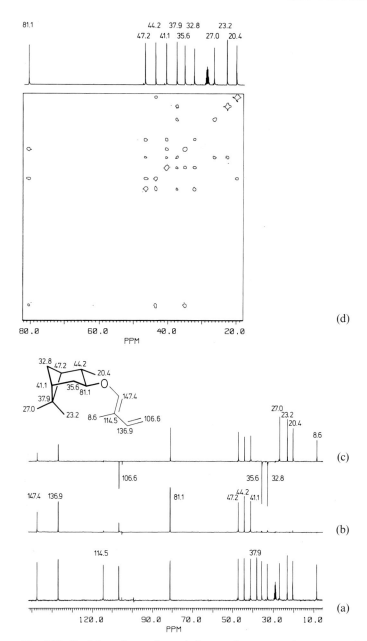

Fig. 2.60. Deriving the carbon skeleton of an organic compound by DEPT and 2 D-IN-ADEQUATE with COSY-like square correlations; sample: 3-(isopinocampheoxy)-2-methyl-1,3-butadiene, 400 mg in 0.4 mL of hexadeuterioacetone, 100.6 MHz; (a) proton-broadband decoupled spectrum (1 scan); (b) DEPT subspectrum of CH carbon nuclei (8 scans); (c) DEPT spectrum with CH, CH$_3$ (positive), and CH$_2$ (negative, 8 scans); (d) 2 D-INADEQUATE experiment of aliphatic carbon nuclei, 256 experiments, 64 scans per experiment. The bicyclic partial structure of the molecule can be derived from the square correlations (e.g. ... −44.1−88.1−35.6− ...). A stacked plot of the carbon-proton shift correlation of this sample is displayed on the cover.

(a) *Proton broadband decoupled ^{13}C NMR spectrum*:
Count nonequivalent carbon atoms (Fig. 2.47(a)).

(b) *J-Modulated spin-echo („APT") or non-selective polarization transfer („DEPT")*:
Determine CH_n multiplicities; generate C, CH, CH_2, and CH_3 subspectra with and without proton coupling (Fig. 2.46; Fig. 2.47(b−e)).

(c) *J-Resolved two-dimensional ^{13}C NMR spectra*:
Separate CH_n multiplets at each carbon-13 shift in the J_{CH} dimension as an alternative to (b) (Fig. 2.52(c)).

(d) *Two-dimensional carbon-proton shift correlation*:
Identify carbon-proton bonds via $^1J_{CH}$ ($C_A H_A$ and $C_X H_X$ substructures, Fig. 2.55) and longer-range carbon-proton via $^2J_{CH}$ and $^3J_{CH}$, particularly for localizing quaternary carbon atoms ($-\overset{|}{\underset{|}{C}}-\overset{H^X}{\underset{|}{C}}-\overset{|}{\underset{H^Y}{C}}-$ substructures by correlation via long-range couplings, „COLOC", as demonstrated in Fig. 5.17).

(e) *Two-dimensional proton-proton shift correlation („COSY")*:
Identify CH_A-CH_X- substructures by shift correlation via proton-proton coupling J_{AX} as a complementary experiment to (d) (Fig. 2.57).

(f) *Two-dimensional carbon-carbon shift correlation („2 D-INADEQUATE")*:
Trace out the entire carbon-carbon bonding network, provided enough sample is available (Fig. 2.59(a−b)). In combination with two DEPT experiments, this method affords the most straightforward and convenient elucidation of the CH skeleton of an organic compound as illustrated in Fig. 2.60.

Fig. 2.61 portrays the practical value of the basic methods of shift correlations for structure elucidation.

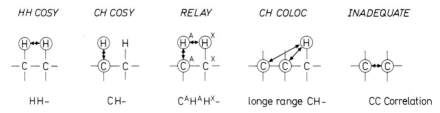

Fig. 2.61. Homo- and heteronuclear shift correlations for the determination of the bonding network (the "connectivities").

3 ^{13}C NMR Spectral Parameters and Structural Properties

3.1 Chemical Shifts

3.1.1 Comparison of ^{13}C and ^{1}H Shifts

Magnetic field strengths used in ^{13}C NMR at present usually are between 1.9 and 11.7 Tesla, corresponding to ^{13}C Larmor frequencies between 20 and 500 MHz (Fig. 2.35). The range Δ of ^{13}C chemical shifts of all organic compounds measured so far and including reactive intermediates such as carbenium ions approaches 400 ppm or 10 kHz at 25 MHz. The average ^{13}C line widths $\Delta v_{1/2}$ now attainable are about 1 Hz, i.e. of the same order as those observed in ^{1}H NMR.

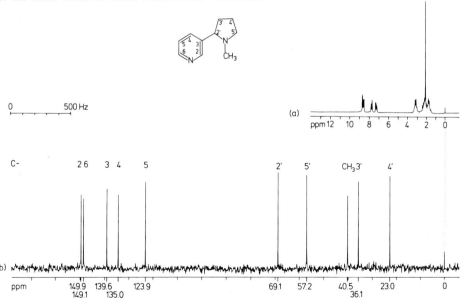

Fig. 3.1. Nicotine, 75% by vol. in hexadeuterioacetone at 25 °C;
(a) 80 MHz ^{1}H NMR spectrum;
(b) 20 MHz ^{13}C NMR spectrum, proton broadband-decoupled.
Both spectra were recorded at the same magnetic field strength with identical frequency scales.

In order to observe good spectral resolution, chemical shifts must provide a high spectral width to line width ratio $\Delta:\Delta v_{1/2}$. This ratio is about 10 kHz:1 Hz or 10^4 for ^{13}C at 25 MHz, or one power of ten higher than that observed in ^{1}H NMR at the same field and 100 MHz with $\Delta v_{1/2} = 1$ Hz and $\Delta = 1$ kHz. A comparison of signal separation in ^{1}H and ^{13}C NMR at the same magnetic field strength is given in Fig. 3.1 (page 107).

Together with the large number of correlations now available between ^{13}C chemical shifts and structures, its high $\Delta:\Delta_{1/2}$ ratio makes ^{13}C the NMR nucleus of choice for structural analysis of organic compounds using chemical shifts. Furthermore, ^{13}C chemical shifts provide more direct information concerning the carbon skeleton than do proton shifts, and ^{13}C NMR spectra are often more easily assignable than homonuclear coupled ^{1}H NMR spectra. This is not only due to the spectrum simplification and sensitivity enhancement by broadband or off-resonance proton decoupling in ^{13}C NMR, but also to the low natural abundance of ^{13}C, which assures that complicating homonuclear ^{13}C–^{13}C couplings are lost in the noise as long as ^{13}C enriched compounds are not measured.

3.1.2 Referencing ^{13}C Chemical Shifts

Any heteronuclear signal of a solvent or an added reference substance can be used for referencing ^{13}C shifts. For example, ^{13}C shifts can be directly measured relative to a deuterium signal of the deuterated solvent usually required for field/frequency stabilization. However, homonuclear shift references such as the ^{13}C signals of tetramethylsilane (TMS), carbon disulfide, benzene, cyclohexane, 1,4-dioxane or the easily localizable multiplet signals of deuterated solvents (Fig. 2.22) are predominantly applied in ^{13}C NMR.

For ^{13}C shift/structure correlations and for tabulations of ppm values one generally accepted reference should be used. Carbon disulfide, which appears in the low field region of ^{13}C spectra, was widely used in the early literature [73a, b]. Later, tetramethylsilane (TMS), known from proton NMR, became the generally accepted carbon-13 shift reference, particularly because of some parallels in the behavior of ^{1}H and ^{13}C shifts.

Neither CS$_2$ nor TMS are ideal standards. The ^{13}C signals of CS$_2$ and carbonyl carbons overlap, as do the ^{13}C signals of cyclopropane and some methyl carbons with TMS (Fig. 3.3). Furthermore, the ^{13}C resonance of TMS has been shown to suffer from solvent shifts of the order of ± 0.1 to ± 1.5 ppm in common NMR solvents, even at infinite dilution [74]. This must be considered if ^{13}C shifts relative to TMS of one compound in different solvents are to be compared. There are two alternative methods to overcome this problem: one is to use cyclohexane as the internal reference; cyclohexane was shown to have ^{13}C solvent shifts lower than ± 0.5 ppm [74]. The other alternative is to use TMS as an external reference (Sections 1.9.3 and 2.8.5) and to make bulk susceptibility shift corrections according to eq. (1.44).

Since medium effects may be expected for any resonance, the solvent must be mentioned when ^{13}C shift values are tabulated. In this work, all ^{13}C shifts are given relative to TMS. Those shifts which were originally reported relative to other references (*e.g.* carbon disulfide, δ_{CS_2}, or 1,4-dioxane, $\delta_{C_4H_8O_2}$), have been converted into shifts relative to TMS (δ_{TMS}) using the known shift difference between common reference substances, as

Table 3.1. ^{13}C Data of Common Deuterated Solvents and of their ^{1}H Isotopes.

Compound	C–D signal multiplicity	C–D coupling constants Hz	^{13}C shifts ppm	Isotopic ^{1}H compound	
				^{13}C shift ppm	C–H coupling constant Hz
Dideuteriomethylene-chloride	Quintet	27	53.1	53.8	177.5
Deuteriochloroform	Triplet	32	77.0±0.5	78.0±0.5	210.5
Deuteriobromoform	Triplet	31.5	10.2	10.3	205
Trideuterionitromethane	Septet	23.5	60.5	61.1	146.5
Trideuterioacetonitrile	Septet *	32 <1	1.3 118.2	1.7 118.2	117.4 <5
Tetradeuteriomethanol	Septet	21.5	49.0	49.9	141
Hexadeuterioethanol	Septet Quintet	19.5 22	15.8 55.4	16.9 56.3	127 140.5
Decadeuteriodiethylether	Septet Quintet	19 21	13.4 64.3	14.5 65.3	127 142
Octadeuterio-1,4-dioxane	Quintet	22	66.5	67.6	146
Octadeuteriotetrahydrofuran	Quintet Quintet	20.5 22	25.2 67.4	26.2 68.2	132 137
Hexadeuterioacetone	Septet *	20 <1	29.8 206.5	30.7 206.7	126 7.5
Hexadeuteriodimethylsulfoxide	Septet	21	39.7	40.9	137
Tetradeuterioacetic acid	Septet *	20 <1	20.0 178.4	20.9 178.8	129 7.5
Heptadeuteriodimethylformamide	Septet Septet Triplet	21 21 30	30.1 35.2 167.7	30.9 36.0 167.9	138 138 188
Octadecadeuterio-hexamethylphosphoramide	Septet	21	35.8	36.9	129.5
Dodecadeuteriocyclohexane	Quintet	19	26.4	27.8	126
Hexadeuteriobenzene	Triplet	24	128.0±0.5	128.5±0.5	161
Pentadeuteriopyridine	Triplet (3,5) Triplet (4) Triplet (2,6)	25 24.5 27.5	123.5 135.5 149.2	124.2 136.2 149.7	166 163 177

* long-range multiplet, not resolved

illustrated in Table 3.1 and outlined by eq. (3.1):

$$\delta_{TMS} = 192.5 + \delta_{CS_2} = 67.6 + \delta_{C_4H_8O_2} = \cdots \qquad (3.1)$$

However, conversion errors due to medium effects may be as high as 1 ppm.

3.1.3 Survey of ^{13}C Chemical Shifts [73 a–k]

The magnetic shielding constant σ_i (eq. 1.39) arises from the influence of chemical effects on the Larmor frequency of a nucleus. It is described in terms of three additive contributions (eq. 3.2):

$$\sigma_i = \sigma^{dia} + \sigma^{para} + \sigma^N. \qquad (3.2)$$

The diamagnetic term σ^{dia} accounts for local electronic circulations around the nucleus induced by the applied field B_0. According to the Lenz rule, the resultant intramolecular field opposes B_0, so that the nucleus will be shielded. As known from the Lamb formula, eq. 3.3 [9],

$$\sigma^{dia} \propto r^{-1}, \qquad (3.3)$$

the diamagnetic term decreases with distance r between nucleus and circulating electrons. Thus, s electrons will cause a stronger shielding than p electrons ($r_s : r_p = 1 : \sqrt{3}$!), and for atoms with s electrons only, such as hydrogen, σ^{dia} is the dominant shielding term.

The contribution σ^N is referred to as the neighbor anisotropy effect. This term accounts for the fields arising from electronic circulations around the atoms surrounding the observed nucleus. It depends on the nature of the neighbor atoms and on molecular geometry. – Sometimes, a medium term must be added in eq. (3.2), correcting for solvent and pH effects.

For nuclei other than hydrogen, such as ^{13}C, the paramagnetic shielding term σ^{para} predominates. It opposes σ^{dia} and therefore deshields. According to eq. 3.4 derived by Karplus and Pople [75],

$$\sigma^{para} = -\frac{e^2 h^2}{m^2 c^2} \Delta E^{-1} r_{2p}^{-3} [Q_{AA} + \Sigma Q_{AA}] \qquad (3.4)$$

the paramagnetic term increases with a decreasing average electronic excitation energy ΔE and with the inverse cube r_{2p}^{-3} of the distance between a $2p$ electron and the nucleus. It further depends on the number of electrons occupying the p orbital (Q_{AA}) and multiple bond contributions (ΣQ_{AX}). These two effects are included in the $[Q_{AA} + \Sigma Q_{AX}]$ factor, also known as the charge density bond order matrix in the MO formalism.

The connection between ^{13}C shielding and electronic excitation agrees with the fact that carbonyl compounds are most deshielded ($\delta_c > 170$ ppm, n → π* transitions with $\Delta E \approx 7$ eV) when compared with alkenes and aromatic compounds ($\delta_c \approx 100–150$ ppm, π → π* transitions with $\Delta E \approx 8$ eV) and alkanes ($\delta_c < 50$ ppm, σ → σ* transitions with $\Delta E \approx 10$ eV).

Fig. 3.2 reflects the dependence of σ^{para} on r_{2p}^{-3}: The ^{13}C shifts of non-benzenoid aromatic ions correlate linearly with the π electron density relative to benzene ($Q_\pi = 1$)

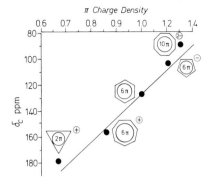

Fig. 3.2. π-Electron density *versus* ^{13}C chemical shift of ring carbons in benzene and non-benzenoid monocyclic aromatic ions [76].

[76]. In the negative ions, for example, an increased π electron density at the carbons causes electronic repulsion. As a result, the bonding orbitals expand, r increases, σ^{para} decreases and the carbons are more shielded. The upfield shift of 160 ppm per additional π-electron, as found in Fig. 3.2, illustrates that the large ^{13}C chemical shift range can be mainly attributed to the r_{2p}^{-3} dependence of σ^{para}.

Generally, ^{13}C chemical shifts correlate with some typical properties of carbon within its molecular environment discussed in the following sections.

3.1.3.1 Carbon Hybridization

The hybridization of a carbon determines to a great extent the range within which its ^{13}C signal is found. As illustrated in Fig. 3.3, sp^3 carbons resonate at highest field, followed by sp carbons, while sp^2 hybridized centers are shifted farthest to low field. The hybridization effect in ^{13}C NMR thus parallels the effect observed in ^{1}H NMR. The ^{13}C resonances of sp^3 carbons are found between — 20 and 100 ppm relative to TMS; sp carbons resonate from 70 to 110 ppm; the low field sp^2 carbon signals occur at 120 to 240 ppm in organic compounds.

3.1.3.2 Electronegativity [73]

A correlation between ^{13}C chemical shifts and substituent electronegativities is found for the α carbons in many classes of compounds, as is illustrated for the halohexanes in Table 3.2.

Table 3.2. ^{13}C Shift Increments $Z = \delta_{RX} - \delta_{RH}$ (in ppm) Induced in Hexane by Halogenation of a Terminal Hydrogen.

	α X—CH$_2$—	β CH$_2$—	γ —CH$_2$—	δ —CH$_2$—	ε —CH$_2$—	>ε —CH$_3$
X = H	δ = 14.2	23.1	32.2	32.2	23.1	14.2
X = I	Z = − 7.2	10.9	−1.5	−0.9	0.0	0.0
X = Br	Z = 19.7	10.2	−3.8	−0.7	0.0	0.0
X = Cl	Z = 31.0	10.0	−5.1	−0.5	0.0	0.0
X = F	Z = 70.1	7.8	−6.8	0.0	0.0	0.0

This is explained in terms of increasing electron withdrawal from the α carbon orbitals with increasing electronegativity of the substituent. Correspondingly, a larger r^{-3} appears in eq. (3.4), and hence a deshielding occurs.

The bond polarization due to electronegative substituents should propagate along the carbon chain and decrease with the inverse cube of the distance. Much smaller inductive effects are thus expected in β and γ position. Table 3.2 shows, however, that the observed β and γ effects do not correlate at all with substituent electronegativities and that the influence of other effects must therefore be involved.

3.1.3.3 Crowding of Alkyl Groups and Substituents [73]

Increasing crowding of alkyl groups or electron withdrawing substituents at a carbon causes a successive downfield shift of its ^{13}C resonance (Table 3.3). Similarly, crowding of shielding heteroatoms such as iodine successively reinforces upfield shifts (Table 3.3).

Table 3.3. ^{13}C Chemical Shifts δ_C (in ppm) of Halogenated and Alkylated Methane Derivatives.

	CH_4	CH_3X	CH_2X_2	CHX_3	CX_4
X = Cl	−2.3	23.8	52.8	77.7	95.5
X = I		−21.8	−55.1	−141.0	−292.5
		CH_3R	CH_2R_2	CHR_3	CR_4
R = CH_3		5.7	15.4	24.3	31.4

3.1.3.4 Unshared Electron Pairs at Carbon

An unshared electron pair which occupies a compact orbital localized at a terminal carbon (e.g. an sp orbital with small r_{2p} in eq. 3.4), causes a downfield shift of its ^{13}C resonance. A comparison of carbon dioxide with carbon monoxide, and nitrile with cyanide anion and isocyanide [77] illustrates this:

$$O=C=O \quad |\overset{\ominus}{C}\equiv\overset{\oplus}{O}| \qquad \text{(ppm)}$$
$$132.0 \qquad 181.3$$

$$\overset{0.3}{CH_3}-C\equiv N \quad |\overset{\ominus}{C}\equiv N|^{\ominus} \quad \overset{29.3}{CH_3}-N\equiv\overset{\ominus}{C}$$
$$117.7 \qquad 168.5 \qquad 158.5$$

On the other hand, odd-numbered carbons in conjugated dienyl anions are shielded [78]:

$$\overset{\ominus}{}\diagup\!\!\diagdown\!\!\diagup^{65.0} \qquad \overset{\ominus}{}\diagup\!\!\diagdown\!\!\diagup\!\!\diagdown^{64.2} \qquad \text{(ppm)}$$
$$73.2 \; 138.7 \qquad\qquad 92.1 \; 146.6$$

This has been explained by MO approximations in terms of a change of the charge density bond order term in eq. (3.4).

3.1.3.5 Electron Deficiency at Carbon

Electron deficiency at a carbon causes drastic deshielding. This is observed for the sp² carbons typical of carbocations [79]. In such systems, the sp² ^{13}C chemical shift range may approach 400 ppm relative to TMS. If the positive charge is dispersed in a carbocation, e.g. by resonance, the electron deficient carbon will be more shielded. The following comparison of *t*-butyl-, dimethylhydroxy- and dimethylphenyl-carbenium ion illustrates this:

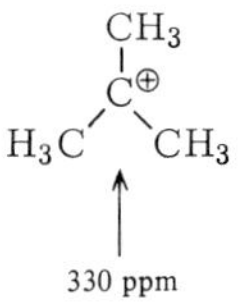

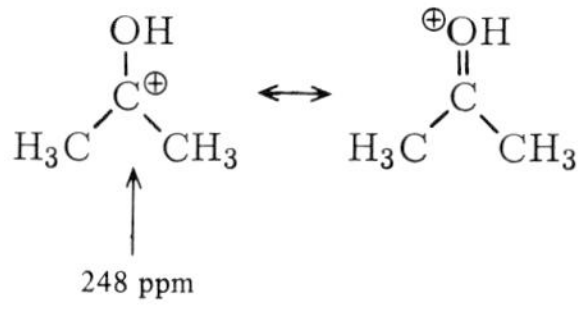

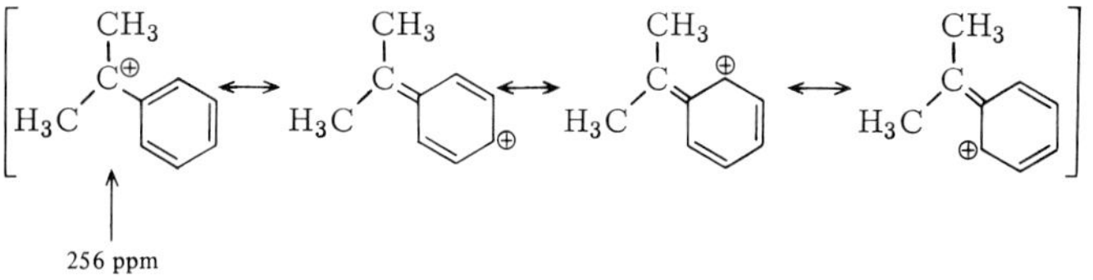

Downfield shifts due to electron deficiency are observed for carbene metal complexes [80a] and for the sp carbons of metal carbonyls [80b].

3.1.3.6 Mesomeric Effects

The ^{13}C chemical shifts of benzenoid carbons largely depend on the mesomeric interaction between substituent und benzene ring. Electron releasing substituents (*e.g.* $-\bar{\text{N}}\text{H}_2$, $-\bar{\text{O}}\text{H}$) will increase the electron density at the *o* and *p* carbons relative to benzene (128.5 ppm), while slight electron deficiencies will be induced by electron withdrawing groups (*e.g.* $-\text{NO}_2$, $-\text{CN}$).

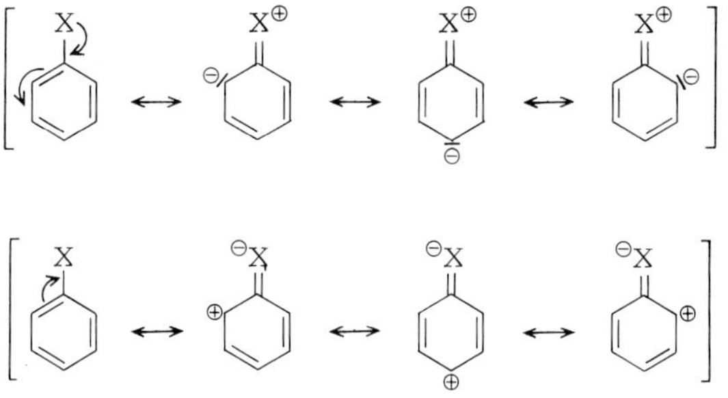

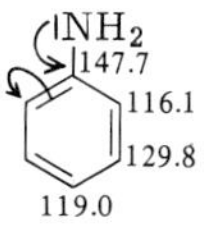

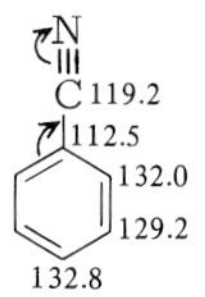

These formulae explain why the *o* and *p* carbons of monosubstituted benzenes are shielded by electron releasing substituents but deshielded by electron acceptors, while the *m* carbons remain almost unaffected by both kinds of substituents [73].

Inductive and electric field effects of the substituents may overlap the *o*-mesomeric effects. However, the *m* and *p* shifts of monosubstituted benzenes generally follow the pattern discussed above.

On the other hand, ^{13}C chemical shifts may reflect the contribution of polar species to the bonding state of a molecule. For example, the methylene sp^2 carbon shielding in diazomethane and ketene indicates that the carbanionic polar states of both molecules are significant [81].

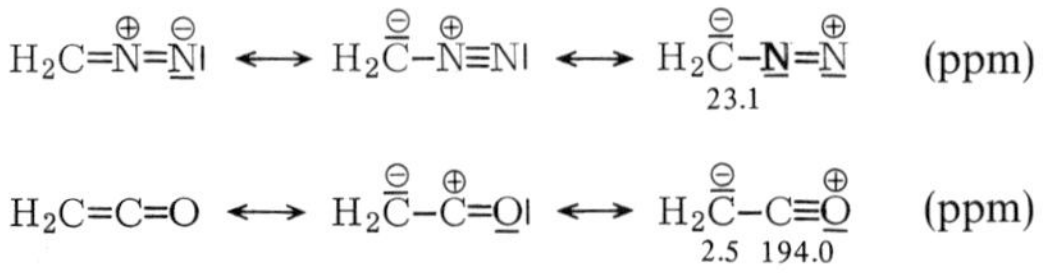

$$H_2C=\overset{\oplus}{N}=\overset{\ominus}{\underline{N}}| \longleftrightarrow H_2\overset{\ominus}{C}-\overset{\oplus}{N}\equiv N| \longleftrightarrow H_2\overset{}{C}-\underline{N}=\overset{\oplus}{N} \quad (ppm)$$
$$\phantom{H_2C=\overset{\oplus}{N}=\overset{\ominus}{\underline{N}}| \longleftrightarrow H_2\overset{\ominus}{C}-\overset{\oplus}{N}\equiv N| \longleftrightarrow H_2\overset{}{C}-} 23.1$$

$$H_2C=C=O \longleftrightarrow H_2\overset{\ominus}{C}-\overset{\oplus}{C}=\underline{O}| \longleftrightarrow H_2\overset{}{C}-C\equiv\overset{\oplus}{O} \quad (ppm)$$
$$\phantom{H_2C=C=O \longleftrightarrow H_2\overset{\ominus}{C}-\overset{\oplus}{C}=\underline{O}| \longleftrightarrow} 2.5 \; 194.0$$

In α, β-unsaturated carbonyl compounds, a distinct deshielding of the β-olefinic carbon clearly responds to the polarized state:

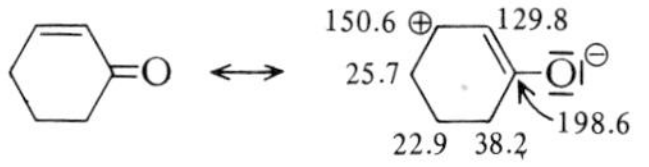

2-Cyclohexenone (ppm): 150.6, 129.8, 25.7, 198.6, 22.9, 38.2

3.1.3.7 Conjugation

The decrease in bonding order arising from delocalization of multiple bonds in conjugated systems results in a shielding of the central carbon atoms. This can be clearly seen by comparison of the pairs 1-butene/1,3-butadiene and *cis*-3-octene/*cis*-*cis*-3,5-octadiene:

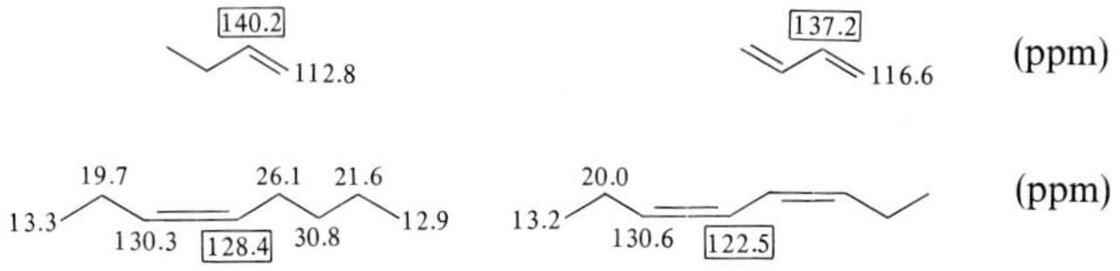

140.2, 112.8   137.2, 116.6   (ppm)

19.7, 26.1, 21.6, 13.3, 130.3, 128.4, 30.8, 12.9   20.0, 13.2, 130.6, 122.5   (ppm)

In a similar manner, carbonyl carbons in α, β-unsaturated aldehydes, ketones and carboxylic acids are shielded by about 10 ppm when compared with those of corresponding saturated carbonyl compounds [73]:

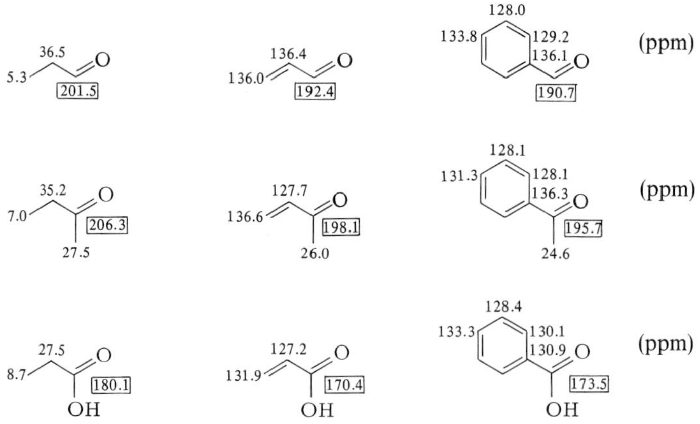

36.5, 5.3, 201.5   136.4, 136.0, 192.4   128.0, 133.8, 129.2, 136.1, 190.7   (ppm)

35.2, 7.0, 206.3, 27.5   127.7, 136.6, 198.1, 26.0   128.1, 131.3, 128.1, 136.3, 195.7, 24.6   (ppm)

27.5, 8.7, 180.1, OH   127.2, 131.9, 170.4, OH   128.4, 133.3, 130.1, 130.9, 173.5, OH   (ppm)

3.1.3.8 Steric Interactions

Steric interactions, mostly arising from touching or overlapping of van der Waals radii of closely spaced hydrogens, usually cause a shielding of the carbons attached to these hydrogens. The steric perturbation of the C—H bond involved causes the charge to drift towards carbon; the bonding orbitals at carbon expand, and a shielding will arise according to eq. (3.4) [82].

In saturated open-chain and cyclic systems [87–89], the steric effect on carbon shielding is observed when two hydrogenated carbons are *γ-gauche* relative to each other. In mobile open-chain alkanes with *gauche* rotamer populations of approximately 30%, averaged shieldings of about − 2 ppm are observed for the γ carbon when a methyl group is introduced in the α position. Other substituents (*e.g.* the halogens in Table 3.2) cause γ effects of up to −7 ppm.

According to a simple model developed by Grant [82], the steric shift δ_{St} depends on the proton-proton repulsive force $F_{HH}(r_{HH})$, which is a function of the proton-proton distance r_{HH}, but also on the angle Θ between the H—H axis and the perturbed C—H bond:

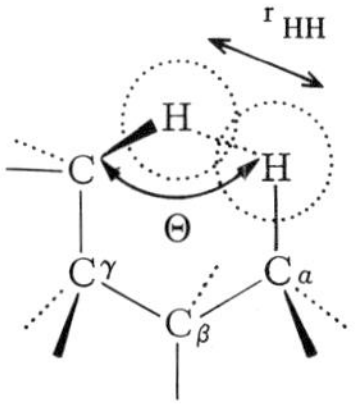

$$\delta_{St} = \text{const.}\ F_{HH}(r_{HH}) \cos \Theta. \tag{3.5}$$

Eq. (3.5) shows that the sign of δ_{St} may change with Θ. In conclusion, steric interactions are not generally associated with upfield shifts.

Typical γ-gauche effects are observed in conformationally rigid systems such as methylcyclohexanes and methylnorbornanes [73, 87], *e.g.*:

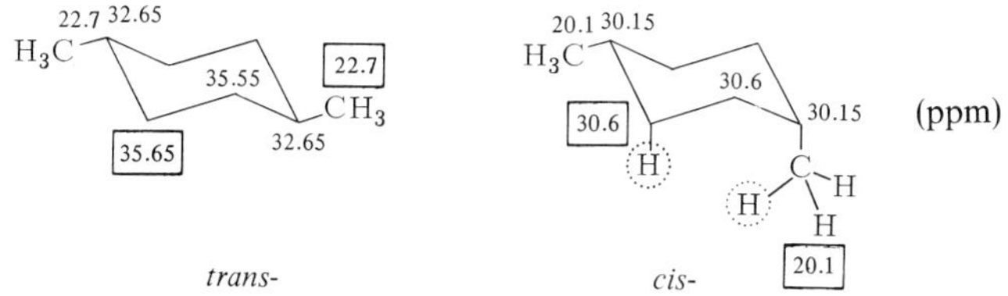

trans-   *cis-*

1,4-Dimethylcyclohexane

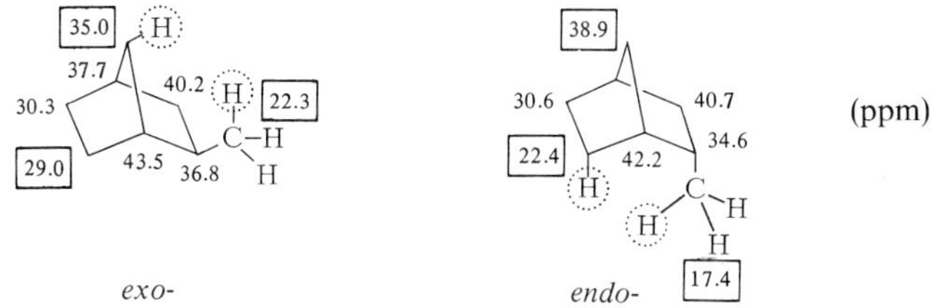

exo-   *endo-*

2-Methylnorbornane

116   3 ^{13}C NMR Spectral Parameters and Structural Properties

Another example of sterically induced upfield shifts is the $\delta_{trans} > \delta_{cis}$ relation characteristic of the carbons α to the double bond in cis- and trans-alkenes [73, 90], e.g.:

[16.5]
H$_3$C   H
123.9 C=C 130.6   22.0
   H   CH$_2$–CH$_2$–CH$_3$ (ppm)
          [34.1]   12.5

(E)-2-Hexene

   H   H
122.8 C=C 129.7 21.4
H–C   C–CH$_2$–CH$_3$
[11.4] H   H [28.2] 12.5

(Z)-2-Hexene

In conjugated systems, steric repulsion between alkyl groups may oppose conjugation and cancel the upfield shift characteristic of the conjugated state. This can be seen for a series of acetophenones, and a dependence on the torsional angle Φ between phenyl and carbonyl bonding planes becomes obvious [73 c], e.g.:

C 195.7 ppm
O CH$_3$

$\Phi = 0°$

      CH$_3$
C 199.0 ppm
O CH$_3$

$\Phi \approx 28°$

H$_3$C   CH$_3$
C 205.5 ppm
O CH$_3$

$\Phi \approx 50°$

3.1.3.9 Electric Fields of Charged Substituents

Electric fields originating from charged substituents may also contribute to carbon shielding. The o carbons in nitrobenzene are an example; they are expected to be deshielded according to the mesomeric effect of the nitro group as outlined in Section 3.1.3.6. In fact, a shielding of −5.3 ppm relative to the benzene carbons is observed. This can partially be explained by the intramolecular electric field of the nitro group which causes the charge of the o C–H bond to move towards the o carbon. A shielding for the latter will be the result.

Intramolecular electric fields also contribute to the pH induced shifts [84] discussed for basic and acidic groups in Section 3.1.4.3.

3.1.3.10 Anisotropic Intramolecular Magnetic Fields

The neighbor anisotropy term σ^N of eq. (3.2) plays an important role in proton shielding, permitting, for example, a distinct differentiation between aromatic and olefinic protons due to the ring current effect. However, this contribution is small in ^{13}C NMR (<2 ppm). A comparison of the methyl carbon shieldings in methylcyclohexene and toluene shows that the ring current effect often cannot be clearly separated from other shielding contributions:

δ_C (in ppm)

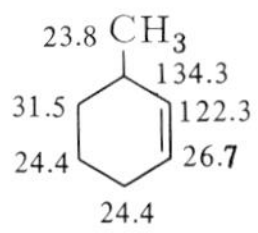

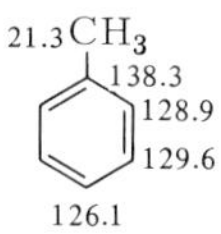

23.8 CH_3
  134.3
31.5 ⟨ring⟩ 122.3
24.4   26.7
   24.4

21.3 CH_3
  138.3
  128.9
  129.6
  126.1

(δ_H (in ppm))

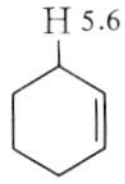

H 5.6

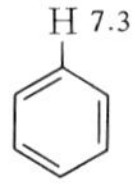

H 7.3

A small ring current was described for 1,4-dodecamethylene benzene [83]. The most shielded methylene carbons of this *ansa* compound shift upfield by 0,7 ppm relative to those in cyclopentadecane:

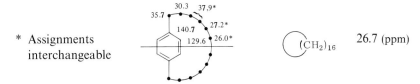

* Assignments
  interchangeable

30.3 37.9*
35.7
   140.7 27.2*
   129.6 26.0*

(CH_2)_{16}   26.7 (ppm)

3.1.3.11 Heavy Atoms

Tables 3.2 and 3.3 illustrated for iodine that heavy atom substitution causes a significant upfield shift of the substituted carbon. This is attributed to increased diamagnetic shielding caused by the large number of electrons introduced by heavy atoms.

3.1.3.12 Isotope Effect

Upfield shifts may also arise from substitution of an atom (e.g. ^{1}H) by its heavier isotope (e.g. ^{2}H). The upfield shift follows from a lower potential energy of the ground state, hence an increased ΔE in eq. (3.4), and a reduction of bond distance. The influence of both effects will be to decrease the paramagnetic shielding term. Deuterium isotope shifts depend on the degree of deuteration and reach values of up to 1.5 ppm (Table 3.1).

3.1.3.13 Intramolecular Hydrogen Bonding

Intramolecular hydrogen bonding, e.g. in salicylaldehyde or *o*-hydroxyacetophones, involves the nonbonding electron of the carbonyl oxygen. As a result, the carbonyl carbon will become more positive [73], so that a deshielding is observed:

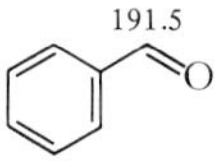

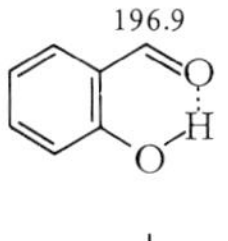

191.5   196.9   (ppm)

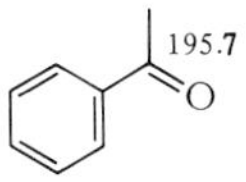

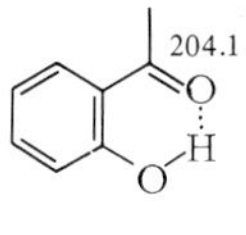

195.7   204.1   (ppm)

118   3 ^{13}C NMR Spectral Parameters and Structural Properties

Table 3.4. Solvent ^{13}C and ^{1}H Shifts of Chloroform and ^{13}C–^{1}H One-Bond Coupling Constants [92] (internal reference: cyclohexane; mole fraction of CHCl$_3$: 0.16; temperature: 45–50 °C).

Solvent	δ_C ppm	δ_H ppm	$J_{^{13}C-^1H}$ Hz
Cyclohexane	0.00	0.000	208.1
Carbon tetrachloride	0.20	0.120	208.4
Neat[a]	0.37	0.147	209.5
Benzene[b]	0.47	−0.747	210.6
Acetic acid	1.02	0.400	—
Nitromethane	1.12	0.382	213.6
Nitrobenzene	1.12	0.410	—
Diethyl ether	1.24	0.605	213.7
Methanol	1.35	0.611	214.3
Acetonitrile[b]	1.63	0.445	214.6
Acetone	1.76	0.812	215.6
Triethylamine	1.85	0.967	214.2
Dimethylformamide	2.55	1.182	217.4
Pyridine	2.63	1.282	215.0
Hexamethylphosphoramide	4.24	1.995	—

[a] with 10% cyclohexane as internal reference.
[b] shifts and coupling constants do not follow the linear correlation found for the other solvents.

In a similar manner, intermolecular hydrogen bonds will cause a deshielding, as is found for chloroform in pyridine (Table 3.4). If the hydrogen bonds are broken, *e.g.* by solvation in dilute solutions with nonpolar solvents (*e.g.* carbon tetrachloride or cyclohexane), upfield shifts have to be expected (Table 3.4).

3.1.3.14 *Substituent Increments and Functional Group Shifts*

Empirical additive substituent increments obtained by analysis of substituted alkanes, alkenes, cycloalkanes, aromatic and heteroaromatic compounds have proved to be useful for the prediction of ^{13}C chemical shifts. These substituent increments will be tabulated for the various classes of organic compounds in Section 4.13.

Many functional groups containing carbon resonate within characteristic shift ranges. A selection of these shift ranges is outlined in Fig. 3.3, which also summarizes some previously disussed substituent effects on aliphatic carbon shielding. Both parallels and significant differences to the trends of proton shifts can be observed here. As in ^{1}H NMR, cyclopropane carbons are more deshielded than the ring carbons of other alicycles. In contrast to proton shifts, however, the ^{13}C signals of aromatic ring carbons overlap with the sp^2 carbon signals of alkenes. This stems from the fact that the induced field attributed to the cyclic delocalization of π electrons in aromatics [2–5] weakens the $\boldsymbol{B}_0$ field in the center and strengthens it at the periphery of the ring, but does not significantly influence the ring carbons themselves.

3.1 Chemical Shifts   119

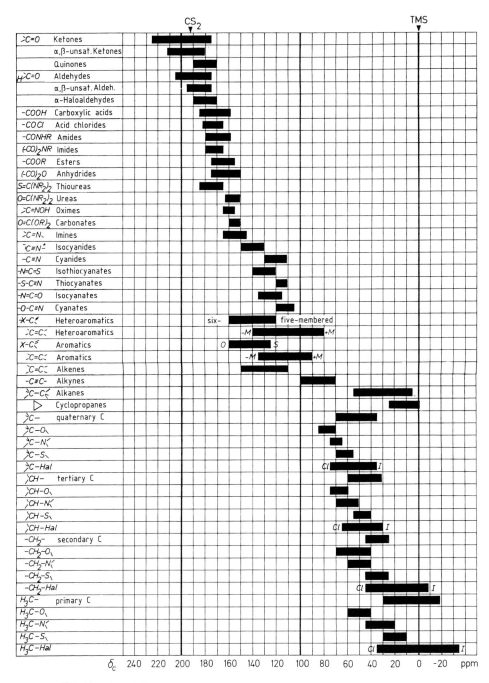

Fig. 3.3. ^{13}C Chemical shift ranges in organic compounds.

3.1.4 Medium Shifts

3.1.4.1 Dilution Shifts

Dilution shifts of ^{13}C signals may reach a magnitude of several ppm. The ^{13}C resonance of methyl iodide dissolved in cyclohexane [91a] or tetramethylsilane [91b] shifts upfield by about 7 ppm upon dilution. A much smaller upfield dilution shift (0.5 ppm) is observed for the ^{13}C signal of chloroform in cyclohexane [92]. A constant shift independent of further dilution may be reached at lower concentrations. In this case, the solution behaves as if it were infinitely diluted in terms of chemical shifts. This was observed for substituted benzenes in trifluoroacetic acid at a solute concentration in moles of 10 to 15% [93].

3.1.4.2 Solvent Shifts

Solvent shifts of ^{13}C signals are often larger than those observed in ^{1}H NMR. Downfield shifts are found for both the ^{13}C and ^{1}H signals of chloroform on going from nonpolar solvents, such as cyclohexane or carbon tetrachloride, to media susceptible to hydrogen bonding, such as pyridine or hexamethylphosphoramide [92] (Table 3.4). The ^{13}C solvent shift range of chloroform is more than 4 ppm, which is about twice as much as that found for the proton signal of that compound (Table 3.4).

Using the data shown in Table 3.4, a linear plot of ^{13}C versus ^{1}H solvent shifts is obtained [92]. Moreover, ^{13}C solvent shifts correlate linearly with the one-bond carbon-13-proton coupling constants [92]. This is attributed to changes in the average distance of bonding electrons in the C−H bond of chloroform due to intermolecular association [92]. Since much smaller solvent shifts of the carbon tetrachloride ^{13}C resonance are found [92], interactions between chlorine and the solvent can be disregarded. Thus,

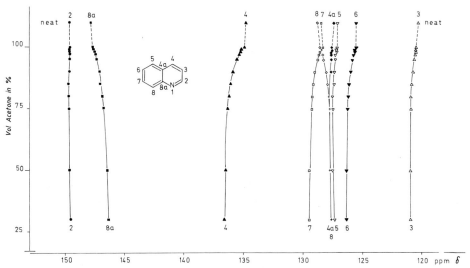

Fig. 3.4. ^{13}C chemical shifts of quinoline in the system acetone-water-quinoline with varying water concentration at 25 °C [94].

3.1 Chemical Shifts   121

having a C–H bond, chloroform appears to associate with polar solvents by hydrogen bonding. Hydrogen bonding between polar solvents and basic substituents in aromatic compounds also generates downfield shifts of the aromatic ring carbons [93].

In solutions of dipolar compounds in water, hydrogen bonding is the most important kind of intermolecular interaction; downfield shifts of the ^{13}C signals are to be expected. In the system acetone-water-quinoline, most of the quinoline carbon signals (C-2 to C-7) follow this pattern on increasing the water concentration, C-2 showing the smallest effect. The carbon nuclei C-8 and C-8 a, however, shift to higher field [94], as shown in Fig. 3.4. Thus, medium shifts may be different in magnitude and direction for nonequivalent nuclei. Closely spaced signals may then invert their sequence or coalesce. Both effects are observed in following ^{13}C resonances of C-4 a, C-7, and C-8 of quinoline on addition of water in Fig. 3.4.

3.1.4.3 pH Shifts

Protonation of amines, carboxylate anions and α-aminocarboxylic acids causes considerable upfield shifts, particularly in the position β to the protonated group [84, 95–98]:

$$-\overset{|}{C}-\overset{|}{C}-\overset{|}{C}-COO^{\ominus} \underset{\longleftarrow}{\overset{+ H^{\oplus}}{\longrightarrow}} -\overset{|}{\overset{\gamma}{C}}-\overset{|}{\overset{\beta}{C}}-\overset{|}{\overset{\alpha}{C}}-COOH$$

$\Delta\alpha \approx -4.5$ ppm
$\Delta\beta \approx -1.5$ ppm
$\Delta\gamma \approx -0.3$ ppm

$$-\overset{|}{C}-\overset{|}{C}-\overset{|}{C}-NH_2 \underset{\longleftarrow}{\overset{+ H^{\oplus}}{\longrightarrow}} -\overset{|}{\overset{\gamma}{C}}-\overset{|}{\overset{\beta}{C}}-\overset{|}{\overset{\alpha}{C}}-\overset{\oplus}{N}H_3$$

$\Delta\alpha \approx -1.5$ ppm
$\Delta\beta \approx -5.5$ ppm
$\Delta\gamma \approx -0.5$ ppm

$$-\overset{|}{C}-\overset{|}{\underset{\overset{|}{\overset{\oplus}{N}H_3}}{C}}-\overset{|}{C}-COO^{\ominus} \underset{\longleftarrow}{\overset{+ H^{\oplus}}{\longrightarrow}} -\overset{|}{\overset{\gamma}{C}}-\overset{|}{\underset{\overset{|}{\overset{\oplus}{N}H_3}}{\overset{\beta}{C}}}-\overset{|}{\overset{\alpha}{C}}-COOH$$

$\Delta\alpha \approx -1.5$ ppm
$\Delta\beta \approx -3.0$ ppm
$\Delta\gamma \approx -1.0$ ppm

The upfield protonation shifts, particularly for C-β, are mainly attributed to the electric field E of the charged group and its gradient $\frac{\partial E}{\partial r}$ at the site of the observed carbon [84a–c], while in the α and γ positions the contributions of inductive and steric effects will be more relevant. A model was developed [84c] in order to estimate and separate the contributions E and $\frac{\partial E}{\partial r}$ to the shielding constant. Under the simplifying assumption of a uniform electric field, no net charge will be induced at a carbon of undistorted tetrahedral bond symmetry, e.g. at the quaternary carbon of neopentylamine. It was concluded that the incremental protonation shift Δ_β upon perturbation of the tetrahedral bonding symmetry (e.g. in ethylamine) is caused by the field gradient $\frac{\partial E}{\partial r}$ [84c]. The difference in protonation shifts (about 3.4 ppm) can thus be considered as a measure for the electric field of $-\overset{\oplus}{N}H_3$ [84b]:

$$
\begin{array}{cc}
\text{CH}_3\\
\text{CH}_3-\overset{|}{\underset{|}{C}}\beta-\text{CH}_2-\text{NH}_2 & \text{H}-\overset{|}{\underset{|}{C}}\beta-\text{CH}_2-\text{NH}_2\\
\text{CH}_3 & \text{H}
\end{array}
$$

$\Delta = \delta_{\beta R - N\oplus H_2} - \delta_{\beta R - NH_2} \approx -5.2 \qquad \Delta' \approx -1.8; \qquad \Delta - \Delta' \approx 3.4 \text{ ppm}$

In nitrogen heteroaromatics, upfield protonation shifts are found for carbons α to nitrogen, while those in β and γ positions are deshielded on protonation [94, 99, 100]. This is shown in Fig. 3.5 for quinoline [94]. The protonation shifts for C-β and C-γ can be rationalized in terms of the cannonical formulae of protonated pyridine [73 d], while the upfield shifts for C-α are probably due to the lower π character of the N – C-α bond. The curves in Fig. 3.5 representing the pH dependence of ^{13}C shifts resemble titration curves. pK values and, in the case of amino acids, the isoelectric points pI can be obtained from the point of inflection of the δ versus pH plot for each individual carbon [84, 94, 98].

According to the Henderson-Hasselbach equation (3.6), an accurate approximation of the pK value is obtained from a semilogarithmic plot

of $\log \dfrac{\delta_{max.} - \delta}{\delta - \delta_{min.}}$ versus pH

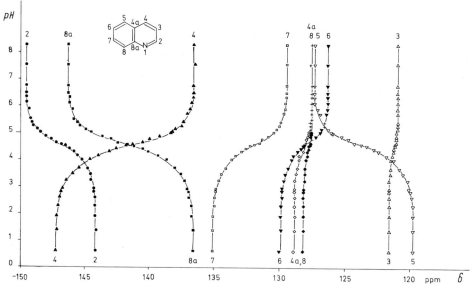

Fig. 3.5. pH dependence of quinoline ^{13}C shifts in water-acetone (vol. ratio 70:30%) at 25 °C [94].

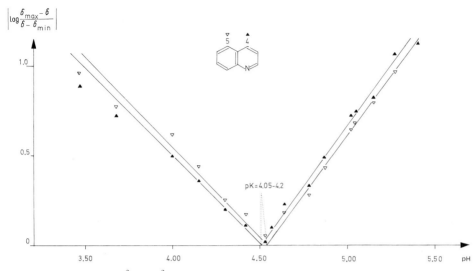

Fig. 3.6. Amount of $\log \dfrac{\delta_{\text{max.}} - \delta}{\delta - \delta_{\text{min.}}}$ versus pH plot for pK determination of quinoline, using the data shown for C-4 and C-5 in Fig. 3.5.

for values of δ near the point of inflection:

$$\text{pH} = \text{pK} + \log \dfrac{\delta_{\text{max.}} - \delta}{\delta - \delta_{\text{min.}}} \quad [101, 102]. \tag{3.6}$$

$\delta_{\text{max.}}$ and $\delta_{\text{min.}}$ are the maximum and minimum shifts of the carbon atom, corresponding to the pH invariant segments of the δ versus pH curves (e.g. $\delta_{\text{max.}} = 147.2$ ppm at pH < 2.0; $\delta_{\text{min.}} = 136.6$ ppm at pH > 6.5 for C-4 of quinoline in Fig. 3.5). Henderson-Hasselbach plots are illustrated in Fig. 3.6 for C-4 and C-5 of quinoline, using the data shown in Fig. 3.5.

Alternatively, chemical shifts δ at pH values near pK can be calculated with eq. (3.6), if pK, $\delta_{\text{max.}}$ and $\delta_{\text{min.}}$ are known.

3.1.5 Isotropic Shifts

In addition to line broadening due to accelerated relaxation, paramagnetic salts and chelates give rise to isotropic shifts when added to samples containing molecules with groups susceptible to coordination with metal ions, e.g. $-\text{OH}$, $-\text{NH}_2$, $-\text{SH}$, $-\text{COOH}$, $>\text{C}=\text{O}$. The isotropic shift Δ_i of a nucleus i in a sample S is the difference between chemical shifts measured before and after addition of the paramagnetic shift reagent SR [103]:

$$\delta_{i(S+SR)} - \delta_{i(S)} = \Delta_i. \tag{3.7}$$

Δ_i arises from two contributions:

$$\Delta_i = \Delta_i^{dipolar} + \Delta_i^{contact}. \qquad (3.7\,a)$$

The pseudocontact term $\Delta_i^{dipolar}$ is due to the valence electrons of the shift reagent SR; they cause an additional intramolecular field in the adduct S + SR, shielding or deshielding the nucleus i in the molecule S. The Fermi contact term $\Delta_i^{contact}$ accounts for interaction between the nucleus i and the field of the unpaired electrons of the paramagnetic additive SR which may be delocalized within the adduct S + SR [103].

The net effect of a shift reagent is an expansion of the NMR spectrum. The magnitude of the isotropic shift increases with the concentration of the paramagnetic additive; coupling constants remain largely unaffected. Shift reagents thus simulate high-field NMR to some degree. However, whereas NMR spectra are expanded linearly and without changes in δ values by raising the magnetic field strength, shift reagents change δ and sometimes signal sequences, since the magnitude of the isotropic shift Δ_i depends on the distance r_i of nucleus i from the paramagnetic metal ion in the adduct S + SR [103].

Since the unpaired 4f electrons in lanthanide ions occupy orbitals of extremely low radial extension, the Fermi contact term $\Delta_i^{contact}$ is small for lanthanide shift reagents, and pseudocontact shifts $\Delta_i^{dipolar}$ predominate. Moreover, the influence of unpaired electrons in lanthanide additives on relaxation of sample nuclei is so small, that practically no line broadening is observed. The pseudocontact term $\Delta_i^{dipolar}$ of nucleus i is a function of the polar coordinates r_i, ϑ_i, and φ_i of i, with the paramagnetic ion at the origin (Fig. 3.7 (a)) [103].

$$\Delta_i^{dipolar} = f(r_i, \vartheta_i, \varphi_i). \qquad (3.8)$$

Crudely approximated, the magnitude of the pseudocontact shift $\Delta_i^{dipolar}$, which predominates in lanthanide shift reagents, decreases with increasing distance r_i between the observed nucleus i and the paramagnetic ion:

$$\Delta_i^{dipolar} \propto \frac{1}{r_i^3}. \qquad (3.8\,a)$$

For expanding NMR spectra of aqueous sample solutions, lanthanide salts such as the chlorides, nitrates, or perchlorates of europium and praseodymium are used [103]. In organic solvents, tris-(β-diketonato)-europium(III) chelates *(1)* are usually applied [103]. Chiral europium(III) chelates such as tris-(3-*tert*-butylhydroxymethylene)-d-camphorato-europium(III) *(2)* separate the signals of enantiomers in the NMR spectra of racemic solutions [103].

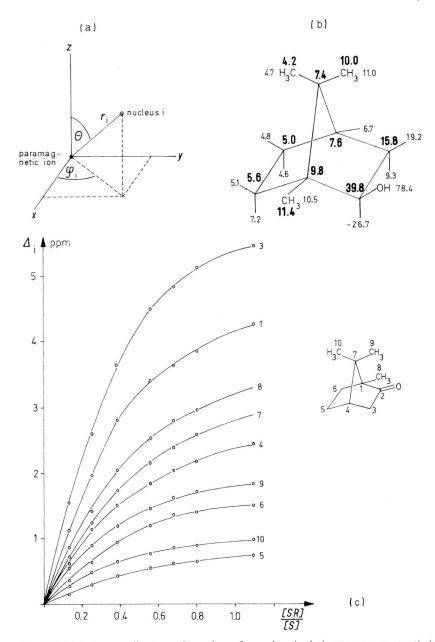

Fig. 3.7. (a) Polar coordinates r_i, ϑ_i, and φ_i of a nucleus i relative to a paramagnetic ion.
(b) Proton (small numbers) and carbon-13 (large numbers) isotropic shifts Δ_i of isoborneol in ppm; shift reagent: tris-(dipivaloylmethanato)-europium(III) (*1*, $R^1 = R^2 = t$-butyl); molar ratio [SR]/[S] = 1 [103, 105].
(c) Carbon-13 isotropic shift of camphor as a function of the molar ratio [SR]/[S] of shift reagent and sample [106]; shift reagent: tris-[4,4,4-trifluoro-(2-thienyl)-1,3-butanedionato]-europium(III) (*1*, R^1 = trifluoromethyl, R^2 = thienyl).

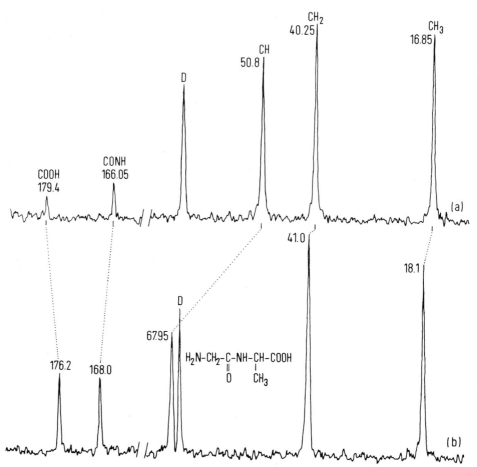

Fig. 3.8. 22.63 MHz PFT ^{13}C{^{1}H} NMR spectrum of glycylalanine; 0.3 mol/L in deuterium oxide; pH = 3.4; 25 °C; 2048 accumulated pulse interferograms; pulse width: 5 µs; pulse interval: 0,4 s; magnitude; internal reference: D = 1,4-dioxane;
(a) without, (b) with praseodymium perchlorate ([SR]/[S] = 0.86).

Lanthanide shift reagents can aid in signal assignments in ^{13}C{^{1}H} experiments, as was reported for the ^{13}C{^{1}H} NMR spectrum of ribose-5-phosphate [104], using europium chloride in aqueous solution, and in the case of isoborneol [105], isotropic shifts being induced by tris-(dipivaloyl methanato)-europium(III) (*1*, $R^1 = R^2 = t$-butyl). As shown for isoborneol in Fig. 3.7 (b), proton and carbon isotropic shifts do not differ significantly in their order of magnitude [103]. Plotting the isotropic shifts Δ_i observed for camphor *versus* the molar ratio between shift reagent and sample, [SR]/[S] [106], as illustrated in Fig. 3.7 (c), demonstrates three characteristics:

(1) The plot $\Delta_i = f([SR]/[S])$ is not linear, but may be approximated by a straight line at small concentrations of the shift reagent.

(2) In accordance with the proportionality (3.8 a), the magnitude of the pseudocontact shift, which predominates for lanthanide shift reagents, decreases with the distance of nucleus i from the paramagnetic ion. Thus, C-9 of camphor is shifted more than C-10 on addition of the europium chelate. Often, the crude approximation (3.8 a) is a valuable assignment aid.

(3) Carbonyl compounds (*e.g.* camphor in Fig. 3.7 (c)) show smaller isotropic shifts than comparable alcohols (*e.g.* isoborneol in Fig. 3.7 (b)).

In aqueous oligopeptide solutions, C_α in the C-terminal amino acid shifts drastically downfield, whereas the carboxy signal of this residue suffers an upfield shift [106]. Fig. 3.8 illustrates this for glycylalanine [106].

3.1.6 Intramolecular Mobility and the Temperature Dependence of ^{13}C Chemical Shifts and Line Widths

3.1.6.1 Introduction

Many molecules show intramolecular mobility: Rotations of groups about σ bonds or inversions of cycloaliphatic rings are representative phenomena. A well-known example is N,N-dimethylformamide, which exists as an equilibrium mixture of *cis* and *trans* isomers due to the partial π character of the N–CO bond: Rotation of the dimethylamino group is restricted at room temperature but occurs upon heating.

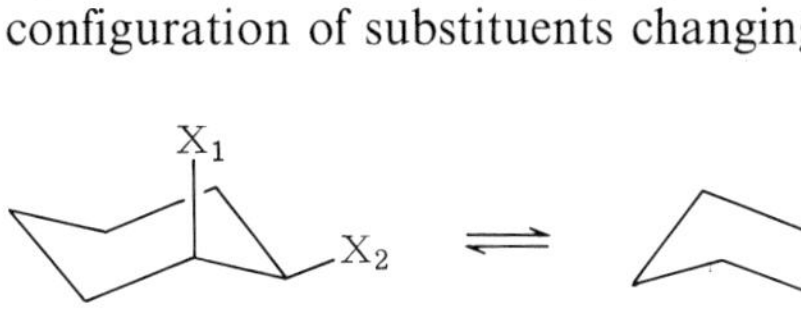

Another example is the ring inversion of cyclohexanes. These are usually nonresolvable equilibrium mixtures of rapidly interconverting conformers at room temperature, the configuration of substituents changing from axial to equatorial.

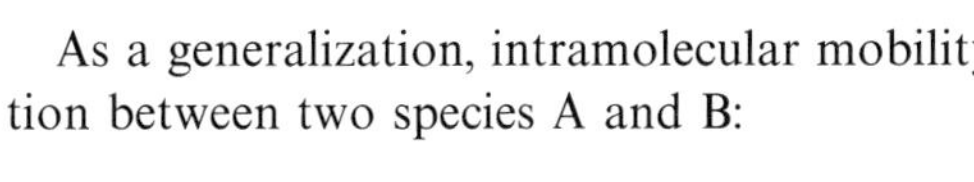

As a generalization, intramolecular mobility may give rise to equilibria of isomerization between two species A and B:

$$A \rightleftharpoons B.$$

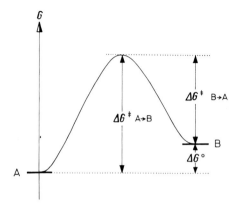

Fig. 3.9. The free enthalpy barrier of an intramolecular rotation or inversion A ⇌ B; ΔG^0 is zero if A and B are degenerated in energy.

Such equilibria depend on the temperature $T(K)$ and the free enthalpy of activation $\Delta G^{\ddagger}$ (Fig. 3.9). The rate constant k_r of isomerization is given by the Eyring equation (3.9) [107]:

$$k_r = \frac{kT}{h} e^{-\Delta G^{\ddagger}/RT}. \tag{3.9}$$

Eq. (3.9) arises from the absolute rate theory and can be expressed in the following logarithmic form, using the numeric values of the Boltzmann constant k, the gas constant R, the Planck constant h, and $\log e$ [108].

$$\Delta G^{\ddagger} = 4.57\, T\left(10.32 + \log \frac{T}{k_r}\right) \cdot 4.19 \text{ kJ/mol} \quad (1 \text{ kcal/mol} = 4.19 \text{ kJ/mol}). \tag{3.9 a}$$

In an NMR experiment, A and B have two separate signals for the same nucleus when the following conditions are met.

(1) The nucleus has different chemical environments in A and B.

(2) The chemical shift difference $(v_A - v_B)$ of the nucleus in A and B is large relative to its line width $\Delta v_{1/2}$ $((v_A - v_B) \gg \Delta v_{1/2})$, contributions of relaxation and field inhomogeneities to $\Delta_{1/2}$ being small.

(3) Interconversion between A and B occurs so slowly that k_r is very small relative to the chemical shift difference.

Due to the third condition and the fact that the rate constant k_r is a function of temperature according to eq. (3.9), the NMR spectra of molecules susceptible to interconversion due to intramolecular mobility are temperature dependent. This behavior, which is discussed in more detail in standard texts (e.g. [5]) and reviews [107, 108], is illustrated in Fig. 3.10.

If the rate constant is very small compared to the chemical shift difference $(v_A - v_B)$ at low temperatures, two separate signals are observed for the corresponding nucleus in A

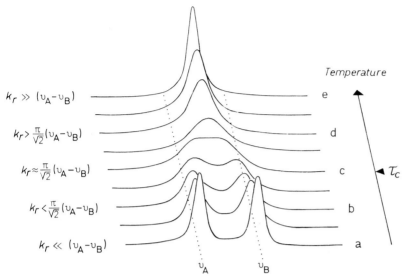

Fig. 3.10. Temperature dependence of chemical shifts and line shapes due to intramolecular mobility.

and B (Fig. 3.10(a)); the line widths are determined by the spin-spin relaxation time and the field inhomogeneity, being of the order of 1 Hz in $^{13}C\{^1H\}$ experiments. As the temperature increases, the life-time $1/k_r$ of A and B, and thus the life-time of a nucleus in a particular spin state, is reduced due to increasing k_r. As a result, the line widths of the A and B signals increase (Fig. 3.10(b)), recalling Sections 1.5.2 and 1.5.3. Furthermore, k_r approaches the magnitude of the chemical shift difference $(v_A - v_B)$. Consequently, the signals of A and B approach each other coalescing at the temperature T_c in Fig. 3.10(c). At the coalescence temperature T_c, the rate constant is related to the chemical shift difference by eq. (3.10).

$$k_r = \frac{\pi}{\sqrt{2}}(v_A - v_B). \tag{3.10}$$

As k_r gradually becomes larger than $(v_A - v_B)$ with increasing temperature, the line width of the averaged signal decreases, finally reaching the "natural" line width $\Delta v_{1/2}$ at high temperatures (Fig. 3.10(e)).

Using the temperature dependence of NMR spectra, thermodynamic data of interconversions due to intramolecular mobility can be determined. If the coalescence temperature is known, the rate constant at T_c is calculated according to eq. (3.10). Often, T_c and $(v_A - v_B)$ are not known exactly. In this case, a rough value for k_r can be calculated by measuring the half maximum intensity line widths $\Delta v_{1/2(T_c)}$ at temperatures near T_c and using eq. (3.11) [107, 108].

$$k_r \cong 2\Delta v_{1/2(T_c)} \quad \text{(near } T_c\text{).} \tag{3.11}$$

Using the k_r values thus obtained, the free enthalpies $\Delta G^{\ddagger}_{T_c}$ of rotation or inversion can be determined according to eq. (3.9a). Other thermodynamic parameters of intramolecular mobility are the energy of activation, ΔE_a, from the Arrhenius equation (3.12),

$$k_r = A\,e^{-\Delta E_a/RT}, \tag{3.12}$$

the enthalpy of activation, $\Delta H^{\ddagger}$,

$$\Delta H^{\ddagger} = \Delta E_a - RT, \tag{3.13}$$

and the entropy of activation, $\Delta S^{\ddagger}$, given by the Gibbs-Helmholtz equation (3.14),

$$\Delta G^{\ddagger} = \Delta H^{\ddagger} - T\Delta S^{\ddagger}. \tag{3.14}$$

The calculation of these data requires the temperature dependence of k_r [107]; the energy of activation can then be obtained as the slope of a $\log k_r$ versus $1/T$ plot according to eq. (3.12). However, the measurement of $k_r = f(T)$ is more difficult and includes more errors than the determination of k_r at or near T_c by using eqs. (3.10) and (3.11) [107, 108].

For this reason it is usually preferable to determine the numerically more accurate free enthalpy of activation, $\Delta G_{T_c}^{\ddagger}$, from the temperature dependent NMR spectra according to eq. (3.9 a).

If energies of activation are required, a convenient and accurate way to obtain rate constants k_r as functions of temperature is to calculate line shapes for different values of k_r. The experimentally obtained line shapes at $T(K)$ are compared with the calculated ones. The best-fitting pairs of experimental and computer-simulated line shapes give corresponding values for rate constant and temperature, so that $k_r = f(1/T)$ can be plotted for the determination of ΔE_a. This is illustrated in Fig. 3.11 for the methyl resonance of N,N-dimethyltrichloroacetamide [109].

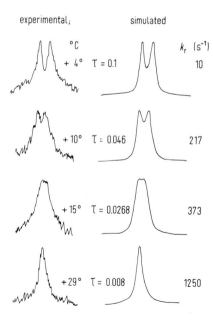

Fig. 3.11. Experimental and computer-simulated methyl ^{13}C signal shapes (22.63 MHz) of N,N-dimethyltrichloroacetamide at various temperatures (measured) fitted to rate constants (simulated) [109]. (Reproduced by permission of the copyright owner from Ref. [109].)

3.1.6.2 Temperature Dependence of ^{13}C NMR Spectra

Studies of intramolecular mobility using ^{1}H NMR often suffer from small shift differences between rotational and inversional isomers, as well as from unresolved splitting due to homonuclear spin-spin coupling. Because of the large spectral width to line width ratio in ^{13}C{^{1}H} NMR (Section 3.1.1), temperature-dependent ^{13}C{^{1}H} experiments should provide a larger signal separation for rotational and inversional isomers than ^{1}H NMR while eliminating the complications of homonuclear spin-spin coupling, at least in samples with naturally abundant ^{13}C. Indeed, this has been reported for the inversion of several cycloaliphatic rings [110-113].

The high-field 251 MHz ^{1}H NMR spectrum of cyclononane (Fig. 3.12 (a)) changes from a single line at room temperature to two broad overlapping signals about 30 Hz apart at $-160\,°$C [110]. At the same field strength B_0, and within the same range of temperatures, the PFT ^{13}C{^{1}H} NMR spectrum changes from a single resonance to two widely spaced signals of intensity ratio 2:1, the shift difference being 570 Hz or about 9 ppm at 63.1 MHz (Fig. 3.12(b)) [110]. From temperatures near coalescence (about 120 K for ^{1}H and 140 K for ^{13}C) an inversion barrier of about 25 kJ/mol is obtained [110].

Cyclononane prefers a twist boat-chair (TBC) conformation (see references cited in [110]), with two classes of CH$_2$ groups differing in symmetry (● and ·). At room temperature, a rapid inversion between two enantiomeric TBC forms probably occurs, involving an intermediate boat-chair form: An averaged spectrum having one signal as in Fig. 3.10(e) is observed (Fig. 3.12) [110]. At very low temperature, this inversion is frozen,

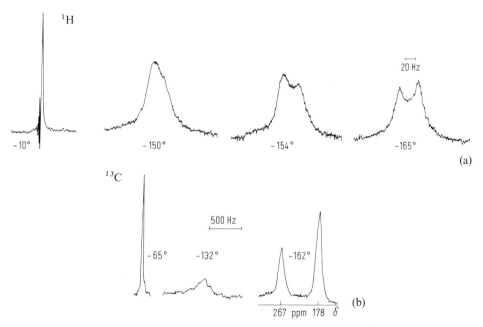

Fig. 3.12. Temperature-dependent ^{1}H and PFT ^{13}C{^{1}H} NMR spectra of cyclononane [110], 251 MHz for ^{1}H and 63.1 MHz for ^{13}C.

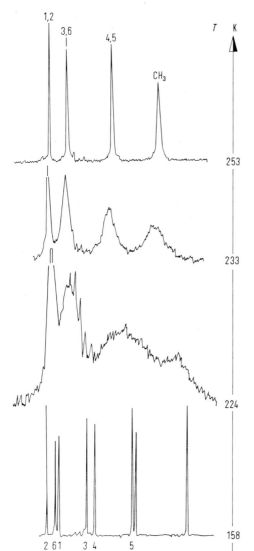

Temperature °C	Carbon	δ_C ppm
−100	1	33.45
	2	35.5
	3	28.7
	4	27.3
	5	21.0
	6	34.05
	1-CH$_3$ (ax.)	11.75
	2-CH$_3$ (eq.)	20.3
40	1,2	35.05
	3,6	32.05
	4,5	24.2
	1,2-CH$_3$	16.25

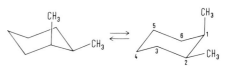

Fig. 3.13. Temperature-dependent 22.63 MHz PFT ^{13}C{^{1}H} NMR spectra of *cis*-1,2-dimethylcyclohexane [112].

and two signals are observed for the nonequivalent nuclei (three • and six ·), those lying on C_2 axes (•) appearing at lower field.

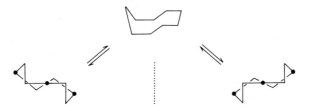

Similarly, ring inversion of methyl cyclohexanes [111], dimethylcyclohexanes [112], and cis-decalin [113] is more easily examined with $^{13}C\{^1H\}$ experiments than with ^1H NMR. As an example, the PFT $^{13}C\{^1H\}$ NMR spectra of cis-1,2-dimethylcyclohexane at various temperatures are shown in Fig. 3.13 [112]. At high temperatures, i.e. $T = 250$ K, sharp, averaged ^{13}C signals for C-1,2, C-3,6, C-4,5, and the axial and equatorial methyl carbons are observed. This was attributed to a fast ring inversion between two chair forms of equal potential energy, the methyl group rapidly exchanging axial and equatorial configurations: $k_r \gg (v_{eq.} - v_{ax.})$. As the temperature decreases, the line widths of the averaged signals increase, as was the case in going from (e) to (a) in Fig. 3.10. At 158 K, a single signal is obtained for each carbon (Fig. 3.13). Using the line widths near the coalescence point ($T_c \approx 220$ K), the free enthalpy barrier of inversion is found to be $\Delta G^{\pm}_{220} \approx 41.4$ kJ/mol for cis-1,2-dimethylcyclohexane and $\Delta G^{\pm}_{220} \approx 38.9$ kJ/mol for the cis-1,4-isomer [112], which agrees well with values known from ^1H NMR.

Intramolecular rotations in carboxylic and carbonic acid amides have been thoroughly investigated by ^1H NMR [108]. The thermodynamic data for rotation in N,N-dimethyltrichloroacetamide as obtained by ^1H and $^{13}C\{^1H\}$ NMR agree quite well (^1H: $\Delta G^{\pm}_{294} = 76.5$ kJ/mol; ^{13}C: $\Delta G^{\pm}_{288} = 73.6$ kJ/mol) [109].
Any interconversion in a sample may give rise to temperature-dependent NMR spectra. For example, the ^{13}C NMR spectrum of dimeric cyclopentadienyl iron dicarbonyl is temperature dependent [114]. This was attributed to intermolecular exchanges of carbonyls and interconversion between cis and trans complexes.

3.2 ^{13}C Coupling Constants

3.2.1 Basic Theoretical Considerations

The mechanism assumed to predominate in nuclear spin-spin coupling was outlined for an AX group in Section 1.10.4: The spin of nucleus A polarizes the bonding electrons of atom A in the molecule AX; spin correlation between the electrons in the AX bond affects the spin of nucleus X. It is therefore reasonable to assume that at least one-bond coupling constants depend on the angular and radial distributions of the bonding electrons occupying molecular orbitals. The angular distribution of bonding electrons is expressed in terms of the s character of hybrid orbitals making up the bond, e.g. 0.25, 0.33, and 0.5 for sp^3, sp^2, and sp orbitals, respectively. In valence bond and molecular orbital approximations, the charge bond-order matrix $P_{s_Cs_X}$ accounts for these hybridization parameters [115]. The radial distribution of valence electrons may be expressed in terms of bond polarity. More precisely, this radial part is related to the carbon 2s and proton 1s orbital densities (for A = ^{13}C and X = ^1H), denoted as $s_C^2(0)$ and $s_H^2(0)$ [115]. According to an

MO treatment of one-bond coupling constants, $^{13}C-X$ couplings are expressed by eq. (3.15).

$$J_{CX} = \left(\frac{4}{3}\right)^2 h\mu_\beta \gamma_C \gamma_X P^2_{s_C s_X} s^2_C(0) s^2_X(0) (\Delta E_e)^{-1}. \tag{3.15}$$

In this equation, ΔE_e is the approximated electronic excitation energy; γ_C and γ_X are the gyromagnetic ratios of the coupled nuclei, ^{13}C and X; μ_β is the Bohr magneton, h the Planck constant.

In this form, eq. (3.15) yields only positive $^{13}C-X$ couplings; however, negative one-bond couplings (e.g. $J_{C^{19}F}$) are also observed [115]. Moreover, interpretation of carbon one-bond coupling constants in terms of carbon s character or hybridization (Fig. 3.14) assumes the factor $s^2_C(0) s^2_X(0) (\Delta E_e)^{-1}$ to be constant. In fact, ΔE_e and $s^2_C(0)$ are found to vary considerably in substituted methanes [116, 117].

Another more successful MO approach, referred to as INDO (intermediate neglect of differential overlap), avoids the average electronic energy approximation [115]. Its concept is a self-consistent field perturbation calculation. The INDO approach permits computation of one-bond carbon-13 coupling constants. The results obtained for J_{CH} agree well with the experimental data for hydrocarbons and molecules with $-F$, $-OR$, $-NR_3$ as substituents, but not for those containing $-C\begin{smallmatrix}\nearrow O\\\searrow X\end{smallmatrix}$, $-NO_2$, or $-CN$. Qualitatively good agreement between computed and experimental carbon couplings is also achieved with INDO for carbon-carbon, but nor for carbon-fluorine and carbon-nitrogen-15 coupling constants. Nevertheless, the INDO approach yields the negative sign of some $^{13}C-^{19}F$ coupling constants as is observed experimentally [115].

For structural elucidation by means of carbon-13 coupling constants, an empirical approach is often sufficient: Carbon-13 one-bond coupling constants, particularly J_{CH} values roughly correlate with carbon hybridization and bond polarity. The latter is greatly affected by electron-withdrawing heteroatoms or substituents. These relations will be outlined in the following sections.

3.2.2 Carbon-Proton Coupling

3.2.2.1 One-Bond Coupling (J_{CH})

One-bond carbon-proton coupling constants J_{CH} range from 120 to 320 Hz (Fig. 3.14). There are two structural features that increase carbon-proton coupling constants. These are

1) increasing s character of the carbon hybrid orbital making up the C−H bond;
2) electron-withdrawing substituents at the coupling carbon and an increased degree of substitution with such groups.

For closely analogous representatives of sp^3, sp^3, and sp carbons, such as the series ethane, ethene, ethyne, or methylamine, formaldimine, hydrogen cyanide, J_{CH} increases

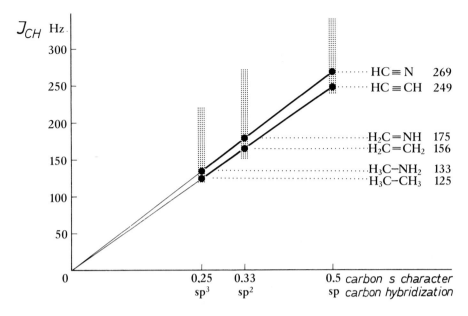

Fig. 3.14. Ranges of one-bond carbon-13-proton coupling constants for sp³, sp², and sp carbons (dotted rectangles) and linear correlation between J_{CH} and carbon s character of comparable compounds (coupling constants from Ref. [115]). The empirical relation is $J_{CH} \approx 500 \cdot s$ (Hz).

linearly with carbon s character as shown in Fig. 3.14. The J_{CH} ranges for the three types of carbon hybridization are narrow and do not overlap, provided that only one heteroatom or substituent is attached to the coupling carbon (Fig. 3.14). Thus, sp³ carbons in CH_3X compounds or $-CH_2X$ groups have J_{CH} values between 120 and 150 Hz (Tables 3.4, 3.5). In the series ethane, methylamine, methanol (Table 3.5), the methyl coupling J_{CH} increases with increasing electron withdrawal by the substituent at the methyl carbon. For terminal $=CH_2$ carbons of vinyl compounds and for benzenoid $>CH$ carbons, one-bond couplings between 150 and 170 Hz are observed. The influence of a substituent, a heteroatom, conjugation or cumulation of π bonds is within the same range (Tables 3.6, 3.7). Much smaller J_{CH} values (25–75 Hz) are reported for the sp² carbons of open-chain carbenium ions [75], since dynamic effects cause an averaging with the two-bond coupling constant $^2J_{CH}$ in this case. A one-bond coupling constant of 250 ± 5 Hz is characteristic of the terminal sp carbon of an ethynyl group (Table 3.6, bottom).

The range of J_{CH} expands considerably if more than one electron-withdrawing substituent is found at the coupling carbon. This trend is exemplified in a series of formaldehyde derivatives (Table 3.6) and in di- and trisubstituted methanes, CH_2X_2 and CHX_3 (Table 3.5). Whereas methyl iodide, bromide, chloride, and fluoride show almost equal J_{CH} values of about 150 Hz, the coupling constants increase almost linearly for each additional substituent at the methyl carbon, but individually for each kind of substituent.

For substituted methanes of type CHXYZ, the coupling constant J_{CHXYZ} can be approximated by adding the coupling constant of methane, J_{CH_4}, and the empirical incre-

Table 3.5. One-Bond Coupling Constants J_{CH} of Methane Derivatives HCH_2X (Left Side) and $HCHX_2$ or HCX_3 (Right Side); Data from Ref. [115].

Compound	Formula	J_{CH} (Hz)	Compound	Formula	J_{CH} (Hz)
Methane	HCH_3	125.0			
Ethane	HCH_2CH_3	124.9			
			Propane	$HCH(CH_3)_2$	119.4
			t-Butane	$HC(CH_3)_3$	114.2
Propene	$HCH_2CH=CH_2$	122.4			
Toluene	$HCH_2C_6H_5$	129.4			
Propyne	$HCH_2C\equiv CH$	132.0			
Methyl iodide	HCH_2I	151.1			
Methyl bromide	HCH_2Br	151.5			
Methyl chloride	HCH_2Cl	150.0			
			Dichloromethane	$HCHCl_2$	178.0
			Chloroform	$HCCl_3$	209.0
Methyl fluoride	HCH_2F	149.1			
			Difluoromethane	$HCHF_2$	184.5
			Trifluoromethane	HCF_3	239.1
Methylamine	HCH_2NH_2	133.0			
Methanol	HCH_2OH	141.0			
			Ethanol	$HCH(OH)CH_3$	140.3
				HCH_2CH_2OH	126.9
			2-Propanol	$HC(OH)(CH_3)_2$	142.8
				$(HCH_2)_2CHOH$	126.9
Dimethyl ether	HCH_2OCH_3	140.0			
			Formaldehyde dimethylacetal	$HCH(OCH_3)_2$	161.8
			Methyl orthoformate	$HC(OCH_3)_3$	186.0
Acetic acid	HCH_2COOH	130.0			
			Malonic acid	$HCH(COOH)_2$	132.0
Acetonitrile	HCH_2CN	136.1			
			Malodinitrile	$HCH(CN)_2$	145.2
Nitromethane	HCH_2NO_2	146.0			
			Dinitromethane	$HCH(NO_2)_2$	169.4
Fluoroacetonitrile	HCHFCN	166.0			
			Difluoroacetonitrile	HCF_2CN	205.5

3.2 ^{13}C Coupling Constants 137

Table 3.6. One-Bond Coupling Constants J_{CH} of sp^2 and sp Carbons.

Class	Compound	Formula	J_{CH} (Hz)	Ref.
sp^2	Ethene	H$_2$C=CH$_2$	156.2	[115]
	Vinyl fluoride	$_a$H, $_b$H / C=C / H^c, F	159.2 (a) 162.2 (b) 200.2 (c)	[115]
	Trimethylethene	H, H$_3$C / C=C / CH$_3$, CH$_3$	148.4	[124]
	1,1-Di-*t*-butylethene	H, H / C=C / C(CH$_3$)$_3$, C(CH$_3$)$_3$	151.9	[124]
	Tri-*t*-butylethene	H, (H$_3$C)$_3$C / C=C / C(CH$_3$)$_3$, C(CH$_3$)$_3$	143.3	[124]
	Methylenecyclobutane	(CH$_2$)$_n$ C=CH$_2$, n = 3	154.9	[124]
	Methylenecyclopentane	n = 4	154.2	[124]
	Methylenecyclohexane	n = 5	153.5	[124]
	Methylenecycloheptane	n = 6	153.4	[124]
	Allene	H$_2$C=C=CH$_2$	168.2	[115]
	cis-Stilbene	H, H$_5$C$_6$ / C=C / H, C$_6$H$_5$	155.0	[126]
	trans-Stilbene	H, H$_5$C$_6$ / C=C / C$_6$H$_5$, H	151.0	[126]
	cis-Diazastilbene*	H, NH$_4$C$_5$ / C=C / H, C$_5$H$_4$N	159.0	[126]
	trans-Diazastilbene*	H, NH$_4$C$_5$ / C=C / C$_5$H$_4$N, H	157.0	[126]
	Acetaldoxime, anti-	H, H$_3$C / C=N / OH	163.0	[115]
	Acetaldoxime, syn-	H, H$_3$C / C=N / OH	177.0	
	Formaldehyde	H, H / C=O	172.0	[115]
	Acetaldehyde	H$_3$C, H / C=O	172.4	[115]
	Formamide	H$_2$N, H / C=O	188.3	[115]

* cis- and trans-di-α-pyridylethene.

Table 3.6 (continued).

Class	Compound	Formula	J_{CH} (Hz)	Ref.
sp	N,N-Dimethylformamide	$(H_3C)_2N\text{-CH=O}$	191.2	[115]
	Formate anion in aq. sol.	$^-O\text{-CH=O}$	194.8	[115]
	Formic acid	$HO\text{-CH=O}$	222.0	[115]
	Methyl formate	$H_3CO\text{-CH=O}$	226.2	[115]
	Formyl fluoride	$F\text{-CH=O}$	267.0	[115]
	Ethyne	HC≡CH	249.0	[115]
	Propyne	HC≡C–CH$_3$	248.0	[115]
	2-Propyn-1-ol	HC≡C–CH$_2$OH	248.0	[115]
	Phenylethyne	HC≡C–C$_6$H$_5$	251.0	[115]
	Hydrogen cyanide	HC≡N	269.0	[115]

ments, j, according to eq. (3.16). The j parameters include j_X, j_Y, j_Z for the substituents [118] and j_{XY}, j_{XZ}, j_{YZ} for combinations of substituent pairs [119].

$$J_{CHXYZ} = J_{CH_4} + j_X + j_Y + j_Z + j_{XY} + j_{XZ} + j_{YZ}. \quad (3.16)$$

The one-bond coupling constants J_{CH} of cycloaliphatic rings, such as cyclohexane, or heteroalicyclics, such as 1,4-dioxane, are almost equal to those found for comparable open-chain compounds (Table 3.7).

In strongly strained systems such as cyclopropane or cyclopropene, very large coupling constants J_{CH} are measured (Table 3.7). This is attributed to the high s character of the ring carbon hybrid orbitals, cyclopropane (161 Hz) having olefinic character, and cyclopropene (220 Hz) being more closely related to an alkyne [2, 120, 121]. Similarly, the three-membered heteroalicyclic compounds aziridine (172 Hz), oxirane (175.5 Hz), and thiirane (170.5 Hz) show much larger J_{CH} values than their higher membered homologues [122].

The J_{CH} values reported for cis and trans isomers of alkenes and azomethenes differ as shown in Table 3.6. The larger coupling constant, found for the acetaldoxime isomer with a proton configuration trans to the hydroxy group, is attributed to the lone electron pair at nitrogen which is cis to the proton [123]. In alkenes, crowding of alkyl groups such as methyl or t-butyl at the sp^2 carbons reduces one-bond C–H coupling constants of the olefinic carbons by 5–10 Hz [124] (Table 3.6).

3.2 ^{13}C Coupling Constants   139

Table 3.7. One-Bond Coupling Constants J_{CH} of some Alicyclic and Aromatic Carbons.

Compound	Structure	J_{CH} (Hz)	Ref.
Cyclopropane		161	[120]
Cyclobutane		136	[120]
Cyclopentane		131	[120]
Cyclohexane		127	[120]
Tetrahydrofuran	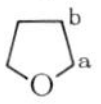	149 (a) 133 (b)	[125]
1,4-Dioxane	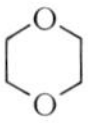	145	[125]
Cyclopropene (sp^2)		220	[121]
Benzene		165	[158]
Mesitylene	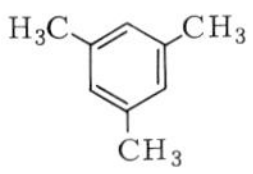	154	[115]
Pyridine	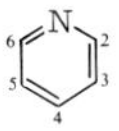	170 (2,6) 163 (3,5) 152 (4)	[115]
2,4,6-Trimethylpyridine	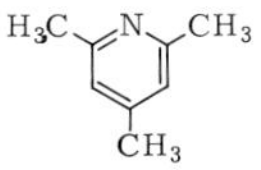	158.5	[115]
Pyrrole	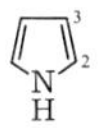	170 (3) 182 (2)	[130]
Pyrazole	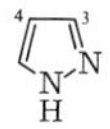	178 (4) 190 (3)	[130]
Imidazole	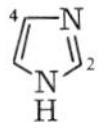	199 (4) 208 (2)	[130]
1,2,3-Triazole	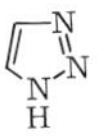	205	[130]
1,2,4-Triazole	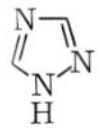	208	[130]
Tetrazole	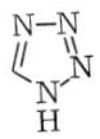	216	[130]

The one-bond coupling constants of halo- and pseudohalobenzenes, $X-C_6H_5$ (X = halogen, CN, NO$_2$) decrease slightly with decreasing electronegativity and distance from the substituent X in the order

$$J_{CH(2,6)} > J_{CH(3,5)} > J_{CH(4)},$$

fluorobenzene being an exception, as shown in Table 3.9. Methyl coupling constants J_{CH} of substituted toluenes, t-butylbenzenes, N,N-dimethylanilines, and anisoles are reported to correlate linearly with the Hammet substituent constants in each series [125].

The one-bond coupling constants J_{CH} of pyridine ring carbons behave similarly to those of benzenes substituted by electron acceptors such as cyano and nitro groups (Table 3.7). In azole and azines, J_{CH} increases with the number of nitrogen atoms in the ring (Table 3.7).

One-bond coupling constants J_{CH} may suffer from slight solvent effects. Table 3.4 shows this behavior for chloroform, whose carbon-proton coupling increases with the polarity of the medium when measured in different solvents, being 208 Hz in cyclohexane and 215 Hz in pyridine [92]. This is attributed to association between chloroform and solvents susceptible to hydrogen bonding.

3.2.2.2 Longer-Range Carbon-Proton Couplings: The J_{CH}/J_{HH} Ratio

Longer-range carbon-proton coupling constants $^2J_{CH}$ and $^3J_{CH}$ correlate with comparable proton-proton couplings [129] (Table 3.8). $^3J_{CH}$ is about 0.5 to 0.7 of $^3J_{HH}$. Exclusive consideration of the gyromagnetic ratios of ^{13}C and ^1H according to eq. (3.15) would predict

$$J_{CH}/J_{HH} = \gamma_C/\gamma_H = 0.25$$

and does therefore not account adequately for the actual J_{CH}/J_{HH} ratio. Nevertheless, the parallel behavior of $^{2,3}J_{CH}$ and $^{2,3}J_{HH}$ can be attributed to similar coupling mechanisms, mainly due to the same spin quantum number and the absence of nonbonding electron pairs in both ^{13}C and ^1H.

Table 3.8. Longer-Range Carbon-Proton Coupling Constants in Relation to Comparable Proton-Proton Couplings.

		J_{CH} Hz	Type	J_{HH} Hz	
Cyclohexane carboxylic acid	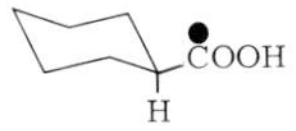	−6.2	$^2J_{gem}$	12.6	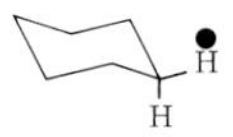 Cyclohexane
Propenoic acid	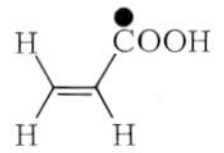	4.1 7.6 14.1	$^2J_{gem}$ $^3J_{cis}$ $^3J_{trans}$	2.5 11.6 19.1	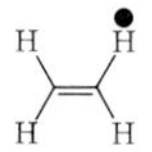 Ethene
Benzoic acid	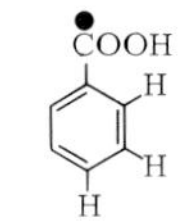	4.1 1.1 0.5	$^3J_{ortho}$ $^4J_{meta}$ $^5J_{para}$	7.7 1.4 0.6	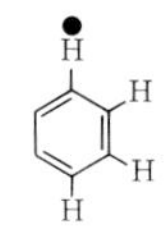 Benzene

Table 3.9. Two Bond Coupling Constants $^2J_{CH}$ of Selected Alkyl, Alkenyl, and Alkynyl Compounds *.

Compound	Formula*	$^2J_{CH}$ (Hz)	Ref.
Ethane	$H_3\mathbf{C}-C\mathbf{H}_3$	−4.5	[125]
1,2-Dichloroethane	$ClH_2\mathbf{C}-C\mathbf{H}_2Cl$	−3.4	[125]
1,1,2,2-Tetrachloroethane	$Cl_2H\mathbf{C}-C\mathbf{H}Cl_2$	1.2	[125]
Ethene	$H_2\mathbf{C}=C\mathbf{H}_2$	−2.4	[125]
1,2-Dichloroethene, *trans*	H\\C=C/Cl ; Cl/C=C\\H	0.8	[125]
1,2-Dichloroethene, *cis*	H\\C=C/H ; Cl/C=C\\Cl	16.0	[125]
Methylenecyclopentane	$-C\mathbf{H}_2$ \\ $\mathbf{C}=CH_2$ / $-CH_2$	4.2	[124]
Methylenecyclohexane		5.2	[124]
Methylenecycloheptane		5.5	[124]
Acetone	$(C\mathbf{H}_3)_2\mathbf{C}=O$	5.5	[126]
3-Aminoacrolein	$H_2N-C\mathbf{H}=C\mathbf{H}-CH=O$	6.0	[126]
Acetaldehyde	$H_3C-C\mathbf{H}=O$ (H on C)	26.7	[127]
2-Ethylbutyraldehyde	$(C_2H_5)_2CH-C\mathbf{H}=O$	22.1	[127]
Acrolein	$H_2C=CH-C\mathbf{H}=O$	26.9	[127]
3-Aminoacrolein	$H_2N-CH=CH-C\mathbf{H}=O$	20.0	[126]
Propynal	$H-C\equiv C-C\mathbf{H}=O$	33.2	[127]
Chloroacetaldehyde	$ClCH_2-C\mathbf{H}=O$	32.5	[127]
Dichloroacetaldehyde	$Cl_2CH-C\mathbf{H}=O$	35.3	[127]
Trichloroacetaldehyde	$Cl_3C-C\mathbf{H}=O$	46.3	[127]
Ethyne	$H-\mathbf{C}\equiv C-\mathbf{H}$	49.3	[125]
Phenoxyethyne	$C_6H_5O-\mathbf{C}\equiv C-\mathbf{H}$	61.0	[125]
1-Phenoxy-1-propyne	$C_6H_5O-\mathbf{C}\equiv C-C\mathbf{H}_3$	10.8	[125]
Typical values:	\\C=C/ ; =C\\	$^2J_{CH}=8$	
	O=C\\H /C\\	$^2J_{CH}=5-8$	
	−C−C−H (sp3)	$^2J_{CH}=1-6$	
	\\C=C/H	$^2J_{CH}=1-16$	
	O=C\\H /C\\ (aldehyde)	$^2J_{CH}=20-25$	
	−C≡C−H	$^2J_{CH}\geq 40$	

* the coupling nuclei are printed in bold type.

3.2.2.3 Two-Bond Coupling ($^2J_{CH}$)

Similar to *geminal* proton-proton coupling, two-bond carbon-proton coupling $^2J_{CH}$ becomes more positive when the C−C−H bond angle increases [129].

Cyclohexane carboxylic acid (109.5°): −6.2 Hz

(Z)-2-Butenoic acid (120°): 3.1 Hz

(E)-2-Butenoic acid (120°): 3.4 Hz

When an oxygen is *synclinal* to a C—C—H coupling path, $^2J_{CH}$ is negative, while an *antiperiplanar* configuration causes geminal coupling to be positive [129]:

$^2J_{CH} < 0$ $\quad\quad\quad$ $^2J_{CH} > 0$

This trend is exemplified not only in the β- and α-anomers of glucopyranose and other monosaccharides ([129], Fig. 2.7 (b)),

β- $\quad\quad$ α-

D-Glucopyranose

but is also apparent in alkenes substituted by electronegative groups. $^2J_{CH}$ of a proton *syn* to the substituent is negative, while a positive $^2J_{CH}$ is observed for the *anti* proton [129] as shown for bromoethane. Correspondingly, cis-1,2-dichloroethene displays a larger $^2J_{CH}$ than the *trans* isomer (Table 3.9).

Bromoethene

In benzenoid compounds, $^2J_{CH}$ of the substituted carbon becomes more positive when the electronegativity of the substituent decreases (Table 3.10) [128]:

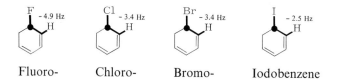

Fluoro- $\quad$ Chloro- $\quad$ Bromo- $\quad$ Iodobenzene

A corresponding behavior is found for the $^2J_{CH}$ couplings in the series furan, pyrrole, and thiophene (Table 3.11, [130]):

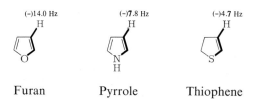

Furan $\quad\quad$ Pyrrole $\quad\quad$ Thiophene

3.2 ^{13}C Coupling Constants 143

Particularly large $^2J_{CH}$ values are characteristic of aldehyde protons, as is illustrated for ethanal, propenal, and benzaldehyde (Table 3.9, [129]). Obviously, the positive polarization of the aldehyde carbon causes $^2J_{CH}$ to increase significantly.

26.7 Hz	26.9 Hz	24.1 Hz
H_3C-CHO	$H_2C=CH-CHO$	C_6H_5-CHO
Ethanal	Propenal	Benzaldehyde

Increasing s character of the hybrid orbitals not only at the coupling carbon but also at that bearing the coupling proton increases the magnitude of two-bond CH coupling. This trend is recognized when comparing the $^2J_{CH}$ values of 1-phenoxy-1-propyne ($-C(sp)-C(sp)-C(sp)-C(sp^3)-H(s)$) and phenoxyethyne ($-C(sp)-C(sp)-H$), as illustrated in Table 3.9.

3.2.2.4 Three-Bond Coupling ($^3J_{CH}$)

In aliphatic and cycloaliphatic compounds, *vicinal* carbon-proton coupling constants $^3J_{CH}$ are related to the dihedral angle Φ, as known from the Karplus-Conroy relation for $^3J_{HH}$. The Fermi-contact contribution to $^3J_{CH}$ as a function of the dihedral angle Φ calculated for propane [131] is displayed in Fig. 3.15, and the Karplus relation given by eq. (3.17) can be derived:

$$^3J_{CH} = 4.26 - \cos\Phi + 3.56 \cos 2\Phi. \qquad (3.17)$$

According to Fig. 3.15 and eq. (3.17), *synclinal* carbon-proton coupling is $^3J_{CH_s} = 2$ Hz, while an *antiperiplanar* carbon-proton pair will couple with $^3J_{CH_a} = 8.8$ Hz. An averaged

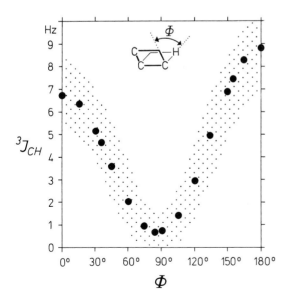

Fig. 3.15. $^3J_{CH}$ versus Φ plot of propane [131]. Calculated values are given by full circles.

three-bond coupling of 4 to 4.5 Hz will be expected for alkyl groups with free rotation, according to $^3J_{CH} = (2\,^3J_{CH_s} + {}^3J_{CH_a})/3$. Relation (3.17) was verified in several cases [129] using dihedral angles obtained from x-ray crystallographic data. As known from *vicinal* proton-proton coupling, electronegative substituents within the coupling path, such as oxygen in glycosides, reduce the magnitude of $^3J_{CH}$.

Couplings of carbon atoms with alkene protons exhibit the $^3J_{trans} > {}^3J_{cis}$ relationship known from proton NMR. The magnitudes of $^3J_{CH}$ increase slightly with increasing s character of the coupling carbon atom, as shown in the series propene, methylene-cyclohexene, and 2-methyl-1-buten-3-yne:

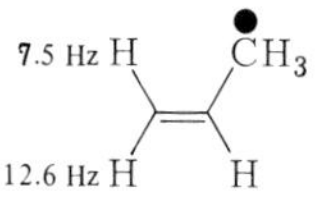

7.5 Hz H CH₃	7.8 Hz H	8.1 Hz H
12.6 Hz H H	11.9 Hz H	14.7 Hz H CH₃
Propene	3-Methylene-cyclohexene	2-methyl-1-buten-3-yne

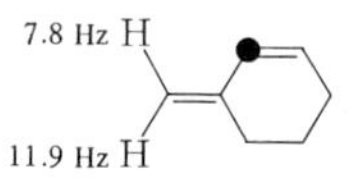

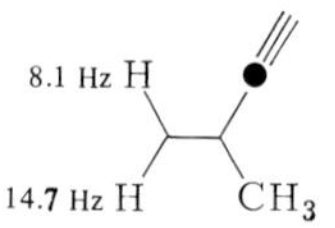

Electronegative substituents within the coupling path decrease while those at the coupling carbon atoms increase $^3J_{CH}$ with *cis* and *trans* protons, as shown for 2- and 3-bromopropene [129].

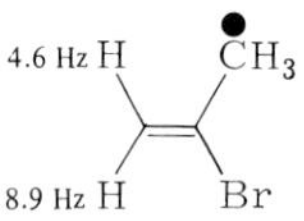

4.6 Hz H CH₃	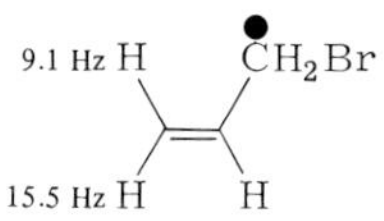 9.1 Hz H CH₂Br
8.9 Hz H Br	15.5 Hz H H
2-Bromopropene	3-Bromopropene

Steric interactions within the coupling path decrease *trans* $^3J_{CH}$ values, as illustrated for propenal derivatives [129].

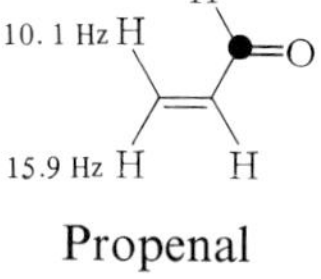

10.1 Hz H H=O	9.5 Hz H H=O	H₃C H=O
15.9 Hz H H	15.1 Hz H CH₃	11.0 Hz H CH₃
Propenal	2-Methylpropenal	(Z)-2,3-Dimethyl-propenal

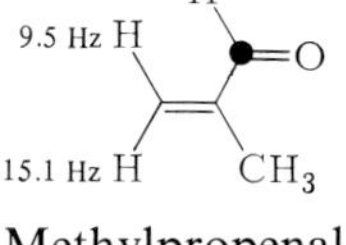

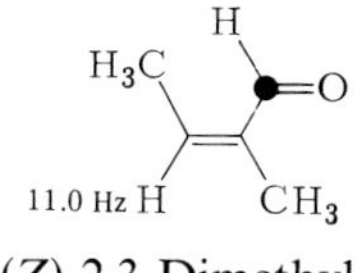

The sterically induced decrease of *trans* $^3J_{CH}$ apparently occurs when the coupling carbon atom is attached to at least one hydrogen atom. Thus, no steric effect will be observed when groups like —COOR, —COOH, and —CN include the coupling carbon.

For benzenoid $^3J_{CH}$ couplings, two essential trends can be recognized from the data given in Table 3.10 [128]:

Table 3.10. One-, Two-, Three-, and Four-Bond Carbon-Proton Coupling Constants (Hz) of Benzene (a), Pyridine (b), Pyrimidine (c), Monosubstituted Benzenes [128] (d) and Five-Membered Heteroaromatics [130] (e).

(a)
$J_{CH} = 158 - 165$
$^2J_{CH_o} = 0 - 5$
$^3J_{CH_m} = 6 - 11$
$^4J_{CH_p} = 0.5 - 2$

(b)
$J_{C-2-2-H} = 172$
$J_{C-3-3-H} = 163$
$J_{C-4-4-H} = 160$
$^2J_{C-2-3-H} = 2$
$^3J_{C-2-4-H} = 7$
$^3J_{C-2-6-H} = 12$
$^2J_{C-3-2-H} = 7$
$^3J_{C-3-4-H} = 2$
$^3J_{C-3-5-H} = 6$
$^3J_{C-4-2-H} = 4$

(c)
$J_{C-2-2-H} = 203$
$J_{C-4-4-H} = 181.5$
$J_{C-5-5-H} = 168$
$^3J_{C-2-4-H} = 10.4$
$^2J_{C-4-5-H} = 3$
$^3J_{C-4-2-H} = 9$
$^3J_{C-4-6-H} = 5.3$
$^2J_{C-5-4-H} = 7.8$
$^4J_{C-5-2-H} = 1.6$

(d)

X	$^2J_{12}$	$^3J_{13}$	$^4J_{14}$	J_{22}	$^2J_{23}$	$^3J_{24}$	$^3J_{26}$	$^4J_{25}$
				C-1			C-2	
F	−4.89	10.95	−1.73	162.55	1.10	8.29	4.11	−1.50
Cl	−3.38	10.93	−1.83	164.92	1.47	8.08	5.06	−1.36
Br	−3.39	11.20	−1.87	165.72	1.46	8.04	5.41	−1.32
I	−2.53	10.79	−1.88	165.62	1.55	7.97	6.07	−1.26
OH	−2.80	9.71	−1.56	158.35	1.20	8.09	4.68	−1.40
OCH$_3$	−2.79	9.22	−1.51	158.52	1.42	8.01	4.80	−1.42
NH$_2$	−0.88	8.62	−1.37	155.71	1.28	7.90	5.38	−1.36
CH$_3$	0.54	7.61	−1.40	155.89	1.19	7.78	6.59	−1.37
Si(CH$_3$)$_3$	4.19	6.34	−1.10	156.14	1.37	7.50	8.63	−1.15
H	1.15	7.62	−1.31	158.43	1.15	7.62	7.62	−1.31
NO$_2$	−3.57	9.67	−1.75	168.12	1.84	8.07	4.45	−1.33
CHO	0.29	7.19	−1.26	160.95	1.40	7.84	6.25	−1.29
CN	0.14	9.01	−1.44	165.45	1.75	7.90	6.13	−1.28

X	J_{33}	$^2J_{32}$	$^2J_{35}$	$^3J_{35}$	$^4J_{36}$	J_{44}	$^2J_{43}$	$^3J_{42}$
		C-3					C-4	
F	161.14	−0.57	1.74	9.02	−0.76	161.37	0.82	7.57
Cl	161.41	0.27	1.57	8.41	−0.87	161.35	0.90	7.51
Br	161.53	0.35	1.62	8.37	−1.05	161.27	0.89	7.51
I	161.26	0.56	1.52	8.20	−1.17	161.04	0.92	7.45
OH	158.99	−0.30	1.66	8.70	−0.74	160.84	0.77	7.42
OCH$_3$	158.35	−0.32	1.77	8.73	−0.75	160.43	0.85	7.52
NH$_2$	156.92	0.20	1.69	8.35	−0.82	160.39	0.82	7.41
CH$_3$	157.58	1.07	1.35	7.91	−1.06	158.83	1.07	7.54
Si(CH$_3$)$_3$	157.70	1.58	1.14	7.26	−1.35	158.10	1.24	7.45
H	158.43	1.15	1.15	7.62	−1.31	158.43	1.15	7.62
NO$_2$	165.12	−0.28	1.50	8.18	−0.74	162.75	1.25	7.65
CHO	161.92	0.76	1.28	7.58	−1.06	160.43	1.46	7.62
CN	163.84	0.43	1.27	7.55	−1.07	162.42	1.02	7.48

Table 3.10 (continued).

(e) Compound	Structure	C	One-bond coupling constants J_{CH} (Hz)	Long-range coupling constants (Hz)			
				$^2J_{CH-2}$	$^2J_{CH-3}$	$^{2,3}J_{CH-4}$	$^{3,4}J_{CH-5}$
Pyrrole		2	182	—	7.6	7.6	7.6
		3	170	7.8	—	4.6	7.8
Furan		2	201	—	7.0	10.8	7.0
		3	175	14.0	—	4.0	5.8
Thiophene		2	185	—	7.35	10.0	5.15
		3	168	4.7	—	5.9	9.5
Selenophene		2		—	7.0	10.0	3.5
		3		4.5	—	6.0	10.4
Pyrazole		3	190	—	—	6.5; 7.5	
		4	178	—	9.5	—	9.5
Imidazole		2	208	—	—	9.4	9.4
		4	189	7.3	—	—	13.0
1,2,4-Triazole		3,5	205	—	—	—	9.4

(1) For substituted benzene carbon atoms, $^3J_{CH}$ decreases with decreasing electronegativity of the substituent:

Phenol H 9.7 Hz   Aniline H 8.6 Hz   Toluene H 7.5 Hz

(2) When the electronegative substituent is located within the coupling route, $^3J_{CH}$ increases as the electronegativity of the substituent decreases:

OH H 4.7 Hz   NH₂ H 5.4 Hz   CH₃ H 6.6 Hz

Carbon-proton couplings of substituent carbons with benzenoid protons attenuate with the coupling distance, as known from proton-proton coupling (Table 3.8). Benzenoid ring carbon atoms coupled to substituent protons behave similarly:

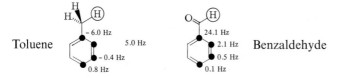

Toluene: −6.0 Hz, 5.0 Hz, −0.4 Hz, 0.8 Hz
Benzaldehyde: 24.1 Hz, 2.1 Hz, 0.5 Hz, 0.1 Hz

Phenolic hydroxy groups display $^3J_{CH}$ couplings through oxygen when the OH proton undergoes intramolecular hydrogen bonding. A slight dependence on *cis* and *trans* configuration of the coupling carbon relative to the chelated hydrogen is observed in salicylaldehyde [129].

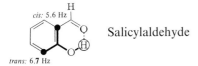
Salicylaldehyde

3.2.3 Carbon-Deuterium Coupling

Carbon-proton and carbon-deuterium coupling constants are related to each other by eq. (3.15a), which follows not only from eq. (3.15) but also from Ramsay's theory [5, 10].

$$J_{CH}/J_{CD} = \gamma_H/\gamma_D = 6.51 . \qquad (3.15\,a)$$

As can be seen in Table 3.1 for commonly used deuterated solvents, deviations of J_{CH}/J_{CD} from the ratio γ_H/γ_D are rare and small, so that only J_{CH} or J_{CD} need be determined experimentally. To conclude, all relations found for J_{CH} can also be derived for J_{CD}. The signal multiplicities, however, are different (3 for CD, 5 for CD_2 and 7 for CD_3, respectively) because of different spin quantum numbers ($I_D = 1$; $I_H = \frac{1}{2}$).

Carbon-deuterium multiplets are weak or even lost in the noise of proton-decoupled ^{13}C NMR spectra, particularly in a larger molecule. A practical use of this fact leads to the unequivocal assignment of specifically deuterated carbons by comparison with the spectrum of the fully protonated compound [132]. This method is advantageous if other assignment techniques such as proton decoupling fail due to equal multiplicities, non-resolved couplings, higher order splittings or multiplet overcrowding.

3.2.4 Carbon-Carbon Coupling

3.2.4.1 One-Bond Coupling (J_{CC})

The carbon skeleton of an organic compound can be derived by measurement of one-bond $^{13}C-^{13}C$ coupling constants, because identical J_{CC} values found for two carbon atoms identify their connectivity (Section 2.9.4.). However, homonuclear carbon multiplets are usually lost in the noise when ^{13}C NMR spectra of samples with naturally abundant ^{13}C are recorded. Only in strong solutions or neat liquid samples of smaller molecules can CC multiplets be observed as weak satellites (0.5% of the normal carbon-13 signal intensity) in $^{13}C\{^1H\}$ NMR [133]. Otherwise, ^{13}C enriched samples must be synthesized or the INADEQUATE technique (Section 2.9.4) has to be applied. However, as illustrated for uniformly labeled and mutarotated D-glucose in (Fig. 3.16(b), homonuclear CC coupling may give rise to $^{13}C\{^1H\}$ NMR spectra of higher than first order. In this case, complete assignment is difficult (Fig. 3.16(b)), and much stronger magnetic fields in addition to two-dimensional techniques (Fig. 3.16(c), Section 2.10.6) can be helpful in order to perform a more detailed analysis.

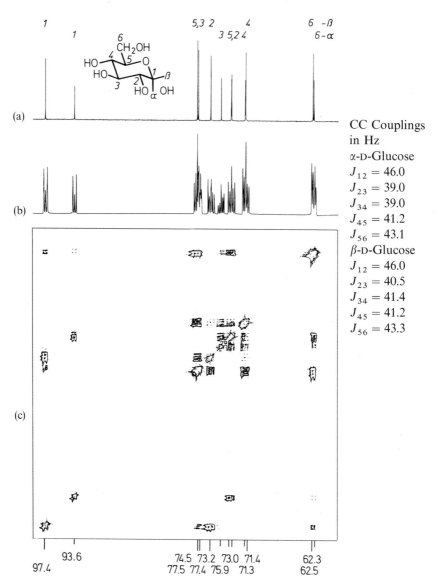

Fig. 3.16. $^{13}C\{^1H\}$ NMR spectra of mutarotated D-glucose; 40 mg/0.4 mL deuterium oxide; 30 °C; 100.576 MHz.
(a) Sample with natural abundance; 128 scans;
(b) ^{13}C-enriched sample (66.1% ^{13}C); 16 scans;
(c) homonuclear CC correlation with proton decoupling (CC COSY) of sample (b) for evaluation of carbon-carbon connectivities and carbon shift assignment as given on top of spectrum (a).

3.2 ^{13}C Coupling Constants   149

Table 3.11 (a). One-Bond Carbon-Carbon Coupling Constants J_{CC}.*

Class	Compound	Formula	J_{CC} (Hz)	Ref.
sp^3-sp^3	Ethane	H$_3$C–CH$_3$	34.6	[115]
	2-Methylpropane	H$_3$C–CH(CH$_3$)$_2$	36.9	[115]
	Ethylbenzene	H$_3$C–CH$_2$–C$_6$H$_5$	34.0	[115]
	Propionitrile	H$_3$C–CH$_2$–CN	33.0	[115]
	1-Propanol	H$_3$C–CH$_2$–CH$_2$OH	34.2	[115]
		H$_3$C–CH$_2$–CH$_2$OH	37.8	[115]
	Ethanol	H$_3$C–CH$_2$OH	37.7	[115]
	2-Propanol	(H$_3$C)$_2$CHOH	38.4	[133a]
	t-Butylamine	(H$_3$C)$_3$CNH$_2$	37.1	[115]
	t-Butylalcohol	(H$_3$C)$_3$COH	39.5	[115]
Cyclopropanes	Methylcyclopropane	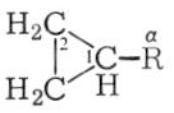 R = CH$_3$	44.0 (1–α)	[133a]
	Dicyclopropylketone	R = –C(O)–cyclopropyl	54.0 (1–α)	[133a]
			10.2 (1–2)	[133a]
	Cyclopropanecarboxylic acid	R = COOH	72.5 (1–α)	[133a]
			10.0 (1–2)	[133a]
	Cyclopropyl cyanide	R = CN	77.9 (1–α)	[133a]
			10.9 (1–2)	[133a]
	Cyclopropyl iodide	R = I	12.9 (1–2)	[133a]
	Cyclopropyl bromide	R = Br	13.3 (1–2)	[133a]
	Cyclopropyl chloride	R = Cl	13.9 (1–2)	[133a]
sp^3-sp^2	2-Butanone	H$_3$C–C(O)–C$_2$H$_5$	38.4	[115]
	Acetaldehyde	H$_3$C–C(O)–H	39.4	[115]
	Acetone	H$_3$C–C(O)–CH$_3$	40.1	[115]
	3-Pentanone	H$_3$C–CH$_2$–C(O)–CH$_2$–CH$_3$	35.7	[133a]
		H$_3$C–CH$_2$–C(O)–CH$_2$–CH$_3$	39.7	[133a]
	Acetophenone	H$_3$C–C(O)–C$_6$H$_5$	43.3	[115]
	Acetate anion (aq.)	H$_3$C–C(O)–O$^\ominus$	51.6	[115]
	N,N-Dimethylacetamide	H$_3$C–C(O)–N(CH$_3$)$_2$	52.2	[115]
	Acetic acid	H$_3$C–C(O)–OH	56.7	[115]
	Ethyl acetate	H$_3$C–C(O)–OC$_2$H$_5$	58.8	[115]

150   3 ^{13}C NMR Spectral Parameters and Structural Properties

Table 3.11 (a) (continued).

Class	Compound	Formula	J_{CC} (Hz)	Ref.
sp^3-sp	t-Butyl cyanide	$(H_3C)_3C-C\equiv N$	52.0	[115]
	iso-Propyl cyanide	$(H_3C)_2CH-C\equiv N$	54.8	[115]
	Propionitrile	$H_3C-CH_2-C\equiv N$	55.2	[115]
	Acetonitrile	$H_3C-C\equiv N$	56.5	[115]
	Propyne	$H_3C-C\equiv C-H$	67.4	[133a]
sp^2-sp^2 alkenic	Ethene	$H_2C=CH_2$	67.6	[115]
	Acrylic acid	$H_2C=CH-COOH$	70.4	[115]
	Acrylonitrile	$H_2C=CH-CN$	70.6	[115]
	Styrene	$H_2C=CH-C_6H_5$	70.0±3	[115]
sp^2-sp^2 aromatic	Benzene	X=H	57.0	[115]
	Nitrobenzene	X=NO$_2$	55.4 (1−2)	[115]
			56.3 (2−3)	
		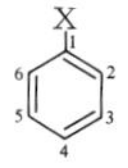	55.8 (3−4)	
	Iodobenzene	X=I	60.4 (1−2)	[133a]
			53.4 (2−3)	
			58.0 (3−4)	
	Anisole	X=OCH$_3$	58.2 (2−3)	[133a]
			56.0 (3−4)	
	Aniline	X=NH$_2$	61.3 (1−2)	[115]
			58.1 (2−3)	
			56.6 (3−4)	
	Pyridine	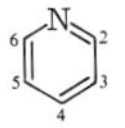	53.8 (2−3)	[115]
			56.2 (3−4)	
	Thiophene	X=S	64.2	[133a]
	Pyrrole	 X=NH	65.9	[133a]
	Furan	X=O	69.1	[133a]
sp^2-sp	Benzonitrile	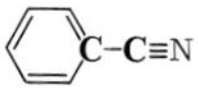 C−C≡N	80.3	[115]
	1,1-Dimethylallene	$(H_3C)_2C=C=CH_2$	99.5	[133a]
sp-sp	Phenylethynyl cyanide	$C_6H_5-C\equiv C-C\equiv N$	155.8	[115]
	Ethyne	$H-C\equiv C-H$	171.5	[115]
	Phenylethyne	$C_6H_5-C\equiv C-H$	175.9	[115]

* the coupling carbons are printed in bold type.

The one-bond coupling constants J_{CC} tabulated so far cover a span of 30 to 180 Hz (Table 3.11 (a)). They increase with increasing s character of the hybrid orbitals contributing to the bond between the coupling carbon nuclei. A rough correlation between J_{CC} and the product $s_1 \cdot s_2$ of the s characters of the interacting carbon hybrid orbitals (C-1, C-2), i.e. the "s" bond order of the C−C bond, is observed (Table 3.11 (a); Fig. 3.17).

Calculated carbon-carbon couplings J_{CC} obtained with the INDO approach agree qualitatively with the experimental data of open-chain hydrocarbons. Considerable deviations are found for molecules containing electron withdrawing substituents and for aromatic compounds [115].

Table 3.11 (b). Longer-Range Carbon-Carbon Coupling Constants $^2J_{CC}$ and $^3J_{CC}$; Data from Ref. [133 a].

Compound	Formula		$^2J_{CC}$ (Hz)
Propyne	$H_3C-C\equiv C-H$		11.8
2-Butanone	$H_3C-\overset{O}{\overset{\|}{C}}-CH_2-CH_3$		15.2
			9.5
Cyclobutanone	$H_2C\overset{CH_2}{\underset{CH_2}{\diagup}}C=O$		9.0
			$^3J_{CC}$ (Hz)
Cyclobutyl bromide	$H_2C\overset{CH_2}{\underset{CH_2}{\diagup}}CH-Br$		
Pyridine		2–5	13.95
Aniline		$X=NH_2$ 2–5	7.9
Iodobenzene		$X=I$ 2–5	8.6
Nitrobenzene		$X=NO_2$ 2–5	7.6

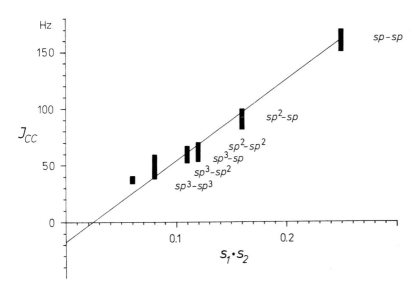

Fig. 3.17. Homonuclear carbon one-bond coupling constants observed for the six possible combinations of carbon nuclei in organic compounds; the ranges of J_{CC} are represented as a function of the product of s characters, $s_1 \cdot s_2$, of the coupling nuclei. The straight line follows the equation $J_{C_1C_2} = 730\, s_1 \cdot s_2 - 17$ (Hz) [133].

3 ^{13}C NMR Spectral Parameters and Structural Properties

Some empirical relations are deducible by inspection of the experimental data. Electron withdrawal increases J_{CC}. However, the effects are weaker than those in $^{13}C-{}^1H$ coupling. Nevertheless, the parallel behavior of J_{CH} and J_{CC} toward hybridization and electron withdrawal is reflected in the linear correlation found between J_{CH} in compounds of the type XYZCH and J_{CC} in compounds of the type $XYZCCH_3$ [133a].

An increase of J_{CC} of only 2.5 to 3 Hz is found in going from ethane to ethanol, or from *t*-butylamine to *t*-butylalcohol (Table 3.11 (a)). Homonuclear carbon couplings of the type sp^3-sp^2 in carbonyl compounds increase more significantly with electron withdrawal, e.g. in going from aldehydes or ketones ($J_{CC} \approx 40$ Hz) to comparable carboxylic acid derivatives ($J_{CC} \approx 50$ Hz) (Table 3.11 (a)). In substituted benzenes and in pyridine, couplings between adjacent carbons, not including the substituted atom, differ by about 2 Hz. Differences in J_{CC} of the same order are reported for the C-1—C-2 couplings of α- and β-D-glucose (Fig. 3.16(b)), the β-anomer with an equatorial hydroxy group at C-1 showing the larger homonuclear coupling [134].

It has been suggested that because the carbon orbitals forming the C—H bonds have more *s* character in cyclopropane than in normal hydrocarbons, the orbitals forming the cyclopropane C—C bonds must have more *p* character [122, 133a]. The small 1—2-couplings (10—15 Hz) for the cyclopropane ring carbons and the large 1-α couplings (44—78 Hz) in substituted cyclopropanes appear to provide evidence for this assumption.

3.2.4.2 Longer-Range Carbon-Carbon Couplings ($^2J_{CC}$, $^3J_{CC}$)

Two-bond carbon-carbon coupling constants ($^2J_{CC}$) are usually smaller than three-bond couplings ($^3J_{CC}$), as shown for hexanoic and (E)-2-hexenoic acid:

Hexanoic acid
2.9 Hz 55.2 Hz
1.0 Hz 1.7 Hz COOH

2-Hexynoic acid
1.5 Hz
19.3 Hz 123.0 Hz
COOH

6.4 Hz 71.6 Hz
0.7 Hz COOH
(E)-2-Hexenoic acid

Large $^2J_{CC}$ couplings are observed if one or both of the coupling carbon atoms have a high *s* character (hexynoic acid, propyne, cyclobutane ring carbons), or when the coupling path includes an electron deficient carbon atom such as a carbonyl function (Table 3.11 (b)).

Three-bond carbon-carbon coupling constants ($^3J_{CC}$) have been compared with corresponding carbon-proton ($^3J_{CH}$) and proton-proton couplings ($^3J_{HH}$) [129]. Some aliphatic couplings averaged due to free rotation and several *trans* couplings of alkenes exemplify the trend:

3.2 ^{13}C Coupling Constants

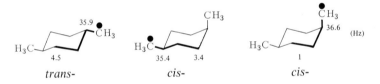

It turns out that substitution of a proton by carbon reduces three-bond coupling by a factor of 0.6–0.8, so that

$$^3J_{CC} = c\,^3J_{CH} = c\,^3J_{HH} \quad \text{where } c = 0.6\text{–}0.8.$$

The consistent behavior of all three coupling constants is attributed to similar coupling mechanisms.

In *aliphatic compounds*, $^3J_{CC}$ is related to the dihedral angle Φ according to a Karplus cosine function as portrayed in Fig. 3.18. Similar to $^3J_{HH}$, the maximum at $\Phi = 180°$ is larger than that at $\Phi = 0°$ for carboxylic acids. A reversed pattern is found for aldehydes and alcohols, due to heteroatom through-space effects [129].

Accordingly, *trans* $^3J_{CC}$ couplings involving cyclohexane ring carbon atoms are larger compared with coupling carbons in a *cis* configuration [129].

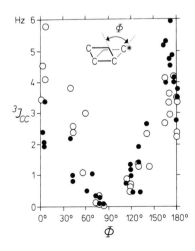

Fig. 3.18. $^3J_{CC}$ versus Φ plot of $C-C-C-C^*$ connections with $C^* = CH_2OH$ (empty circles) and $C^* = COOH$ (full circles) [129].

When the coupling path includes a cyclopropane ring, however, *cis* coupling is larger, as known from $^2J_{HH(cis)}$ in proton NMR [129].

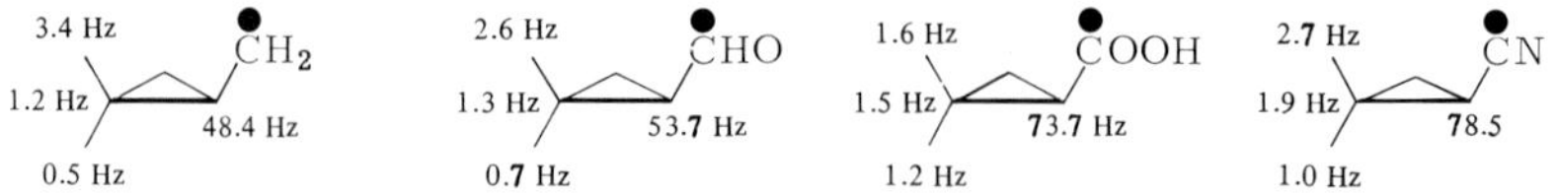

3.4 Hz — CH₂, 1.2 Hz, 48.4 Hz, 0.5 Hz	2.6 Hz — CHO, 1.3 Hz, 53.7 Hz, 0.7 Hz	1.6 Hz — COOH, 1.5 Hz, 73.7 Hz, 1.2 Hz	2.7 Hz — CN, 1.9 Hz, 78.5, 1.0 Hz
1-Hydroxymethyl-	1-Formyl-	1-Carboxy-	1-Cyano-

2,2-dimethylcyclopropane

In alkenes, $^3J_{CC(trans)}$ is significantly larger than $^3J_{CC(cis)}$, a fact which is also analogous to the behavior known from $^3J_{HH}$ [129].

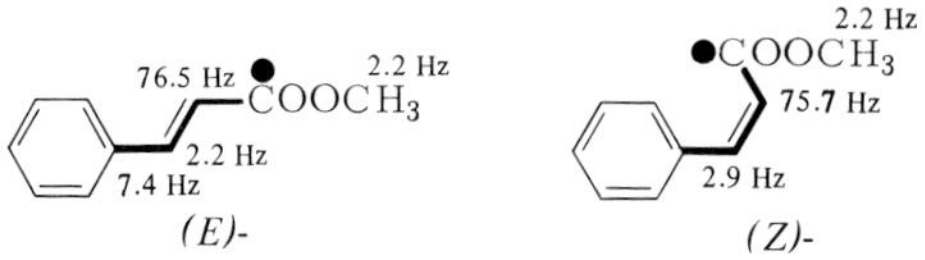

(E)-: 76.5 Hz, 2.2 Hz COOCH₃, 2.2 Hz, 7.4 Hz

(Z)-: 2.2 Hz COOCH₃, 75.7 Hz, 2.9 Hz — Methyl cinnamate

(E)-: 72.6 Hz COOH, 1.4 Hz, 6.6 Hz

(Z)-: COOH, 72.0 Hz, 2.0 Hz — Cinnamic acid

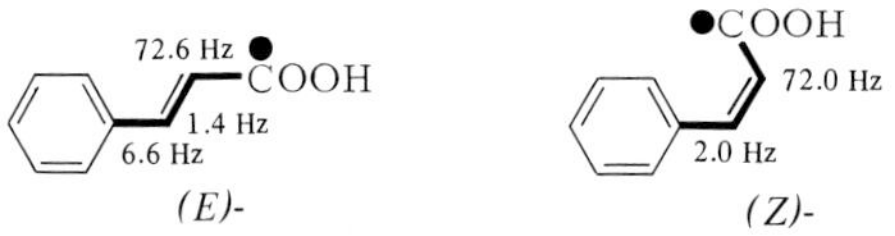

(E)-: H₃C—, 1.3 Hz COOH, 7.3 Hz

(Z)-: H₃C—, 0 Hz COOH, 2.4 Hz

2-Butenoic acid

H₃C— 2.2 Hz COOH, 1.5 Hz, CH₃ 7.5 Hz — 3-Methyl-2-butenoic acid

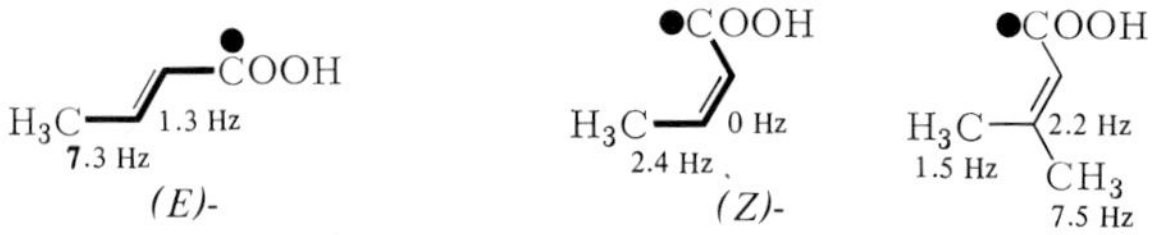

Following this pattern, $^3J_{CC}$ couplings of *transoid* carbon atoms in aromatic compounds are larger than those in a *cisoid* coupling path [129].

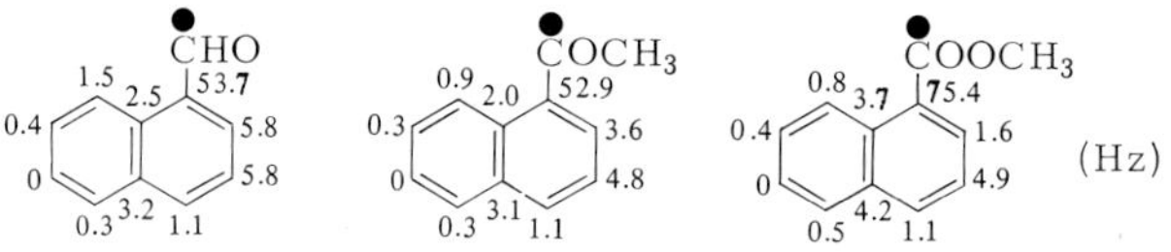

1-Formyl-: CHO 53.7, 1.5, 2.5, 0.4, 0, 0.3, 3.2, 1.1, 5.8, 5.8

1-Acetyl-naphthalene: COCH₃ 52.9, 0.9, 2.0, 0.3, 0, 0.3, 3.1, 1.1, 3.6, 4.8

1-Methoxycarbonyl-: COOCH₃ 75.4, 0.8, 3.7, 0.4, 0, 0.5, 4.2, 1.1, 1.6, 4.9 (Hz)

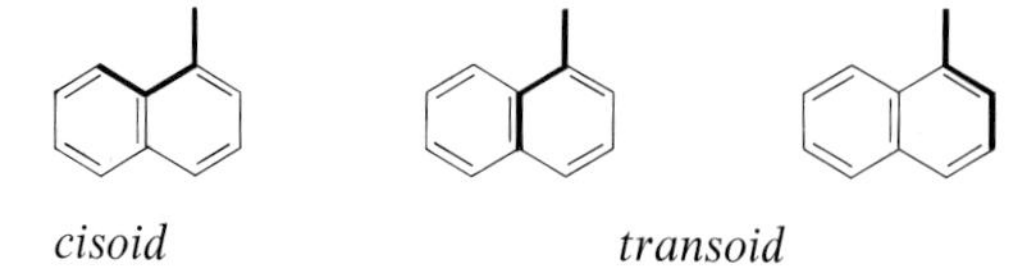

cisoid
$^3J_{CC}$ small

transoid
$^3J_{CC}$ large

$^3J_{CC}$ also reflects delocalized π bonding in benzene (π bond order $p = 0.67$) relative to an alkene with a localized π bond ($p = 1.0$), as shown for (E)-butenoic acid and benzoic acid. Correspondingly, the carboxy carbons of 1- and 2-naphthoic acid display larger three-bond carbon-carbon couplings to the ring positions with the larger p value. On the other hand, more extensive π bond delocalization in pyrene-1-carboxylic acid results in "benzenoid" $^3J_{CC}$ values of about 4.5 Hz [129].

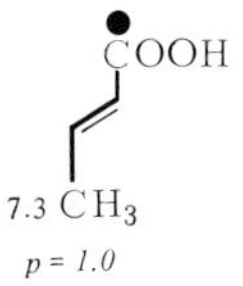

[Structures shown:

(E)-2-Butenoic acid: 7.3 CH₃, $p = 1.0$

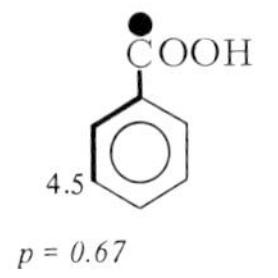

Benzoic acid: 4.5, $p = 0.67$

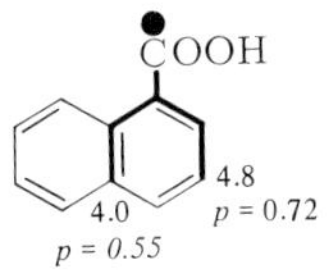

1-Naphthoic acid: 4.8, 4.0, $p = 0.72$, $p = 0.55$

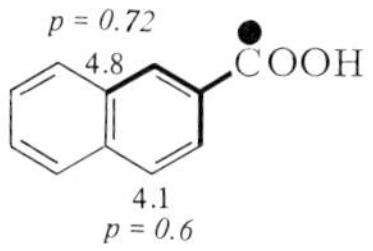

2-Naphthoic acid: $p = 0.72$, 4.8, 4.1, $p = 0.6$

Pyrene-1-carboxylic acid: $p = 0.59, 0.67$; 4.5, 4.5 (Hz)]

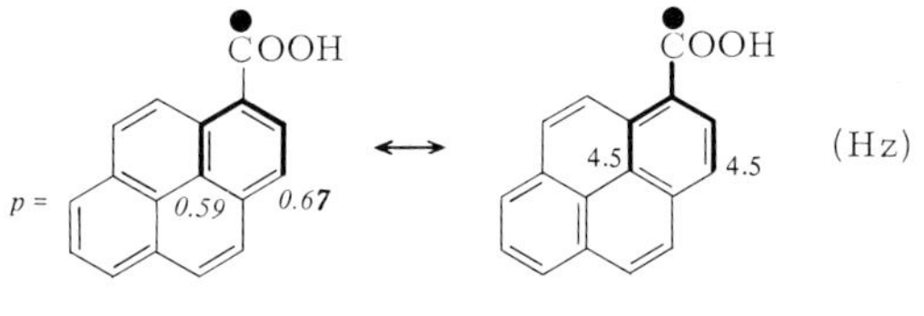

3.2.5 $^{13}C - ^{15}N$ Coupling Constants

The magnetic moment of ^{15}N is small compared with protons and carbon-13 nuclei ($\mu_H : \mu_C : \mu_N = 10 : 2.5 : 1$) and negative. Thus, $^{13}C - ^{15}N$ coupling constants have small magnitudes and their signs are, for the most part, reversed relative to comparable carbon-proton or carbon-carbon couplings. Due to the low natural abundance of both isotopes ^{13}C (1.1%) and ^{15}N (0.37%), the probability of a $^{13}C - ^{15}N$ connection is even smaller than the occurrence of a $^{13}C - ^{13}C$ linkage. As a result, $^{13}C - ^{15}N$ coupling constants are lost in the noise of carbon-13 and nitrogen-15 spectra under usual measuring conditions. They are observable, however, when one of the two isotopes is enriched and the NMR of the other is observed. Since ^{13}C is the more sensitive NMR probe, convenient investigation of $^{13}C - ^{15}N$ coupling constants involves recording of ^{13}C NMR spectra with ^{15}N-enriched samples. Couplings of carbon-13 with the more abundant nitrogen isotope ^{14}N are usually not resolvable in ^{13}C NMR spectra due to their small magnitudes and quadrupolar broadening.

3.2.5.1 One-Bond Couplings (J_{CN})

According to eq. (3.15), the magnitude of one-bond carbon-13–nitrogen-15 coupling constants correlates with the product of s characters of the hybrid orbitals overlapping in the C–N bond [135]:

$$|J_{CN}| = 125 \, s_C \, s_N . \tag{3.15 b}$$

Therefore, $|J_{CN}|$ increases in the series propylamine, N-methylaniline, N-methyl-N-propynylaniline due to increasing s character of carbon hybrid orbitals,

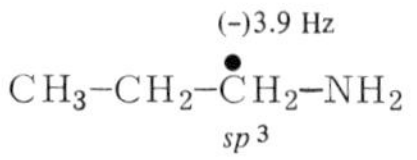

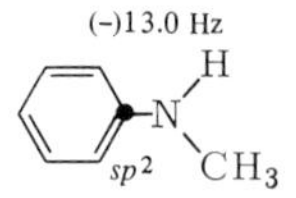

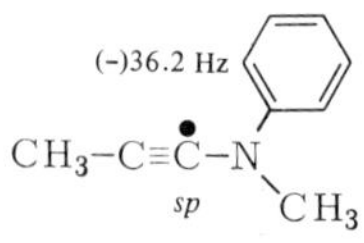

$$\underset{sp^3}{CH_3-CH_2-\overset{(-)3.9\ Hz}{\overset{\bullet}{C}}H_2-NH_2} \qquad \underset{sp^2}{\underset{}{\text{PhN}(CH_3)}}\ (-)13.0\ Hz \qquad \underset{sp}{CH_3-C\equiv\overset{(-)36.2\ Hz}{\overset{\bullet}{C}}-N(CH_3)\text{Ph}}$$

as well as in the series methylamine, nitromethane, diazomethane with increasing s character of nitrogen hybrid orbitals [135].

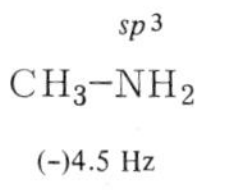

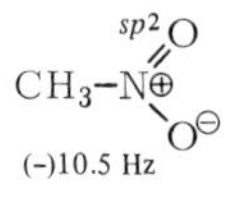

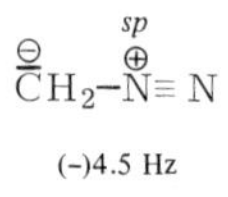

$$\underset{(-)4.5\ Hz}{\underset{sp^3}{CH_3-NH_2}} \qquad \underset{(-)10.5\ Hz}{\underset{sp^2}{CH_3-NO_2}} \qquad \underset{(-)4.5\ Hz}{\underset{sp}{CH_2-N\equiv N}}$$

A *positive charge* at the coupling nitrogen induced by protonation, inductive or mesomeric effects also causes the magnitude of J_{CN} to increase. Thus, the azaadamantane cation displays a larger $|J_{CN}|$ than the free base [135].

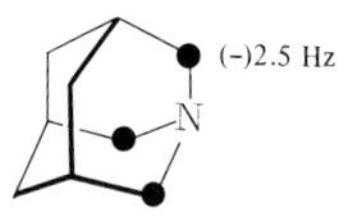

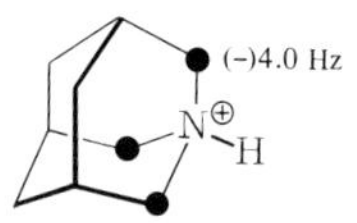

(azaadamantane: (−)2.5 Hz; azaadamantane cation: (−)4.0 Hz)

A larger $|J_{CN}|$ of urea relative to acetic amide is essentially attributed to the $(-)$-I-effect of the second amino group,

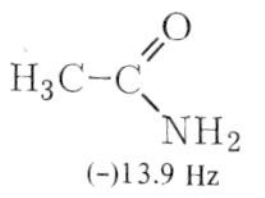

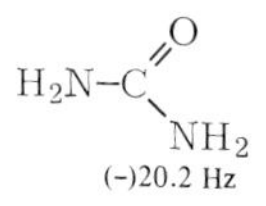

$$\underset{(-)13.9\ Hz}{H_3C-C(=O)NH_2} \qquad \underset{(-)20.2\ Hz}{H_2N-C(=O)NH_2}$$

and the electron-withdrawing $(-)$-M-effect of a *p*-nitro group enhances $|J_{CN}|$ of *p*-nitroaniline relative to aniline [135].

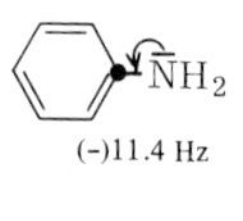

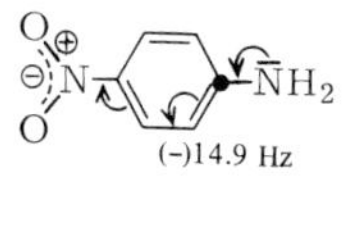

(aniline: (−)11.4 Hz; *p*-nitroaniline: (−)14.9 Hz)

A *nonbonding electron pair* at the coupling nitrogen located in the plane of the α-CH bond of an imine makes up a positive contribution to J_{CN}. This is recognized when comparing J_{CN} of pyridine with the data of its cation and its N-oxide [135],

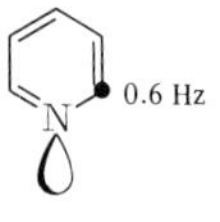

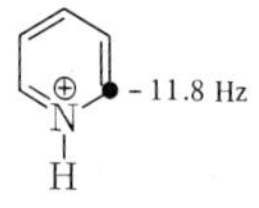

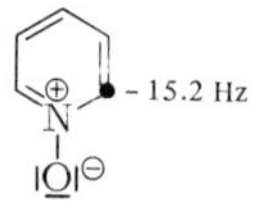

(pyridine: 0.6 Hz; pyridinium: −11.8 Hz; pyridine N-oxide: −15.2 Hz) and can be used to distinguish (E)- from (Z)-isomers of aldoximes:

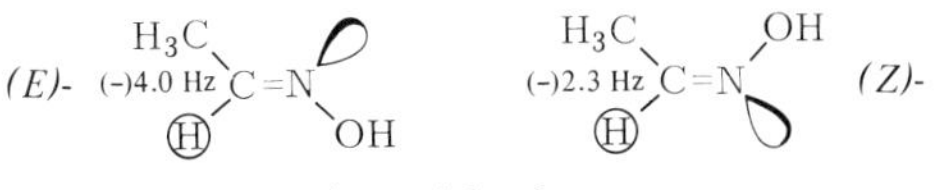

$$(E)\text{-} \quad \underset{(-)4.0\,\text{Hz}}{\overset{H_3C}{\underset{(H)}{\diagdown}}}C=N\overset{}{\diagdown}_{OH} \qquad \underset{(-)2.3\,\text{Hz}}{\overset{H_3C}{\underset{(H)}{\diagdown}}}C=N\overset{OH}{\diagdown} \quad (Z)\text{-}$$

Acetaldoxime

3.2.5.2 Longer-Range Couplings ($^2J_{CN}$, $^3J_{CN}$)

Two-bond carbon-13–nitrogen-15 couplings ($^2J_{CN}$) are largely determined by the orientation of the nonbonding electron pair at the coupling nitrogen. In α, β-unsaturated imines, for example, a nonbonding electron pair at nitrogen closely spaced to the coupling carbon causes a negative contribution to $^2J_{CN}$. Thus, N–C-8 coupling in quinoline is $^2J_{CN} = -9.3$ Hz due to the adjacent electron pair, while a $^2J_{CN}$ of 2.7 Hz is observed for carbon C-3, similarly to the $^2J_{CN}$ couplings (1–2.5 Hz) displayed by the quinolinium and pyridinum ions as well as by pyridine itself. In contrast, pyrrole, without an electron pair outside the π electron sextet, is characterized by a negative $^2J_{CN}$ [135].

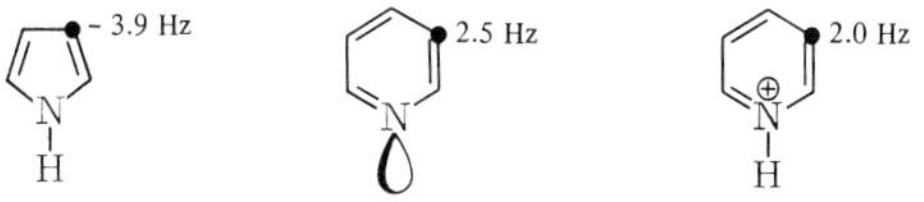

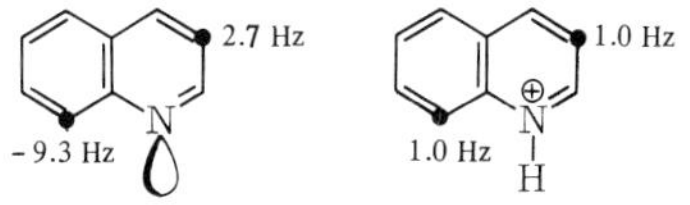

The negative contribution of a closely spaced nonbonding electron pair permits one to distinguish (E)- from (Z)-isomers of α, β-unsaturated oximes and of azoxybenzenes [135].

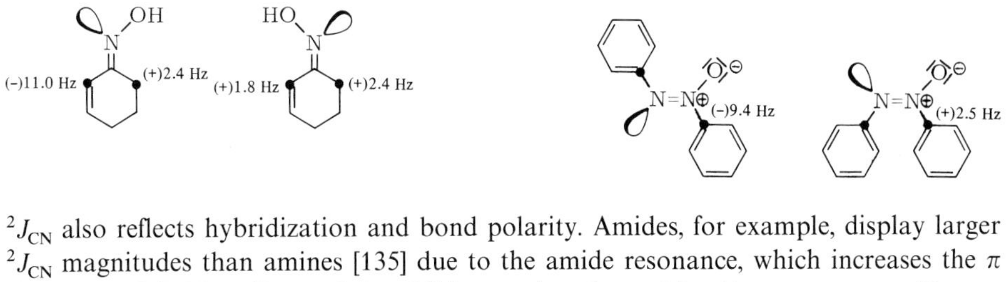

$^2J_{CN}$ also reflects hybridization and bond polarity. Amides, for example, display larger $^2J_{CN}$ magnitudes than amines [135] due to the amide resonance, which increases the π character of CN bonding and, in addition, makes the amide nitrogen more positive.

Propylamine $CH_3-CH_2-CH_2-NH_2$ 1.2 Hz

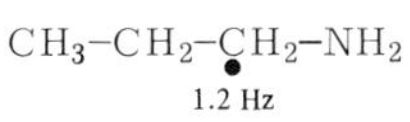

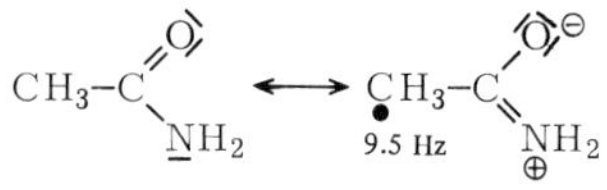

$CH_3-C\overset{O}{\underset{NH_2}{\diagdown}} \longleftrightarrow CH_3-C\overset{O^{\ominus}}{\underset{\overset{\oplus}{NH_2}}{\diagdown}}$ Acetamide
9.5 Hz

158   3 ^{13}C NMR Spectral Parameters and Structural Properties

The magnitudes of *three-bond carbon-13 – nitrogen-15 couplings* ($^3J_{CN}$) in aliphatic compounds are small, so that configurational assignments derived from Karplus relationships are rare [135]. A positive charge at the coupling nitrogen and the location of its nonbonding electron pair have a more significant influence on $^3J_{CN}$. Protonation of the coupling nitrogen induces a negative increment, while a closely spaced nonbonding electron pair makes up a positive contribution to $^3J_{CN}$. This is exemplified by the $^3J_{CN}$ data of both the free bases and the cations of aniline, pyridine and quinoline [135].

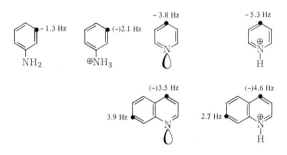

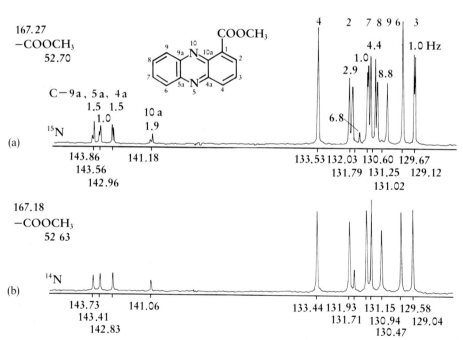

Fig. 3.19. ^{13}C Chemical shift assignments of 1-Methoxycarbonylphenazine (a) in deuteriochloroform at 30 °C based on long-range ^{13}C – ^{15}N couplings of the 10-^{15}N-enriched derivative (b).

Table 3.12. Carbon-13–Nitrogen-15 Coupling Constants of Selected Organic Compounds.*

Compound	Formula	J_{CN} (Hz)	Ref.
Methylamine	$H_3\textbf{C}-\textbf{N}H_2$	135	[135]
Methylammonium ion	$H_3\textbf{C}-\overset{\oplus}{\textbf{N}}H_3$	< 8.0	[137]
Tetramethylammonium ion	$\overset{\oplus}{\textbf{N}}(\textbf{C}H_3)_4$	5.8	[138]
Tetraethylammonium ion	$\overset{\oplus}{\textbf{N}}(\textbf{C}H_2CH_3)_4$	4.0	[138]
Glycine	$H_3\overset{\oplus}{\textbf{N}}-\textbf{C}H_2-COO^{\ominus}$	6.2	[139]
	$H_2\textbf{N}-\textbf{C}H_2-COO^{\ominus}$	4.9	[139]

DL-Alanine	$H_3\overset{\oplus}{\textbf{N}}-\textbf{C}H-COO^{\ominus}$
	          $|$
	         CH_3
		5.6	[139]

Dimethylnitrosamine	$(H_3\textbf{C})_2\textbf{N}-{}^{15}NO$	7.5 C_{anti} $(^2J_{CN})$ 1.4 C_{syn}	[139]
1,1-Dimethylhydrazine	$(H_3\textbf{C})_2-{}^{15}\textbf{N}H_2$	< 1.0 $(^2J_{CN})$	[139]
trans-N-Methylbenzaldimine	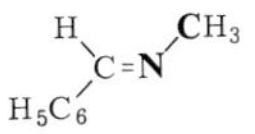 $\begin{array}{c} H \diagdown \diagup CH_3 \\ \textbf{C}=\textbf{N} \\ H_5C_6 \diagup \end{array}$	< 3.0	[137]
	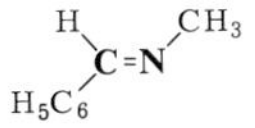 $\begin{array}{c} H \diagdown \diagup CH_3 \\ \textbf{C}=\textbf{N} \\ H_5C_6 \diagup \end{array}$	7.1	[137]
Acetanilide	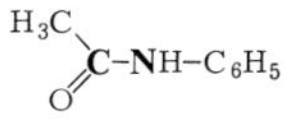 $\begin{array}{c} H_3C \diagdown \\ \textbf{C}-\textbf{N}H-C_6H_5 \\ \diagup\!\!\!\!O \end{array}$	9.3	[139]
	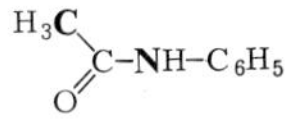 $\begin{array}{c} H_3\textbf{C} \diagdown \\ C-\textbf{N}H-C_6H_5 \\ \diagup\!\!\!\!O \end{array}$	13.0	[139]
N,N-Dimethylformamide	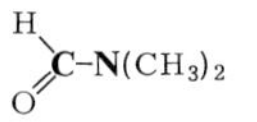 $\begin{array}{c} H \diagdown \\ \textbf{C}-\textbf{N}(CH_3)_2 \\ \diagup\!\!\!\!O \end{array}$	13.4	[139]
Cyanide anion	$^{\ominus}\textbf{C}\equiv\textbf{N}$	± 5.9	[115]
Acetonitrile	$H_3C-\textbf{C}\equiv\textbf{N}$	−17.5	[140]
	$H_3\textbf{C}-C\equiv\textbf{N}$	3.0 $(^2J_{CN})$	[140]
Propionitrile	$H_5C_2-\textbf{C}\equiv\textbf{N}$	−16.4	[115]
t-Butyl cyanide	$(H_3C)_3C-\textbf{C}\equiv\textbf{N}$	−15.4	[115]
Methylisocyanide	$H_3\textbf{C}-\overset{\oplus}{\textbf{N}}\equiv\overset{\ominus}{C}$	−10.6	[141]
Methyl isothiocyanate	$H_3\textbf{C}-\textbf{N}=C=S$	13.4	[137]

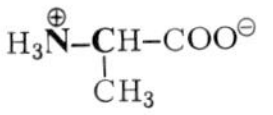

* the coupling nuclei are printed in bold type.

160   3 ^{13}C NMR Spectral Parameters and Structural Properties

Accordingly, C-8 of α-tetralone oxime displays a larger $^3J_{CN}$ compared with C-4 a, which is more removed from the electron pair, and the (E)- and (Z)-isomers of cyclohexenone oxime can be characterized by their different $^3J_{CN}$ couplings [135].

^{13}C–^{15}N couplings of selected compounds are summarized in Table 3.12. They are useful aids in assigning the ^{13}C signals of organonitrogen compounds. 1-Methoxycarbonylphenazine, for example, exhibits ^{13}C resonances for protonated carbon atoms between 129 and 133.5 ppm (Fig. 3.19(a)), which cannot be assigned unequivocally due to small shift differences. In the ^{13}C NMR spectrum of the 10–^{15}N-labeled sample (Fig. 3.19(b), page 158), however, long-range carbon-13–nitrogen-15 splittings decrease with distance from the nonbonding electron pair at ^{15}N ($|J_{CN}|$: C-1,9 > C-2,8 > C-3,7 > C-4,6), so that a straightforward assignment is possible [136].

3.2.6 Coupling between Carbon and Other Heteronuclei X (X ≠ C, H, D)

Couplings between carbon and heteronuclei X (X ≠ C, H) may lead to an observable splitting of the ^{13}C signals in ^{13}C{^{1}H} experiments. A knowledge of the magnitude of heteronuclear coupling constants J_{CX} helps to assign signals in ^{13}C{^{1}H} NMR spectra.

The largest heteronuclear carbon couplings known so far are the J_{CTl} data of organothallium compounds. They may approach the order of 10 kHz for one-bond and 1 KHz for longer-range couplings (Table 3.13(a, b), [133, 142]). Carbon-13–nitrogen-15 coupling constants lie at the other extreme (Section 3.2.5).

The relative signs of one-bond coupling constants may be negative, as is the case for all J_{CF} data (Table 3.14, [115]). Variable signs, depending on the types of compounds, are observed for J_{CP}: A small negative J_{CP} is measured for triphenylphosphine with trivalent phosphorus, and a larger positive one for tetramethylphosphonium salts (Table 3.13(a)).

Couplings between carbon and many heteronuclei, e.g. the J_{CF} data shown in Table 3.14, do not depend simply on carbon s character as is the case for J_{CH} and J_{CC}. Nevertheless, electron withdrawal at the coupling carbon often increases J_{CX}. This trend is obvious when the J_{CF} values of mono- di-, and trifluoroethanol, or those of mono- and trifluoroacetic acid are compared (Table 3.14).

Relatively large longer-range C–X splittings are found not only for X = mercury (Table 3.13(b)), but also for fluorine and phosphorus(III,V).

The magnitudes of two-bond carbon-fluorine couplings $^2J_{CF}$ in trifluoroacetic acid and hexafluoroacetone hydrate are 44 and 34 Hz, respectively (Table 3.14). Five-bond couplings of more than 20 Hz are reported for 1,4,8-trimethyl-5-fluorophenanthrene derivatives; their magnitudes decrease on saturation of the 9,10 bond in the phenanthrene

Table 3.13 (a). One-Bond Coupling Constants J_{CX} in Methyl Derivatives $X(CH_3)_n$ of Heteroatoms; Data from Ref. [142] if not otherwise indicated.

Hetero-atom X	Compound $X(CH_3)_n$	J_{CX} (Hz)	Ref.
^1H	HCH_3	125.0	
^{13}C	$C(CH_3)_4$	36.9	
^{14}N	$^\oplus N(CH_3)_4$	10.0	
^{15}N	$^\oplus N(CH_3)_4$	5.8	[128]
^{19}F	FCH_3	−157.5	
^{29}Si	$Si(CH_3)_4$	− 52.0	
$^{31}P(III)$	$P(CH_3)_3$	− 13.6	
	$^\oplus P(CH_3)_4$	55.5	
$^{31}P(V)$	$OP(CH_3)(OCH_3)_2$	141.5	[126]
		5.0 ($^2J_{CP}$)	
^{77}Se	$Se(CH_3)_2$	− 62.0	
	$^\oplus Se(CH_3)_3$	− 50.0	
^{113}Cd	$Cd(CH_3)_2$	−537.0	
^{119}Sn	$Sn(CH_3)_4$	−340.0	
^{125}Te	$Te(CH_3)_2$	162.0	
^{199}Hg	$Hg(CH_3)_2$	687.4	
	$O_2NOHgCH_3$	1800.0	
^{205}Tl	$Tl(CH_3)_3$	1930	

Table 3.13 (b). One-Bond and Longer-Range Couplings J_{CX} and $^2J_{CX}$, $^3J_{CX}$, $^4J_{CX}$, of Selected Phenyl Derivatives $X(C_6H_5)_n$; Data from Ref. [142].

Hetero-atom X	Compound $X(C_6H_5)_n$	J_{CX} (Hz)	$^2J_{CX}$ (Hz)	$^3J_{CX}$ (Hz)	$^4J_{CX}$ (Hz)
^1H	HC_6H_5	157.5	1.0	7.4	−1.1
^{11}B	$^\ominus B(C_6H_5)_4$	49.5	—	2.6	—
^{19}F	FC_6H_5	−245.3	21.0	7.7	3.3
^{31}P	$P(C_6H_5)_3$	12.4	19.6	6.7	0.0
	$^\oplus P(C_6H_5)_4$	88.4	10.9	12.8	2.9
^{199}Hg	$Hg(C_6H_5)_2$	1186.0	88.0	101.6	17.8

moiety [143]. These large five-bond splittings provide evidence that a through-space mechanism might be operative in carbon-13 coupling [143].

In alkyl phosphonates, one-bond coupling constants J_{CP} range from 125 to 165 Hz; negative two-bond couplings between 6 and 7 Hz and positive three-bond splittings of about 5 to 6 Hz are observed [144]. Four-membered cyclic phosphines such as 2,2,3,4,4-pentamethylphosphetanes show two-bond coupling constants $^2J_{CP}$ which are larger (25 Hz) for *trans* and smaller (5 Hz) for *cis* configuration of the exocyclic substituent X (X = −Cl, −CH$_3$, −C$_6$H$_5$) at phosphorus [145]. A similar stereospecificity of

Table 3.14. Carbon-Fluorine Coupling Constants of Selected Organic Compounds; Data from Ref. [115] if not otherwise indicated.

Compound	Formula	$-J_{CF}$ (Hz)	Ref.
Fluoromethane	FCH_3	157.5	
Difluoromethane	$FCFH_2$	234.8	
Trifluoromethane	FCF_2H	274.3	
Tetrafluoromethane	FCF_3	259.2	
1,1,1-Trifluoroethane	FCF_2CH_3	271.0	
Benzyl fluoride	$FCH_2C_6H_5$	165.0	
t-Butyl fluoride	$FC(CH_3)_3$	167.0	
2-Fluoroethanol	FCH_2CH_2OH	167.0	
2,2-Difluoroethanol	$FCFHCH_2OH$	240.5	
2,2,2-Trifluoroethanol	FCF_2CH_2OH	278.0	
Hexafluorodiethyl ether	FCF_2OCF_3	265.0	
Trifluoroacetone	$FCF_2C(O)CH_3$	289.0	
Hexafluoroacetone hydrate	$(FCF_2)_2C(OH)_2$	285.5	[126]
		$^2J_{CF}$: 34.0	[126]
Monofluoroacetic acid	FCH_2COOH	181.0	
Trifluoroacetic acid	FCF_2COOH	283.2	
		$^2J_{CF}$: 44.0	[126]
p-Fluoroanisole	$p\text{-}FC_6H_4OCH_3$	237.0	
p-Fluorotoluene	$p\text{-}FC_6H_4CH_3$	241.0	
Fluorobenzene	FC_6H_5	244.0	
p-Trifluoromethylfluorobenzene	$p\text{-}FC_6H_4CF_3$	252.0	
p-Fluoroacetophenone	$p\text{-}FC_6H_4C(O)CH_3$	253.0	
p-Nitrofluorobenzene	$p\text{-}FC_6H_4NO_2$	257.0	
1,1-Difluoroethene	$F_2C=CH_2$	287.0	
Fluorophosgene	$F(H_3C)C=O$	308.4	
Acetyl fluoride		353.0	
Formyl fluoride	$F(H)C=O$	369.0	

carbon-phosphorus three-bond coupling has been reported for nucleotides [146]. Two- and three-bond coupling constants in this class of compounds cover the range of 1 to 10 Hz. These longer-range splittings are valuable aids for assigning the ^{13}C signals of pyranose, furanose and polyalcoholic carbons near phosphorus in nucleotides. An illustrative example is the assignment achieved for C-5' and C-4' of ribose and ribitol in the enzyme cofactor flavin adenine dinucleotide as the result of carbon-phosphorus long-range coupling [147] (Fig. 5.9).

3.3 Spin-Lattice Relaxation Times

3.3.1 Mechanism of ^{13}C Spin-Lattice Relaxation [7, 148]

In spin-lattice relaxation, the excited nuclei transfer their excitation energy to their environment. They do so *via* interaction of their magnetic vectors with fluctuating local fields of sufficient strengths and a fluctuation frequency of the order of the Larmor frequency of the nuclear spin type. Depending upon the atomic and electronic environment of a nucleus in a molecule and the motion of that molecule, there are five potential mechanisms contributing to spin-lattice relaxation of the nucleus.

3.3.1.1 Relaxation Resulting from Chemical Shift Anisotropy (CSA Mechanism)

The magnetic shielding of a nucleus arising from the surrounding electrons can be anisotropic, *e.g.* in derivates of benzene, ethyne, and carbonyl compounds (*chemical shift anisotropy*). The shielding constant σ then possesses spatially oriented components, which can change with time if the molecule moves relative to the field $\boldsymbol{B}_0$. There arise fluctuating local fields which permit spin-lattice relaxation of the anisotropically shielded nucleus. A contribution of the CSA mechanism is apparent from a proportionality of the T_1 values measured to the square of the magnetic field strength B_0 applied. This contribution is usually negligible for the ^{13}C nuclei of organic molecules.

3.3.1.2 Relaxation by Scalar Coupling (SC Mechanism)

The spins of two proximate nuclei A and X in a molecule undergo coupling, *i.e.* the signals of A and X are split, but only if the life-times of these nuclei in their magnetic energy levels (Fig. 1.3) are sufficiently large (*scalar coupling*).

If nucleus X relaxes very much faster than nucleus A then no signal splitting is observed. However, the fast relaxation of X generates fluctuating fields which in turn contribute to the relaxation of nucleus A. Quadrupole nuclei having $I \geqslant 1$, *i.e.* nuclei whose charge distribution is not spherically symmetrical, relax so rapidly that, for instance, they accelerate the relaxation of neighboring nuclei. The contribution of the SC mechanism, which can be recognized from a frequency and temperature dependence of T_1, is particularly large when the coupling nuclei precess with similar Larmor frequencies. This applies to ^{13}C and the quadrupole ^{79}Br. As a consequence, ^{13}C nuclei bound to Br relax relatively fast ($CHCl_3$: $T_1 = 32.4$ s; $CHBr_3$: $T_1 = 1.65$ s).

3.3.1.3 Relaxation by Spin Rotation (SR Mechanism)

If a molecule or molecular segment rotates, then the magnetic vectors of the bonding electron spins will also rotate. This gives rise to fluctuating local fields which can contribute to spin-lattice relaxation of the nuclei of a rotating molecule or a rotating alkyl group. This SR mechanism plays an important role in small symmetrical molecules (methane, cyclopropane) or in small segments of larger molecules (methyl groups). If the ^{13}C nuclei are not protonated (as in CS_2), spin rotation can even become the predominant relaxation mechanism. In such cases T_1 is found to decrease significantly with increasing temperature.

3.3.1.4 Relaxation by Internuclear Dipole-Dipole Interaction (DD Mechanism)

Each nuclear spin generates a local magnetic field upon molecular motion. If two magnetic nuclei such as ^{13}C and ^1H are linked by a bond, then each of the nuclei will experience not only the external field B_0, but also the local field of the other nuclear spin (*internuclear dipole-dipole interaction*). Now molecular motion is very rapid in liquids. Rotations of C—H bonds are correspondingly fast in dissolved organic molecules, and the orientations of ^1H and ^{13}C relative to B_0 will be constantly changing. The resulting constantly imposed magnetic reorientation of the two nuclear spins generates fluctuating local fields which contribute to the relaxation of the nuclei. Most of the ^{13}C nuclei in organic molecules, especially those linked to hydrogen (CH, CH_2, CH_3), are relaxed predominantly by internuclear dipole-dipole interaction.

T_1 values of ^{13}C are usually measured with proton noise decoupling. Thus, just one signal, not a multiplet, and consequently only one T_1 value results for each C atom of a molecule. During this decoupling, the protons transfer their excitation energy to the "lattice" primarily by internuclear DD interaction with ^{13}C and, consequently, forced ^{13}C relaxation. As a result, the population of the energetically more favorable ^{13}C spin states (Fig. 1.3) increases, and the ^{13}C signal intensities are enhanced on proton-decoupling more than would be expected from the multiplet intensities in the spectra recorded without decoupling (*nuclear Overhauser enhancement*, NOE [34]).

If ^{13}C relaxation proceeds exclusively by the DD mechanism, the NOE factor η_C, indicating the enhancement of the ^{13}C signal intensity on decoupling, is determined by the gyromagnetic ratios of ^1H and ^{13}C according to eq. (3.18) [34]:

$$\eta_C = \frac{\gamma_H}{2\gamma_C} = 1.988. \tag{3.18}$$

NOE factors of this magnitude are indeed measured for numerous ^{13}C signals of organic molecules, especially for signals of CH and CH_2 carbon atoms, as shown for formic acid in Fig. 2.20 [33].

NOE factors smaller than 1.988 indicate participation of other mechanisms in the spin-lattice relaxation of protonated ^{13}C nuclei. The percentange contribution of the DD mechanism can be ascertained from the measured NOE factor according to eq. (3.19):

$$\% \text{ DD} = \frac{\eta_C}{1.988} \times 100. \tag{3.19}$$

Accordingly, the time constant $T_{1(DD)}$ of the DD mechanism can be calculated from the measured η_C and T_1 values [eq. (3.20)]:

$$T_{1(DD)} = T_1 \frac{1.988}{\eta_C}. \tag{3.20}$$

3.3 Spin-Lattice Relaxation Times

Since in the presence of several relaxation mechanisms the relaxation rates $1/T_1$ can be assumed to be additive, as shown in eq. (3.21),

$$\frac{1}{T_1} = \frac{1}{T_{1(DD)}} + \frac{1}{T_{1(SR)}} + \frac{1}{T_{1(SC)}} + \frac{1}{T_{1(CSA)}} + \cdots \quad (3.21)$$

the measured T_1 value and the amount of $T_{1(DD)}$, accessible from the NOE factor η_C via eq. (3.20), can be used to calculate the time constant of a second relaxation mechanism, provided that no further mechanisms are operative. For instance, in benzene the DD mechanism is accompanied only by the SR mechanism. Benzene contains no quadrupole nucleus, and the contribution of the CSA mechanism to ^{13}C relaxation of liquid benzene is so small that the T_1 value can be shown to be independent of the magnetic field strength [149]. Then $T_{1(DD)}$ and $T_{1(SR)}$ can be determined from the experimental values of T_1 and η_C according to Scheme 3.1 [148, 150–152].

Experimental data: Spin-lattice relaxation time $T_1 = 29.3$ s
NOE factor $\eta_C = 1.6$

Percentage of DD relaxation:

% DD = $\eta_C \times 100/1.988 = 80\%$

Contribution of DD mechanism:

$T_{1(DD)} = T_1 \times 1.988/\eta_C = 29.3 \times 1.988/1.6 = 36.4$ s

Contribution of SR mechanism:

from $\dfrac{1}{T_1} = \dfrac{1}{T_{1(DD)}} + \dfrac{1}{T_{1(SR)}}$

it follows

$$T_{1(SR)} = \frac{T_1 \, T_{1(DD)}}{T_{1(DD)} - T_1} = \frac{29.3 \times 36.4}{36.4 - 29.3} = 150 \text{ s}$$

Scheme 3.1. Determination of $T_{1(DD)}$ and $T_{1(SR)}$ from T_1 and η_C measured for benzene.

3.3.1.5 Electron Spin-Nucleus Interactions and Consequences

During molecular motion, unpaired electrons generate much stronger fluctuating local fields than the nuclear spins, owing to their considerably larger magnetic moments. As a result, relaxation in the presence of unpaired electrons is dominated by electron spin-nucleus dipole-dipole interactions analogous to internuclear dipole-dipole interactions, especially for nonprotonated ^{13}C nuclei. Thus, smaller T_1 values are observed for paramagnetic compounds themselves, or in their presence, e.g. in solvents containing oxygen. Paramagnetic contaminants, including the oxygen nearly always dissolved in liquids, must therefore be removed if characteristic and reproducible T_1 values are to be measured.

Acceleration of relaxation of a nuclear spin by paramagnetic compounds is the greater the slower the nucleus relaxes. For example, the nonprotonated carbon nuclei of phenylethyne in a normally prepared sample (20% in $(CD_3)_2CO$ as solvent) display T_1 values lower by a factor of two (numbers in parentheses) than those for a degassed, i.e. largely oxygen-free, solution [151]. In contrast, the faster-relaxing protonated carbon atoms ($T_1 < 20$ s) are affected much less.

```
           132.0
          (53.0)
  8.2        C≡C-H      Phenylethyne
 (9.0)  107.0    9.3
  14.0 14.0 (56.0) (8.5)   (s)
 (13.3) (13.2)
```

The pronounced relaxation acceleration observed for slowly relaxing nuclei in the presence of paramagnetic compounds is exploited in Fourier transform ^{13}C NMR spectroscopy. On use of the fast pulse sequences that are frequently necessary, the spin-lattice relaxation of "slow" ^{13}C nuclei can no longer follow excitation, and the corresponding ^{13}C signals have low intensities. In such cases, addition of small amounts of relaxation accelerators, such as radicals or transition metal salts, to the sample amplifies these signals [153].

It ist quite reasonable that a paramagnetic central ion should accelerate the relaxation of the C atoms of a ligand, and that the magnitude of the acceleration should depend upon the distance between the ion and the C atom being observed.

In such cases, T_1 measurements provide information about the distances between the paramagnetic ion and the ligand. Thus, the spin-lattice relaxation times of the ^{13}C nuclei in α- and β-methyl D-glucopyranoside become selectively smaller on binding to the Mn^{2+} complex of the protein concanavalin A [154]. The experimental T_1 values show that the two glucopyranoside anomers retain their $^4C_1$ chair conformation, but adopt different orientation on binding to the protein complex [154]. The nonreducing end of the α-anomer (C-3, C-4) approaches the Mn^{2+} ion most closely (average separation approx. 1 nm) [154].

3.3.2 Influence of Molecular Motion on Dipole-Dipole Relaxation

The majority of ^{13}C NMR measurements are performed with solutions or liquid samples with proton decoupling. Under these conditions, the ^{13}C spin-lattice relaxation times T_1 depend mainly upon the speed of molecular motion relative to ^{13}C Larmor precession. The overall tumbling of molecules in liquids cannot easily be resolved into components such as rotation, vibration, or translation. In order to derive a useful relation with the spin-lattice relaxation, the average time required by a molecule between two reorientations is taken as a measure of molecular motion. This time is referred to the unit circle, i.e. divided by 2π, and then designated the *effective molecular correlation time* τ_C [7, 155–158]. If the molecules are rotating or vibrating, one revolution or vibration per second (1 Hz) is equal to 2π rad/s. τ_C then corresponds to the averge time required by the molecules to rotate through 1 radian.

Only those molecular motions whose "frequencies" lie in the region of the ^{13}C Larmor precession lead to rapid ^{13}C dipolar relaxation. The ^{13}C nuclei are known to precess with a frequency of $v_0 \approx 2.26 \times 10^7$ Hz or $\omega_0 = 2\pi v_0 \approx 1.42 \times 10^8$ rad/s at a field strength of

3.3 Spin-Lattice Relaxation Times

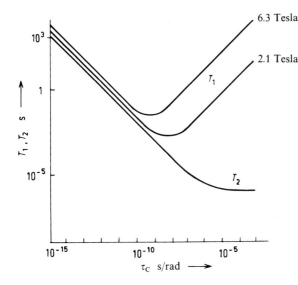

Fig. 3.20. Relation between the relaxation times T_1 and T_2 and the effective correlation time τ_C. In stronger magnetic fields the T_1 minimum shifts towards smaller correlation times, i.e. higher "frequencies" of molecular motion.

$B_0 \approx 2.1$ Tesla. Most effective DD relaxation then becomes possible owing to molecular motions having correlation times of

$$\tau_C \approx \frac{1}{1.42 \times 10^8} \approx 7 \times 10^{-9} \text{ s/rad.}$$

The correlation function $T_1 = f(\tau_C)$ shown in Fig. 3.20 gives a minimum for such motion; however, this is true only for a field strength of 2.1 Tesla. At higher fields the minimum shifts to shorter correlation times, i.e. efficient DD relaxation requires faster molecular motion (Fig. 3.20).

Very slow molecular motion ($\tau_C > 10^{-9}$ s/rad at $B_0 \approx 2.1$ Tesla) leads to an increase in T_1, while T_2 decreases (Fig. 3.20). The signals then broaden (line width at half-maximum intensity $\Delta v_{1/2} \sim 1/T_2$). Therefore, the more sluggish macromolecules usually give poorly resolved ^{13}C NMR spectra having a bandlike shape.

If the viscosity is sufficiently low, small and medium-sized molecules tumble very fast. The "frequencies" of their motion often even exceed the Larmor frequency $\omega_0 = 2\pi v_0$, and the correlation time consequently becomes shorter than the value $\tau_{C(\min)}$, leading to the most effective DD relaxation ($\omega_0 \tau_C \ll 1$). This situation corresponds to the declining branch of the correlation function (Fig. 3.20): T_1 and T_2 increase, are approximately equal, and the signals become sharper owing to $\Delta v_{1/2} \sim 1/T_2$ (*motional narrowing*). Under these circumstances the dipolar relaxation time $T_{1(DD)}$ of a ^{13}C nucleus is shown in eq. (3.22) to be inversely proportional to the number N of directly adjacent protons and to the effective correlation time τ_C [155–158].

$$\frac{1}{T_{1(DD)}} = \hbar^2 \gamma_C^2 \gamma_H^2 r_{CH}^{-6} N \tau_C. \tag{3.22}$$

$\hbar \equiv h/2\pi$; h is Planck's constant; γ_C and γ_H are the gyromagnetic ratios of ^{13}C and ^{1}H respectively; r_{CH} is C—H internuclear distance, usually about 0.109 nm.

3.3.3 Information Content of ^{13}C Spin-Lattice Relaxation Times

3.3.3.1 Degree of Alkylation and Substitution of C Atoms

Within rigid molecules the correlation time τ_C is equal for all carbon atoms. Moreover, the C–H bond lengths in nearly all organic compounds are approximately 0.109 nm, except for alkynes ($r_{CH} \sim 0.106$ nm). Under these conditions, the spin-lattice relaxation time T_1 of a ^{13}C nucleus relaxing by the DD mechanism is shown by eq. (3.23) to depend solely upon the number N of directly bonded H atoms:

$$T_{1\,(DD)} = \text{const.}/N. \tag{3.23}$$

Neglecting the methyl carbons, which do not belong to the rigid skeleton of a molecule owing to internal rotation of the CH$_3$ groups (Section 3.3.4), the T_1 values for the CH and CH$_2$ groups of the same molecule give a ratio of 2:1; quaternary C atoms having $N = 0$ relax much more slowly:

$$T_{1\,(C)} \gg T_{1\,(CH)} > T_{1\,(CH_2)} \quad \text{and} \quad T_{1\,(CH)}: T_{1\,(CH_2)} = 2:1. \tag{3.24}$$

This relation (3.24) can be verified for simple rigid molecules, *e.g.* for adamantane having $T_{1\,(CH)} = 17.0$ s and $T_{1\,(CH_2)} = 7.8$ s [159 *]. It is a valuable aid in the assignment of the ^{13}C NMR spectra of large molecules, particularly when signal crowding precludes clear distinction between singlets, doublets, and triplets in coupled or proton off-resonance decoupled ^{13}C NMR spectra.

3.3.3.2 Molecular Size and Relaxation Mechanisms

The T_1 values of the ^{13}C nuclei generally decrease with increasing molecular size if the molecules are rigid. Thus values in the ms or ns range are observed for the ^{13}C nuclei of the rigid backbone of macromolecules. Medium-sized molecules (C$_{10}$–C$_{50}$) typically show T_1 values between 0.1 s and 20 s. For very small molecules of high symmetry (dumbbells, tetrahedra, pyramids, small regular polyhedra), however, values of up to 100 s and beyond are possible. This applies, *e.g.* to methane derivatives except for bromides (effective SC relaxation, *cf.* Section 3.3.1.2), especially when the molecule contains no H atoms (^{13}CCl$_4$: $T_1 \approx 160$ s). The reason is that the ^{13}C nuclei of very mobile small molecules relax partly, and sometimes even predominantly, by the less effective spin-rotation mechanism. The considerable contribution of spin rotation to ^{13}C spin-lattice relaxation has already been demonstrated for benzene (Section 3.3.1.4).

The relation between molecular size, T_1, and the relaxation mechanisms is apparent from the homologous series of cycloalkanes. T_1 and NOE data as well as the contributions of the DD and SR mechanisms accessible from these experimental values *via* eqs. (3.20) and (3.21) are listed in Table 3.15. The ^{13}C nuclei of cyclopropane relax almost as fast by spin-rotation as by the DD mechanism ($T_{1\,(SR)} \approx T_{1\,(DD)}$). However, in cyclobutane the DD mechanism predominates ($T_{1\,(DD)} < T_{1\,(SR)}$), and the carbon atoms of cyclohexane and its higher homologs relax exclusively by the DD mechanism ($T_{1\,(DD)} \ll T_{1\,(SR)}$) [160].

---

* See also Fig. 2.29 and compare the results.

Table 3.15. Spin-Lattice Relaxation Times T_1 and NOE Factors η_C of the ^{13}C Nuclei of Cycloalkanes Containing n Carbon Atoms [160].

n	T_1 s	η_C	DD %	$T_{1(DD)}$ s	$T_{1(SR)}$ s
3	36.7	1.0	50.2	72.2	74.6
4	35.7	1.4	70.5	50.7	121
5	29.2	1.52	76.5	38.2	124
6	19.6	1.9	95.5	20.5	447
7	16.2	1.96	98.5	16.4	>1000
8	10.3	2.0	100	10.3	>1000
10	4.7	2.0	100	4.7	>1000

The $T_{1(DD)}$ values themselves decrease fairly regularly with increasing ring size (Table 3.15).

3.3.3.3 Anisotropy of Molecular Motion

Eq. (3.21) discussed in Section 3.3.2 is only valid if the motion of the molecules under study has no preferential orientation, *i.e.* is not anisotropic. Strictly speaking, this applies only for approximately spherical bodies such as adamantane. Even an ellipsoidal molecule like *trans*-decalin performs anisotropic motion in solution: it will preferentially undergo rotation and translation such that it displaces as few as possible of the other molecules present. This anisotropic rotation during translation is described by the three diagonal components R_1, R_2, and R_3 of the rotational diffusion tensor. If the principal axes of this tensor coincide with those of the moment of inertia – as can frequently be assumed in practice – then R_1, R_2, and R_3 indicate the speed at which the molecule rotates about its three principal axes.

The connection between anisotropic molecular motion and nuclear relaxation was derived by Woessner as early as 1962 [161]. Accordingly, the dipole-dipole relaxation time of a carbon nucleus is a function of the diagonal components R_1, R_2, and R_3 of the rotational diffusion tensor and the cosines λ, μ, and ν of the angles assumed by the C–H bonds relative to the principal axes of this tensor:

$$1/T_{1(DD)} = f(R_1, R_2, R_3, \lambda, \mu, \nu). \tag{3.25}$$

If the position of the principal axes of the rotational diffusion tensor were known with respect to the molecular coordinates, then the motion of the molecule could be calculated from the measured relaxation times. With simple molecules, however, it is possible to interpret the T_1 values qualitatively in terms of an anisotropic motion.

Thus, the C–H nuclei in the *para* position of monosubstituted benzene derivatives relax faster than those in the *ortho* or *meta* positions (Table 3.16 [151]). The reason for this behavior lies in a preferred rotation about the molecular axis passing through the substituent X and the *p*-carbon. During this motion, the *para* C–H bond does not change its direction relative to the field $\boldsymbol{B}_0$; fluctuating local fields can only arise at the *p*-C nucleus by rotations of the molecule perpendicular to the preferred axis. However,

Table 3.16. Spin-Lattice Relaxation Times T_1 of the ^{13}C Atoms in Benzene Derivatives[a].

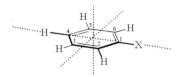

X	C Atom	T_1 (s)	Ref.
H	1–6	29.3[b]	[151]
CH_2-CH_3	1	36.0	[126]
	2,6	18.0	
	3,5	18.0	
	4	13.0	
	$-CH_2$	13.0	
	$-CH_3$	19.0	
$CH=CH_2$	1	75.0	[151]
	2,6	14.8	
	3,5	13.5	
	4	11.9	
	$-CH=$	17.0	
	$=CH_2$	7.8	
$C\equiv CH$	1	107.0[b]	[151]
	2,6	14.0	
	3,5	14.0	
	4	8.2	
	$-C\equiv$	132.0	
	$\equiv CH$	9.3	
CH_3	1	58.0	[151]
	2,6	20.0	
	3,5	21.0	
	4	15.0	
	CH_3	16.3	
C_6H_5	1	61.0	[151]
	2,6	5.9	
	3,5	5.9	
	4	3.2	
NO_2	1	56.0	[151]
	2,6	6.9	
	3,5	6.9	
	4	4.8	
OH	1	21.5	[151]
	2,6	4.4	
	3,5	3.9	
	4	2.4	

[a] for an additional example see Fig. 2.27.
[b] degassed.

these less preferred rotations still have sufficiently large frequencies to effectively relax the p-C atom. The preferred rotation itself continuously changes the orientation of the *ortho* and *meta* C−H bonds with respect to B_0, but too fast for effective dipolar relaxation of the *ortho* and *meta* carbon nuclei.

Preferred rotation is also conceivable in 3-methyl-5,6,7,8-tetrahydroquinoline, i.e. about the axis passing through C-7, C-3, and the methyl group [162]. Accordingly, the C-7 methylene group exhibits a smaller T_1 value than all the other CH_2 groups of this molecule.

3-Methyl-5,6,7,8-tetrahydroquinoline

The first quantitative estimate of the rotational diffusion tensor for simple molecules was accomplished by Grant et al. [163]. By solving the Woessner equations, they were able to show e.g. for *trans*-decalin that the molecule rotates preferentially like a propeller, i.e. about the axis perpendicular to the plane of the molecule. The values given as a measure of the rotational frequencies do not correlate with the moments of inertia, but instead with the "ellipticities" of the molecule as defined [163]. They are accessible from the ratios of the interatomic distances perpendicular to the axes of rotation, and can be adopted as a measure of the number of solvent molecules that have to be displaced on rotation about each of the three axes.

The greater the differences between the interatomic distances perpendicular to an axis, the greater will be the ellipticity ε, and the more slowly will the molecule rotate about that axis.

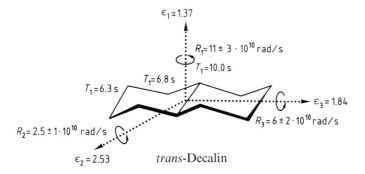

trans-Decalin

For less symmetric molecules one has to resort to computer programs [164] to solve the Woessner equations. The orientation of the rotational diffusion tensor is usually defined by assuming that its principal axes coincide with those of the moment of inertia tensor. This assumption is probably a good approximation for molecules of low polarity containing no heavy atoms, since under these conditions the moment of inertia tensor roughly represents the shape of the molecule.

A rather sophisticated application of Woessner's theory has been accomplished for all-*trans*-retinal and its isomers [165]. After determination of the components of the rotational diffusion tensor in retinal for various dihedral angles between the olefinic chain

172   3 ^{13}C NMR Spectral Parameters and Structural Properties

and the cyclohexene ring, the T_1 values could be calculated for each one of these conformations. The best agreement between experimental and calculated ^{13}C T_1 values results when the 5,6–7,8 dihedral angle is 60°, in other words when the cyclohexene ring is bent by 60° towards the plane of the olefinic chain [165]. This is an example of a conformation being determined by measurement of ^{13}C relaxation times.

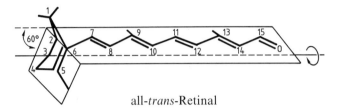

all-*trans*-Retinal

Woessner's equations thus permit prediction of spin-lattice relaxation times for the dipole-dipole mechanism, which can be of help in the assignment of ^{13}C NMR spectra. Moreover, the calculations described can be applied to the problem of internal molecular motion.

3.3.3.4 Internal Molecular Motion

Rotation of Methyl Groups

While the skeleton of large molecules is often relatively rigid, the methyl groups bonded to the backbone are highly mobile. Their rotation is thus much faster than the overall motion of the molecule ($\tau_{C(CH_3)} \ll \tau_{C(skeleton)}$). The methyl groups in proton-decoupled ^{13}C NMR spectra therefore exhibit NOE factors that are smaller than the typical value for pure dipolar relaxation (1.988), *i.e.* the signals are relatively weak. At the same time, the spin-lattice relaxation times of the methyl carbon nuclei are usually much longer than permitted by the ratio expected for the DD mechanism according to eq. (3.22), *i.e.*

$$T_{1\,(CH)}: T_{1\,(CH_2)}: T_{1\,(CH_3)} = 6:3:2.$$

The experimental values found for all the methyl carbons in 3-methyl-5,6,7,8-tetrahydroquinoline [162], 8,9,9-trimethyl-5,8-methano-5,6,7,8-tetrahydroquinazoline [162], and cholesteryl chloride [166] may be cited as examples. Lower intensities and longer T_1 values relative to the methylene and methine carbon nuclei thus frequently facilitate detection of methyl resonances in ^{13}C NMR spectra.

8,9,9-Trimethyl-5,8-methano-5,6,7,8-tetrahydroquinazoline

Cholesteryl-chloride

3.3 Spin-Lattice Relaxation Times

Intramolecular Steric Interactions

Steric interactions can hinder the rotation of methyl groups and thus accelerate methyl relaxation. It can be deduced, *e.g.* from the T_1 values of butanone oxime, that the CH_3 group *syn* to the OH group in the more stable *anti* isomer rotates faster ($T_1 = 6.1$ s) than that *anti* to the OH group in the more labile *syn* isomer ($T_1 = 2.8$ s) [167]. As an explanation it is assumed that van der Waals interactions between the methyl and hydroxyimino groups, and between the methyl and methylene groups are of the same order of magnitude in the *anti* isomer. In contrast, the methyl group in the *syn* isomer adopts an energetically more favorable conformation involving one-sided interaction with the methylene group, with the result that its rotation is hindered [167].

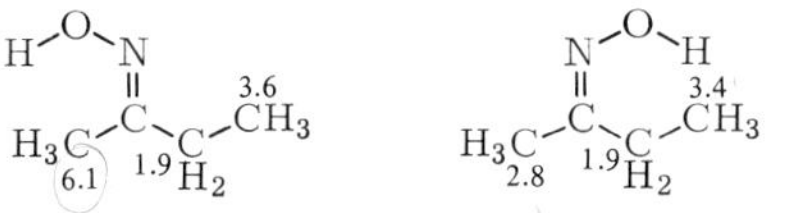

anti   Butanone oxime   syn

The T_1 values of the methyl carbon nucleus in 1-methylnaphthalene and 9-methylanthracene are interpreted accordingly: in 1-methylnaphthalene the *peri* proton forces a preferred conformation of the methyl group and thereby inhibits its rotation. On the other hand, in 9-methylanthracene two energetically equivalent *peri* H $\cdots$ CH_3 interactions occur, so that methyl rotation is less hindered because there is no preferred conformation [148].

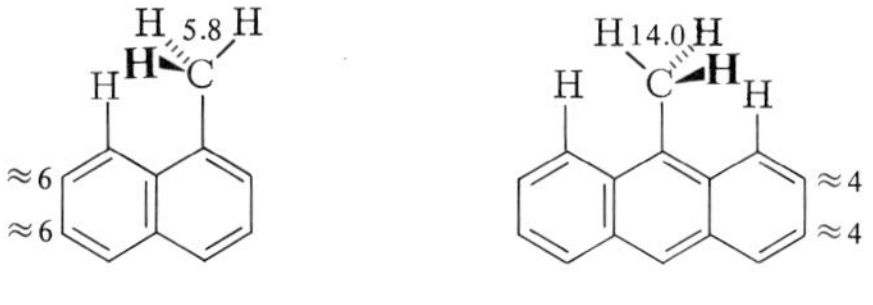

1-Methylnaphthalene   9-Methylanthracene

In the series 6,7-dihydrolinalool, linalool, and 6,7-dehydrolinalool the methyl C atoms of the terminal 1,1-dimethylvinyl group which are located *trans* to the alkyl group show widely differing absolute values of T_1; however, the ratio $T_{1\,(trans)} : T_{1\,(cis)}$ is always about 2:1. The rotation of the *trans* methyl groups therefore appears to be less hindered than that of the *cis* methyl groups [126].

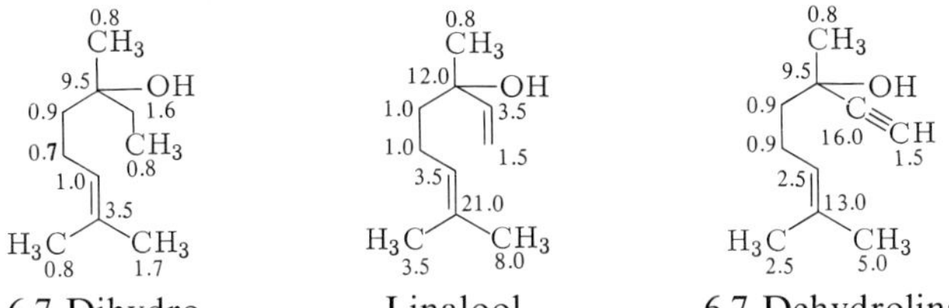

6,7-Dihydro-linalool   Linalool   6,7-Dehydrolinalool

174   3 ^{13}C NMR Spectral Parameters and Structural Properties

The T_1 values of the methyl carbon atoms of methylated phosphetanes (in s) reveal that the methyl groups attached to C-2 and C-4, as well as to C-3, are subject to greater rotational hindrance than the P–CH$_3$ group [168]. Surprisingly, the pseudoaxial methyl groups sometimes relax more slowly than their pseudoequatorial counterparts, in spite of the stronger van der Waals repulsion expected for the former [168]. Shorter ^{13}C spin-lattice relaxation times thus do not always constitute proof of hindered rotation. Instead, the effects of methyl mobility found from relaxation time measurements must always be considered in relation to the motion of the entire molecule, which can vary markedly during a change of configuration.

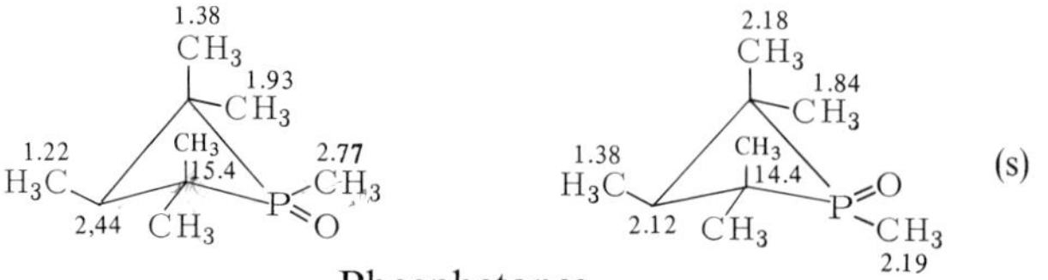

Phosphetanes

A more quantitative interpretation of methyl relaxation requires a knowledge of the motional anisotropy of the entire molecule. Thus the activation energy of methyl rotation can be estimated from T_1 data if the rotational diffusion tensor of the molecule, mentioned in Section 3.3.3.3, is known [164].

In biphenyl derivatives and related compounds, substituents can hinder rotation of the phenyl groups. Since phenyl rotation is anisotropic (Section 3.3.3.3, $T_{1\,(o,m)} > T_{1\,(p)}$), the ratio $T_{1\,(o,m)} : T_{1\,(p)}$ of the phenyl carbon nuclei will decrease on rotational hindrance. Hence, a smaller ratio $T_{1(m)} : T_{1\,(p)}$ is found in 2,2',6,6'-tetramethylbiphenyl [169] than in biphenyl itself, owing to the methyl groups in the o- and o'-positions (Table 3.16). In 3-bromobiphenyl and phenyl benzoate, the ratio $T_{1\,(o,m)} : T_{1\,(p)}$ is significantly larger for the unsubstituted phenyl and phenoxy ring than for the 3-bromophenyl and benzoyl ring [148].

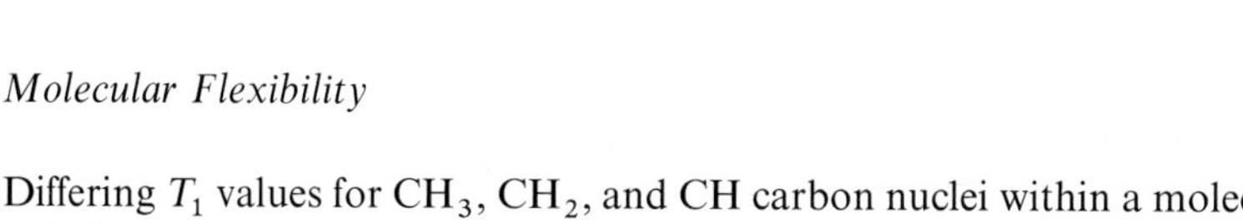

3-Bromobiphenyl   Phenyl-   2,2',6,6'-Tetramethyl-
                      benzoate   biphenyl

Molecular Flexibility

Differing T_1 values for CH$_3$, CH$_2$, and CH carbon nuclei within a molecule can arise not only by methyl rotation or anisotropic molecular motion, but also from the segmental mobility of partial structures, even when the dipolar mechanism predominates. Thus the spin-lattice relaxation times of methylene carbon atoms in long alkane chains pass through a minimum at the middle of the chain. In the presence of heavy nonassociating substituents, the minimum is displaced somewhat towards the heavier end of the molecule, as is apparent for decane [170] and 1-bromodecane [171]. The molecular periphery is accordingly more flexible than the center.

$$\underset{\text{Decane}}{\overset{8.74\quad 6.64\quad 5.71\quad 4.95\quad 4.36\quad 4.36\quad 4.95\quad 5.71\quad 6.64\quad 8.74}{CH_3-CH_2-CH_2-CH_2-CH_2-CH_2-CH_2-CH_2-CH_2-CH_3}} \quad (s)$$

$$\underset{\text{1-Bromodecane}}{\overset{2.8\quad 2.7\quad 1.9\quad 2.0\quad 2.1\quad 2.1\quad 2.2\quad 3.1\quad 3.9\quad 5.3}{Br-CH_2-CH_2-CH_2-CH_2-CH_2-CH_2-CH_2-CH_2-CH_2-CH_3}} \quad (s)$$

In the side chain of cholesteryl chloride C atoms with the same number of attached protons are found to behave similarly [166]: the methyl carbon atoms C-26 and C-27 relax more slowly than C-21, the methylene carbon C-24 slower than C-23, and the methine carbon atom C-25 slower than C-20. The mobility of the side chain thus increases with increasing distance from the steroidal skeleton.

In the case of such flexible molecules the correlation time τ_C can be different for each C atom. $N T_1$ is then no longer a constant as for rigid molecules according to eq. (3.23), but inversely proportional to the correlation time, as can be seen from eq. (3.22) (N is the number of protons bonded to a carbon atom).

$$N T_1 \propto 1/\tau_C \tag{3.26}$$

The product $N T_1$ can therefore be interpreted as an internal molecular mobility parameter, although only qualitatively and with caution, as was apparent for a correlation of carbon-13 T_1 values with diffusion coefficients [170]. The increase in T_1 or $N T_1$ with increasing internal mobility is not only a valuable aid in the signal assignment of carbons having the same degree of substitution, but also offers the possibility of determining the geometry and internal dynamics in the liquid and dissolved states, particularly in the case of large molecules. ^{13}C-T_1 measurements with this aim have been performed on phospholipids, peptides, proteins, and synthetic polymers.

It has been proposed to use $N T_1$ values for sugar sequence determination in oligosaccharides. The sequence is reflected by a progressive increase in the average $N T_1$ times of the pyranose moiety carbon nuclei when going from the monosaccharide connected with the aglycone to the terminal sugar, as illustrated for the steroidal oligoglycoside k-strophanthoside [172].

In dipalmitoyllecithin the carbon nuclei relax increasingly slowly going from the central glycerol group to the ends of the two fatty acids, and thence to the tetraalkylammonium end of the choline group [173]. Thus, the mobility becomes successively higher, starting from the glycerol skeleton and proceeding along the fatty acid and choline chains to the molecular periphery. The terminal propyl groups of the fatty acid chains appear to undergo particularly rapid motion. If the fatty acid is shortened or a double bond is introduced, the chain assumes more internal flexibility. This may be seen from a comparison of data for dipalmitoyl-, dioctanoyl-, and dioleoyllecithin *: the "central" methylene carbon atoms in dioctanoyl- and dioleoyllecithin not only relax more slowly than in dipalmitoyllecithin, their T_1 values also differ more significantly [174].

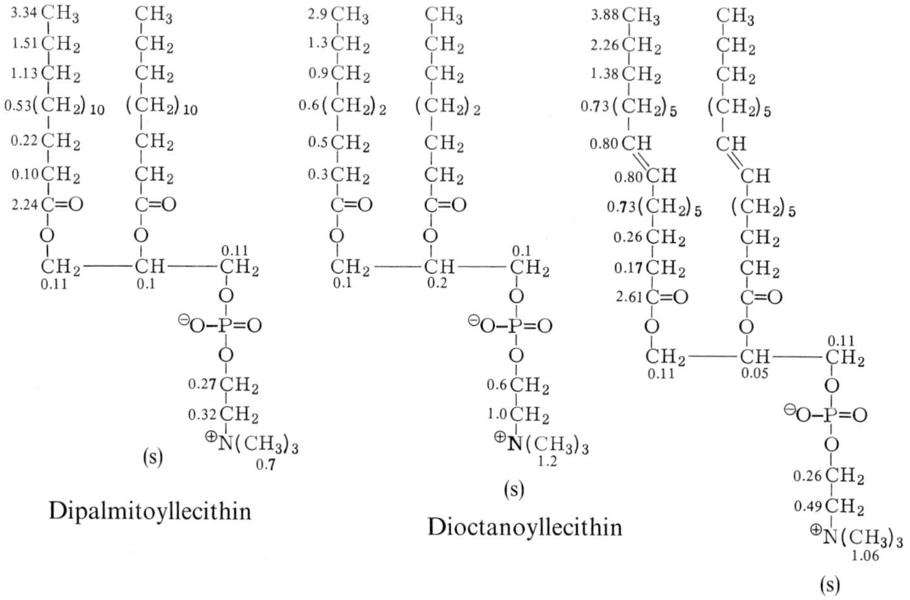

Dipalmitoyllecithin

Dioctanoyllecithin

Dioleoyllecithin

Together with proteins, phospholipids are the most important structural components of biological membranes. Since mobility of the lipid segments favors molecular transport through a membrane and thereby increases its permeability, a marked increase in T_1 along a lipid-fatty acid chain also reflects a more efficient molecular diffusion through the lipid layer of a membrane [175].

In methanol, all the α carbons of the cyclic decapeptide antibiotic grammicidin S relax at about the same rate ($T_1 \approx 135$ to 150 ms) [176]. Therefore, this molecule undergoes approximately isotropic motion in solution, and the motion of the α carbons within the peptide ring is slower than the overall molecular motion itself. The mobility parameters NT_1 increase significantly in the order C-α, C-β, C-γ. Thus the side chains become increasingly mobile with increasing length. It is only in the case of the relatively rigid

---

* Measured at 52 °C in D_2O solutions that had been treated with ultrasound.

Table 3.17. ^{13}C Spin-Lattice Relaxation Times T_1 of Ribonuclease A in Aqueous solutions (conc. 0.019 mol/L; 45 °C; 15.08 MHz; maximum deviation $\pm 30\%$ [177]).

C Atoms	Native T_1 (s)	Denaturated T_1 (s)
Carbonyl	0.416	0.539
α-C	0.042	0.120
β-C (Thr)	0.040	0.099
ε-C (Lys)	0.330	0.306

proline skeleton that all carbon nuclei have similar T_1 value. The phenylalanine side chain displays the anisotropy of rotation ($T_{1\,(o,m)} > T_{1\,(p)}$) usually encountered in phenyl derivatives. Since the axis of rotation passes through C-β and C-p, these carbons should show the same $N\,T_1$ values. The differences measured in practice ($N\,T_{1\,(\beta)} > N\,T_{1\,(p)}$) are possibly due to somewhat different C–H bond lengths (C-β: sp^3 C–H bond; C-p: sp^2 C–H bond) [176].

In ribonuclease A ^{13}C spin-lattice relaxation of the carbonyl and α and β carbon atoms is slower in the denaturated protein than in the native sample [177]. Apparently, the skeleton of this macromolecule becomes more flexible on denaturation, probably owing to conformational changes. However, the ε carbons of lysine in the native protein exhibit relatively large T_1 values which change only insignificantly on denaturation [177]. This behavior is attributed to a considerable segmental mobility of the lysine side chain (Table 3.17 [177]).

Indeed, ^{13}C spin-lattice relaxation times can also reflect conformational changes of a protein, i.e. helix to random coil transitions. This was demonstrated with models of polyamino acids [178–180], in which definite conformations can be generated, e.g. by addition of chemicals or by changes in temperature. Thus effective molecular correlation times τ_C determined from spin-lattice relaxation times and the NOE factors were 24–32 ns/rad for the α carbons of poly-(β-benzyl L-glutamate) in the more rigid helical form and about 0.8 ms/rad for the more flexible "random coil" form [180].

A certain segmental mobility of the chains can also be deduced from ^{13}C-relaxation measurements on synthetic polymers. For instance, the ratio of the spin-lattice relaxation times for the methine and methylene carbon atoms of polystyrene is $T_{1\,(CH)} : T_{1\,(CH_2)} = 2:1$. The NOE factors η_C lie between 0.8 and 1.1, depending upon the solvent. Both parameters T_1 and η_C only slightly depend on tacticity. These results indicate that internuclear dipole-dipole interaction predominates in the ^{13}C relaxation of polystyrene, but that the polymer chain is also relatively flexible [181]. The carbons in amorphous *trans*-polyisoprene relax more slowly and display larger NOE factors than the corresponding atoms of amorphous *cis*-polyisoprene [182]. Apparently, the "frequency" of internal motion of these molecules lies in the region of the minimum of the correlation function (Fig. 3.20): the *trans* chain shows slightly shorter and the *cis* chain slightly longer correlation times τ_C; hence the *trans* isomer is more flexible [182]. If carbon black is added as filler to *cis*-polyisoprene (natural rubber), then no significant decrease in T_1 is observed; however, individual line broadening occurs for each carbon atom. The filler therefore appears to impair the mobility of some carbon atoms in the *cis* chain [182]. A situation can then be reached in which segmental motion in the chain becomes so slow that T_1 increases again, while the spin-spin relaxation time T_2 continues to decrease (Fig. 3.20, to right of minimum) so that the signals become broader ($\Delta v_{1/2} \propto 1/T_2$).

Natural Rubber

$$\left. \begin{array}{c} \underset{0.35}{H_3C} \diagdown \underset{0.095}{H} \\ {}_{0.7}C=C \\ \underset{0.05}{-H_2C} \diagup \underset{0.055}{CH_2-} \end{array} \right\} \eta_C = 1.2$$

Guttapercha

$$\left. \begin{array}{c} \underset{0.45}{H_3C} \diagdown \underset{0.16}{CH_2-} \\ {}_{0.95}C=C_{0.16} \\ \underset{0.085}{-H_2C} \diagup H \end{array} \right\} \eta_C = 1.5$$

3.3.3.5 Association and Solvation of Molecules and Ions

Strong intermolecular interactions such as hydrogen bonds or ion-dipole pairs restrict the motion of molecules and pertinent molecular segments. These interactions increase the correlation time τ_C and accelerate the ^{13}C spin-lattice relaxation. Shorter ^{13}C relaxation times can therefore also indicate the presence of such interactions. The T_1 values of the C atoms of carboxylic acids, phenols, alcohols, and solvated molecular ions behave in this way.

The ^{13}C nuclei in formic and acetic acids relax faster and show larger nuclear Overhauser enhancements than the ^{13}C nuclei of the methyl esters (Table 3.18) [183]. Hence a methyl ester is more mobile than its parent carboxylic acid, which is dimerized *via* hydrogen bonding. It may also be seen that a pure DD mechanism operates in formic acid, owing to the directly bonded proton ($\eta_C = 2.0$), whereas spin rotation also contributes to ^{13}C relaxation in methyl formate ($\eta_C = 1.55$).

Owing to hydrogen bonding, the ^{13}C nuclei of phenol relax faster than the corresponding nuclei of nonassociating benzenes (Table 3.16). – In decanoic acid [171] and 1-decanol [171, 184], ^{13}C spin-lattice relaxation becomes progressively slower on going from the associating COOH or CH$_2$OH group to the hydrophobic end of the molecule. While the

Table 3.18. T_1 and η_C Values of ^{13}C Nuclei in Formic and Acetic Acids, and their Methyl Esters [183].

		T_1 (s) at 15 MHz	T_1 (s) at 25 MHz	η^C
Formic acid	$-$CH$=$O	10.3$\pm$1	10.1$\pm$1	2.0 $\pm$0.15
Methyl formate	$-$CH$=$O	15.1$\pm$2	14.5$\pm$1.5	1.55$\pm$0.15
	$-$OCH$_3$	16.8$\pm$2	18.0$\pm$2	0.8 $\pm$0.1
Acetic acid	$>$C$=$O	29.1$\pm$2.2	30.0$\pm$3.5	1.08$\pm$0.04
	$-$CH$_3$	10.5$\pm$0.5	9.8$\pm$0.6	1.4 $\pm$0.13
Methyl acetate	$>$C$=$O	35.0$\pm$5	29.0$\pm$5	0.25$\pm$0.08
	$-$CH$_3$	16.3$\pm$1.3	17.4$\pm$1.5	0.63$\pm$0.08
	$-$OCH$_3$	17.0$\pm$1.5	18.3$\pm$2.1	0.93$\pm$0.04

3.3 Spin-Lattice Relaxation Times

hydrocarbon part becomes increasingly mobile towards the exterior, the polar end is rooted by hydrogen bonding. This behavior is also observed in branched alcohols and those having shorter chains such as 3,7-dimethyl-1-octanol [126] and 1-butanol [156], although it is less pronounced in such cases. In very small alcohols like methanol [185] relaxation is considerably slower ($T_1 = 17.5$ s) as a result of the much faster molecular motion.

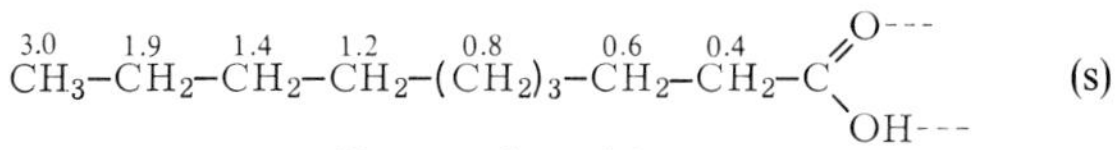

$$\underset{3.0}{CH_3}-\underset{1.9}{CH_2}-\underset{1.4}{CH_2}-\underset{1.2}{CH_2}-\underset{0.8}{(CH_2)_3}-\underset{0.6}{CH_2}-\underset{0.4}{CH_2}-C\underset{OH---}{\overset{O---}{\diagup}} \quad (s)$$

Decanoic acid

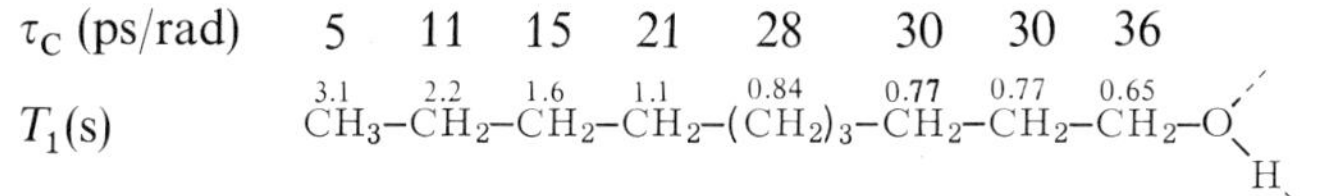

τ_C (ps/rad)	5	11	15	21	28	30	30	36
T_1(s)	3.1	2.2	1.6	1.1	0.84	0.77	0.77	0.65

$$CH_3-CH_2-CH_2-CH_2-(CH_2)_3-CH_2-CH_2-CH_2-O\diagdown_H$$

Decanol

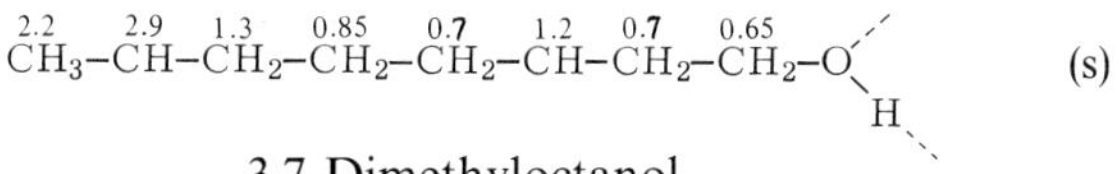

$$\underset{2.2}{CH_3}-\underset{2.9}{CH}-\underset{1.3}{CH_2}-\underset{0.85}{CH_2}-\underset{0.7}{CH_2}-\underset{1.2}{CH}-\underset{0.7}{CH_2}-\underset{0.65}{CH_2}-O\diagdown_H \quad (s)$$

3,7-Dimethyloctanol

$$\underset{4.2}{CH_3}-\underset{3.6}{CH_2}-\underset{3.9}{CH_2}-\underset{3.0}{CH_2}-O\diagdown_H \quad (s)$$

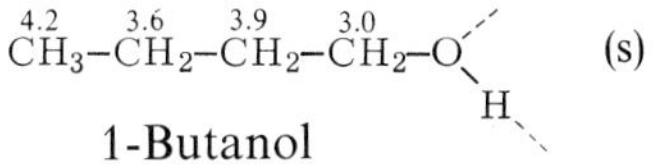

1-Butanol

In prostaglandin $F_{2\alpha}$, the T_1 values of ^{13}C nuclei in the alkyl side chain increase more significantly towards the periphery than in the associating carboxylic acid chain, thus confirming the assignment of the closely crowded CH_2 signals of the chains in the ^{13}C NMR spectrum [171]. Generally, the increased mobility on going from the associating

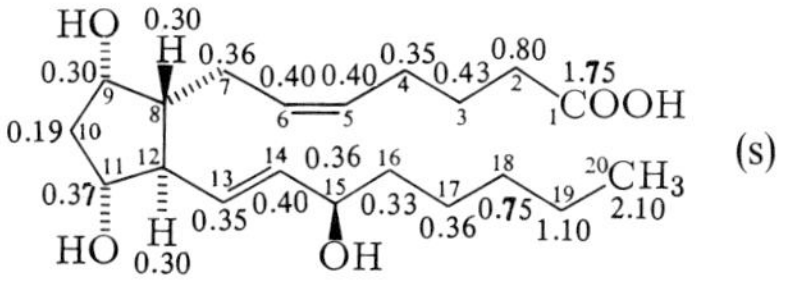

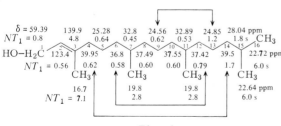

Prostaglandin $F_{2\alpha}$

Phytol

end of the molecule, and the concomitant increase in T_1 or $N T_1$ [eq. (3.26)] can be utilized in signal assignment. Thus in phytol, the diterpene alcohol component of chlorophyll, the signals of the carbon atom pairs 4 and 14, 6 and 12, and 9 and 13 defy unequivocal assignment, owing to very similar shifts and equal multiplicities (CH_2 groups). However, the C atoms 12, 13, and 14, which are more removed relative to the associating end of the molecule, undergo faster motion, and the $N T_1$ values are therefore larger. Signal assignment then follows readily [186].

In weakly solvating solvents interionic interactions between organic molecular ions can lead to fixation of the ionic end of the molecule. The T_1 values of pertinent and neighboring ^{13}C nuclei become smaller. In contrast, strongly solvating solvents such as water and alcohols inhibit interionic interaction and lead to an enhanced mobility of the ions solvated by ion-dipole interactions; ^{13}C spin-lattice relaxation is consequently slower in such solvents. Thus the T_1 values of n-butylammonium trifluoroacetate increase with the polarity of the solvent, as shown in Table 3.19 [148].

Table 3.19. Solvent Dependence of Spin-Lattice Relaxation Times of the ^{13}C Atoms in n-Butylammonium Trifluoroacetate [148].

Solvent	Conc. (% by wt.)	T_1 (s) CH_3—	—CH_2—	—CH_2—	—CH_2—	—$NH_3^{\oplus} CF_3COO^{\ominus}$
1,4-Dioxane	20	3.35	1.95	1.54	0.88	
CH_2Cl_2/acetone	24	3.90	2.41	1.67	0.91	
CF_3COOH	15.4	3.98	3.12	2.30	1.54	
	28.2	3.46	2.13	1.50	0.97	
CD_3OD	20	6.00	5.35	4.52	3.10	
D_2O	20	5.00	5.00	4.26	3.75	

3.3.3.6 Determination of Quadrupole Relaxation Times and Coupling Constants from ^{13}C Spin-Lattice Relaxation Times

While the nuclei ^1H and ^{13}C relax predominantly by the DD mechanism, relaxation of a quadrupole nucleus such as deuterium essentially involves fluctuating fields arising from interaction between the quadrupole moment and the electrical field gradient at the quadrupole nucleus [16]. If the molecular motion is sufficiently fast (decreasing branch of the correlation function, Fig. 3.20), the ^2H spin-lattice relaxation time is inversely proportional to the square of the quadrupole coupling constant $e^2qQ/\hbar$ of deuterium and the effective correlation time [16]:

$$\frac{1}{T_1} = \frac{3}{8}\left(\frac{e^2qQ}{\hbar}\right)^2 \tau_C. \qquad (3.27)$$

Sine ^{13}C and ^2H relaxation are the consequence of intramolecular interactions, the relaxation times T_1 of ^{13}C and ^2H within a given molecule may be compared directly with the help of eqs. (3.22) and (3.27):

$$\frac{T_1(^{13}C)}{T_1(^2\overline{H})} = \frac{3}{8} \frac{r^6 (e^2qQ/\hbar)^2}{N\hbar^2 \gamma_C^2 \gamma_H^2}. \qquad (3.28)$$

By using the numerical values for the constants (e.g. $r_{CH} \approx 0.109$ nm and $e^2 q Q/\hbar \approx 170$ kHz for an sp^3 C–H bond) a simple linear relation between the ^{13}C and ^{2}H relaxation times is obtained:

$$T_1(^{13}C)/T_1(^2H) = 19.9/N \quad (N = 1, 2, 3 \text{ for CD, CD}_2, \text{CD}_3). \tag{3.29}$$

Knowledge of the T_1 values for deuterium thus allows prediction of the ^{13}C relaxation times and *vice versa* [187]. On the other hand, if these data are known, the deuterium quadrupole coupling constants can be determined according to eq. (3.28). Agreement between calculated and experimental quadrupole coupling constants is good, any pronounced deviations indicating that alternative relaxation mechanisms are operative. For instance, the quadrupole coupling constant calculated for the terminal ethyne deuteron of phenylethyne according to eq. (3.28) is much too large (calc. 254 kHz; exp. 215 ± 5 kHz). This is explained in terms of the chemical shift anisotropy in phenylethyne [187].

3.3.4 Medium and Temperature Effects

Any change in the medium, i.e. the solvent, the concentration, the pH value, and in the temperature will affect the mobility of the molecules and hence also the spin-lattice relaxation. However, few systematic studies have so far been performed on the concentration dependence or on the precise influence of the macroscopic viscosity, the main reason, in the case of ^{13}C, lying in the need for highly protracted measurements in concentration studies. Moreover, little is known about the pH dependence of ^{13}C relaxation [188]. Nevertheless, the concentration dependence of ^{13}C relaxation is apparent in the case of saccharose (Table 3.20) [166], and intramolecular hydrogen bonds can be detected by measuring the concentration dependence of T_1 [189].

Solvating solvents appear to facilitate molecular motion, as has already been shown for n-butylammonium trifluoroacetate (Table 3.19) [148].

The ^{13}C nuclei of molecules whose protons undergo exchange relax at different rates in H$_2$O and D$_2$O solutions. Thus T_1 for the carbonyl carbon of acetic amide is 72 s in D$_2$O, but only 37 s in H$_2$O [189]. In this context, the influence of paramagnetic impurities should again be mentioned. Traces of paramagnetic ions (e.g. Cu^{2+}) no longer detectable by the usual analytical techniques can drastically lower the ^{13}C relaxation times of complexing substrates (e.g. amino acids).

Publications on the temperature dependence of ^{13}C relaxation [190–199] are concerned with simple molecules such as carbon disulfide, iodomethane, and acetonitrile [190–194]. Any interpretation of the temperature dependence of relaxation times requires a knowledge of the relative contributions of dipole-dipole and spin-rotation relaxation, as the former becomes progressively slower and the latter steadily faster with increasing temperature. In special cases, such as that of cyclopropane [200], the effects of both contributions can almost cancel each other.

The temperature dependence of dipole-dipole relaxation is that of the correlation time, which is usually written in the form of an Arrhenius equation (3.30):

$$\tau_C = \tau_{C_0} \times e^{\frac{\Delta E}{RT}}. \tag{3.30}$$

Plotting the logarithm of the dipole-dipole relaxation time $T_{1(DD)}$ *versus* the reciprocal temperature therefore gives an activation energy ΔE for molecular reorientation, which is of the order of 8.4 kJ/mol for most of the molecules hitherto studied. In the case of 4,4'-dimethylbiphenyl a value of 16.8 kJ/mol was found from the temperature dependence of T_1 for C-2 and C-3 (Fig. 3.21) [169]. For T_1 of the methyl carbon atom the Arrhenius plot is curved at lower temperatures, since the internal rotation of this group is then probably faster than the overall motion of the molecule.

The temperature dependence of the SR component is more complicated. An Arrhenius relation can again be formulated for the correlation time of spin rotation; however, this function additionally contains a linear temperature term.

Table 3.20. ^{13}C Spin-Lattice Relaxation Times T_1 (s) of Saccharose in H$_2$O and D$_2$O at 42 °C [166].

	C Atom	0.5 mol/L in D$_2$O	0.5 mol/L H$_2$O	2.0 mol/L H$_2$O
C	2	6.0	7.8	2.6
CH	1'	0.54	0.69	0.16
	3	0.51	0.64	0.16
	4	0.52	0.49	0.16
	5	0.58	0.61	0.16
	3'	0.61	0.73	0.15
	2'	0.53	0.69	0.17
	5'	0.60	0.70	0.16
	4'	0.59	0.60	0.17
CH$_2$	1	0.48	0.40	0.12
	6	0.35	0.30	0.15
	6'	0.34	0.37	0.13

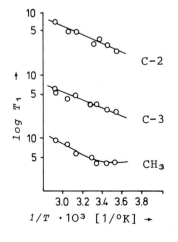

Fig. 3.21. Temperature dependence of spin-lattice relaxation times T_1 of protonated ^{13}C nuclei in 4,4'-dimethylbiphenyl [169]

4 ^{13}C NMR Spectroscopy of Organic Compounds

4.1 Saturated Hydrocarbons

4.1.1 Alkanes

From the ^{13}C chemical shift data collected in Table 4.1, Grant and Paul [85] deduced their additivity rule for the ^{13}C chemical shifts of alkanes. The signal assignments for the alkanes, also given in Table 4.1, are based on signal intensities and proton decoupling experiments.

Table 4.1. ^{13}C Chemical Shifts of Alkanes (δ_C in ppm) [85, 91a].

Compound	C-1	C-2	C-3	C-4	C-5
Methane	− 2.3				
Ethane	6.5				
Propane	16.1	16.3			
n-Butane	13.1	24.9			
n-Pentane	13.7	22.6	34.6		
n-Hexane	13.7	22.8	31.9		
n-Heptane	13.8	22.8	32.2	29.3	
n-Octane	13.9	22.9	32.2	29.5	
n-Nonane	13.9	22.9	32.3	29.7	30.0
n-Decane	14.0	22.8	32.3	29.8	30.1
Isobutane	24.6	23.3			
2-Methylbutane	21.9	29.7	31.7	11.4	
Neopentane	27.4	31.4			
2,2-Dimethylbutane	28.7	30.2	36.5	8.5	
2,3-Dimethylbutane	19.1	33.9			
2-Methylpentane	22.4	27.6	41.6	20.5	14.0
3-Methylpentane	11.1	29.1	36.5	18.4 (C-6)	
3,3-Dimethylpentane	7.7	33.4	32.2	25.6 (C-6)	

^{13}C chemical shift values of n-alkanes can be predicted using the additivity relationship given in eq. (4.1) [85]:

$$\delta_C(k) = B + \sum_l A_l n_{kl} + \sum S_{kl} \qquad (4.1)$$

$\delta_C(k)$ = chemical shift of the k-th C-atom;
B = a constant given by the ^{13}C chemical shift of methane (-2.3 ppm);
n_{kl} = number of C atoms in position l with respect to C atom k;
A_l = additive shift parameter of C atom l.

The values of A_l given in Table 4.2 were determined for linear alkanes (standard deviation of predicted chemical shifts: ± 0.10 ppm, $B = -2.3$).

Additional parameters S_{kl} are required in order to calculate the chemical shift values of branched chain alkanes (standard deviation of predicted chemical shifts ± 0.3 ppm, $B = -2.3$ ppm) (Table 4.3). The values symbolized by Greek letters indicate the change in chemical shift due to substitution of hydrogen by a methyl group at the α to ε carbon atoms. The remaining 8 correction parameters S_{kl} account for the effect of branching.

Primary, secondary, tertiary, and quarternary carbon atoms are symbolized by 1°, 2°, 3°, and 4°, respectively. Using this notation, the connectivity symbol 3° (2°) in Table 4.3 indicates the observed carbon k to be tertiary and attached to a secondary carbon l.

For illustration of the increment system, the ^{13}C chemical shift of C-3 in 2-methylhexane is calculated below according to eq. (4.1) using the shift parameters listed in Tables 4.2 and 4.3 [85]:

2-Methylhexane

21.1	28.5	39.1	29.7	23.4	13.4	ppm (calculated)
22.95	28.55	39.45	30.3	23.5	14.3	ppm in CDCl$_3$
CH$_3$—	CH—	CH$_2$—	CH$_2$—	CH$_2$—	CH$_3$	
1	2	3	4	5	6	
β	α		α	β	γ	

CH$_3$
1
β

$$\delta_3 = B + 2A_\alpha + 3A_\beta + A_\gamma + S_{2°(3°)}$$
$$= -2.3 + 18.2 + 28.2 - 2.5 - 2.5$$
$$= 39.1 \text{ ppm}$$

The calculated shift of C-3 is 39.1 ppm as compared to an observed value of 39.45 ppm. Prediction of carbon chemical shifts using the Grant-Paul relation (4.1) is a practical aid in assigning the carbon signals of larger alkyl groups, *e.g.* in cholestane derivatives (Section 5.2.2). – Other increment systems have been proposed [201, 202], as well as an absolute scale for carbon shielding [203].

Valence-bond calculations of ^{13}C chemical shift parameters of alkanes permit the following general conclusions [204].

Table 4.2. A_l Parameters of Equation (4.1) (n-alkanes).

Carbon position l	A_l in ppm
α	9.1 ± 0.10
β	9.4 ± 0.10
γ	−2.5 ± 0.10
δ	0.3 ± 0.10
ε	0.1 ± 0.10

Table 4.3. Corrective Terms S_{kl} for Equation (4.1) (Branched Chain Alkanes).

	S_{kl} in ppm
1° (3°)	− 1.10 ± 0.20
1° (4°)	− 3.35 ± 0.35
2° (3°)	− 2.50 ± 0.25
2° (4°)	− 7.5
3° (2°)	− 3.65 ± 0.15
3° (3°)	− 9.45
4° (1°)	− 1.50 ± 0.10
4° (2°)	− 8.35

(1) The paramagnetic screening term is mainly responsible for ^{13}C chemical shifts.

(2) Low-lying excited states cause downfield shifts of the ^{13}C signals.

(3) Increasing electron charge density at a carbon atom causes increased shielding and vice versa.

(4) Parallel-spin pairing at a carbon atom causes a decrease in the paramagnetic shielding term and, consequently, the carbon resonates at higher field.

(5) γ-Substitution effects can be explained in terms of sterically induced polarizations of the electrons of the C−H bond.

(6) Bulky β-substituents force carbons to resonate at lower field due to orbital contraction, which causes an increase in the paramagnetic shielding term.

As shown in Table 4.2, substitution of a hydrogen by carbon causes a deshielding in the α and β positions ($A_\alpha = 9.1$; $A_\beta = 9.4$ ppm), while a shielding is found as the γ effect ($A_\gamma = -2.5$ ppm). The γ effect is attributed to Van der Waals repulsion between hydrogens (Section 3.1.3.8) which causes the electrons of the CH σ bond to move away from hydrogen (deshielding) towards carbon (shielding). When free rotation is neither hindered nor frozen, the observed γ effect is averaged, as illustrated for butane which has three stable conformers.

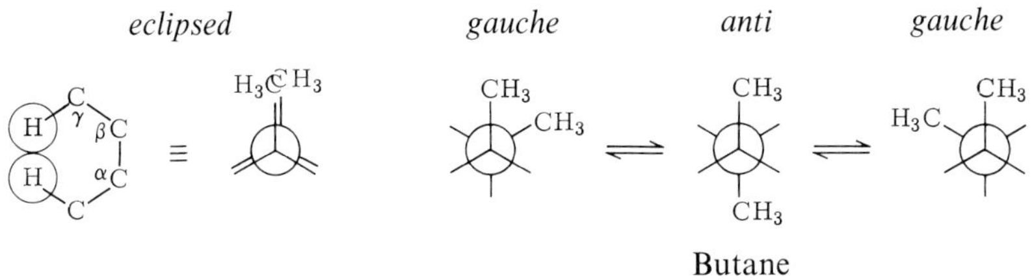

Butane

Stronger γ shieldings will be found when rigid conformations involve *gauche* or *eclipsed* interactions. – Energetically favored conformations of alkanes can be identified by low-temperature ^{13}C NMR, as shown for 2,3-dimethylbutane which, at $-188\,°$C (85 K), prefers the *gauche* conformation [205].

```
      H₃C     CH₃              H₃C
34.2                                       31.5     (ppm)
      H₃C     CH₃              H₃C    CH₃
              20.8                    20.4
                                CH₃
                                12.7
       anti                    gauche
```

2,3-Dimethylbutane at $-188\,°$C

4.1.2 Cycloalkanes

Carbon-13 shifts and one-bond carbon-proton couplings of parent cycloalkanes are summarized in Table 4.4 [206]. Cyclopropane carbons are remarkably shielded and parallel the behavior known from cyclopropane protons in ^{1}H NMR. This is explained in terms of a ring current model [206]. Some ring strain is also reflected by a slight shielding of cyclobutane carbons. Shieldings of medium-sized rings ($C_{10}-C_{14}$, Fig. 4.1) are essentially attributed to steric repulsions of inner hydrogen atoms. Very large rings ($>C_{20}$) approach the carbon shifts of CH_2 groups in long-chain, unbranched alkanes (29–30 ppm [207]).

Table 4.4. ^{13}C Chemical Shifts and One-Bond Coupling Constants J_{C-H} of Cycloalkanes (δ_C in ppm) [206].

Compound	Chemical shift (ppm)	J_{CH} (Hz)
Cyclopropane	−2.80	162±2
Cyclobutane	22.40	136±1
Cyclopentane	25.80	131±2
Cyclohexane	27.60	127±2
Cycloheptane	28.20	126±2
Cyclooctane	26.60	127±2
Cyclononane	25.80	125±2
Cyclodecane	25.00	126±2
Cycloundecane	26.30	126±2
Cyclododecane	23.20	123±3
Cyclotridecane	26.20	127±2
Cyclotetradecane	24.60	126±2
Cyclopentadecane	27.00	126±2
Cyclohexadecane	26.50	126±3
Cycloheptadecane	26.70	126±2

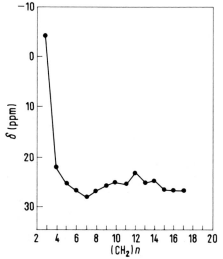

Fig. 4.1. Plot of ^{13}C chemical shifts of cycloalkanes *versus* ring size [206].

Table 4.5. ^{13}Chemical Shifts of Methylcycloalkanes (δ_C in ppm).

Compound	C-1	C-2	C-3	C-4	C-5	C-6	CH$_3$	Ref.
Methylcyclopropane	4.9	5.6					19.4	[208]
1,1-Dimethylcyclopropane	11.3	13.7					25.5	
cis-1,2-Dimethylcyclopropane	9.3		13.1				12.8	
trans-1,2-Dimethylcyclopropane	13.9		14.2				18.9	
Methylcyclobutane	31.2	30.2	18.3				22.1	[209]
1,1-Dimethylcyclobutane	35.9	34.9	14.8				29.4	
cis-1,2-Dimethylcyclobutane	32.2		26.6				15.4	
trans-1,2-Dimethylcyclobutane	39.2		26.6				20.5	
cis-1,3-Dimethylcyclobutane	26.9	38.5					22.5	
trans-1,3-Dimethylcyclobutane	26.1	36.4					22.0	
Methylcyclopentane	35.8	35.8	26.4				21.4	[210]
1,1-Dimethylcyclopentane	40.1	42.3	25.8				30.0	
cis-1,2-Dimethylcyclopentane	38.6		34.2	24.2			16.1	
trans-1,2-Dimethylcyclopentane	43.7		36.0	24.3			19.7	
cis-1,3-Dimethylcyclopentane	36.3	46.0		35.3			22.1	
trans-1,3-Dimethylcyclopentane	34.2	48.2		36.2			22.4	
Methylcyclohexane	33.1	35.8	26.6	26.4			22.7	[87, 211]
1,1-Dimethylcyclohexane	30.0	39.8	22.6	26.7			28.8	
cis-1,2-Dimethylcyclohexane	34.4		31.5	23.7			15.7	
trans-1,2-Dimethylcyclohexane	39.6		36.0	26.9			20.2	
cis-1,3-Dimethylcyclohexane	32.8	44.7		35.5	26.5		22.8	
trans-1,3-Dimethylcyclohexane	27.1	41.5		33.9	20.8		20.5	
cis-1,4-Dimethylcyclohexane	32.6	35.6					22.6	
trans-1,4-Dimethylcyclohexane	30.1	30.9					20.1	
Methylcycloheptane	34.9	37.5	26.9	28.9			24.4	[212]
1,1-Dimethylcycloheptane	33.3	42.6	23.8	30.8			30.8	
cis-1,2-Dimethylcycloheptane	37.5		34.0	26.6	29.2		17.9	
trans-1,2-Dimethylcycloheptane	41.3		35.8	26.7	29.7		22.6	
cis-1,3-Dimethylcycloheptane	34.2	46.9		37.4	26.5		24.9	
trans-1,3-Dimethylcycloheptane	31.1	44.8		37.5	29.1		24.3	
cis-1,4-Dimethylcycloheptane	34.2	33.5			38.4	27.0	24.3	
trans-1,4-Dimethylcycloheptane	32.2	36.6^a			36.8^a	24.1	24.3	

^{a} assignments interchangeable.

The large CH coupling constants J_{CH} of cyclopropane and – to a smaller extent – of cyclobutane are attributed to a higher s character of the carbon bond orbitals.

Carbon-13 shifts of some methylcycloalkanes are given in Table 4.5. Methyl substitution increments have been derived for methylated cyclopentanes (Table 4.6 [210]) and cyclohexanes (Table 4.7 [87]) in order to predict carbon shift values. While γ effects in methylated cyclopentanes are small (Table 4.6), shieldings of carbon atoms in γ position

Table 4.6. Methyl Substituent Effects on ^{13}C Chemical Shifts of Cyclopentane and Methylcyclopentane Resonances [210]

Compound	α effect	β effect	γ effect
Methylcyclopentane	9.3	9.3	−0.1
trans-1,2-Dimethyl-cyclopentane	7.9	7.9 9.6	0.2 −2.1
cis-1,3-Dimethyl-cyclopentane	10.0	8.9 10.2	0.6 −0.5
1,1-Dimethyl-cyclopentane	4.3	6.5	−0.5
cis-1,2-Dimethyl-cyclopentane	2.8	2.8 7.8	−1.6 −2.2
trans-1,3-Dimethyl-cyclopentane	8.1	8.3 9.8	0.4 −1.3

Table 4.7.(a) Carbon Shift Increments A_l for Methyl Substitution in Position l to the Observed Carbon of Cyclohexane and Corrections S_i for Geminal (e.g. $α_eα_a$) Vicinal (e.g. $α_e β_e$) and other Substitution Patterns [87].

$$\delta_C(k) = 26.5 + \sum A_l \cdot n_{kl} + \sum S_i$$

l	A_l	Type i	S_i
$α_e$	5.95	$α_aα_e$	−3.8
$α_a$	1.4	$β_aβ_e$	−1.25
$β_e$	9.0	$γ_aγ_e$	2.0
$β_a$	5.4	$α_eβ_e$	−2.45
$γ_e$	0.05	$α_eβ_a$	−2.9
$γ_a$	−6.35	$α_aβ_e$	−3.4
$δ_e$	−0.2	$β_eγ_a$	−0.8
$δ_a$	−0.05	$β_aγ_e$	1.55

Table 4.7.(b) Increments A_l for Prediction of Axial and Equatorial Methyl Carbon Shieldings [87].

$$\delta_{CH_3}(k) = B + \sum A_i n_{kl}$$

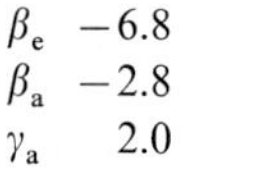

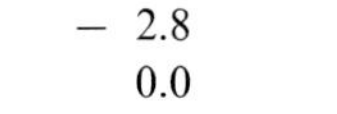

B = 18.8 ppm A_l		B = 23.1 ppm A_l	
α	6.4		10.4
$β_e$	−6.8		−2.8
$β_a$	−2.8		−2.8
$γ_a$	2.0		0.0
$γ_e$	0.0		0.0

of an axial methyl group in cyclohexane is − 5.4 ppm (Table 4.7) [87] due to *1,3-diaxial steric interaction*, as illustrated for methylcyclohexane:

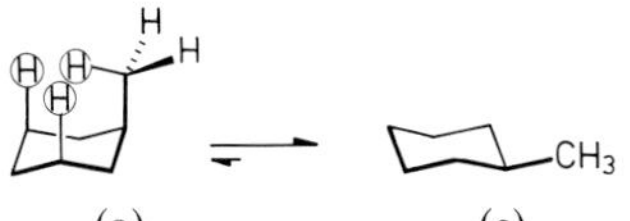

(a) ⇌ (e)

Ring inversion of methylcyclohexane, which exchanges equatorial and axial methyl groups at room temperature, is frozen at − 100 °C [213]. In this situation, the methyl and

C-3,5 carbons in (a) are shielded by about 6 ppm relative to (e). Similar shift differences have been obtained by low-temperature ^{13}C NMR of 1,2- and 1,4-dimethylcyclohexane [112] (Fig. 3.13).

4.1.3 Polycycloalkanes

Carbon-13 shifts of bicyclo[1.1.0]butane [213], bicyclo[2.1.0]hexanes [214], bicyclo-[2.2.1]heptanes (norbornanes) [215, 216], bicyclo[2.2.2]octanes [217], bicyclo-[3.3.0]octanes [218], bicyclo[3.3.1]nonanes [219], bicyclo[n.1.0]alkanes [220], bicyclo-[4.n.0]alkanes [221], spiro[4.5]alkanes [222], [1.1.1]propellane [223], [3.1.1]propellane [224] and adamantane derivatives [225] have been analyzed. Chemical shifts range from −4 ppm in tetracyclo[4.1.0.0^{2,4}0^{3,5}]heptane [227], due to high s character of bond obitals and steric repulsions, up to 66.9 ppm in dodecahedrane at the other extreme, due to many deshielding α and β carbon atoms. Another special case is tetra-t-butyl-tetrahedrane [228] with a quaternary t-butyl carbon shielded relative to the methyl carbons due to high s character of bond orbitals and angle strain.

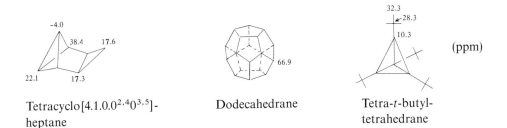

Tetracyclo[4.1.0.0^{2,4}0^{3,5}]-   Dodecahedrane   Tetra-t-butyl-
heptane   tetrahedrane

^{13}C chemical shifts of norbornane and some derivatives are collected in Table 4.8. The values are predominantly influenced by steric repulsions. In 2-methyl-norbornene, shieldings are observed for C-6 (−6.6 ppm) and the methyl carbon (−4.9 ppm) when the methyl group changes from *exo* to *endo* configuration, while the signal of C-7 experiences a deshielding (3.9 ppm) [229]. Shieldings are attributed to sterically induced CH bond polarizations as discussed for alkanes and cycloalkanes. This γ_{syn} effect can generally be used for configurational assignments of norbornane derivatives, but not for tricyclo-[3.2.1.0^{2,4}]octanes with *endo* and *exo* cyclopropane rings fused to the bicyclo-[2.2.1]heptane system [227]. Deshielding of the *endo* methano carbon is explained in terms of an electron withdrawal, induced by interaction of the unoccupied Walsh orbitals at the bridge carbons with the occupied orbitals of the adjacent carbon atoms [227].

Tricyclo[3.2.1.0^{2,4}]octane

Table 4.8. ^{13}C Chemical Shifts of Norbornane and Related Compounds (δ_C in ppm) [215, 229].

Compound	C-1	C-2	C-3	C-4	C-5	C-6	C-7	CH$_3$
Quadricyclane	22.9	14.7	14.7	22.9	14.7	14.7	31.9	
Tricyclene	9.6	9.6	32.9	29.4	32.9	9.6	32.9	
Norbornane	36.5	29.8	29.8	36.5	29.8	29.8	38.4	
1-Methylnorbornane	43.8	36.8	31.3	37.9	31.3	36.8	45.3	20.7
exo-2-Methylnorbornane	43.2	36.5	39.9	37.0	30.0	28.7	34.7	22.0
endo-2-Methylnorbornane	41.9	34.3	40.4	37.9	30.3	22.1	38.6	17.1
7-Methylnorbornane	40.7	26.9	26.9	40.7	30.7	30.7	44.0	12.4
exo-exo-2,3-Dimethyl-norbornane	39.9	39.9	44.5	29.7	29.7	44.5	31.9	15.9
endo-exo-2,3-Dimethyl-norbornane	45.2	44.0	42.0	21.2	30.3	44.4	36.9	21.3 exo 15.9 endo
endo-endo-2,3-Dimethyl-norbornane	34.8	34.8	43.1	21.8	21.8	43.1	39.7	11.8

Sterically induced CH bond polarizations are also reflected by the shieldings of carbon atoms in *cis*-decalin relative to its *trans* isomer [230]. In the 9-methyl derivatives, steric repulsion between the methyl- and the *syn*-axial 2,4-hydrogens shields the attached carbon atoms relative to the parent *trans*-decalin. In the *cis* isomer, shielding of the bridgehead carbons relative to those in the *trans* compound clearly reflects steric interaction of hydrogen atoms in the 2,4- und 6,8-positions.

trans- 27.1 34.7 44.2   (ppm)   29.6 36.9 24.6 cis-

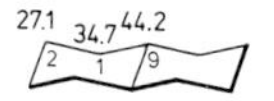

Decalin

trans- 22.2 15.7 CH$_3$ H 42.4 34.8 27.4 29.4 46.2   (ppm)   36.4 28.2 CH$_3$ 32.9 22.8 41.8 24.5 28.1 cis-

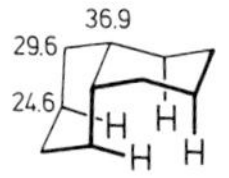

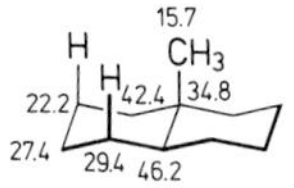

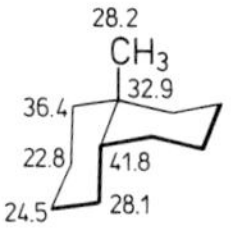

9-Methyldecalin

An increment system has been derived for methyldecalins [230] and methylated perhydrophenanthrenes [231] by regressional analysis, because these compounds are steroid models and the increments, in fact, have proved to be of help in carbon-13 shift assignments of steroids (Section 5.2). Eq. (4.2) permits prediction of δ_C using the increments A_l of structural elements in position l to the carbon atom k to be considered (Table 4.9). These increments clearly indicate the configurational influence of the γ substituent, decreasing for example from *eclipsed* via *gauche* to *trans* interactions ($V_e > V_g > V_t$ and

Table 4.9. Increments A_l of Structural Elements in Position l Relative to the Carbon Atom k in order to Predict its Carbon-13 Shift $\delta_C(k)$ according to eq. (4.2).

$$\delta_C(k) = B + \sum A_l n_{kl} \quad (n_{kl}: \text{number of structural elements}) \tag{4.2}$$

		Methyldecaline	Perhydroanthracene Perhydronaphthacene Perhydrophenanthrene Perhydropyrene
Reference shift B		-3.07	-2.49
Structural element		A_l	A_l
α		9.94	9.43
β		8.49	8.81
T		-2.91	-1.13
Q		-9.04	
V_g		-3.50	-3.38
V_{tr}			-0.79
V_e			-8.19
$\beta_g \gamma_{tr}$	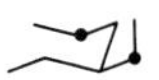	-1.91	
$\gamma_{H \cdots H}$	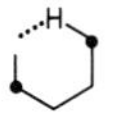	$+1.91$	-5.53
$\gamma_{2H \cdots H}$	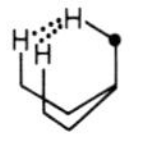	-4.56	-3.01
γ_p	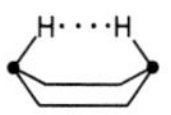	-3.95	-11.01
δ_{Synaxial}	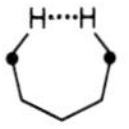		-3.65

similarly $\gamma_p > \gamma_{H \cdots H} > \gamma_{2H \cdots H}$). γ Effects due to eclipsed (V_e) or flagpole interactions (γ_p) in boat conformations may approach the magnitudes of α and β effects and may cause methine and methylene carbons to be more shielded than methylene and methyl carbons, respectively, in polycyclic systems such as triterpenes and steroids (Chapter 5).

4.2 Alkenes

4.2.1 Open-Chain Alkenes and Dienes

Carbon-13 chemical shifts of some representative alkenes [90, 232–236] are collected in Table 4.10. Inspection of the data shows that the shift value increases with increasing alkylation of the observed olefinic carbon atom.

$$\delta_{=CH_2} < \delta_{=CHR} < \delta_{=CR_2}$$
$$<125 \quad >125 \quad >135 \text{ ppm}$$

The α alkylation effect on an alkene carbon is similiar to the values reported for alkanes (~ 10 ppm), but two β effects must be considered: Across a σ bond, the β effect deshields

Table 4.10. ^{13}C Chemical Shifts of Alkenes [90, 232–235], Dienes [233] and Methylene-cycloalkanes [236] (δ_C in ppm).

Compound	C-1	C-2	C-3	C-4	C-5	C-6	C-7	C-8
Ethene	123.5							
Propene	115.9	133.4	19.4					
1-Butene	113.5	140.5	27.4	13.4				
1-Pentene	114.5	139.0	36.2	22.4	13.6			
1-Hexene	114.2	139.2	33.8	31.5	22.5	14.0		
1-Heptene	114.2	139.2	34.0	29.0	31.7	22.8	14.1	
1-Octene	114.2	139.2	34.0	29.1	29.3	32.0	22.8	14.0
2-Methylpropene	111.3	141.8	24.2					
2-Methylbutene	109.1	147.0	31.1	12.5	22.5			
(E)-2-Butene	16.8	125.4						
(Z)-2-Butene	11.4	124.2						
(E)-2-Pentene	17.3	123.5	133.2	25.8	13.6			
(Z)-2-Pentene	12.0	122.8	132.4	20.3	13.8			
(E)-2-Hexene	17.9	124.9	131.7	35.0	23.0	13.7		
(Z)-2-Hexene	12.8	124.0	130.8	29.2	23.0	13.9		
(E)-2-Heptene	17.9	124.7	131.8	32.6	32.1	22.5	14.0	
(Z)-2-Heptene	12.7	123.7	131.0	26.8	32.1	22.6	14.1	
(E)-2-Octene	17.7	124.7	131.9	33.0	29.8	31.9	22.9	14.0
(Z)-2-Octene	12.5	123.6	131.0	27.1	29.7	31.9	22.9	14.0
(E)-3-Hexene	14.1	25.9	131.2					
(Z)-3-Hexene	14.4	20.7	131.2					
(E)-3-Octene	13.9	25.9	132.6	129.6	32.6	32.3	22.5	14.0
(Z)-3-Octene	14.3	20.7	131.6	129.4	27.1	31.8	22.6	13.9
(E)-4-Octene	13.6	23.1	33.1	130.6				
(Z)-4-Octene	13.7	23.2	29.6	130.0				

Table 4.10 (continued).

Compound	C-1	C-2	C-3	C-4	C-5	C-6	C-7	C-8
1,3-Butadiene	116.6	137.2						
(E)-1,3-Pentadiene	114.4	137.8	129.5	133.2	17.2			
(Z)-1,3-Pentadiene	116.5	132.5	130.9	126.4	12.8			
(E, E)-2,3-Hexadiene	17.5	126.2	132.5					
(Z, Z)-2,4-Hexadiene	12.9	124.9	125.3					
(E, Z)-2,4-Hexadiene	18.0	128.3	130.2	127.4	123.1	13.0		
2-Methyl-1,3-butadiene	113.0	142.9	140.3	116.4	17.6			
(E)-2-Methyl-1,3-pentadiene	114.5	142.2	135.5	125.1	18.8	18.3		
(E)-3-Methyl-1,3-pentadiene	110.3	142.1	135.5	127.1	11.1	13.6		
2,3-Dimethyl-1,3-butadiene	113.0	143.8			20.3			
(E, E)-2,6-Octadiene	17.8	125.1	131.4	33.1				
(Z, Z)-2,6-Octadiene	12.6	124.1	130.5	27.1				
(E, Z)-2,6-Octadiene	17.8	125.1	131.1	32.9	27.3	130.5	124.1	12.6
1,7-Octadiene	114.5	138.9	34.1	27.8				
Methylenecyclopropane	103.5	131.0	3.0					
Methylenecyclobutane	105.2	148.8	32.3	17.2				
Methylenecyclopentane	105.2	152.3	33.2	27.1				
Methylenecyclohexane	106.9	149.2	35.8	28.8	26.9			
Methylenecycloheptane	110.9	151.2	36.6	29.0	30.0			

as known from alkanes ($\beta_\sigma \sim 7$ ppm). A shielding, however, is observed when the β effect is transmitted through a π bond ($\beta_\pi \sim -7$ ppm).

$A_\alpha \sim 10$ $A_{\beta_\rho} \sim 7$ $A_{\beta_\pi} \sim -7$ ppm

Similarly to alkanes, an increment system has been proposed in order to predict carbon-13 shifts of olefinic carbons from the reference value of ethene (122.1 ppm) [237] and the increments A_l for α, β, γ, and δ alkylation, as well as multiple substitution corrections S (Table 4.11). Accuracy is about 1 ppm.

In (Z)-(E)-isomeric pairs of alkenes, the olefinic carbons of the *cis* isomers are slightly shielded when compared with those in the *trans* compounds. A much stronger shielding is experienced by the carbon atoms α to the double bond in the *cis* isomers due to the γ effect, as outlined in Section 3.1.3.8 and as demonstrated for the α carbons of several (E)-(Z) pairs in Table 4.10. The relation $\delta_{\alpha(Z)} < \delta_{\alpha(E)}$ permits straightforward configurational assignments in alkenes, also includung dienes (Table 4.10), e.g.

Table 4.11. Alkylation Increments for Prediction of Olefinic Carbon Shifts in Alkenes [237] According to the Equation
$\delta_C(k) = 122.1 + \sum A_l n_{kl} + S$.

Position l	Increment A_l (ppm)	Substitution	Correction S (ppm)
α_π	11.0	(Z)	−1.2
β_π	−7.1	gem_α	−4.9
β_σ	6.0	gem_β	1.2
γ_π	−1.9	$mult_\sigma$	1.3
γ_σ	−1.0	$mult_\pi$	−0.7
δ_π	1.1		
δ_σ	0.7		
ε	0.2		

gem: geminal alkylation at the double bond
mult: branching at the α carbon
n_{kl}: number of carbons in position *l* to carbon *k*

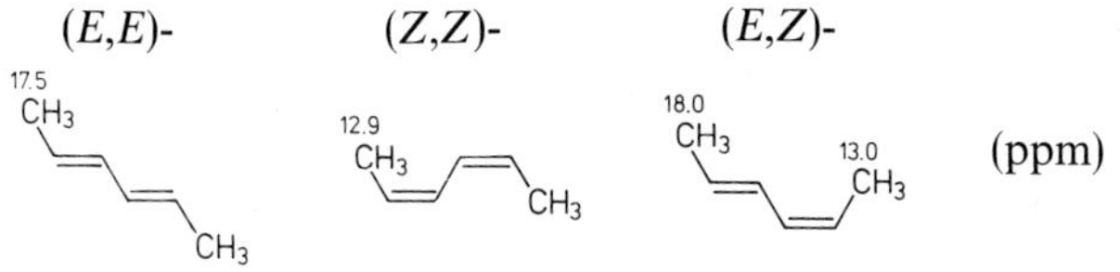

2,4-Hexadiene

Conjugation does not significantly change olefinic carbon shifts in 1,3-dienes relative to those of an isolated double bond. Shieldings observed for the central carbons are usually caused by steric repulsions.

4.2.2 Cycloalkenes

^{13}C chemical shifts of cycloalkenes given in Table 4.12 [229, 233, 238–241] again reflect the special bonding state of three-membered rings, characterized by the smallest shift values in the series. As shown for cyclooctene in Table 4.12, the relation $\delta_{\alpha(Z)} < \delta_{\alpha(E)}$ can also be applied to distinguish *cis-trans*-isomeric cycloalkenes.

A successive polarization of the exocyclic double bond, as monitored by the carbon shifts in the 7-methylene derivatives of norbornane, norbornene and norbornadiene (Table 4.12), is attributed to *homoconjugative interaction of π bonds*, pushing π electrons towards the *exo* methylene carbon [239].

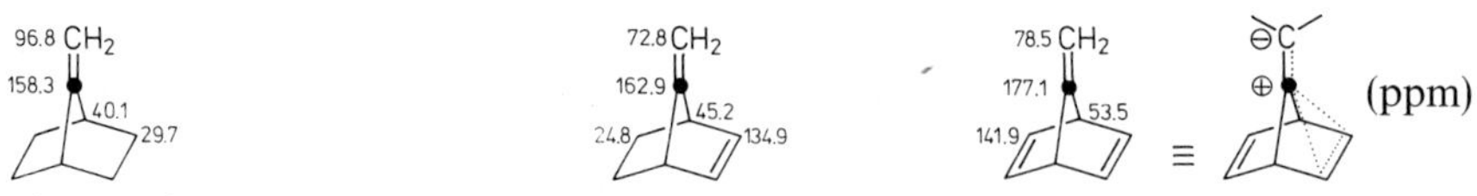

7-Methylenenorbornane   7-Methylenenorbornene   7-Methylenenorbornadiene

Table 4.12. ^{13}C Chemical Shifts of Cycloalkenes [229, 233, 238–241] and Cope Systems [243–245] (δ_C in ppm).

Compound	C-1	C-2	C-3	C-4	C-5	C-6	C-7	C-8	CH$_3$
Cyclopropene	108.7		2.3						
Cyclobutene	137.2		31.4						
Cyclopentene	130.8		32.8	23.3					
Cyclohexene	127.4		25.4	23.0					
Cycloheptene	130.4		26.0	27.0	29.8				
(Z)-Cyclooctene	130.2		25.8	26.8	29.6				
(E)-Cyclooctene	132.8		34.3	34.3	28.7				
1-Methylcyclopropene	116.5	98.8	6.2						12.5
1-Methylcyclopentene	140.2	124.8	33.2	24.3	37.3				16.6
1-Methylcyclohexene	133.0	120.9	25.0	22.8	22.2	29.7			23.4
1-Methylcycloheptene	140.3	126.2	32.9	28.7^a	27.9^a	26.8^a			26.4
Cyclopentadiene	132.2	132.8			41.6				
1,3-Cyclohexadiene	126.1	124.6			22.3				
1,3-Cycloheptadiene	126.6	134.2			33.0	26.9			
1,3-Cyclooctadiene	126.5	132.1			23.4	28.5			
1,4-Cyclohexadiene	125.5		26.0						
1,4-Cyclooctadiene	128.5		28.5						
1,8-Tetradecadiene	130.4		26.9	28.7	28.1				
Fulvene	124.9	134.3			152.6	123.4			
1,3,5-Cycloheptatriene	120.4	126.8	131.0				28.1		
1,5,9-Cyclododecatriene	132.7		32.9						
Cyclooctatetraene	131.5								
Norbornene	41.9	135.2			24.6		48.5		
1-Methylnorbornene	49.6	139.7	135.5	43.0	27.7	32.3	54.7		17.7
exo-5-Methylnorbornene	42.4	136.9	135.9	48.4	32.7	34.7	44.7		21.4
endo-5-Methylnorbornene	43.3	136.9	132.2	47.5	32.7	33.9	50.2		19.2
syn-7-Methylnorbornene	47.5	132.1			25.6		54.4		12.2
anti-7-Methylnorbornene	45.7	137.5			21.5		53.0		14.1
7-Methylenenorbornene	45.2	134.9			24.8		162.9	72.8	
Norbornadiene	50.9	143.5					75.5		
7-Methylenenorbornadiene	53.5	141.9					177.1	78.5	
Homotropilidene	19.4	129.8	127.7	28.8			20.1		(-37°C)
Semibullvalene	42.2	121.7	131.8	53.1			50.0		(-160°C)
	86.5	120.4		50.0			53.1		(25°C)
Bullvalene	21.0	128.3	128.5	31.0					(-60°C)
	86.4								(141°C)

^{a} assignments interchangeable.

4 ^{13}C NMR Spectroscopy of Organic Compounds

Even stronger polarizations of double bonds in alkenes are induced by electron withdrawing substituents, as present in enol ethers, enones, and enamines (Sections 4.6.2, 4.7, and 4.9.2). Deshielding of C-7 in norbornadiene (75.5 ppm, Table 4.12) is understood as arising from interaction of antibonding π orbitals at the olefinic carbon atoms with σ orbitals of the bridgehead bonds [214, 216]. Spiroconjugation in spiro[4.4]nonatetraene is interpreted similarly [242].

Norbornadiene 143.5 75.5 50.9 ≡ (ppm)

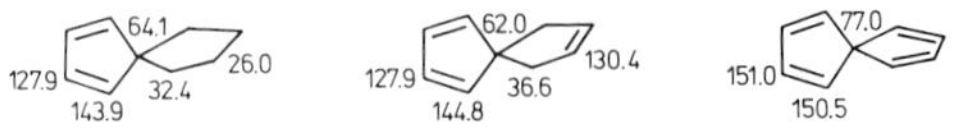

Spiro[4.4]nona- -diene (127.9, 64.1, 26.0, 32.4, 143.9) -triene (127.9, 62.0, 130.4, 36.6, 144.8) -tetraene (151.0, 77.0, 150.5) (ppm)

Relatively balanced carbon shifts of some fulvenes [241] indicate that polar resonance formulae contribute to the ground state of these crossed conjugated systems only slightly.

(ppm) 130.7, 120.6, 142.8, 149.3, 22.9   ↔   6,6-Dimethylpentafulvene

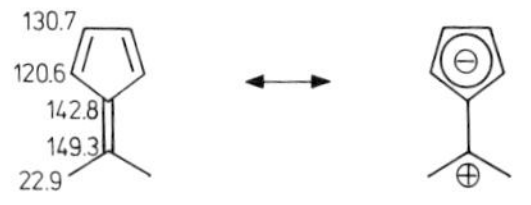

Temperature-dependent ^{13}C NMR has been used to investigate Cope systems such as homotropilidene [243], semibullvalene [244], and bullvalene [245] (Table 4.12). Activation parameters have been determined.

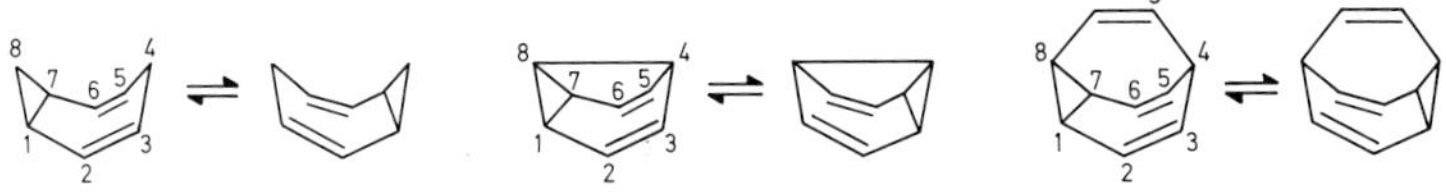

	Homotropilidene	Semibullvalene	Bullvalene	
$\Delta G^{\ddagger}$	59 (25 °C)	23 (−143 °C)	52 (0 °C)	J/mol
$\Delta H^{\ddagger}$	52	20	53	J/mol

4.3 Alkynes and Allenes

Carbon-13 shifts of alkynes (Table 4.13) [246–250] are found between 60 and 95 ppm. To conclude, alkyne carbons are shielded relative to olefinic but deshielded relative to alkane carbons, also paralleling the behavior of protons in proton NMR. Shielding relative to alkenes is attributed to the higher electronic excitation energy of alkynes which decreases the paramagnetic term according to eq. (3.4), and to the anisotropic effect of the triple bond. An increment system can be used to predict carbon shieldings in alkynes

Table 4.13. ^{13}C Chemical Shifts of Alkynes [246, 247], Polyalkynes [248], Alkenynes [249] and Cycloalkynes [250] (δ_C in ppm).

Compound	C-1	C-2	C-3	C-4	C-5	C-6	C-7	C-8
Ethyne	71.9							
1-Pentyne	68.2	83.6	20.1	22.1	13.1			
1-Hexyne	68.6	86.3	18.6	31.1	22.4	14.1		
1-Heptyne	68.4	83.9	18.7	29.1	31.7	23.4	15.0	
1-Octyne	70.0	85.0	19.4	30.2	30.2	33.0	24.4	15.9
2-Hexyne	2.7	74.7	77.9	20.6	22.6	13.1		
2-Heptyne	3.3	75.2	78.6	18.7	32.2	22.8	14.5	
2-Octyne	3.9	75.8	79.5	18.4	30.5	32.4	23.7	15.5
3-Hexyne	15.4	13.0	80.9					
3-Heptyne	14.7	13.0	79.3	81.2	21.2	23.5	14.1	
3-Octyne	16.1	14.0	80.7	82.0	19.7	32.9	23.8	15.6
1,3-Pentadiyne	64.7	68.8	65.4	74.4	3.9			
2,4-Hexadiyne	4.0	72.0	64.8					
2,4,6-Octatriyne	4.4	74.8	65.0	60.0				
1-Buten-3-yne	129.2	117.3	82.8	80.0				
(E)-3-Penten-1-yne	75.8	82.5	110.1	141.3	18.6			
(Z)-3-Penten-1-yne	82.1	80.3	109.4	140.3	15.9			
Cyclooctyne	94.4		20.8	34.7[a]	29.7[a]			

[a] assignments interchangeable.

Table 4.14. Increments A_l for Prediction of sp Carbon Shifts in Alkynes According to $\delta_C(k) = 71.9 + \sum A_l n_{kl}$ [247].

$$\overset{\delta}{-C}-\overset{\gamma}{C}-\overset{\beta}{C}-\overset{\alpha}{C}-C_k \equiv C-\overset{\alpha_\pi}{C}-\overset{\beta_\pi}{C}-\overset{\gamma_\pi}{C}-\overset{\delta_\pi}{C}-.$$

Position l	α	β	γ	δ	α_π	β_π	γ_π	δ_π	
Increment A_l	6.9	4.8	−0.1	0.5	−5.7	2.3	−1.3	0.6	(ppm)[a]

[a] rounded to the next tenth of a ppm.

[247]. The signs of the shift increments A_α, A_{β_σ}, A_{β_π}, etc., (Table 4.14) are the same as reported for alkenes, but their magnitudes are reduced to about 2/3 of the alkene increments.

Due to cooperating anisotropic effects, shieldings of central sp carbon atoms increase with the number of conjugated triple bonds in polyalkynes (Table 4.14).

Cycloalkynes display a deshielding with increasingly strained rings, as reported for a series of cyclooctynes [250]; an outstanding alkyne carbon shift is found in cyclooctadienyne.

198   4 ^{13}C NMR Spectroscopy of Organic Compounds

Table 4.15. ^{13}C Chemical Shifts of Selected Allenes [251, 252]. (δ_C in ppm)

Compound	C-1	C-2	C-3	C-4	C-5	C-6	C-7	C-8
Allene	73.3	212.4						
1,2-Butadiene	73.2	209.2	84.0	13.0				
1,2-Pentadiene	74.5	208.6	91.4	21.4	13.0			
1,2-Hexadiene	74.4	209.3	89.7	30.3	19.1	13.5		
1,2-Heptadiene	73.6	208.8	89.4	30.9	27.4	21.6	13.0	
1,2-Octadiene	74.8	209.2	90.2	31.7	29.3	28.6	23.0	14.4
3-Methyl-1,2-butadiene	70.8	206.7	92.8	18.9	18.9			
2,3-Pentadiene	14.4	84.7	206.3	84.7	14.4			
2-Methyl-2,3-pentadiene	20.5	103.9	203.3	83.3	20.5	14.8		
2,4-Dimethyl-2,3-pentadiene	19.8	91.9	199.8	91.9	19.8	19.8	19.8	
1,2,6,7-Cyclononatetraene	90.2	208.3		26.8				

1,5-Cyclooctenyne: 19.2, 100.4, 30.8, 131.4

1,4-Cyclooctenyne: 20.0, 95.8, 26.9, 96.0, 26.8, 20.2, 133.1, 129.8

1,3-Cyclooctenyne: 21.6, 114.5, 32.9, 94.5, 25.3, 111.8, 35.5, 149.1

1,5-Cyclooctadien-3-yne: 112.7, 116.3, 152.0, 32.6   (ppm)

The large carbon-13 shifts of the central sp carbons in allenes (200–210 ppm) relative to the terminal ones (70–90 ppm, Table 4.15) are attributed to an increased paramagnetic shielding due to two localized π bonds originating from the central carbon. The β_π effect of methyl substitution to the central carbon is considerably smaller than that found for alkenes and alkynes.

4.4 Halides

4.4.1 Alkyl Halides

Carbon-13 chemical shifts of haloalkanes collected in Table 4.16 and displayed in Fig. 4.2 range from 119.9 ppm for tetrafluormethane to −292.2 ppm for tetraiodomethane. Substituent effects on alkyl carbon atoms induced by the different halogens are given in Table 3.2 (Section 3.1.3.2). It was pointed out that the α *effect* correlates with substituent *electronegativity* ((−)-*I effect*). Deshieldings relative to methane and alkanes are found for fluorine, chlorine, and bromine, while iodine causes a shielding due to the heavy atom influence (Section 3.1.3). As shown in Fig. 4.2, substituent effects observed for monosubstitution in Table 3.2 do not simply add to produce the shifts of multiply halogenated carbon atoms. Charge polarizations, changes in bond lengths caused by steric effects, and

4.4 Halides   199

Table 4.16. ^{13}C Chemical Shifts (δ_C in ppm) and Coupling Constants ($^nJ_{CH}$ in Hz) of Haloalkanes, Haloalkenes and Haloalkynes [86, 91 a, 255–261].

Compound	^{13}C chemical shifts (ppm)	$^1J_{CH}$ (Hz)	$^2J_{CH}$ (Hz)	Compound	^{13}C chemical shifts (ppm)	$^1J_{CH}$ (Hz)	$^2J_{CH}$ (Hz)
CH_3F	71.6	149		$(CH_3)_2CHI$	31.85		
CH_3Cl	25.6	150			20.95		
CH_3Br	9.6	152		$(CH_3)_3CCl$	33.45		
CH_3I	−24.0	151			165.15		
CH_2F_2	111.2	185		$(CH_3)_3CBr$	36.35		
CH_2Cl_2	54.4	178			60.55		
CH_2ClBr	38.75			$(CH_3)_3CI$	40.45		
CH_2Br_2	21.6				41.85		
CH_2I_2	−53.8	173		$CH_2=CHCl$	116.00	198	4.3
CHF_3	118.8	238			124.90	162	9.3; 6.7
$CHFCl_2$		220		$CH_2=CHBr$	120.80		6.25
CHF_2Cl		231			114.30		6.75; −8.2
$CHCl_3$	77.70	209		$CH_2=CHI$	129.20		4.0
$CHCl_2Br$	56.05				84.10		4.15; −7.5
$CHClBr_2$	33.25			$CHCl=CHCl$ (*trans*)	119.90	199	<1.8
$CHBr_3$	12.3	206		$CHCl=CHCl$ (*cis*)	118.10	198	16.5
CHI_3	−139.7			$CHBr=CHBr$ (*cis*)	115.10		
CCl_4	96.7			$CHBr=CHBr$ (*trans*)	108.10		
CCl_3Br	66.50			$CHI=CHI$ (*cis*)	95.20		
CCl_2Br_2	35.40			$CHI=CHI$ (*trans*)	78.10		
$CClBr_3$	3.80			$CH_2=CCl_2$	112.10	167	
CBr_4	−28.5				125.90		1.3
CH_3-CH_2F	13.30			$CHCl=CCl_2$	116.40	201	
	78.00				123.90		8.5
CH_3-CH_2Cl	17.50	128	2.6	$CCl_2=CCl_2$	120.10		
	38.70	150	4.2	$CH_3-CH=CHJCl$ (*cis*)	125.40		
CH_3-CH_2Br	18.95				118.80		
	26.95			$CH_3-CH=CHCl$ (*trans*)	128.10		
CH_3-CH_2I	21.85				116.40		
	−0.25			$CH_3-CH=CHBr$ (*cis*)	129.60		
CH_3-CHCl_2	31.30	130	<1.2		110.10		
	68.00	179	5.1	$CH_3-CH=CHBr$ (*trans*)	133.40		
CH_3-CHBr_2	34.75				104.40		
CH_3-CCl_3	45.10	132		n-$C_4H_9-CH=CHI$ (*cis*)	140.90		
	95.00		5.9		82.90		
CH_2Cl-CH_2Cl	43.90	155	<2	n-$C_4H_9-CH=CHI$ (*trans*)	146.40		
$CH_2Cl-CHCl_2$	50.40	158	<2		75.30		
	71.10	181	2.5	$C_2H_5-CH=Cl-C_2H_5$ (*cis*)	134.70		
$CH_2Cl-CCl_3$	59.50	158			110.80		
	96.40		3.2	$C_2H_5-CH=Cl-C_2H_5$ (*trans*)	141.50		
$CHCl_2-CHCl_2$	74.30	179	<2		104.80		
$CHCl_2-CCl_3$	80.00	182		n-$C_4H_9-C\equiv CCl$	67.50		
	99.70		1.5		55.40		
CCl_3-CCl_3	104.90			n-$C_4H_9-C\equiv CBr$	78.50		
$(CH_3)_2CHCl$	26.85				37.10		
	53.75			n-$C_4H_9-C\equiv CI$	95.50		
$(CH_3)_2CHBr$	28.05				−6.60		
	44.05						

200   4 ^{13}C NMR Spectroscopy of Organic Compounds

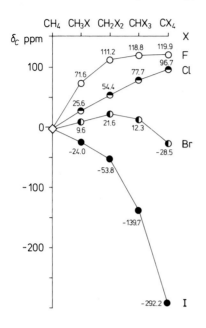

Fig. 4.2. ^{13}C chemical shifts of substituted methanes *versus* the degree of substitution.

deviations from normal electron pairing have been shown to be responsible for these ^{13}C chemical shift anomalies of the halomethanes [86]. The expressions (4.3–4.6) have been developed for the calculation of ^{13}C chemical shifts of halomethanes [86] from the carbon shift value of methane, δ_0:

$$\delta_C(CH_3X) = \delta_0 + \Delta(X) \tag{4.3}$$

$$\delta_C(CH_2XX') = \delta_0 + \Delta(X) + \Delta(X') + \Delta(X - X') \tag{4.4}$$

$$\delta_C(CHXX'X'') = \delta_0 + \Delta(X) + \Delta(X') + \Delta(X'') + \Delta(X - X') + \Delta(X - X'') \\ + (X' - X'') \tag{4.5}$$

$$\delta_C(CXX'X''X''') = \delta_0 + \Delta(X) + \Delta(X') + \Delta(X'') + \Delta(X''') \\ + \Delta(X - X') + \Delta(X - X'') + \Delta(X - X''') + \Delta(X' - X'') \\ + \Delta(X' - X''') + \Delta(X'' - X'''). \tag{4.6}$$

Substitution parameters Δ, calculated from a least-squares regression analysis, are collected in Table 4.17. By applying eqs. (4.3)–(4.6), the ^{13}C chemical shifts of halomethanes can be predicted with a standard deviation of about 4 ppm.

The α carbon shifts of haloalkanes depend on temperature and solvent. Strong solvent effects are observed for the iodinated carbon atoms in iodoalkanes, as shown in Table 4.18 [253]. As expected from theory [254], carbon-13 solvent shifts are linearly dependent on $(\varepsilon - 1)/(2\varepsilon + n^2)$ (ε: dielectric constant; n: refractive index) [253].

No correlation with substituent properties can be recognized for the β effect of halogens in Table 3.2. The magnitude of shielding due to the γ effect, however, decreases with increasing size of the halogen (Table 3.2). A reduced population of the *gauche* conformer

Table 4.17. Substitution Parameters of Halomethanes Obtained from Least-Squares Regression Analysis [86].

Δ(Cl)	33.98 ± 3.04
Δ(Br)	24.18 ± 3.04
Δ(I)	-7.83 ± 4.15
Δ(Cl–Cl)	-5.44 ± 1.87
Δ(Cl–Br)	-11.29 ± 1.52
Δ(Br–Br)	-18.88 ± 1.87
Δ(I–I)	-35.67 ± 3.94

Table 4.18. ^{13}C Chemical Shifts of Alkyl Iodides in Different Solvents (δ_C in ppm) [253].

Compound	C-1	C-2	C-3	C-4	C-5	C-6	C-7	Solvent
1-Iodoethane	−5.6	19.6						Cyclohexane
	−1.4	20.5						Neat liquid
	−1.2	19.8						Nitromethane
	−0.4	20.3						N,N-Dimethylformamide
1-Iodopentane	3.1	33.1	32.4	21.2	12.8			Cyclohexane
	5.9	33.2	32.5	21.4	13.6			Neat liquid
	7.0	33.2	32.5	21.3	12.9			Nitromethane
	7.5	33.3	32.5	21.3	13.3			N,N-Dimethylformamide
1-Iodoheptane	3.3	33.5	30.3	28.0	31.5	22.2	13.1	Cyclohexane
	5.8	33.5	30.3	28.0	31.5	22.3	13.7	Neat liquid
	7.1	33.6	30.3	28.0	31.5	22.3	13.2	Nitromethane
	7.7	33.7		28.0	31.6	22.3	13.5	N,N-Dimethylformamide
2-Iodopropane	30.6	16.9						Cyclohexane
	31.4	20.2						Neat liquid
	30.5	22.4						Nitromethane
	30.9	22.8						N,N-Dimethylformamide
2-Iodo-2-methyl-propane	39.9	38.3						Cyclohexane
	40.8	41.7						Neat liquid
	39.9	45.4						Nitromethane

(Section 4.1.1) with increasing steric hindrance in going from fluorine *via* chlorine and bromine to iodine obviously attenuates the shielding contribution of the γ gauche effect.

A systematic investigation of chloro-substituted ethanes $\overset{1}{C}H_{3-x}Cl_x - \overset{2}{C}H_{3-y}Cl_y$ shows that for $y = 0$, the C-1 resonance is shifted to lower field by about 30 ppm per hydrogen-chlorine substitution on C-1; with increasing chlorine substitution at C-2 the relative deshielding for C-1 decreases [256].

As the range of chemical shifts is an order of magnitude larger for ^{13}C than for ^{1}H, CMR is better suited for investigating stereoisomers. This has been demonstrated by ^{13}C NMR measurements on 2,4-dichloropentane and 2,3-dichlorobutane, which are models

Table 4.19. ^{13}C Chemical Shifts of 2,4-Dichloropentanes, 2,3-Dichlorobutanes and Polyvinyl Chloride (δ_C in ppm) [262].

Compound	CHCl	CHCl (syndiotactic)	CHCl (heterotactic)	CHCl (isotactic)	CH$_2$	CH$_3$
meso-2,4-Dichloropentane	54.2				50.2	24.4
racem-2,4-Dichloropentane	55.3				50.4	25.2
meso-2,3-Dichlorobutane	61.3					21.8
racem-2,3-Dichlorobutane	60.2					19.7
Polyvinyl chloride		56.8	55.7	54.7	46.6	45.6

for polyvinyl chloride [262]. Table 4.19 gives the ^{13}C chemical shift values for *meso-* and *racem-*2,4-dichloropentane and 2,3-dichlorobutane and for polyvinyl chloride.

In *racem-*2,4-dichloropentane (Fig. 4.3) the methyl groups are in *anti* position to the methine carbons, and in the *meso* isomer, these two groups are partly *syn* and partly *anti* to each other. This different spatial arrangement of groups in the *racem* and *meso* isomers is reflected in a shielding difference of about 1 ppm for the methyl and methine carbons. The ^{13}C chemical shift difference of the methylene carbons of *racem-* and *meso-*2,4-dichloropentane is only 0.2 ppm. Though in both conformations the methylene carbons are not chemically equivalent, their shielding is very similar because no interaction is possible with an alkyl group in the *anti* or *syn* position.

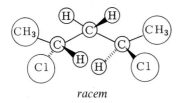

racem

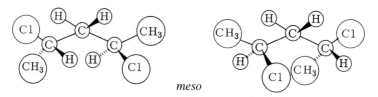

meso

Fig. 4.3. Racem- and meso-2,4-dichloropentane.

Comparison of the ^{13}C NMR spectrum of commercial polyvinyl chloride with the CMR spectra of *racem-* and *meso-*2,4-dichloropentane permits signal identification of the syndiotactic, heterotactic and isotactic methine carbons of the polymer [262].

4.4.2 Cycloalkyl Halides

Halogenation of cycloalkanes causes α, β, and γ substituent increments which essentially follow the pattern outlined for alkanes. Detailed investigations have been achieved for substituted cyclohexanes, as the conformational and configurational relations are clearly defined by the equilibrium of ring inversion (Fig. 4.4). Empirical increments of equatorial and axial substituents have been determined either by freezing ring inversion of substituted cyclohexanes or on the basis of the shift increments of *cis-trans*-isomeric *t*-butylcyclohexanes [263–265]. The bulky *t*-butyl group is assumed to adopt an equatorial configuration without distorting the geometry of the cyclohexane ring. Empirical increments $(Z = \delta_{R-X} - \delta_{R-H})$ of equatorial and axial halogens, also including other substituents, are summarized in Table 4.79 (Section 4.16). The relations found for chlorocyclohexane are exemplified in Fig. 4.4: Substituents in *axial configuration* induce a shielding in β and particularly *in γ positions* due to the γ *gauche effect*, when compared with those in equatorial configuration. If ring inversion is not frozen, averaged increments $(Z_{av.} = (Z_a + Z_e)/2)$ are observed.

In 2,2-difluoronorbornanes, *exo* methyl groups in the 3, 5, and 6 position cause a shielding of C-7 while an *endo* methylation in the 3 and 5 position shields C-5 and C-3, respectively, due to γ *gauche* interaction *via* the γ CH bond (Table 4.20(a) [229]). In *endo*-6-methyl-2,2-difluoronorbornane, however, the fluorines at the 2 position attenuate the γ *gauche* shielding as expected for C-2.

exo-   34.2   28.6 CH_3 11.9   (ppm)   20.1 CH_3 9.0   *endo*-

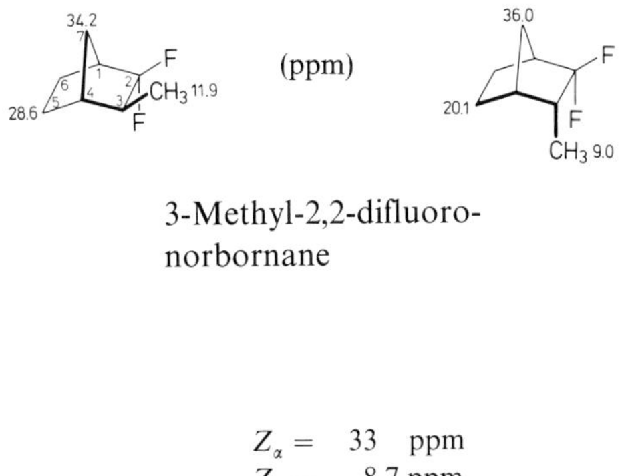

3-Methyl-2,2-difluoro-
norbornane

(a)
$Z_\alpha = 33$ ppm
$Z_\beta = 8.7$ ppm
$Z_\gamma = -3.4$ ppm
$Z_\delta = -1.4$ ppm

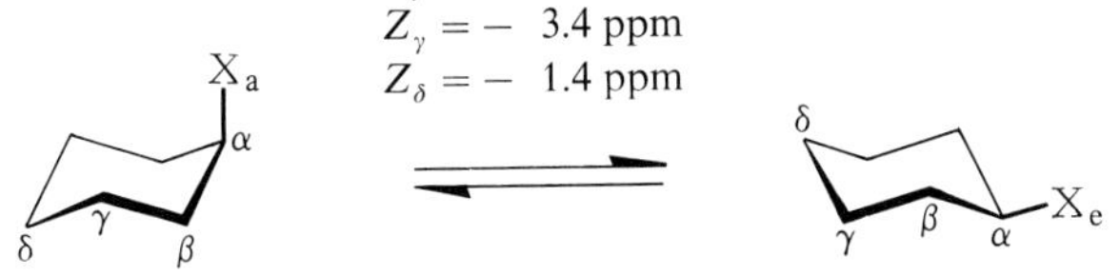

(b) $Z_{\alpha a} = 33$ ppm $Z_{\alpha e} = 33$ ppm
$Z_{\beta a} = 7$ ppm $Z_{\beta e} = 10.5$ ppm
$Z_{\gamma a} = -6.5$ ppm (γ_{gauche}) $Z_{\gamma e} = 0$ ppm (γ_{anti})
$Z_{\delta a} = -1$ ppm $Z_{\delta e} = -2$ ppm

Fig. 4.4. Carbon-13 shift increments of chlorocyclohexane for rapid (a) and frozen (b) cyclohexane ring inversion (X = Cl).

Table 4.20.(a) ^{13}C and ^{19}F Chemical Shifts of Methyl-2,2-difluoronorbornanes [229] (δ_C in ppm; ^{19}F shifts relative to CFCl$_3$ = 0 in ppm).

-norbornane	C-1	C-2	C-3	C-4	C-5	C-6	C-7	CH$_3$	F_{exo}	F_{endo}
2,2-Difluoro-	44.9	131.1	42.7	36.2	27.4	20.8	36.9		86.2	109.3
1-Methyl-2,2-difluoro-	48.7	130.2	43.3	35.3	29.6	28.8	43.0	12.2	94.3	116.7
exo-3-Methyl-2,2-difluoro-	44.9	130.9	46.8	43.6	28.6	21.0	34.2	11.9	104.8	109.6
endo-3-Methyl-2,2-difluoro-	44.0	129.3	45.6	41.2	20.1	20.9	36.0	9.0	84.6	120.6
exo-5-Methyl-2,2-difluoro-	45.6	131.0	43.5	42.5	34.6	30.6	33.4	21.5	86.2	108.0
endo-5-Methyl-2,2-difluoro-	46.0	131.0	35.4	41.3	32.2	29.2	38.5	16.1	83.7	110.1
exo-6-Methyl-2,2-difluoro-	51.4	131.0	41.9	36.5	37.8	27.1	33.4	20.5	85.5	110.0
endo-6-Methyl-2,2-difluoro-	49.5	131.5	43.5	36.1	36.3	32.2	38.7	17.8	74.1	101.8
syn-7-Methyl-2,2-difluoro-	48.8	131.4	40.3	41.1	27.7	22.0	45.2	12.2	81.4	97.8
anti-7-Methyl-2,2-difluoro-	49.1	130.6	44.1	40.1	24.5	17.9	42.1	11.5	86.5	106.5
exo-2-Fluoro-	42.1	95.6	39.8	34.6	28.0	22.3	35.0		158.9	

Table 4.20.(b) ^{13}C – ^{19}F Coupling Constants (in Hz) of Methyl-2,2-difluoronorbornanes and *exo*-2-Fluoronorbornane [229].

-norbornane	C-1	C-2	C-3	C-4	C-6	C-7	CH$_3$	
2,2-Difluoro-	24.1; 22.0	253.9; 253.9	24.3; 22.0	5.4; 2.8	5.8; 5.8	4.9		
1-Methyl-2,2-difluoro-	21.9; 21.9	256.5; 256.5	24.7; 22.7	4.5; 2.6	6.3; 4.2	5.3	4.0	
exo-3-Methyl-2,2-difluoro-	24.0; 21.8	258.0; 258.0	24.0; 20.3	5.6; 1.2	5.8; 5.8	4.7	14.1; 2.7	
endo-3-Methyl-2,2-difluoro-	24.2; 22.4	259.9; 252.6	22.4; 22.1	4.4; 1.9	6.9; 5.9	5.8	1.5; 9.7	
exo-5-Methyl-2,2-difluoro-	23.0; 21.8	255.0; 250.2	23.6; 21.6	3.2; 3.2	6.0; 6.0	5.0	<1	
endo-5-Methyl-2,2-difluoro-	23.5; 21.3	254.0; 251.6	24.6; 21.8	4.2; 2.2	6.0; 6.0	4.0	<1	
exo-6-Methyl-2,2-difluoro-	22.3; 20.9	255.0; 250.2	23.1; 21.1	4.6; 2.0	5.8; 5.8	5.0	≈1	
endo-6-Methyl-2,2-difluoro-	22.3; 19.1	257.3; 250.3	21.8; 21.8		3.4; 3.4	5.2	<1;	7.0
syn-7-Methyl-2,2-difluoro-	20.7; 20.7	253.0; 253.0	25.4; 21.6	4.0; 3.0	8.1; 6.8	5.5	4.5; <1	
anti-7-Methyl-2,2-difluoro-	22.6; 20.6	254.5; 254.5	24.1; 21.7	3.5; 2.6	5.8; 5.8	4.3	<1	
exo-2-Fluoro-	20.2	182.0	20.4	2.3	9.8	<1		

4.4.3 Alkenyl Halides

In alkenyl halides, electron donation $((+)$-M effect) of the halogen attenuates the inductive effect in the α and increases electron densitiy in the β position:

Thus, substitution of an alkenyl hydrogen by halogen causes much smaller deshieldings in the α position when compared with the α effects induced in alkyl halides. Further, the β_π effect is a net shielding relative to comparable alkene hydrocarbons, as shown for representative alkenyl halides in Table 4.16 [258], due to the negative charge induced in the β position by the halogen. Polarization decreases from iodine via bromine to chlorine as substituent. CC triple bonds respond similiarly upon halogen substitution (Table 4.16 [260]). Finally, the change from trans to cis configuration of the halogen at the double bond initiates more significant shift differences (α: $\delta_{trans} > \delta_{cis}$; β: $\delta_{cis} > \delta_{trans}$) when compared with alkyl groups (Table 4.16; Table 4.10). Again, shift differences decrease in the order iodine (6–8 ppm), bromine (4–6 ppm), chlorine (1–3 ppm).

4.4.4 Carbon-Fluorine Coupling Constants in Alkyl and Cycloalkyl Fluorides

The gyromagnetic ratio of fluorine is positive. Nevertheless, one-bond carbon-fluorine couplings are negative because fluorine predominantly uses p orbitals for bond formation [115]. Similarly to J_{CH}, J_{CC}, and J_{CN}, one-bond carbon-fluorine couplings J_{CF} correlate with the s character of carbon hybrid orbitals, as is verified by some of the data given in Table 3.14. More importantly, the magnitude of J_{CF} increases significantly with increasing positive polarization of the coupling carbon, induced, for example, by crowding of fluorine atoms in cyclobutane (Table 4.21). J_{CF} also clearly responds to ring strain, as demonstrated in Table 4.21 for fluorinated cycloalkanes.

The magnitude of carbon-fluorine coupling attenuates with coupling distance in alkyl and cycloalkyl fluorides, as shown in Tables 4.20(b) and 4.21. Analogously to vicinal proton-proton, carbon-proton and carbon-carbon coupling (Sections 3.2.2.4, 3.2.4.2), $^3J_{CF}$ depends on the dihedral angle Φ enclosed by the C–C–C–F connection, and Karplus cosine relations such as eq. (4.7) have been derived empirically [266] and theoretically [267].

$$^3J_{CF} = 5.5 \cos 2\Phi + 5.5 \text{ (Hz)} \tag{4.7}$$

Thus, trans-4-t-butylfluorocyclohexane displays a larger $^3J_{CF}$ than the cis isomer, and in the conformers of fluorocyclohexane, frozen at 180 K, $^3J_{CF}$ is larger for equatorial than for axial fluorine (Table 4.21 [266]). Similarly, $^3J_{CF}$ for exo- is larger than for endo-fluorine in 2,2-difluoronorbornanes (Tables 4.20(b) and 4.21 [229]). Further, when compared with exo-2-fluoronorbornane, 2,2-difluoronorbornane attenuates its $^3J_{CF}$ to the exo-fluorine

Table 4.21. Survey of Carbon-Fluorine Coupling Constants (in Hz) in Representative Alkyl and Cycloalkylfluorides [229, 266].

	CH₃-CH₂-CH₂-F	cyclohexyl-F	cyclopentyl-F	cyclopropyl-F	1,1,2,2-tetrafluoro	1,1,1-trifluoro-t-butyl
J_{CF}	(−) 163.5	(−) 170	(−) 173	(−) 215	(−) 298	(−) 329
$^2J_{CF}$	19.5	19	22	19		
$^3J_{CF}$	6.7	5	< 1.5	19		

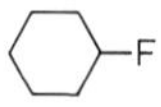

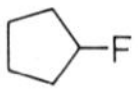

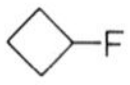

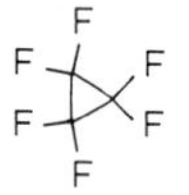

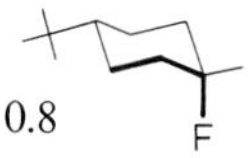

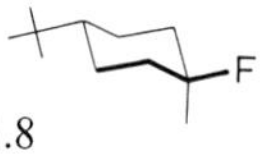

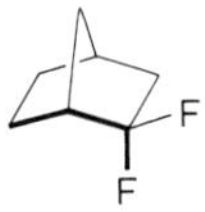

$^3J_{CF}$ and Configuration

| < 0.8 (axial F, cyclohexane) | 11.0 (equatorial F, cyclohexane) | 10.0 (norbornyl) |
| < 0.8 (t-Bu axial, F axial) | 11.8 (t-Bu axial, F equatorial) | 2.8 (norbornyl endo) | 5.4 (norbornyl exo) |

(Table 4.21) due to two electronegative fluorine atoms at the coupling carbon. Attenuation of ^3J by electronegative substituents is also known from proton NMR.

4.5 Alcohols

4.5.1 Alkanols

Two empirical increment systems (Table 4.22) derived from experimental data as collected in Table 4.23 permit prediction of alkanol carbon-13 shifts. One is related to the shift value B of the hydrocarbon $R-H$ and involves, as usual, addition of the increments $Z_i = \delta_{i(R-OH)} - \delta_{i(R-H)}$ according to eq. (4.8a) [268]. The other employs a linear equation (4.8b), correlating the shifts of an alkanol $R-OH$ and the corresponding methylalkane $R-CH_3$ by a constant b_k and a slope a_k, which is 0.7–0.8 for α and about 1 for β and γ positions [269]. Specific parameter sets characterize primary, secondary, and tertiary alcohols (Table 4.22). The magnitudes of Z_γ increments in eq. (4.8a) decrease successively from primary to tertiary alcohols (Table 4.22), obviously as a result of reduced populations of conformers with γ_{gauche} interactions in the conformational equilibrium when the degree of alkylation increases.

Isopropyl methyl groups in 3-methyl-2-butanol (Table 4.23) are *diastereotopic* ($\delta_C = 18.1$ and 18.3 ppm) due to the adjacent asymmetric carbon atom. It was shown that

4.5 Alkohols

Table 4.22. Increment Sets for Prediction of Alkanol Carbon-13 Shifts According to the Equations (4.8 a) [268] and (4.8 b) [269].

$$\delta_C(k) = \delta_C(k, R-H) + \sum Z_i \quad (4.8\,\text{a}) \qquad \delta_C(k) = b_k + a_k \delta_C k, R-CH_3 \quad (4.8\,\text{b})$$

Alkohol	Position i	Z_i (ppm)		b_k (ppm)	a_k
Primary	α	48.3		46.45	0.709
	β	10.2		1.81	0.963
	γ	−5.8		−2.28	0.963
	δ	0.3		0	1.0
Secondary	α	44.5		45.65	0.786
2-alkanols	β	9.7 (C-1) 7.4 (C-3)	all	2.07	0.958
	γ	−3.3		−0.48	0.982
	δ	0.2		0	1.0
Others	α	40.8			
	β	7.7			
	γ	−3.7			
	δ	−0.7			
Tertiary	α	39.7		48.37	0.755
	β	7.3 (1°) 5.0 (2°)		−0.06	1.029
	γ	−1.8		−1.03	1.137
	δ	0		0.07	0.995

diastereotopic shift differences grow with increasing size of the alkyl group replacing methyl (C-1) in 3-methyl-2-butanol [269].

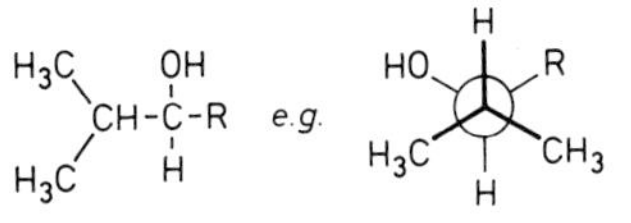

$R = CH_3,\ i\text{-}C_3H_7;\ t\text{-}C_4H_9$
$\delta = 0.2 \quad 2.7 \quad 6.9\ \text{ppm}$

A slightly negative γ effect of the hydroxy group is found when comparing the carbon shifts of propene and allyl alcohol, and (E)-(Z)-isomeric pairs of 2-en-1-ols can be distinguished following the $\delta_{(E)} > \delta_{(Z)}$ rule for the carbons α to the double bond (Table 4.23). Positive γ effects of the OH group, however, are identified when going from alkynes to alkynols, indicating that γ effects do not exclusively arise from steric interactions [270].

Concentration effects on ^{13}C shifts of alcohols are small [271]. Solvent-induced shifts are enhanced on going from primary to tertiary alcohols and may be as high as 2 ppm [271]. Protonation shifts are much larger. The methanol carbon, for example, is shielded by −14.6 ppm relative to the neat liquid value when dissolved in magic acid [272]. *Protonation shifts* of 1-alkanols in trifluoroacetic acid are shieldings for C-1 and alternating deshieldings for all other carbon atoms, e.g. $\delta_{C\text{-}2} > \delta_{C\text{-}4} > \delta_{C\text{-}3}$ for 1-butanol [272]. They may be useful for signal assignments, as are *acetylation shifts* [210]. O-acetylation

Table 4.23. ^{13}C Chemical Shifts of Acyclic Alcohols (δ_C in ppm) [268].

Compound	C-1	C-2	C-3	C-4	C-5	C-6	C-7	C-8	C-9	C-10
Methanol	49.9									
Ethanol	57.0	17.6								
1-Propanol	63.6	25.8	10.0							
1-Butanol	61.4	35.0	19.1	13.6						
1-Pentanol	61.8	32.5	28.2	22.6	13.8					
1-Hexanol	61.9	32.8	25.8	32.0	22.8	14.2				
1-Heptanol	61.9	32.9	26.1	29.4	32.1	22.8	13.9			
1-Octanol	61.9	32.9	26.1	29.7	29.6	32.1	22.8	13.9		
1-Nonanol	62.0	32.9	26.2	29.8	29.9	29.6	32.2	22.9	14.0	
1-Decanol	61.9	32.9	26.1	29.8	29.8	29.9	29.6	32.2	22.8	14.0
2-Propanol	25.1	63.4								
2-Butanol	22.6	68.7	32.0	9.9						
2-Pentanol	23.3	67.0	41.6	19.1	14.0					
2-Hexanol	23.3	67.2	39.2	28.3	22.9	13.9				
2-Heptanol	23.3	67.2	39.5	25.8	32.3	22.9	13.9			
2-Octanol	23.4	67.2	39.6	26.1	29.7	32.2	22.8	14.0		
2-Decanol	23.4	67.2	39.6	26.2	30.1	30.0	29.6	32.2	22.9	14.0
3-Pentanol	9.8	29.7	73.8							
3-Hexanol	9.9	30.3	72.3	39.4	19.4	14.0				
3-Heptanol	10.0	29.7	72.6	36.9	28.2	23.0	14.0			
3-Octanol	10.0	30.3	72.6	37.2	25.7	32.3	22.9	13.9		
4-Heptanol	14.1	19.1	40.0	70.6						
4-Octanol	14.0	19.1	40.0	70.9	37.5	28.2	23.0	14.0		
5-Nonanol	14.0	23.0	28.3	37.5	71.1					
2-Methyl-1-propanol	68.9	30.8	18.9							
2-Methyl-2-propanol	31.3	68.4								
2,2-Dimethyl-1-propanol	72.6	32.6	26.3							
2-Methyl-1-butanol	66.9	37.5	25.9	11.1	16.0					
3-Methyl-1-butanol	60.2	41.8	24.8	22.5						
3-Methyl-2-butanol	19.7	72.0	35.1	18.1	18.3					
4-Methyl-2-pentanol	24.0	65.2	48.9	24.8	22.8					
3,3-Dimethyl-1-butanol	58.9	46.4	29.7	29.8						
2,3-Dimethyl-2-butanol	26.3	72.2	38.8	17.5						
3,3-Dimethyl-2-butanol	17.9	74.8	35.0	25.5						
2,3,3-Trimethyl-2-butanol	25.4	74.1	37.5	25.6						
2,2,4,4-Tetramethyl-3-pentanol	28.5	36.9	84.7							
2-Chloroethanol	62.9	46.6								
2-Bromoethanol	63.3	35.1								
2-Iodoethanol	64.0	10.5								
2-Propen-1-ol	62.5	136.7	114.0							
(E)-2-Buten-1-ol	62.9	132.1	126.0	17.3						
(Z)-2-Buten-1-ol	57.9	131.4	125.3	12.7						
3-Buten-1-ol	61.3	36.9	134.7	117.2						
2-Propyn-1-ol	50.4	82.0	73.8							
2-Butyn-1-ol	50.4	77.7	81.6	3.6						
3-Butyn-1-ol	60.7	22.9	80.7	70.5						
2,4-Pentadiyn-1-ol	50.8	74.7	69.7	67.5	68.6					

causes a deshielding of a α carbon which increases from primary (2 ppm) *via* secondary (3 ppm) to tertiary alcohols (10 ppm). Carbon atoms in the β position are shielded by about −3 to −4 ppm upon O-acetylation; much smaller shieldings (≤ 1 ppm) are induced in the γ and δ position.

4.5.2 Cycloalkanols

Substitution of hydrogen by a hydroxy function in a cycloalkane induces deshieldings in the α and β position and shieldings in the position γ to the OH group, following the pattern outlined for alkanols (Tables 4.22, 4.24). In rings with well defined conformational and configurational relations, *cis-trans* isomerism is clearly reflected by carbon-13 shifts, as illustrated in Fig. 4.5 for *cis-* and *trans*-4-*t*-butylcyclohexanol and in Table 4.24 for a series of isomeric cycloalkanols [210, 268] and bicycloalkanols [229, 273], including *cis*- and *trans*-9-methyl-*trans*-decalins. These are useful models for configurational assignments of 3α- and 3β-hydroxy-triterpenes and -steroids.

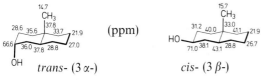

trans- (3 α-)   cis- (3 β-)

3-Hydroxy-9-methyl-*trans*-decalin

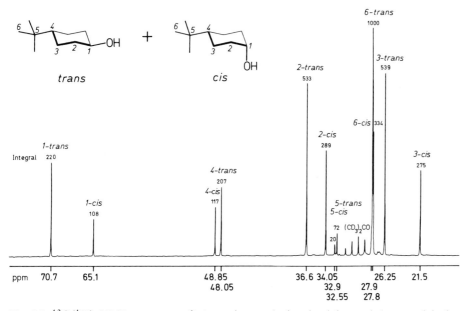

Fig. 4.5. ^{13}C{^{1}H} NMR spectrum of *cis-* and *trans*-4-*t*-butylcyclohexanol (saturated in hexadeuterioacetone; 28 °C; 22.63{90} MHz). Evaluation of the integrals gives 66±2% *trans-* and 33±2% *cis*-isomer. Deviations from the shift value given in Table 4.24 are solvent induced.

Table 4.24. ^{13}C Chemical Shifts of Representative Cycloalkanols [210, 268], Norbornanols [229] and Hydroxydecalins [273] (δ_C in ppm).

Compound	C-1	C-2	C-3	C-4	C-5	C-6	C-7	CH_3	$-\overset{\mid}{\underset{\mid}{C}}-$	$(CH_3)_3$
Cyclopropanol	45.7	6.8								
Cyclobutanol	70.8	33.4	12.2							
Cyclopentanol	73.3	35.0	23.4							
Cyclohexanol	69.5	35.5	24.4	25.9						
Cycloheptanol	72.4	37.7	23.3	28.6						
Cyclooctanol	71.9	34.7	23.0	25.5	27.8					
1-Methylcyclobutanol	72.2	37.1	11.4					26.5		
1-Methylcyclopentanol	79.7	41.2	24.3					28.2		
1-Methylcyclohexanol	69.8	39.7	22.9	26.0				29.5		
1-Methylcycloheptanol	72.2	42.4	20.0	29.2				30.5		
cis-2-Methylcyclopentanol	76.4	41.0	35.7	23.3	32.2			14.9		
trans-2-Methylcyclopentanol	81.0	43.4	35.1	22.7	33.0			19.5		
cis-2-Methylcyclohexanol	71.1	35.8	29.3	24.2	21.5	31.8		16.2		
trans-2-Methylcyclohexanol	76.6	39.7	34.0	25.8	25.4	35.1		18.1		
cis-2-Methylcycloheptanol	73.2	38.4	29.4	25.5	27.7	21.9	34.3	17.3		
trans-2-Methylcycloheptanol	77.3	41.7	31.7	26.3	24.1	21.8	36.0	20.2		
cis-3-Methylcyclopentanol	73.2	44.1	33.0	32.4	35.4			21.1		
trans-3-Methylcyclopentanol	73.2	44.3	31.9	32.7	35.2			20.1		
cis-3-Methylcyclohexanol	70.5	44.0	31.7	34.8	24.4	34.4		22.5		
trans-3-Methylcyclohexanol	66.5	41.2	26.6	34.4	20.2	32.8		20.2		
cis-4-Methylcyclohexanol	65.9	31.4	28.7	30.6				20.9		
trans-4-Methylcyclohexanol	69.7	33.1	31.1	31.4				21.7		
cis-4-t-Butylcyclohexanol	65.7	33.4	20.9	48.1					32.2	27.5
trans-4-t-Butylcyclohexanol	70.9	36.0	25.6	47.2					32.2	27.6

Table 4.24 (continued).

Compound	C-1	C-2	C-3	C-4	C-5	C-6	C-7	CH$_3$	$-\overset{\vert}{\underset{\vert}{C}}-$	(CH$_3$)$_3$
endo-2-Hydroxy-norbornane	42.8	72.2	39.3	37.4	30.0	20.1	37.5			
exo-2-Hydroxy-norbornane	44.2	74.1	42.1	35.5	28.5	24.6	34.3			
1-Methyl-2-norbornanol (*endo*)	47.9	76.8	40.8	37.1	31.5	27.1	44.6	18.5		
1-Methyl-2-norbornanol (*exo*)	47.6	76.6	43.5	36.1	30.6	33.4	40.3	16.3		
endo-3-Methyl-2-norbornanol (*endo*)	43.9	71.4	37.1	42.6	21.9	19.8	36.8	10.2		
endo-3-Methyl-2-norbornanol (*exo*)	45.6	a	46.4	41.1	21.2	25.2	37.0	14.6		
exo-3-Methyl-2-norbornanol (*endo*)	43.5	a	45.7	44.1	30.2	19.9	34.3	19.5		
exo-3-Methyl-2-norbornanol (*exo*)	45.1	a	43.5	43.3	29.4	24.7	31.9	19.6		

	C-1	C-2	C-3	C-4	C-5	C-6	C-7	C-8	C-9	C-10
1α-Hydroxy-*trans*-decalin	70.4	34.3	20.0	33.8	33.9	26.8	26.4	29.6	47.4	35.6
1β-Hydroxy-*trans*-decalin	74.6	35.8	24.1	33.5	33.7	26.4	26.2	29.1	50.4	41.1
2α-Hydroxy-*trans*-decalin	43.0	70.2	35.6	32.0	33.8	26.5	26.3	33.3	41.2	42.3
2β-Hydroxy-*trans*-decalin	40.4	66.6	33.9	27.6	33.8	26.7	26.7	32.9	36.4	43.2
cis-3-Hydroxy-9-methyl-*trans*-decalin	40.0 15.7 (CH$_3$)	31.2	71.0	38.1	28.8	26.7	21.9	41.1	33.0	43.1
trans-3-Hydroxy-9-methyl-*trans*-decalin	35.6 14.7 (CH$_3$)	28.6	66.6	36.0	28.8	27.0	21.9	41.5	33.7	37.8

a not unequivocally identified separately from the resonances of other 3-methyl isomers [229].

Table 4.25. Carbon-13 Shift Increments Z_i Induced by Hydroxylation of Cyclohexane, trans-Decalin and Bicyclo[2.2.1]heptane [274] According to Eq. (4.8a). (Reference shifts B are 27.0 ppm for cyclohexane, 34.7 (C-1,4,5,8), 27.1 (C-2,3,6,7), 44.2 ppm (C-9,10) for trans-decalin, and 36.5 (C-1,4), 29.8 (C-2,3,5,6), 38.7 ppm (C-7) for norbornane, respectively.)

Ring	Substitution	Z_α	Z_β	Z_γ	Z_δ
Cyclohexane	e-OH	43.7	8.5	− 2.8	− 1.8
	a-OH	38.9	5.6	− 7.5	− 0.9
trans-Decalin	e-1-OH	40.3	8.9 (C-2)	− 2.8 (C-3)	− 0.8 (C-4)
			6.7 (C-9)	− 2.5 (C-5)	− 0.6 (C-6)
				− 5.2 (C-9)	− 0.7 (C-8)
	a-1-OH	36.1	7.4 (C-2)	− 6.9 (C-3)	− 0.4 (C-4)
			3.7 (C-9)	− 8.1 (C-5)	− 0.4 (C-6)
				− 4.7 (C-9)	− 0.5 (C-8)
	e-2-OH	43.3	8.7	− 2.4	− 1.2
	a-2-OH	39.7	6.1 (C-1)	− 6.7 (C-4)	− 0.5 (C-5)
			7.0 (C-3)	− 7.3 (C-10)	− 1.4 (C-9)
Bicyclo[2.2.1]heptane	1-OH	46.4	5.5 (C-2)	− 1.6 (C-4)	0.4 (C-3)
			5.6 (C-7)		
	exo-2-OH	45.1	7.9 (C-1)	− 0.9 (C-4)	− 1.5 (C-5)
			12.6 (C-3)	− 5.2 (C-6)	
				− 3.9 (C-7)	
	endo-2-OH	43.3	6.2 (C-1)	− 0.9 (C-4)	0.3 (C-5)
			9.8 (C-3)	− 9.6 (C-6)	
				− 0.7 (C-7)	
	7-OH	41.4	4.1	− 2.8	
	Δ^2-anti-OH	33.5	3.8	− 3.2 (C-5,6)	
	Δ^2-syn-OH	38.4	5.6	− 2.4 (C-5,6)	

Increments Z_i ($i = \alpha, \beta, \gamma, \delta$) derived from the shift values in Table 4.24 ($Z_i = \delta_{i(\text{ROH})} - \delta_{i(\text{RH})}$) for axial and equatorial OH in cyclohexane and trans-decalin as well as for endo- and exo- OH in norbornane are summarized in Table 4.25 [274]. Increments of equatorial OH groups are more positive than those of axial ones with the exception of Z_δ values. γ shieldings in cyclohexanes and trans-decalins are significantly larger for axial compared with equatorial OH groups due to 1,3-diaxial γ-gauche type interactions between OH and axial hydrogen atoms in γ positions. Endo-exo-increments (Table 4.25) of OH in norbornanols follow the trend discussed for methyl- and fluoronorbornanes [229], exo-2-OH shielding C-7 and endo-2-OH particularly shielding C-6, again as a result of γ-gauche sterically induced polarization of CH bonds. For norbornanols it was pointed out early that carbon-13 shifts predicted by increment calculations differ considerably from experimental shifts for carbon atoms involved in

4.6 Ethers

multiple substitution, such as C-2 and C-3 in 3-methyl-2-norbornanols [229]. In these cases, application of shift reagents, as demonstrated for methylcyclopentanols [210], or two-dimensional shift correlations (Sections 2.10.3 and 2.10.4) permit more straightforward assignments.

4.6.1 Dialkyl Ethers

When comparing alcohols with corresponding ethers, the α carbon atom of the ether is deshielded by about 10–12 ppm while the carbon in the β position experiences a shielding of about 3 ppm (Table 4.26):

```
14.6  20.1  36.0  62.4              15.0  20.5  32.9  73.4  59.1
CH₃-CH₂-CH₂-CH₂-OH                  CH₃-CH₂-CH₂-CH₂-OCH₃        (ppm)
 δ    γ    β    α                    δ    γ    β    α
                                     ε'   δ'   γ'   β'   α'

    Butanol                              Butyl-
                                         methyl-
                                         ether
```

Using the increment formalism, this is rationalized as a β effect of the O-alkyl carbon to the α and a γ effect to the β-carbon transmitted through the ether oxygen [210, 275–277]. Corresponding changes are observed for methoxy relative to hydroxy shift increments for α and β carbons of cyclohexane [263–265].

Cyclohexanol (ppm)   [structures with shifts]   Cyclohexyl methylether (ppm)

Successive enhancement of γ shieldings in the series ethyl, iso-propyl and t-butyl methyl ether (Table 4.26) is attributed to an increased number of *gauche* interacting methyl groups:

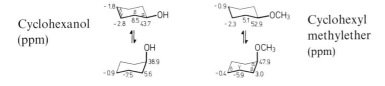

Dimethyl　　Ethyl-　　Methyl-　　t-Butyl-
　　　　　　methyl　　iso-propyl　methyl-
　　　　　　　　　　　　　　　　　ether

Table 4.26. ^{13}C Chemical Shifts of Ethers [210, 275–277] and Enol Ethers [278–280] (δ_C in ppm).

Compound	C-1	C-2	C-3	C-4	OCH$_3$	OCH$_2-$	$-$CH$_3$
Dimethylether					61.2		
Ethylmethylether	69.2	16.1			59.1		
Methylpropylether	75.4	24.0	11.7		59.1		
Butylmethylether	73.4	32.9	20.5	15.0	59.1		
Methyl-*iso*-propylether	74.1	22.9			56.4		
Methyl-*t*-butylether	73.6	28.2			50.1		
Diethylether	66.9	16.4					
Dipropylether	73.7	24.4	11.8				
Dibutylether	71.6	33.5	20.8	15.1			
Di-*iso*-propylether	69.2	24.3					
Di-*t*-butylether	73.6	31.7					
1,2-Dimethoxyethane	72.3				58.4		
2,2'-Dimethoxydiethyl ether	72.0	70.6			58.9		
Allylmethylether	73.1	134.4	116.4		57.4		
Diallylether	71.1	135.3	116.5				
Methoxyethene	153.2	84.1			52.5		
(*E*)-1-Methoxypropene	147.5	96.0	12.5		54.9		
(*Z*)-1-Methoxypropene	146.4	100.2	8.8		58.5		
(*E*)-1-Trimethylsilyloxypropene	138.5	103.8	10.3				0.3[a]
(*Z*)-1-Trimethylsilyloxypropene	136.8	103.4	7.2				0.2[a]
1,1-Dimethoxyethene	167.9	54.7			55.0		
(*E*)-1,2-Dimethoxyethene	135.2				57.9		
(*Z*)-1,2-Dimethoxyethene	130.3				59.6		
Tetramethoxyethene	141.9				57.7		
(*E*)-1-Ethoxy-2-methyl-1,3-butadiene	147.8	114.7[b]	136.9	107.8		68.5	15.3
Ethoxyethyne	89.6	23.4				72.6	10.0
1-Methoxypropyne	88.6	28.2	$-$4.2		62.6		
Methoxyallene	123.1	202.0	90.3		56.1		

[a] $-$OSi(CH$_3$)$_3$.
[b] C$-$2$-$CH$_3$: 8.8 ppm

Conversion of diethyl ether into triethyloxonium tetrafluoroborate deshields the α while shielding the γ carbon atom. The reason is not only the positive charge at oxygen but also an additional β and γ effect introduced by the third ethoxy group.

$$16.4 \quad 66.9 \qquad\qquad 12.1 \quad 84.1 \qquad \text{ppm}$$

$$(CH_3-CH_2)_2O \qquad (CH_3-CH_2)_3\overset{\oplus}{O} \ \overset{\ominus}{BF_4}$$

4.6.2 Enol Ethers and Alkynyl Ethers

An alkoxy group at the double or triple bond polarizes the π system in a similar manner but more strongly than that described for halogens (Section 4.4.3). The α effect is predominantly inductive, while the $β_π$ effect results from electron donation induced by the nonbonding pairs at the alkoxy oxygen:

Thus, strong shieldings are observed for β carbons of enol ethers and alkynyl ethers, as shown for 1,1-dimethoxyethene (54.7 ppm) and ethoxyethyne (23.4 ppm) in Table 4.26. In 1-alkoxy-1,3-butadienes, transmission of the electron releasing effect along the conjugated double bonds affects alternate carbons similarly, shielding the carbons in β and δ position as illustrated for 1-ethoxy-2-methyl-1,3-butadiene.

(ppm) 136.9 147.8 68.5
       107.8 114.7   15.3
              8.8

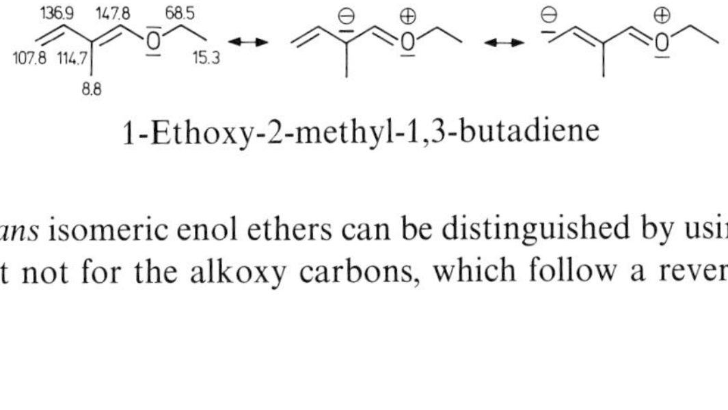

1-Ethoxy-2-methyl-1,3-butadiene

Cis-trans isomeric enol ethers can be distinguished by using the $δ_{trans} > δ_{cis}$ relation for the α but not for the alkoxy carbons, which follow a reverse pattern (Table 4.26).

4.7 Carbonyl Compounds

4.7.1 General

Carbonyl carbon-13 shifts of aldehydes, ketones and carboxylic acids, including all derivatives, occur between 150 and 220 ppm [281]. Within this range, carboxy carbons are shielded (150–180 ppm) relative to carbonyl carbons in aldehydes and ketones (190–220 ppm). This is attributed to an electron releasing effect of the additional hetero

216   4 ^{13}C NMR Spectroscopy of Organic Compounds

atom X (X = O, N, halogen) attached to the carbonyl group in carboxylic acids and derivatives which, in part, compensates for electron withdrawal of the carbonyl oxygen according to the following canonical formulae.

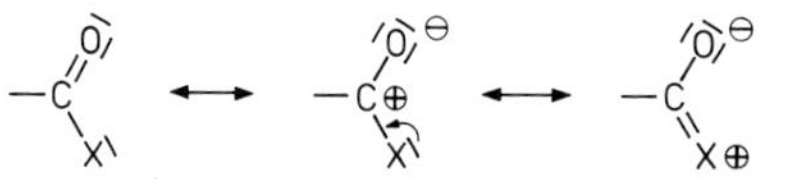

Carboxylic acids and derivatives:
150–180 ppm

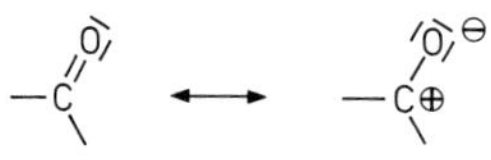

Aldehydes and ketones:
190–220 ppm

In α, β-unsaturated carbonyl compounds [282, 283], carbonyl carbons are shielded relative to those in comparable saturated compounds. Electron withdrawal of the carbonyl oxygen at the carbonyl carbon is attenuated by electron donation of the double bond. This, in turn, generates a positively charged β carbon atom which is deshielded as a result:

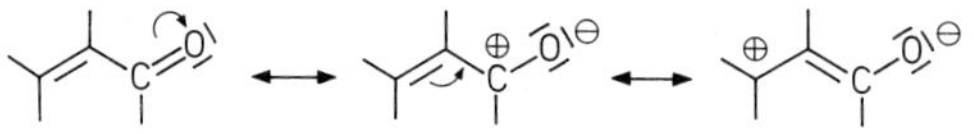

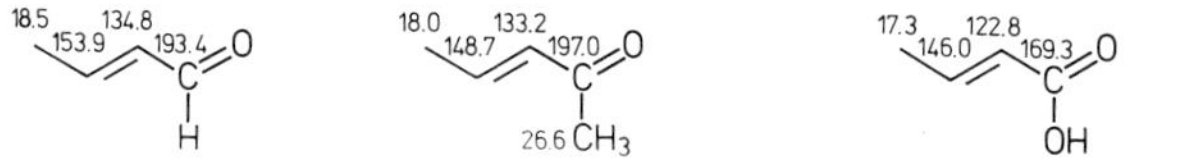

18.5 134.8	18.0 133.2	17.3 122.8	
153.9 193.4 O	148.7 197.0 O	146.0 169.3 O	ppm
H	26.6 CH$_3$	OH	
(E)-2-Butenal	(E)-3-Penten-2-one	(E)-2-Butenoic acid	

13.8 45.9	13.5 45.2	13.7 36.2	
15.8 202.5 O	17.5 206.6 O	18.7 179.3 O	ppm
H	29.3 CH$_3$	OH	
Butanal	2-Pentanone	Butanoic acid	

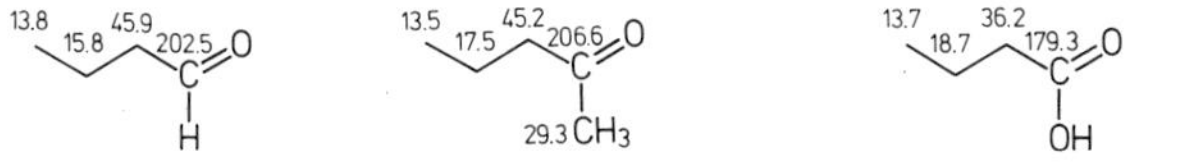

4.7.2   Aldehydes and Ketones

Carbon-13 chemical shifts of representative aldehydes [284] and ketones [285–288] are collected in Tables 4.27 and 4.28. Inspection of the data shows that α, β, and γ effects are up to 7, 2, and − 1 ppm, respectively. These increments are significantly smaller compared with those reported for alkyl carbons. Obviously, the electron releasing effect of alkyl groups ((+)-I-effect) slightly attenuates positive polarization of the carbonyl carbons.

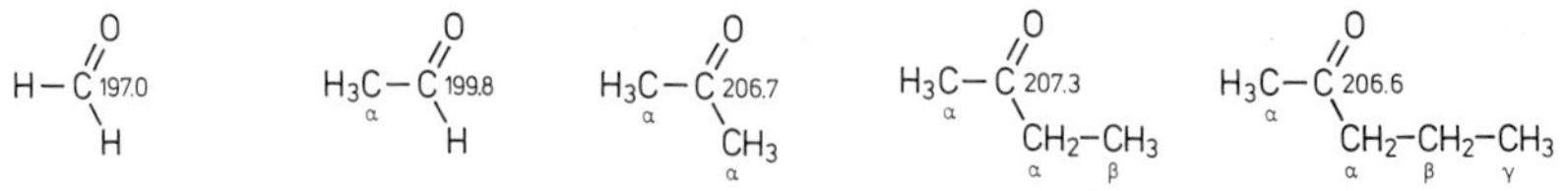

H–C 197.0	H$_3$C–C 199.8	H$_3$C–C 206.7	H$_3$C–C 207.3	H$_3$C–C 206.6	ppm
H	α H	α CH$_3$	α CH$_2$–CH$_3$	α CH$_2$–CH$_2$–CH$_3$	
		α	β	α β γ	
Methanal	Ethanal	Propanone	2-Butanone	2-Pentanone	

Table 4.27. ^{13}C Chemical Shifts of Selected Aldehydes (δ_C in ppm) [284].

Compound	C-1	C-2	C-3	C-4	C-5	C-6	C-7
Methanal	197.0						
Ethanal	199.8	30.9					
Propanal	202.8	37.3	6.0				
Butanal	202.5	45.9	15.8	13.8			
Pentanal	202.3	43.7	24.3	22.3	13.8		
2-Methylpropanal	204.7	41.2	15.5				
2-Methylbutanal	204.9	47.9	23.2	11.4	12.9		
3-Methylpentanal	202.4	50.9	29.8	29.8	11.4	19.6	
2-Ethylbutanal	205.0	55.0	21.5	11.4			
2,2-Dimethylpropanal	205.6	42.4	23.4				
Propenal	193.3	136.0	136.4				
2-Methylpropenal	194.9	146.9	134.4	14.0			
(E)-2-Butenal	193.4	134.8	153.9	18.5			
(E)-3-Phenylpropenal	193.5	128.2	152.5	134.1 α	129.0 o	128.5 m	131.0 p
Propynal	176.8	81.8	83.1				
Formylcyclohexane	204.7	50.1	26.1	25.2	25.2		

Table 4.28. ^{13}C Chemical Shifts of Representative Ketones (δ_C in ppm) [283, 285–288].

Compound	C-1	C-2	C-3	C-4	C-5	C-6	C-7	C-8
Propanone (Acetone)	30.7	206.7						
Butanone	27.5	207.3	35.2	7.0				
2-Pentanone	29.3	206.6	45.2	17.5	13.5			
2-Hexanone	29.4	206.8	43.5	31.9	23.8	14.8		
3-Methylbutanone	27.5	212.5	41.6	18.2				
3,3-Dimethylbutanone	24.5	212.8	44.3	26.5				
3,3,4,4-Tetramethyl-2-pentanone	23.5	210.0	49.4	32.6	19.1	16.1		
3-Pentanone	8.0	35.5	210.7					
2,4-Dimethyl-3-pentanone	17.8	38.0	215.1					
2,2,4-Trimethyl-3-pentanone	24.9	43.4	217.4	32.5	19.0			
2,2,4,4-Tetramethyl-3-pentanone	28.6	45.6	218.0					
Butenone	25.7	197.5	137.1	128.0				
(E)-3-Penten-2-one	26.6	197.0	133.2	142.7	18.0			
4-Methyl-3-penten-2-one	31.4	197.7	124.5	154.3	27.5	20.5 (4-CH$_3$)		
3,5-Heptadien-2-one	26.7	197.0	128.3	143.3	130.0	139.7	18.3	
6-Methyl-3,5-heptadien-2-one	27.4	198.5	128.0	139.4	124.1	147.5	19.0 [a]	26.6 [b]

4 ^{13}C NMR Spectroscopy of Organic Compounds

Table 4.28 (continued).

Compound	C-1	C-2	C-3	C-4	C-5	C-6	C-7	C-8
2,3-Butanedione (Diacetyl)	22.4	197.1						
2,4-Pentanedione (Acetylacetone)	30.9	203.5	58.7 (keto)					
	24.8	192.6	101.1 (enol)					
2,5-Hexanedione	29.6	206.9	37.0					
Fluoropropanone	84.9	203.5	25.1					
Chloropropanone	49.4	200.1	27.2					
Bromopropanone	35.5	199.0	27.0					
3-Hydroxy-2-butanone	24.9	211.2	73.1	19.4				
4-Hydroxy-2-butanone	30.2	209.2	46.1	57.5				
Cyclobutanone	208.9	47.8	9.9					
Cyclopentanone	219.6	37.9	22.7					
Cyclohexanone	209.7	41.5	26.6	24.6				
Cycloheptanone	215.0	43.9	24.4	30.6				
2-Methylcyclohexanone	212.9	45.4	36.4	25.3	28.0	42.0	14.8	
3-Methylcyclohexanone	211.6	50.0	34.2	33.4	25.3	41.1	22.1	
4-Methylcyclohexanone	208.7	39.5	33.8	30.1			19.8	
4-t-Butylcyclohexanone	211.5	41.1	27.5	46.6			32.3	27.5
2-Cyclopentenone	209.8	134.2	165.3	29.1	34.0			
2-Cyclohexenone	199.0	129.9	150.6	25.8	22.9	38.2		
2-Methyl-2-cyclopentenone	209.5	141.8	158.0	26.5	34.2	10.1		
3-Methyl-2-cyclopentenone	209.5	130.5	178.9	33.1	35.8	19.3		
3-Methyl-2-cyclohexenone	198.4	126.5	162.3	30.9	24.2	37.1	22.8	
4,4-Dimethyl-2,5-cyclohexadienone	185.8	127.3	156.7	37.9		26.7		

^{a} (Z)-CH$_3$
^{b} (E)-CH$_3$.

Particularly large carbonyl shifts (215–218 ppm) are measured for 2,4-dimethyl-, 2,2,4-trimethyl- and 2,2,4,4-tetramethyl-3-pentanone (di-t-butyl ketone). Strong steric repulsion of the bulky alkyl groups which spreads the bond angle from 115° in acetone to 130° in di-t-butyl ketone is assumed to be responsible for these deviations from increment additivity [285]; the IR and UV spectra of di-t-butyl ketone also exhibit anomalies.

2,4-Dimethyl- 215.1   2,2,4,-Trimethyl- 217.4   2,2,4,4-Tetramethyl-3-pentanone 218.0   ppm

4.7 Carbonyl Compounds

It has been demonstrated for a series of cyclic and acyclic ketones that there is a correlation between the ^{13}C chemical shifts of carbonyl groups and their n → π* transition energy, as is expressed by eq. (4.9) [289].

$$\delta_{CO} = a \lambda^{n \to \pi^*}_{max.} + b \qquad (4.9)$$

δ_{CO}: ^{13}C chemical shift of the carbonyl group in cyclic ketones;

$\lambda^{n \to \pi^*}_{max.}$: absorption maxima of n → π* transitions in cyclic ketones;

a, b: regression coefficients of the least-squares linear fit.

The chemical shift dependence of the carbonyl resonances on ring size in cycloalkanones is particularly remarkable: In the series of cycloalkanones, cyclopentanone is found to have the largest carbonyl shift (219.6 ppm). The CO signals of cyclobutanone and cyclohexanone are both observed at higher field (≈ 209 ppm). The carbonyl carbons of cyclooctanone and cyclononanone are much more deshielded than those of cyclohexanone, cycloheptanone, cyclodecanone and cycloundecanone. The carbonyl resonances of the twelve to seventeen membered ring ketones occur at δ values similar to those of acyclic ones [282, 288].

An empirical increment system permits prediction of charge distribution in α,β-unsaturated carbonyl compounds, assuming additivity of electronic effects and neglecting the conformational dependence of carbon-13 chemical shifts [290]. Moreover, carbonyl and alkenyl carbon shifts of α, β-unsaturated ketones may be used to differentiate between planar and twisted conjugated systems, as shown in Table 4.29 [291] and outlined for phenones in Section 3.1.3.8.

Bicycloalkanones display carbonyl carbon-13 shifts (215 ± 5 ppm) similar to those reported for alkanones with bulky substituents [229]. Double bonds in β, γ position to the carbonyl function in 2-norbornanone (bicyclo[2.2.1]heptan-2-one) induce a shielding of the carbonyl carbon of up to 7 ppm, as shown for the 5,6-dimethylene derivative. This is attributed to homo- and hyperconjugative interactions [292]. An even larger shielding of the carbonyl carbon induced by homoconjugation is found for 7-norbornenone relative to 7-norbornanone [292].

2-Norbornanone: 37.1, 23.7, 26.7, 49.3, 34.8, 215.8, 44.7

2-Norbornenone: 50.8, 130.4, 142.7, 55.8, 40.0, 214.7, 37.0

5,6-Dimethylene-2-norbornanone: 38.6, 106.1, 102.5, 142.4, 148.8, 60.3, 43.5, 210.0, 43.8

7-Norbornenone: 216.2, 24.3, 37.9

7-Norbornanone: 205.1, 21.1, 45.4, 133.3

(ppm)

Table 4.29. ^{13}C Chemical Shifts of Carbons in α, β-Unsaturated Ketones (δ_C in ppm) [291].

Carbon atom	Planar	Nonplanar
Carbonyl	197.5-195.0	205.2-198.0
C-α	142.5-122.5	136.5-120.5
C-β	178.5-127.5	142.5-136.5

The substituent effects of carbonyl groups on alkyl carbon shifts are

$$Z_\alpha = 30.0, \quad Z_\beta = -0.6, \quad Z_\gamma = -2.7 \text{ ppm} \quad \text{for the aldehyde function, and}$$
$$Z_\alpha = 23.0, \quad Z_\beta = 3.0, \quad Z_\gamma = -3.0 \text{ ppm} \quad \text{for the acetyl group,}$$

respectively. Thus, α and γ effects of both groups work in the same direction; but a negative β effect characterizes the aldehyde function. Another unique feature of aldehydes is the large $^2J_{CH}$ coupling (20–25 Hz) of the α carbon with the aldehyde proton, as outlined in Section 3.2.2.3.

The potential of carbon-13 NMR in the analysis of keto-enol tautomerism has been demonstrated for 2,4-pentanedione (acetylacetone) and dimedone [293]. Quantitative evaluation of equilibrium concentrations is possible by application of the inverse gated decoupling technique illustrated in Fig. 2.23.

15 % 85 %
30.9
58.7
203.5
O O
101.1
24.8
192.6
O H O

Oxo- 2,4-Pentanedione *Enol-*

(ppm)

O
203.4
57.0
32.4
O
53.4
27.9
OH
190.2
102.5
30.4
O
46.0
27.9

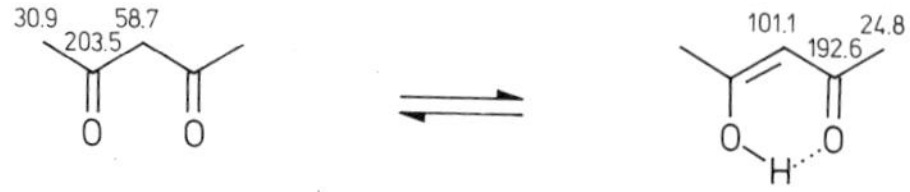

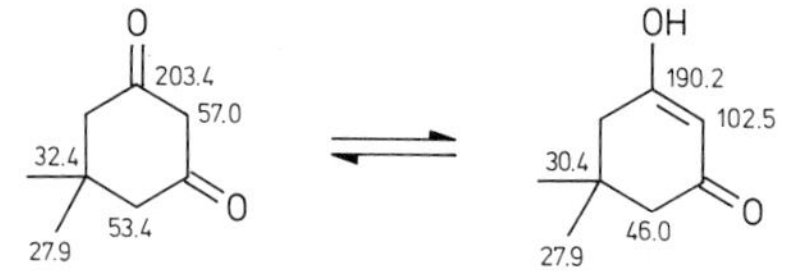

Oxo- Dimedone *Enol-*

Acetal and ketal carbon nuclei are shielded relative to the parent functions. Their shifts occur at 105 ± 5 ppm, reflecting the change from sp^2 carbonyl to sp^3 dialkoxy carbons. Increasing the number of alkoxy groups successively enhances the associated carbon shift. But the effects are not additive when comparing the data of formaldehyde dimethylacetal, triethyl orthoformate and tetramethyl orthocarbonate.

109.9
CH2(OCH3)2
53.7

99.6
CH3CH(OCH2CH3)2
19.9 60.7 15.4

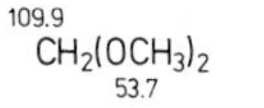

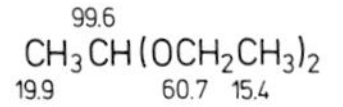

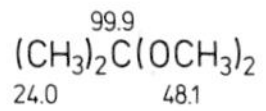

99.9
(CH3)2C(OCH3)2
24.0 48.1

Formaldehyde dimethylacetal Acetaldehyde diethylacetal Acetone dimethylketal

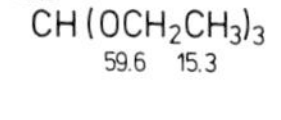

112.9
CH(OCH2CH3)3
59.6 15.3

Triethyl orthoformate

(ppm)

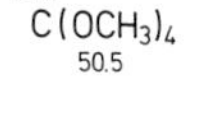

121.2
C(OCH3)4
50.5

Tetramethyl orthocarbonate

Representative benzaldehyde derivatives [73 k] and phenones [294] (Tables 4.30 and 4.31) display carbonyl shifts which are essentially influenced by steric repulsions and intramolecular hydrogen bonding. Steric repulsions by bulky alkyl groups in *o*, *o'* position of the carbonyl group prevent coplanarity of carbonyl double bond and phenyl ring,

Table 4.30. ^{13}C Chemical Shifts of Selected Benzaldehyde Derivatives in Deuteriochloroform as Solvent (δ_C in ppm) [73k].

Compound	−CH=O	C-1	C-2	C-3	C-4	C-5	C-6	Others
Benzaldehyde	192.0	136.7	129.7	129.0	134.3	129.0	129.7	
2-Hydroxy-	196.7	120.9	161.6	117.6	136.9	119.9	133.8	
2-Nitro-	188.6	131.3	149.5	124.5	143.3*	134.1*	129.7	
2-Amino-	195.2	119.4	151.1	116.7	136.5	116.8	136.0	
3-Hydroxy-	194.8	139.4	116.5	159.4	123.2	131.4	123.2	
4-Hydroxy-	192.0	130.1	133.5	117.4	164.9	117.4	133.5	
4-Methoxy-	190.4	130.3	131.9	114.5	164.7	114.5	131.9	55.4 (OCH$_3$)
3-Nitro-	190.4	140.3	130.6	124.4	151.4	124.4	130.6	
4-Dimethyl-amino-	191.0	126.3	133.0	112.5	155.6	112.5	133.0	40.9 (N(CH$_3$)$_2$)
4-Cyano-	190.8	138.9	129.9	133.0	117.7	133.0	129.9	110.9 (CN)
4-Bromo-	190.9	135.2	130.9	132.4	129.7	132.4	130.9	
4-Trimethylsilyl-	192.3	136.6	128.7	133.8	149.0	133.8	128.7	1.4 (Si(CH$_3$)$_3$)
2-Hydroxy-3-methoxy-	196.6	121.0	148.4	151.4	118.3	119.6	124.6	56.3 (OCH$_3$)
4-Hydroxy-3-methoxy-	192.4	130.5	110.1	148.4	153.2	115.5	128.3	56.5 (OCH$_3$)
3,4-Dimethoxy-	190.7	130.3	110.7	149.8	154.6	109.4	126.5	56.0 (OCH$_3$) 56.1 (OCH$_3$)
3,4-Methylenedioxy-	190.0	132.1	108.4	148.8	153.2	107.0	128.4	102.2 (O−CH$_2$−O)

* assignments interchangeable.

Table 4.31. ^{13}C Chemical Shifts of Arylalkylketones (Phenones) (δ_C in ppm) [294].

Compound	C=O	C-1	C-2	C-3	C-4	C-5	C-6	C-α	C-β	Others
Acetophenone	195.7	136.3	128.1	128.1	131.3			24.6		
o-Methyl-	199.2	137.8	138.3	129.5	131.4	125.8	132.1	28.9		21.5
o-Ethyl-	199.4	135.5	142.7	128.5	130.0	123.8	130.0	28.2		26.0 14.9
o-iso-Propyl-	200.5	137.4	145.9	124.4	129.5	124.4	126.2	28.1		28.1 22.5
2,6-Dimethyl-	205.4	141.7	131.5	126.9	126.9			30.5		17.7
2,4,6-Trimethyl-	205.9	139.4	131.7	128.1	137.1			30.8		18.5
2,6-Diethyl-	205.5	141.4	137.7	125.3	128.1			31.0		25.4 15.1
2,6-Di-iso-propyl-	206.0	141.2	143.0	123.2	129.2			31.7		34.0 25.7
2,4,6-Tri-iso-propyl-	206.3	138.8	142.7	120.1	148.6			31.7		32.9 23.6
m-Methyl-	195.2	136.6	127.0	136.6	132.2	127.0	124.2	24.8		20.2
p-Methyl-	195.7	133.4	128.2	128.2	142.7			25.6		20.9

Table 4.31 (continued).

Compound	C=O	C-1	C-2	C-3	C-4	C-5	C-6	C-α	C-β	Others
o-Chloro-	198.4	136.8	131.3	126.2	129.9	126.2	126.2	29.2		
m-Chloro-	195.1	137.7	128.8	133.8	132.2	128.8	126.5	25.5		
p-Chloro-	195.5	134.3	128.8	128.8	137.9			24.5		
o-Bromo-	199.0	139.9	122.0	131.5	131.5	127.4	127.4	29.4		
m-Bromo-	195.1	137.5	129.6	121.3	135.2	129.6	126.5	25.8		
p-Bromo-	196.3	134.4	130.5	130.5	126.3			26.2		
o-Amino-	200.0	117.2	149.8	114.7	133.3	116.8	131.4	26.8		
m-Amino-	198.3	137.8	113.0	149.0	118.7	129.1	116.1	26.6		
p-Amino-	196.9	126.8	132.2	114.3	155.1			27.2		
o-Hydroxy-	204.1	118.9	161.5	118.3	135.5	118.3	129.7	25.4		
m-Hydroxy-	197.4	138.6	113.9	157.4	129.9	119.6	129.9	25.0		
p-Hydroxy-	197.1	129.6	130.9	115.2	161.8			24.8		
o-Methoxy-	197.5	127.4	158.2	111.1	132.8	119.4	129.3	30.7		54.3
m-Methoxy-	195.8	137.4	111.4	158.5	118.5	128.4	118.5	25.2		54.0
p-Methoxy-	195.0	129.5	129.8	113.0	162.8			25.2		54.6
o-Nitro-	198.0	135.4	144.7	122.5	129.8	133.0	126.3	28.1		
p-Nitro-	195.8	140.8	129.0	123.8	150.1			26.6		
Propiophenone	198.1	136.0	126.9	126.9	130.8			30.3	6.7	
Isobutyrophenone	202.8	135.8	127.9	127.9	132.4			34.6	18.1	
Pivalophenone	206.9	137.8	127.3	127.3	130.1			43.5	27.9	
Benzophenone	196.4	137.6	130.0	128.3	132.3					

as outlined in Section 3.1.3.8. Therefore, o alkylated acetophenones display larger carbonyl shifts ($\geq$ 200 ppm) than acetophenone with undistorted coplanarity (195 ppm), as shown for several examples in Table 4.31. Intramolecular hydrogen bonding also deshields the carbonyl carbon, as illustrated for the o-amino- and o-hydroxy- derivatives of benzaldehyde and acetophenone (Tables 4.30 and 4.31).

4.7.3 Quinones and Annulenones

Carbonyl resonances of quinones occur between 170 and 190 ppm [295–299] (Table 4.32). Carbonyl carbons of o-quinones are shielded relative to those of the p-isomers. Substitution of ring hydrogens by chlorine or other electron donors such as N,N-dimethylamino and methoxy groups deshield the α while shielding the β carbon due to (−)-I electron withdrawal in the α and (+)-M electron donation in the β position. This

4.7 Carbonyl Compounds

parallels the usual behavior of an electron donor attached to a CC double bond, as outlined for alkenyl halides and enol ethers in Sections 4.4.3 and 4.6.2.

4,5-Dimethylamino-
o-benzoquinone

2,5-Dimethylamino-
p-benzoquinone

2-Methoxy-8-methyl-
1,4-naphthoquinone

Table 4.32. ^{13}C Chemical Shifts of Representative o- and p-Benzoquinones [295–297], Naphthoquinones [298] and Anthraquinones [298, 299] (δ_C in ppm).

o-Benzoquinone   p-Benzoquinone   1,4-Naphthoquinone   Anthraquinone

Quinone	C-1	C-2	C-3	C-4	C-5	C-6	C-7	C-8	C-9	C-10
p-Benzoquinone	187.0	136.4								
2-Methyl-	187.5	145.9	133.3	187.7	136.6	136.5	15.8			
2,6-Dimethyl-	187.6	145.8	133.8	188.3			15.8			
Tetramethyl-	187.4	140.4					12.4			
2,5-Di-t-butyl-	188.2	154.2	133.5				34.5	29.0		
2,6-Di-t-butyl	187.7	157.7	130.1	188.6			35.5	29.3		
2,5-Diphenyl-	187.0	147.7	132.6				133.2	129.5	130.1	128.6
2,6-Diphenyl-	186.0	146.4	132.5	187.5			133.1	129.9	129.4	128.3
Tetraphenyl-	186.9	143.3					132.9	130.9	128.4	127.7
2-Chloro-	179.2	144.1	133.7	184.9	136.8	136.0				
2,5-Dichloro-	176.7	143.8	132.8							
2,6-Dichloro-	172.7	143.6	133.9	182.5						
Tetrachloro-	169.4	139.4								
2,5-Dimethyl- amino	181.3	152.0	101.8				42.7			
o-Benzoquinone	180.4		130.8	139.4						
4-Methyl-	179.2	180.1	127.7	152.0	143.4	129.5	22.8			
3,6-Di-t-butyl-	180.9		149.5	134.0			34.8	29.0		
4,6-Di-t-butyl-	179.5	180.4	121.4	162.7	133.0	149.2	35.5	28.7	34.9	27.4
Tetrachloro-	168.8		143.8	131.9						
4,5-Dimethyl- amino	178.4		103.8	157.1			41.7			
1,4-Naphthoquinone	184.7	138.5			126.2	133.8			131.8	
Anthraquinone	127.2	134.1			133.5(C-a)				183.1	

224   4 ^{13}C NMR Spectroscopy of Organic Compounds

The relatively narrow shift range of sp^2 carbons in unsubstituted o and p-benzoquinones, including naphthoquinone (Table 4.32) is attributed to a more balanced charge distribution in quinones compared with other α, β-unsaturated carbonyl compounds in accordance with MO theory. This also holds true for [n]-annulenones (n = 3, 5, 7,...), some of which are collected in Table 4.33 [300–304]. The considerable shielding observed for the carbonyl carbon in [3]annulenones arises from the geometry of the cyclopropenone system. ^{13}C shifts of [5]annulenones are predominantly influenced by the (−)-I and (−)-M electron withdrawing effects of the carbonyl group towards the double bonds, as is reflected in large carbonyl shifts and strong polarization of the sp^2 carbons.

Cylopentadienones

Carbon-13 shifts of fluorenone reflect a weaker polarization, indicating that this molecule is a benzene derivative rather than a cyclopentadienone. Similarly, olefinic carbon

Table 4.33. ^{13}C Chemical Shifts of Selected [n] Annulenones [300–302], Tropones [303] and H$_2$C$_n$O$_n$ Acid Derivatives [304] (δ_C in ppm).

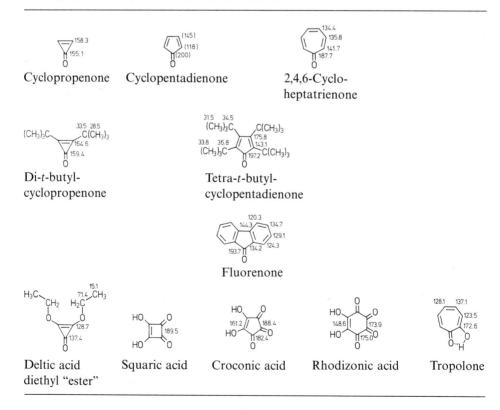

atoms of tropone ([7]annulenone) display balanced shifts (134–142 ppm). Thus, the contribution of the tropylium oxide state becomes more significant.

2,4,6-Cycloheptatrienone
(tropone)

Carbon-13 NMR spectra of the $H_2C_nO_n$ series, such as deltic, squaric, croconic and rhodizonic acids, obtained in anhydrous solvents [304] display carbonyl shifts similar to those reported for quinones (Table 4.33). Considerable shielding of the carbonyl carbon of deltic acid diethyl ester is not only attributed to the three-membered ring but also to an electron releasing effect of the ethoxy groups.

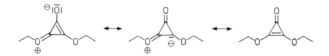

Table 4.34. ^{13}C Chemical Shifts of Representative Carboxylic Acids (δ_C in ppm) [305–309]

Acid	C-1	C-2	C-3	C-4	C-5	C-6
Methanoic (formic)	166.5					
Ethanoic (acetic)	175.7	20.3				
Propanoic (propionic)	179.8	27.6	9.0			
Butanoic (butyric)	179.3	36.2	18.7	13.7		
Pentanoic (valeric)	180.6	34.8	27.7	22.7	14.2	
Hexanoic (caproic)	180.6	34.2	31.4	24.5	22.4	13.8
2-Methylpropanoic (isobutyric)	184.1	34.1	18.8			
2,2-Dimethylpropanoic (tiglic)	185.9	38.7	27.1			
Chloroethanoic	173.7	40.7				
Dichloroethanoic	170.4	63.7				
Trichloroethanoic	167.0	88.9				
Trifluoroethanoic	163.0	115.0				
Hydroxyethanoic (glycolic)	177.2	60.4				
2-Hydroxypropanoic (lactic)	176.8	66.0	19.9			
Methoxyethanoic	173.9	69.3	59.3 (OCH_3)			
Propenoic (acrylic)	168.9	129.2	130.8			
2-Methylpropenoic	170.8	136.3	126.2	17.5		
(E)-2-Butenoic (crotonic)	169.3	122.8	146.0	17.3		
(Z)-2-Butenoic	169.8	121.0	146.2	15.0		
(E)-2-Pentenoic	171.9	119.5	152.5	24.6	11.1	
(Z)-2-Pentenoic	171.5	118.5	154.0	22.3	12.7	

Table 4.34 (continued).

Acid	C-1	C-2	C-3	C-4	C-5	C-6
(E)-Chloropropenoic	166.4	125.2	137.0			
(Z)-Chloropropenoic	166.2	121.4	133.2			
(E)-Bromopropenoic	166.5	129.1	128.3			
(Z)-Bromopropenoic	166.3	124.7	122.0			
(E)-Iodopropenoic	166.5	137.4	100.3			
(Z)-Iodopropenoic	167.5	130.6	97.4			
Ethanedioic (oxalic)	160.1					
Propanedioic (malonic)	169.2	40.9				
Butanedioic (succinic)	173.9	28.9				
Pentanedioic (glutaric)	176.6	33.2	20.6			
Hexanedioic (adipic)	174.9	33.5	24.9			
Heptanedioic (pimelic)	175.8	33.8	25.0	29.2		
2-Hydroxybutanedioic	175.2	68.0	39.4	172.6		
2,3-Dihydroxybutanedioic	172.7	73.8 (tartaric)				
	175.3	72.8 (meso-tartaric)				
(E)-Butenedioic (fumaric)	166.6	134.2				
(Z)-Butenedioic (maleic)	166.1	130.4				
(Z)-2-Methylbutenedioic	175.5	147.0	121.7	166.3	20.9	
(Z)-2,3-Dichlorobutenedioic	162.3	130.9				
Propynoic	156.5	74.0	78.6			
Cyclopropanecarboxylic	180.7	12.0	9.2			
Cyclobutanecarboxylic	179.4	38.4	25.6	18.8		
Cyclopentanecarboxylic	183.8	43.3	29.5	25.3		
Cyclohexanecarboxylic	182.1	43.1	29.0	25.6	26.0	

4.7.4 Carboxylic Acids and Derivatives

Alkyl carbon atoms attached in a position α, β, or γ to a carboxy function give rise to α, β, and γ effects which shift the carboxy carbon resonances by 10, 4 and -1 ppm, respectively. This trend is illustrated by the ^{13}C shift values collected for formic, acetic, propionic, and butyric acids among others in Table 4.34 [305–309]. Further, carboxy carbons of α halo acids and dicarboxylic acids with closely spaced carboxy groups are shielded relative to those of parent alkanoic acids (Table 4.34). On the other hand, the α, β and γ carboxy increments $Z_i = \delta_{i(RCOOH)} - \delta_{i(RH)}$ for the carbon shifts of an alkyl chain are

$$Z_\alpha = 22 \pm 1, \ Z_\beta = 3 \pm 1, \ \text{and} \ Z_\gamma = -2 \pm 1,$$

4.7 Carbonyl Compounds

respectively, following the overall pattern outlined for the other carbonyl groups. Similiar effects are observed for carboxylic acid derivatives such as esters [210, 310–312] and amides [313–315] (Table 4.35).

Equilibration of carboxylic acid dimers and monomers in the sample solution depends on the extent to which the carboxy group is involved in hydrogen bonding with the solvent:

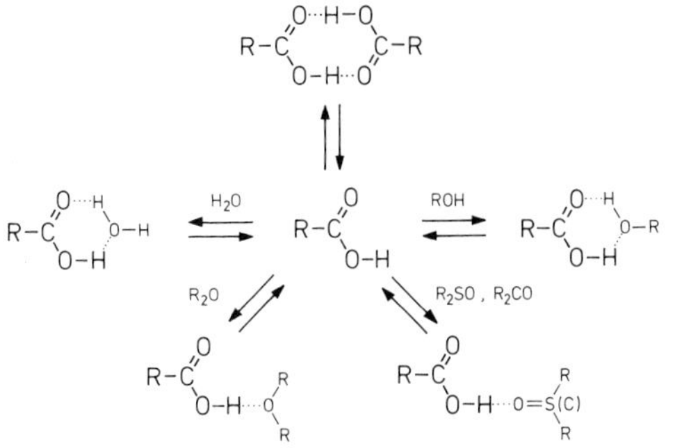

Therefore, solvent shifts of up to 4 ppm must be expected, as illustrated for acrylic acid in Table 4.36 [282].

Dissociation of the carboxy proton causes deshieldings decreasing from the carboxy (4.7 ppm) to the γ carbon (0.6 ppm) [305, 307] which are attributed to electric fields (Section 3.1.4.3). Intramolecular electric fields are also made responsible for the different sp^2 carbon shifts in long-chain carboxylic acids [309].

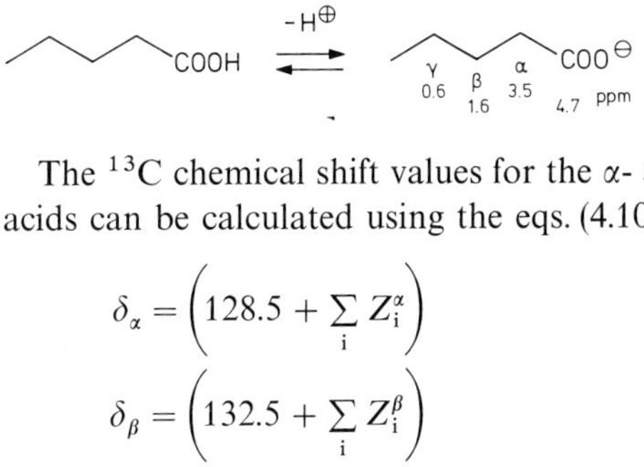

The ^{13}C chemical shift values for the α- and β-carbons in α, β-unsaturated carboxylic acids can be calculated using the eqs. (4.10) [306]:

$$\delta_\alpha = \left(128.5 + \sum_i Z_i^\alpha\right) \quad (4.10\,\text{a})$$

$$\delta_\beta = \left(132.5 + \sum_i Z_i^\beta\right) \quad (4.10\,\text{b})$$

$\sum_i Z_i$ = sum of increments for the substituents R_{gem}, R_{cis} and R_{trans};

Z_i^α, Z_i^β = increments (see Table 4.37);

$R_{gem}, R_{cis}, R_{trans}$ = substituents at the olefinic carbons (configuration relative to the carboxy group):

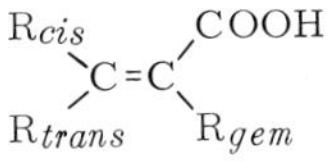

Deviations between calculated and experimental values are generally smaller than 4 ppm [306]. Differences may be attributed partly to varying molecular ground states and partly to varying excited states of α, β-unsaturated acids [306].

Table 4.35. ^{13}C Chemical Shifts of Carboxylic Acid Esters [210, 310–312], Tetramethylammonium Salts [305], Anhydrides [316], Chlorides [317], and Amides [313–315] (δ_C in ppm).

Compound	C-1	C-2	C-3	C-4	C-1'	C-2'	C-3'	C-4'
Esters								
Methyl methanoate	160.9				49.1			
Ethyl methanoate	160.4				58.8	13.0		
Propyl methanoate	160.3				64.3	21.3	9.1	
iso-Propyl methanoate	159.3				66.3	20.7		
Methyl ethanoate	171.3	20.6			51.5			
Ethyl ethanoate	170.6	21.0			60.3	14.2		
Propyl ethanoate	169.0	19.6			64.6	21.6	9.3	
Butyl ethanoate	169.1	19.4			63.1	30.4	18.6	12.7
iso-Propyl ethanoate	170.4	21.4			67.5	21.8		
t-Butyl ethanoate	170.4	22.5			80.1	28.1		
Methyl propanoate	173.3	27.2	9.2		51.0			
Ethyl propanoate	172.4	27.4	9.0		59.7	14.2		
Methyl butanoate	172.2	35.6	18.9	13.8	51.9			
Ethyl butanoate	171.6	36.2	18.9	13.8	59.7	14.5		
Methyl 2-methylpropanoate	173.5	31.0	16.2		49.8			
Methyl 2,2-dimethylpropanoate	178.8	38.7	27.3		51.5			
Ethyl 2,2-dimethylpropanoate	178.8	38.7	27.2		60.2	14.2		
iso-Propyl 2,2-dimethylpropanoate	177.9	38.6	27.2		67.1	21.7		
t-Butyl 2,2-dimethylpropanoate	177.8	39.2	27.2		79.4	27.9		
Ethanediol diethanoate	170.6	20.8			62.9			
Propanetriol triethanoate (glycerol triacetate)	169.7 / 169.2	20.8 / 20.5			62.4 / 69.2			
Ethenyl ethanoate (vinyl acetate)	167.9	20.5			141.5	97.5		
Methyl propenoate (methyl acrylate)	166.0	128.7	129.9		50.9			
Methyl 2-methylpropenoate	166.7	136.2	125.0	17.9	51.3			
(E)-Methyl 2-butenoate	166.0	122.3	144.1	17.1	50.3			
(Z)-Methyl 2-butenoate	166.2	120.4	144.5	15.6	50.6			
Methyl propynoate	153.4	74.8	75.6		53.0			
Dimethyl ethanedioate	158.4				53.1			
Diethyl ethanedioate	156.7				62.1	13.2		
Dimethyl propanedioate	167.6	41.2			52.3			
Diethyl propanedioate	165.7	41.5			61.0	14.1		
Dimethyl butanedioate	173.1	29.1			51.3			
Diethyl butanedioate	171.1	29.2			60.3	14.3		
Dimethyl pentanedioate	173.2	33.1	20.7		51.3			
Dimethyl (E)-butenedioate	165.7	129.9			52.1			
Dimethyl butynedioate	152.3	74.6			53.6			
Tetramethylammonium Salts								
methanoate	171.1							
ethanoate	181.4	23.7						
propanoate	184.8	31.0	10.5					
butanoate	184.0	39.9	19.6	13.6				
pentanoate	184.1	37.7	28.4	22.3	13.4 (C-5)			

Table 4.35 (continued).

Compound	C-1	C-2	C-3	C-4	C-1'	C-2'	C-3'	C-4'
Anhydrides								
Ethanoic (acetic) anhydride	167.4	21.8						
Propanoic (propionic) anhydride	170.9	27.4	8.5					
Butanoic (butyric) anhydride	169.6	37.2	18.2	13.3				
Butanedioic (succinic) anhydride	172.5	28.2						
Pentanedioic (glutaric) anhydride	168.5	30.1	16.7					
Butenedioic (maleic) anhydride	165.9	137.4						
2-Methylbutenedioic anhydride	167.4	150.2	130.2	165.3	11.3 (2-CH$_3$)			
Dimethylbutenedioic anhydride	167.0	141.0						
Dibromobutenedioic anhydride	159.8	131.4						
Chlorides								
Ethanoic acid chloride	170.5	33.6						
2-Methylpropanoic acid chloride	177.8	46.5	19.0					
2,2-Dimethylpropanoic acid chloride	180.0	49.4	27.1					
Ethanedioic acid dichloride	159.3							
Dimethylpropanedioic acid dichloride	171.9	69.4	23.2					
Butanedioic acid dichloride	172.3	41.4						
Hexanedioic acid dichloride	173.2	46.4	23.7					
Amides								
Methanoic acid amide	167.6							
N-methylamide, *syn*	165.5				25.4			
anti	168.7				29.4			
N,N-dimethylamide	162.6				31.5 (*syn*)			
					36.5 (*anti*)			
N,N-diethylamide	162.2				36.7	12.8 (*syn*)		
					41.9	14.9 (*anti*)		
N,N-dipropylamide	162.8				43.8	20.6	11.3 (*syn*)	
					49.2	21.9	10.9 (*anti*)	
N,N-di-*iso*-propylamide	161.7				43.4	20.0 (*syn*)		
					46.3	23.4 (*anti*)		
Ethanoic acid amide	178.1	22.3						
N,N-dimethylamide	170.4	21.5			35.0 (*syn*)			
					38.0 (*anti*)			
N,N-diethylamide	169.6	21.4			40.0	13.1 (*syn*)		
					42.9	14.2 (*anti*)		
Propanoic acid								
N,N-dipropylamide	173.5	26.4	9.7		47.6	21.0	11.4 (*syn*)	
					49.6	22.3	11.3 (*anti*)	
Butanoic acid								
N,N-diethylamide	172.2	35.1	18.9	14.0	40.1	13.2 (*syn*)		
					42.0	14.4 (*anti*)		

Table 4.36. Solvent-Induced Shifts of Acrylic Acid Carbons (δ_C in ppm) [282].

$$H_2C=CH-COOH$$
$$321$$

Solvent	C-1	C-2	C-3
Neat	171.2	128.3	132.9
H_2O	170.6	128.9	133.4
CH_3OH	168.9	129.2	130.8
$(CH_3)_2SO$	168.4	130.7	131.6
1,4-Dioxane	167.8	129.1	131.2
C_6H_6	172.1	128.3	132.9

Table 4.37. Substitution Increments Z_i for Calculation of the Chemical Shifts of α, β-Unsaturated Acids (see eqs. (4.10a) and (4.10b)) [306].

Substituent R	Z_i^β			Z_i^α		
	gem	cis	trans	gem	cis	trans
H	0	0	0	0	0	0
Cl	−8	−1	4	2	−6	4
Br	−8	−13	−1	−8	−5	1
I	−9	−39	−35	−38	2	11
CH_3	−6	12	10	7	−6	−4
n-C_2, C_3	−6	21	19	16	−9	−8
α-Branched C_3, C_4	−6	26	24	24	−10	−9
CH_2OH	−4			18		
CH_2Br	6			13		
CHO		22			−4	
COOH		−2	2		2	6

Due to reduction or lack of hydrogen bonding, carbonyl carbon nuclei of amides [313–315], anhydrides [316], esters [310–312], and halides [317] display smaller shift values relative to the parent acids (Tables 4.34 and 4.35). *Methyl esterification shieldings* are about −6±1 ppm for mono- and −1.5±1 ppm for diesters, reflecting weaker hydrogen bonding in dicarboxylic acids [316]: Taking the methoxy carbon as a β effect to the carboxy nucleus ($-\overset{\alpha}{O}-\overset{\beta}{CH_3}$), which is about 4 ppm as mentioned above, the estimated contribution of hydrogen bonding to the carboxy shift is about 10 ppm for mono- and 5 ppm for dicarboxylic acids. Alkyl carbon shifts of dicarboxylic acid dimethyl esters can be predicted by using a selective increment system [318]. Different *acetylation shifts* of primary, secondary, and tertiary alcohols, as outlined in Section 4.5.2, are useful for identification and signal assignments. *Chloroacetylation shifts* are even larger. They increase with the number of chloro atoms in the series mono-, di-, and trichloroacetylated alcohols [319].

4.7 Carbonyl Compounds 231

Separate signals for N-alkyl groups *syn* or (Z) and *anti* or (E) to the carbonyl oxygen are observed for N-N-dialkylamides (Table 4.35) [313–315] due to the *amide resonance*. But the rotational barrier of the CN partial double bond decreases with increasing size of the carboxylic acid residue R [315].

N,N-Dialkyl amides

A particularly large α effect of about 33 ppm characterizes the carboxylic acid chloride function [317], while the β and γ effects are similar to those of other carboxylic acid derivatives.

Carboxy carbons of methyl benzoates are shielded by electron-withdrawing substituents in the *o*-position of the benzene ring [320] (Table 4.38). Carbonyl shifts of phthalic acid diesters and phthalimide are larger than those of phthalic anhydride [321]. β effects of the O-alkyl group in the esters and hydrogen bonding of the imide are the obvious reasons.

Phthalic acid diethylester   Phthalic anhydride   Phthalimide

Table 4.38. ^{13}C Chemical Shifts of Substituted Methyl Benzoates (δ_C in ppm) [320].

Compound	CH$_3$O	C=O	C-1	C-2	C-3	C-4	C-5	C-6
Methyl benzoate	51.8	166.8	130.3	129.5	128.3	132.8		
Methyl 4-methylbenzoate	51.6	166.8	127.9	129.7	129.3	143.4		
Methyl 4-methoxybenzoate	51.6	166.7	122.8	131.7	113.7	163.5		
Methyl 4-dimethylaminobenzoate		166.5	116.1	131.0	110.5	153.1		
Methyl 4-nitrobenzoate	53.0	164.9	135.8	130.4	123.3	149.9		
Methyl 3-methylbenzoate	50.4	165.3	129.8	126.9	136.9	130.5	126.9	125.3
Methyl 2-methylbenzoate	51.3	166.9	129.7	140.2	131.8	132.0	125.8	130.7
Methyl 2,3-dimethylbenzoate	50.3	166.8	129.6	136.7	136.7	132.1	126.8	133.9
Methyl 2,6-dimethylbenzoate	51.7	170.3	133.7	134.8	127.4	129.2		
Methyl 2,4,6-trimethylbenzoate	51.3	170.4	131.3	135.3	128.4	129.2		
Methyl 2,6-diethylbenzoate	49.9	168.7	133.7	140.2	125.2	128.6		
Methyl 2-nitrobenzoate	52.0	164.1	125.9	146.8	122.6	131.4	131.4	128.7
Methyl 2-chlorobenzoate	51.5	164.6	129.5	136.1	126.1	130.7	126.1	126.1
Methyl 2-bromobenzoate	51.9	165.1	131.7	121.0	131.4	133.3	126.3	126.3
Methyl 2-iodobenzoate	52.4	165.6	134.5	94.1	141.3	132.1	127.5	130.5
Methyl 2,6-dichlorobenzoate	52.0	163.7	130.8	132.6	126.6	130.2		
Methyl 2,4,6-tribromobenzoate	53.3	165.3	136.6	120.3	133.4	123.7		
Methyl 2,4,6-trinitrobenzoate		161.5	129.7	144.9	124.2	144.9		

4 ^{13}C NMR Spectroscopy of Organic Compounds

The ^{13}C NMR spectrum of phthalic dichloride shows additional signals for a five-membered cyclic pseudo halide, characterized by an sp^3 carbon shift of 104 ppm. This shift value is also observed for laevulic acid chloride, which exists in the cyclic form only. Similarly, phthalaldehyde acid (o-formylbenzoic acid) turns out to be hydroxyphthalide, as indicated by a hemiacetal carbon shift of 98.4 ppm.

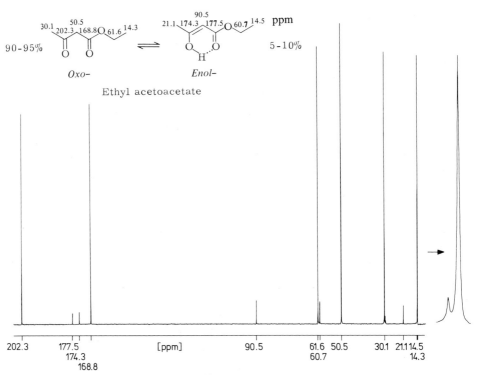

Keto-enol tautomerism of β-keto esters can be analyzed by carbon-13 NMR as illustrated by an inverse gated proton-decoupled ^{13}C NMR spectrum of ethyl acetoacetate in Fig. 4.6 [73i].

Fig. 4.6. Inverse gated proton-decoupled ^{13}C NMR spectrum of ethyl acetoacetate, 95% in hexadeuterioacetone (30.1 and 206.7 ppm) at 100.576 MHz and 25 °C [73i].

4.8 Aliphatic Organosulfur Compounds

Some ^{13}C shift increments (Table 4.39) [322] derived from the values collected in Table 4.40 [322–327] indicate that γ effects are of similar magnitude to those determined for other substituents. Negative β effects, however, are found for methyl sulfoxides and sulfones, and α effects of SH, S$^-$, and SCH$_3$ are significantly lower than those reported for oxygen analogs (Tables 4.22 and 4.39). This reflects not only the low electronegativity of bivalent sulfur but also the heavy atom effect. Methylation at sulfur causes the increments to increase by 9–10 ppm for each methyl group due to the β effect. Further, α increments increase with the oxidation state of sulfur, but not continuously (Table 4.39).

Polarization of double bonds due to the electron releasing effect of thioalkyl groups in thioenol ethers is much weaker than that reported for enol ethers (Table 4.26), as can be verified for the (E) and (Z) isomers of methyl propenyl sulfide in Table 4.40 [326, 327].

Thioacetal and thioketal carbon shieldings (45–65 ppm) [328, 329] also reflect the heavy atom influence when compared with acetal and ketal carbon shifts (100 ± 5 ppm).

```
      S            25.2              S           25.2              S
  21.3 ⌐S⌐                       30.8 ⌐S⌐                       30.4 ⌐S⌐  13.7   (ppm)
      41.9   30.6                     45.1  27.0                     54.7   23.2
```

2-Methyl-1,3-dithiane   2,2-Dimethyl-1,3-dithiane   Acetone
(cyclic thioacetals and thioketals)   diethylthioketal

Thiocarbonyl carbon nuclei resonate at much lower fields than comparable carbonyl groups (Table 4.41) [330–334]. This deshielding is attributed to the lower average electronic excitation energy ΔE of the thiocarbonyl group which increases the paramagnetic shift contribution according to eq. (3.4). The trend is also apparent when comparing the shift values reported for carbon dioxide and its sulfur and seleno analogs [335].

  125.1   153.8   156.6   193.1   200.8   209.9 ppm
  O=C≡O   O=C=S   O=C=Se   S=C=S   Se=C=S   Se=C=Se

A linear relation (eq. (4.11)) permits prediction of thiocarbonyl shifts from the C=O shift values of the parent carbonyl compounds [330].

$$\delta_C(C=S) = a + b\,\delta_C(C=O) \text{ (ppm)}; \quad a = 46.5-47.5; \quad b = 1.45-1.5. \quad (4.11)$$

Table 4.39. Substituent Increments ($Z_i = \delta(RX) - \delta(RH)$ in ppm) of Selected Functional Groups Containing Sulfur [322].

Substituent X	Z_α	Z_β	Z_γ	Z_δ
SH	11.1	11.8	−2.9	0.7
S$^-$	11.7	14.0	−2.3	0.4
SCH$_3$	21.1	6.4	−3.0	0.5
$^+$S(CH$_3$)$_2$	30.1	1.1	−3.0	0.7
SOCH$_3$	41.8	−0.3	−2.7	0.5
SO$_2$CH$_3$	41.0	−0.2	−2.9	0.3
SO$_2$Cl	54.5	3.4	−3.0	0.0

Table 4.40. ^{13}C Chemical Shifts of Selected Aliphatic Organosulfur Compounds (δ_C in ppm) [322–327].

Class/Compound	C-1	C-2	C-3	C-5	Others
Thiols					
Methanethiol	6.5				
Ethanethiol	19.8	17.3			
1-Propanethiol	26.4	27.6	12.6		
1-Butanethiol	23.7	35.7	21.0	12.0	
2-Propanethiol	29.9	27.4			
2-Butanethiol	24.7	36.7	33.5	11.3	
2-Methyl-2-propanethiol	41.1	35.1			
Cyclohexanethiol	38.5	38.5	26.8	25.9	
Thiolates					
1-Propanethiolate	27.5	30.2	13.6		
1-Butanethiolate	24.7	38.8	22.5	13.6	
Thioethers (Sulfides)					
Dimethylsulfide	19.3				
Diethylsulfide	25.5	14.8			
Dipropylsulfide	34.3	23.2	13.7		
Dibutylsulfide	34.1	31.4	22.0	13.7	
Di-*iso*-propylsulfide	33.4	23.6			
Di-*t*-butylsulfide	45.6	33.2			
Diallylsulfide	33.3	134.4	116.9		
(*E*)-Methylpropenylsulfide	125.5	121.7	18.4		
(*Z*)-Methylpropenylsulfide	128.5	122.7	14.4		
Sulfonium Ions (as Iodides)					
Trimethylsulfonium ion	27.5				
Dimethylethylsulfonium ion	38.2	8.3			24.3 (SCH$_3$)
Dimethylpropylsulfonium ion	45.4	17.7	12.7		24.8 (SCH$_3$)
Dimethylbutylsulfonium ion	43.2	25.7	21.4	13.7	25.3 (SCH$_3$)
Disulfides					
Dimethyldisulfide	22.0				
Diethyldisulfide	32.8	14.5			
Di-*t*-butyldisulfide	45.6	30.5			
Sulfoxides					
Dimethylsulfoxide	40.0				
Diethylsulfoxide	44.9	6.8			
Dipropylsulfoxide	54.4	16.3	13.4		
Diallylsulfoxide	54.3	125.8	123.5		
Sulfones					
Dimethylsulfone	42.6				
Diethylsulfone	40.4	6.6			
Dipropylsulfone	54.5	15.5	13.2		
Methylpropylsulfone	56.3	16.3	13.0		40.3 (SOCH$_3$)
Methylbutylsulfone	54.5	24.4	21.7	13.5	40.4 (SOCH$_3$)
Diallylsulfone	56.5	125.0	125.4		
Sulfinic Acids					
2,2-Dimethylpropanesulfinic acid	72.0	30.9	29.8		
Sulfonic Acids					
Ethanesulfonic acid	49.0	9.2			
Propanesulfonic acid	55.4	18.8	13.7		
Butanesulfonic acid	54.0	26.9	22.9	14.7	
2,2-Dimethylpropanesulfonic acid	63.4	30.9	29.4		
Sulfonic Acid Derivatives					
Ethanesulfochloride	60.2	9.1			
Sodium methanesulfonate	41.1				
Sodium 3-trimethylsilylpropane-sulfonate	57.2	21.1	17.4		0.0 (Si(CH$_3$)$_3$)

Table 4.41. ¹³C Chemical Shifts of Selected Thio- and Thiono Acids and Esters [73m, 331], Thioamides and Thioureas [332, 333], and Thioketones [334] in Comparison to the Corresponding Carbonyl Compounds (δ_C in ppm).

Carbonyl compound	Thiocarbonyl derivative		
H₃C–C(=O)OH 20.3, 175.7	H₃C–C(=O)SH 32.6, 194.5	Thioacetic acid	
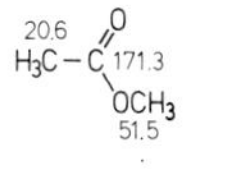 H₃C–C(=O)OCH₃ 20.6, 171.3, 51.5	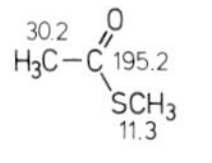 H₃C–C(=O)SCH₃ 30.2, 195.2, 11.3	Thioacetic acid S-methyl ester	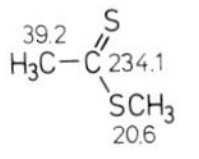 H₃C–C(=S)SCH₃ 39.2, 234.1, 20.6 — Dithioacetic acid methylester
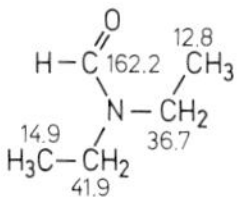 H–C(=O)N(CH₂CH₃)₂ 162.2, 12.8, 36.7, 41.9, 14.9	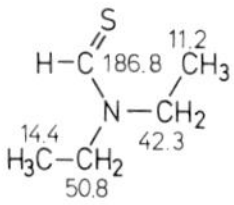 H–C(=S)N(CH₂CH₃)₂ 186.8, 11.2, 42.3, 50.8, 14.4	Thioformic acid N,N-diethylamide	
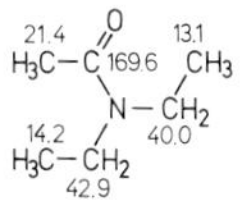 H₃C–C(=O)N(CH₂CH₃)₂ 21.4, 169.6, 13.1, 40.0, 42.9, 14.2	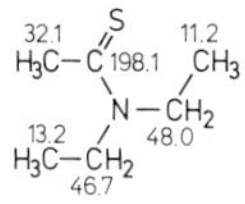 H₃C–C(=S)N(CH₂CH₃)₂ 32.1, 198.1, 11.2, 48.0, 46.7, 13.2	Thioacetic acid N,N-diethylamide	
(H₃C)₂N–C(=O)–N(CH₃)₂ 165.7, 38.6	(H₃C)₂N–C(=S)–N(CH₃)₂ 194.0, 43.2	Tetramethylthiourea	
H₃C–C(=O)CH₃ 30.1, 206.7	H₃C–C(=S)CH₃ 41.8, 252.7	Propanethione (Thioacetone)	
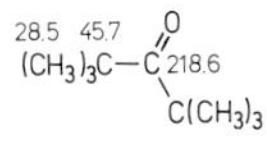 (CH₃)₃C–C(=O)–C(CH₃)₃ 28.5, 45.7, 218.6	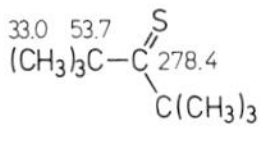 (CH₃)₃C–C(=S)–C(CH₃)₃ 33.0, 53.7, 278.4	2,2,4,4,-Tetramethylpentanethione (Di-*t*-butylthioketone)	
cyclopropyl–C(=O)–cyclopropyl 10.4, 20.7, 210.4	cyclopropyl–C(=S)–cyclopropyl 18.3, 33.3, 259.7	Dicyclopropylthioketone	
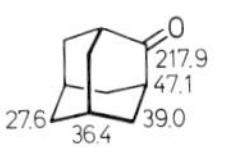 Adamantanone 217.9, 47.1, 39.0, 27.6, 36.4	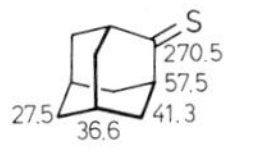 Adamantanethione 270.5, 57.5, 41.3, 27.5, 36.6	Adamantanethione	

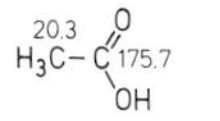

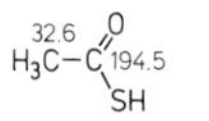

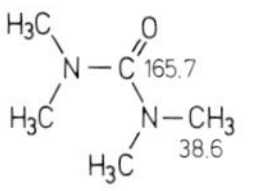

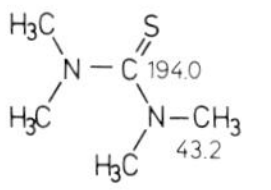

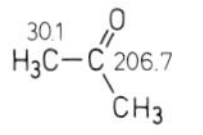

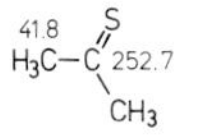

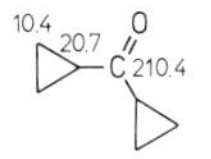

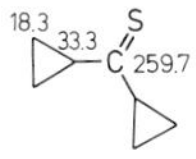

In contrast to N,N-dimethylaminoacrolein, restricted rotation of the thioaldehyde group is observed in the ¹³C NMR spectrum of N,N-dimethylaminothioacrolein [336]. The C-2 carbon shift of the thioaldehyde (118.5 ppm) also indicates that electron release by the dimethylamino group is attenuated when compared with the aldehyde (101.3 ppm).

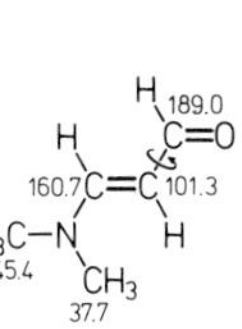

N,N-Dimethylaminoacrolein: H 189.0, C=O; 160.7 C=C 101.3; H₃C–N (45.4, 37.7) CH₃

(ppm)

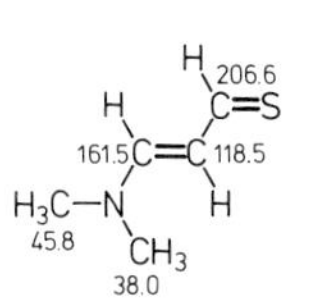

S-*trans* N,N-Dimethylaminothioacrolein: H 206.6, C=S; 161.5 C=C 118.5; H₃C–N (45.8, 38.0) CH₃

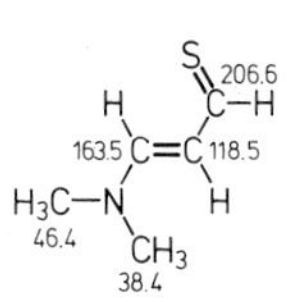

S-*cis*: H 206.6, C–H; 163.5 C=C 118.5; H₃C–N (46.4, 38.4) CH₃

4.9 Aliphatic Organonitrogen Compounds

4.9.1 Amines

Carbon-13 shift values of aliphatic amines [337–339] collected in Table 4.42 indicate that the γ effect is about -4.5 ppm for primary, secondary, and tertiary amines. The α effect, however, increases while the β effect decreases with an increasing degree of alkylation at the amino nitrogen. Analogously to alkanols, two increment systems can be used to predict alkyl carbon shifts of amines according to eq. (4.12 a, b), based on the reference shifts of corresponding alkanes (R–H) or the methyl homologs (R–CH$_3$) [337, 338] (Table 4.43).

$$\delta_C(k) = \delta_{C(k, R-H)} + \sum Z_i \qquad (4.12\,\text{a})$$

$$\delta_C(k) = b_k + a_k \delta_{C(k, R-CH_3)} \qquad (4.12\,\text{b})$$

Following the trend reported for other substituents (Section 4.5.2), carbon shift increments for equatorial amino groups at noninverting cyclohexane are more positive than those of axial ones, particularly in the γ position [340].

Cyclohexane as reference: $\delta_C = 27.0$ ppm

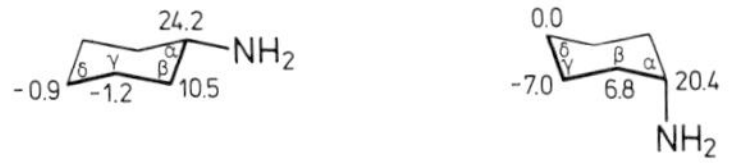

Similarly, an *exo*-amino group at C-2 in norbornane shields C-7 while *endo*-2-NH$_2$ particularly shields C-6 [229] due to γ-gauche type interaction with CH bonds.

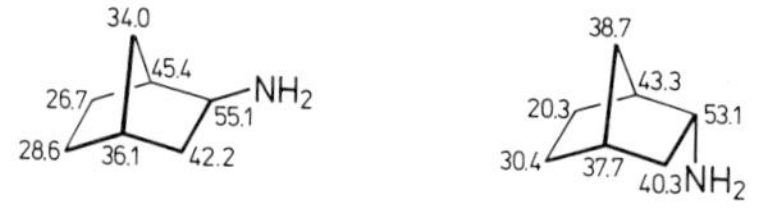

exo-   endo-   2-Aminonorbornane

Protonation of the amino group usually induces a shielding of all carbon atoms, explained in terms of electric field effects [84] and CH bond polarizations [338], and follows the pattern observed on protonation of carboxylate anions (Section 4.7.4).

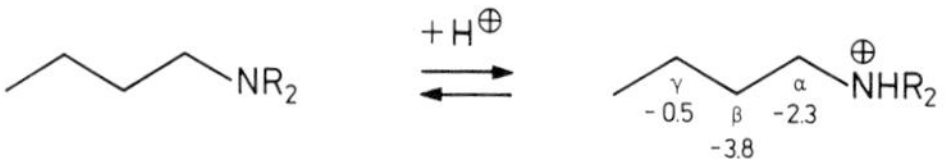

In α-branched amines, however, protonation induces considerable deshieldings α to the amino function, as demonstrated in the series *t*-butyl-, di-*iso*-propyl-, and *t*-butyldimethylamine, while shieldings are found for β carbons as usual [338].

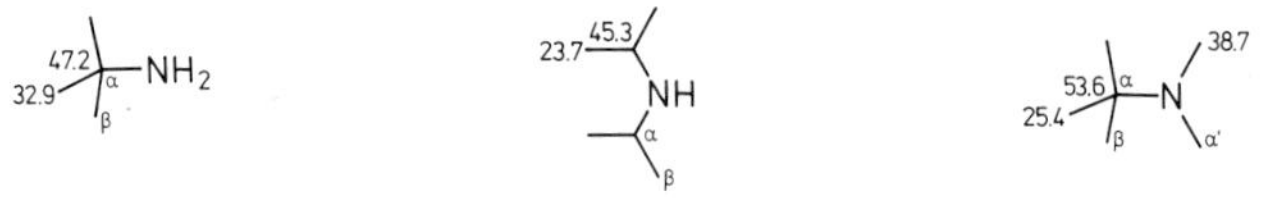

t-Butyl-   Di-*iso*-propyl-   *t*-Butyldimethylamine

$\delta_{NH^+} - \delta_N$:

Table 4.42. ^{13}C Chemical Shifts of Representative Amines [337–338] and Alkylammonium Ions [339] (δ_C in ppm).

Compound	C-1	C-2	C-3	C-4	C-5	C-6	C-7	C-8
-amine								
Methyl-	28.3							
Ethyl-	36.9	19.0						
Propyl-	44.6	27.4	11.5					
Butyl-	42.3	36.75	20.45	14.15				
Pentyl-	42.65	34.3	29.7	23.1	14.3			
Hexyl-	42.65	34.6	27.1	32.35	23.15	14.2		
Heptyl-	42.55	34.45	27.35	29.7	32.35	23.05	14.2	
Octyl-	42.55	34.45	27.4	30.0	29.85	32.35	23.1	14.2
iso-Propyl-	42.95	26.45						
sec-Butyl-	48.8	33.4	10.8	23.9 (2')				
t-Butyl-	47.2	32.85						
1-Methylbutyl-	46.85	42.8	19.85	14.3	24.0 (2')			
2-Methylbutyl-	48.35	38.5	27.2	11.55	17.2 (3')			
3-Methylbutyl-	40.6	43.7	25.9	22.9				
2,2-Dimethylpropyl-	54.45	32.1	27.0					
Dimethyl-	38.2							
Diethyl-	44.45	15.7						
Dipropyl-	52.35	23.95	12.0					
Dibutyl-	50.1	33.1	20.9	14.2				
Dipentyl-	50.45	30.6	30.1	23.1	14.25			
Dihexyl-	50.4	30.35	27.55	32.35	23.1	14.2		
Trimethyl-	47.55							
Triethyl-	51.4	12.85						
Tripropyl-	56.75	21.25	12.0					
Tributyl-	54.3	30.3	20.95	14.25				
Tripentyl-	54.5	27.65	30.05	23.0	14.25			
Trihexyl-	54.55	27.95	27.5	32.25	23.05	14.2		
Allyl-	44.8	139.9	113.6					
Allylmethyl-	54.5	136.9	115.8	35.9				
Allyldimethyl-	62.9	136.0	117.4	45.2				
-ammonium ion								
Tetramethyl- (iodide)	56.5							
Tetraethyl- (iodide)	54.4	9.5						
Tetrapropyl- (bromide)	60.4	16.0	10.9					
Tetrabutyl- (bromide)	59.1	24.3	19.8	13.8				
Allyl- (chloride)	42.2	130.2	120.9					
Diamino-								
-ethane	44.3							
-propane	39.6	36.1						
-butane	41.8	30.4						
-pentane	42.1	33.8	24.2					
-ethane, N,N'-dimethyl-	52.0	36.4 (NCH$_3$)						
-ethane, N,N,N',N'-tetramethyl-	58.7	46.0 (NCH$_3$)						
Hydroxyethylamine	44.6	64.2						
Tris-(hydroxyethyl)-amine	57.4	60.3						

Table 4.43. Increment Sets for Prediction of Alkylamine Carbon-13 Shifts According to the Equations (4.12a, b) [337, 338].

Amine	Position i	Z_i (ppm)	b_k (ppm)	a_k
Primary	α	28.5	23.09	0.846
	β	11.43	3.0	0.955
	γ	−4.6	0.07	0.941
	δ	0.6		
Secondary	α	36.5	22.88	0.9
	β	7.6	2.07	0.942
	γ	−4.5	−0.68	0.951
	δ	0		
Tertiary	α	41.0	22.62	0.914
	β	5.0	0.45	0.999
	γ	−4.5	−0.43	0.934
	δ	0		

Quaternization of tertiary amines causes β and γ effects which overlap with the shieldings induced by the positively charged nitrogen, as shown for some tetraalkylammonium salts in Table 4.42 [339]. Tetraalkylammonium ions display triplet splittings of 4 Hz due to $^{13}C-^{14}N$ coupling ($I_{^{14}N} = 1$). Examples are cholin and acetylcholin [341].

$$\underset{\underset{\text{Cholin (Chloride)}}{68.3\,(t)\quad 54.8\,(t)}}{\overset{56.6}{HO-CH_2-CH_2-\overset{\oplus}{N}(CH_3)_3}\,Cl^{\ominus}\text{ (ppm)}}$$

$$\underset{\underset{\text{Acetylcholin (Chloride)}}{\qquad\qquad\qquad\qquad 65.2\,(t)\quad 54.8\,(t)}}{\overset{21.3\quad 173.8\qquad 59.2}{H_3C-CO-O-CH_2-CH_2-\overset{\oplus}{N}(CH_3)_3}\,Cl^{\ominus}\text{ (ppm)}}$$

4.9.2 Enamines, Enaminoaldehydes, Cyanines

Carbon-13 shifts of enamines [342] follow the behavior described for other donor substituted alkenes (Sections 4.4.3 and 4.6.2). Electron release by the dialkylamino group has two consequences: The inductive electron withdrawal at the α alkene carbon is reduced ($Z_\alpha \sim 10-15$ ppm) compared with the α increments of aliphatic amines (Table 4.43). Further, electron density at the β olefinic carbon increases, as indicated by considerable shieldings in pyrrolidino- and morpholinoalkenes.

The configuration of the dialkylamino group is monitored by the shift values of the α alkyl carbon nuclei at the other side of the double bond, similar to the manner described for alkenes ($\delta_{(E)} > \delta_{(Z)}$) and shown for the (E)- and (Z)-pairs of 1-pyrrolidino- and 1-morpholinopropene. − ^{13}C shifts of 1-dimethylamino-1,3-dienes clearly show that elec-

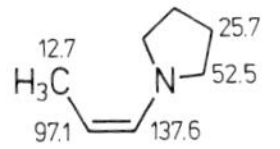

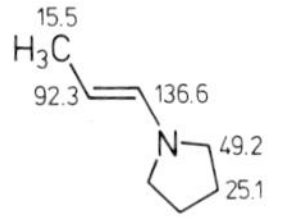

4.9 Aliphatic Organonitrogen Compounds

(Z)- (E)-1-Pyrrolidinopropene

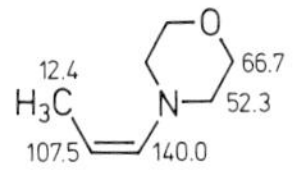

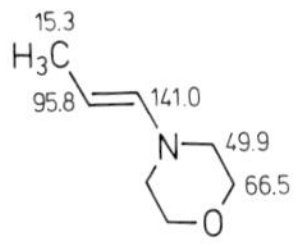

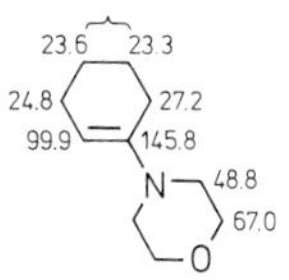

(Z)- (E)-1-Morpholinopropene 1-Morpholino-cyclohexene

trons released by the dimethylamino group are transmitted to the β and the δ carbon [126].

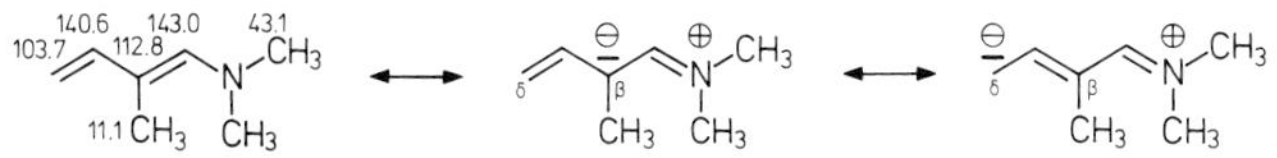

1-Dimethylamino-2-methyl-1,3-butadiene

Electron withdrawing substituents such as aldehyde or imonium functions at the other end of the alkene or the 1,3-diene transform enamines and 1,3-dienamines into push-pull systems characteristic of dyes. Electron release (push) of the donor group shields the β, δ, ε, ... carbons and electron withdrawal (pull) of the acceptor group (carbonyl oxygen or imonium nitrogen) deshields in the α, γ, ... position [343].

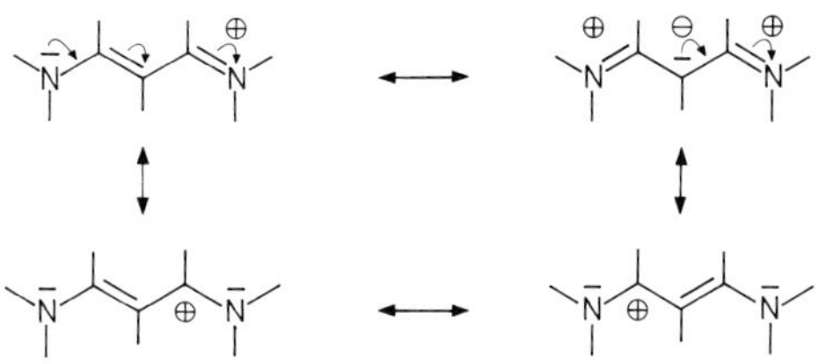

Thus, deshieldings and shieldings alternate along the polyene chain, as demonstrated by the carbon-13 shifts of N,N-dimethylaminoacrolein, N,N,N',N'-tetramethyltrimethinecyanine and their vinyl analogs [343].

3-N,N-Dimethylaminoacrolein

H₃C (37.7) \
   N (101.3) — CH=CH—CHO \
H₃C (43.4)   160.7 189.0

N,N,N′,N′-Tetramethyltrimethinecyanine

H₃C (38.5) \
   N (90.8) \
H₃C (46.3) 163.7 163.7 N—CH₃ / CH₃   ClO₄⁻

(ppm)

H₃C (41.3) 155 156 Hz \
   N 97.3 118.6 \
H₃C 157.7 154.6 191.1 =O \
     163   146   162 Hz

H₃C (38.6) 158 158 Hz CH₃ \
  N 103.7 103.7 N⁺ \
H₃C (46.1) 162.1 162.6 162.1 CH₃ \
     167   150   167 Hz ClO₄⁻

5-N,N-Dimethylamino-2,4-pentadienal

N,N,N′,N′-Tetramethylpentamethine-cyanine

Due to the partial π character of the CN bonds which are, in fact, vinylogous formamide or amidine connections in keeping with the resonance formulae, different methyl shift values are observed for some N,N-dimethylamine groups. Moreover, one-bond carbon-proton coupling constants decrease from the end to the center of the polymethine chain. This is explained by CNDO/2 calculations, in which corrected bond orders and bond angles alternating between 125° (α, γ, ε) and 117.5° (β, δ) are taken into account [343].

4.9.3 Imines

Carbon-13 shifts of imino functions $>$C=N$<$ present in Schiff bases (aldimines, ketimines) [344, 345], hydrazones [346], oximes [284, 347], amidines [348] and guanidines [349] occur between 150 and 180 ppm. Parallel to the trend known from alkenes, (Z) configuration of alkyl groups relative to *vicinal* substituents at the CN double bond results in a shielding particularly at the α carbon due to γ *gauche* type interactions. Using the $\delta_{(E)} > \delta_{(Z)}$ relation for the carbon α to the CN double bond, straightforward configurational assignments can be achieved for oximes, as shown for aldoximes and ketoximes in Table 4.44 [284, 347]. (E), (Z) pairs of hydrazones [346] follow the same trend as illustrated for the N,N-dimethylhydrazones of acetone and 2-pentanone. Obviously, the nonbonding electron pair of the N,N-dimethylamino group makes up the predominant contribution to shielding of the *cis*-α carbon; the N,N-dimethylamino carbon nuclei are almost independent of configuration.

H₃C (18.0) \
   164.6 C=N—N(CH₃)₂ (47.1) \
H₃C (25.1)

(E)- \
16.2 H₃C \
165.5 C=N—N(CH₃)₂ (46.5) \
40.5 H₂C \
20.1 CH₂ \
13.7 CH₃

(Z)- \
22.6 H₃C \
164.6 C=N—N(CH₃)₂ (47.0) \
33.1 H₂C \
19.7 CH₂ \
14.2 CH₃

(ppm)

Propanone N,N-dimethylhydrazone

2-Pentanone N,N-dimethylhydrazone

4.9 Aliphatic Organonitrogen Compounds

Table 4.44. ^{13}C Chemical Shifts of Selected Aldoximes and Ketoximes [284, 347] (δ_C in ppm).

Compound	C-1	C-2	C-3	C-4	C-5
(E)-Ethanaldoxime	148.2	15.0			
(Z)-Ethanaldoxime	147.8	11.2			
(E)-Propanaldoxime	153.1	23.1	10.9		
(Z)-Propanaldoxime	153.7	18.6	10.4		
(E)-Butanaldoxime	152.1	31.5	20.1	13.6	
(Z)-Butanaldoxime	152.6	27.0	19.5	13.9	
(E)-Pentanaldoxime	152.3	29.2	28.8	22.3	13.7
(Z)-Pentanaldoxime	152.8	24.9	28.4	22.6	13.7
(E)-2-Methylpropanaldoxime	156.9	29.4	20.0		
(Z)-2-Methylpropanaldoxime	157.8	24.5	19.7		
(E)-2-Methylbutanaldoxime	156.4	36.1	27.7	11.4	17.7 (2-CH_3)
(Z)-2-Methylbutanaldoxime	157.0	31.2	27.7	11.4	17.1 (2-CH_3)
Propanone oxime	15.0	155.4	21.7		
(E)-Butanone oxime	13.0	159.1	28.9	10.7	
(Z)-Butanone oxime	18.9	159.5	21.7	9.6	
3-Pentanone oxime	10.1	21.0	163.5	27.1	10.7
(E)-3-Methylbutanone oxime	10.8	162.2	34.3	19.5	
(Z)-3-Methylbutanone oxime	15.1	162.7	25.7	18.7	
(E)-2-Methyl-3-pentanone oxime	20.0	33.6	166.4	20.0	10.7
(Z)-2-Methyl-3-pentanone oxime	19.0	26.5	165.8	23.3	10.8

Due to an α effect of the additional alkyl group, the imino carbon nuclei of ketoximes are deshielded relative to those of comparable aldoximes (Table 4.44). Aldimines and ketimines [344, 345] (Schiff bases) behave similiarly (Table 4.45), but (E), (Z) configurational assignments are not as unequivocal as reported for oximes and hydrazones. – Conjugation of the CN double bond with a phenyl ring, for example, shields the imino carbon nucleus as expected. Conversion of an imine into the corresponding imonium salt, on the other hand, increases electron deficiency at the imino carbon and deshields as a result [350, 351] (Table 4.45). However, this cannot be generalized for all types of CN double bonds. Protonation of guanidines [349], for example, slightly shields the imino carbon, obviously arising from a better distribution of the positive charge by the other amino groups.

2-N-Phenyliminoimidazolidine

(ppm)

2-N-Phenyliminopyrrolidine

Acetamidinium chloride

242   4 ^{13}C NMR Spectroscopy of Organic Compounds

Table 4.45. ^{13}C Chemical Shifts of Representative Aldimines, Ketimes [344, 345], Guanidines [349] and Comparable Imonium Salts [350, 351] (δ_C in ppm).

[Structures with chemical shift values:]

N-Methylbenzaldimonium fluorosulfonate: 139.2, 130.6, 131.7, 125.7, 172.7 C=N⊕, SO₃F⊖, 39.5 CH₃

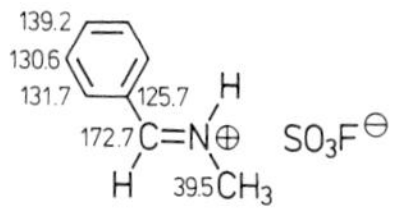

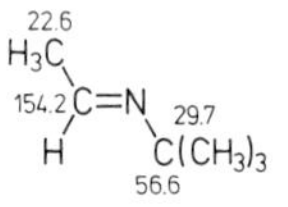

N-t-Butylacetaldimine: 22.6 H₃C, 154.2 C=N, 29.7 C(CH₃)₃, 56.6

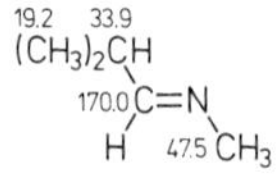

N-Methyl-isobutyraldimine: 19.2, 33.9 (CH₃)₂CH, 170.0 C=N, 47.5 CH₃

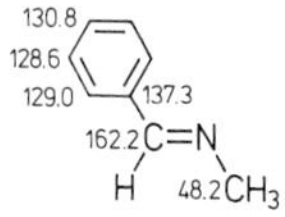

N-Methylbenzaldimine: 130.8, 128.6, 129.0, 137.3, 162.2 C=N, 48.2 CH₃

N-Methylacetonimine: 29.1 H₃C, 168.0 C=N, 18.0 H₃C, 38.6 CH₃

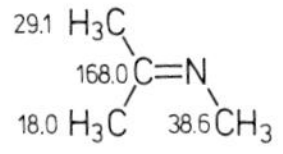

(E)-N-Methylbutanonimine: 10.7, 36.9 H₃C—CH₂, 172.5 C=N, H₃C 16.5, CH₃ 35.5

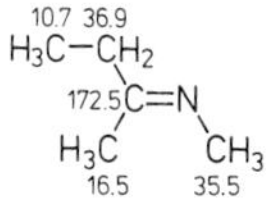

(Z)-N-Methylbutanonimine: 10.7 H₃C, 175.0 C=N, H₃C—CH₂ 7.8 35.4, CH₃ 38.2

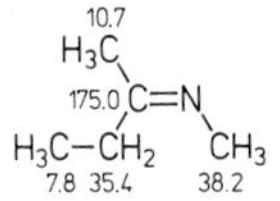

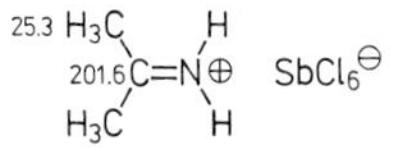

Acetonimonium hexachloroantimonate: 25.3 H₃C, 201.6 C=N⊕, H₃C, SbCl₆⊖

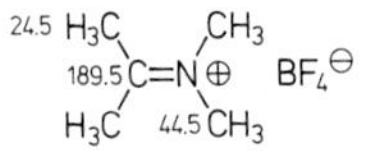

N,N-Dimethylacetonimonium tetrafluoroborate: 24.5 H₃C, CH₃, 189.5 C=N⊕, H₃C 44.5 CH₃, BF₄⊖

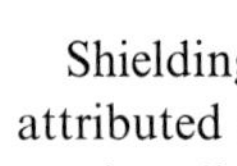

Tetramethylguanidine: 39.6 (CH₃)₂N, 167.8 C=N, (CH₃)₂N

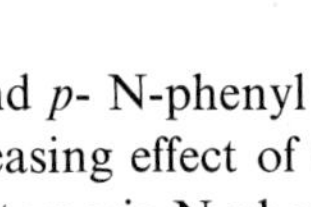

Tetramethylphenylguanidine: 119.4, 128.2, 114.4, 151.3, 39.4 (CH₃)₂N, 159.1 C=N, (CH₃)₂N

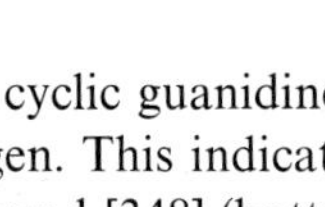

Tetramethylphenylguanidinium iodide: 124.9, 130.0, 116.3, 147.9, 41.4 (CH₃)₂N, 157.6 C=N⊕, (CH₃)₂N, I⊖

Shieldings observed for the o- and p- N-phenyl ring carbons in cyclic guanidines are attributed to a (+)-M electron releasing effect of the imino nitrogen. This indicates an N-phenylimino rather than the tautomeric N-phenylamino compound [348] (bottom of page 241).

4.9.4 Nitriles and Isonitriles

Carbon-13 shifts of cyano groups in nitriles are found between 110 und 125 ppm (Table 4.46) [77a, 352]. Shift values close to 125 ppm are characteristic of nitriles with α branched alkyl groups (Table 4.46). Similiar to the isoelectronic ethynyl group, the nitrile function shields the α carbon owing to the anisotropy effect. Alkenyl carbon shifts of

Table 4.46. ^{13}C Chemical Shifts of Representative Nitriles [77a, 353] and Isonitriles [353, 354] (δ_C in ppm). Where Available, One-Bond ^{13}C–^{14}N Couplings (in Hz) are given in Parentheses.

Class/Compound	C-1	C-2	C-3	C-4	C-5
Cyanides (Nitriles)					
Hydrogen cyanide	110.9				
Methyl cyanide	117.7	0.3			
Ethyl cyanide	120.8	10.8	10.6		
Propyl cyanide	119.8	19.3	19.0	13.3	
iso-Propyl cyanide	123.7	19.8	19.9		
t-Butyl cyanide	125.1	28.1	28.5		
Cyclohexyl cyanide (*a*-CN)	122.0	26.4	27.4	22.9	25.0
(*e*-CN)	122.7	27.7	29.2	24.5	24.4
Malodinitrile	110.5	8.6			
Succinodinitrile	118.0	14.6			
Adipodinitrile	119.3	24.3	16.4		
Acrylonitrile	117.3	107.9	137.3		
(*E*)-1-Cyanopropene	117.5	101.2	151.4	19.1	
(*Z*)-1-Cyanopropene	117.5	100.8	150.2	17.5	
(*E*)-1,2-Dicyanoethene	114.2	119.3			
Isocyanides (Isonitriles)					
Methyl isocyanide	158.2 (5.8)	26.8 (7.5)			
Ethyl isocyanide	156.8 (5.3)	36.4 (6.5)	15.3		
Propyl isocyanide	156.3	43.4	22.9	11.0	
iso-Propyl isocyanide	155.6 (4.8)	45.5 (5.6)	23.4		
t-Butyl isocyanide	154.5 (3.7)	54.0 (5.0)	30.7		
Cyclohexyl isocyanide (*a*–NC)	155.3 (4.5)	50.3 (5.8)	30.5	21.1	25.2
(*e*–NC)	153.8	51.9	33.7	24.4	25.2
Ethenyl isocyanide	165.7 (5.0)	119.4 (11.7)	120.6		
(*Z*)-1-Propenyl isocyanide	166.2 (5.7)	113.3 (11.5)	131.7	13.1	
2-Propenyl isocyanide	158.3	44.1	128.7	117.5	
Allenyl isocyanide	166.2 (4.7)	86.1 (14.3)	210.1	86.1	

acrylonitrile and other α, β-unsaturated nitriles reflect the considerable polarization of the CC double bond induced by $(-)$-*M* electron withdrawal of the cyano group (α: 100–110, β: 137–150 ppm). (*E*), (*Z*) isomeric alkenyl cyanides can be distinguished as usual by means of the $\delta_{(E)} > \delta_{(Z)}$ relation for the carbon nuclei α to the double bond (Table 4.46).

244 4 ^{13}C NMR Spectroscopy of Organic Compounds

Carbon nuclei of isonitriles are considerably deshielded (> 150 ppm) compared with those of nitriles (Table 4.46) [353, 354]. This arises from the lone pair which occupies a compact *sp* orbital at the terminal carbon and enhances the paramagnetic screening term according to eq. (3.4). Isonitrile groups display α effects similiar to those of amino groups and do not polarize attached double bonds as much as nitrile functions (Table 4.46). Another feature characteristic of isonitriles is a triplet splitting of both carbon nuclei attached to nitrogen due to one-bond ^{13}C–^{14}N coupling. J_{CN} couplings of α alkyl carbon atoms respond, as usual, to the increasing *s* character of the hybrid orbitals overlapping to the CN bond (Table 4.46).

4.9.5 Azacumulenes

Carbon-13 shifts of azacumulenes, including ketenimines, diazoalkanes, carbodiimides, isocyanates and isothiocyanates, can be readily interpreted in terms of mesomeric effects described by cannonical formulae.

Central *sp* carbon nuclei of ketenimines [355, 356] resonate at 190–195 ppm, while the terminal sp^2 carbons appear at 30–60 ppm, depending on the number of alkyl groups attached to them (α and β effects). Shielding of the terminal carbon reflects a significant contribution of cannonical formulae with carbanionic carbons.

$$\diagup\!\!\!\!\diagdown C = C = N\diagdown \quad \longleftrightarrow \quad \ominus\diagup\!\!\!\!\diagdown C - C \diagup\!\!\!\!\diagdown^{N-}_\oplus \quad \longleftrightarrow \quad \ominus\diagup\!\!\!\!\diagdown C - C \equiv N - \overset{\oplus}{}$$

N-Phenylketenimine: 37.0, 189.2 (C=C=N), aromatic 127.1, 129.8, 123.3, 140.8

Methyl-N-phenylketenimine: 48.0, 193.4 (C=C=N), H$_3$C 8.4, aromatic 126.8, 129.7, 123.2, 142.9

Dimethyl-N-phenylketenimine: 58.2, 195.5 (C=C=N), H$_3$C 15.8, aromatic 126.5, 129.2, 123.2, 144.1 (ppm)

Diazoalkane carbon shifts behave similiarly to those of the isoelectronic ketenimines [357, 358]. The contribution of resonance formulae with carbanionic carbons predominates, as indicated by a considerable shielding of the carbon nucleus α to the diazonium residue.

$$\diagup\!\!\!\!\diagdown C = \overset{\oplus}{N} = \overset{\ominus}{N} \mid \quad \longleftrightarrow \quad \ominus\diagup\!\!\!\!\diagdown C - \overset{\oplus}{N} \equiv N \mid \quad \longleftrightarrow \quad \ominus\diagup\!\!\!\!\diagdown C - N \diagup\!\!\!\!\diagdown^{N\oplus}_{} \quad \longleftrightarrow \quad \oplus\diagup\!\!\!\!\diagdown C - N \diagup\!\!\!\!\diagdown^{N\ominus}_{}$$

Diazomethane: 23.1 (CN$_2$)

Di-*t*-butyl-diazomethane: (CH$_3$)$_3$C 30.7, 31.3; (CH$_3$)$_3$C 28.2; CN$_2$

Diazoacetic acid ethyl ester: H$_3$C–CH$_2$ 14.6, 61.0; O–C 167.1; CN$_2$ 46.2

Diazomalonic acid diethyl ester: H$_3$C–CH$_2$ 14.5, 61.7; O–C 161.1; CN$_2$ 65.4

Diphenyl-diazomethane: aromatic 125.4, 129.9, 124.9, 129.3; CN$_2$ 62.5

4.9 Aliphatic Organonitrogen Compounds   245

Carbodiimide *sp* carbon atom resonances occur at much higher fields (~ 140 ppm) [359] than those of ketenimines, probably due to electron release of the additional nitrogen which induces a cyanamide-like polarization of both nitrogen atoms.

$$\bar{N}=C=\underline{N} \longleftrightarrow -N\equiv C-\underline{\ddot{N}}^{\ominus}$$

This explanation is supported by the presence of an equilibirum between carbodiimide and cyanamide, as found for the ethyltrimethylsilyl derivative [360].

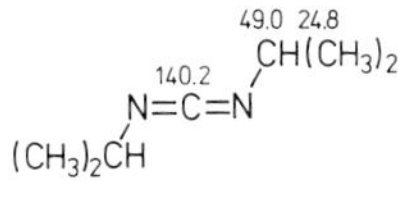

Di-*iso*-propylcarbodiimide 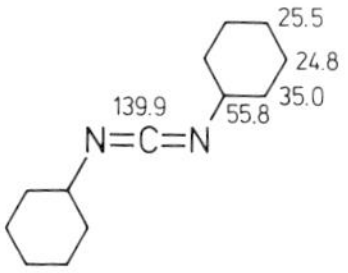   Dicyclohexylcarbodiimide

                                                                                                                (ppm)

55% $(CH_3)_3Si-N=C=N-CH_2-CH_3$ ⇌ $N\equiv C-N\begin{matrix}CH_2-CH_3\\Si(CH_3)_3\end{matrix}$ 45%

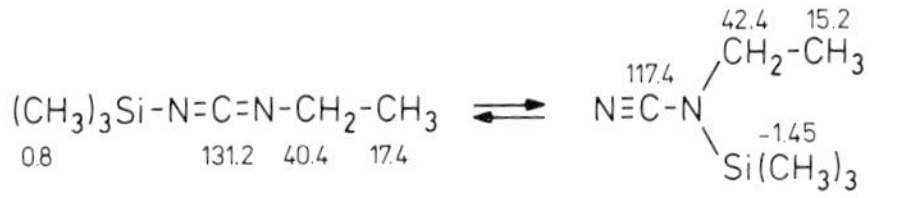

Ethyltrimethylsilylcarbodiimide

Isocyanate and isothiocyanate carbon nuclei resonate between 120 and 130 ppm [77a]. Isothiocyanate carbons are slightly deshielded relative to comparable isocyanates. Typical nitrile shift values (110–115 ppm) are characteristic of thiocyanates (rhodanides), while the carbon nuclei of rhodanide and cyanate anions shift to lower field due to significant contributions of heterocumulene-type resonance formulae.

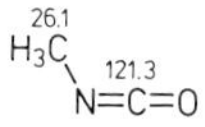

Methyl-isocyanate   Methyl-isothiocyanate   Allyl isothiocyanate

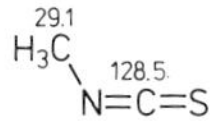

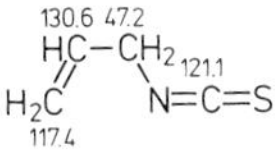

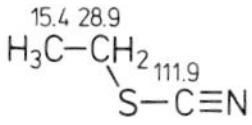

Methyl-thiocyanate

Cyanate anion:
   139.8 ppm $:O=C=\ddot{N}:^{\ominus} \leftrightarrow :\ddot{O}=C=N:^{\ominus} \leftrightarrow {}^{\ominus}:O-C\equiv N: \leftrightarrow {}^{\ominus}:O\equiv C-N:^{2\ominus}$

Thiocyanate anion:
   134.4 ppm $:S=C=\ddot{N}:^{\ominus} \leftrightarrow :\ddot{S}=C=N:^{\ominus} \leftrightarrow {}^{\ominus}:S-C\equiv N: \leftrightarrow {}^{\ominus}:S\equiv C-N:^{2\ominus}$

4.9.6 Nitroso and Nitro Compounds

The NN connection of N-nitrosamines is a partial double bond described by two resonance formulae.

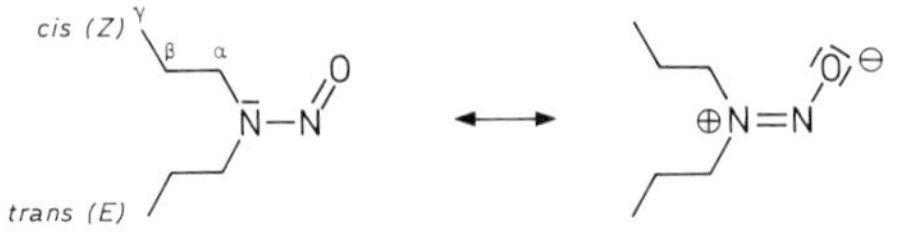

Therefore, N-alkyl substituents in (E)- and (Z)-configuration relative to the nitroso oxygen may be distinguished by their different carbon-13 shift values; in fact, for a given position $\delta_{(E)}$ is generally larger than $\delta_{(Z)}$ (Table 4.47) [361]. The N-methyl carbon shift of N-methyl-N-nitrosoaniline indicates that the molecule exists in the (E) configuration; separate signals are observed for the o- and m-carbon pairs of the phenyl ring due to hindered rotation [361]. While N-alkyl carbon shifts of N-nitrosamines occur in the same range as those of amino N-alkyl groups, alkyl nitrite shift values are more closely related to those of O-alkyl groups [362].

$$\begin{array}{cccc} 22.7 & 25.8 & 38.5 & 67.4 \\ CH_3-CH-CH_2-CH_2-ONO \\ | \\ CH_3 \end{array} \qquad \begin{array}{cccc} 11.3 & 26.7 & 35.6 & 73.7 \quad \text{ppm} \\ CH_3-CH_2-CH-CH_2-ONO \\ | \\ 16.6 \; CH_3 \end{array}$$

3-Methylbutylnitrite   2-Methylbutylnitrite

α-effects of nitro groups in nitroalkanes [363], obtained by comparing the C-1 shifts in Table 4.48 with those of the parent alkanes (Tables 4.1 and 4.4), are about 60 ppm. Of all substituents investigated so far, only fluorine induces a larger deshielding in the α position (Table 3.2). The β effect is relatively small (~ 3 ppm), and γ shieldings are in the range of those known for other organonitrogen functions (~ − 4.5 ppm). Carbon shifts of primary but not of secondary nitroalkanes ($R-CH_2-NO_2$) correlate with those of the parent methyl hydrocarbons ($R-CH_2-CH_3$), following linear equations (4.8b and 4.12b) given for alcohols and amines [363].

Table 4.47. ^{13}C Chemical Shifts of N-Alkyl Carbons in N-Nitrosamines (δ_C in ppm) [361].

Compound	α position		β position		γ position	
	cis (Z)	trans (E)	cis (Z)	trans (E)	cis (Z)	trans (E)
N-Nitrosodimethylamine	32.6	40.5				
N-Nitrosodiethylamine	38.4	47.0	11.5	14.5		
N-Nitrosodi-n-propylamine	45.2	54.2	20.3	22.5	11.3	11.8
N-Nitrosodiisopropylamine	45.4	51.1	19.1	23.7		
N-Nitrosodi-n-butylamine	41.8	51.9	28.9	31.1	20.3	20.9
N-Nitrosodiisobutylamine	50.4	59.9	26.5	27.5	20.1	20.5

Table 4.48. ^{13}C Chemical Shifts of Representative Nitroalkanes [363] and Nitroalkenes (δ_C in ppm).

Compound	C-1	C-2	C-3	C-4	C-5	C-6
Nitromethane	61.4					
Nitroethane	70.8	12.3				
Nitropropane	77.4	21.2	10.8			
Nitrobutane	75.6	29.6	19.8	13.3		
Nitrohexane	75.8	26.2	27.6	31.3	22.5	13.9
2-Nitropropane	20.1	79.1				
2-Nitrobutane	28.6	85.0	18.7	10.1		
2-Methyl-2-nitropropane	26.9	85.2				
2-Methyl-1-nitropropane	82.9	28.2	19.6			
3-Methyl-1-nitrobutane	74.4	36.2	25.8	22.0		
Nitrocyclohexane	84.6	31.4	24.7	25.5		
Nitroethene	145.6	139.0				
(E)-1-Nitropropene	139.0	140.9	13.9			
1-Nitrocyclohexene	149.9	134.5	24.1[a]	22.0[a]	20.9[a]	25.0[a]

[a] assignments interchangeable.

Electron withdrawal arising from the inductive effect in the α and from the resonance (mesomeric) effect in the β position of nitroalkenes (Table 4.48) induces deshieldings of similiar magnitudes and large two-bond carbon-proton coupling constants (up to 8 Hz).

Carbon-13 NMR spectra of all nitro compounds are characterized by quadrupolar broadening of the α carbon signal.

4.10 Organophosphorus Compounds

4.10.1 Survey of Carbon-13 Shifts

Carbon-13 shifts of representative phosphines [364], phosphonium salts [365], phosphonium ylides [365, 366], diphosphines [367], phosphonates [368], phosphorous and phosphoric acid derivatives [369] are summarized in Table 4.49. ^{13}C shift data of some group V organoelement compounds are compared in Table 4.50. It turns out that α sp^3 carbon nuclei of phosphines and arsines are shielded (0–25 ppm) relative to those of amines (30–60 ppm), as expected from the heavy atom effect. sp^2 carbons of CC double bonds behave correspondingly, as shown for the triphenyl derivatives in Table 4.50, with

Table 4.49. ^{13}C Chemical Shifts of Representative Phosphines [364], Phosphonium Salts [365], Phosphonium Ylides [365, 366], Diphosphines [367], Phosphonates [368], Phosphorous and Phosphoric Acid Derivatives [369] (δ_C in ppm).

Class/Compound		C-1	C-2	C-3	C-4
Phosphines R_3P, *Phosphine oxides* $R_3P=O$					
Methylphosphine		−4.4			
Dimethylphosphine		7.1			
Trimethylphosphine		14.3			
Triethylphosphine		19.5	10.3		
Tributylphosphine		29.3	28.6	25.4	14.7
Triphenylphosphine		137.8	134.0	128.8	128.5
Triphenylphosphine oxide		135.6	132.3	128.8	132.2
Phosphonium salts $R_4P^+\ X^-$					
Tetramethylphosphonium iodide		11.3			
Tetraethylphosphonium bromide		11.9	5.9		
Tetrabutylphosphonium iodide		19.5	23.6	24.1	13.5
Methyltriphenylphosphonium iodide	CH_3	10.3			
	C_6H_5	119.3	133.5	130.7	135.4
Acetonyltriphenylphosphonium bromide	CH_2COCH_3	40.1	201.4	32.4	
	C_6H_5	119.0	123.2	130.3	134.9
Ethoxycarbonylmethyl-triphenylphosphonium bromide	$CH_2COOC_2H_5$	32.9	164.5	62.8	13.7
	C_6H_5	118.0	134.1	130.7	135.6
Phosphonium ylides $R_3P=CHR'$					
Trimethylphosphonium methylide	CH_2	2.3			
	CH_3	18.9			
Triethylphosphonium methylide	CH_2	−14.2			
	C_2H_5	20.3	6.6		
Triphenylphosphonium methylide	CH_2	5.3			
	C_6H_5				
Triphenylphosphonium acetonylide	$CHCOCH_3$	51.3	190.5	28.4	
	C_6H_5	127.4	131.3	128.7	131.4
Triphenylphosphonium ethoxycarbonylmethylide	$CHCOOC_2H_5$	29.6	168.8	57.6	14.6
	C_6H_5	126.2	131.0	126.9	132.3
Diphosphines R_2P-PR_2					
Tetramethyldiphosphine		10.3			
Tetraethyldiphosphine		16.9	12.1		
1,2-Diethyl-1,2-dimethyldiphosphine	C_2H_5	20.1	11.7		
	CH_3	6.9			
Phosphorous acid derivatives PX_3					
Trimethylphosphite		48.9			
Triethylphosphite		58.0	17.4		
Phosphorous acid tri-(N,N-dimethyl-)amide		37.9			
Phosphonic acid esters $RPO(OR')_2$					
Diethyl methylphosphonate	CH_3	11.8			
	OC_2H_5	51.8	17.2		
Diethyl ethylphosphonate	C_2H_5	19.9	7.5		
	OC_2H_5	52.1	17.0		
Phosphoric acid derivatives $O=PX_3$					
Trimethylphosphate		54.2			
Triethylphosphate		63.4	15.9		
Phosphoric acid tri-(N,N-dimethyl-)amide		36.5			

Table 4.50. ^{13}C Chemical Shifts of Group V Organoelement Compounds (δ_C in ppm) [370, 371].

Class/Compound		C-1	C-2	C-3	C-4
(CH$_3$)$_3$X	Trimethylamine	47.5			
	Trimethylphosphine	17.3			
	Trimethylarsine	11.4			
(CH$_3$)$_3$X=CH$_2$	Trimethylphosphonium methylide	18.9	2.3 (P=CH$_2$)		
	Trimethylarsonium methylide	15.6	7.6 (As=CH$_2$)		
X(C$_6$H$_5$)$_2$	Triphenylamine	147.9	124.2	129.2	122.7
	Triphenylphosphine	137.8	134.0	128.8	128.5
	Triphenylarsine	140.5	134.3	129.3	129.0
	Triphenylstibine	139.3	136.8	129.4	129.1
	Triphenylbismuthine	131.1	138.1	131.0	128.3
	Pyridine		149.6	124.2	136.2
	Phosphabenzene		154.1	133.6	128.8
	Arsabenzene		167.7	133.1	128.2
	Stibabenzene		178.0	134.0	127.4

an exceptional shielding of C-1 in triphenylphosphine, arising from the propeller conformation of this molecule. If the sp^2 carbon belongs to a CX double bond (X = N, P, As, Sb), however, heavy atom substitution causes a deshielding, as shown for pyridine and its group V analogs in Table 4.50 [370], following the trend discussed for the carbonyl-thiocarbonyl series in Section 4.8. Both heavy atom effects – shielding of sp^3-C-X and deshielding of $sp^2-C=X$ carbon nuclei – can be verified for the ylides of phosphorus and arsenic [371].

As expected, carbon-13 shifts of phosphines (Table 4.49) are enhanced by increasing alkylation (α effect) and increasing number of β alkyl carbons (β effect). In contrast to amines, however, quaternization of phosphines causes shieldings in α and β posititions (Table 4.49). A particularly large α shielding is induced by quaternization of triphenylphosphine.

$\delta_{R_4P^+} - \delta_{R_3P}$: C−1: − 18.5 ppm
C−2: − 0.5 ppm
C−3:   1.9 ppm
C−4:   6.9 ppm

Conversion of phosphines via phosphonium salts into phosphonium ylides induces a deshielding of the α alkyl and aryl carbons (Table 4.49). Considerable shielding of the ylide $sp^2-C=P$ carbon nucleus reflects the presence of a large electron density, thus indicating a significant contribution of the ylide dipole to the actual state of the molecule.

250   4 ^{13}C NMR Spectroscopy of Organic Compounds

Ylene $\diagdown\!\!\!\!P=C\!\!\!\diagup$ ⟷ $\diagdown\!\!\!\!\overset{\oplus}{P}-\overset{\ominus}{C}\!\!\!\diagup$ Ylide

P-Alkyl carbon shifts of phosphonic acid esters (Table 4.49) are similiar to those of phosphines, while the O-alkyl carbons display shift values closely related to those of other alkoxy groups.

4.10.2 Carbon-Phosphorus Couplings

Phosphorus occurs as the pure isotope ^{13}P with spin quantum number $I = \frac{1}{2}$. Therefore the signals in proton broadband-decoupled ^{13}C NMR spectra of organophosphorus compounds split into doublets due to carbon-13–phosphorus-31 coupling, as shown for *o*-aminophenylphosphine in Fig. 4.7. Similarly to fluorine (Section 4.4.4), phosphorus is assumed to use *p* orbitals for bond formation in phosphines. This explains why negative one-bond carbon-phosphorus couplings are observed in the phosphine series (Table 4.51) in spite of a positive gyromagnetic ratio. The other organophosphorus compounds display positive J_{CP} couplings (Table 4.51). For one-bond and longer-range carbon-phosphorus couplings a characteristic dependence on the bonding states of both nuclei is observed. Table 4.52 gives a brief survey.

One-bond carbon-phosphorus couplings increase with increasing *s* character of carbon hybrid orbitals, as shown for some phosphine oxides [366, 372], and with growing number of electronegative substituents attached to phosphorus (or carbon) as demonstrated for some phosphines [376] (Table 4.52). In the 1-phenyl-1-phosphacycloalkane series, J_{CP} becomes more positive with decreasing ring strain, obviously due to changing hybridization of orbitals overlapping to bonds [377]. As illustrated for *cis-trans*-isomeric cyclohexylphosphonates, J_{CP} is also dependent on the configuration of the CP bond and, further, is enhanced by electronegative substituents such as OH attached to the coupling carbon [376] (Table 4.52).

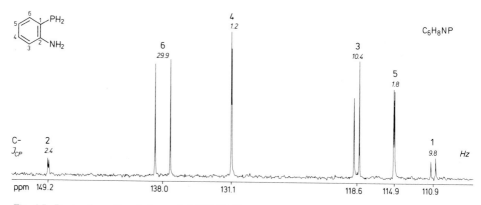

Fig. 4.7. Proton broadband-decoupled ^{13}C NMR spectrum of *o*-aminophenylphosphine (50 mg in 1 mL deuteriochloroform; 500 scans; 20.115 MHz).

Table 4.51. One-Bond (J and Longer-range (^2J, ^3J, ^4J) Carbon-Phosphorus Coupling Constants of Representative Phosphines [372], Phosphonium Salts [365], Phosphine oxides [366, 372, 373], Phosphonium Ylides [365, 373], Phosphonates [368, 372], Phosphites [374], and Phosphates [375] (in Hz).

Class/Compound		J_{CP}	$^2J_{CP}$	$^3J_{CP}$	$^4J_{CP}$
Phosphines R_3P					
Trimethylphosphine		−13.6			
Tributylphosphine		(−)10.9	11.7	12.5	0
Triphenylphosphine		−12.5	19.7	6.8	0.3
Tripropynylphosphine		8.8	10.9	−1.2	
Phosphonium salts $R_4P^+ \, X^-$					
Tributylphosphonium bromide		47.6	(−)4.3	15.4	0
Methyltriphenylphosphonium iodide	CH_3	57.1			
	C_6H_5	88.6	10.7	12.9	3.0
Phosphine oxides $R_3P=O$					
Trimethylphosphine oxide		68.3			
Triphenylphosphine oxide		104.4	9.8	12.1	2.8
Diphenylpropynylphosphine oxide	C_6H_5	121.6	11.3	13.4	2.9
	C_3H_3	174.4	31.4	3.2	
Phosphonium ylides $R_3P=CHR'$					
Trimethylphosphonium methylide	CH_3	56.0			
	CH_2	90.5			
Triphenylphosphonium methylide	C_6H_5	83.6	9.8	11.6	2.4
	CH_2	100.0			
Triphenylphosphonium 2-propylide	C_6H_5	81.2	8.5	11.6	2.4
	$C(CH_3)_2$	121.5	13.4		
Triphenylphosphonium methoxy-carbonylmethylide	C_6H_5	91.9	10.1	12.2	3.0
	$CHCOOCH_3$	126.7	12.7	2.6	
Phosphonates $RPO(OR')_2$					
Dimethyl methylphosphonate	CH_3	142.2			
	OCH_3		−6.0		
Diethyl butylphosphonate	C_4H_9	140.9	5.1	16.3	1.2
	OC_2H_5		−6.0	5.8	
Diethyl propynylphosphonate	C_3H_3	299.8	53.5	4.8	
	OC_2H_5		−6.3	5.9	
Phosphites $P(OR)_3$					
Trimethyl phosphite			10.1		
Triphenyl phosphite			3.0	4.9	0
Phosphates $O=P(OR)_3$					
Triethyl phosphate			−5.8	6.8	
Tributyl phosphate			−6.1	7.2	0

Table 4.52. Structural Features and Carbon-Phosphorus Couplings [370–380] (in Hz).

J_{CP} *and Carbon hybridization*

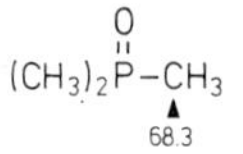

(CH$_3$)$_2$P(=O)–CH$_3$ 68.3
Trimethyl-

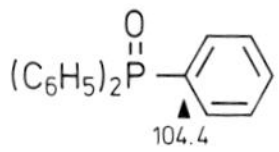

(C$_6$H$_5$)$_2$P(=O)–C$_6$H$_5$ 104.4
Triphenyl-

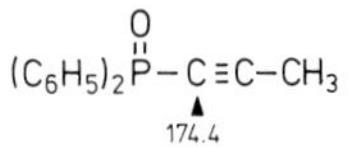

(C$_6$H$_5$)$_2$P(=O)–C≡C–CH$_3$ 174.4
Diphenylpropynylphosphine oxide

Electronegativity

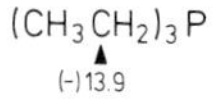

(CH$_3$CH$_2$)$_3$P   (−)13.9
Triethyl-

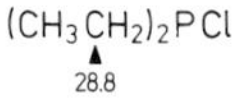

(CH$_3$CH$_2$)$_2$PCl   28.8
Chlorodiethyl-

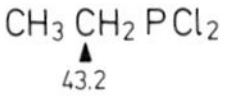

CH$_3$CH$_2$PCl$_2$ 43.2
Dichloroethylphosphine

Ring Strain

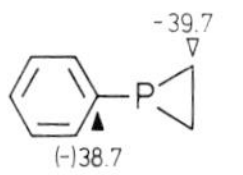

Ph–P(cyclopropane): −39.7, (−)138.7
1-Phenyl-1-phospha-cyclopropane

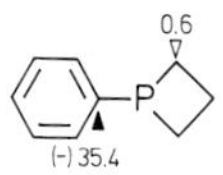

Ph–P(cyclobutane): 0.6, (−)35.4
-cyclobutane

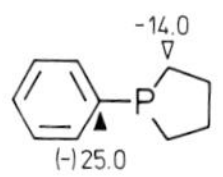

Ph–P(cyclopentane): −14.0, (−)25.0
-cyclopentane

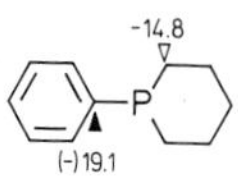

Ph–P(cyclohexane): −14.8, (−)19.1
-cyclohexane

Configuration

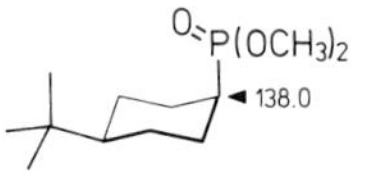

cis-: O=P(OCH$_3$)$_2$, 138.0

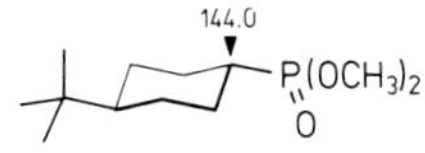

trans-: 144.0, P(OCH$_3$)$_2$=O

Dimethyl 4-*t*-butylcyclohexylphosphonate

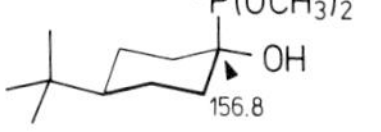

cis-: O=P(OCH$_3$)$_2$, OH, 156.8

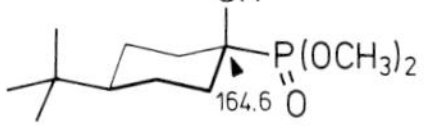

trans-: OH, P(OCH$_3$)$_2$=O, 164.6

Dimethyl 1-hydroxy-4-*t*-butylcyclohexylphosphonate

$^2J_{CP}$ *and Lone-pair orientation*

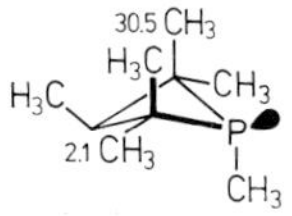

endo-: 30.5 CH$_3$, 2.1 CH$_3$

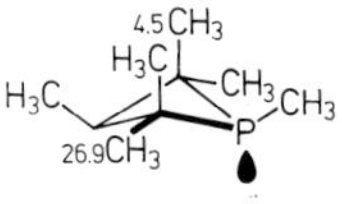

exo-: 4.5 CH$_3$, 26.9 CH$_3$

Hexamethylphosphetane

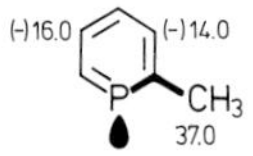

2-Methylphosphabenzene: (−)16.0, (−)14.0, 37.0

Table 4.52. (continued).

| $^3J_{CP}$ and Lone-pair orientation | Configuration |

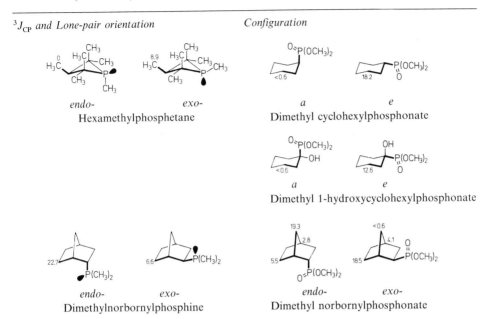

Two types of one-bond carbon-phosphorus couplings occur in phosphonium ylides: sp^3-C-P couplings are similiar to those observed for phosphonium salts, while $sp^2-C=P$ couplings are much larger, indicating a high degree of s character in the anionic ylide carbon [365, 371, 373] (Table 4.51). The outstandingly small J_{CP} measured for the ylide carbon of triphenylphosphonium cyclopropylide is attributed to the pyramidal geometry of the carbanionic carbon [373], in accordance with the results of X-ray diffraction.

Triphenylphosphonium 2-propylide

Triphenylphosphonium cyclopropylide

Two-bond carbon-phosphorus couplings in phosphines are essentially determined by the lone-pair geometry. Analogously to $^2J_{CN}$ (Section 3.2.5.2), $^2J_{CP}$ is considerably enhanced when the nonbonding electron pair at phosphorus is *cis* to the coupling carbon, as shown for the isomers of hexamethylphosphetane [378] and 2-substituted phosphabenzenes [370, 379].

Three-bond carbon-phosphorus couplings $^3J_{CP}$ depend on configuration, as expected, *transoid* couplings being *larger* than *cisoid* ones, as demonstrated for cyclohexyl- and 2-norbornylphosphonates in Table 4.52. The Karplus relation (4.13) can be derived from the data for 2-norbornyl derivatives [376].

$$^3J_{CP} = 8.6 \cos 2\Phi - 4.7 \cos \Phi + 7.7 \text{ (Hz)} \qquad 4.13$$

Parallel to $^3J_{HH}$ and $^3J_{CH}$, electronegative substituents within the coupling path reduce $^3J_{CP}$ as shown for cyclohexyl- in comparison to 1-hydroxycyclohexylphosphonate in Table 4.52 [376]. Relations similiar to eq. (4.13) can be applied to other organophosphorus compounds with tetra- and pentavalent but not with trivalent phosphorus. In the latter compounds, orientation of the nonbonding electron pair at phosphorus again plays the dominant role: A larger $^3J_{CP}$ coupling is observed when the coupling carbon comes closer to the lone pair and this may reverse the $^3J_{CP(exo)} > {^3J}_{CP(endo)}$ pattern, as shown for the *exo- endo-* pair of 2-norbornylphosphine [380] (Table 4.52).

2,4,6-Triphenyl-   2-Methyl-   Phosphabenzene

Four-bound carbon-phosphorus couplings can also be resolved in some cases (Table 4.51). Carbon-phosphorus couplings transmitted by more than four bonds are detectable in π electron systems and have been reported for biphenylmethylphosphonates [368] and triphenylphosphabenzene, in which a $^7J_{CP}$ of 1 Hz can be resolved [370, 379].

4.11 Aromatic Compounds

4.11.1 General

In contrast to proton shifts, carbon-13 shifts cannot be used as criteria for aromaticity (Section 3.1.3.10). No difference exists between aromatic (128.5 ppm for benzene) and comparable alkene carbon nuclei (127.5 ppm for cyclohexene). Aromatic ring carbon nuclei are practically not influenced by the ring current (Section 3.1.3.4), which makes up a deshielding of about 2 ppm and thus is small compared with other (*e.g.* steric) effects on carbon-13 shifts.

A linear correlation between ^{13}C chemical shifts and local π electron densities has been reported for monocyclic $(4n + 2)$ π electron systems such as benzene and nonbenzenoid aromatic ions [76] (Section 3.1.3, Fig. 3.2). In contrast to theoretical predictions (86.7 ppm per π electron [75]), the experimental slope is 160 ppm per π electron (Fig. 3.2), so that additional parameters such as σ electron density and bond order have to be taken into account [381]. Another semiempirical approach based on perturbational MO theory predicts alkyl-induced ^{13}C chemical shifts in aromatic hydrocarbons by means of a two-parameter equation; parameters are the atom-atom polarizability π_{ij}, obtained from HMO calculations, and an empirically determined substituent constant [382].

Chemical shifts of aromatic compounds occur between 120 and 150 ppm. On inclusion of electron releasing and electron withdrawing substituents as well as multiple substitution, this shift range may expand considerably (90–185 ppm).

4.11 Aromatic Compounds   255

In aromatic compounds carbon-13 shifts are largely determined by *mesomeric (resonance)* and *inductive effects*. *Field effects* arising from through-space polarization of the π system by the electric field of a substituent, and the influences of *steric (γ) effects* on the *ortho* carbon nuclei should also be considered. Substituted carbon (C-1) shifts are further influenced by the *anisotropy effect* of triple bonds (alkynyl and cyano groups) and by *heavy atom shielding*.

4.11.2 Substituted Benzenes

Carbon-13 shift values of a small selection of monosubstituted benzenes [383] are collected in Table 4.53. Signal assignments are based on conventional techniques such as proton off-resonance and gated decoupling as well as comparative measurements of specifically deuterated compounds [384].

Substituted carbon shifts (C-1) in *alkylbenzenes* grow with increasing number of β carbon atoms, e.g. in the series toluene (no β), ethylbenzene (1 β), iso-propylbenzene (2 β), and t-butylbenzene (3 β effects). They diminish with increasing γ alkylation, e.g. from ethyl- via propyl- to iso-butylbenzene, or due to heavy atom substitution in an α position, e.g. on going from t-butyl- to trimethylsilylbenzene (Table 4.53). Alkenyl groups cause deshieldings similiar to those of alkyl substituents, while alkynyl residues induce shieldings arising from the anisotropy effect of the CC triple bond (Table 4.53).

Alkyl substitution has practically no influence on *meta* (C-3,5) carbon shifts which occur at 128.5±0.5 ppm (Table 4.53). The *ortho* (C-2,6) carbons are shielded upon α branching of the alkyl group. This kind of γ effect is clearly recognized when comparing the C-2,6 shift values of toluene, ethylbenzene (1 γ), iso-propylbenzene (2 γ), and t-butylbenzene (3 γ CH$_3$ groups) in Table 4.53. All alkyl groups induce similiar shieldings of the *para* (C-4) carbons; these generally occur at 126±0.5 ppm. Slight deshieldings of the *p* carbon nuclei in benzyl derivatives relative to toluene as parent compound (Table 4.53) are explained in terms of an electric field effect denoted as F_π or π *inductive effect* [385]; it involves through-space polarization of the aromatic π system by an electric field of the α substituent.

π inductive effect
in benzyl halides and
related compounds

^{13}C chemical shifts of *non-alkyl substituted benzenes* (electron donors and acceptors) are predominantly determined by mesomeric (resonance) and inductive effects. Substituent-induced carbon-13 shifts of substituted benzenoid carbons (C-1) are essentially inductive and comparable to the α effects known from alkenes. They may be correlated with substituent electronegativities as shown in Fig. 4.8 [386] and for the halobenzene series in Table 4.53. Extreme shielding of C-1 in iodobenzene arises from the heavy atom influence. Shielding of C-1 in benzonitrile originates from an anisotropy effect of the CN triple bond. The large C-1 carbon shift of phenyllithium, at the other extreme, is attributed to paramagnetic deshielding induced by excess electron density concentrated in a nonbonding-type orbital orthogonal to the plane of the benzene ring.

Table 4.53. ^{13}C Chemical Shifts of Representative Monosubstituted Benzenes (δ_C in ppm) [383].

Class/Compound	C-1	C-2	C-3	C-4	C-α	C-β	C-γ	C-δ (others)
Alkyl-, alkenyl-, alkynylbenzenes								
Benzene	128.5							
Toluene	137.8	129.2	128.4	125.5	21.3			
Ethylbenzene	144.3	128.1	128.6	125.9	29.7	15.8		
Propylbenzene	142.5	128.6	128.3	125.8	38.3	25.0	13.9	
Butylbenzene	143.3	129.0	128.2	125.7	36.0	34.0	22.9	14.9
iso-Propylbenzene	148.8	126.6	128.6	126.1	34.4	24.1		
sec-Butylbenzene	148.4	127.9	129.3	126.8	42.3	31.7	12.2	22.2 (β')
iso-Butylbenzene	141.1	128.7	127.6	125.3	45.3	30.1	22.2	
t-Butylbenzene	150.9	125.4	128.3	125.7	34.6	31.4		
Trimethylsilylbenzene	140.2	134.4	127.8	128.2		−1.1		
Cyclopropylbenzene	143.7	125.5	128.1	125.2	15.4	9.1		
Cyclohexylbenzene	147.9	127.1	128.6	126.2	50.6	35.1	27.2	26.6
Diphenylmethane	141.3	129.0	128.5	126.2	42.1			
Triphenylmethane	143.8	129.4	128.2	126.2	56.8			
Tetraphenylmethane	146.9	131.2	127.5	126.0	65.0			
Styrene	137.6	126.1	128.3	127.6	137.1	113.3		
(E)-Propenylbenzene	138.0	126.8	128.5	125.9	131.2	125.4	18.4	
Allylbenzene	140.1	128.8	128.6	126.3	40.4	137.7	115.7	
(E)-Stilbene	137.6	126.8	128.9	127.8	129.0			
(Z)-Stilbene	137.5	129.1	128.4	127.3	130.5			
Allenylbenzene	131.3	125.6	128.6	126.6	94.0	209.6	78.8	
Ethynylbenzene	122.7	131.4	128.6	128.9	84.0	77.8		
Diphenylethyne	123.5	131.7	128.4	128.2	89.6			
Biphenyl	140.6	126.7	128.4	126.9				
Benzyl derivatives								
Benzyl fluoride	136.6	127.6	128.6	128.7	84.5			
Benzyl chloride	137.9	128.9	128.8	128.6	47.0			
Benzyl bromide	138.2	129.5	128.8	128.7	33.5			
Benzyl iodide	139.0	128.5	128.5	127.6	5.9			
Benzyl alcohol	141.8	127.7	129.1	128.1	65.1			
Benzylamine	143.4	129.9	128.4	126.6	46.4			
Benzylmercaptane	141.0	127.9	128.5	126.9	28.8			
Benzyl methyl ether	139.5	129.0	128.1	128.0	75.8		56.9	
Benzyldimethylamine	139.6	129.3	128.3	127.0	64.4		45.3	
Benzyl methyl sulfide	138.3	128.9	128.4	126.9	38.2		14.7	
Dibenzyl ether	139.0	128.0	129.0	128.0	72.5			
Dibenzylamine	140.4	128.0	128.2	126.8	53.1			
Dibenzyl sulfide	139.0	128.8	129.4	127.0	37.0			
Benzenes substituted by donors ($\delta_{C-4} < 128.5$ ppm)								
Fluorobenzene	163.6	114.2	129.4	124.1				
Chlorobenzene	134.9	128.7	129.5	126.5				
Bromobenzene	122.6	131.5	130.0	127.0				
Iodobenzene	96.2	138.4	131.1	128.1				

Table 4.53. (continued).

Class/Compound	C-1	C-2	C-3	C-4	C-α	C-β	C-γ	C-δ (others)
O-substituted								
Phenol	155.1	115.7	130.1	121.4				
Phenyl acetate	150.9	121.4	128.1	125.3	169.7	23.9		
Anisol	159.9	114.1	129.5	120.7	54.8			
Trimethylsilyloxybenzene	155.3	120.1	129.4	121.4				0.2
Diphenyl ether	157.7	119.1	129.9	123.2				
Phenyl cyanate	153.1	115.3	130.7	127.0	108.7 (OCN)			
Phenolate anion	168.1	120.3	130.4	114.9				
S-substituted								
Thiophenol	130.5	129.1	128.7	125.2				
Methyl phenyl sulfide	138.6	126.8	128.8	125.0	15.9			
Diphenyl sulfide	135.8	130.9	129.1	126.9				
Diphenyl disulfide	136.0	127.2	129.3	127.4				
Phenyl thiocyanate	124.8	130.0	129.7	129.7	111.9 (SCN)			
N-substituted								
Aniline	148.7	114.4	129.1	116.3				
Acetanilide	138.2	120.4	128.7	124.1	169.5	24.1		
N-Methylaniline	150.4	112.1	129.1	115.9	29.9			
N,N-Dimethylaniline	150.7	112.7	129.0	116.7	40.3			
Diphenylamine	143.2	117.8	129.4	118.0				
Phenylhydrazine	152.3	112.8	129.9	119.9				
Diphenylcarbodiimide	139.3	124.9	130.2	126.3	136.0 (NCN)			
Phenyl isocyanate	133.9	124.8	129.7	125.9	120.8 (NCO)			
Phenyl isothiocyanate	145.9	126.3	130.5	128.1	131.9 (NCS)			
Benzenes substituted by acceptors ($\delta_{C-4} > 128.5$ ppm)								
Trimethylphenyl-ammonium ion	148.0	121.2	131.0	130.9				
(E)-Azobenzene	152.7	123.0	129.7	131.8				
Phenyldiazonium ion	115.8	134.5	134.2	144.5				
Nitrobenzene	149.1	124.2	129.8	134.7				
Phenyl isocyanide	126.7	126.3	129.9	129.4	165.7 (NC)			
Methyl phenyl sulfoxide	145.1	122.6	129.6	130.0	39.9			
Methyl phenyl sulfone	140.8	127.1	129.3	133.6	40.5			
Diphenyl sulfoxide	145.7	124.7	129.2	131.0				
Diphenyl sulfone	140.8	127.1	129.3	133.6				
Benzenesulfonic acid	145.7	131.8	128.4	134.4				
Benzenesulfonyl chloride	144.3	129.9	126.9	135.5				
Benzonitrile	112.8	132.1	129.2	132.8	119.5 (CN)			
Benzaldehyde	136.7	129.7	129.0	134.3	192.0			
Acetophenone	136.3	128.1	128.1	131.3	195.7	24.6		
Benzophenone	137.6	130.0	128.3	132.3	196.4			
Benzil	133.0	129.7	128.9	134.7	194.3			
Benzoic acid	131.4	129.8	128.9	133.1	168.0			
Benzoate anion	138.8	131.3	130.7	133.6	177.9			
Benzamide	134.4	127.4	128.2	131.2	168.2			

Table 4.53. (continued).

Class/Compound	C-1	C-2	C-3	C-4	C-α	C-β	C-γ	C-δ (others)
N,N-Dimethylbenzamide	136.5	127.0	128.3	129.5	171.4		39.5 and 35.2	
Methyl benzoate	130.5	129.7	128.4	132.8	166.8		51.0	
Ethyl benzoate	131.0	129.5	128.0	132.4	166.3		60.8	14.4
Phenyllithium	186.6	143.7	124.7	133.9				
Phenylmagnesium bromide	164.3	139.9	131.2	132.5				
Diphenylmercury	172.5	139.7	129.4	128.7				

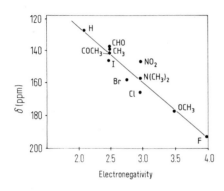

Fig. 4.8. Substituent electronegativities E_x versus ^{13}C shifts (δ_C in ppm) of substituted benzenoid carbons, corrected for magnetic anisotropy effects [386].

Meta carbon atoms (C-3,5) remain almost unaffected by all kinds of substituents. In monosubstituted benzenes they usually resonate at 129 ± 1 ppm (Table 4.53). Recalling Section 3.1.3.6, *electron releasing substituents* ((+)-*M substituents, electron donors*) increase π electron densities in *ortho* and *para* positions and thereby induce a shielding relative to benzene ($\delta_{C(o,p)} < 128.5$ ppm). *Electron withdrawing groups* ((−)-*M substituents, electron acceptors*) decrease π electron densities in *o* and *p* positions; thus, a deshielding relative to benzene is observed ($\delta_{C(o,p)} > 128.5$ ppm).

Electron releasing: $\delta_{C(o,p)} < 128.5$ ppm

Electron withdrawing: $\delta_{C(o,p)} > 128.5$ ppm

Shieldings induced by donors and deshieldings arising from acceptors are clearly indicated by the *p* carbon nuclei of monosubstituted benzenes selected in Table 4.53. In the *ortho* positions, inductive and electric field effects overlap with the resonance effects. In nitrobenzene, for example, an intramolecular electric field of the nitro group increases σ electron densities at the *o, o'* carbon nuclei, so reducing the electron density at the attached protons. This effect, in fact, overcompensates electron withdrawal of the nitro group, and a net shielding is observed (Table 4.53). For similiar reasons, (−)-*M* electron withdrawal of carbonyl groups in the *o, o'* positions is reduced.

Electric field effect of the nitro group in nitrobenzene

Para carbon shieldings, however, clearly follow the pattern described by the cannonical formulae. They may be correlated with the total charge densities Δq, obtained by CNDO calculations, according to eq. 4.14 [387], and with the Hammet σ constants, as shown in Fig. 4.9 [386, 388]

$$\delta_{C_6H_6} - \delta_{C-4(C_6H_5X)} = 166 \Delta q \tag{4.14}$$

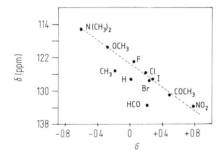

Fig. 4.9. Hammet σ constants *versus* ^{13}C shifts (δ_C in ppm) of *para* carbon nuclei in monosubstituted benzenes [386].

Substituent increments Z_i obtained from ^{13}C shifts of numerous monosubstituted benzenes according to eq. (4.15) have been tabulated [383]. They permit prediction of benzene ring carbon shifts in multisubstituted benzenes according to eq. (4.16). These increments and their practical application will be summarized in Section 4.16.

$$Z_i = \delta_{i(C_6H_5X)} - 128.5 \text{ (ppm)} \tag{4.15}$$

$$\delta_{i(C_6H_{6-n}X_n)} = 128.5 + \sum Z_i \text{ (ppm)} \tag{4.16}$$

Originating from monosubstituted benzenes, these shift increments do not take any kind of substituent interaction into account. To conclude, deviations of experimental data from the shifts predicted according to eq. (4.16) are to be expected, if substituents display *intramolecular interactions* such as hydrogen bonding, *strong resonance or inductive effects of opposite signs*, or *steric repulsions* which induce bond angle deformations in saturated substituents and torsional distortions of groups containing double bonds in conjugation to the benzenoid π system.

Nevertheless, eq. (4.16) predicts the signal sequence of multisubstituted benzenes. Straightforward assignment can be achieved frequently, provided signals are separated by at least 2 ppm. – Experimental data of several *o*-, *m*-, *p*-disubstituted isomeric series and higher substituted benzene derivatives are collected in Table 4.54 [73 k, l, m].

Steric repulsions between substituents with π bonds and other groups attached to the benzene ring may twist the planes of both π systems as outlined in Sections 3.1.3.8 and 4.7.2 for *o*, *o'*-dialkylated acetophenones. This sterically induced hindrance of conjugation is not only monitored by an increase in the carbonyl shifts but also by a shielding of the *para* benzenoid carbon due to an attenuated mesomeric effect of the carbonyl group. This

Table 4.54. ^{13}C Chemical Shifts of Representative Polysubstituted Benzenes (δ_C in ppm) [73 k, l, m, 126].

Class/Compound	C-1	C-2	C-3	C-4	C-5	C-6	C-α	C-β and (or) others
Polyalkylbenzenes								
o-Xylene	136.2		129.7	125.9			19.6	
m-Xylene	137.6	130.0		126.3	128.3		21.3	
p-Xylene	134.6	129.1					20.9	
o-Diethylbenzene	141.0		128.0	125.7			25.3	15.3
o-Di-*t*-butylbenzene	148.8		129.5	125.5			37.6	34.9
m-Di-*t*-butylbenzene	150.6	127.6		122.4	122.2		34.8	31.5
p-Di-*t*-butylbenzene	147.8	124.8					34.1	31.4
p-Divinylbenzene	136.9	126.1					136.5	113.2
1,2,3-Trimethylbenzene	136.1	134.8		127.9	125.5		20.4	15.0 (C-2-CH$_3$)
1,2,4-Trimethylbenzene	133.4	136.3	130.5	135.2	126.7	129.8	19.1	19.5 (C-2-CH$_3$)
1,3,5-Trimethylbenzene (Mesitylene)	137.6	127.4					21.2	20.9 (C-4-CH$_3$)
1,3,5-Tri-*t*-butylbenzene	149.8	119.3					34.9	31.6
1,2,4,5-Tetramethylbenzene	133.8		131.2				19.2	
1,2,3,5-Tetramethylbenzene	136.0	131.6		128.9	134.3		20.3	14.6 (C-2-CH$_3$) 20.9 (C-5-CH$_3$)
1,2,3,4-Tetramethylbenzene	133.5	134.4			127.3		20.6	15.5 (C-2-CH$_3$)
Pentamethylbenzene	133.0	132.1	134.5			131.5	20.7	15.8 (C-2-CH$_3$) 16.1 (C-3-CH$_3$)
Hexamethylbenzene	132.3						16.9	
Dihalobenzenes								
o-Difluorobenzene	150.3		124.6	117.0				
m-Difluorobenzene	163.1	103.3		110.9	130.7			
p-Difluorobenzene	159.1	116.5						
o-Dichlorobenzene	132.6		130.5	127.7				
m-Dichlorobenzene	134.9	128.5		126.9	130.8			
p-Dichlorobenzene	132.7	130.3						
o-Dibromobenzene	124.7		133.5	128.2				
m-Dibromobenzene	123.1	134.0		130.4	131.5			
o-Diiodobenzene	108.6		139.6	129.6				
m-Diiodobenzene	95.8	145.0		137.0	132.2			
Diphenols								
Catechol	147.1		119.3	124.1				
Resorcinol	159.3	105.3		110.3	133.3			
Hydroquinone	151.5	118.5						
Diamines								
o-Phenylenediamine	135.0		117.0	120.5				
m-Phenylenediamine	147.9	102.2		106.2	130.4			
p-Phenylenediamine	138.8	116.9						
Dicarboxylic acids								
Phthalic acid	120.7		133.6	138.6				181.0 (protonated)
Isophthalic acid	121.0	134.9		140.9	130.9			178.0 (protonated)
Terephthalic acid	127.9	131.5						179.4 (protonated)
Dialdehydes								
Phthalaldehyde	137.2		133.1	135.8				196.1
Isophthalaldehyde	137.9	132.6		136.4	131.5			193.8
Terephthalaldehyde	141.2	131.7						201.9

Table 4.54. (continued).

Class/Compound	C-1	C-2	C-3	C-4	C-5	C-6	C-α	C-β and (or) others
Mixed disubstituted benzenes								
o-Chlorotoluene (Cl, CH$_3$)	136.0	134.5	129.1	127.1	126.6	131.0	19.9	
m-Chlorotoluene	139.9	129.3	134.2	125.7	129.5	127.3	21.1	
p-Chlorotoluene	136.2	130.4	128.5	131.3			20.3	
o-Cresol (OH, CH$_3$)	153.5	124.0	131.0	121.4	127.7	115.9	16.7	
m-Cresol	154.9	116.1	139.3	122.2	130.3	112.7	20.9	
p-Cresol	152.6	115.3	130.2	130.5			20.6	
o-Toluidine (NH$_2$, CH$_3$)	145.2	122.2	130.4	118.3	127.0	115.0	17.0	
m-Toluidine	147.2	116.1	138.8	119.1	129.2	112.5	21.3	
p-Toluidine	144.1	115.0	129.5	126.7			20.3	
o-Nitrotoluene (NO$_2$, CH$_3$)	133.4	149.5	124.5	127.1	133.1	132.8	20.1	
p-Nitrotoluene	146.2	130.0	123.5	146.2			21.4	
o-Chlorophenol (OH, Cl)	151.8	120.3	129.3	121.5	128.5	116.3		
m-Chlorophenol	156.3	116.1	135.1	121.5	130.5	113.9		
p-Chlorophenol	154.1	116.8	129.7	125.9				
o-Chloroaniline (NH$_2$, Cl)	143.0	119.1	129.3	118.9	127.6	115.9		
m-Chloroaniline	148.0	114.8	134.7	118.2	130.4	113.3		
p-Chloroaniline	144.9	115.9	128.7	128.6				
o-Nitrophenol (OH, NO$_2$)	155.3	133.8	125.2	120.4	137.7	120.1		
p-Nitrophenol	161.5	115.9	126.4	141.7				
o-Nitroaniline (NH$_2$, NO$_2$)	144.5	131.7	125.6	118.5	135.2	116.5		
m-Nitroaniline	151.4	109.0	150.4	111.7	131.1	121.6		
p-Nitroaniline	157.1	114.1	127.8	137.6				
o-Toluic acid (COOH, CH$_3$)	129.2	137.7	128.9	130.0	124.3	130.0	21.3	167.0 (COOH)
m-Toluic acid	129.5	128.5	136.3	125.2	126.9	131.9	20.7	165.7 (COOH)
p-Toluic acid	128.5	129.7	129.4	143.3			21.4	167.8 (COOH)
o-Nitrobenzoic acid	126.1	146.8	122.3	131.5	130.8	128.5		164.0 (COOH)
m-Nitrobenzoic acid	131.0	122.2	146.1	125.7	128.9	133.7		163.6 (COOH)
o-Aminobenzoic acid	111.5	152.9	116.4	135.3	118.0	132.8		169.1 (COOH)
p-Aminobenzoic acid	118.7	132.8	114.3	154.6				171.3 (COOH)
o-Methoxybenzoic acid	119.9	156.3	111.1	131.5	118.7	129.2		165.5 (COOH) 55.0 (OCH$_3$)
m-Methoxybenzoic acid	133.0	115.6	160.8	120.2	131.0	123.1		168.7 (COOH) 56.6 (OCH$_3$)
o-Hydroxybenzoic acid (Salicylic acid)	114.4	162.9	118.4	136.7	120.3	131.7		173.6 (COOH)
Mixed tri- and higher substituted benzenes								
2,3-Dichlorophenol	152.9	119.2	132.8	122.3	128.2	114.6		
2,4-Dichlorophenol	150.3	120.6	128.7	125.8	128.7	117.3		
2,5-Dichlorophenol	152.0	118.5	129.7	121.8	133.9	116.8		
2,6-Dichlorophenol	148.1	121.3	128.5	121.3				
2,4,5-Trichlorophenol	150.4	118.9	129.6	124.3	132.0	117.7		
2,4,6-Trichlorophenol	147.1	121.8	128.2	125.5				
Pentachlorophenol	148.3	119.9	131.6	125.2				
2,4-Dinitrophenol	159.2	132.8	122.0	140.4	131.8	121.4		
2,4,6-Trinitrophenol	153.2	139.2	126.2	138.2				
2,4-Dichloronitrobenzene	146.3	128.2	131.6	139.2	128.0	126.7		
2,4,5-Trichloronitrobenzene	146.0	126.2	132.9	137.9	132.1	127.1		

can be verified by comparing the C-4 carbon shifts of acetophenone and its o-ethyl- and 2,6-diethyl derivatives [294] in Table 4.31. Following an approach known from UV spectroscopy, the torsional angle Φ can be obtained from the experimental shift value δ_X, and the reference data of the parent acetophenone ($\delta_{0°}$) and the 2,6-di-t-butyl derivative ($\delta_{90°}$) according to eq. (4.17) [389, 390]. Using this method results have been obtained for acetophenones and benzoic acid derivatives [390], diphenyl ethers [389], and azomethines [391].

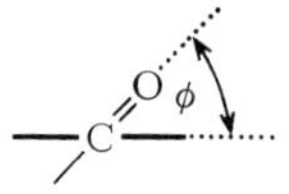

$$\cos^2 \Phi = \frac{\delta_X - \delta_{90°}}{\delta_{0°} - \delta_{90°}} \tag{4.17}$$

Benzene ring plane

4.11.3 Substituted Naphthalenes

Carbon-13 shifts of monosubstituted naphthalenes [392–394] (Table 4.55) and anthracenes [395] follow the overall pattern described for monosubstituted benzenes. Deshieldings arising from acceptors (*e.g.* CN) and shieldings induced by electron donors (*e.g.* OCH_3, Table 4.55) can be rationalized by writing down the cannonical formulae of naphthalenes substituted in a 1- and 2-position. Positive charges on ring carbons due to acceptors X have to be exchanged for negative ones when the substituent is an electron donor.

1-substituted naphthalene

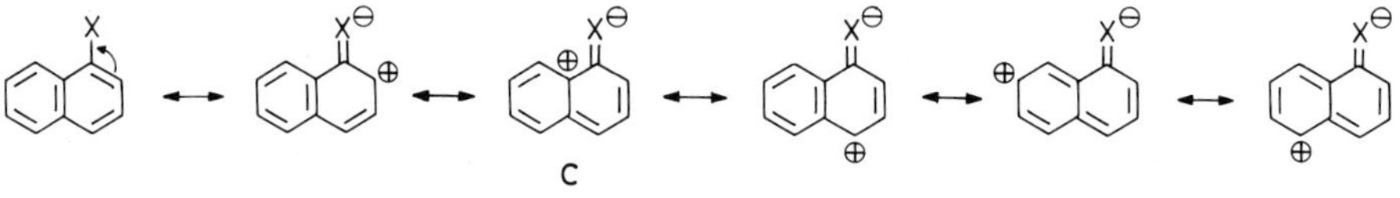

2-substituted naphthalene

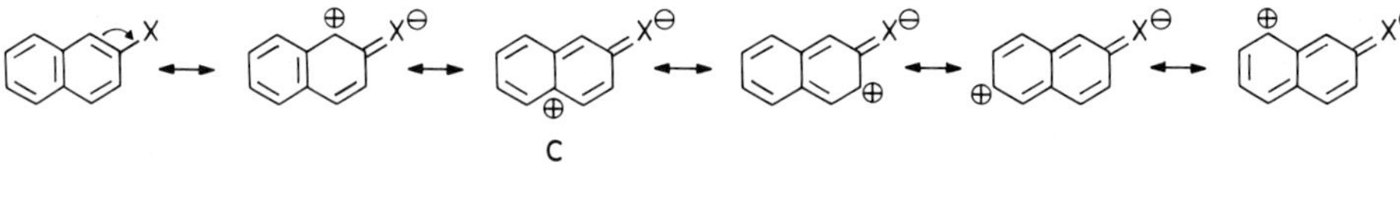

Substituent increments are obtained as usual by subtracting the reference shifts of naphthalene (C-1,4,5,8: 127.7; C-2,3,6,7: 125.6; C-9,10: 133.3 ppm) from the individual data of C-1 to C-10 in the substituted derivatives given in Table 4.55. It turns out that comparable Z_1, Z_{ortho}, Z_{meta}, and Z_{para} increments in naphthalene and benzene differ substantially in magnitude, as exemplified in Table 4.56. In 1-substituted naphthalenes, C-9 increments are attenuated in favor of C-2 relative to comparable *ortho* effects known from benzene; C-3 and C-1 in 2-substituted naphthalenes behave correspondingly (Table 4.56). This can be explained by the cannonical formulae c, which do not contribute so much to the actual molecular state due to disrupted π electron sextets. Full inter-

Table 4.55. ^{13}C Chemical Shifts of Representative 1- and 2-Monosubstituted Naphthalenes (δ_C in ppm) [394].

-Naphthalene	C-1	C-2	C-3	C-4	C-5	C-6	C-7	C-8	C-9	C-10
1-F	159.4	110.2	126.6	124.7	128.5	127.8	172.2	120.8	124.3	135.9
1-Cl	132.1	126.9	126.6	128.2	129.1	127.6	128.0	124.6	131.4	135.5
1-Br	122.9	130.7	127.1	128.9	129.2	127.5	128.2	127.3	132.6	135.5
1-I	99.6	138.2	127.7	129.8	129.4	127.5	128.5	132.4	134.9	134.9
1-OH	151.5	108.7	125.8	120.7	127.6	126.4	125.2	121.4	124.3	134.6
1-OCH$_3$	155.3	103.6	125.7	120.1	127.3	126.2	125.0	121.9	125.5	134.4[a]
1-NH$_2$	142.0	109.4	126.2	118.7	128.3	125.6	124.6	120.7	123.4	134.3
1-N(CH$_3$)$_2$	151.7	114.7	126.6	123.4	129.2	126.3	125.6	124.8	129.7	135.0[b]
1-NO$_2$	146.5	123.8	123.9	134.5	128.5	127.2	129.3	122.9	124.9	134.2
1-CN	110.6	133.5	126.0	134.2	129.7	128.4	129.5	125.3	132.9	133.8[c]
1-CHO	132.2	137.4	125.9	135.8	129.3	127.6	129.6	125.3	131.0	134.5[d]
1-COCH$_3$	136.1	129.6	125.3	133.5	129.2	127.0	128.4	126.7	130.9	134.8[e]
1-COOH	128.0	131.2	125.5	134.1	129.4	127.0	128.3	126.7	132.3	134.9[f]
1-Si(CH$_3$)$_3$	137.8	133.1	125.5	129.7	129.2	125.1	125.2	128.1	137.4	133.8[g]
2-F	111.4	161.4	116.7	131.4	128.7	126.0	127.8	128.1	135.1	131.5
2-Cl	127.2	131.9	127.2	130.5	128.5	127.0	127.8	127.8	134.9	132.6
2-Br	130.6	120.1	129.8	130.7	128.6	127.2	127.8	127.8	135.4	132.8
2-I	137.2	91.8	134.9	130.3	128.5	127.2	127.4	127.4	135.7	132.8
2-OH	109.4	153.2	117.6	129.8	127.7	123.5	126.4	126.3	134.5	128.5
2-OCH$_3$	105.7	157.5	118.7	129.3	127.6	123.4	126.3	126.7	134.6	129.0[h]
2-NH$_2$	108.4	144.0	118.1	129.0	127.6	122.3	126.2	125.6	134.8	127.8
2-N(CH$_3$)$_2$	106.9	149.5	117.1	129.2	128.2	122.5	126.7	126.9	135.0	127.7[i]
2-NO$_2$	125.1	146.3	119.7	130.4	128.7	130.5	128.7	130.8	132.7	136.6
2-CN	134.8	110.1	127.0	130.1	128.8	129.8	128.4	129.2	133.1	135.9[j]
2-CHO	135.1	135.2	123.1	129.8	128.8	129.8	127.9	130.3	133.5	137.1[k]
2-COCH$_3$	130.9	135.4	124.4	128.9	128.4	129.1	127.5	130.3	133.4	136.3[l]
2-COOH	131.8	128.8	126.2	129.0	128.6	129.1	127.6	130.1	133.5	136.4[m]
2-Si(CH$_3$)$_3$	133.1	137.8	129.8	127.0	128.0	126.2	125.7	128.1	133.8	135.8[n]

[a] OCH$_3$: 55.3; [b] N(CH$_3$)$_2$: 45.2; [c] CN: 118.1; [d] CH=0: 194.2; [e] CO: 201.7; CH$_3$: 30.0; [f] COOH: 169.0; [g] Si(CH$_3$)$_3$: −1.0; [h] OCH$_3$: 55.0; [i] N(CH$_3$)$_2$: 40.7; [j] CN: 119.6; [k] CHO: 192.7; [l] CO: 197.8; CH$_3$: 26.6; [m] COOH: 168.0; [n] Si(CH$_3$)$_3$: −1.1.

pretation of substituent increments additionally requires consideration of anisotropy, electric field, and steric effects.

Steric effects (γ *shieldings*) are operative for C-8 in 1-substituted naphthalenes. Table 4.55 illustrates this and also shows that shielding decreases with increasing size of the halogen [392], probably arising from an expanded X−C-1−C-9 bond angle.

Carbon-13 shifts of multisubstituted naphthalenes can be assigned by addition of substituent increments similar to eq. (4.16), provided no intramolecular interactions between substituents take place. This applies for 1,4-, 1,5-, 2,6-, and 2,7-disubstituted deriv-

Table 4.56. Comparison of Methoxy- and Cyano Substituent Effects on Naphthalene and Benzene Carbon-13 Shifts (Z_i in ppm). Benzoid Shift Increments are Printed in *Italic Type*.

-Naph-thalene	C-1	C-2	C-3	C-4	C-5	C-6	C-7	C-8	C-9	C-10
1-OCH$_3$	27.6	−22.0	0.1	−7.6	−0.4	0.6	−0.6	−5.8	−7.8	1.1
	31.4	*−14.4*	*1.0*	*−7.7*					*−14.4*	*1.0*
2-OCH$_3$	−22.0	31.9	−6.9	1.6	−0.1	−2.2	0.7	−1.0	1.3	−4.3
	−14.4	*31.4*	*−14.4*	*1.0*					*1.0*	*−7.7*
1-CN	−17.9	7.0	−0.5	5.6	1.1	1.8	3.0	−3.2	−1.5	−0.5
	−15.7	*3.6*	*0.7*	*4.3*					*3.6*	*0.7*
2-CN	6.3	−18.5	0.5	1.6	0.3	3.3	1.9	0.7	−1.3	1.1
	3.6	*−15.7*	*3.6*	*0,7*					*0.7*	*4.3*

atives, e.g. 1,5-dibromonaphthalene. Considerable deviations, however, must be expected when the substituents are connected in *ortho* and particularly in 1,8-positions (e.g. 1,8-diiodonaphthalene). Bond angle distortions are the obvious reasons. In such cases, selective proton decoupling, two-dimensional shift correlations, or deuterium labeling may help in assigning the signals.

1-5-Dibromo

Br
132.4
132.9 122.9 122.8
       130.8 130.2
       127.2 127.0
       127.1 126.5
Br

1,8-Diido-naphthalene (predicted values in *italic* type)

I   I
136.5
132.7 96.6 104.3
       144.7 141.1
       127.5 129.6
136.4 131.7 131.5
136.5

4.11.4 Benzocycloalkenes and Hydroaromatic Compounds

The ^{13}C chemical shifts of benzocycloalkenes [396] and other hydroaromatic hydrocarbons [73e] are listed in Table 4.57. Relative to *o*-diethylbenzene (Table 4.54) as reference compound, small deviations are observed for benzocyclooctene and benzocycloheptane. On going from benzocyclohexene (tetralin) to benzocyclopropene, however, an increasing deshielding is measured with increasing ring strain for the carbons *meta* to the bridgeheads, while the *ortho* carbons become successively shielded as the ring strain increases [396]. Additionally, the C−H couplings of the *o* and α carbons increase with decreasing carbon shielding. A corresponding trend is apparent in the cycloalkane series (Table 4.4).

The ^{13}C shifts of the other hydroaromates (Table 4.57) are closely related to those of tetralin. Angular fusion of the rings usually causes a slight sterically induced shielding of the *o, o'* or α, α' carbons, as can be seen for the hydrogenated phenanthrene derivatives.

Table 4.57 ^{13}C Chemical Shifts of Selected Benzocycloalkenes [396]^a, Hydroaromatic Compounds [73e, 394, 397] and [2,2]Cyclophanes [398] (δ_C in ppm).

Benzocyclopropene	Benzocyclobutene	Benzocyclopentene	Benzocyclohexene
168.5, 114.7, 159, 128.8, 125.2, 170, 18.4	162, 122.1, 157.5, 126.5, 145.2, 138, 29.5	155.5, 124.0, 157, 125.8, 143.3, 127, 128, 32.7, 25.2	155, 126, 128.8, 29.3, 159, 125.2, 136.4, 23.3

Benzocycloheptene	Benzocyclohepta-triene	Benzocyclooctene	Biphenylene
155, 123, 128.7, 36.6, 161, 142.2, 28.2, 125.7, 32.6	130.3, 127.7, 137.2, 130.8, 125.9, 26.6	155, 32.1, 128.7, 140.6, 25.8, 160, 126.0, 32.1	118.0, 151.9, 128.8

Fluorene	1,2,3,4,5,6,7,8-Octa-hydroanthracene	9,10-Dihydro-anthracene	1,2,3,4,5,6,7,8-Octa-hydrophenanthrene
119.7, 141.6, 126.5, 126.5, 143.2, 124.8, 36.8	130.8, 30.7, 137.3, 25.3	37.7, 128.0, 137.3, 128.0	134.5, 27.1, 24.3, 126.2, 134.5, 24.3, 30.9

9,10-Dihydrophen-anthrene	Dodecahydrotri-phenylene	[2,2]-*m*-Cyclophane	[2,2]-*p*-Cyclophane
137.3, 123.9, 127.5, 127.5, 29.2, 134.7	25.2, 28.4, 132.9	41.4, 136.3, 138.6, 125.1, 128.6	132.8, 139.5, 35.7

^{a} when available, one-bond carbon-proton couplings (Hz) are printed in *italic*.

4.11.5 Fused and Bridged Aromatic Rings

A ^{13}C NMR study of non-alternant fused hydrocarbons (Table 4.58) led to the conclusion that these systems may be distinguished from alternant ones by larger chemical shift ranges of the carbon signals (14–22 ppm in comparison to values smaller than 10 ppm in alternating systems) [399]. According to this criterion, acepleiadiene and acepleiadylene belong to the class of non-alternant hydrocarbons [400]. A plot of ^{13}C chemical shifts *versus* π or total charge densities derived from CNDO/2-SCF-MO calculations yields a poor correlation for naphthalene, azulene, acenaphalene and acenaphthene [399].

Steric interactions generally predominate relative to ring current effects in carbon-13 NMR. Proton shifts of bridged [10]- and [14]annulenes, for example, clearly identify these compounds to be aromatic due to the typical deshieldings of about 2 ppm relative to comparable non-aromatic cyclopolyenes [401a]. In contrast, carbon-13 shifts of bridged annulenes (Table 4.58) are more closely related to those of comparable non-aromatic cyclopolyenes with similar steric interactions than to those of other annulenes [401b].

Table 4.58. ^{13}C Chemical Shifts of Representative Fused Aromatic Compounds [381, 394, 399, 400] and Bridged Annulenes [401 b] (δ_C in ppm).

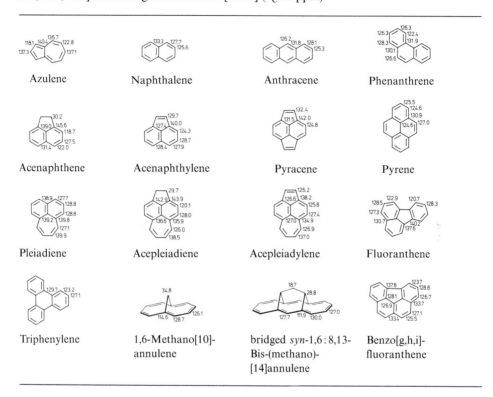

4.11.6 Typical Coupling Constants

Recalling the relation $^3J_{CH(m)} > {}^2J_{CH(o)} > {}^4J_{CH(p)}$ outlined for CH couplings in Section 3.2.2.4 (Table 3.10), benzenoid carbon nuclei without coupling hydrogen atoms in any *meta* position can generally be recognized as strong one-bond doublets for unsubstituted and as singlets in the case of substituted carbons. Typical examples are vanilline (Fig. 4.10) and 1,8-diiodonaphthalene (Fig. 4.11), illustrating the assignment of benzenoid carbon signals by means of CH couplings in multisubstituted benzenes and naphthalenes. Fig. 4.10 exemplifies the large $^2J_{CH}$ coupling with an aldehyde proton (Section 3.2.2.2), and Fig. 4.11 illustrates that *transoid* three-bond CH couplings are larger than *cisoid* ones (C-4 with 2-H: $^3J_{CH} = 7.3$ Hz, *transoid*; C-4 with 5-H: $^3J_{CH} = 5.5$ Hz, *cisoid*). – The high-field part of a multiplet arising from one-bond and longer-range couplings frequently differs from the low-field half. This asymmetry of multiplets indicates that the spectrum is second order. Thus, first-order analysis of single frequency ^{13}C NMR spectra as illustrated in Figs. 4.10 and 4.11 is not generally possible. Even if the proton spectrum of a compound is first-order, the coupled carbon spectrum may be second-order, and a simulation is required for a correct analysis.

4.11 Aromatic Compounds

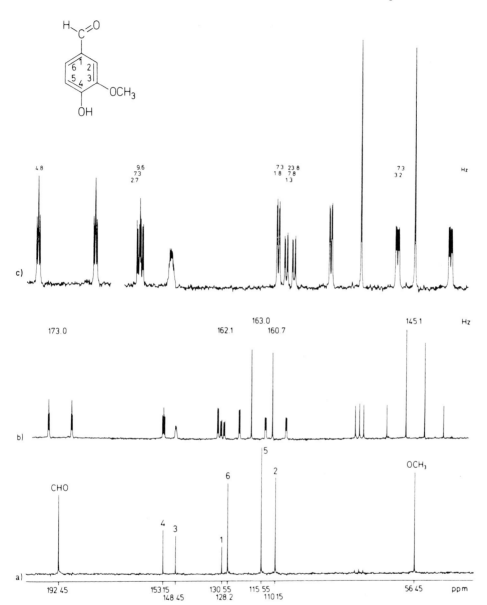

Fig. 4.10. ^{13}C NMR spectrum of vanilline in deuteriochloroform (150 mg/1 mL), 28 °C, 22.63 MHz; (a) proton broadband-decoupled (200 scans); (b) gated-decoupled (1600 scans); (c) expanded output of (b). The CHO carbon splits into a doublet of long-range triplets (4.8 Hz), the latter arising from 2,6-H. C-4 shows *m* coupling with 2-H and 6-H (9.6 and 7.3 Hz) and *o* coupling with 5-H (2.7 Hz). C-3 splits into a multiplet due to *m* coupling with 5-H, *o* coupling with 2-H and three-bond coupling with the methoxy protons. C-1 couples with the formyl proton (23.8 Hz), with 5-H (7.8 Hz) and with 2-H or 6-H (1.3 Hz). The long-range splittings of C-6 originate from 2-H (7.3 Hz) and CHO (1.8 Hz), those of C-2 from 6-H (7.3 Hz) and CHO (3.2 Hz), while C-5 does not have any hydrogen in a *m* position and does not show any longer-range splitting at all.

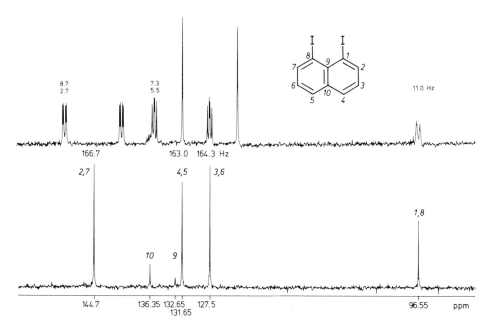

Fig. 4.11. ^{13}C NMR spectrum of 1,8-diiodonaphthalene (saturated in deuteriochloroform, 25 °C, 20.115 MHz); (a) proton broadband-decoupled (200 scans); (b) gated-decoupled (675 scans). Carbon C-1 (8) splits into a doublet ($^3J = 11$ Hz) due to coupling with 3(6)-H; no coupling proton is in a *meta* position relative to C-3(6) which, therefore, appears as a simple one-bond CH doublet; C-4(5) splits into a ^3J doublet of doublets with *transoid* 2(7)-H (7.3 Hz) and *cisoid* 5(4)-H (5.5 Hz); fine structures of the C-2(7) doublet arise from $^2J = 2.7$ Hz with 3(6)-H and $^3J = 8.7$ Hz with 4(5)-H. Quaternary carbon multiplets are weak (C-9) or overlap (C-10).

Table 4.59. One-Bond and Longer-Range ^{13}C–^{13}C Coupling Constants (Hz) of ^{13}C-7-Labeled Monosubstituted Benzene Derivatives [133].

Compound	$J_{1,7}$	$^2J_{2,7}$	$^3J_{3,7}$	$^4J_{4,7}$	$^2J_{COC}$
$C_6H_5-^{13}CH_3$	44.19	3.10	3.84	0.86	
$C_6H_5-^{13}CH_2OH$	47.72	3.45	3.95	0.73	
$C_6H_5-^{13}CH_2Cl$	47.78	3.69	4.23	0.69	
$C_6H_5-^{13}CO_2^-Na^+$	65.90	2.23	4.11	0.8	
$C_6H_5-^{13}CO_2H$	71.87	2.54	4.53	0.90	
$C_6H_5-^{13}CO_2CH_3$	74.79	2.38	4.56	0.90	2.63
$C_6H_5-^{13}COCl$	74.35	3.53	5.46	1.18	
$C_6H_5-^{13}CN$	80.40	2.61	5.75	1.59	

Some *carbon-carbon* couplings measured for ^{13}C-7 labeled monosubstituted benzenes [133] and collected in Table 4.59 summarize the trends outlined in Section 3.2.4.2: One-bond carbon-carbon couplings increase with the *s* character of the hybrid orbitals originating from, and with the electronegativity of the substituent attached to C-7; $^3J_{CC}$ values follow a similiar pattern.

4.11 Aromatic Compounds

Table 4.60. Carbon-Fluorine Couplings in Fluorobenzene and Substituted Fluorobenzenes (Hz) [402].

Compound	$-J_{C-1,F}$	$^2J_{C-2,F}$	$^3J_{C-3,F}$	$^4J_{C-4,F}$	$^3J_{C-5,F}$	$^2J_{C-6,F}$	Others
Fluorobenzene	245.3	21.0	7.7	3.3	7.7	21.0	
o-Difluorobenzene	248.8	14.1	−3.0	5.2	5.2	20.5	
o-Chlorofluorobenzene	248.7	17.5	0	4.05	7.2	20.8	
o-Bromofluorobenzene	247.0	20.7	0	3.35	7.1	21.95	
o-Iodofluorobenzene	245.6	25.2	1.46	3.5	7.2	23.4	
o-Nitrofluorobenzene	264.4		4.25	2.8	8.7	20.6	
o-Aminofluorobenzene	237.5	12.7	3.8	3.6	6.7	18.4	
o-Hydroxyfluorobenzene	238.8	13.7	1.94	3.78	6.6	18.0	
o-Methylfluorobenzene	243.9	17.0	4.8	3.7	7.9	22.1	3.8 (CH$_3$)
o-Fluorobenzaldehyde	257.7	8.2	1.86	3.75	9.1	20.45	6.4 (CHO)
o-Fluoroacetophenone	254.2	12.8	2.54	3.4	9.0	23.7	3.2 (CO) < 0.4 (CH$_3$)
2-Fluoropyridine	236.7		14.7	4.2	7.75	37.4	
m-Difluorobenzene	245.4	25.3	12.1	3.6	9.8	21.2	
m-Chlorofluorobenzene	249.5	24.6	10.0	3.4	8.9	21.3	
m-Bromofluorobenzene	250.4	24.5	9.3	3.4	8.4	21.1	
m-Iodofluorobenzene	249.0	23.2	7.8	3.3	8.1	20.8	
m-Nitrofluorobenzene	250.9	26.5	8.3	3.3	8.2	21.5	
m-Aminofluorobenzene	241.4	24.6	11.0	2.3	10.2	21.3	
m-Hydroxyfluorobenzene	244.5	24.8	11.3	3.0	10.2	21.2	
m-Methylfluorobenzene	243.6	21.1	7.2	2.2	8.5	21.2	1.75 (CH$_3$)
m-Fluorobenzaldehyde	248.2	21.7	6.3	2.9	7.8	21.8	2.4 (CHO)
m-Fluoroacetophenone	246.3	22.2	5.9	2.9	7.75	21.6	1.9 (CO) 0.68 (CH$_3$)
p-Difluorobenzene	242.0	24.3	8.5	3.8	8.5	24.3	
p-Chlorofluorobenzene	245.5	23.1	8.2	3.1	8.2	23.1	
p-Bromofluorobenzene	246.7	23.7	8.0	3.3	8.0	23.7	
p-Iodofluorobenzene	247.4	22.2	7.6	3.4	7.6	22.2	
p-Nitrofluorobenzene	256.6	24.0	10.2		10.2	24.0	
p-Aminofluorobenzene	233.2	22.4	7.5	1.86	7.5	22.4	
p-Hydroxyfluorobenzene	237.4	23.0	7.9	2.14	7.9	23.0	
p-Methylfluorobenzene	243.5	21.1	7.75	2.9	7.75	21.1	0 (CH$_3$)
p-Fluorobenzaldehyde	255.0	22.4	9.7	2.6	9.7	22.4	0 (CHO)

Carbon-fluorine coupling constants of fluorobenzene and selected substituted derivatives are collected in Table 4.60 [402]. Benzenoid J_{CF} couplings are about 245 ± 15 Hz. They depend on both type and position of the substituents: Electron withdrawing groups increase while electron releasing ones decrease one-bond carbon-fluorine coupling in fluorobenzene, particularly when they are ortho and para to fluorine. These observations can be explained by cannonical resonance formulae which take (+)- and (−)-M effects into account. The data of fluoroanilines ((+)-M) and fluorobenzaldehydes ((−)-M) provide typical examples (Table 4.60).

270   4 ^{13}C NMR Spectroscopy of Organic Compounds

In contrast to the $^3J > {}^2J$ relation known from carbon-proton couplings of benzene, benzenoid two-bond carbon-fluorine coupling constants are larger ($^2J_{CF} \sim 15-25$ Hz) than $^3J_{CF}$ values ($\sim 7 \pm 2$ Hz, Table 4.60) due to the strong positive polarization of the fluorinated carbon within the coupling path. Electronegative substituents at the coupling carbon, however, reduce $^2J_{CF}$. Fluorobenzene and its *o*-halogenated derivatives illustrate this effect.

Benzenoid three-bond carbon-fluorine couplings follow overall trends described for $^3J_{CH}$ (Section 3.2.2.4). *Transoid* couplings are larger than *cisoid* ones, as shown for the CH carbon nuclei of 1-fluoronaphthalene [394]. But this example also shows that *transoid* $^3J_{CF}$ is reduced for bridgehead carbons. Further, electronegative substituents within the coupling path attenuate $^3J_{CF}$ as shown for the *o*-substituted fluorobenzenes in Table 4.60. Attached to the coupling carbon, however, electronegative groups increase $^3J_{CF}$, for example, in the *m*-substituted fluorobenzenes (Table 4.60).

2-Fluoronaphthalene (Hz)

Longer-range coupling constants up to $^8J_{CF}$ can be resolved in fluorinated fused aromatic compounds such as isomeric fluoropyrenes [394]. They are particularly large for closely spaced coupling nuclei. A $^5J_{CF} = 24$ Hz found for the 8-methyl carbon of 4,5,8-trimethyl-1-fluorophenanthrene is a striking example which indicates the presence of through-space mechanisms in carbon-fluorine long-range coupling [403].

1- Fluoropyrene   2-   4,5,8-Trimethyl-1-fluorophenanthrene (Hz)

4.11.7 ^{13}C- and ^{12}C-Enriched Aromatic Compounds for Structural Assignments and Mechanistic Studies

^{13}C NMR spectroscopy on specifically ^{13}C-enriched compounds may lead to the unequivocal elucidation of doubtful structures, as has been shown for the dimer of [α-^{13}C]-triphenylmethyl [404a]. According to general chemical shift rules, the two resonances at 138.6 and 63.9 ppm are in accordance with the shift ranges of sp^2 and sp^3 carbon atom signals, thus confirming structure I instead of II for the dimer. The dimer of 9-phenyl-[9-^{13}C]-fluorenyl displays one resonance only at 94.35 ppm, thus indicating a hexaarylethane structure III [404b].

[1-^{13}C]Naphthalene, investigated before and after heating in Al$_2$O$_3$ with benzene, gives identical ^{13}C NMR spectra [405a]. Thus no automerization (equal ^{13}C enrichment on all

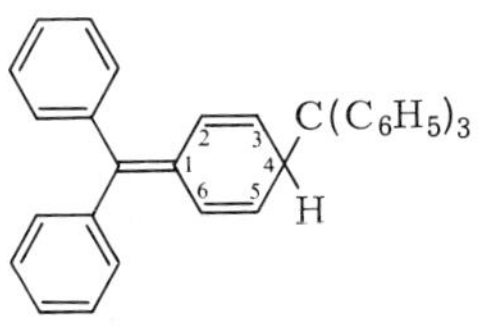

4.12 Aromatic Compounds

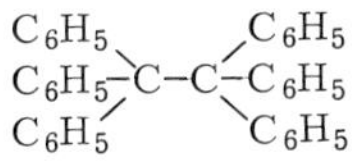

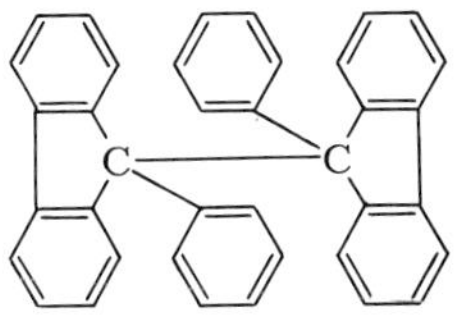

[Structures I, II, III shown at top of page]

carbon positions) of naphthalene occurs under these conditions as was earlier reported [405b].

^{13}C NMR measurements of 6-dimethylamino-6-methylfulvene, synthesized from ^{13}C-7 labeled phenyldiazomethane *via* fulvenallene and ethynylcyclopentadiene, supported the proposed mechanism of the ring contraction of phenylcarbene to fulvenallene. The uniform distribution of ^{13}C in the product fulvenallene may be explained in terms of a "preequilibrium": Before ring contraction, phenylcarbene, a bicyclic intermediate and cycloheptatrienylidene interconvert rapidly [406]:

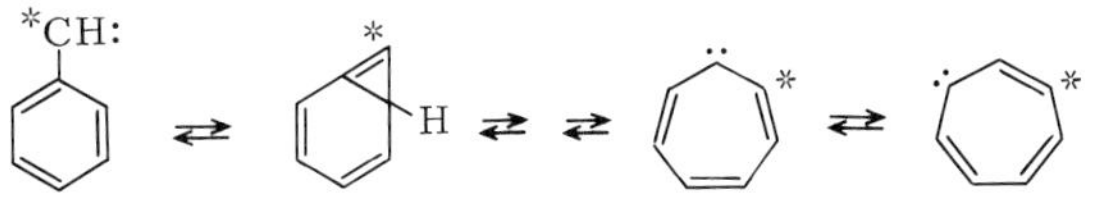

[Equilibrium scheme showing phenylcarbene ⇌ bicyclic intermediate ⇌ cycloheptatrienylidene structures with * denoting labeled carbon]

Instead of ^{13}C labeling, ^{12}C-enriched precursors with a ^{13}C abundance of less than 1.1% can be used for mechanistic studies. In this case, the spectra are more easily analyzed as no homonuclear carbon-carbon coupling occurs. An illustrative example is the base-catalyzed dehydrohalogenation of 7,7-dichloronorcarene to benzocyclopropene [407]. For this reaction, two pathways A and B have been proposed, mechanism B involving a skeletal rearrangement:

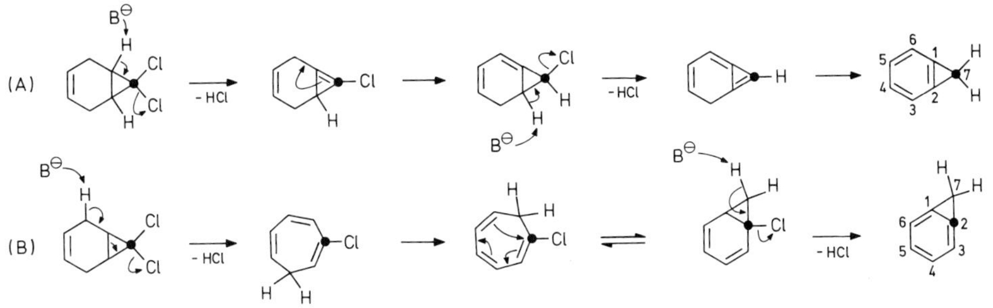

[Schemes (A) and (B) showing reaction mechanisms]

● : ^{12}C-labeled carbon

In order to distinguish between the two alternatives, [7-^{12}C]-7,7-dichloronorcarene was prepared by reaction of 1,4-cyclohexadiene with [^{12}C]-dichlorocarbene originating from [^{12}C]-deuteriochloroform:

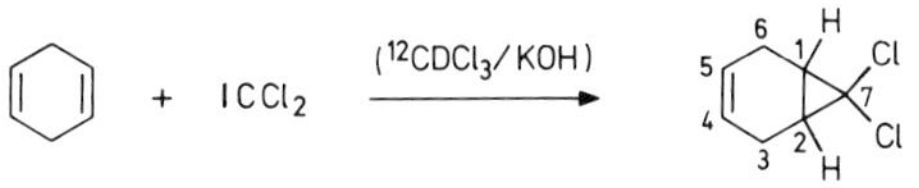

[Reaction scheme: cyclohexadiene + :CCl$_2$ →(^{12}CDCl$_3$/KOH) labeled dichloronorcarene]

After dehydrohalogenation of the labelled dichloronorcarene, a benzocyclopropene was obtained whose CH$_2$ signal (denoted as C-7 with 18.4 ppm, Table 4.57) was lost in the noise [407]. As a result, it was concluded the reaction proceeds *via* mechanism A.

4.12 Heterocyclic Compounds

4.12.1 Heterocycloalkanes

Carbon-13 shift values of parent heterocycloalkanes [408] collected in Table 4.61 are essentially determined by the heteroatom electronegativity, in analogy to the behavior of open-chain ethers, acetals, thioethers, thioacetals, secondary and tertiary amines. Similarly to cyclopropanes, three-membered heterocycloalkanes (oxirane, thiirane, and azirane derivatives) display outstandingly small carbon-13 shift values due to their particular bonding state. Empirical increment systems based on eq. (4.1) permit shift predictions of alkyl- and phenyl-substituted oxiranes [409] and of methyl-substituted tetrahydropyrans, tetrahydrothiapyrans, piperidines, 1,3-dithianes, and 1,3-oxathianes [408], respectively. Methyl increments of these heterocycloalkanes are closely related to those derived for cyclohexane (Table 4.7) due to common structural features of six-membered rings.

Low-temperature ^{13}C NMR indicates piperidine to prefer the conformation with equatorial hydrogen, while the N-methyl derivative occurs exclusively with an equatorial methyl group [410].

The shift values of some methyl-substituted piperidines (Table 4.62) [410] illustrate how to distinguish cis- from trans-isomeric methylated piperidines by using 1,3-diaxial shieldings arising from the γ-effect.

In contrast to cyclohexane, the γ-anti-effect generates a shielding in a 4-position to the heteroatom in heterocyclohexanes, as shown for the 3,5-dimethyl derivatives [410]. Shielding increases with the electronegativity of the heteroatom.

3,5-Dimethylcyclohexane   -tetrahydrothiapyran   -piperidine   -tetrahydropyran

Protonation of piperidines shields α, β, and γ to nitrogen provided these positions are unsubstituted. This behavior corresponds to that found for open-chain amines (Section 4.9.1). Quaternization also shields β and γ to nitrogen but deshields in the α position due to a β effect of the N-alkyl group [410].

Table 4.61. ^{13}C Chemical Shifts of Parent Heterocycloalkanes (δ_C in ppm) [408].

Compound	C-2	C-3	C-4	C-5	C-6	NCH$_3$
Oxirane	40.6					
Oxetane	72.6	22.9				
Tetrahydrofuran	67.9	25.8				
Tetrahydropyran	69.5	27.7	24.9			
Oxepane	70.1	31.0	27.0			
Thiirane	18.1					
Thietane	26.1	28.1				
Tetrahydrothiophene	31.7	31.2				
Tetrahydrothiapyran	29.1	27.8	26.6			
Thiepane	33.8	31.6	27.1			
Azirane (Aziridine)	18.2					
Azetane (Azetidine)	48.1	19.0				
Pyrrolidine	47.1	25.7				
Piperidine	47.9	27.8	25.9			
Azepane	49.3	31.5	27.2			
1-Methylazirane	28.5					48.6
1-Methylazetane	57.7	17.5				46.1
1-Methylpyrrolidine	56.7	24.4				42.7
1-Methylpiperidine	57.0	26.6	24.6			48.0
1,3-Dioxolane	95.0		64.5			
1,3-Dioxane	94.8		67.5	27.5		
1,3-Dioxepane	94.7		67.2	30.1		
1,4-Dioxane	67.6					
1,3-Dithietane	18.6					
1,3-Dithiolane	34.4		38.1			
1,3-Dithiane	31.9		29.8	26.6		
1,4-Dithiane	29.1					
1,2-Dimethyltetrahydropyridazine		54.1	23.4			40.9
1,3-Dimethyltetrahydropyrimidine	79.8		54.3	23.8		42.9
1,4-Dimethylpiperazine	54.7					46.5
Piperazine	47.9					
Thiazolidine	55.7		53.1	34.1		
1,3-Oxathiane	69.2		69.7	26.4	27.9	
1,4-Oxathiane	27.0	68.5				
Morpholine	68.1	46.7				
N-Methylmorpholine	66.8	55.9				46.5
Thiomorpholine	28.3	47.9				
1,3,5-Trioxane	93.7					
2,4,6-Trimethyl-1,3,5-trithiane	47.6				20.3 (CH$_3$)	
1,3,5-Trimethylhexahydra-1,3,5-triazine	78.1					40.6
5-Methyl-5-aza-1,3-dioxane	95.7		85.2			39.0
2-Methyl-2-aza-1,4-dioxane		87.4		64.8	66.9	39.8

274   4 ^{13}C NMR Spectroscopy of Organic Compounds

Table 4.62. ^{13}C Chemical Shifts of Methylated N-Methylpiperidines (δ_C in ppm) [410].

Substituent	C-2	C-3	C-4	C-5	C-6	C–CH$_3$[a]	C–CH$_3$[b]	N–CH$_3$
2-Methyl-	59.4	34.8	24.7	26.4	57.2	20.4		43.3
3-Methyl-	64.1	31.2	32.5	25.6	56.0	19.7		46.5
4-Methyl-	56.0	34.4	30.2			21.8		46.4
cis-2,3-Dimethyl-	60.7	34.8	29.3	23.5	52.9	10.8	15.7	43.4
trans-2,3-Dimethyl-	66.1	37.2	34.0	25.7	57.5	17.1	19.7	43.5
cis-2,4-Dimethyl-	59.0	43.7	31.4	34.8	57.3	20.9	22.1	42.9
trans-2,4-Dimethyl-	53.5	40.7	25.2	33.2	49.5	14.6	20.1	43.0
cis-2,5-Dimethyl-	56.4	30.8	28.6	30.1	59.2	14.1	18.9	43.4
trans-2,5-Dimethyl-	59.1	34.9	33.6	31.6	65.3	20.6	19.6	43.2
cis-2,6-Dimethyl-	59.5	35.1	24.7			21.6		38.0
trans-2,6-Dimethyl-	52.9	33.6	19.3			15.1		40.0
cis-3,4-Dimethyl-	61.1	33.9	30.7	31.8	54.0	14.3	15.8	46.7
trans-3,4-Dimethyl-	64.1	37.8	37.0	34.8	56.4	17.1	19.3	46.4
cis-3,5-Dimethyl-	63.5	31.1	41.6			19.5		46.2
trans-3,5-Dimethyl-	63.3	27.3	38.4			19.2		47.0

[a] at lower-numbered carbon.   [b] at higher-numbered carbon.

Piperidine → N-Methylpiperidine   (ppm)

Similarly to N-methyl groups, N-nitroso functions induce deshieldings of the α but shieldings of the β and γ piperidine ring carbon nuclei. Due to the partial π character of the NN bond, N-nitrosopiperidine displays different shifts for the C-2,6 and C-3,5 carbon pairs, and substituted derivatives exist as syn-anti-isomers. 2-Methyl-N-nitrosopiperidine, for example, occurs as a 1:2 syn:anti mixture; carbon-13 shifts indicate the syn isomer to exist as one conformer with an axial methyl group, while both conformers (a and e) are present in the case of the anti isomer [411].

N-Nitrosopiperidine

a syn-   a 60%   e 40%
         anti
2-Methyl-N-nitrosopiperidine

Conversion of thiacycloalkanes into sulfoxides and sulfones deshield the α ring carbon nuclei due to the enhanced (−) inductive effects of SO and SO$_2$ groups. Stronger deshieldings are induced by SO$_2$ in four- and six-membered rings, and by SO in three- and five-membered rings [412]. Carbon nuclei in a β and γ position are shielded, particu-

4.12 Heterocyclic Compounds   275

Table 4.63. ^{13}C Chemical Shifts of Representative 2-Substituted 1,3-Dithianes (δ_C in ppm) [413].

1,3-Dithiane	C-2	C-4,6	C-5	Others			
2-Methyl-	42.2	30.9	25.5	21.3 (α)			
2-Ethyl-	49.3	30.5	25.9	28.9 (α)	11.5 (β)		
2-Butyl-	47.6	30.4	26.1	35.2 (α)	28.9 (β)	22.3 (γ)	14.0 (δ)
2-*iso*-Propyl-	56.2	30.7	26.3	33.5 (α)	20.0 (β)		
2-*t*-Butyl-	61.8	31.2	26.0	35.6 (α)	27.9 (β)		
cis-2,5-Dimethyl-	41.6	36.3	25.2	21.6 (α)	17.4 (5-α)		
trans-2,5-Dimethyl-	41.8	38.1	30.8	20.2 (α)	22.3 (5-α)		
cis-2-*t*-Butyl-5-methyl-	61.8	37.2	24.3	36.0 (α)	27.9 (β)	16.4 (5-α)	
trans-2-*t*-Butyl-5-methyl-	61.3	38.1	31.7	35.4 (α)	28.1 (β)	22.2 (5-α)	
(*E*)-3-Oxopropenyl-	44.6	28.4	25.0	152.1 (α)	133.1 (β)	193.0 (CH=O)	
2-Phenyl-	51.3	31.9	25.0	139.1 (α)	128.7 (*o*)	128.4 (*m*)	127.7 (*p*)
2-(*p*-Methoxyphenyl-)	50.7	32.1	25.1	131.4 (α)	128.9 (*o*)	114.1 (*m*)	159.7 (*p*) 55.2 (OCH$_3$)
2-Formyl-	47.7	25.4	24.9	188.5 (CH=O)			
2-Methoxycarbonyl-	40.0	26.1	25.1	170.3 (α)	52.5 (OCH$_3$)		

larly when the oxygen attached to sulfur occupies an axial configuration. For this reason, two conformers of thiacyclohexane-S-oxide can be distinguished by low-temperature carbon-13 NMR [412].

Thiacyclohexane: 26.6, 27.8, 29.1 (S)

-S-oxide: 24.7, 18.9, 47.5 (S=O); 24.7, 15.5, 45.1 (S=O axial) ⇌ 24.7, 23.3, 52.1 (S=O)

-S,S-dioxide: 22.8, 24.1, 51.2

(ppm)

Carbon-13 shifts of some 2-substituted 1,3-dithianes as protected carbonyl derivatives and nucleophilic carbonyl equivalents are collected in Table 4.63 [413]. Methyl group increments for shift predictions according to eq. (4.1) are available, also taking contributions of the twist in addition to the chair conformation into account [413].

1,3-Dithiane   Chair   2,5-Twist

Some heteropolycycloalkanes, e.g. indolizidine, quinolizidine, tropane, and decahydroquinoline, are parent skeletons of alkaloids. Carbon-13 shifts of these and other parent polycycloalkanes [408] (Tables 4.64 and 4.65) are predominantly determined by heteroatom electronegativity, ring size, stereoisomerism, and dynamic effects. Aza- and thiadecalins are representative examples (Table 4.64). As described for *cis*- and *trans*-decalin (Section 4.1.3), carbon nuclei in the *cis* isomers of aza- and thiadecalins are shielded

Table 4.64. ^{13}C Chemical Shifts of Isomeric 1-Aza- and 1-Thiadecalins (δ_C in ppm) [414, 415].

X	°C	Conformer	C-2	C-3	C-4	C-5	C-6	C-7	C-8	C-9	C-10	NCH$_3$
cis- NH	−68	10% (a)	39.7	n.o.	24.0	31.7	n.o.	n.o.	n.o.	53.9	35.6	
		90% (b)	47.7	21.2	30.6	25.0	26.3	20.3	32.8	54.9	35.1	
	~30		40.6	22.7	29.8	26.6	25.6	21.6	31.9	55.0	35.9	
	65		46.8	23.1	30.1	26.9	25.8	21.9	32.2	55.4	36.3	
trans-NH	~30		47.6	27.6	32.9	32.9	26.5	25.9	34.1	62.5	43.5	
cis- NCH$_3$	~30	71% (a)	55.8	23.4	28.4	28.4	25.0	21.9	25.9	62.9	36.9	42.9
trans-NCH$_3$	~30		57.9	25.8	32.6	33.1	26.0	25.9	30.5	69.3	41.8	42.6
cis- S	60	58% (b)	27.6	24.6	30.0	29.0	23.9	24.2	30.9	42.5	37.1	
trans-S	~30		30.0	28.2	34.4	34.6	26.3	26.8	32.6	47.0	44.3	

n.o.: not observable.

relative to those in the *trans* derivatives due to γ effects of closely spaced axial hydrogen atoms. Moreover, spectra of the *cis* isomers are temperature dependent due to ring inversion, and preferred conformations are detectable at low temperatures. Thus, the ^{13}C NMR spectrum of *cis*-decahydroquinoline indicates conformer (b) to predominate at −68 °C [414], while thiadecalin prefers conformer (a) at −70 °C [415] (Table 4.64).

4.12.2 Non-Aromatic Heterocycles with sp² Ring Carbons

Carbon-13 shift of common non-aromatic heterocycles with endo- and exocyclic double bonds are reviewed in Table 4.66 [416–432]. – Deshieldings of β-carbons induced by carbonyl groups in heterocyclic α, β-enones due to (−)-*M* electron withdrawal (*e.g.* 2-pyrones, coumarins) and shieldings of β carbons in cyclic enol ethers arising from (+)-*M* electron release (*e.g.* 2,3-dihydrofuran and oxepine derivatives in Table 4.66) fully correspond to the effects described for the open-chain analogs. Outstandingly large shift values are observed for the lithiated carbon in cyclic α-lithium enol ethers (Table 4.66). In terms of its α and β carbon-13 shifts, 2,7-dimethyloxepine is also a typical enol ether [420]. Further, 2,6-dimethyl-4-pyrone [421] and flavone [422] display similiar shift values for the α, β-enone substructure.

Surprisingly, imino carbon shifts of strained 2*H*-azirines [431] do not differ substantially from those of open-chain imines (Section 4.9.3). On the other hand, the alkene carbon shifts of 1-azabicyclo[2.2.2]oct-2-ene do not reflect the polarization characteristic of other enamines (Section 4.9.2); (+)-*M* electron donation of the bridgehead enamino nitrogen cannot take place since the lone-pair orbital and the *p* orbitals of the π bond are

Table 4.65. ^{13}C Chemical Shifts of Representative Heteropolycycloalkanes (δ_C in ppm) [408].

Compound	C-1	C-2	C-3	C-4	C-5	C-6	C-7	C-8	C-9 (NCH$_3$)
2-Oxabicyclo[3.1.0]hexane	57.0			27.3	18.4				
2-Oxabicyclo[4.1.0]heptane	52.2			24.7	19.6				
2-Thiabicyclo[4.1.0]heptane	36.7			25.9	19.5				
2-Azabicyclo[4.1.0]heptane	28.8			25.1	20.5				
1-Azabicyclo[2.2.2]octane		47.8	26.8	20.8					
2-Azabicyclo[2.2.2]octane	42.7		46.9	24.2	24.4	27.2			
2-Methyl-2-azabicyclo-[2.2.2]octane	50.9		57.2	25.5	23.3	24.3			(42.5)
2,3-Dimethyl-2,3-diazabicyclo-[2.2.2]octane	52.2				22.8				(43.5)
8-Azabicyclo[3.2.1]octane	54.7	32.9	17.2			29.0			
8-Methyl-8-azabicyclo[3.2.1]-octane (Tropane)	61.2	29.2	15.9			25.6			(40.4)
endo-3-Oxatricyclo[3.2.1.0^{2,4}]-octane	37.7	62.0				25.5		50.4	
exo-3-Oxatricyclo[3.2.1.0^{2,4}]-octane	36.8	51.0				25.3		26.3	
1-Azabicyclo[4.3.0]nonane (Indolizidine)		53.9	20.3	30.1	64.1	30.7	24.2	25.3	52.7
1,5-Diazabicyclo[4.3.0]nonane		55.0	22.6			54.1	25.0		
1-Aza-3-oxabicyclo[4.3.0]nonane		86.0		68.4	60.1	25.1	21.9	24.4	46.9
9-Oxabicyclo[4.2.1]nonane	77.7	36.2	24.4				31.6		
9-Methyl-9-azabicyclo-[4.2.1]nonane	64.7	35.4	24.6				30.2		(42.9)
9-Oxabicyclo[3.3.1]nonane	66.6	29.4	18.9						
9-Thiabicyclo[3.3.1]nonane	33.2	32.1	21.6						
9-Methyl-9-azabicyclo-[3.3.1]nonane	52.3	26.4	20.4						(40.9)
1-Azabicyclo[4.4.0]decane (Quinolizidine)		56.2	25.6	24.4	33.2	62.9			
1,6-Diazabicyclo[4.4.0]decane		58.1	25.4						
2-Oxaadamantane	67.8	36.1	26.5	35.8					
2-Thiaadamantane	33.7	38.5	27.4	36.9					
2-Azaadamantane	47.2	37.6	27.5	37.0					
2-Methyl-2-azaadamantane	53.3	32.4	26.8	36.9					(41.0)
1-Azaadamantane		59.7	33.7	37.2					
1,3,5,7-Tetraazaadamantane		73.7							
2-Oxaspiro[2,4]heptane	65.1		52.1	32.6	25.4				
2-Oxaspiro[2,5]octane	58.8		54.3	33.8	25.0	25.5			
2,6-Dioxaspiro[4.4]nonane	114.6		66.8	25.0	34.9				
2,6-Dioxaspiro[4.5]decane	105.5		66.9	24.2	33.9		61.4	25.8	20.8^a
2-Thiaspiro[4.5]decane	58.4		32.1	29.1	44.2	41.1	25.0	25.8	
2,7-Dioxaspiro[5.5]undecane	94.9		60.3	25.4	18.6	35.8			
2-Thia-7-oxaspiro[5.5]undecane	82.6		25.1	27.2	21.1	41.6		61.6	25.9^b

^{a} C-10: 38.2; ^{b} C-10: 25.9; C-11: 41.6 ppm.

4 ¹³C NMR Spectroscopy of Organic Compounds

Table 4.66. ¹³C Chemical Shifts of Heterocycles with sp² Ring Carbons (δ_C in ppm). Part 1: Oxygen and Sulfur as Heteroatoms [416–424].

Lactones [416]

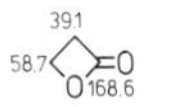

Propio-   γ-Butyro-   δ-Valerolactone

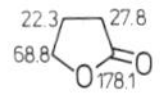

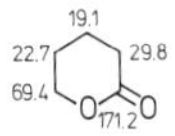

Unsaturated Lactones [417]

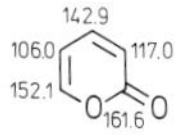

2-Pyrone 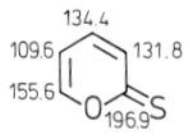   2-Pyranthione 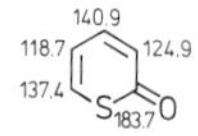   2-Thiapyranone 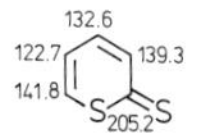   2-Thiapyranthione

Coumarins [418]

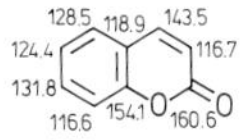

Coumarin 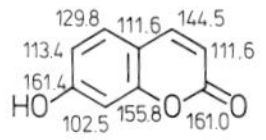   7-Hydroxycoumarin 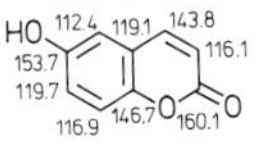   6-Hydroxycoumarin 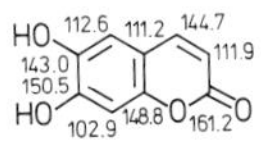   6,7-Dihydroxycoumarin

Cyclic Enol Ethers and Related Compounds [419, 420]

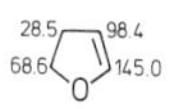

2,3-Dihydrofuran 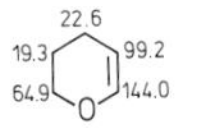   3,4-Dihydro-2H-pyran 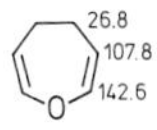   4,5-Dihydrooxepine 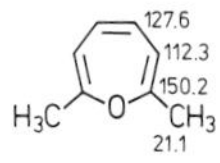   2,7-Dimethyloxepine

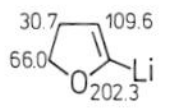

2-Lithium-4,5-dihydrofuran   6-Lithium-3,4-dihydro-2H-pyran

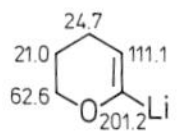

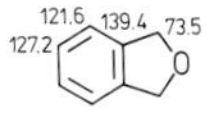

1,3-Dihydro-isobenzofuran   1,2-Methylenedioxy-benzene

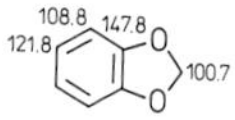

4-Pyrone Derivatives [421]

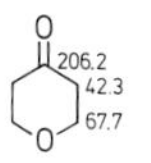

Tetrahydro-pyran-4-one   Tetrahydro-thiapyran-4-one   2,6-Dimethyl-4-pyrone   2,6-Dimethyl-4-pyranthione

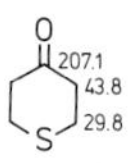

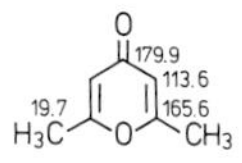

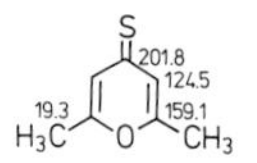

4.12 Heterocyclic Compounds   279

Table 4.66. Part 1 (continued).

Chromanone and Chromone Derivatives [422–424]

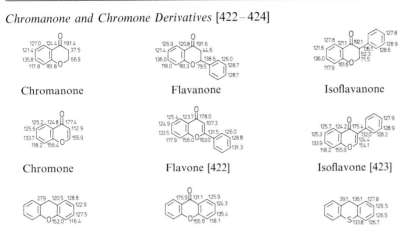

Table 4.66. Part 2: Nitrogen as Heteroatom [425–432].

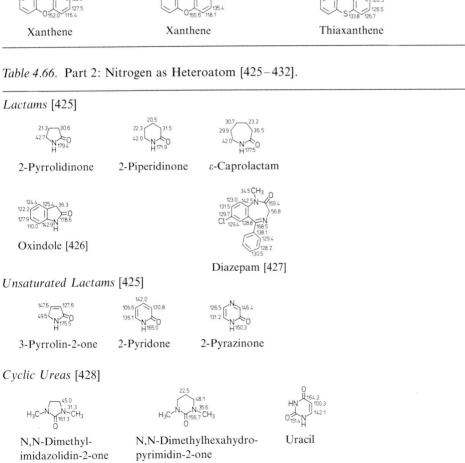

280   4 ¹³C NMR Spectroscopy of Organic Compounds

Table 4.66. Part 2 (continued).

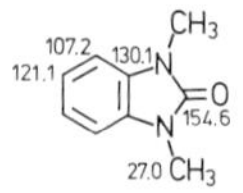

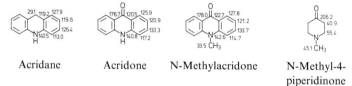

N,N-Dimethyl-1,3-dihydro-benzimidazol-2-one

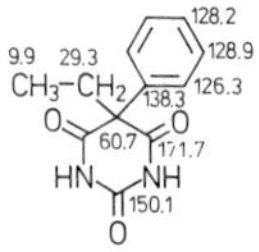

Phenobarbital

Cyclic Benzylamines [429]

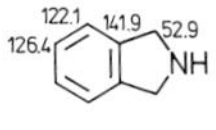

1,3-Dihydro-isoindole

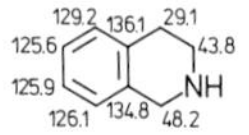

1,2,3,4-Tetrahydro-isoquinoline

Acridane and Related Compounds [430]

Acridane

Acridone

N-Methylacridone

N-Methyl-4-piperidinone

Cyclic Imines [431]

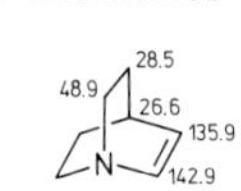

2-Phenyl-2H-azirine

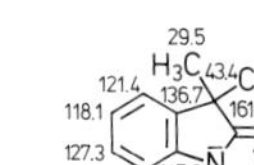

3-Phenyl-2H-azirine

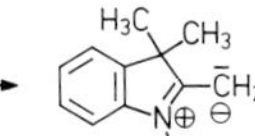

2,3,4,5,6,7,8a-Octahydroquinoline

Cyclic Enamines

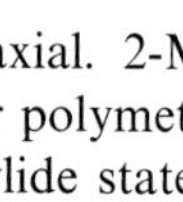

1-Azabicyclo-[2.2.2]oct-2-ene

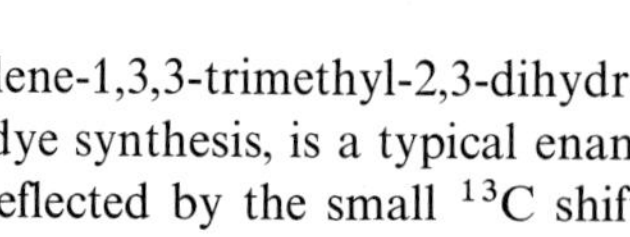

2-Methylene-1,3,3-trimethyl-2,3-dihydroindole

---

not coaxial. 2-Methylene-1,3,3-trimethyl-2,3-dihydroindole, better known as Fischer base for polymethine dye synthesis, is a typical enamine with a significant contribution of the ylide state, as reflected by the small ¹³C shift of the exocyclic sp^2 carbon [432] (Table 4.66).

4.12.3 Heteroaromatic Compounds

Carbon-13 signals of parent heteroaromatic compounds [433–464] (Table 4.67) occur between 105 and 170 ppm. CNDO calculations performed for six-membered heteroaromatic compounds such as pyridine and azines [456, 458, 465] give a linear correlation between carbon-13 shift and total electron density at the individual carbon. The experimental slope, 160 ppm per electron, is the same as that found for monocyclic aromatic compounds (Section 4.11.1). The ^{13}C shift values of pyrrole (107–117 ppm) and pyridine (123–150 ppm, Table 4.67) clearly reflect the chemical classification into π-*excessive* or pyrrole-type and π-*deficient* or pyridine-type heteroaromatic compounds, in accordance with the cannonical formulae:

π-*excessive*

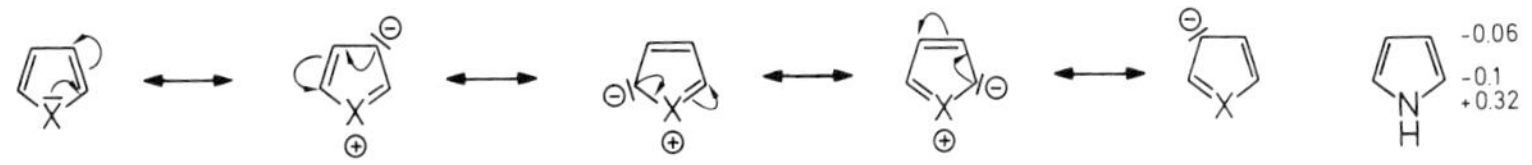

π charge density

π-*deficient*

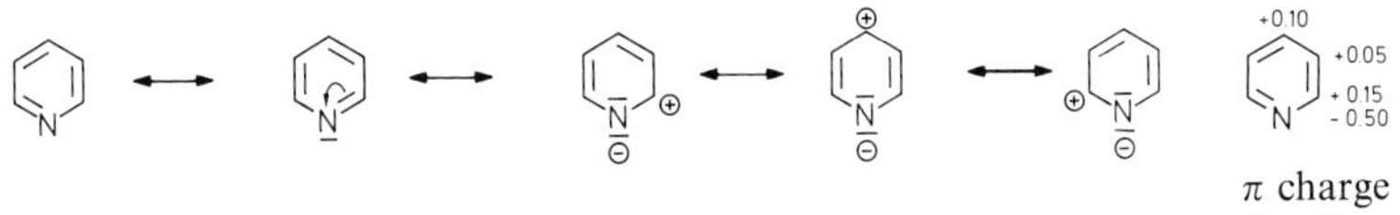

π charge density

Increments for a larger number of substituents in various positions permit ^{13}C shift predictions of pyrroles [435], thiazoles [437], indoles [445], and pyridines [452]. Similarly to benzene, substituted carbon shifts essentially are related to substituent electronegativity, while *ortho*- and *para*-like carbons are predominantly affected by mesomeric (resonance) effects of donor ((−)-M) or acceptor ((−)-M) substituents as shown for methoxy and formyl substituted thiophenes and pyridines.

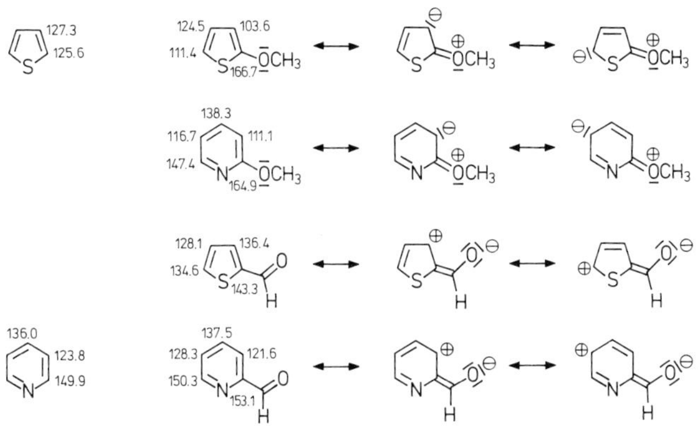

2-Methoxythiophene (+)-M

2-Methoxypyridine (+)-M

Thiophene-2-aldehyde (−)-M

Pyridine-2-aldehyde (−)-M

Table 4.67. ^{13}C Chemical Shifts of Parent Heteroaromatic Compounds with References [433–463] for Substituted Derivatives (δ_C in ppm).

Class/Compound	C-2	C-3	C-4	C-5	C-6	C-7	C-8	C-9	C-10
Five-membered, monocyclic									
Furan [433]	143.6	110.4							
Thiophene [434]	125.6	127.3							
Selenophene	131.0	128.8							
Tellurophene	126.2	137.6							
Pyrrole [435]	117.3	107.6							
N-Methyl-	121.1	107.8			35.2 (NCH$_3$)				
2,2'-Bipyrrole	126.4	103.2	108.7	116.7					
P-Phenylphosphole	135.1	136.7			129.6	133.3	128.3	129.0 (C$_6$H$_5$)	
Isoxazole [436]		150.0	100.5	158.9					
Oxazole [436]	150.6		125.4	138.1					
Isothiazole [437]		157.0	123.4	147.8					
Thiazole [437]	153.6		143.3	119.6					
Pyrazole [438]		134.6	105.8	134.6					
N-Methyl-		139.0	105.4	129.8	38.2 (NCH$_3$)				
anion		138.5	103.4						
cation		135.0	108.4						
Imidazole [439]	135.4		121.9						
N-Methyl	138.3		129.6	120.3	32.6 (NCH$_3$)				
anion	145.1		126.7						
cation	134.6		120.0						
1,2,3-Thiadiazole [440]			147.3	135.8					
1,2,3-2*H*-Triazole [441]			131.6						
N-Methyl-			134.7		41.5 (NCH$_3$)				
1,2,4-1*H*-triazole		147.4							
N-Methyl-		150.7			35.0 (NCH$_3$)				
1,2,4-4*H*-triazole		147.9							
N-Methyl-		144.1			30.7 (NCH$_3$)				
Tetrazole [442]				143.3					
N-Methyl-				144.2	33.7 (NCH$_3$)				
Five-membered, benzo-fused									
Benzo[b]furan [443]	145.0	106.9	121.6	123.2	124.6	111.8	155.5	127.9	
Benzo[b]thiophene [444]	126.4	124.0	123.8	124.3	124.4	122.6	139.9	139.8	
Indole [445]	124.1	102.1	120.5	121.7	119.6	111.0	135.5	127.6	
Benzo[d]isoxazole		147.1	124.3	123.0	130.6	109.9	122.2	162.7	
Benzoxazole [446]	154.1		121.1	126.2	125.2	111.6	150.7	141.1	
Benzo[c]isothiazole		146.4	123.1	124.7	129.2	122.1	162.2	135.4	
Benzothiazole [447]	155.1		124.0	126.6	126.0	122.6	134.5	154.2	
Indazole [448]		133.4	120.4	120.1	125.8	110.0	139.9	122.8	
Benzimidazole [449]	141.5		115.4	122.9			137.9		
2,1,3-Benzoxadiazole			116.9	132.7			149.8		
2,1,3-Benzothiadiazole			121.9	130.0			155.3		
1,2,3-Benzothiadiazole			124.1	129.9	127.7	120.3	141.4	158.9	
1-Methylbenzotriazole [441]			119.3	123.4	126.8	108.8	133.1	145.5	33.7 (NCH$_3$)

Table 4.67. (continued).

Class/Compound	C-1	C-2	C-3	C-4	C-5	C-6	C-7	C-8	C-9 (C-4a)	C-10 (C-9a)
Five-membered, dibenzo-fused										
Dibenzofuran [450]	120.6	122.6	127.0	111.6					156.2	124.2
Dibenzothiophene [450]	121.9	124.6	127.0	122.9					138.5	134.9
Carbazole [441, 450]	120.0	118.4	125.4	110.8					139.6	122.6
N-methyl-	120.1	118.6	125.5	109.9	29.8 (NCH$_3$)				140.5	121.9
Six-membered, monocyclic										
Pyrylium ion (BF$_4^-$) [451]		169.3	127.7	161.2						
Thiapyrylium ion (BF$_4^-$)		158.8	138.3	150.8						
Selenapyrylium ion (BF$_4^-$)		170.7	137.3	149.5						
Pyridine [452]		149.9	123.8	136.0						
cation		142.5	129.0	148.4						
-N-oxide		139.1	125.6	126.1						
2,2'-Bipyridyl		157.5	121.8	137.3	124.4	150.1				
Phosphabenzene		154.1	133.6	128.8						
Arsabenzene		167.7	133.2	128.2						
Stibabenzene		178.3	134.4	127.4						
Pyridazine [99, 453]			152.8	127.6						
cation			151.7	137.7						
Pyrimidine [454]		158.8			156.4	121.4				
cation		152.2			158.8	125.1				
dication		151.0			159.0	128.1				
Pyrazine [455]		145.6								
cation		142.9								
dication		144.0								
1,3,5-Triazine [73j, m]		166.6								
1,2,4,5-Tetrazine [73j, m]			160.9							
Six-membered, benzo-fused										
Quinoline [456]		150.9	121.7	136.1	128.5	127.0	129.9	130.5	149.3	128.9
Isoquinoline [456]	152.5		143.0	120.4	126.4	130.2	127.2	127.6	128.6	135.7
Cinnoline [456]			146.1	124.7	127.9	132.3	132.1	129.5	151.0	126.8
Phthalazine [456]	152.0				126.7	133.2			126.7	
Quinazoline [456]		160.7		155.9	127.6	128.1	134.4	128.6	150.3	125.4
Quinoxaline [456]	145.7				130.0	130.1			143.4	

284   4 ^{13}C NMR Spectroscopy of Organic Compounds

Table 4.67. (continued).

Class/Compound	C-1	C-2	C-3	C-4	C-5	C-6	C-7	C-8	C-9 (C-4a)	C-10 (C-9a)
Six-membered, dibenzo-fused										
Dibenzodioxine [457]	116.2	123.6							142.2	
Phenoxathiine [457]	127.4	124.2	126.5	117.5					151.9	119.9
Phenoxazine [457]	114.5	120.0	123.0	112.8					131.8	142.7
Phenothiazine [457]	126.7	121.3	125.6	113.8					141.7	116.8
Acridine [458]	129.6	128.4	125.5	130.3					149.2	126.6
Phenazine [459]	131.0	130.3							144.0	
Fused heterocyclic rings										
Indolizine [458]	99.5	114.1	113.0		125.6	110.5	117.2	119.6	133.4	
1-Aza-		134.1	113.4		127.0	112.2	124.6	117.6	145.6	
2-Aza-	120.0		128.4		122.8	112.7	119.4	118.2	130.7	
3-Aza-	96.3	141.3			128.1	110.9	112.4	117.4	139.5	
1,8-Diaza-		135.2	112.7		136.1	109.4	151.0		148.9	
1,2,3-Triaza-					125.6	116.8	132.1	115.9	148.6	
Pyrrolo[2,3-b]-pyridine		125.5	100.5	129.0	115.6	142.1			120.7	148.9
Pyrrolo[2,3-d]-pyrimidine [406]		127.2	99.5	148.9		150.9			118.2	151.3
Purine [461		146.1 (C-8)		145.6 (C-6)		152.2 (C-2)			130.5 (C-5)	154.7 (C-4)
1,5-Naphthyridine [462]		151.6	124.2	137.3					143.9	
1,6-Naphthyridine		154.8	122.5	135.6	152.9		147.4	123.8	150.0	122.1
1,7-Naphthyridine		151.9	125.1	134.6	119.8	143.7		154.2	143.3	131.1
1,8-Naphthyridine		156.1	122.8	153.5					153.5	122.1
2,6-Naphthyridine	152.1		145.9	119.6					131.0	
2,7-Naphthyridine	153.4		147.2	119.6					124.3	138.5
Pyrido[3,4-b]-pyrazine		148.4	150.2		122.8	147.8		155.8	139.0	146.3
Pteridine [463]		163.7		159.0		147.8	152.4		152.4	135.0
1,10-Phenanthroline [452]		150.6	123.4	136.3	126.8				129.0	146.5
Azaazulenes										
5-Azaazulene [464]	121.9	138.4	123.6	152.9		155.0	118.1	140.3	139.5	133.5
1,3-Diazaazulene		164.8		139.2	133.8	136.1			162.5	

Indolizine   Pyrrolo[2,3-]pyridine   Pyrrolo[2,3-d]pyrimidine   Purine

Naphthyridines   Pteridine   5-Azaazulene   1,3-Diazaazulene

4.12 Heterocyclic Compounds   285

Electron withdrawal of the aldehyde function at position 3 in pyridine-2-aldehyde is overcompensated by the electric field effect of the negatively charged aldehyde oxygen. – Cooperating resonance effects may induce extreme shieldings or deshieldings. 5-Hydroxy-3-methyl-pyrazole, for example, displays a rather shielded sp^2 carbon for C-4 (89.2 ppm) due to $(+)$-M electron release of both the 5-OH group and the ring nitrogen:

5-Hydroxy-3-methylpyrazole

The lone-pair electrons of bridgehead nitrogens in indolizine and its aza analogs [458] are delocalized, as concluded from carbon-13 shifts and in accordance with CNDO calculations: All ring carbons of the parent indolizine except C-5 and C-9 (Table 4.67) are shielded (99–120 ppm) due to the $(+)$-M electron releasing effect of the bridgehead nitrogen.

Aromatic nitrogen heterocycles display considerable medium shifts (Section 3.1.4). Carbon-13 shifts of pyridine decrease in α but increase in β and γ position upon addition of water (Fig. 4.12). The dilution effect is explained in terms of intermolecular hydrogen bonding between pyridine and water [99].

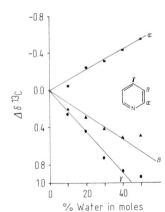

Fig. 4.12. ^{13}C shift changes of pyridine carbon nuclei induced by dilution with water [99].

Protonation shifts of pyridine (Table 4.67) are much stronger but follow the trend of dilution shifts portrayed in Fig. 4.12. Shielding in the α position is attributed to a change of the N–C-α bond order. Deshieldings at C-3 and particularly at C-4 arise from an increased electron withdrawal of the positively charged nitrogen. Shift changes induced

286   4 ^{13}C NMR Spectroscopy of Organic Compounds

by quaternization to the N-methiodide display the same signs. Carbon-13 shifts *versus* pH titration plots are illustrated in Fig. 3.5 [94] (Section 3.1.4.3). – The signs of protonation shifts in heteroaromatic compounds with two dissociation steps are not generally the same for both steps, as shown for pyrazole and pyrazine in Table 4.67. – Shift changes of pyridine induced by *N-oxidation* clearly reflect an increase of negative charge α and γ to nitrogen, as given by the cannonical formulae and as used for activation of electrophilic pyridine substitution.

Pyridine

protonated form   +12.4 +5.5 –7.4   ←   136.0 123.8 149.9   →   +11.1 +6.9 –2.2   N-Methiodide

↓

–10.7 +2.0 –11.3

N-oxide   (ppm)

Fig. 4.13. illustrates how to apply carbon-13 NMR for analysis of tautomerism in heterocyclic chemistry: 3-Methyl-5-oxo-1-phenyl-4,5-dihydropyrazole (the "Knorr-pyrazolone") is shown to exist as the CH tautomer B with a CH_2 carbon at 43.1 ppm in chloroform solution (Fig. 4.13(a)), while the OH tautomer A predominates (90%) in hexadeuteriodimethyl sulfoxide (Fig. 4.13(b)) [73i].

4.12.4 Typical Coupling Constants

One-bond carbon-proton coupling constants J_{CH} of heterocycloalkanes [466] increase with decreasing ring size and, in α position, with increasing electronegativity of the heteroatom (Table 4.68). Thus, oxirane displays the largest J_{CH} in the series. One-bond CH couplings of unsaturated and aromatic heterocycles (Table 4.68) also reflect heteroatom electronegativities, as shown for the flavone/isoflavone isomeric pair as well as the series furan, thiophene, pyrrole, azoles, pyridine, and azines [115, 130, 416–464]. Values of J_{CH} larger than 200 Hz characterize heteroaromatic carbon nuclei attached to two electronegative heteroatoms. Oxazole, thiazole, imidazole, pyrimidine, and triazine exemplify this trend in Table 4.68. These data may help to identify differently substituted heterocycles; 2,4,6- and 4,5,6-triaminopyrimidine [73i] are typical examples: In the 2,4,6-isomer, C-5 appears as a doublet with J_{CH} = 167.7 Hz, strongly shielded (75.5 ppm) due to three electron releasing amino groups. The 4,5,6-isomer, however, displays a doublet signal with stronger CH coupling (202.5 Hz) at 144.3 ppm, characterizing an unsubstituted C-2 of the pyrimidine ring.

4.12 Heterocyclic Compounds   287

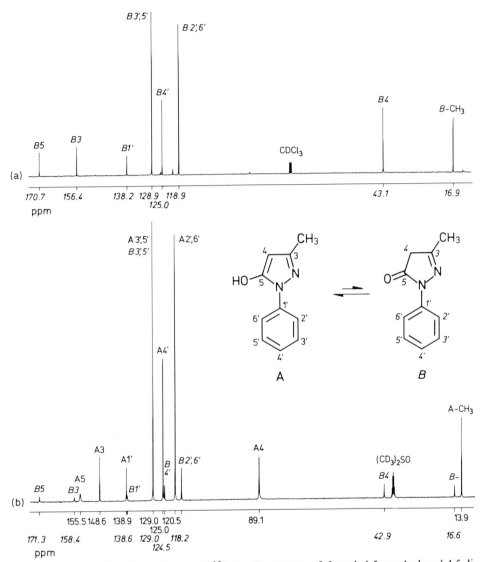

Fig. 4.13. Proton broadband-decoupled ^{13}C NMR spectra of 3-methyl-5-oxo-1-phenyl-4,5-dihydroxypyrazole (saturated solutions, 30 °C, 100.576 MHz, 8 scans, 2 s as relaxation delay); (a) in deuteriochloroform; (b) in hexadeuteriodimethylsulfoxide.

288   4 ^{13}C NMR Spectroscopy of Organic Compounds

Table 4.68. One-Bond Carbon-Proton Coupling Constants of Heterocyclic Compounds (in Hz) [115, 130, 416–464, 466, 467].

Heterocycloalkanes

Heterocycle	X = O		X = S		X = NH	
	C-α	C-β	C-α	C-β	C-α	C-β
(3-ring)	175.7		170.6		168.0	
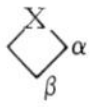 (4-ring)	149.5	137.4	146.5	134.6	140.0	134.0
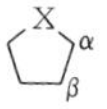 (5-ring)	144.6	133.2	142.1	126.3	139.1	131.4
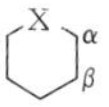 (6-ring)	139.4	128.0	135.0	127.7	133.7	125.6

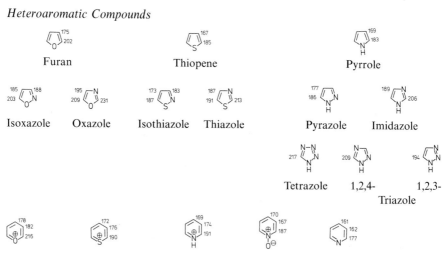

Unsaturated Heterocycles

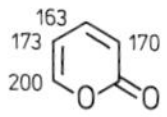

2-Pyrone, 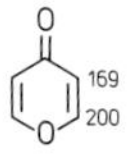 4-Pyrone, 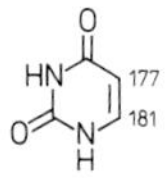 Uracil

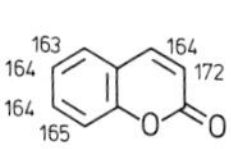

Coumarin, 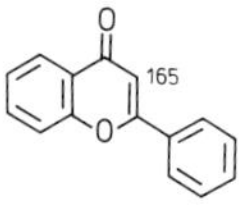 Flavone, 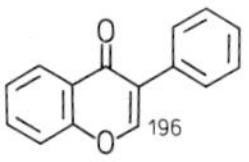 Isoflavone

Heteroaromatic Compounds

Furan, Thiopene, Pyrrole

Isoxazole, Oxazole, Isothiazole, Thiazole, Pyrazole, Imidazole

Tetrazole, 1,2,4-Triazole, 1,2,3-Triazole

Pyrylium-, Thiapyrylium-, Pyridinium-ion, Pyridine N-oxide, Pyridine

4.12 Heterocyclic Compounds   289

Table 4.68. (continued).

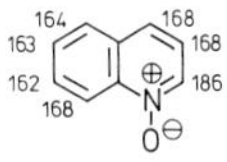

Quinoline N-oxide 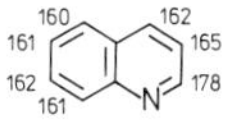   Quinoline 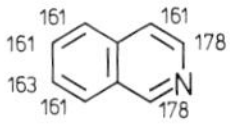   Isoquinoline

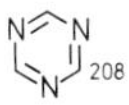

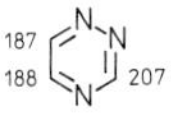

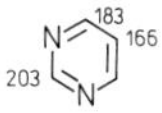

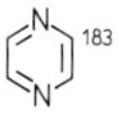

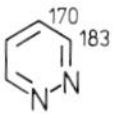

1,3,5-Triazine   1,2,4-Triazine   Pyrimidine   Pyrazine   Pyridazine

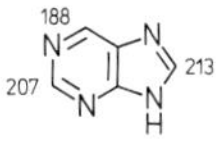

Purine

Table 4.69. Two- and Three-Bond Carbon-Proton Coupling Constants of Representative Heterocyclic Compounds (in Hz) [115, 130, 416–464, 466, 467].

Two-Bond Carbon-Proton Coupling Constants
Unsaturated Heterocyclic Compounds

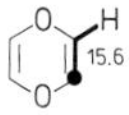

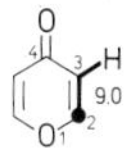

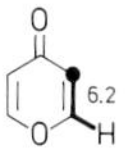

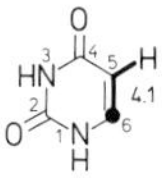

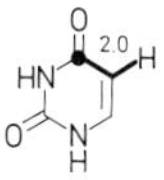

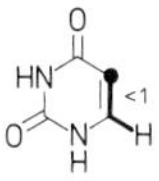

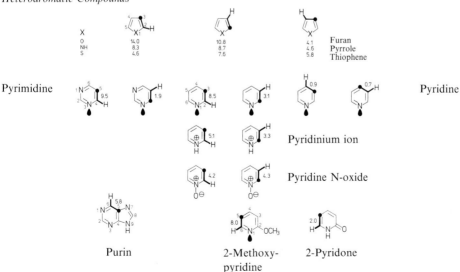

1,4-Dioxine   4-Pyrone   Uracil

Heteroaromatic Compounds

Pyrimidine   Furan / Pyrrole / Thiophene   Pyridine

Pyridinium ion

Pyridine N-oxide

Purin   2-Methoxy-pyridine   2-Pyridone

Table 4.69. (continued).

Three-Bond Carbon-Proton Coupling Constants

X	Furan / Pyrrole / Thiophene (H at 2, coupling to C-4,3)	Position 2-H	Position (with H at ring)	
O	6.9	7.0	6.1	Furan
NH	6.6	7.5	7.4	Pyrrole
S	5.0	10.0	9.8	Thiophene

X	Pyridine series				
N	11.7	6.7	6.4	5.7	Pyridine
N⊕–H	6.4	6.9	7.2	6.1	Pyridinium ion
N⊕–O⊖	4.1	8.1	8.9	6.8	Pyridine N-oxide

Pyrimidine: H 10.4

(pyrazine-type): 5.3

Purine:
8.1 (8-H)
10.3 (2-H) ▸
10.4 (6-H) ▸
10.9 (2-H)
9.2 (8-H)
4.8 (6-H)

Similarly, the large α-CH couplings of pyridine, quinoline, and isoquinoline can be used to locate substituents or fused rings in these heterocycles [73 d, i]. N-Oxidation and protonation further enhance J_{CH} of the α-CH bond in pyridine (Table 4.68) due to the positive charge at nitrogen. To conclude, the sites of oxidation and protonation can be derived from the magnitudes of J_{CH} in pyridines, azines, and their fused derivatives [467].

Two-bond carbon-proton coupling constants $^2J_{CH}$ of heterocyclic compounds increase with the electronegativity of the heteroatom attached to the coupling CCH substructure (O > N > S). Table 4.69 verifies this for unsaturated heterocycles and the parent five-membered heteroaromatic compounds [130]. In pyridine and azines, $^2J_{CH}$ is considerably enhanced when the nonbonding electron pair at nitrogen is *cis* to the CH bond of the coupling hydrogen [467]. The influence of the lone-pair is attributed to hyperconjugative interaction: Charge transfer from the nonbonding electron pair at nitrogen to the attached HCC substructure induces a positive contribution to $^2J_{CH}$. If the lone-pair at nitrogen is not available because of protonation or N-oxidation, two-bond coupling with the *cisoid* proton is attenuated to values close to those with *transoid* hydrogens (3–4 Hz, Table 4.69). The lone-pair effect can be applied to identify the existent tautomers of heterocycles. 2-Hydroxypyridine, for example, displays a two-bond CH coupling of 2 Hz with the proton at C-6 (Table 4.69). No lone-pair at nitrogen is available in the plane of the ring. Thus, the compound exists as the 2-pyridone tautomer. In contrast, 2-methoxy- and 2-aminopyridine display a $^2J_{CH} = 7$–8 Hz of C-5 with 6-H, locating the lone-pair at nitrogen [468].

Three-bond carbon-proton coupling constants $^3J_{CH}$ of five-membered heteroaromatic compounds exhibit trends opposite to those outlined for substituted benzenes in Section 3.2.2.4: When the heteroatom is located within the coupling path, $^3J_{CH}$ increases with the heteroatom electronegativity; if the heteroatom is attached to the coupling carbon but outside of the three-bond coupling path, increasing heteroatom electronegativity reduces $^3J_{CH}$ (Table 4.69). – In pyridines and azines, particularly large $^3J_{CH}$ couplings

4.12 Heterocyclic Compounds

Table 4.70. Carbon-Fluorine Coupling Constants of Isomeric Fluoropyridines and of 2-Fluoropurine (in Hz) [469]. If available, Couplings of the Protonated Heterocycles are Printed in Parentheses.

Compound	J_{CF}	$^2J_{CF}$	$^3J_{CF}$		$^4J_{CF}$
			C-4	C-6	
2-fluoropyridine	(−) 236.3 (263.4)	37.6 (24.4)	7.5 (9.8)	14.9 (1.5)	4.2 (3.4)
3-fluoropyridine	(−) 255.1 (255.3)	22.5 (34.5)	4.3 (7.0)		3.7 (3.7)
4-fluoropyridine	(−) 261.8 (280.2)	16.1 (21.4)	6.4 (12.2)		
2-fluoropurine	(−) 207.5		17.1	15.9	

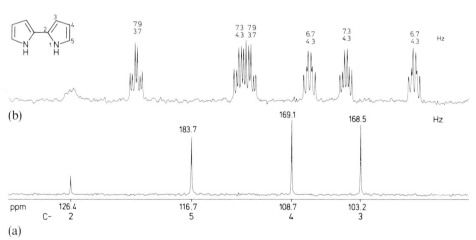

Fig. 4.14.(a) Proton broadband- and (b) gated-decoupled ^{13}C NMR spectrum of 2,2′-bipyrrole (50 mg in 1 mL deuteriochloroform, 20.115 MHz, 1000 scans). One α-CH coupling of 183.7 Hz (not two!) identifies the 2-substituted pyrrole ring; CH coupling constants of unsubstituted carbon nuclei are assigned as follows:
C-5: $J = 183.7$ Hz; $^2J = 7.9$ Hz with 4-H; $^2J = 3.7$ Hz with NH; $^3J = 3.7$ Hz with 3-H;
C-4: $J = 169.1$ Hz; $^2J = 7.3$ Hz with 5-H; $^2J = 4.3$ Hz with 3-H; $^3J = 4.3$ Hz with NH;
C-3: $J = 168.5$ Hz; $^2J = 4.3$ Hz with 4-H; $^3J = 6.7$ Hz with 5-H; $^3J = 4.3$ Hz with NH.

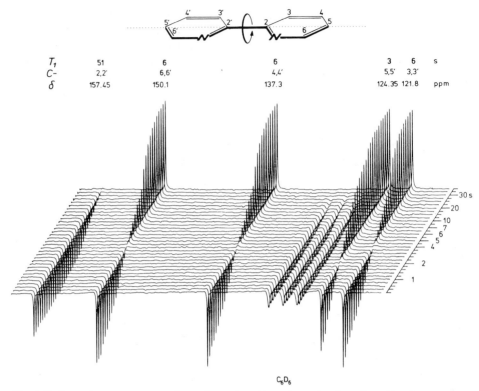

Fig. 4.15. Inversion-recovery experiment for carbon-13 T_1 determination of 2,2'-bipyridine (400 mg in 1 mL hexadeuteriobenzene, 30 °C; 15.08 MHz; 16 scans for a single experiment [73 i]). The principal axis of (the fastest) molecular rotation passes C-2 (2') and C-5 (5'). This rotation is too fast for optimum dipolar relaxation of C-3, C-4, and C-6 but does not influence the C-5-H bond which is affected by rotation about other axes. These rotations are slower and more effectively contribute to dipolar spin-lattice relaxation of C-5 (5') according to Section 3.3.3.3. To conclude, C-5 (5') relaxes faster (3 s) than all other CH carbons (6 s) and can be clearly distinguished from C-3 (3') with similar shift.

(~ 12 Hz) are observed for the CNCH substructure including the ring nitrogen and its lone-pair within the coupling route [467] (Table 4.69); a lone-pair blockade arising from protonation or N-oxidation reduces $^3J_{CH}$ via nitrogen. Other $^3J_{CH}$ couplings of pyridine and azines are slightly smaller than those of benzenoid carbons (5–7 Hz, Table 4.69). Using the $^3J_{CNCH} > {}^3J_{CCCH}$ relation, three-bond coupling of C-4 with 6-H in pyrimidine (5.3 Hz) and purine (4.8 Hz, Table 4.69) permits an unequivocal assignment of the corresponding carbon signals.

Carbon-fluorine coupling constants of fluoropyridines and 2-fluoropurine [469] follow the trends of carbon-proton coupling: The lone-pair at nitrogen induces a positive contribution to all CF couplings of 2-fluoropyridine and 2-fluoropurine which is eliminated upon protonation. The large J_{CF} of 4-fluoropyridine arises from (−)-M electron withdrawal of the ring nitrogen at C-4 (Table 4.70).

Finally, a coupled and decoupled ^{13}C NMR spectrum of 2,2'-bipyrrole (Fig. 4.14(a,b)) and an inversion-recovery series of 2,2'-bipyridine (Fig. 4.15 [73 i]) illustrate signal assignments of heteroaromatic compounds by means of carbon-proton couplings and spin-lattice relaxation times. These spectra also exemplify the characteristic shift differences between π-*excessive* (2,2'-bipyrrole) and π-*deficient* heteroaromatic compounds (2,2'-bipyridine).

4.13 Organometallic Compounds

4.13.1 General

Detailed reviews [470–475] have dealt with numerous investigations of organometallic compounds and transition metal complexes using carbon-13 as NMR probe. Looking at the data selected in Tables 4.71 and 4.72, it is seen that some important features must be considered when recording and evaluating carbon-13 NMR spectra of organometallic compounds.

Table 4.71. ^{13}C Chemical Shifts (δ_C in ppm) and Carbon-Metal Coupling Constants (J_{CM} in Hz) of Representative Methyl-Metal Compounds [474]. Coupling Constants refer to the more Abundant Metal Isotope.

Main Group Metals			Transition Metals		
Compound	δ_C (ppm)	J_{CM} (Hz)	Compound	δ_C (ppm)	J_{CM} (Hz)
CH_3Li	−16.6	15	$[(CH_3)_8Cr_2]^{4-}$	9.0	
CH_3MgI	−14.5		$[(CH_3)_8Mo_2]^{4-}$	4.7	
			$[(CH_3)_8W_2]^{4-}$	2.2	
$(CH_3)_3B$	14.8	46.5	$(CH_3)_6W$	83.8	400
$(CH_3)_4B^-Li$	6.2	39.4			
$(CH_3)_4Al_2$	− 5.6 (bridge)	110.0	$[(CH_3)_8Re]^{2-}$	16.1	
	− 8.2 (term.)		$CH_3Re(CO)_5$	−38.0	
$(CH_3)_3Tl$		1930.0	$[CH_3Fe(CO)_4]^-$	−16.4	
$(CH_3)_3SiH$	− 2.6	− 50.8	$[(CH_3)_6Pt]^{2-}$	−10.6	434
$(CH_3)_4Si$	0	− 50.3	$[(CH_3)_3PtI]_4$	13.2	686
$(CH_3)_6Si_2$	− 2.1	− 43.6			
$(CH_3)_3SiF$	− 0.3	− 60.5	$(CH_3)_2Zn$	− 4.2	
$(CH_3)_3SiCl$	4.1	− 57.4	$(CH_3)_2Cd$	1.0	−537.5
$(CH_3)_3SiBr$	4.6	− 56.0	$(CH_3)_2Hg$	22.1–23.7	687–692
$(CH_3)_3SiI$	6.5	− 54.0	$(CH_3)HgCN$	6.0	1235
$(CH_3)_4Ge$	− 1.4	− 18.7	$(CH_3)HgCl$	8.4	1674
$(CH_3)_4Sn$	− 9.3	−337.2	$(CH_3)HgBr$	11.7	1631
			$(CH_3)HgI$	17.1	1540
$(CH_3)_4Pb$	− 4.2	250.2			

Table 4.72. ^{13}C Chemical Shifts (δ_C in ppm) of Selected Metal Carbonyls and Metallocenes [474].

Metal Carbonyls		Metallocenes	
Compound	δ_C (ppm)	Compound	δ_C (ppm)
$Cr(CO)_6$	212.5	$(\pi\text{-}C_5H_5)_2V$	−790
$Mo(CO)_6$	202.0	$(\pi\text{-}C_5H_5)_2Cr$	−570
$W(CO)_6$	192.1	$(\pi\text{-}C_5H_5)_2Fe$	69.2 [b]
		$(\pi\text{-}C_5H_5)_2Fe^+PF_6^-$	−314.5
$Mn_2(CO)_{10}$	212.9 (*cis*) 223.1 (*trans*)	$(\pi\text{-}C_5H_5)_2Co$	549
$Re_2(CO)_{10}$	192.7 (*cis*) 183.7 (*trans*)	$(\pi\text{-}C_5H_5)_2Ni$	1430
$Fe(CO)_5$	211.0 [a]	$(\pi\text{-}C_5H_5)_2Ti(C_6H_5)_2$	192.9 (C-1)
$Ru_3(CO)_{12}$	188.7 (*cis*) 198.8 (*trans*)		136.0 (C-2,6)
			127.0 (C-3,5)
$Co_2(CO)_8$	203.2		124.4 (C-4)
$Rh_4(CO)_{12}$	190.3		116.8 (C_5H_5)
$Ni(CO)_4$	191.6		

[a] $J_{C^{57}Fe} = 23.5$ Hz; [b] $J_{C^{57}Fe} = 5.0$ Hz.

The usual *carbon-13 chemical shift range* of organic compounds (∼ 250 ppm) considerably *expands* in organometallic compounds. Paramagnetic metals in *metallocenes* (Table 4.72) induce particularly large ^{13}C shift values [476].

Coupling of carbon with attached metals having a spin quantum number of $I > 0$ *distributes signal intensity among* $(2I + 1)$ *multiplet lines. Carbon-metal coupling constants cover a range of several KHz* and may cause difficulties in signal assignment. The magnitudes of carbon-metal coupling constants can be predicted according to eqs. (4.18) [477].

$$K_{AB} = K_{AB}^1 + K_{AB}^2 + K_{AB}^3 \quad (4.18\,a)$$

K_{AB}^1: orbital term

K_{AB}^2: dipolar term

$$K_{AB}^3: \tfrac{16}{9}\pi\mu_0\mu_B^2(^3\Delta E)^{-1}s_A^2(0)s_B^2(0)P_{s_As_B}^2 \quad (4.18\,b)$$

μ_0: permeability in vacuum

μ_B: Bohr magneton

$^3\Delta E$: average triplet excitation energy

$s_A^2(0)$: electron density at nucleus A

$P_{s_As_B}$: s electron bond order of the AB bond

$$J_{AB} = \frac{h\gamma_A\gamma_B}{4\pi^2}K_{AB} \quad (4.18\,c)$$

Agreement of predicted and measured carbon-metal coupling constants for metals without lone-pairs is satisfactory or poor, *e.g.*

^{183}W(CO)$_6$: J_{CW} = 126 Hz (calc.: 137 Hz)
^{199}Hg(CH$_3$)$_2$: J_{CHg} = 689 Hz (calc.: 808 Hz)
^{207}Pb(CH$_3$)$_4$: J_{CPb} = 251 Hz (calc.: 432 Hz).

Quadrupolar nuclei with $I \geq 1$ and large natural abundance, such as $^{10(11)}$B, ^{27}Al, ^{51}V, ^{55}Mn, ^{59}Co, or ^{93}Nb, give rise to *line-broadenings of attached and remote carbon signals* due to scalar relaxation. Signals may be unobservable when the usual high-resolution techniques are applied. Line-broadening can be reduced in some cases by lowering the measuring temperature.

Quaternary carbon nuclei (*e.g.* CO carbons in metal carbonyls) display particularly large spin-lattice relaxation times, so that relaxation reagents such as chromium acetylacetonate must be added to the sample solution in order to obtain sufficient signal:noise within reasonable accumulation times.

4.13.2 Group I Organometallic Compounds

Since alkali metals are electropositive, they induce significant shieldings of carbon nuclei α to the metal; methyllithium in Table 4.71 is a typical example. Covalent bond contributions are small. Organometallic compounds with heavier alkali metals predominantly exist as contact ion pairs (Section 4.14.2). Due to oligomerization and carbon-metal bond exchange within the oligomers, ^{13}C NMR spectra of organolithium compounds depend on temperature. Coupling constants of carbon with the quadrupolar isotopes ^{6}Li ($I = 1$; 7.4%) and ^{7}Li ($I = 3/2$; 92.6%) can be resolved at low temperature or by enrichment of ^{6}Li, which has the lower quadrupolar moment. Coupling constants of the two isotopes are related to each other by their gyromagnetic ratios according to eq. (4.19):

$$J_{C^7Li} = \frac{\gamma_{^7Li}}{\gamma_{^6Li}} J_{C^6Li} = 2.64 J_{C^6Li} \tag{4.19}$$

Doubly labeled (^{13}C–^{6}Li) butyllithium, for example, displays a quintet splitting with J_{C^6Li} = 7.8 Hz at −90°C in tetrahydrofuran [478]. Following the $(2nI + 1)$ multiplicity rule, the number of lithium nuclei ^{6}Li with $I = 1$ coupling to each carbon is 2. Thus, butyllithium occurs as a dimer, similarly to cyclopropyl-, vinyl- and phenyllithium [478].

Butyllithium
J_{C^6Li} (*Multiplicity*)

13.9 31.4 31.9 11.8 ppm
CH$_3$–CH$_2$–CH$_2$–^{13}CH$_2$–^{6}Li

7.8 Hz (5)
(−90°C, THF)

124.9, 144.0, 186.8 ppm
123.3 ^{13}C–^{6}Li

Phenyllithium

8.0 Hz (5)
(−118°C, THF)

4.13.3 Group II Organometallic Compounds

Carbon-13 shifts of Grignard reagents (Table 4.73) [479]) in ether solutions provide evidence that the Schlenck equilibrium is shifted in favor of the dialkylmagnesium compound: There is practically no shift difference between dibenzylmagnesium and benzylmagnesium chloride.

Table 4.73. ^{13}C Chemical Shifts of Aliphatic Grignard Reagents in Diethyl Ether (δ_C in ppm) [479].

Compound	C-1	C-2	C-3	C-4	C-5
CH_3MgI	−13.6				
C_2H_5MgBr	− 2.9	12.2			
C_3H_7MgBr	11.3	22.1	22.1		
C_4H_9MgBr	5.9	31.6	30.6	13.2	
$C_5H_{11}MgBr$	7.4	29.1	40.3	22.7	13.4
iso-C_3H_7MgBr	8.9	22.9			
$C_6H_{11}MgCl$	24.4	34.0	30.9	28.6	

$$2\ RMgX \rightleftarrows R_2Mg + MgX_2$$

Benzylmagnesium chloride

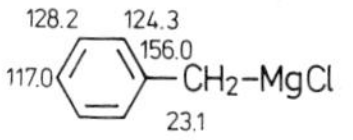

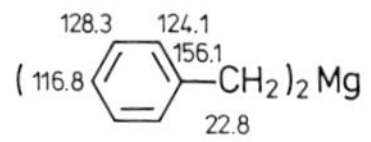

128.2 124.3
117.0 ⟨⟩ 156.0 —CH_2—MgCl
         23.1

128.3 124.1
(116.8) ⟨⟩ 156.1 —(CH_2)$_2$ Mg   (ppm)
           22.8   Dibenzylmagnesium

The equilibrium favors the alkylmagnesium halide, however, upon addition of hexamethylphosphoramide, as monitored by a deshielding of the benzylic carbon and shieldings of benzenoid C-1 and C-4 in benzylmagnesium chloride [479].

In the dibenzyl compounds of group II metals [480], the benzylic carbon is deshielded with increasing electropositive character of the metal due to an increased paramagnetic contribution to the screening constant. Carbon nuclei in ortho- and para- positions, however, are shielded, the extent depending on the amount of negative charge delocalization in the benzene ring.

$(4\langle\stackrel{3\ 2}{\ \ }\rangle1{-}CH_2)_2 M$

	CH_2	C-1	C-2	C-3	C-4
M = Ca	42.1	161.0	119.8	128.6	109.4 ppm
Sr	47.2	158.8	117.4	129.7	107.0 ppm
Ba	57.3	155.4	114.6	131.6	103.6 ppm

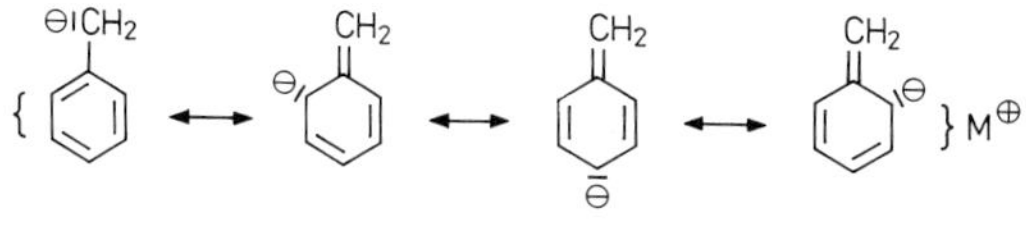

{ $^{\ominus}$|CH_2⟨⟩ ↔ CH_2⟨⟩$^{\ominus}$ ↔ CH_2⟨⟩$_{\ominus}$ ↔ CH_2⟨⟩$^{\ominus}$ } $M^{\oplus}$

4.13.4 Group III Organometallic Compounds

In order to suppress quadrupolar broadening and to quench couplings, carbon-13 NMR spectra of organoboron compounds are conveniently recorded with simultaneous ^1H and ^{11}B decoupling (triple resonance). Carbon-13 shifts thus obtained from trialkylboranes characteristically reflect the bonding state of boron, as shown in the series trimethylborane, lithium tetramethylboranate (Table 4.71), and methylaminodimethylborane: Donor substituents at boron induce considerable shieldings of the α alkyl carbons. Moreover, the partial double bond character of the boron-donor bond segregates α alkyl carbon nuclei into (E)- and (Z)-signals [481].

$$\begin{array}{cc} & \quad 1.7 \quad 29.6 \\ (Z) & H_3C \quad CH_3 \\ & \quad \backslash \quad / \\ & \quad B-N \quad \longleftrightarrow \quad \ominus B=N\oplus \quad (ppm) \\ (E) & \quad / \quad \backslash \\ & H_3C \quad H \\ & \quad 6.9 \end{array}$$

14.8   6.2
$(CH_3)_3B$ $(CH_3)_4B^{\ominus} Li^{\oplus}$

Trimethyl-   Lithium
borane   tetramethyl-   Methylamino-
              boranate   dimethylborane

Carbon-boron coupling constants with both boron isotopes ^{10}B ($I = 3$; 18.8%) and ^{11}B ($I = 3/2$; 81.2%) are not always resolvable due to quadrupolar broadening. Usually, couplings with the more abundant ^{11}B are reported [481], and $J_{C^{10}B}$ can then be calculated by means of the gyromagnetic ratios of the two isotopes using an equation analogous to eq. (4.19). One-bond couplings essentially reflect carbon hybridization and substituent electronegativities, as shown in the series trimethyl-, trivinyl-, and diethylaminodiethynylborane; longer-range couplings are generally not resolvable.

$J_{C^{11}B}$ $(CH_3)_3B$ $(H_2C=CH-)_3B$ $(HC\equiv C-)_2BNC_2H_5$
                  46.7   65.0   132.0 Hz

                Trimethyl-   Trivinyl-   Diethylamino-
                borane   borane   diethynylborane

Outstandingly large one-bond and long-range coupling constants characterize organothallium compounds, as shown for phenylthallium(III)bis-trifluoroacetate [482]. Splitting arises from the isotopes ^{205}Tl (70.5%) and ^{203}Tl (29.5%). Both nuclei possess the same spin quantum number ($I = 1/2$) and similar gyromagnetic ratios. As a result, the different magnitudes of $J_{C^{203}Tl}$ and $J_{C^{205}Tl}$ frequently cannot be resolved, and average values are reported. *Transoid* carbon-thallium couplings can be clearly separated from *cisoid* ones in norbornyl derivatives [482].

2-(3-Acetoxynorbornyl)-thallium (III)bis-acetate   Phenylthallium (III)-bis-trifluoroacetate (Hz)

4.13.5 *Group IV Organometallic Compounds*

Carbon-13 shifts of group IV organometallic compounds (Table 4.71) [483–485] decrease with increasing metal atomic weight due to *heavy atom shielding*; organolead compounds are the exception. Parallel to haloalkanes (Table 3.2), methyl carbon shifts in trimethyl halosilanes increase with decreasing electronegativity of the halogen (Table 4.71). The trimethylstannyl group in organotin compounds $(CH_3)_3SnR$ (Table 4.74) [484], however, shields the α alkyl carbons compared with those of the alkane RH [484]; α alkyl carbon shifts are reported to correlate linearly with those of alkanes RH [486]. Successive substitution of methyl by chlorine steadily deshields the α methyl carbon (Table 4.74). Transmission of substituent effects *via* Si, Sn, or Pb qualitatively follows the trends known from comparable carbon compounds.

Table 4.74. ^{13}C Chemical Shifts (δ_C in ppm) and ^{13}C–^{119}Sn Coupling Constants (in Hz) of Organotin Compounds RSn(CH$_3$)$_n$Cl$_{3-n}$ (n = O, R = n-Alkyl) [484].[a]

Compound	CH$_3$	R C-1	C-2	C-3	C-4	C-5	CH$_3$ J_{CSn}	R J_{CSn}	$^2J_{CSn}$	$^3J_{CSn}$
(CH$_3$)$_3$SnR										
R = CH$_3$	−9.6	−9.6 (−7.5)					350	350		
C$_2$H$_5$	−11.0	2.8 (−3.1)	10.7 (4.8)				319	373	24	
C$_3$H$_7$	−10.4	13.9 (−1.7)	20.5 (4.4)	18.7 (3.1)			319	369	21	
C$_4$H$_9$	−10.5	11.1 (−2.1)	29.2 (4.2)	27.3 (2.3)	13.8 (0.6)		319	368	21	53
C$_5$H$_{11}$	−10.5	11.1 (−2.6)	26.6 (4.0)	36.5 (2.0)	22.5 (−0.1)	14.2 (0.5)	319	368	21	53
(CH$_3$)$_2$SnRCl										
R = C$_4$H$_9$	−1.6	18.7 (5.5)	27.7 (2.7)	26.5 (1.5)	13.7 (0.5)		350	409	25	68
C$_5$H$_{11}$	−1.6	18.9 (5.2)	25.2 (2.6)	35.7 (1.2)	22.2 (−0.4)	14.0 (0.3)	347	410	25	66
CH$_3$SnRCl$_2$										
R = C$_4$H$_9$	6.1	26.0 (12.8)	26.8 (1.8)	26.1 (1.1)	13.5 (0.3)		406	490	37	82
C$_5$H$_{11}$	5.8	26.1 (12.4)	24.5 (1.9)	35.1 (0.5)	22.2 (−0.4)	13.9 (0.2)	405	490	37	84
RSnCl$_3$										
R = C$_4$H$_9$		33.7 (20.5)	26.7 (1.7)	25.4 (0.4)	13.1 (−0.1)			645	60	110
C$_5$H$_{11}$		33.3 (19.6)	24.4 (1.8)	34.3 (−0.2)	21.8 (−0.8)	13.8 (−0.1)		644	60	107

[a] values in parentheses are substituent effects induced upon replacement of a hydrogen in the alkane RH by the corresponding organotin moiety.

Due to the quadrupolar moment of ^{73}Ge (I = 9/2), carbon-germanium couplings are only resolved in molecules with symmetric charge distributions (e.g. (CH$_3$)$_4$Ge: J_{CGe} = −18.7 Hz). All other group IV NMR isotopes have spin quantum number I = 1/2. Coupling constants of less abundant isotopes are obtained by measuring the ^{13}C–M satellite signals. These are weak for M = ^{29}Si (4.7%) so that splittings due to longer-range carbon-silicon couplings may be lost in the tail of the central signal.

Carbon couplings with non-quadrupolar group IV metal nuclei collected in Table 4.75 [483–485] refer to ^{29}Si, the heavier and more abundant tin isotope ^{119}Sn, and to ^{207}Pb.

Table 4.75. Carbon-Metal Coupling Constants of Representative Organo-Silicon [483], Organotin [484], and Organolead Compounds [485] (in Hz).

One-Bond Couplings

sp³-CM	sp²-CM		sp-CM
Si(CH₃)₄ −50.3	(CH₃)₃Si−CH=CH₂ −64.0	(CH₃)₃Si−C₆H₅ (−)66.5	(CH₃)₃Si−C≡C−C₆H₅ (−)83.6
Sn(CH₃)₄ −337.2	(CH₃)₃Sn−CH=CH₂ −478.4	(CH₃)₃Sn−C₆H₅ (−)474.4	(CH₃)₃Sn−C≡C−CH₃ (−)476.5
	Sn(CH=CH₂)₄ −519.3		Sn(C≡C−CH₃)₄ (−)1168.0
Pb(CH₃)₄ 250.2	(CH₃)₃Pb−CH=CH₂ 279.0	(CH₃)₃Pb−C₆H₅ 348.0	(CH₃)₃Pb−C≡C−CH₃ 110.0
	Pb(CH=CH₂)₄ 454.1		Pb(C≡C−CH₃)₄ 1625.0

Longer-Range Couplings

Compound	J_{CM}	$^2J_{CM}$	$^3J_{CM}$	$^4J_{CM}$
Sn(CH₂−CH₂−CH₂−CH₃)	(−)313.7	19.6	52.0	0
Pb(CH₂−CH₂−CH₂−CH₃)	189.2	26.9	74.5	0
(CH₃)₃Sn−C₆H₅	(−)474.4 (−)347.5 (CH₃)	36.6	47.4	10.8
(CH₃)₃Pb−C₆H₅	348.0 274.0 (CH₃)	63.0	66.0	16.0
Sn(C≡C−CH₃)₄	1168.0	241.3	19.5	−
Pb(C≡C−CH₃)₄	1625.0	347.2	30.5	−

Gyromagnetic constants of ^{29}Si and ^{119}Sn are negative, in contrast to ^{207}Pb. One-bond carbon-13-group IV metal coupling constants display corresponding signs, with the exception of lead [483–485] (Table 4.75). A correlation between J_{CM} and the s character of carbon hybrid orbitals is found for organosilicon but not for organotin and organolead compounds (Table 4.75). Moreover, J_{CM} couplings with M = Sn, Pb are tremendously enhanced upon increasing substitution of methyl by alkenyl-, phenyl-, and alkynyl goups. Obviously, mechanisms other than the Fermi contact interaction contribute to carbon-metal coupling in these cases. One-bond carbon-lead coupling constants increase significantly with the coordination number of lead and with the donor strength of the solvent [485].

Two- and three-bond carbon-metal couplings follow patterns known from carbon-proton coupling ($^3J_{CM(transoid)} > {}^3J_{CM(cisoid)}$; $^3J_{meta} > {}^2J_{ortho}$; Table 4.75). Karplus-type equations have been reported for organotin compounds [484, 487]; a clear example of the trend is given by 7-norbornyltrimethylstannane:

7-Norbornyltrimethylstannane
(Hz)

Sn(CH₃)₂
11.9 (cisoid)
(transoid) 67.5

4.13.6 Organo-Transition-Metal Compounds

Carbon-13 chemical shifts of *transition metal carbonyls* [471, 473] decrease as one proceeds down a given group of the periodic table, as demonstrated for the triad $Cr(CO)_6$, $Mo(CO)_6$, and $W(CO)_6$ in Table 4.71. Substitution of CO by cyclopentadienide and other organic ligands deshields the carbonyl carbon nuclei. This is exemplified by iron pentacarbonyl (211 ppm) in comparison to cyclopentadienylirondicarbonyl-dimer, which displays one avaraged carbonyl signal ($\sim$ 240 ppm) at room temperature [488] due to rapid *cis-trans* isomerization and intramolecular bridged-terminal carbonyl interconversion:

Cyclopentadienylirondicarbonyl-Dimer

Temp. °C	$\delta_{term.}$ ppm	$\delta_{av.}$ ppm	δ_{bridge} ppm
55		239.4	
$-$ 40	210.2 (*cis*)	241.9 (*trans*)	273.2 (*cis*)
$-$ 70	210.2 (*cis*)	242.4 (*trans*)	274.1 (*cis*)
-126	210.4 (*cis*)	not obs.	275.9 (*cis*)

At $-$ 40 °C and below, two additional signals for terminal ($\sim$ 210 ppm) and bridge ($\sim$ 275 ppm) carbonyl groups are assigned to the frozen *cis* isomer. The signal at 242 ppm belongs to the *trans* isomer, which still opens its bridges at $-$ 40 °C so that rotation about the Fe$-$Fe bond takes place and only one averaged carbonyl signal is observed. This rotation becomes frozen at lower temperatures, finally resulting in a broadened, undetectable signal at $-$ 126 °C [488].

π-*Complexation of transition metals to alkenes* considerably shields the olefinic carbon nuclei, while carbon-proton couplings J_{CH} remain almost unaffected [475]. Coordination shieldings of *tris*-(alkene)-nickel, -palladium, and -platinum are about $-$ 50 to $-$ 60 ppm, and fixed conformations of *tris*-(bicyclo[2.2.1]heptene)-nickel are detectable by low-temperature ^{13}C NMR [475, 489]. *Bis*-(η^3-allyl)-metal complexes exist as *syn-anti* pairs of *cis* and *trans* isomers when the symmetry of the allyl ligand is distorted by terminal alkylation [490]. Fig. 4.16 illustrates how carbon-13 NMR can be used to detect all four possible isomers of (η^3-1,1-dimethylallyl)-nickel [490].

Transition metal compounds with covalent carbon-metal bonds include organo-zinc, organo-cadmium, and organo-mercury compounds. Carbon-13 shifts of the methyl derivatives (Table 4.71) indicate a heavy atom deshielding. Diphenylmercury displays carbon shifts similar to those of phenyllithium and phenylmagnesium bromide (Table 4.53).

Carbon-13-mercury coupling constants (^{199}Hg with $I = 1/2$ and 16.8% natural abundance) follow relations known from other carbon couplings. One-bond carbon-mercury

4.13 Organometallic Compounds

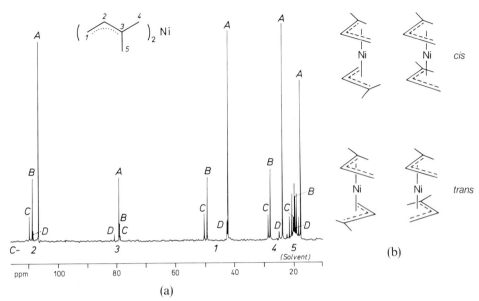

Fig. 4.16. $^{13}C\{^1H\}$ NMR spectrum of bis-(1,1-dimethylallyl)-nickel (25.2 MHz; octadeuteriotoluene; $-20\,°C$ (a), detecting all of the four possible isomers (b) [490].

coupling constants increase with the electronegativity of the halogen attached to mercury (Table 4.71) and with the *s* character of carbon hybrid orbitals [491].

sp³	sp²	sp
$(CH_3-CH_2)_2Hg$	$(CH_2=CH)_2Hg$	$(C_6H_5-C\equiv C)_2Hg$
24 648	37 1161	2584 (Hz)
Diethyl-	Divinyl-	Bis-(phenylethynyl)-mercury

Three-bond carbon-mercury couplings are larger for *transoid* in comparison to *cisoid* coupling routes, as shown for the *endo-exo* pair of 2-norbornylmercury acetate, and *meta*

exo- 2-Norbornylmercuric acetate   *endo*   Diphenylmercury   Phenylmercuric chloride

couplings ($^3J_{CHg}$) are stronger than *ortho* couplings ($^2J_{CHg}$) in phenylmercury compounds [491].

4.14 Ions

4.14.1 Carbocations

Carbocation samples for NMR measurements in solutions are prepared by dissolving appropriate precursors (*e.g.* haloalkanes) in anhydrous strong Lewis acid systems such as *"magic acid"* (SbF_5/FSO_3H in SO_2ClF) at low temperatures ($-40°C$ to $-150°C$). Carbon-13 shifts are referenced to external TMS. Table 4.76 collects ^{13}C shift values of representative *carbenium ions* R_3C^+ [79, 492] in the order of increasing shielding of the positively charged carbon. It turns out that carbenium ion carbon nuclei without attached electron releasing substituents, such as trialkylcarbenium ions, display shift values larger than 300 ppm. Stabilization of the cation increases with the strength and the number of donor substituents (Table 4.76). The C^+ carbon shifts of trimethyl-, triphenyl-, and trihydroxycarbenium ions clearly verify this trend. Due to the "Y" arrangement of donor substituents D, centrosymmetric delocalization of the positive charge in D_3C^+ type ions is referred to as *Y stabilization*. In this context, the guanidinium ion with $R^1 = R^2 = R^3 = NH_2$ and its derivatives belong to the most stabilized carbenium ions, as demonstrated by their particularly small carbon-13 shift values (≥ 160 ppm, Table 4.76).

Y Stabilization

$\bar{D}$: $NH_2 > OH > C_6H_5 >$ alkyl
δ_C: ~160   ~170   ~200   ~300 ppm

Nonplanar carbenium ions, such as the 1-adamantyl cation, display shielded C^+ carbon nuclei compared with open-chain trialkylcarbenium ions (*e.g.* the triethylcarbenium ion in Table 4.76). Deshielding of α and β carbons relative to the parent adamantane is attributed to charge delocalization *via* σ bonds [493].

Adamantane   37.8, 28.4   Adamantyl cation   300.6, 66.6, 87.6, 34.6

The positive charge of allyl cations essentially distributes itself among C-1 and C-3 of the allyl skeleton. Terminal alkylation concentrates the positive charge at the alkylated carbon, as shown for the 1-methylcyclopentenyl cation [494].

1-Methylcyclopentenyl: 147.8, 48.6, 219.0, 261.6, CH_3
Cyclopentenyl: 137.6, 218.6
Cyclohexenyl cation: 146.6, 235.6 (ppm)

Aromatic carbenium ions such as cyclopropenium and cycloheptatrienium ion (Fig. 3.2) [76, 495] are efficiently stabilized as the positive charge is delocalized within the $(4n + 2)$ π electron system. Carbon-13 shifts are smaller than 180 ppm.

Table 4.76. ^{13}C Chemical Shifts of Selected Carbenium Ions in Magic Acid (SbF$_5$/FSO$_3$H/SO$_2$ClF) between $-40\,°C$ and $-80\,°C$ [79, 492] (δ_C in ppm). (Guanidinium chloride and formamidinium chloride (last entries) are measured in hexadeuteriodimethyl sulfoxide at room temperature.)

R^1	R^2	R^3	$\diagdown\!\!\underset{\diagup}{\overset{}{-\text{C}+}}$	R^1	R^2	R^3
C$_2$H$_5$	C$_2$H$_5$	C$_2$H$_5$	336.8	51.8 (α) 8.6 (β)		
C$_2$H$_5$	CH$_3$	CH$_3$	336.0	58.4 (α) 10.3 (β)	45.4	
CH$_3$	CH$_3$	CH$_3$	335.7	48.3		
CH$_3$	CH$_3$	H	320.6	51.5	—	
CH$_3$	CH$_3$	Br	319.8		—	
CH$_3$	CH$_3$	Cl	312.8		—	
CH$_3$	CH$_3$	F	282.9		—	
▷	▷	▷	272.0	33.4 (α) 31.7 (β)		
C$_6$H$_5$	CH$_2$CH$_2$CH$_3$	CH$_2$CH$_2$CH$_3$	257.4	139.1 (1) 141.9 (2,6) 133.4 (3,5) 155.7 (4)	47.8 (α) 31.5 (β) 15.8 (γ)	
C$_6$H$_5$	CH$_3$	CH$_3$	254.1	139.3 (1) 141.1 (2,6) 131.8 (3,5) 154.8 (4)	32.4	
▷	▷	H	253.7	45.7 (α) 38.7 (β)	—	
CH$_3$	CH$_3$	OH	249.5	31.8	—	
C$_6$H$_5$	▷	CH$_3$	245.3	138.9 (1) 133.8 (2,6) 130.3 (3,5) 144.6 (4)	44.9 (α) 44.1 (β)	22.5
OH	H	H	223.8	—	—	—
C$_6$H$_5$	C$_6$H$_5$	C$_6$H$_5$	211.9	140.9 (1) 144.3 (2,6) 131.3 (3,5) 144.1 (4)		
C$_6$H$_5$	C$_6$H$_5$	OH	209.2	131.5 (1) 140.2 (2,6) 132.5 (3,5) 145.5 (4)		—
C$_6$H$_5$	C$_6$H$_5$	H	191.1	138.6 (1) 143.8 (2,6) 134.1 (3,5) 151.2 (4)		—
OH	OH	H	177.6	—	—	—
OH	OH	OH	166.3	—	—	—
NH$_2$	NH$_2$	NH$_2$	158.3	—	—	—
N(CH$_3$)$_2$	N(CH$_3$)$_2$	H	156.4	45.8		—

4 ^{13}C NMR Spectroscopy of Organic Compounds

Acyl cations [496] are stabilized by electron release of the alkyl groups and by interaction with adjacent π orbitals, as shown for benzoyl, acroyl and propiolyl cations. Ketenoid resonance formulae contribute to the states of these ions.

7.5 H_3C 150.3⊕C=O Acetyl

149.4 / 132.9 / 141.3 / 87.7 / 154.8⊕C=O Benzoyl

177.1 92.7 147.1⊕C=O Acroyl

110.7 \\ 48.9 124.1⊕C=O Propiolyl cation   (ppm)

Rapidly equilibrating (degenerate) carbenium ions are characterized by averaged carbon-13 shifts and carbon-proton coupling constants. 2,3-Dimethylbutyl in comparison with 2-propyl cation is an example [497]. Only two instead of four signals are observed at $-80\,°C$ due to a rapid 1,2-hydride shift.

2-Propyl cation: H_3C–⊕C–H –CH_3

$H_3C\underset{H}{\overset{CH_3}{\underset{|}{\overset{|}{C}}}}\underset{CH_3}{\overset{CH_3}{\underset{|}{\overset{|}{C⊕}}}}$ ⇌ $H_3C\underset{H_3C}{\overset{H_3C}{\underset{|}{\overset{|}{⊕C}}}}\underset{H}{\overset{CH_3}{\underset{|}{\overset{|}{C}}}}–CH_3$ 2,3-Dimethylbutyl cation

CH^+: 319.6 ppm (J_{CH} = 171.3 Hz)   C-2: 197.2 ppm (J_{CH} = 67.5 Hz)
CH_3: 61.8 ppm (J_{CH} = 131.7 Hz)   C-1: 42.0 ppm (J_{CH} = 132.2 Hz)

The coupling constant of the carbenium ion carbons is small, since it is an average from one- and two-bond CH couplings of the $sp^2 \rightleftarrows sp^3$ interchanging carbon nuclei C-2 and C-3 according to eqs. (4.20):

$$\tilde{J}_{CH} = \tfrac{1}{2}(\tfrac{1}{2}J_{sp^2-CH} + \tfrac{1}{2}J_{sp^3-CH})$$
$$^2\tilde{J}_{CH} = \tfrac{1}{2}(\tfrac{1}{2}\,^2J_{sp^2-CH} + \tfrac{1}{2}\,^2J_{sp^3-CH})$$
$$J_{obs.} = \tfrac{1}{2}(\tilde{J}_{CH} + {}^2\tilde{J}_{CH}) \qquad (4.20)$$

Other examples of rapidly equilibrating carbenium ions include cyclopropylcarbinyl, cyclopentyl, decalyl, and benzenonium ion [79, 492]. Unlike methylcyclopentyl cation, in which the positive charge is localized at the substituted carbon, the parent cyclopentyl carbenium ion [79] exhibits only one signal at 98.4 ppm. Neglecting longer-range carbon-proton couplings, the averaged value of $J_{CH} = 28.5$ Hz arises from eight sp^3-CH bonds with $J_{CH} \cong 130$ Hz (Table 4.4) and one sp^2-$\overset{+}{C}H$ bond with $J_{CH} = 171$ Hz (2-propyl cation as reference). Thus, a total of nine CH bonds and a five-fold degenerate carbenium ion have to be considered when calculating the average J_{CH}:

Methylcyclo-pentyl cation

27.2 / 64.1 / 337.3 / 37.5 CH_3

Cyclopentyl cation

(five equilibrating structures)

$\delta_C = 98.4$ ppm
$J_{CH} = 28.5$ Hz

$$\tilde{J}_{CH} = \tfrac{1}{5}\tfrac{1}{9}(J_{sp^2-CH} + 8\,J_{sp^3-CH}) = \tfrac{1}{45}(171 + 8 \cdot 130) \cong 26.9\ \text{Hz}$$

The benzenonium ion, obtained by dissolving benzene in "magic acid", behaves correspondingly: Only one signal with $\delta_C = 144.5$ ppm is recorded at $-80\,°C$ due to rapid proton migration. An averaged one-bond carbon-proton coupling of 26.5 Hz is expected

4.14 Ions

when taking five sp^2-CH and two sp^3-CH bonds into account, *i.e.* a total of seven CH connections and a six-fold degenerate cyclohexadienyl cation [498]. Individual signals for the nondegenerate cyclohexadienyl cation are observed at $-140\,°C$.

$$\tilde{J}_{CH} = \tfrac{1}{6}\tfrac{1}{7}(5\,J_{sp^2\text{-}CH} + 2\,J_{sp^3\text{-}CH}) = \tfrac{1}{42}(5 \cdot 171 + 2 \cdot 128) \cong 26.5 \text{ Hz}$$

Benzenonium ion

$\delta_C = 144.5$ ppm
$J_{CH} = 26.5$ Hz
$(-80\,°C)$

(ppm) $(-140\,°C)$

The norbornyl cation, generated by dissolving *exo*-2-chloronorbornane in the Lewis acid system $SbF_5/SO_2ClF/SO_2F_2$ at $-78\,°C$, displays only three signals at $-80\,°C$, providing evidence for a symmetrically bridged non-classical structure with pentacoordinate carbon (R_5C^+) [499], in accordance with the results of other methods [500]. However, individual carbon-13 signals for all carbon nuclei are observed for the 2-methylnorbornyl cation [501].

Norbornyl cation

C-1,2,6: 91.7 ppm (Q, 55.1 Hz)
C-3,5,7: 30.8 ppm (T, 139.1 Hz)
C-4: 37.7 ppm (D, 153,1 Hz)

2-Methyl-2-norbornyl cation

4.14.2 Carbanions

Organometallic compounds with strongly electropositive group I and group II metals form carbanionic species (Sections 4.13.2 and 4.13.3). Their ^{13}C NMR spectra depend on the degree of oligomerization which, itself, is influenced by solvent, concentration, temperature and the nature of both the carbanion and its counter cation. Relatively free carbanions are detected in hexamethylphosphoramide (HMPT), while contact ion pairs or solvated ions appear to occur in ethers [502]. General conclusions concerning the structure of carbanions by means of carbon-13 shifts have to be drawn with care. Allyllithium (Table 4.77) and allylmagnesium bromide, for example, present averaged carbon-13 shifts for the terminal carbon nuclei C-1 and C-3; their ppm values do not differ substantially [503]. This may arise either due to a symmetrical π structure or as a consequence of rapidly equilibrating degenerate σ structures. An application of the deuterium isotope effect identifies allyllithium as a π structure; allylmagnesium bromide,

Allyllithium
π structure

Allylmagnesium bromide
σ structure

(ppm)

Table 4.77. ^{13}C Chemical Shifts of Representative Carbanions (δ_C in ppm).

-Carbanion [Reference]	R$^-$		M$^+$	C-1	C-2	C-3	C-4	C-5	C-6
Allyl- [503-504]	CH$_2$=CH–CH$_2$		Li	51.2	147.2				
	CH$_2$=CH–CH$_2$		K	52.8	144.0				
5,5-Dimethyl-2-hexenyl- [504]	(CH$_3$)$_3$C–CH$_2$–CH=CH–CH$_2$		Li	31.0	142.5	81.9	43.2	32.8	30.1
	(CH$_3$)$_3$C–CH$_2$–CH=CH–CH$_2$		Na	35.7	138.8	72.3	43.8	33.1	30.5
	(CH$_3$)$_3$C–CH$_2$–CH=CH–CH$_2$		K	45.0	137.5	67.5	44.6	32.0	30.3
	(CH$_3$)$_3$C–CH$_2$–CH=CH–CH$_2$		Rb	47.4	138.2	67.5	45.1	31.7	30.4
	(CH$_3$)$_3$C–CH$_2$–CH=CH–CH$_2$		Cs	51.4	139.5	69.0	45.8	31.7	30.5
Pentadienyl- [505]	CH$_2$=CH–CH=CH–CH$_2$		Li	66.2	142.8	86.9			
	CH$_2$=CH–CH=CH–CH$_2$		K	79.9	137.6	78.6			
Benzyl- [506]	C$_6$H$_5$–CH$_2$		Li	36.7	160.6	116.4 o	128.2 m	104.2 p	
	C$_6$H$_5$–CH$_2$		K	52.7	152.7	110.7 o	130.6 m	95.7 p	
Benzhydryl- [506, 507]	(C$_6$H$_5$)$_2$CH		Li	76.8	147.3	117.5 o	128.0 m	107.4 p	
	(C$_6$H$_5$)$_2$CH		K	78.5	145.4	116.7 o	129.4 m	108.4 p	
Triphenylmethyl- [506]	(C$_6$H$_5$)$_3$C		Li	90.2	149.6	123.8 o	127.9 m	113.0 p	
	(C$_6$H$_5$)$_3$C		K	88.3	148.6	123.5 o	128.8 m	114.2 p	
Cyclohexadienyl- [508]			Li$^+$	78.0	131.8	75.8	30.0		
Cycloheptadienyl-			Li$^+$	71.3	134.5	98.9	35.6		
Cyclopentadienyl			Li$^+$	102.8					
Indenyl- [509]			Li$^+$	91.8	115.0		119.4	113.9	128.1 (8
Cyclooctatetraenediyl-			2 K$^+$	85.5					
1,3-Dithianyl [510]			Li$^+$		26.7		34.1	31.2	
Cyanobenzyl [502]	⟨C$_6$H$_5$⟩–CH$^-$–CN		Na$^+$	34.0	150.6	116.9 o	127.5 m	108.7 p	134.4 (CN)
Dicyanomethyl- (Malonitrile)	(NC)$_2$CH$^-$		Na$^+$	–2.1	130.3				
Tricyanomethyl-	(NC)$_3$C$^-$		Na$^+$	71.8	58.4				
Dimethylmalonate	(CH$_3$OOC)$_2$CH$^-$		Na$^+$	61.3	171.9	47.8			
Acetylacetonate	(CH$_3$CO)$_2$CH$^-$		Na$^+$	99.2	190.4	28.0			
Diformylmethyl- (Malonaldehyde)	(CHO)$_2$CH$^-$		Na$^+$	107.4	187.5				
Triformylmethyl-	(CHO)$_3$C$^-$		Na$^+$	118.8	186.4				

4.14 Ions 307

however, selects a two-fold degenerate σ structure [503]; it displays two signals for C-1 and C-3 upon terminal deuteration, separated by the deuterium isotope effect.

In Table 4.77 a small selection of carbanionic species has been compiled [502–510]. The carbanionic carbon shift of methyllithium is − 16.6 ppm (Table 4.71) in comparison to − 23.1 ppm, which is predicted for an sp^2 carbon containing two electrons in a p orbital, following the empirical carbon-13 shift to charge density correlation [76, 507]. Carbanion carbon shifts become progressively more positive with increasing delocalization of the negative charge by resonance (mesomeric) effects, as shown for *allyl* and *pentadienyl* anions [503–505] in Table 4.77.

$$\left[\underset{\ominus}{\overset{137.6}{\underset{78.6}{\diagup}}\diagdown\overset{79.9}{\diagup}} \longleftrightarrow \underset{\ominus}{\diagup\diagdown\diagup} \longleftrightarrow \underset{\ominus}{\diagup\diagdown\diagup} \right] K^{\oplus} \quad \text{(ppm)}$$

Pentadienyl potassium

Charge delocalization of benzyl type carbanions is clearly monitored by the *ortho* and *para* carbon shifts [506–507]. Delocalizaton includes a second phenyl ring in benzhydryl anions, as indicated by an increased deshielding of the carbanionic carbon [507] (Table 4.77). Additional deshielding introduced by a third phenyl ring, however, is only about 10 ppm (Table 4.77) due to steric hindrance of coplanarity in the triphenylmethyl carbanion.

Benzyl- ($R' = R^2 = H$), Benzhydryl- ($R^1 = H$, $R^2 = C_6H_5$), and Triphenylmethyl carbanion ($R^1 = R^2 = C_6H_5$)

Similarly, ^{13}C shift values of *cyclohexadienyl* anions are related to the distribution of negative charge among ring carbons, as portrayed for the parent cyclohexadienate and its dimethyl derivative [508]. – Substitution of the cyclohexadienate anion by typical electron acceptors such as nitro groups results in additional distribution of the negative charge. Significantly deshielded carbanionic carbons in 1-acetonyl-2,4-dinitrocyclohexadienate [511] relative to the 1,1-dimethyl derivative exemplify this trend. Taking 2,4-dinitrophenylacetone as an aromatic reference compound, however, an increased negative charge density of the 2,4,6-carbon nuclei is clearly apparent in the Meisenheimer anion.

1,1-Dimethylcyclohexadienate

Tetramethylammonium 1-acetonyl-2,4-dinitrocyclohexadienate

2,4-Dinitrophenylacetone

Y attachment of electron donors efficiently stabilizes carbenium ions (Section 4.14.1). Correspondingly, Y arrangement of electron acceptors at the anionic carbon induces efficient *Y stabilization*, as indicated by large carbanion shifts and shielded acceptor carbons. Representative examples include the trivinylmethyl carbanion [512] as well as the sodium salts of triformylmethane and cyanoform [502]. The symmetry of the charge distribution is not only reflected by an outstandingly small nitrile carbon shift but also by a resolvable carbon-13-nitrogen-14 coupling of $J_{CN} = 1.5$ Hz in the tricyanomethyl anion.

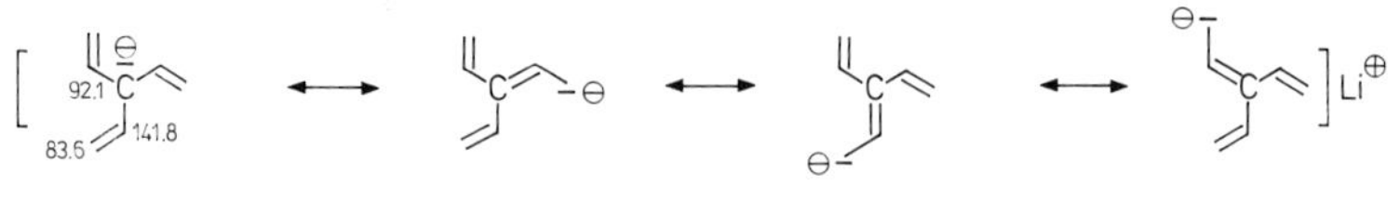

Lithium trivinylmethide (ppm)

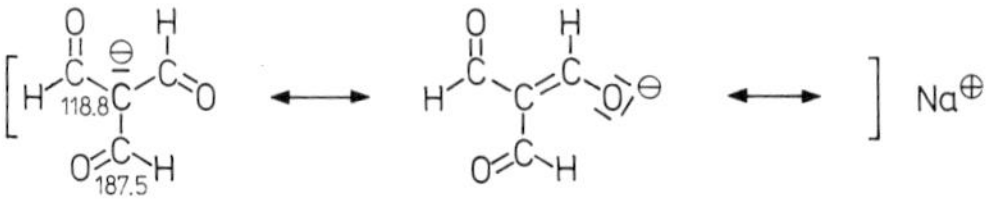

Sodium triformylmethide (ppm)

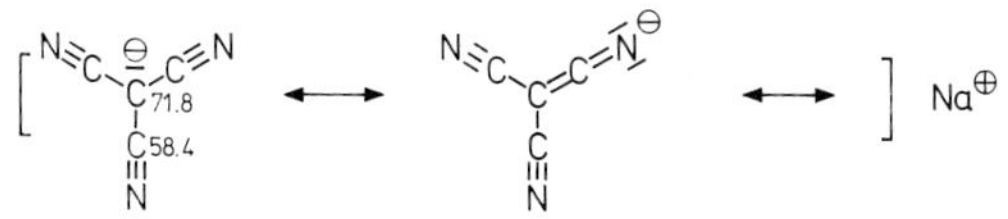

Sodium tricyanomethide (ppm)

4.15 Synthetic Polymers

Compared with the proton, carbon-13 offers some advantages as an NMR probe for investigation of polymers in solution [513–517]. *Firstly*, shift differences of carbon nuclei from structural units with identical constitution but with different steric environment are larger than those of the attached protons. *Secondly*, unlike ^{1}H NMR spectra, proton broadband-decoupled carbon-13 NMR spectra are not complicated by spin-spin coupling. *Thirdly*, dipolar line-broadening induced by the slower motion of macromolecules in solution, which may be more than 10 Hz for proton signals, increases with the square of the gyromagnetic ratio γ. This is small for carbon-13 ($\gamma_C \cong \frac{1}{4} \gamma_H$), so that line broadening of polymer carbon signals is smaller than that of proton resonances.

Important applications of ^{13}C NMR in polymer analysis include subjects such as *stereoregularity (tacticity)* of vinyl polymers, *configurational (cis, trans) isomerism* of diene polymers and the *mobility* of polymer chain segments. Due to poor solubility, however, many polymers cannot be investigated by ^{13}C NMR in solution. In these cases, high resolution solid phase NMR techniques such as *cross-polarization magic angle spinning* (CP-MAS) [518] must be applied as described for several synthetic polymers [519] and biopolymers [520].

4.15.1 Tacticity

Polymerization of unsymmetric vinyl monomers generates asymmetric carbon centers within the polymer chain

4.15 Synthetic Polymers

$$R-CH=CH_2 \quad \rightarrow \quad (-\overset{*}{C}H-CH_2-).$$
$$\phantom{R-CH=CH_2 \quad \rightarrow \quad (-\overset{*}{C}H-}|$$
$$\phantom{R-CH=CH_2 \quad \rightarrow \quad (-\overset{*}{C}}R$$

In order to describe the *stereosequence* of such chains, *meso* or *m* diads with identical and *racemic* or *r* diads with opposite configuration of succeeding asymmetric carbon atoms are defined:

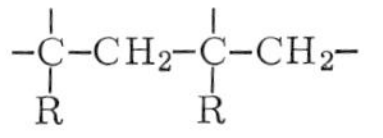

```
   |       |                    R
  -C-CH₂-C-CH₂-           |
   |       |             -C-CH₂-C-CH₂-
   R       R              |       |
                          R

   m diad                   r diad
```

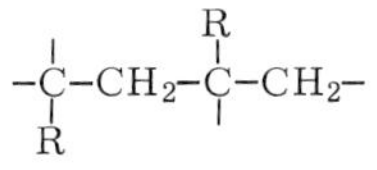

Longer sequences are referred to as triads, tetrads, pentads, ... Taking the pentad as partial sequence of a vinyl polymer with head to tail connections of units, three typical stereoisomers can be denoted by the *m*, *r* symbols: If all asymmetric carbons display identical configurations, the sequence is referred to as an *mmmm* or *isotactic pentad*. A regularly alternating configuration is denoted as an *rrrr pentad* or a *syndiotactic* sequence. Sequences with irregularly alternating configurations such as an *rrrm pentad* are defined as *atactic* or *heterotactic*.

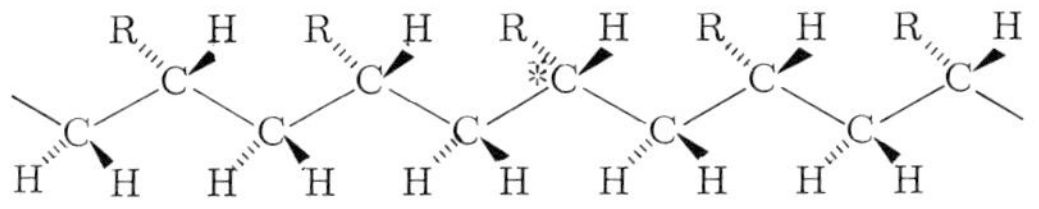

isotactic (mmmm)

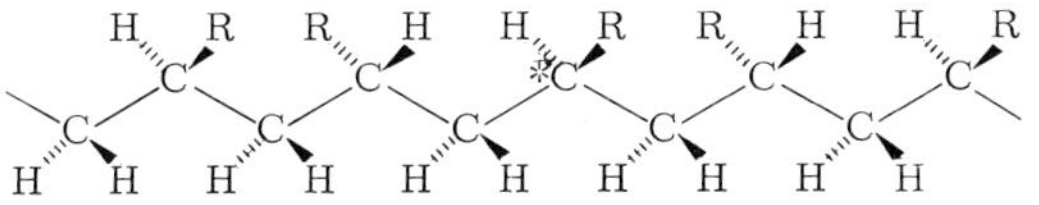

syndiotactic (rrrr)

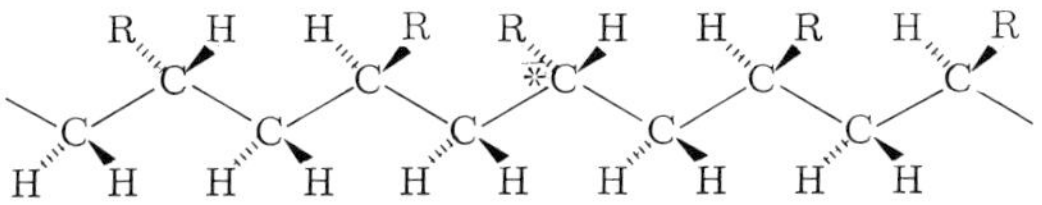

atactic (rrrm)

Seven other possible pentad sequences (*mmmr, rrmr, mmrm, mmrr, mrmr, mrrm,* and *rmmr*) are atactic. Residues R may be CH_3 for polypropylene, C_6H_5 for polystyrene, Cl for polyvinyl chloride, CN for polyacrylonitrile, and CO_2CH_3 for polyacrylic acid methyl ester.

Fig. 4.17 illustrates the potential of carbon-13 NMR to detect the presence of isotactic (a), syndiotactic (b), and atatactic (c) vinyl polymers with polypropylene as sample [521]. The spectrum of atactic polypropylene (Fig. 4.17(c)) displays the signals of all possible stereosequences including iso- and syndiotactic ones. Using the empirical increment systems for alkane carbon shift prediction [85, 201, 202] and including γ effects of $Z_\gamma = -5$ ppm specifically obtained by analysis of stereoisomeric polypropylene partial sequences between 3,5-dimethylheptane and 3,5,7,9,11,13,15-heptamethylheptadecane as a heptad model, the methyl carbon-13 shifts of all 36 possible heptads can be calculated

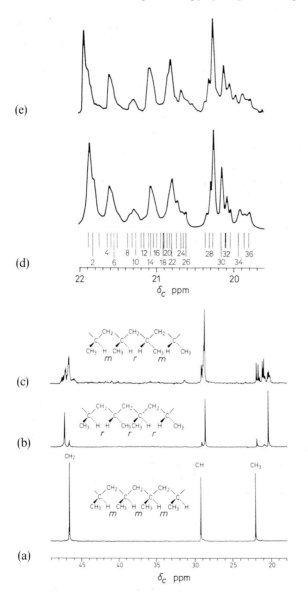

Fig. 4.17. Proton broadband-decoupled ^{13}C NMR spectra of polypropylene ((a–c) 25 MHz; 200 mg/mL 1,2,4-trichlorobenzene at 140 °C; (d–e) 90.52 MHz; 200 mg/mL heptane at 67 °C); (a) isotactic; (b) syndiotactic; (c) atactic sample; (d) methyl carbon spectrum, simulated for calculated carbon shifts and Lorentzian signals of < 0.1 Hz width at half-height; (e) experimental spectrum [521]. Numbers in (d) refer to the 36 possible heptads:

1: *mmmmmm*;	2: *mmmmmr*;	3: *rmmmmr*;	4: *mmmmrr*;	5: *mmmmrm*;	6: *rmmmrr*;
7: *mrmmmr*;	8: *rrmmrr*;	9: *mrmmrr*;	10: *mrmmrm*;	11: *mmmrrm*;	12: *mmmrrr*;
13: *rmmrrm*	14: *rrrmmr*;	15: *mmmmrm*;	16: *mmmmmm*;	17: *rmrmmr*;	18: *mmrmmr*;
19: *rrmrrm*;	20: *rrrmrr*;	21: *mrmrrm*;	22: *rrrmrr*;	23: *rrmrmr*;	24: *mmmrmr*;
25: *mrmrmr*;	26: *mmrmrm*;	27: *mrrrrm*;	28: *rrrrrm*;	29: *rrrrrr*;	30: *rmrrrm*;
31: *rrrrmr*;	32: *mmrrrm*;	33: *rrrrmm*;	34: *rmrrmr*;	35: *rmrrmm*;	36: *mmrrmm*.

[522] (Fig. 4.17(d, bottom)). Adjustment of this line spectrum to Lorentzian signals generates a simulated spectrum (Fig. 4.17(d)) which is practically identical with the experimental spectrum (Fig. 4.17(e)).

Stereosequences of *polyvinyl chloride* (PVC) are assignable by comparing its carbon shifts with those of *meso* and *racem* 2,4-dichloropentane [262] as outlined in Section 4.4.1. 2,4,6-Trichloroheptane can also be used as a model in which the three triads (*mm, rr, mr*) building up the stereoisomers of PVC may be observed [521].

Polyacrylonitrile, as another example, displays well separated nitrile and methine carbon signals for isotactic, syndiotactic and atactic sequences [523]. ^{13}C NMR analysis of polyacrylonitrile samples obtained by different methods of polymerization clearly indicates the isotactic sequence to predominate in the material obtained by γ irradiation of the acrylonitrile-urea channel adduct [523].

$$\begin{array}{c} \overset{CN}{|}\overset{CN}{|}\overset{CN}{|} \\ -CH_2-CH-CH_2-CH-CH_2-CH- \end{array} \qquad mm = \text{isotactic} \qquad : \delta_{CN} = 120.1 \text{ ppm}$$

$$\begin{array}{c} \overset{CN}{|}\overset{CN}{|} \\ -CH_2-CH-CH_2-CH-CH_2-CH- \\ \underset{CN}{|} \end{array} \qquad rr = \text{syndiotactic} : \delta_{CN} = 119.4 \text{ ppm}$$

$$\begin{array}{c} \overset{CN}{|} \\ -CH_2-CH-CH_2-CH-CH_2-CH- \\ \underset{CN}{|}\underset{CN}{|} \end{array} \qquad rm = \text{atactic} \qquad : \delta_{CN} = 119.8 \text{ ppm}$$

The benzenoid C-1 resonance of styrene units in *acrylonitrile-styrene copolymers* is particularly sensitive to the sequence of the chain; relative configurations of triad sequences can be determined by quantitative evaluation of carbon-13 signals [524]. Microstructures of other vinyl polymers such as polystyrene [525], polypropylene oxide [526], and polyalkyl acrylates [527] have also been investigated by ^{13}C NMR.

Besides determination of stereosequences, high-field carbon-13 NMR can be used to trace out *chain branches* and *end groups* by precise signal area measurements with an accuracy of 0.01 to 0.02%, as verified for polyethylene samples [528].

4.15.2 Configurational Isomerism

Similarly to monomeric alkenes, *cis-trans* isomeric polyalkenes obtained by polymerization of 1,3-dienes can be distinguished on the basis of γ shielding experienced by alkyl groups *cis* to a double bond (Section 4.2.1). Natural rubber and guttapercha, the *cis-trans* isomeric pair of 1,4-polyisoprene are typical examples, samples of which have been investigated in the bulk and in solution [529]. Recalling the relation $\delta_{trans} > \delta_{cis}$ for alkyl carbon nuclei α to double bonds, it is found that guttapercha displays a shielded methyl (7.7 ppm) but one deshielded methylene carbon (35.0 ppm) relative to natural rubber, as portrayed in Fig. 4.18(a, b) [73i]. – Shift differences in the order of 0.5 to 1 ppm are also observed for the olefinic carbons ($\delta_{trans} > \delta_{cis}$) as shown in Fig. 4.18. This difference, also, may be useful for configurational assignments as demonstrated for polybutadiene in

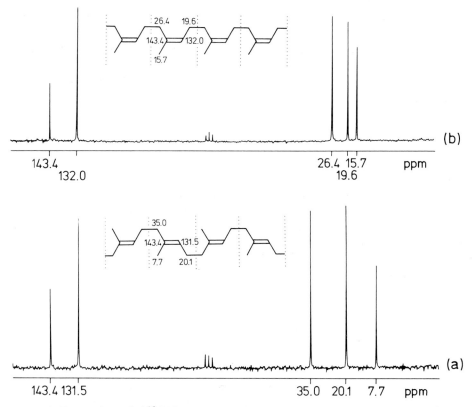

Fig. 4.18. Proton-decoupled ^{13}C NMR spectra (22.63 MHz; 100–150 mg/mL deuteriochloroform) of guttapercha (*trans*-) (a) and natural rubber (*cis*-polyisoprene) (b) [73 i].

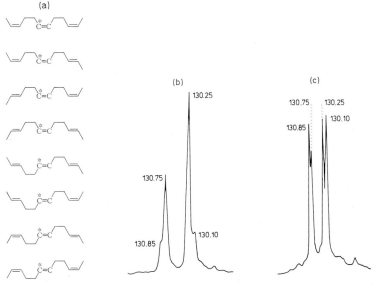

Fig. 19. (a) *Cis-trans* isomers related to an olefinic double bond of polybutadiene; (b,c) proton-decoupled olefinic carbon-13 NMR subspectra (25.2 MHz) of (b) 51% *cis*: 49% *trans* linkages in heptane; (c) 57.5% *cis*: 38% *trans*: 4.5% 1,2-connections in hexadeuteriobenzene [530].

Fig. 4.19. Eight *cis-trans* isomers relative to a central CC double bond may exist (Fig. 4.19(a)). Four resonances can be resolved (Fig. 4.19(b, c)) which are assigned to the most similiar pairs (Fig. 4.19(a)) [530]. Signal intensities clearly respond to the different *cis:trans* ratios of both samples.

4.15.3 Segmental Mobility

Carbon-13 spin-lattice relaxation times T_1 (Section 3.3) are relatively insensitive to the chain length of polymers [531]. The influence of local segmental motions predominates, as shown for low-density polyethylenes in which T_1 values are one to two seconds for the main chain but up to seven seconds for peripheral side-chain carbon nuclei at 120 °C [532] due to segmental mobility (Section 3.3.3.4). To conclude, quantitative evaluation of polymer carbon-13 spectra as necessary for side-chain determination requires the knowledge of spin-lattice relaxation times.

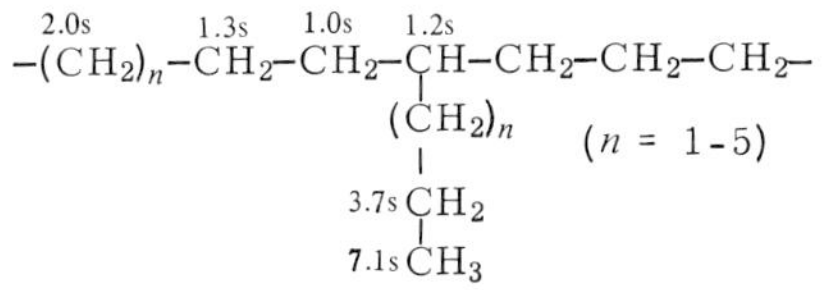

$$\begin{array}{c}
\overset{2.0s}{}\ \overset{1.3s}{}\ \overset{1.0s}{}\ \overset{1.2s}{} \\
-(CH_2)_n-CH_2-CH_2-CH-CH_2-CH_2-CH_2- \\
\underset{3.7s\ CH_2}{\underset{|}{(CH_2)_n}}\quad (n = 1\text{-}5) \\
\underset{7.1s\ CH_3}{|}
\end{array}$$ Low-density polyethylene

Spin-lattice relaxation times of carbon-13 in different polypropylene stereosequences differ slightly while nuclear Overhauser enhancements are almost identical (1.8–2.0) [533]; isotactic sequences display larger T_1 values than the syndiotactic stereoisomers. Other vinyl polymers behave correspondingly [534]. Carbon-13 spin-lattice relaxation times further indicate that dynamic properties in solution depend on configurational sequences longer than pentads. The ratio $T_1(CH):T_1(CH_2)$ varies between 1.6 to 1.9; thus, relaxation can be influenced by anisotropic motions of chain segments or by unusual distributions of correlation times [181].

4.16 Substituent Increments: Summary and Application

In the following sections, substituents, increments for carbon shielding in basic organic skeletons are summarized. Their practical use according to eq. (4.21) is illustrated by a few representative examples. Most of the tabulated increments are not based on regressional analyses, and deviations may be as large as ±0.5 ppm.

$$\delta_k = \delta_{k(RH)} + \sum Z_{ik} (+ \sum S_i) \qquad (4.21)$$

δ_k: ^{13}C shift of carbon in position k with respect to the substituent X;
   ($k = \alpha, \beta, \gamma, \ldots$ in aliphatic compounds and 1, 2, 3, 4, ... or 1, o m p in heteroaromatic and aromatic compounds);

$\delta_{k(RH)}$: ^{13}C shift of corresponding carbon k in parent hydrocarbon RH;

314   4 ^{13}C NMR Spectroscopy of Organic Compounds

Z_{ik}: ^{13}C shift increment of the substituent X in position i to the considered carbon k;
S_i: steric correction(s) in *cis*- and *trans*-cyclohexanes and alkenes only;

4.16.1 Substituted Alkanes

It can be seen from Table 4.78 that the α increments can be reasonably rationalized in terms of inductive effects (Pauling electronegativities), while the shielding of carbons γ to the substituent is generally attributed to a steric polarization of the γ C−H bond. Inductive and electric field effects contribute to the β increments. As electric fields can be evaluated only in rare cases, no general trend for the β effect has been recognized so far. Frequently, the α and β increments depend on whether a substituent X is terminal (*n*) or central (*iso*). If available, the *iso* increments are also given in Table 4.78.

The increments of Table 4.78 permit a prediction of carbon shifts in substituted alkanes. Thereby, they may provide an assignment aid which is particularly useful when proton decoupling experiments (off-resonance, gated, selective) and the application of modern pulse sequences fail because of equal multiplicities $((-CH_2-)_n$ chains) or signal overcrowding in the ^{13}C and ^{1}H NMR spectra. The practical use of the increments is illustrated for 1,4-dibromopentane and lysine.

1,4-Dibromopentane

			5	4	3	2	1	
Reference compound:		pentane	CH$_3$—	CH$_2$—	CH$_2$—	CH$_2$—	CH$_3$	
		δ (ppm)	13.7	22.6	34.6	22.6	13.7	(Table 4.1)
Increments	Z_{nBr}	(ppm)	0	−0.5	−4	10	20	(Table 4.78)
	Z_{isoBr}	(ppm)	10	26	10	−4	−0.5	(Table 4.78)
Predicted shifts		δ (ppm)	23.7	48.1	40.6	28.6	33.1	
Measured shifts		δ (ppm)	26.5	50.0	39.2	30.7	32.8	(in CDCl$_3$)

$$\underset{5}{CH_3}-\underset{4}{\underset{|}{CH}}-\underset{3}{CH_2}-\underset{2}{CH_2}-\underset{1}{CH_2}-Br$$
$$\qquad\quad Br$$

Lysine (2,6-Diaminohexanedioic acid)

			5	4	3	2	1	
Reference compound:		pentane	CH$_3$—	CH$_2$—	CH$_2$—	CH$_2$—	CH$_3$	
		δ (ppm)	13.7	22.6	34.6	22.6	13.7	(Table 4.1)
Increments	Z_{COOH}	(ppm)	0	0	−3	2	20	(Table 4.78)
	Z_{nNH_2}	(ppm)	28.5	11.5	−5	0	0	(Table 4.78)
	Z_{isoNH_2}	(ppm)	0	0	−5	10	24	(Table 4.78)
Predicted shifts		δ (ppm)	42.2	34.1	21.6	34.6	57.7	
Measured shifts		δ (ppm)	40.0	30.7	22.4	27.2	55.3	(in D$_2$O)

$$H_2N-CH_2-CH_2-CH_2-CH_2-\underset{|}{CH}-COOH$$
$$\qquad\qquad\qquad\qquad\qquad\qquad\quad NH_2$$

4.16 Substituent Increments: Summary and Application

Table 4.78. Empirical Increments of Substituents X Replacing H in Alkanes (in ppm) [73].

−X	Z_α		Z_β		Z_γ	Z_δ	Z_ε
	n	iso	n	iso			
−F	70	63	8	6	−7	0	0
−Cl	31	32	10	10	−5	−0.5	0
−Br	20	26	10	10	−4	−0.5	0
−I	−7	4	11	12	−1.5	−1	0
−O−	57	51	7	5	−5	−0.5	0
−OCOCH$_3$	52	45	6.5	5	−4	0	0
−OH	49	41	10	8	−6	0	0
−SCH$_3$	20.5	−	6.5	−	−2.5	0	0
−S−	10.5	−	11.5	−	−3.5	−0.5	0
−SH	10.5	11	11.5	11	−3.5	0	0
−NH$_2$	28.5	24	11.5	10	−5	0	0
−NHR	36.5	30	8	7	−4.5	−0.5	−0.5
−NR$_2$	40.5	−	5	−	−4.5	−0.5	0
−$\overset{+}{\text{N}}$H$_3$	26	24	7.5	6	−4.5	0	0
−$\overset{+}{\text{N}}$R$_3$	30.5	−	5.5	−	−7	−0.5	−0.5
−NO$_2$	61.5	57	3	4	−4.5	−1	−0.5
−NC	27.5	−	6.5	−	−4.5	0	0
−CN	3	1	2.5	3	−3	0.5	0
>C=NOH syn	11.5	−	0.5	−	−2	0	0
>C=NOH anti	16	−	4.5	−	−1.5	0	0
−CHO	30	−	−0.5	−	−2.5	0	0
>CO	23	−	3	−	3	0	0
−COCH$_3$	29	23	3	1	−3.5	0	0
−COCl	33	28	2	2	−3.5	0	0
−COO$^-$	24.5	20	3.5	3	−2.5	0	0
−COOCH$_3$ (and C$_2$H$_5$)	22.5	17	2.5	2	−3	0	0
−CONH$_2$	22	−	2.5	−	−3	−0.5	0
−COOH	20	16	2	2	−3	0	0
−Phenyl	23	17	9	7	−2	0	0
−CH=CH$_2$	20	−	6	−	−0.5	0	0
−C≡CH	4.5	−	5.5	−	−3.5	0.5	0

Significant differences between predicted and measured ^{13}C chemical shifts as found for some of the lysine carbons indicate that shift increments are not additive when interactions between substituents are involved. Although deviations from increment additivity are frequently observed, the sequence of ^{13}C signals can be qualitatively evaluated in many cases.

4.16.2 Substituted Cyclohexanes and Bicyclo[2.2.2]heptanes

Table 4.79 summarizes the increments of some substituents in cyclohexanes. The general trends are similar to those observed for alkyl derivatives (deshielding in α and β and shielding in γ position). In addition, the δ carbons are slightly shielded by substituents. Moreover, axial substituents generally cause a larger shielding of the carbons in α, β and γ position ($\delta_{eq} > \delta_{ax}$). Concerning the γ effect, the γ gauche type shielding (Chapter 3.1.3.8) observed for axial substituents is significantly larger than the γ trans effect of equatorial groups (Table 4.79).

The application of these increments is illustrated for 1-trans-2-cis-3-trimethylcyclohexane and cis-3-methylcyclohexanol (data from Table 4.24).

Table 4.79. (a) Empirical Increments of Equatorial and Axial Substituents in Cyclohexane [263–265] (in ppm); (Reference Compound: Cyclohexane with $\delta_{RH} = 27.6\ (\pm 0.5)$ ppm). (b) Correction Terms S for 1,1- and 1,2-Dimethylcyclohexanes.

(a) X	$Z_{\alpha e}$	$Z_{\alpha a}$	$Z_{\beta e}$	$Z_{\beta a}$	$Z_{\gamma e}$	$Z_{\gamma a}$	$Z_{\delta e}$	$Z_{\delta a}$
CH$_3$	6	1.5	9	5.5	0	−6.5	−0.5	0
C$_2$H$_5$	13	8.5	6	3	−0.5	−5.5	0	0
i-C$_3$H$_7$	17.5	14.1	3	3	0	−5.5	0.5	0
t-C$_4$H$_9$	21.5			0.5	0		0	
CH=CH$_2$	15	10	5.5	3	−1	−6	0	0
C≡CH	1.5	1	5	3	−2	−6	−2.5	−1.5
CH=O	23	19.5	−1.5	−2.5	−1.5	−4.5	0	0
COOCH$_3$	16.5	12	2.5	0.5	−1	−4	−0.5	−0.5
CN	0.5	−0.5	2	0.5	−2.5	−5	−2.5	−2
NC	25	23.5	7	4	−3	−7	−2	−2
NCS	28.5	26	7	4.5	−2.5	−6.5	−2	−2
F	64	61	6	3	−3	−7	−3	−2
Cl	33	33	11	7	0	−6	−2	−1
Br	25	28	12	8	1	−6	−1	−1
I	3	11	13	9	2	−4	−2	−1
OH	43	39	8	5	−3	−7	−2	−1
OCH$_3$	52	47	4	2	−3	−7	−2	−1
OSi(CH$_3$)$_3$	43.5	39	9	6	−2.5	−7	−2	−2
OCOCH$_3$	46	42	5	3	−2	−6	−2	0
OCOC$_6$H$_5$	46	42	4.5	2.5	−3	−6.5	−2.5	−2.5
SH	11	9	10.5	6	−0.5	−7.5	−2.5	−1.3
NH$_2$	24	20	10.5	7	−1	−7	−1	0

4.16 Substituent Increments: Summary and Application   317

1-*trans*-2-*cis*-3-Trimethylcyclohexane

	C-1,3		C-2		C-4,6		C-5
δ_{RH}	27.6		27.6		27.6		27.6
$Z_{\alpha e}$	6.0						
$Z_{\beta e}$	9.0	$Z_{\alpha e}$	6.0	$Z_{\beta e}$	9.0	$2Z_{\gamma e}$	0.0
$Z_{\gamma e}$	0.0	$2Z_{\beta e}$	18.0	$Z_{\gamma e}$	0.0	$Z_{\delta e}$	−0.5
$S_{\alpha e \beta e}$	−2.5	$2S_{\alpha e \beta e}$	−5.0	$Z_{\delta e}$	−0.5		
calculated δ	40.1		46.6		36.1		27.1 ppm
measured δ	39.15		46.25		36.45		26.45 ppm

cis-3-Methylcyclohexanol

	C-1	C-2	C-3	C-4	C-5	C-6
δ_{RH}	27.6	27.6	27.6	27.6	27.6	27.6
Z_{OH}	43 ($Z_{\alpha e}$)	8 ($Z_{\beta e}$)	−3 ($Z_{\gamma e}$)	−2 ($Z_{\delta e}$)	−31 ($Z_{\gamma e}$)	8 ($Z_{\beta e}$)
Z_{CH_3}	0 ($Z_{\gamma e}$)	9 ($Z_{\beta e}$)	6 ($Z_{\alpha e}$)	9 ($Z_{\beta e}$)	0 ($Z_{\gamma e}$)	−0.5 ($Z_{\delta e}$)
calculated δ	70.6	44.6	30.6	34.6	24.6	35.1 ppm
measured δ	70.5	44.0	31.7	34.8	24.4	34.4 ppm

Table 4.80. Empirical Increments of 2-*endo* and 2-*exo* Substituents in Norbornanes [73f, 215] (in ppm).

Reference compound: Norbornane

X	Z_α	Z_β	$Z_{\beta'}$	Z_γ	$Z_{\gamma'}$	Z_δ	$Z_{\gamma''}$
endo-CH₃	4.5	5.4	10.6	−7.7	1.4	0.5	0.2
exo-CH₃	6.7	6.7	10.1	−1.1	0.5	0.2	−3.7
endo-CN	0.1	3.4	5.5	−4.9	0.2	−0.7	0.0
exo-CN	1.0	5.5	6.3	−1.6	−0.3	−1.5	−1.3
endo-COOH	16.2	4.2	2.1	−4.8	0.9	−0.6	1.9
exo-COOH	16.7	4.6	4.4	−1.0	−0.2	−0.3	−1.8
endo-COOCH₃	15.9	4.0	2.2	−5.0	0.7	−0.7	1.7
exo-COOCH₃	16.4	5.1	4.2	−1.4	−0.4	−1.1	−2.1
endo-OH	42.4	6.3	9.5	−9.7	0.9	0.2	−0.9
exo-OH	44.3	7.7	12.3	−5.2	−1.0	−1.3	−4.1
endo-NH₂	23.2	6.8	10.5	−9.5	1.2	0.6	0.3
exo-NH₂	25.3	8.9	12.4	−3.1	−0.4	−1.2	−4.4

4 ^{13}C NMR Spectroscopy of Organic Compounds

γ gauche and γ trans type interactions as discussed for cyclohexanes are also apparent from the shift increments of selected *endo-* and *exo-*substituted norbornanes (Table 4.80). The increased shieldings of the γ carbon in the *endo* isomers and the γ" carbon (C-7) in the *exo* isomers are typical γ gauche effects. The shielding differences are useful for configurational assignments in this class of compounds.

4.16.3 Substituted Alkenes

Table 4.81 shows the empirical increments obtained for substituents in position *i* relative to an olefinic carbon denoted as *k*. Correction terms *S* are required for *cis* and *geminal* substituents. As an application of the increments in Table 4.81 the olefinic carbon shifts of (Z)-4-methyl-2-pentenoic acid are predicted:

Table 4.81. (a) Empirical Increments Z for Substituents in Position *i* (*i'*) Relative to the Olefinic Carbon *k* (in ppm) [233].
(b) Correction Terms $S_{ii'}$ for Alkyl Groups *cis*, *trans* and *geminal* to Each Other (in ppm) [233]. The Reference Compound is Ethene with $\delta_{RH} = 123.5 \pm 0.5$ ppm.

(a) X_i	γ'— $Z_{\gamma'}$	β'— $Z_{\beta'}$	α'— $Z_{\alpha'}$	C=C k	α— Z_α	β— Z_β	γ— Z_γ
C	1.5	−1.8	−7.9		10.6	7.2	−1.5
C$_6$H$_5$			−11		12		
C(CH$_3$)$_3$			−14		25		
Cl		2	−6		3	−1	
Br		2	−1		−8	0	
I			7		−38		
OH		−1	—		—	6	
OR		−1	−39		29	2	
OCOCH$_3$			−27		18		
CHO			13		13		
COCH$_3$			6		15		
COOH			9		4		
COOR			7		6		
CN			15		−16		
(b)	$S_{\alpha\alpha'(trans)}$:	0.0					
	$S_{\alpha\alpha'(cis)}$:	−1.1					
	$S_{\alpha\alpha(gem)}$:	−4.8					
	$S_{\alpha'\alpha'(gem)}$:	2.5					
	$S_{\beta\beta}$:	2.3					

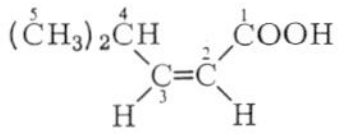

4.16 Substituent Increments: Summary and Application   319

(CH₃)₂ĊH ĊOOH   (Z)-4-Methyl-2-pentenoic acid

	C-3	C-2	
δ_{RH}	123.5	123.5	δ_{RH}
$Z_{\alpha'COOH}$	9	4	$Z_{\alpha COOH}$
$Z_{\alpha C}$	10.6	−7.9	$Z_{\alpha' C}$
$2 Z_{\beta C}$	14.4	−3.6	$2 Z_{\beta' C}$
$S_{\alpha\alpha'(cis)}$	−1.1	−1.1	$S_{\alpha\alpha'(cis)}$ (values for C)
predicted δ	156.4	114.9 ppm	
measured δ	158.3	116.4 ppm	

4.16.4 Substituted Benzenes

Most of the substituent increments presented in Table 4.82 can be derived from ^{13}C shifts of benzenoid carbons in monosubstituted benzenes as listed in Table 4.53. Additional substituent increments are available for fused aromatic rings such as naphthalene and

Table 4.82. Empirical Substituent Increments in Monosubstituted Benzenes [383].

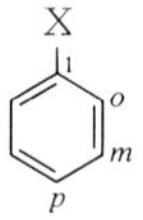

$\delta_{C_6H_6} = 128.5$ ppm

X	Z_1	Z_o	Z_m	Z_p
CH₃	9.3	−0.1	0.7	−3.0
CH₂CH₃	14.9	−1.3	−0.7	−3.3
CH₂CH₂CH₃	14.1	−2.0	0.1	−2.7
CH(CH₃)₂	20.3	−0.2	−0.1	−2.6
C(CH₃)₃	21.8	−3.9	−0.9	−3.6
Cyclopropyl (C₃H₅)	15.1	−3.3	−0.6	−3.6
CH=CH₂	9.1	−2.4	−0.2	−0.9
Phenyl (C₆H₅)	13.0	−1.1	0.5	−1.0
C≡CH	−5.8	3.9	0.1	0.4
CF₃	−9.0	−2.2	0.3	3.2
CH₂F	8.1	−0.9	0.1	0.2
CH₂Cl	9.4	0.3	0.4	0.2
CH₂Br	9.7	1.0	1.3	0.6
CH₂OH	13.3	−0.8	0.6	−0.4
CH₂OCH₂C₆H₅	10.5	−0.5	0.5	−0.5

Table 4.82. (continued).

X	Z_1	Z_o	Z_m	Z_p
CH_2NH_2	15.5	− 1.1	0.0	− 1.9
$CH_2NHCH_2C_6H_5$	11.9	− 0.5	−0.3	− 1.7
$CH_2SCH_2C_6H_5$	10.5	0.3	0.9	− 1.5
typical electron donors				
F	35.1	−14.4	0.9	− 4.4
Cl	6.4	0.2	1.0	− 2.0
Br	− 5.9	3.0	1.5	− 1.5
I	−32.3	9.9	2.6	− 0.4
OH	26.6	−12.8	1.6	− 7.1
OCH_3	31.4	−14.4	1.0	− 7.8
OC_6H_5	29.2	− 9.4	1.4	− 5.3
$OCOCH_3$	23.0	− 6.4	1.3	− 2.3
OCN	24.6	−13.2	2.2	− 1.5
SH	2.0	0.6	0.2	− 3.3
SCH_3	10.1	− 1.7	0.3	− 3.5
SC_6H_5	7.3	2.4	0.6	− 1.6
NH_2	20.2	−14.1	0.6	− 9.6
$NHCH_3$	21.9	−16.4	0.6	−12.6
$N(CH_3)_2$	22.2	−15.8	0.5	−11.8
NHC_6H_5	14.7	−10.6	0.8	− 7.6
$N(C_6H_5)_2$	19.0	− 4.6	0.9	− 5.8
$NHCOCH_3$	11.1	− 9.9	0.2	− 5.6
$N=CHC_6H_5$	24.5	− 6.9	1.0	− 1.9
$N=C=NC_6H_5$	10.8	− 3.6	1.7	− 2.2
$N=C=O$	5.4	− 3.7	1.2	− 2.6
$N=C=S$	17.4	− 2.2	2.0	− 0.4
$P(C_6H_5)_2$	9.3	5.5	0.3	0.0
typical electron acceptors				
NO_2	20.6	− 4.3	1.3	6.2
$N=NC_6H_5$	24.2	− 5.5	1.2	3.3
$N(CH_3)_3^+I^-$	8.8	− 8.2	0.2	3.0
COCl	4.8	2.9	0.6	6.9
COO^-Na^+	10.3	2.8	2.2	5.1
COOH	2.9	1.3	0.4	4.6
$COOCH_3$	2.1	1.2	0.0	4.4
$COOC_2H_5$	2.4	1.2	−0.1	4.3
$CONH_2$	5.8	− 1.1	−0.3	2.7
CHO	8.2	1.2	0.5	5.8
$COCH_3$	8.9	0.1	−0.1	4.5
COC_6H_5	9.1	1.5	−0.2	3.8
CN	−15.5	4.1	1.4	5.0

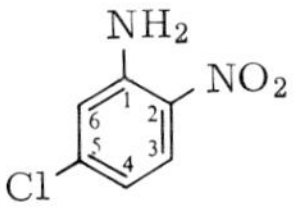

4.16 Substituent Increments: Summary and Application

5-Chloro-2-nitroaniline [73k]

	C-1	C-2	C-3	C-4	C-5	C-6
$\delta_{Benzene}$	128.5	128.5	128.5	128.5	128.5	128.5
Z_{NH_2}	20.2 (Z_1)	−14.1 (Z_o)	0.6 (Z_m)	−9.6 (Z_p)	0.6 (Z_m)	−14.4 (Z_o)
Z_{NO_2}	−4.3 (Z_o)	20.6 (Z_1)	−4.3 (Z_o)	1.3 (Z_m)	6.2 (Z_p)	1.3 (Z_m)
Z_{Cl}	1.0 (Z_m)	−2.0 (Z_p)	1.0 (Z_m)	0.2 (Z_o)	6.4 (Z_1)	0.2 (Z_o)
predicted δ	145.4	133.0	125.8	120.4	141.7	115.6 (ppm)
measured δ in $(CD_3)_2SO$	146.5	131.7	125.4	122.6	136.9	120.1 (ppm)

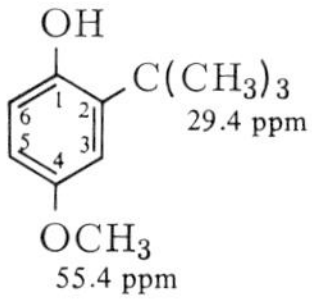

2-t-butyl-4-methoxyphenol [73 k]

	C-1	C-2	C-3	C-4	C-5	C-6
$\delta_{Benzene}$	128.5	128.5	128.5	128.5	128.5	128.5
Z_{OH}	26.6 (Z_1)	−12.8 (Z_o)	1.6 (Z_m)	−7.1 (Z_p)	1.6 (Z_m)	−12.8 (Z_o)
$Z_{C(CH_3)_3}$	−3.9 (Z_o)	21.8 (Z_1)	−3.9 (Z_o)	−0.9 (Z_m)	−3.6 (Z_p)	−0.9 (Z_m)
Z_{OCH_3}	−7.8 (Z_p)	1.0 (Z_m)	−14.4 (Z_o)	31.4 (Z_o)	−14.4 (Z_1)	+1.0 (Z_m)
predicted δ	143.4	138.5	111.8	151.9	112.1	115.8 (ppm)
measured δ in $CDCl_3$	147.7	136.9	110.2	152.3	113.3	116.2 (ppm)

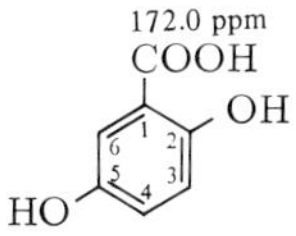

2,5-Dihydroxybenzoic acid [73 k]

	C-1	C-2	C-3	C-4	C-5	C-6
$\delta_{Benzene}$	128.5	128.5	128.5	128.5	128.5	128.5
Z_{COOH}	2.9 (Z_1)	1.3 (Z_o)	0.4 (Z_m)	4.6 (Z_p)	0.4 (Z_m)	1.3 (Z_o)
Z_{2-OH}	−12.8 (Z_o)	26.6 (Z_1)	−12.8 (Z_o)	1.6 (Z_m)	−7.1 (Z_p)	1.6 (Z_m)
Z_{5-OH}	1.6 (Z_m)	−7.1 (Z_p)	1.6 (Z_m)	−12.8 (Z_o)	26.6 (Z_1)	−12.8 (Z_o)
predicted δ	120.2	149.3	117.7	121.9	148.4	118.6 (ppm)
measured δ in $(CD_3)_2SO$	112.8	154.5	114.1	123.9	149.5	117.9 (ppm)

anthracene [394]. It was pointed out (Sections 4.11.2 and 3.1.3.6 that the α effect (Z_1) is predominantly inductive. The o increment (Z_o) arises from inductive and mesomeric effects, while the mesomeric effect predominates in p position (Z_p). Thus the p increments reflect a shielding for $+ M$ substituents ($Z_p < 0$) and a deshielding for $- M$ substituents ($Z_p > 0$). This behavior parallels the activation and deactivation of the p position towards electrophilic substitution (Fig. 4.9) by those substituents. The m carbons are only slightly influenced by substituents ($Z_m < 2$ ppm, mostly $Z_m < 1$ ppm). The practical use of the increments listed in Table 4.82 is illustrated for the benzenoid carbons of 5-chloro-2-nitroaniline, 2-t-butyl-4-methoxyphenol and 2,5-dihydroxybenzoic acid. Clearly, prediction of ^{13}C shifts by increment calculations is not accurate if interactions between substituents are involved, e.g. hydrogen bonding between OH and COOH in 2,5-dihydroxybenzoic acid.

4.16.5 Substituted Pyridines

Substituent increments have been compiled for several heterocyclic systems, such as six-membered heteroalicyclic rings [408], pyrrole [435], indole [445], thiazole [437], and pyridine [452].

Increments of selected substituents in positions 2, 3 and 4 of the pyridine ring are listed in Table 4.83(b). The ^{13}C shifts of C-2,6 (149.7 ppm), C-3,5 (124.2 ppm) and C-4 (136.2 ppm) have been used as references. As can be seen in Table 4.67, N-oxidation causes a significant shielding ($- 10$ ppm) of the carbons α and γ to nitrogen, while those in the β position are slightly shielded. This analogy to the behavior of pyridine-N-oxide towards electrophilic substitution can be rationalized by remembering the known cannonical formulae of this molecule in contrast to pyridine itself.

As is shown below for 2-amino-3-methylpyridine, a reasonable prediction of the signal sequence is possible using the substituent increments presented in Table 4.83.

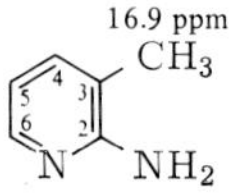

	2-Amino-3-methylpyridine [73k]				
	C-2	C-3	C-4	C-5	C-6
$\delta_{Pyridine}$	149.7	124.2	136.2	124.2	149.7
Z_{2NH_2}	11.3	-14.7	$+2.3$	-10.6	-0.9
Z_{3CH_3}	1.3	9.0	0.2	-0.8	-2.3
predicted δ	162.3	118.5	138.7	112.8	146.5 ppm
measured δ in CDCl$_3$	157.9	116.5	137.5	113.8	145.4 ppm

Table 4.83. Nitrogen Increments (a) and Substituent Effects of Selected Groups (b) in Pyridine [452] (in ppm).

(a)

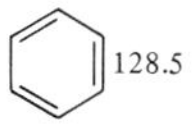

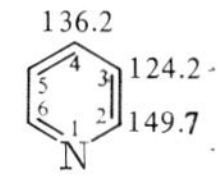

$\delta_2 = 149.7 + Z_{i2}$
$\delta_3 = 124.2 + Z_{i3}$
$\delta_4 = 136.2 + Z_{i4}$
$\delta_5 = 124.2 + Z_{i5}$
$\delta_6 = 149.7 + Z_{i6}$

$Z_{2,6} = 149.7 - 128.5 = 21.2$ ppm
$Z_{3,5} = 124.2 - 128.5 = -4.3$ ppm
$Z_4 = 136.2 - 128.5 = 7.7$ ppm

(b)

2-Substituent	Z_{22}	Z_{23}	Z_{24}	Z_{25}	Z_{26}
CH$_3$	8.8	−0.6	0.2	−3.0	−0.4
CH$_2$CH$_3$	13.6	−1.8	0.4	−2.9	−0.7
Si(CH$_3$)$_3$	17.6	3.8	−2.9	−2.0	−0.5
F	14.4	−13.1	6.1	−1.5	−1.5
Cl	2.3	0.7	3.3	−1.2	0.6
Br	−6.7	4.8	3.3	−0.5	1.4
OH	15.5	−3.5	−0.9	−16.9	−8.2
OCH$_3$	15.3	−7.5	2.1	−13.1	−2.2
NH$_2$	11.3	−14.7	2.3	−10.6	−0.9
NO$_2$	8.0	−5.1	5.5	6.6	0.4
CHO	3.5	−2.6	1.3	4.1	0.7
COCH$_3$	4.3	−2.8	0.7	3.0	−0.2
CN	−15.9	5.0	1.6	3.6	1.4

3-Substituent	Z_{32}	Z_{33}	Z_{33}	Z_{35}	Z_{36}
CH$_3$	1.3	9.0	0.2	−0.8	−2.3
CH$_2$CH$_3$	−0.4	15.5	−0.6	−0.4	−2.7
Si(CH$_3$)$_3$	1.9	8.3	2.5	−3.1	−0.9
F	−11.5	36.2	−13.0	0.9	−3.9
Cl	−0.3	8.2	−0.2	0.7	−1.4
Br	2.1	−2.6	2.9	1.2	−0.9
OH	−10.7	31.4	−12.2	1.3	−8.6
NH$_2$	−11.9	21.5	−14.2	0.9	−10.8
CHO	2.4	7.9	0.0	0.6	5.4
COCH$_3$	3.5	8.6	−0.5	−0.1	0.0
CN	3.6	−13.7	4.4	0.6	4.2

Table 4.83. (continued).

4-Substituent	Z_{42}	Z_{43}	Z_{44}
CH_3	0.5	0.8	10.8
CH_2CH_3	−0.1	− 0.4	17.0
$Si(CH_3)_3$	−3.6	1.6	11.4
F	2.7	−11.8	33.0
Cl	0.9	0.5	8.2
Br	3.0	3.4	− 3.0
OCH_3	0.6	−14.2	29.3
NH_2	0.9	−13.8	19.6
$N(CH_3)_2$	−0.2	−17.1	18.3
NO_2	−1.3	− 3.0	4.1
CHO	1.7	− 0.6	5.5
$COCH_3$	1.6	− 2.6	6.8
CN	2.1	2.2	−15.7

4.16.6 Nitrogen Increments in Fused Heterocycles

Based on the carbon shifts of naphthalene as reference, nitrogen increments can be derived for quinoline and isoquinoline as is shown in Table 4.84. These increments are

Table 4.84. Nitrogen Increments in Quinoline and Isoquinoline Relative to Naphthalene [73] (a); Application of these increments to Quinazoline (b) and 2,7-Naphthyridine (c).

(a)

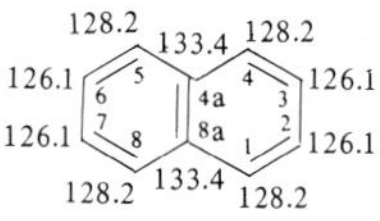

Naphthalene (Reference compound): 128.2, 133.4, 128.2, 126.1, 126.1, 126.1, 126.1, 128.2, 133.4, 128.2

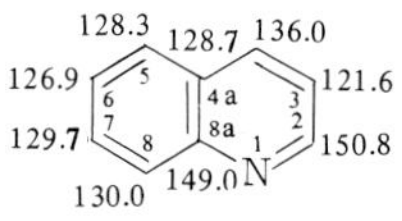

Quinoline (ppm): 128.3, 128.7, 136.0, 126.9, 121.6, 129.7, 150.8, 130.0, 149.0

Isoquinoline (ppm): 127.0, 136.2, 121.0, 130.7, 144.0, 127.7, 128.1, 129.3, 153.3

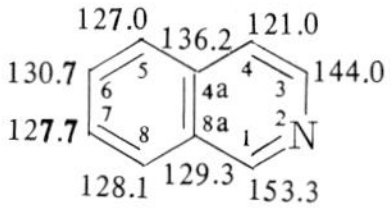

	Quinoline ppm	Isoquinoline ppm
Z_1	–	25.1
Z_2	24.7	–
Z_3	−4.5	17.9
Z_4	7.8	−7.2
Z_5	0.1	−1.2
Z_6	0.8	4.6
Z_7	3.6	1.6
Z_8	1.8	−0.1
Z_{4a}	−4.7	2.8
Z_{8a}	15.6	−4.1

(b) Quinazoline

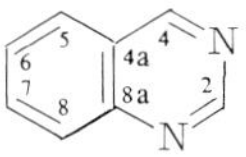

	Calculated ppm	Measured ppm
$\delta_2 = 126.1 + Z_{2q} + Z_{3i} = 126.1 + 24.7 + 17.9 = 168.7$		160.5
$\delta_4 = 128.2 + Z_{4q} + Z_{1i} = 128.2 + 7.8 + 25.1 = 161.1$		155.8
$\delta_5 = 128.2 + Z_{5q} + Z_{8i} = 128.2 + 0.1 - 0.1 = 128.2$		127.4
$\delta_6 = 126.1 + Z_{6q} + Z_{7i} = 126.1 + 0.8 + 1.6 = 128.5$		127.9
$\delta_7 = 126.1 + Z_{7q} + Z_{6i} = 126.1 + 3.6 + 4.6 = 134.3$		134.1
$\delta_8 = 128.2 + Z_{8q} + Z_{5i} = 128.2 + 1.8 - 1.2 = 128.8$		128.6
$\delta_{4a} = 133.4 + Z_{4aq} + Z_{8ai} = 133.4 - 4.7 - 4.1 = 124.6$		125.2
$\delta_{8a} = 133.4 + Z_{8aq} + Z_{4ai} = 133.4 + 15.6 + 2.8 = 151.8$		150.1

(c) Naphthyridine

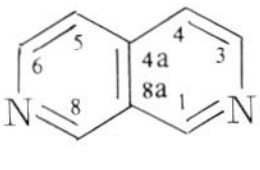

	Calculated	Measured
$\delta_{1,8} = 128.2 + Z_1 + Z_8 = 128.1 + 25.1 - 0.1 = 153.1$		153.4
$\delta_{3,6} = 126.1 + Z_3 + Z_6 = 126.1 + 17.9 + 4.6 = 148.6$		147.2
$\delta_{4,5} = 128.2 + Z_4 + Z_5 = 128.2 - 7.2 - 1.2 = 119.8$		119.6
$\delta_{4a} = 133.9 + 2Z_{4a} \quad = 133.9 + 5.6 \quad = 139.5$		138.5
$\delta_{8a} = 133.9 + 2Z_{8a} \quad = 133.9 - 8.1 \quad = 125.8$		124.3

useful aids in assigning the ^{13}C shifts of diazanaphthalenes such as quinazoline (Table 4.84(b) or 2,7-naphthyridine (Table 4.84(c)). Significant deviations between increment prediction and measurement are found for C-2 and C-4 of quinazoline due to additional interactions between closely attached nitrogens. The values predicted for benzenoid carbons, however, agree well with the experimental data.

5 ^{13}C NMR Spectra of Natural Products

5.1 Terpenes

The different types of terpenoid carbons resonate according to the chemical shift rules discussed in Chapter 3. In several communications the ^{13}C chemical shifts of a large series of open-chain terpenes have been reported [535, 536]. Off-resonance decoupling and measurements in the presence of lanthanoid shift reagents were used as additional aids for the signal assignments. Measurements of T_1 values may be useful to differentiate between the terminal methyls of the 1,1-dimethylvinyl group, as was demonstrated for a

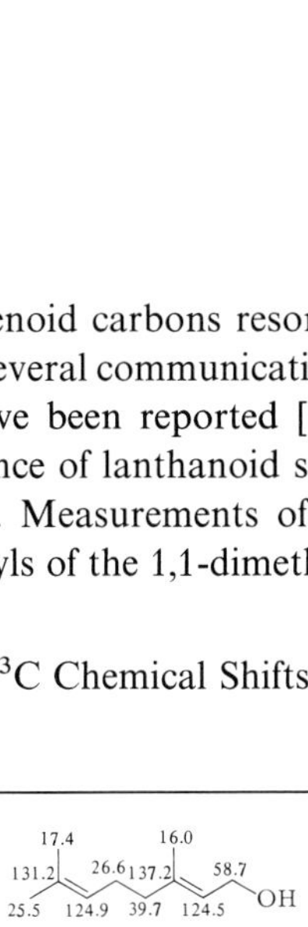

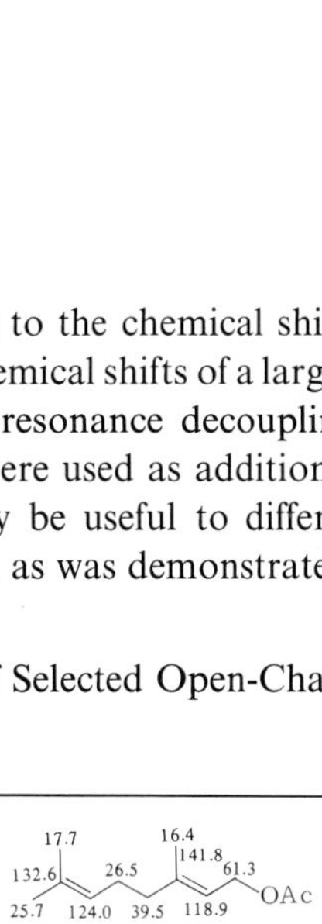

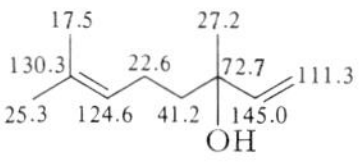

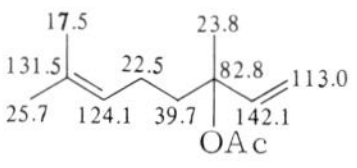

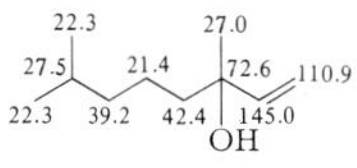

Table 5.1. Structures and ^{13}C Chemical Shifts (δ_C in ppm) of Selected Open-Chain Terpenes .

Dimethyloctanol [535]

Geraniol [535]

Geraniol acetate [535]

Linalool [535]

Linalool acetate [535]

Dihydrolinalool [535]

Isolavandulol [535]

cis-Citral [536]

trans-Citral [536]

Farnesol [536]

Myrcene [536]

Squalene [536]

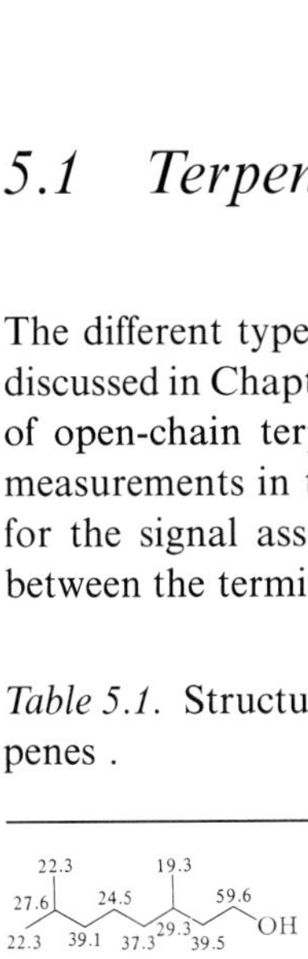

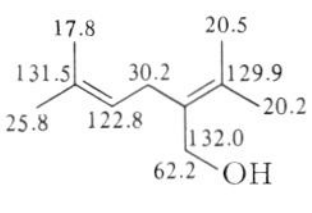

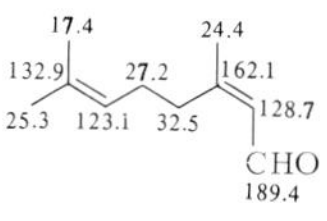

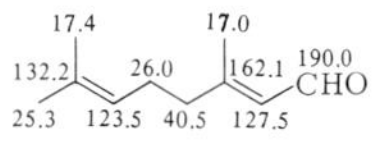

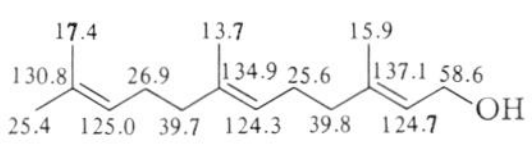

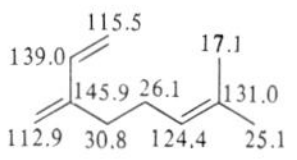

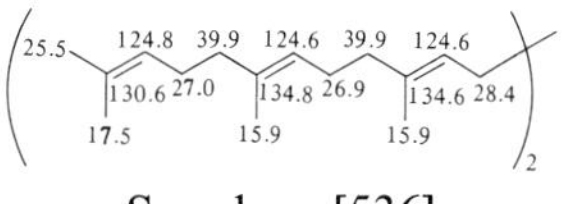

5 ¹³C NMR Spectra of Natural Products

Table 5.2. Structures and ¹³C Chemical Shifts (δ_C in ppm) of Selected Menthane Terpenes.

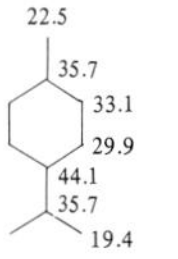

Menthane [536] 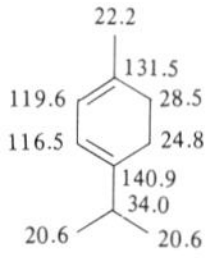
α-Terpinene [536] 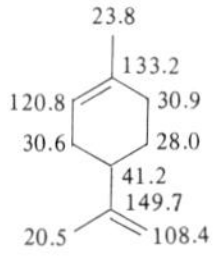
Limonene [536] 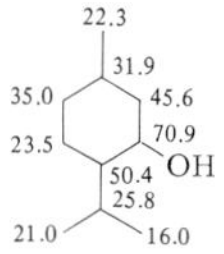
Menthol [536] 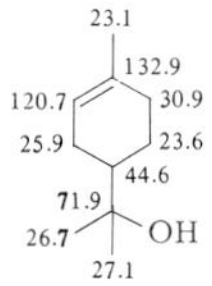
α-Terpinol [536]

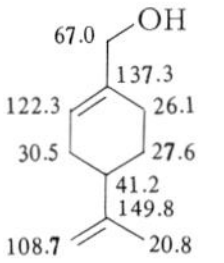

Perillaalcohol [535]
Cuminalcohol [535] 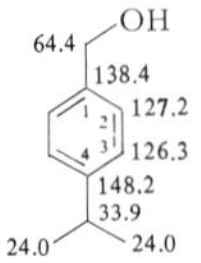
Pulegone [536] 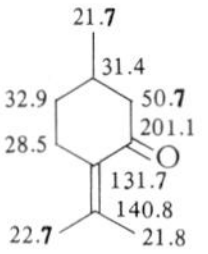
Carvone [536] 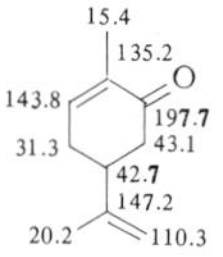
Cineol [535] 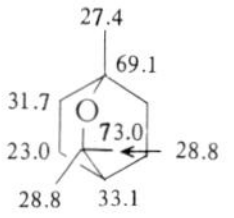

series of linalools where the ratio $T_{1(trans)}:T_{1(cis)}$ was found to be always about 2:1 [126]. In Tables 5.1 and 5.2 structures and ¹³C chemical shifts of selected open-chain and menthone terpenes are given.

For the bicyclic monoterpenes the originally reported and used signal assignments [535–544] could be revised using the following methods:

1. Graphical analyses of series of single-frequency off-resonance decoupled spectra,

2. one-bond and long-range C–H couplings (Table 5.3),

3. silver/lanthanide induced shifts and chiral splittings [545],

4. empirically computed shifts, calculated from $\delta_{C-4} = 48.3 - 3.5\,F + 4.1\,R$ of seven 1-substituted camphenes or from shift increments [546] and

5. from ¹³C NMR spectra of α- and β-pinene enriched at position C-10 [547, 548].

Structures and ¹³C chemical shifts (in ppm) of selected bicyclic monoterpenes are collected in Table 5.4.

The ¹³C chemical shifts of the parent compounds of camphor and its derivatives are given with the following structures:

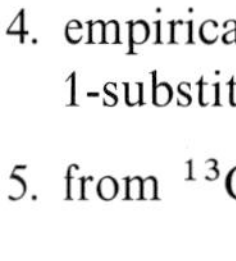

Norbornane
Bornane 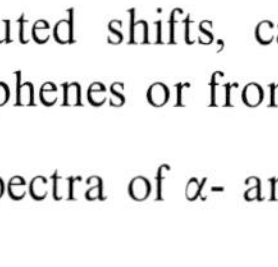
Bornylene

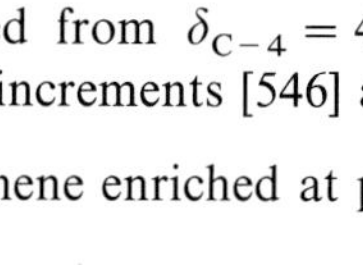

Table 5.3. C–H Coupling Constants (in Hz) of Camphene, α-Pinene and Bornene [548].

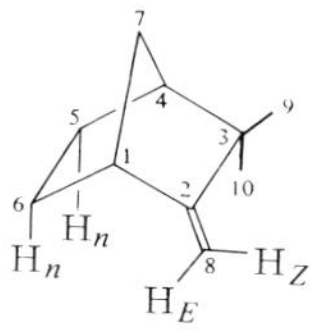

Camphene *1* 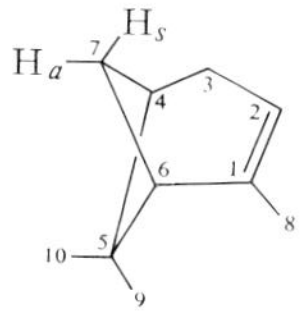   α-Pinene *2* 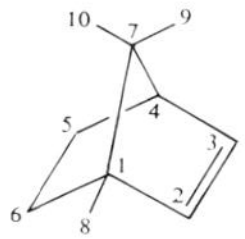   Bornene *3*

Compound	C-1	C-2	C-3	C-4	C-5	C-6	C-7	C-8	C-9	C-10
1	d 143 (H-1) d 10 (H-8Z) t 5 (H-4, 8E)	s, b	s, b	d 140 (H-4) b	t 133 (H-5) t 9 d 6 d 3	t 134 (H-6) t 9	t 133 (H-7) t 7 (H-5,6n)	d 155 (H-8E) d 153 (H-8Z) d 2 (H-1)	qa 125 (H-9) qa 5 (H-10) d 4 (H-4)	qa 126 (H-10) qa 5 (H-9)
2	sx 5 d 2	d 160 (H-2) sx 6 (H-4,6,8) d 2	t 126 (H-3) t 5 (H-2) (H-7a)	d 144 (H-4)	s, b	d 141 (H-6 d 6 (H-2) t 4	t 127 (H-7) 6	qa 125 (H-8) t 5 (H-2,6)	qa 124 (H-9) sx 5 (H-4,6,10)	qa 123 (H-10) sx 4 (H-4,6,9)
3	s, m	d 165 (H-2) b	d 168 (H-3) d 6	d 144 (H-4) m	t 133 (H-5) qa 3	t 132 (H-6) qa 4 (H-8)	s, m	qa 124 (H-8) t 2	qa 125 (H-9) qa 5 (H-10)	qa 123 (H-10) qa 5 (H-9)

s: singlet, d: doublet, t: triplet, qa: quartet, qi: quintet, sx: sextet, m: multiplet, b: broad line.

Table 5.4. Structures and ^{13}C Chemical Shifts (δ_C in ppm) of Selected Bicyclic Monoterpenes.

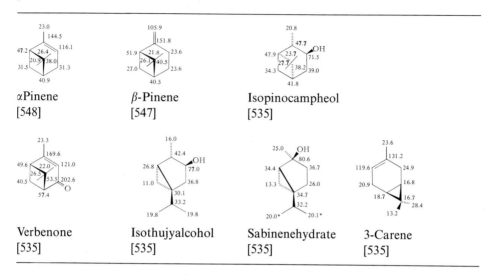

Several rules found for the spectra of a series of camphor derivatives (bicyclo[2,2,1]heptanones) are worthy of mention (Table 5.5) [548, 549]:

(1) The signals of methylene carbons in the 7-position (the one-carbon bridge) are deshielded by about 10 ppm relative to those of methylene groups in the two-carbon bridges.

(2) Changing the bromine atom in 3-bromo-norbornanones from *endo* to *exo* position causes the signal of C-3 to shift upfield about 6 ppm.

Based on off-resonance decoupling, lanthanoid shift studies, spectral comparison with different analogues and decalin models, signal assignments of pimaradienes [550], kauranoid diterpenes [551, 552], podocarpane derivatives [553] and labdanic diterpenes [554] could be achieved (Table 5.6).

Using single-frequency and noise-modulated resonance and off-resonance proton decoupling, T_1 relaxation time measurements, relaxation reagents like Gd (fod)$_3$ and specifically deuterated compounds, all the carbons in retinal isomers, the model compounds α- and β-ionone, and vitamin A and its isomers [165, 555–557] were assigned. The olefinic ring carbons (C-5 and C-6) could be identified on the assumption that the ^{13}C relaxation times are dominated by intramolecular dipole-dipole interactions with neighboring protons and that the same rotational correlation time characterizes the interactions for both carbons. Consequently the ratio of T_1's for C-5 and C-6 can be estimated from eq. (5.1),

$$\frac{T_1(\text{C-6})}{T_1(\text{C-5})} = \frac{\sum_{^1H_i} r_{i,\,\text{C-5}}^{-6}}{\sum_{^1H_j} r_{j,\,\text{C-6}}^{-6}} \tag{5.1}$$

Table 5.5. Structure and ^{13}C Chemical Shifts (δ_C in ppm) of Camphor and Related Compounds.

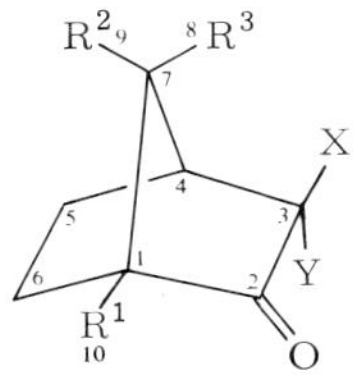

Norbornanone: $R^1 = R^2 = R^3 = X = Y = H$
3-Bromonorbornanone (endo): $R^1 = R^2 = R^3 = X = H, Y = Br$
3-Bromonorbornanone (exo): $R^1 = R^2 = R^3 = Y = H, X = Br$
3-Methylnorbornanone (endo): $R^1 = R^2 = R^3 = X = H, Y = Me$
1-Methylnorbornanone: $R^1 = Me, R^2 = R^3 = X = Y = H$
Camphor: $R^1 = R^2 = R^3 = Me, X = Y = H$
3-Bromocamphor (endo): $R^1 = R^2 = R^3 = Me, X = H, Y = Br$
9,3-Dimbromocamphor (endo): $R^1 = R^3 = Me, R^2 = CH_2Br, X = H, Y = Br$
Fenchone: $R^1 = X = Y = Me, R^2 = R^3 = H$
Camphene: $R^1 = R^2 = R^3 = H, X = CH_3(9), Y = CH_3(10), O = CH_2(8)$

Solvent Ref. C Atom	Norbornanone CCl$_4$ [549]	3-Bromonorbornanone (endo) CCl$_4$ [549]	3-Bromonorbornanone (exo) CCl$_4$ [549]	3-Methylnorbornanone (endo) CCl$_4$ [549]	1-Methylnorbornanone CCl$_4$ [549]	Camphor CCl$_4$ [549]	3-Bromocamphor (endo) CCl$_4$ [549]	9,3-Dibromocamphor (endo) CHCl$_3$ [549]	Fenchone CCl$_4$ [549]	Camphene CDCl$_3$ [548]
1	49.6	48.2	49.2	50.0	53.1	57.0	57.1	58.3	53.8	47.0
2	213.8	207.8	208.3	216.2	214.8	214.7	209.3	209.9	219.1	165.8
3	45.0	55.9	50.1	48.1	45.0	43.2	53.9	52.4	47.1	41.3
4	35.6	42.6	44.8	40.7	34.1	43.2	49.8	47.8	45.5	48.3
5	24.4	23.1	24.1	21.2	29.0	27.4	22.9	22.1	25.3	23.9
6	27.7	25.4	26.6	25.6	31.4	30.1	30.7	30.4	32.0	28.9
7	37.7	35.3	35.3	37.2	43.9	46.6	45.8	50.1	41.8	37.4
8				10.7		19.5	20.2	17.2	21.9	99.3
9						20.0	20.2		23.6	29.4
10					13.8	9.7	10.0			25.9

332   5 ^{13}C NMR Spectra of Natural Products

Table 5.6. Structures and ^{13}C Chemical Shifts (δ_C in ppm) of Selected Diterpenes.

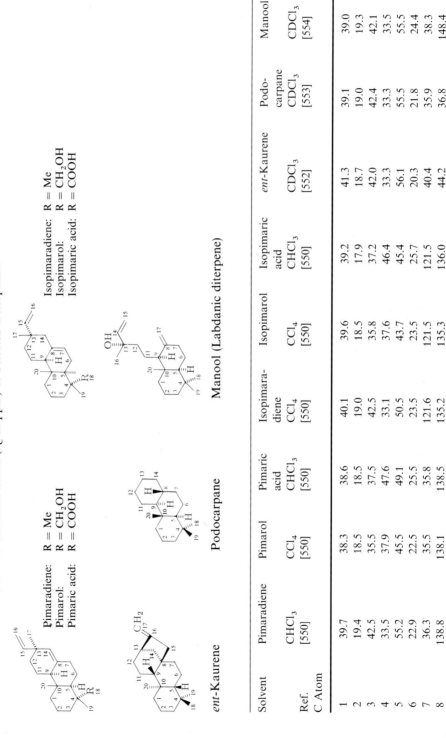

Solvent	Pimaradiene CHCl$_3$ [550]	Pimarol CCl$_4$ [550]	Pimaric acid CHCl$_3$ [550]	Isopimara-diene CCl$_4$ [550]	Isopimarol CCl$_4$ [550]	Isopimaric acid CHCl$_3$ [550]	ent-Kaurene CDCl$_3$ [552]	Podo-carpane CDCl$_3$ [553]	Manool CDCl$_3$ [554]
Ref.									
C Atom									
1	39.7	38.3	38.6	40.1	39.6	39.2	41.3	39.1	39.0
2	19.4	18.5	18.5	19.0	18.5	17.9	18.7	19.0	19.3
3	42.5	35.5	37.5	42.5	35.8	37.2	42.0	42.4	42.1
4	33.5	37.9	47.6	33.1	37.6	46.4	33.3	33.3	33.5
5	55.2	45.5	49.1	50.5	43.7	45.4	56.1	55.5	55.5
6	22.9	22.5	25.5	23.5	23.5	25.7	20.3	21.8	24.4
7	36.3	35.5	35.8	121.6	121.5	121.5	40.4	35.9	38.3
8	138.8	138.1	138.5	135.2	135.3	136.0	44.2	36.8	148.4

Table 5.6. (continued).

Solvent	Pimaradiene CHCl$_3$	Pimarol CCl$_4$	Pimaric acid CHCl$_3$	Isopimara-diene CCl$_4$	Isopimarol CCl$_4$	Isopimaric acid CHCl$_3$	ent-Kaurene CDCl$_3$	Podo-carpane CDCl$_3$	Manool CDCl$_3$
Ref.	[550]	[550]	[550]	[550]	[550]	[550]	[552]	[553]	[554]
C Atom									
9	51.8	51.5	51.9	52.2	52.0	52.4	56.1	56.3	57.2
10	38.8	38.8	38.1	35.6	35.4	35.5	39.3	36.9	39.8
11	19.1	19.3	19.5	20.3	20.5	20.5	18.1	25.1	17.6
12	36.3	36.0	36.0	36.4	36.5	36.0	33.3	27.1	41.3
13	38.8	39.0	39.0	37.0	36.9	37.5	44.2	26.4	73.4
14	128.1	128.1	128.2	46.3	46.4	46.5	39.9	35.5	144.9
15	147.7	147.0	147.8	149.9	150.0	150.7	49.2		111.4
16	112.9	113.1	113.2	109.5	109.5	109.7	156.0		27.9
17	29.8	29.8	29.9	21.8	21.8	21.9	102.8		106.2
18	34.5	71.7	185.7	33.9	71.9	183.9	33.7	33.6	33.5
19	22.5	18.3	17.6	22.6	18.5	17.5	21.7	22.0	21.7
20	14.9	15.6	15.4	15.2	15.9	15.7	17.6	14.3	14.4

334   5 ^{13}C NMR Spectra of Natural Products

where $r_{i,C-k}$ is the *i*-proton to *k*-carbon distance. The shifts in 9-*cis* and 13-*cis*-retinal follow the rules discussed in Section 4.2.1 that α-carbons of *cis* isomers resonate about 6 ppm upfield compared to *trans* isomers. In all *cis* isomers with new steric interactions between hydrogen atoms downfield shifts of up to 0.5 ppm were observed in the ^{1}H NMR spectra of isomeric vitamin A and related compounds compared to their *trans* isomers. According to the theory by Cheney and Grant [204, 558], a downfield shift of the ^{1}H NMR signal should be accompanied by an upfield shift of the attached carbon atom. A strict reversal of ^{1}H and ^{13}C shifts could, however, not be observed for all cases of a large series of *cis-trans* isomeric vitamins A and carotenoids because a quantitative separation of steric from substituent effects is not always possible [557]. ^{13}C chemical shifts of ionones, retinals and vitamin A acetate are given with the corresponding structures in Table 5.7.

On a similar basis the ^{13}C NMR spectra of squalene [559], functionalized squalene-like compounds [559], and carotenoids like β-carotene [577, 560, 561], ε-carotene [557], astaxanthin [557], astacene [557], crustaxanthin tetraacetate [557], isocryptoxanthin [561], isozeaxanthin [561], zeaxanthin [561], lutein [561], violaxanthin [561], violeoxanthin [561], alloxanthin [561], isomytiloxanthin [561], capsorubin [561], canthaxanthin [561], fucoxanthin [561], all *trans*-phytoene [562], *cis*-phytoene [562], methyl azafrin [561], all *trans*-methyl bixin [561], methyl bixin [561], crocetindial [561], flavoxanthin [563], chrysanthemaxanthin [563], xanthophyll [563], and its epoxide [563] were determined. The assignment of the ^{13}C NMR spectrum of β-carotene was achieved by measurements of a series of β-apocarotenoates [561] and β-apocarotenals [560], and it was observed that the polarization of the polyene chain by the carbonyl group causes strong

Table 5.7. Structures and ^{13}C Chemical Shifts ($δ_C$ in ppm) of Ionones, Retinals, and Vitamin A Acetate.

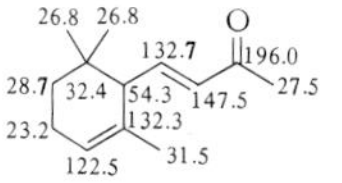

α-Ionone [536]

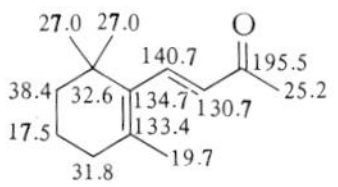

β-Ionone [555]

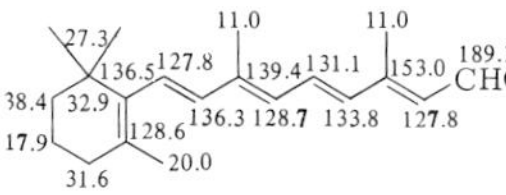

all *trans*-Retinal [555]

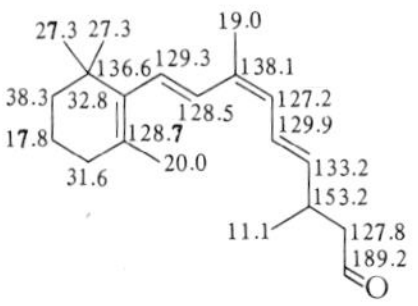

9-*cis*-Retinal [555]

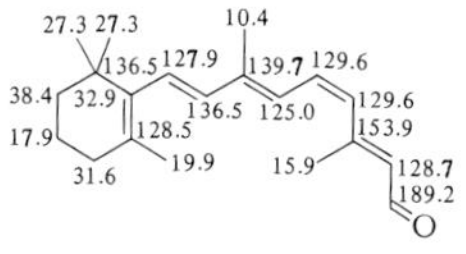

11-*cis*-Retinal [555]

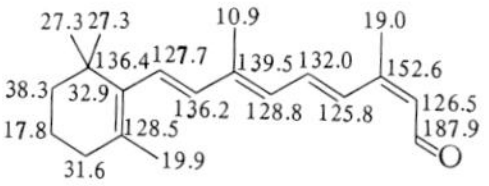

13-*cis*-Retinal [555]

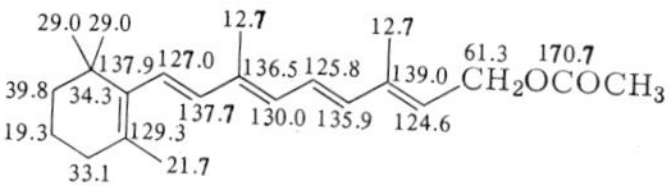

Vitamin A acetate [557]

Table 5.8. Structures and ^{13}C Chemical Shifts (δ_C in ppm) of Selected Carotenoids [561].

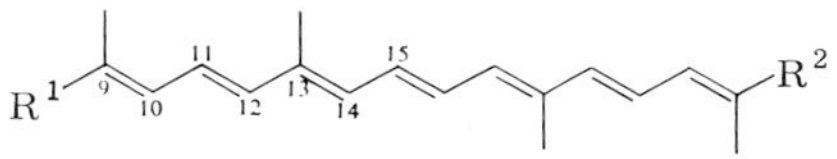

I: β-Carotene: $R^1 = R^2 = a$
II: Isocryptoxanthin: $R^1 = a, R^2 = c$
III: Isozeaxanthin: $R^1 = R^2 = c$
IV: Zeaxanthin: $R^1 = R^2 = b$
V: Lutein: $R^1 = b, R^2 = f$
VI: Violaxanthin: $R^1 = R^2 = d$
VII: Alloxanthin: $R^1 = R^2 = e$

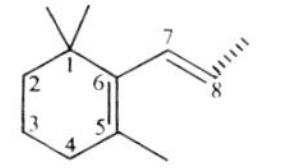

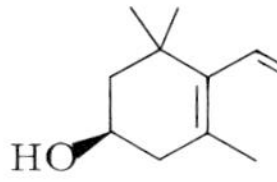

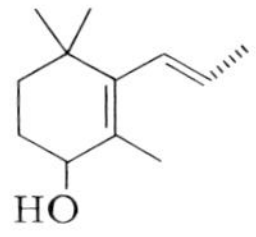

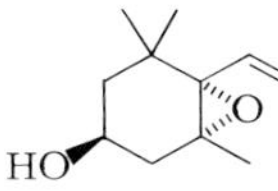

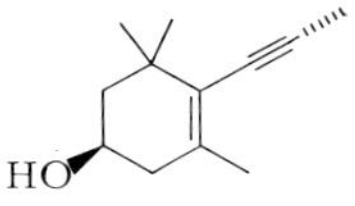

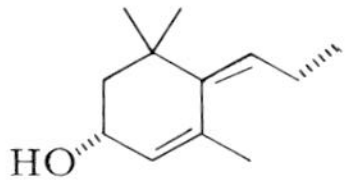

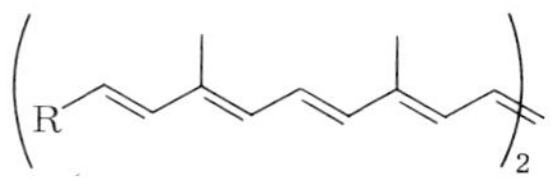

VIII: Capsorubin: R = h
IX: Canthaxanthin: R = g
X: All trans methyl bixin: R = CO_2Me

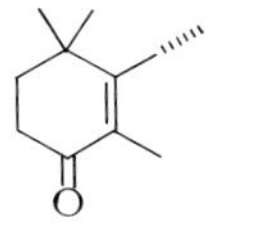

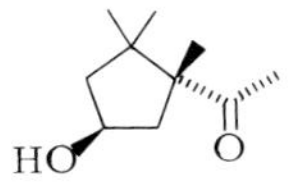

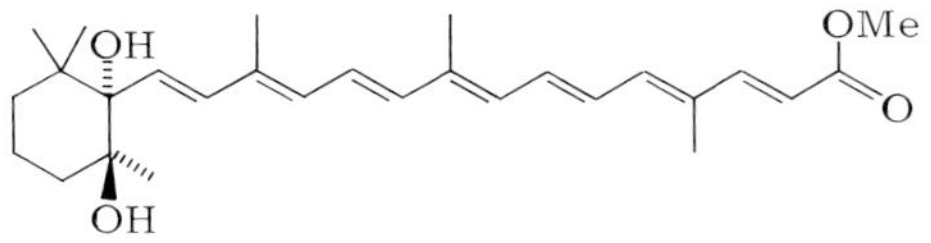

XI: Methyl azafrin

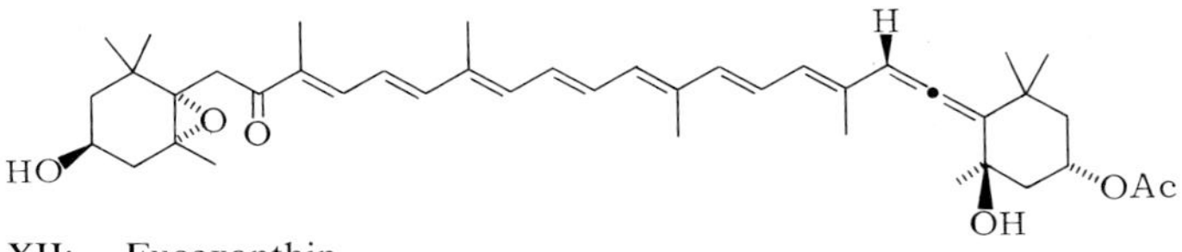

XII: Fucoxanthin

Table 5.8. (continued).

C Atom Compound	1	2	3	4	5	6	7	8	9	10	11	12	13	14	15	1-Me's	5-Me	9-Me	13-Me
I	34.3	39.7	19.3	33.2	129.3	138.0	126.7	137.8	136.0	130.8	125.0	137.3	136.4	132.4	130.0	29.0	21.7	12.7	12.7
Ia	34.2	39.7	19.3	33.1	129.4	138.0	126.8	137.8	136.3	130.8	125.5	137.4	137.1	126.8	125.5	29.0	21.8	12.7	12.5
Ib	34.7	40.3	20.2	33.7	129.6	138.3	127.3	138.3	135.5	131.1	127.3	137.2	146.4	111.6	99.3	29.3	22.1	13.0	15.5
II (β end)	34.6	39.7	19.3	33.2	129.5	137.9	126.8	137.9	136.2	131.0	125.3	137.3	136.4	132.5	130.0	29.0	21.8	12.8	12.8
(OH end)	34.6	35.0	28.6	70.5	129.8	142.1	125.7	139.0	135.5	131.8	124.8	137.9	136.7	132.9	130.3	27.6 29.0	18.8	12.8	12.8
III	34.6	34.6	28.5	70.2	129.8	142.0	125.6	138.9	135.5	131.7	124.8	137.8	136.5	132.7	130.1	27.5 29.1	18.7	12.7	12.7
IV	37.1	48.2	65.1	42.4	126.1	137.6	125.5	138.5	135.7	131.3	124.9	137.8	136.5	132.6	130.0	28.7 30.2	21.7	12.8	12.8
V (β end)	37.1	48.4	65.1	42.5	126.2	137.6	125.6	138.5	135.6	131.3	124.9	137.6	136.5	132.6	130.0	28.7 30.2	21.6	12.7	12.7
(α end)	34.0	44.7	65.9	125.6	137.8	55.0	128.6	137.8	135.0	130.8	124.5	137.6	136.5	132.6	130.0	24.3 29.5	22.8	13.2	12.7
VI	35.4	47.2	64.1	40.9	67.3	70.4	123.9	137.2	134.2	132.4	124.6	138.2	136.4	132.9	130.2	24.7 29.7	20.0	12.8	13.1
VIa (trans end)	35.4	47.3	64.2	41.0	67.2	70.3	123.8	137.3	134.2	132.3	124.6	138.2	136.3	132.8	130.7	24.9 28.7	20.0	12.8	13.1
(cis end)	35.4	47.3	64.2	41.0	67.8	70.6	126.0	129.2	132.6	130.8	123.6	137.3	136.7	132.6	130.0	24.9 29.7	20.0	21.1	13.1
VII	36.9	46.7	64.8	41.5	137.2	124.3	89.1	98.8	119.1	138.0	124.3	135.2	136.4	133.4	130.3	28.8 30.5	22.4	18.0	12.7
VIII	44.0	50.8	70.3	45.3	59.0	203.0	121.1	146.8	134.0	140.6	124.7	141.8	136.9	134.9	131.2	25.1 25.9	21.3	12.8	12.8
IX	35.7	37.7	34.3	198.7	129.9	160.9	124.2	141.1	134.8	134.3	124.7	139.3	136.6	133.6	130.5	27.7	13.7	12.5	12.7
X						167.9	115.9	149.0	133.5	139.4	124.4	141.7	136.8	134.8	131.1			12.6	12.6[a]
Xa (trans end)						167.9	115.9	149.1	133.5	139.5	124.3	141.9	136.6	135.0	131.3			12.6	12.8[a]
(cis end)						168.1	117.6	140.5	131.6	138.0	123.3	140.5	137.0	134.3	130.8			20.3	13.1[a]
XI (diol end)	38.7	36.4	18.0	36.4	75.4	79.6	130.6	134.7	135.8	131.4	126.2	137.4	138.9	131.9	133.9	26.6 27.1	25.2	13.3	13.0
(apo end)										168.0	116.0	149.0	133.7	139.4	128.8				12.8[a]
XII (keto end)	35.2	47.3	64.2	40.9	67.2	72.6	41.8	197.9	134.5	139.2	123.4	145.1	135.4	136.7	129.5	25.0 29.2	21.5	11.9	14.0
(allenic end)	35.8	45.4	68.2	45.5	66.2	117.6	202.4	103.4	132.6	128.5	125.7	137.1	138.1	132.2	132.6	31.3 32.2	28.2	13.0	12.7[b]

Ia: 15,15′-cis-β-carotene; Ib: 15,15′-dehydro-β-carotene; VIa: violeoxanthin (9-cis VI); Xa: methyl bixin (9-cis X).

[a] δ_C (MeO) 51.5.
[b] δ_C (acetate) 21.2 (Me), 170.5 (CO).

shielding for the α-carbons and strong deshielding for the β-carbons. In Table 5.8 ^{13}C NMR shifts of selected cartenoids are collected.

5.2 Steroids

Complete signal assignments have been achieved for a large number of steroids [564]. A series of assignment aids is used for the signal identification of this class of compounds:

1. *Chemical shifts.* The resonances of different carbons of steroids are found in the following ranges:

primary alkane carbons:	12 to 24 ppm
secondary alkane carbons:	20 to 41 ppm
tertiary alkane carbons:	35 to 57 ppm
quaternary alkane carbons:	27 to 43 ppm
alcohol carbons:	65 to 91 ppm
olefinic carbons:	119 to 172 ppm
carbonyl carbons:	177 to 220 ppm
carbons bearing fluorine atoms:	88 to 102 ppm.

Alcohol carbons are identified by spectral comparison of the alcohol with the corresponding acetate: Upon acetylation of cyclohexanols the α-carbons shift about 3 to 4 ppm downfield and the β-carbons move 2–3 ppm upfield. Esterification of axial hydroxy groups also causes a downfield shift of 1 ppm of the γ-carbons [65a].

2. *Comparison of closely related compounds.*

3. *Isotopic labeling.* Replacing hydrogens attached to carbons by deuterium atoms causes spin-spin splitting, quadrupole broadening and – in ^{13}C{^{1}H} experiments – reduction of signal:noise due to lack of Overhauser effects. Therefore resonances of specifically deuterated carbons in a proton broad-band-decoupled ^{13}C NMR spectrum are usually lost in the noise. Recording the spectra before and after specific deuteration allows unequivocal assignment of the corresponding carbon atoms [65a, 565, 566].

Due to the isotopic effect the resonances of neighboring carbons of the site of deuteration are shifted upfield by 0.05 to 0.1 ppm [566].

^{13}C-enriched steroids are helpful in studying biosynthetic pathways [567].

4. *Lanthanide shift reagents.* Lanthanide shift reagents, especially containing ytterbium (causing downfield shifts) and praseodymium (causing upfield shifts), have been used in a series of communications for unambiguous signal assignments of steroids [568–575].

5. *Proton off-resonance decoupling* (see also Section 2.6.5). Proton off-resonance decoupling is used mainly for identifying primary and quaternary carbons of steroids, as the

closely spaced –CH and >CH₂ carbon resonances often do not split into assignable doublets or triplets in off-resonance experiments [65a].

6. *Spin-lattice relaxation (T_1) times.* The information deduced from measurements of spin-lattice relaxation times of steroids [166, 570, 576, 577] is discussed in detail in Section 3.3.3, in which the T_1 values for cholesteryl chloride [166] are given as an example.

7. *Modern techniques.* J-Modulated spin-echo ("APT"), non-selective polarization transfer ("DEPT") experiments and two-dimensional carbon-proton, proton-proton ("COSY") and carbon-carbon ("2 D-INADEQUATE") shift correlations are in use (see the following section) and will be used in order to clarify doubtful signal assignments (see Sections 2.9 and 2.10).

5.2.1 Androstanes, Pregnanes and Estranes

The ^{13}C chemical shifts of selected androstanes, pregnanes and estranes are given in Table 5.10. The formulae of the parent compounds and some related steroid hormones with numbering of the carbon atoms are given below:

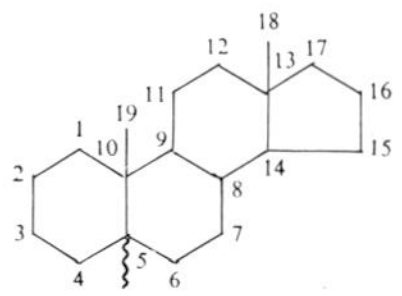

5α, β-Androstane 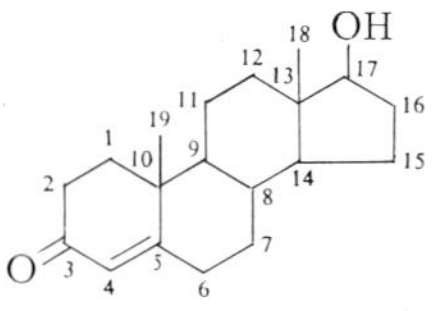   Testosterone 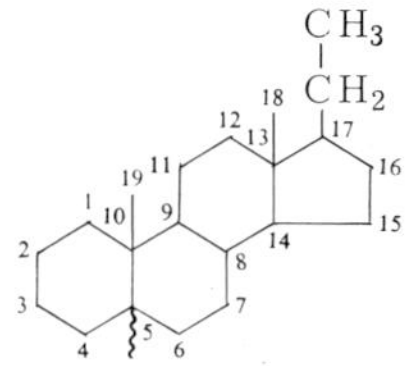   5α, β-Pregnane

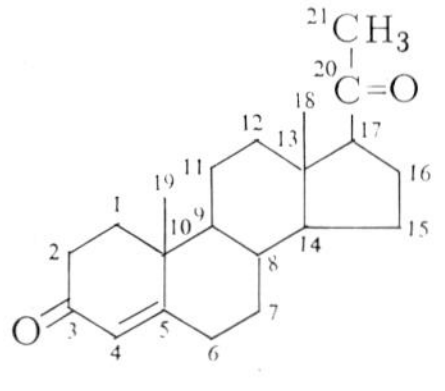

Progesterone 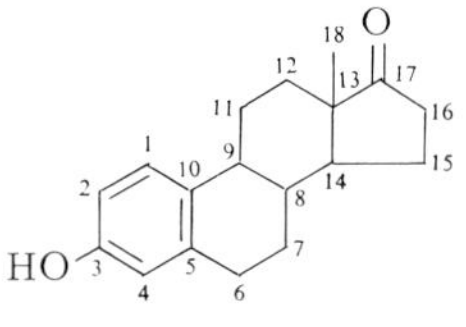   Estrone

The signal assignments of the parent compounds 5α- and 5β-androstane, 5α- and 5β-pregnane and estrane [564] were performed by comparing the spectra with closely related derivatives, using substituent effects, off-resonance decoupled spectra and specifically deuterium-labeled analogues [564]. The influence of different structural environments on the ^{13}C chemical shift of the carbonyl carbon in keto steroids is illustrated by the values given in Table 5.10 [566]. The carbonyl carbon atom frequency in cyclopentanone moieties is shifted about 5 ppm downfield relative to that in cyclohexanone residues, and the effect of alkyl substitution on the carbonyl signal is relatively small. For steroids with the carbonyl group in positions 2 to 11, constant chemical shift values of 212 ($\pm$ 1) ppm are found. If the carbonyl group is neighbored by a quaternary carbon atom, downfield shifts of 3–4 ppm are observed. The signals of methylene and quaternary carbon atoms in β-position to the carbonyl group are shifted by 16 and those of methine carbon atoms by 12 ppm (average values) downfield. The values are reduced by 2–3 ppm in cyclopentanone residues. The influence of a keto group on the chemical shifts of γ-carbon atoms varies from + 6.3 to − 9.5 ppm [566].

For monohydroxylated androstanes the chemical shifts could be rationalized in terms of α, β and γ substituent effects [575]. From investigations on cyclohexanols [584, 585, 268] it was concluded that the chemical shift of the hydroxylated carbon depends markedly on the configuration of the hydroxy group. For a series of monohydroxylated androstanes, however, it was found that the chemical shift of the hydroxylated carbon is not primarily dependent on the stereochemistry of the hydroxy group but, instead, is related to the number (n) of γ-gauche carbons with hydrogen atoms interacting with the hydroxy group and to the number of skew pentane interactions (p) of the hydroxy group with carbon atoms. The α-substituent effect can be expressed by eq. (5.2) [575]:

$$\Delta_\alpha \text{ (ppm)} = 45.0 + 3.5\, p - 3.5\, n. \tag{5.2}$$

The hydroxy group in androstanols influences with large variation the chemical shifts of β-carbons in a range from 13.5 to 2.3 ppm. Though the origin of the β-substituent effect is still not clearly understood, the suggestion was made that γ-gauche interactions of the axial hydroxy group should be responsible for the difference in β-substitution effects between axial and equatorial hydroxy groups [268]. For monohydroxy steroids eq. (5.3) was found for the hydroxy β-substituent effect Δ_β [575]:

$$\Delta_\beta \text{ (ppm)} = 9.3 - 2.4\, q, \tag{5.3}$$

where q is the number of γ-gauche interactions of the hydroxy group with γ carbon atoms connected to the β carbon atom in question.

The ^{13}C NMR spectra of monohydroxylated steroids demonstrate that the γ-gauche substituent effects (the hydroxy group and the γ-carbon are gauche to each other) are different for secondary (average value − 6.5 ppm) and tertiary (average value − 7.8 ppm) γ-carbons. For the γ-trans shifts much smaller values in the range of + 1.3 to − 3.0 ppm are found [575].

For a series of dihydroxy steroids it was found that observed and calculated (see eqs. 5.2 and 5.3) shifts differ significantly for 1,2- and 1,3-dihydroxy steroids. Steric interactions are made responsible for this effect [586].

The ^{13}C NMR spectra of progesterone, deoxycorticosterone and cortisole were also analyzed in terms of α, β and γ effects [565]. Complete signal assignments were achieved on a basis similar to that discussed for the androstane derivatives. Replacing the 5α-hydrogen of pregnanes by amino or azido groups causes the signal of C-9 to shift downfield by more than 10 ppm. This was attributed to a 1,3-diaxal interaction between the axial amino or azido group at position 5 and the axial proton attached to C-9. Thus, a downfield shift of the C-9 resonance can be considered as a proof of the axial configuration of the nitrogen-containing substituents in these pregnane derivatives [578].

Table 5.9. $^{13}C-^{19}F$ Coupling Constants in Hz of Selected Steroids in Hz [579].

	C-1	C-2	C-3	C-5	C-8	C-9	C-10	C-11	C-19
16α-Methyl-11β,21-dihydroxy-9α-fluoropregna-1,4-diene-3,20-dione	–	–	–	–	18.1	174.9	22.6	40.7	4.5
16α-Methyl-11β,17α,21-trihydroxy-9α-fluoropregna-1,4-diene-3,20-dione	–	–	–	–	18.1	174.9	22.6	40.7	4.5
17β-Hydroxy-2α-fluoroandrost-4-en-3-one	34.7	182.5	12.8	–	–	–	4.5	–	–
2-Fluoroestrone	19.6	229.2	19.6	3.8	–	–	3.8	–	–

The assignments of the aromatic carbons in ring A of estrone were performed by comparing the ^{13}C NMR spectrum of the steroid hormone with that of its acetate. A correlation with the acetylation shifts of phenol (C-1 (3.7 ppm); C-2 (6.2 ppm); C-3 (0.6 ppm); C-4 (4.4 ppm)) resulted in the signal assignments outlined in Table 5.10.

The $^{13}C-^{19}F$ coupling constants of fluorinated steroids are helpful for the signal identification of these fluorine derivatives and nonfluorinated analogues. In Table 5.9 the $^{13}C-^{19}F$ coupling constants of four fluorinated steroids are listed [579].

As can be seen from Table 5.9, the $^{13}C-^{19}F$ coupling constants decrease drastically with increasing distance of the carbon atoms from the CF group in the molecule. With the exception of C-19, fluorine substitution causes the γ-carbon signals to shift upfield (cf. resonances of C-1, C-5, C-7, C-12 and C-14 of the compounds in Table 5.9).

5.2.2 Cholestanes and Cholanes

All alkane carbons of cholestane are found in the range 57 to 12 ppm [587]. The signal range of primary, secondary, tertiary and quaternary carbons was outlined at the beginning of this chapter. Due to strong steric interactions of C-11 and C-15 with C-18, C-19 and C-20, the two methylene signals at highest field must be assigned to C-11 and C-15. Substituting hydrogen atoms by double bonds, hydroxy, carbonyl or fluorine groups causes the above mentioned downfield shifts of the corresponding carbon signals. For the signal identification of cholestanes the ^{13}C NMR spectra of the following compounds were used for correlation: cyclohexane, trans-1,2-dimethylcyclohexane, cyclohexene, 2,6-dimethyloctane [65a], 2-methyl-2-heptene, $\Delta^{8,9}$-sandaracopimaradiene [587], 2,2,5,6-tetramethyloctane, and 2,5-dihydroxy-2,6-dimethyloctane [588].

The ^{13}C NMR spectrum of 2,6-dimethyloctane can be used for assigning the resonances of the cholestane side chain. Applying the rules proposed by Grant and Paul [85], all signals of the model compound 2,6-dimethyloctane can be identified. The Grant-Paul parameters also helped in identifying the side chain resonances of ergosterol [65a].

Inversion of ring A causes significant shifts of the carbon signals of rings A and B in steroids. In 5β-cholestane, the C-19 methyl carbon resonates about 12 ppm towards

Table 5.10. ^{13}C Chemical Shifts (δ_C in ppm) of Androstanes, Pregnanes, and Estranes.

Solv. Ref. C-	5α-Androstane CDCl$_3$ [273]	5α-Androstan-1-one CDCl$_3$ [566]	5α-Androstan-2-one CDCl$_3$ [65b]	5α-Androstan-3-one CDCl$_3$ [566]	5α-Androstan-4-one CDCl$_3$ [566]	5α-Androstan-6-one CDCl$_3$ [566]	5α-Androstan-7-one CDCl$_3$ [65b]	5α-Androstan-11-one CDCl$_3$ [566]	5α-Androstan-12-one CDCl$_3$ [566]
1	38.8	215.8	54.1	38.7	37.8	38.3	38.4	37.8	38.3
2	22.2	38.2	210.7	38.1	22.7	21.5	21.8	21.9	21.9
3	26.9	28.0		211.0	41.2	25.3	26.5	26.8	26.6
4	29.1	28.0	28.0	44.6	212.6	20.4	29.2	28.6	28.8
5	47.1	49.8	45.2	46.7	59.3	58.8	47.2	46.9	47.0
6	29.1	28.0	29.2	29.0	20.5	211.8		28.5	28.8
7	32.6	31.5	32.1	32.1	30.9	47.1	211.2	33.2	31.7
8	36.0	36.2	35.2	35.7	35.5	38.3	50.4	37.4	35.0
9	55.1	47.2	54.2	54.1	54.8	55.1	56.5	64.9	56.5
10	36.4	52.0	40.6	35.7	42.6	41.8	36.7	36.0	36.9
11	20.9	22.7	21.0	21.5	21.8	21.2	21.3	210.7	37.5
12	39.1	38.8		38.8	38.9	38.5		56.9	215.3
13	40.8	41.0	40.6	40.8	40.8	41.2	40.7	44.9	54.9
14	54.7	54.4	54.4	54.3	54.5	54.7	48.9	54.2	54.6
15	25.5	25.5	25.5	25.5	25.5	25.3		24.9	24.8
16	20.6	20.4	20.5	20.5	20.5	20.5	20.6	20.9	19.5
17	40.5	40.4		40.3	40.4	40.2		39.3	31.9
18	17.6	17.8	17.4	17.4	17.6	17.5	17.4	18.2	17.7
19	12.3	12.3	12.5	11.4	13.8	13.1	11.7	12.1	11.9

5.2 Steroids 341

Table 5.10. (continued).

Solv. Ref. C-	5α-Andro- stan-15-one CDCl$_3$ [566]	5α-Andro- stan-16-one CDCl$_3$ [566]	5α-Andro- stan-17-one CDCl$_3$ [566]	5α-Andro- stan-1α-ol CDCl$_3$ [575]	5α-Andro- stan-1β-ol CDCl$_3$ [575]	5α-Andro- stan-2α-ol CDCl$_3$ [575]	5α-Andro- stan-2β-ol CDCl$_3$ [575]	5α-Andro- stan-3α-ol CDCl$_3$ [575]	5α-Andro- stan-3β-ol CDCl$_3$ [575]
1	38.7	38.4	38.6	71.5	78.8	48.2	45.3	32.4	37.1
2	22.2	22.1	22.1	29.0	33.4	68.0	68.1	29.1	31.6
3	26.8	26.8	26.7	20.3	24.7	36.3	33.9	66.6	71.2
4	29.0	28.8	28.8	28.6	28.7	27.7	23.9	36.0	38.2
5	47.3	47.0	47.0	39.0	46.3	46.4	47.4	39.2	44.9
6	28.6	29.0	29.0	29.0	28.9	28.2	28.9	28.7	28.8
7	30.8	32.4	31.0	32.2	32.5	32.4	32.5	32.5	32.5
8	32.5	35.0	35.1	35.9	36.3	35.3	35.4	36.0	35.9
9	55.0	54.7	54.9	47.5	55.4	55.0	55.9	54.7	54.7
10	36.5	36.5	36.4	40.2	42.6	37.6	36.1	36.3	35.6
11	20.4	20.4	20.1	20.1	24.7	21.1	21.0	20.9	21.3
12	39.4	38.4	31.7	38.8	39.3	39.0	39.1	39.0	38.9
13	39.2	39.2	47.7	40.8	40.2	40.9	40.9	40.9	40.8
14	63.4	51.9	51.6	54.5	54.6	54.6	54.7	54.7	54.5
15	216.1	39.3	21.7	25.6	25.8	25.6	25.5	25.6	25.5
16	35.1	218.3	35.7	20.5	20.4	20.5	20.5	20.6	20.5
17	35.4	55.9	220.4	40.5	40.6	40.5	40.5	40.5	40.4
18	18.3	18.1	13.8	17.5	17.4	17.6	17.7	17.6	17.6
19	12.2	12.3	12.2	12.9	6.7	13.4	14.8	11.2	12.4

Table 5.10. (continued).

Solv. Ref. C-	5α-Androstan-4α-ol CDCl$_3$ [575]	5α-Androstan-4β-ol CDCl$_3$ [575]	5α-Androstan-6β-ol CDCl$_3$ [575]	5α-Androstan-11α-ol CDCl$_3$ [575]	5α-Androstan-11β-ol CDCl$_3$ [575]	5α-Androstan-12α-ol CDCl$_3$ [575]	5α-Androstan-12β-ol CDCl$_3$ [575]	5α-Androstan-15α-ol CDCl$_3$ [575]	5α-Androstan-15β-ol CDCl$_3$ [575]
1	38.1	38.7	40.5	40.8	38.9	38.8	38.7	38.8	38.8
2	20.5	17.1	22.2	22.6	22.0	22.3	22.2	22.2	22.3
3	36.4	33.9	27.1	26.7	26.6	26.9	26.8	26.7	26.9
4	70.5	72.5	26.1	29.7	28.6	29.2	29.0	29.0	29.1
5	54.3	50.2	49.8	47.0	47.8	47.2	47.1	46.9	47.4
6	22.8	26.1	72.5	29.7	28.6	29.2	29.0	28.9	29.1
7	32.1	32.9	40.0	32.8	32.9	32.4	32.2	32.5	31.9
8	35.6	36.0	30.7	35.4	31.7	36.1	34.9	35.5	31.9
9	55.1	55.9	55.0	61.2	59.0	48.3	53.9	55.0	55.7
10	37.7	36.3	36.4	38.4	36.5	36.1	36.3	36.4	36.6
11	21.0	20.3	20.7	69.2	68.6	28.4	29.9	20.7	20.8
12	39.0	39.0	39.0	50.5	47.8	72.7	79.7	39.4	40.6
13	40.8	40.9	40.9	41.2	39.9	45.3	46.3	41.7	40.6
14	54.7	54.8	54.4	53.7	56.4	46.4	53.3	61.9	59.6
15	25.5	25.6	25.5	25.6	25.4	25.2	25.2	75.7	72.5
16	20.5	20.5	20.5	20.6	20.1	20.2	20.7	32.9	34.0
17	40.5	40.5	40.5	40.2	40.8	33.0	38.2	38.3	40.4
18	17.6	17.6	17.6	18.4	20.0	18.7	11.8	18.8	20.0
19	13.5	14.7	15.8	12.8	15.5	12.2	12.2	12.3	12.3

Table 5.10. (continued).

Solv. Ref. C-	5α-Androstan-16α-ol CDCl$_3$ [575]	5α-Androstan-16β-ol CDCl$_3$ [575]	5α-Androstan-17α-ol CDCl$_3$ [575]	5α-Androstan-17β-ol CDCl$_3$ [575]	5α-Androstan-3,17-dione CDCl$_3$ [564]	5α-Androstane-3α,17β-diol Pyridine [273]	5α-Androstane-3β,17β-diol Pyridine [273]	Androst-4-ene-3,17-dione CDCl$_3$/Dioxane [65a]
1	38.8	38.8	38.8	38.8	38.5	32.1	37.5	35.0
2	22.2	22.3	22.2	22.2	38.0	29.8	32.0	33.4
3	26.9	26.9	26.8	26.8	210.9	65.5	70.5	197.9
4	29.1	29.1	29.1	29.1	44.6	36.9	39.2	124.3
5	47.2	47.2	47.0	47.2	46.6	39.5	45.3	169.7
6	29.1	29.1	29.1	29.1	28.7	29.1	29.1	32.0
7	32.5	32.5	32.5	31.8	30.5	32.1	32.3	31.2
8	35.5	35.5	35.8	35.8	35.0	36.0	35.8	34.9
9	55.1	55.0	54.6	55.0	53.9	55.0	54.9	54.4
10	36.4	36.5	36.4	36.4	35.8	36.6	35.8	38.3
11	20.5	20.6	20.3	20.5	20.8	20.9	21.3	20.7
12	39.0	39.3	31.6	36.9	31.5	37.6	37.5	30.6
13	41.9	40.3	45.3	43.1	47.7	43.5	43.5	47.5
14	52.3	54.3	48.9	51.3	51.3	51.5	51.4	51.2
15	37.3	37.3	24.6	23.4	21.8	23.8	23.8	22.0
16	71.8	71.9	32.5	30.6	35.8	31.0	31.0	35.4
17	52.2	51.5	80.0	82.1	220.2	81.3	81.2	218.7
18	18.8	19.1	17.2	11.2	13.8	11.9	11.8	13.7
19	12.3	12.3	12.3	12.3	11.5	11.6	12.5	17.3

Table 5.10. (continued).

Solv. Ref. C-	17-β-Hydroxyandrost-4-en-3-one[a] CDCl$_3$/Dioxane [65a]	Androsta-1,4-diene-3,17-dione CDCl$_3$ [580]	Androsta-1,4-diene-3,11,17-trione CDCl$_3$ [564]	5β-Androstane CDCl$_3$ [564]	5β-Androstane-3,17-dione CDCl$_3$ [564]	5α-Pregnane CDCl$_3$ [564]	5-Amino-5α-pregnane CDCl$_3$ [581]	5-Azido-5α-pregnane CDCl$_3$ [581]	5-α-Pregnan-11-one CDCl$_3$ [564]
1	36.1	155.2	154.3	37.8	37.1	38.9	31.2	31.5	37.8
2	34.1	127.6	127.7	21.4	37.0	22.3	20.8	20.7	21.8
3	198.0	186.0	185.6	27.0	212.1	26.9	20.8	21.1	26.8
4	124.2	124.0	124.8	27.3	42.2	29.2	34.9	20.7	28.7
5	170.4	168.2	165.3	43.8	44.2	47.2	52.6	69.3	43.0
6	32.8	32.3	32.1	27.5	24.8	29.2	34.9	30.6	28.5
7	32.2	31.2	31.9	27.1	26.4	32.4	26.3	26.8	32.9
8	36.1	35.0	36.0	36.3	35.2	35.7	35.4	34.7	37.3
9	54.6	52.3	60.9	40.6	41.1	55.3	46.4	46.5	65.0
10	39.0	43.4	42.2	35.5	35.1	36.4	38.9	38.7	36.0
11	21.2	22.0	206.9	20.8	20.6	20.7	20.4	20.7	211.0
12	37.1	32.5	50.4	39.2	31.8	38.3	38.1	37.9	56.2
13	43.2	47.6	50.1	41.0	47.8	42.3	42.2	42.1	46.6
14	51.1	50.4	49.8	54.7	51.4	56.2	55.9	55.5	54.4
15	23.8	21.8	21.7	25.6	21.8	24.6	24.3	24.3	24.0
16	30.7	35.5	35.9	20.6	35.8	28.2	28.2	28.0	28.5
17	81.3	219.6	215.7	40.6	220.2	53.2	52.9	52.8	52.0
18	11.3	13.8	14.8	17.5	13.9	12.6	12.6	12.6	13.1
19	17.3	18.7	19.0	24.3	22.6	12.3	15.6	15.4	12.1
20						23.3	23.0	23.0	23.0
21						13.4	13.3	13.3	13.1

[a] Testosterone

346 5 ^{13}C NMR Spectra of Natural Products

Table 5.10. (continued).

Solv. Ref. C-	5α-Pregnan-11α-ol CDCl$_3$ [564]	5α-Pregnan-11β-ol CDCl$_3$ [564]	5α-Pregnane-3,20-dione CDCl$_3$ [564]	3β-Hydroxy-5α-pregnan-20-one CDCl$_3$ [564]	5α-Pregnane-3β,20-diol Pyridine [564]	Pregn-4-ene-3,20-dione[b] CDCl$_3$ [564]	20-Carboxypregna-1,4-dien-3-one (methyl ester) CDCl$_3$ [582]
1	40.8	38.9	38.5	37.0	37.5	35.6	155.4
2	22.5	21.9	38.0	32.0	32.5	33.8	127.5
3	26.7	26.7	212.3	71.0	70.5	198.5	185.9
4	29.7	28.6	44.5	38.1	39.2	123.7	123.8
5	47.1	47.8	46.6	44.8	45.3	169.8	168.7
6	29.7	28.6	28.8	28.6	29.2	32.6	35.5
7	32.5	32.7	31.6	31.4	32.5	31.8	32.7
8	35.0	31.5	35.4	35.4	35.7	35.4	35.5
9	61.4	59.2	53.7	54.2	54.8	53.5	52.8
10	38.3	36.6	35.6	35.4	35.8	38.5	43.4
11	69.2	68.4	21.4	21.2	21.5	21.0	22.7
12	50.0	47.0	38.9	39.0	40.5	38.5	27.0
13	42.7	41.3	44.1	44.2	43.0	43.7	42.2
14	55.1	57.7	56.4	56.6	56.3	55.9	55.1
15	24.9	24.5	24.3	24.3	24.9	24.2	24.3
16	28.3	27.8	22.9	22.8	26.2	22.8	39.2
17	52.9	53.7	63.6	63.7	58.9	63.3	52.3
18	13.4	15.1	13.4	13.4	12.7	13.2	17.0
19	12.9	15.5	11.4	12.3	12.5	17.3	18.7
20	23.2	23.1	210.0	209.2	69.5	208.3	42.6
21	13.3	13.3	31.4	31.4	24.4	31.0	12.2
CO (ester)							176.7
CH$_3$ (ester)							51.1

[b] Progesterone

Table 5.10. (continued).

Solv. Ref. C-	21-Hydroxy-pregn-4-ene-3,20-dione[c] DMSO [565]	17,21-Dihydroxypregn-4-ene-3,20-dione[d] DMSO [565]	11β,21-Dihydroxypregn-4-ene-3,20-dione[e] DMSO [565]	11β,17,21-Trihydroxypregn-4-ene-3,20-dione[f] DMSO [565]	5β-Pregnane CDCl₃ [564]	5-Amino-5β-pregnane CDCl₃ [581]	5-Azido-5β-pregnane CDCl₃ [581]	5β-Pregnan-11-one CDCl₃ [564]	5β-Pregnan-11α-ol CDCl₃ [564]
1	35.0	35.2	34.2	34.0	37.7	31.3	31.8	36.7	40.9
2	33.4	33.4	33.5	33.4	21.4	21.1	20.8	22.1	22.4
3	197.5	197.4	197.7	197.7	27.1	21.9	22.1	27.0	27.5
4	123.0	123.0	121.5	121.4	27.3	32.9	29.1	27.1	27.7
5	170.4	170.4	171.9	172.0	43.9	52.9	67.7	44.6	45.5
6	31.8	31.9	32.6	32.7	27.6	37.5	31.8	27.3	28.0
7	31.5	32.1	31.5	31.3	26.8	28.2	28.3	27.3	26.8
8	34.8	35.2	31.2	31.1	36.0	35.0	34.9	37.6	35.3
9	52.9	53.0	55.7	55.5	41.1	42.9	43.2	52.0	48.3
10	38.0	38.1	38.9	38.8	35.5	39.6	39.8	35.0	36.9
11	20.4	20.3	66.3	66.4	20.6	20.5	20.2	211.3	69.4
12	37.6	30.1	46.8	39.0	38.5	38.1	38.0	56.4	50.2
13	43.5	47.0	43.2	46.2	42.3	41.8	42.0	46.7	42.7
14	55.3	49.9	57.1	51.5	56.2	56.2	55.9	55.2	55.1
15	23.9	23.2	24.2	23.3	24.6	24.5	24.5	24.1	24.7
16	22.3	33.4	22.0	32.9	28.3	28.2	28.2	28.5	28.3
17	57.4	88.4	58.2	88.3	53.2	52.9	52.9	52.0	53.0
18	13.1	14.5	15.7	16.8	12.5	12.4	12.4	13.1	13.5
19	16.7	17.0	20.5	20.4	24.3	17.2	18.0	24.1	24.7
20	209.6	211.3	209.8	211.4	23.1	23.0	23.1	23.0	23.2
21	68.6	65.9	68.6	65.8	13.3	13.3	13.3	13.1	13.3

[c] Deoxycorticosterone, [d] Cortexolone, [e] Corticosterone, [f] Cortisol

Table 5.10. (continued).

Solv. Ref. C-	5β-Pregnan-11β-ol CDCl₃ [564]	5β-Pregnane-3,20-dione CDCl₃ [564]	3α-Hydroxy-5β-pregnan-20-one CDCl₃ [570]	Estra-1,3,5(10)-triene Dioxane [583]	3-Hydroxy-estra-1,3,5(10)-trien-17-one[g] Dioxane [583]	Estra-1,3,5(10)-triene-3,17α-diol Dioxane [583]	Estra-1,3,5(10)-triene-3,17β-diol Dioxane [583]
1	37.6	37.1	35.7	126.0	126.9	127.2	126.9
2	21.8	37.0	30.8	126.2	113.5	113.7	113.5
3	27.2	212.3	71.6	126.2	155.8	155.7	155.6
4	27.3	42.3	36.7	129.7	115.9	116.1	115.9
5	44.6	44.2	42.4	137.4	138.2	138.7	138.4
6	27.0	25.8	27.5	30.0	30.2	30.4	30.2
7	26.8	26.6	26.7	28.6	27.4	28.9	28.0
8	31.7	35.6	36.1	39.5	39.3	40.1	39.8
9	46.4	40.8	40.7	45.1	45.0	44.6	44.8
10	35.8	34.9	34.9	141.5	131.9	132.6	132.3
11	68.4	21.2	21.2	26.9	26.4	27.1	27.1
12	47.2	39.2	39.5	39.2	32.5	33.1	37.6
13	41.3	44.2	44.4	41.4	48.3	46.2	43.9
14	57.7	56.6	56.9	54.2	51.1	48.4	50.8
15	24.6	24.4	24.7	25.5	22.2	24.9	23.7
16	27.7	23.0	22.9	20.9	35.9	32.4	31.0
17	53.8	63.9	64.0	40.9	219.3	79.7	81.9
18	15.1	13.5	13.4	17.6	13.9	17.5	11.5
19	27.8	22.6	23.4				
20	23.1	208.8	209.7				
21	13.3	31.4	31.5				

[g] Estrone

5.2 Steroids 349

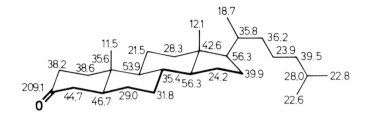

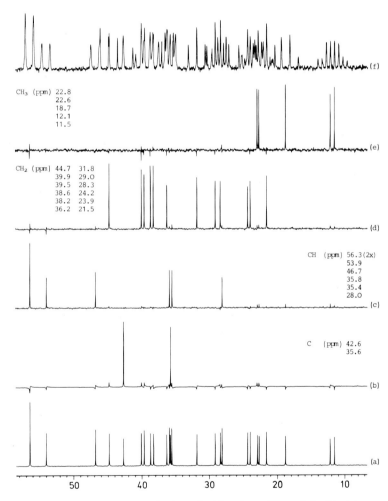

Fig. 5.1. ^{13}C NMR spectra of 5α-cholestan-3-one in deuteriochloroform (50 mg/0.5 mL); (a) proton broadband-decoupled, 400 scans; (b) *J*-modulated spin-echo experiment for quaternary carbon selection, 1000 scans; (c–e) CH, CH$_2$, and CH$_3$ subspectra generated from linear combination of three DEPT experiments (see Section 2.9.3.2), 200 scans per experiment; (f) gated proton-decoupled experiment for comparison.

350   5 ¹³C NMR Spectra of Natural Products

(A)

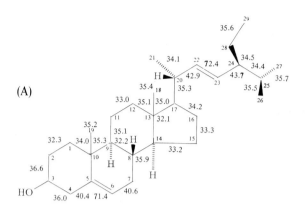

(B)

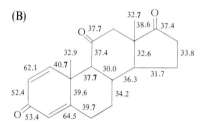

Fig. 5.2. One-bond ¹³C–¹³C coupling constants (in Hz) of stigmasterol (A) and 1,4-androstadiene-3,11,17-trione (B) measured at 100.6 MHz and 70 °C in pyridine-d_5 (700 mg/2.5 mL). (A): 1D and 2D INADEQUATE experiments were performed with $\tau = 6$ msec, digital resolution: 0.21 Hz/point. $^1J(C_7-C_8)$ cannot be observed because the $\Delta\delta$ value is only 0.1 ppm. The long T_1 value for C-10 resulted in a poor signal to noise ratio which prevented the observation of $^1J(C_{10}-C, ^5)$. (B): 1D INADEQUATE experiments: $\tau = 5$ msec, digital resolution: 0.33 Hz/point. $^1J(C_6-C_7)$ and $^1J(C_{12}-C_{13})$ cannot be observed because the $\Delta\delta$ values are only 0.22 and 0.01 ppm, respectively [595].

lower field than in the corresponding 5α-steroid. Conversely, the resonances of C-10, C-9, C-7 and C-5 are shifted 5 to 10 ppm upfield in going from 5α-cholestane to 5β cholestane (cf. Table 5.11) [589].

Fig. 5.1 demonstrates for cholestane-3-one the utility of recording proton broadband-decoupled spectra, performing J-modulated spin echo experiments, and generating subspectra from linear combinations of DEPT experiments for the unequivocal identification of primary, secondary, tertiary and quaternary carbon atoms.

Recent advances in pulsed NMR methods and the availability of high-field spectrometers with enhanced sensitivity have extended the scope and utility of ¹³C–¹³C coupling constant measurements at natural abundance to structural elucidation of medium molecular weight molecules [590–592]. The first complete ¹³C–¹³C coupling patterns for steroidal molecules have been reported [593–595]. The determination of carbon-carbon coupling constants on complex natural products and at natural abundance requires carefully designed experiments. Most important is to ensure long T_2 values for proton-bearing carbon atoms. Only in this way can the closely spaced satellite lines be made sharp enough for unambiguous analysis. Sometimes the consecutive use of 1D

Table 5.11. ^{13}C Chemical Shifts (δ_C in ppm) of Cholestanes and Cholanes.

Solv. Ref. C-	5α-Cholestane Dioxane/ CHCl$_3$ [65a]	5β-Cholestane CDCl$_3$ [589]	5α-Ergostane CDCl$_3$ [589]	Cholestan- 3β-ol Dioxane/ CHCl$_3$ [65a]	Cholestan-3α-yl acetate Dioxane/ CHCl$_3$ [65a]	Cholestan- 3β-yl acetate Dioxane/ CHCl$_3$ [65a]	Cholesta- 3,5-diene Dioxane/ CHCl$_3$ [65a]	Cholesterol Dioxane/ CHCl$_3$ [65a]	Vitamin D$_3$ CDCl$_3$ [588a]
1	38.9	37.4	38.5	37.3	33.1	36.9	34.0	37.5	32.0
2	22.3	21.1	20.3	31.8	26.2	27.6	23.1	31.6	35.3
3	27.0	26.4	26.4	70.4	69.5	73.2	129.6	71.3	69.3
4	29.3	26.9	28.9	38.6	33.1	34.2	124.3	42.4	46.0
5	47.3	40.4	46.9	45.2	40.3	44.8	141.4	141.2	135.2
6	29.3	27.1	28.9	29.0	28.4	28.8	122.9	121.3	122.5
7	32.3	28.2	32.0	32.3	32.1	32.1	31.9	32.0	117.7
8	35.8	35.6	35.3	35.8	35.7	35.7	32.0	160.5	142.3
9	55.2	45.6	54.6	54.8	54.6	54.6	48.6	50.5	29.1
10	36.5	27.8	36.1	35.6	35.8	35.6	35.2	36.5	145.2
11	21.1	20.7	18.6	21.4	21.0	21.4	21.2	21.2	22.3
12	28.3	27.8	28.0	28.3	28.4	28.2	28.3	28.3	40.6
13	42.8	42.5	42.3	42.8	42.7	42.7	42.6	42.4	45.9
14	56.9	56.3	56.3	56.7	56.8	56.7	57.2	56.9	56.4
15	24.3	24.1	24.1	24.3	24.2	24.2	24.2	24.3	23.6
16	40.4	40.2	40.0	40.4	40.3	40.2	40.2	40.0	27.7
17	56.8	56.1	56.0	56.7	56.8	56.7	56.6	56.5	56.8
18	12.2	11.9	12.3	12.1	12.0	12.0	12.0	12.0	12.1
19	12.1	24.1	12.1	12.0	11.2	12.0	18.6	19.4	112.4
20	36.0	35.8	36.1	35.9	36.0	35.9	36.0	35.8	36.2
21	18.8	18.5	14.2	18.6	18.7	18.7	18.8	18.8	18.9
22	36.4	36.0	33.6	36.4	36.4	36.4	36.5	36.4	36.2

Table 5.11. (continued).

Solv. Ref. C-	5α-Cholestane Dioxane/CHCl$_3$ [65a]	5β-Cholestane CDCl$_3$ [589]	5α-Ergostane CDCl$_3$ [589]	Cholestan-3β-ol Dioxane/CHCl$_3$ [65a]	Cholestan-3α-yl acetate Dioxane/CHCl$_3$ [65a]	Cholestan-3β-yl acetate Dioxane/CHCl$_3$ [65a]	Cholesta-3,5-diene Dioxane/CHCl$_3$ [65a]	Cholesterol Dioxane/CHCl$_3$ [65a]	Vitamin D$_3$ CDCl$_3$ [588a]
23	24.2	23.7	30.6	24.1	24.1	24.1	24.2	24.1	23.9
24	39.7	39.4	39.0	39.7	39.7	39.6	39.8	39.6	39.6
25	28.1	27.4	17.4	28.0	28.0	28.0	28.1	28.0	28.1
26	22.5	22.7	31.2	22.4	22.5	22.4	22.5	22.5	22.6
27	22.7	22.3	21.9	22.6	27.0	22.6	22.7	22.8	22.9
28			20.6						
CH$_3$ (acetyl)					20.7	20.6			
CO (acetyl)					169.3	169.4			

5.2 Steroids 353

Table 5.11. (continued).

Solv. Ref. C-	Cholesteryl acetate Dioxane/CHCl$_3$ [65a]	Cholesteryl methyl ether Dioxane/CHCl$_3$ [65a]	7-Dehydro-cholesteryl acetate Dioxane/CHCl$_3$ [65a]	Ergosterol Dioxane/CHCl$_3$ [65a]	Lanosterol CDCl$_3$ [587]	Dihydro-lanosterol CDCl$_3$ [587]
1	37.3	37.4	38.1	38.6	35.0	35.1
2	28.2	28.3	28.2	32.2	27.4	27.4
3	73.7	80.3	72.6	69.7	78.3	77.9
4	38.4	39.0	36.8	41.1	38.1	38.2
5	139.9	141.0	141.0	140.7	49.8	49.8
6	122.6	121.3	120.4	119.4	20.3	20.5
7	32.2	32.0	116.7	116.7	27.4	27.4
8	32.2	160.5	138.5	140.6	133.9	133.0
9	50.4	50.5	46.2	46.5	133.9	133.0
10	36.7	36.9	37.2	37.2	36.3	36.5
11	21.3	21.2	21.1	21.2	17.6	17.6
12	32.5	28.2	28.2	28.3	25.8	25.8
13	42.5	42.5	43.0	43.0	43.7	43.8
14	57.0	57.0	54.5	54.6	49.1	49.1
15	24.6	24.3	23.0	23.1	30.2	30.3
16	40.1	40.1	39.4	39.4	30.2	30.3
17	56.6	56.6	56.2	56.0	49.8	49.8
18	12.0	11.9	11.8	11.8	15.3	15.2
19	19.3	19.3	16.0	16.0	18.0	18.1
20	36.1	36.0	36.3	40.5	35.1	35.8
21	18.9	18.8	18.9	19.4	18.4	18.5
22	36.7	36.4	36.3	132.2	35.1	35.8
23	24.3	24.1	24.0	136.0	24.2	23.6
24	39.8	39.7	39.6	43.0	124.7	38.9

Table 5.11. (continued).

Solv. Ref. C-	5α-Cholestane Dioxane/CHCl$_3$ [65a]	5β-Cholestane CDCl$_3$ [589]	5α-Ergostane CDCl$_3$ [589]	Cholestan-3β-ol Dioxane/CHCl$_3$ [65a]	Cholestan-3α-yl acetate Dioxane/CHCl$_3$ [65a]	Cholestan-3β-yl acetate Dioxane/CHCl$_3$ [65a]	Cholesta-3,5-diene Dioxane/CHCl$_3$ [65a]	Cholesterol Dioxane/CHCl$_3$ [65a]	Vitamin D$_3$ CDCl$_3$ [588a]
25	28.2		28.0	28.0		33.2		130.2	27.4
26	32.7		22.5	22.5		19.7		24.9	22.3
27	22.9		22.7	22.7		21.0		16.9	22.0
28			55.1			17.4		23.5	23.6
CH$_3$ (methoxyl)									
CH$_3$ (acetyl)	20.9				20.9				
CO (acetyl)	169.6				169.6				
29								27.4	27.4
30								14.8	15.0

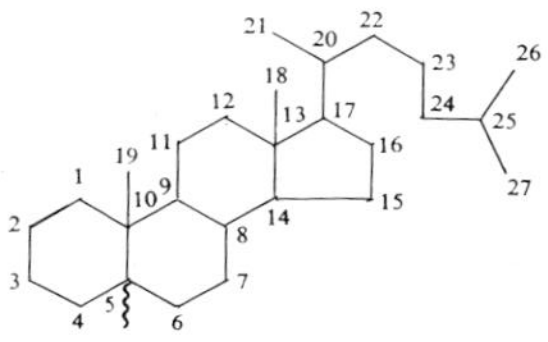

5α, β-Cholestane

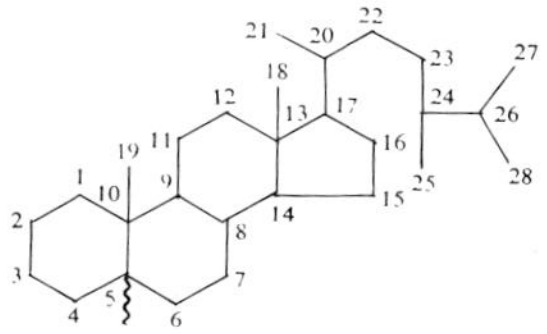

5α, β-Ergostane

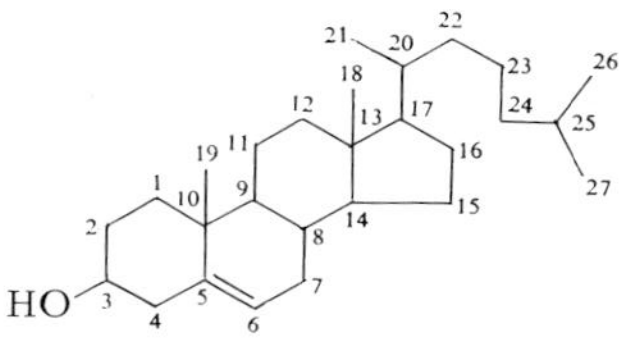

Cholesterol

Table 5.11. (continued).

Solv. Ref. C-	Cholestan-3-one Dioxane/CHCl$_3$ [65a]	Cholesta-3,5-dien-7-one Dioxane/CHCl$_3$ [65a]	Cholest-5-en-7-on-3β-yl acetate Dioxane/CHCl$_3$ [65a]	2β,3β,14α-Trihydroxy-5β-cholest-7-en-6-one Pyridine-d$_5$ [588]	α-Ecdysone Pyridine-d$_5$ [588]	α-Ecdysone triacetate Pyridine-d$_5$ [588]
1	38.6	33.0	36.1	37.2	37.2	33.9
2	37.9	23.5	27.5	66.8		68.2
3	209.1	136.1	72.3	66.8	66.9	66.6
4	44.5	124.4	37.7	31.7	31.7	28.8
5	46.7	160.0	163.2	50.3	50.3	50.7
6	29.1	127.9	126.7	175.7	175.7	175.7
7	31.9	200.4	200.1	119.5	119.5	119.4
8	35.7	46.0	45.3	163.0	162.9	162.9
9	54.1	49.8	49.9	33.9	33.9	33.9
10	35.6	36.2	38.3	37.9	37.9	37.8
11	21.6	21.4	21.3	20.7	20.7	20.9
12	28.3	28.7	28.6	26.9	26.1	25.9
13	42.7	43.6	43.2	46.3	46.7	46.8
14	56.5	50.9	50.2	82.8	82.5	82.4
15	24.3	26.5	26.4	31.2	31.3	31.3
16	40.2	39.2	39.0	30.8	30.8	30.8
17	56.6	55.3	55.2	50.2	47.4	47.3
18	12.0	12.0	11.8	15.6	15.5	15.4
19	11.2	16.5	16.9	24.0	24.0	23.7
20	35.9	35.9	35.9	35.4	41.6	39.1
21	18.7	19.0	18.9	18.9	13.3	13.6
22	36.4	36.4	36.4	36.1	72.8	76.5

Table 5.11. (continued).

Solv. Ref. C-	5α-Cholestane Dioxane/CHCl₃ [65a]	5β-Cholestane CDCl₃ [589]	5α-Ergostane CDCl₃ [589]	Cholestan-3β-ol Dioxane/CHCl₃ [65a]	Cholestan-3α-yl acetate Dioxane/CHCl₃ [65a]	Cholestan-3β-yl acetate Dioxane/CHCl₃ [65a]	Cholesta-3,5-diene Dioxane/CHCl₃ [65a]	Cholesterol Dioxane/CHCl₃ [65a]	Vitamin D₃ CDCl₃ [588a]
23	24.0		24.1		24.1	24.0	25.0		22.5
24	39.6		39.6		39.6	39.1	42.2		40.7
25	28.0		28.0		28.0	27.7	68.6		68.2
26	22.5		22.6		22.4	22.6	29.7		29.7
27	22.7		22.8		22.6	22.3	29.4		29.2
28									
CH₃ (acetyl)					20.5				20.5
CO (acetyl)					169.3				162.9
									167.4

Table 5.11. (continued).

Solv. Ref. C-	Methyl 5β-cholan-24-oate CD$_2$Cl$_2$ [570]	3α-Hydroxy-5β-cholan-24-oic acid CD$_3$OD [570]	3α,7α-Dihydroxy-5β-cholan-24-oic acid CD$_3$OD [570]	3α,12α-Dihydroxy-5β-cholan-24-oic acid CD$_3$OD [570]	3α,7α,12α-Trihydroxy-5β-cholan-24-oic acid CD$_3$OD [570]	5β-Cholane-3α,7α,24-triol CD$_3$OD [570]	5β-Cholane-3α,12α,24-triol CD$_2$Cl$_2$ [570]
1	37.9	35.6	35.8	35.7	35.7	35.8	35.7
2	21.6	30.4	31.0	30.5	30.6	31.3	30.5
3	27.4	71.8	71.4	71.7	71.6	72.3	71.4
4	27.8	36.5	39.7	36.1	39.6	39.7	36.3
5	44.1	42.5	42.1	42.7	42.0	42.1	42.7
6	27.5	27.5	35.4	27.6	35.1	35.4	27.7
7	26.9	26.7	67.8	26.6	68.0	68.8	26.6
8	36.2	36.1	39.9	36.4	39.9	40.0	36.3
9	40.8	40.6	33.3	33.9	26.7	33.3	33.9
10	35.6	34.8	35.5	34.5	35.1	35.5	34.5
11	21.1	21.1	21.1	29.0	28.5	20.9	29.0
12	40.6	40.5	39.9	72.8	72.5	39.9	73.1
13	43.0	43.0	43.0	46.8	46.8	43.0	46.9
14	56.9	56.7	50.9	48.4	42.0	50.9	48.4
15	24.4	24.4	24.0	24.2	23.5	24.0	24.2
16	28.4	28.4	28.5	26.5	28.0	28.5	26.5
17	56.4	56.5	56.3	47.4	47.4	56.5	47.6
18	12.1	11.8	11.9	12.9	12.8	11.9	12.9
19	24.2	23.2	23.0	23.4	22.8	23.0	23.4
20	35.6	35.6	35.8	35.7	35.7	35.9	35.9
21	18.3	18.1	18.3	17.5	17.6	18.6	18.0
22	31.2	31.2	31.2	31.3	31.3	32.3	32.1
23	31.1	31.1	31.1	31.2	31.2	29.5	29.5
24	175.0	174.5	174.6	177.2	174.5	62.9	63.8
OCH$_3$	51.3						

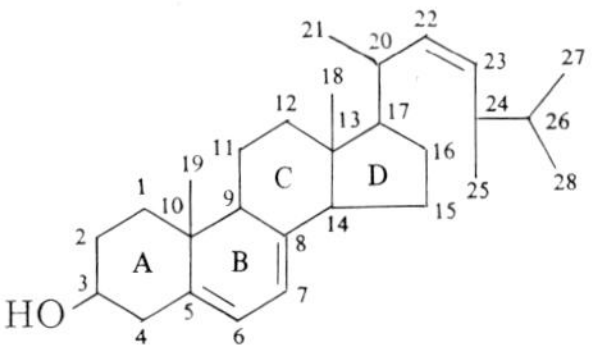

Ergosterol

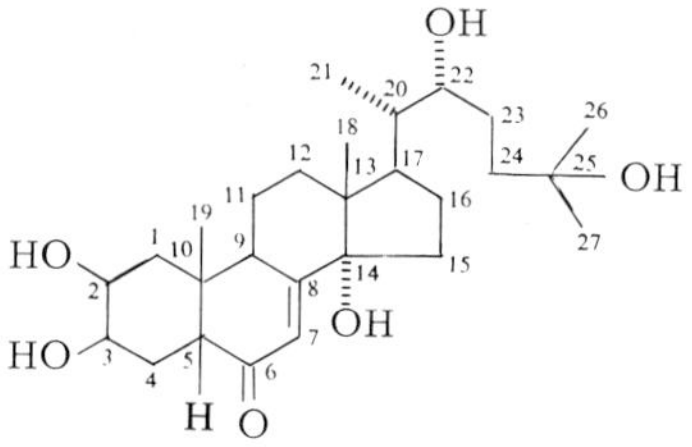

α-Ecdysone

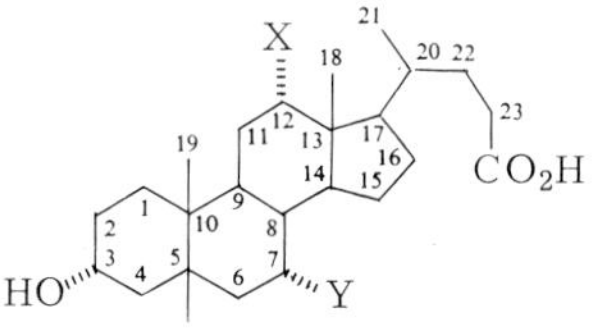

Cholic acid (X = Y = OH)
(3α, 7α, 12α-Trihydroxy-5β-cholan-24-oic acid)

and 2 D experiments is the only remedy for interpretation difficulties caused by accidental overlapping. From a comparison of the one-bond $^{13}C-^{13}C$ coupling constants of stigmasterol with 1,4-androstadiene-3,11,17-trione (Fig. 5.2) several conclusions can be drawn. The most interesting coupling information displayed by the two steroidal systems is attributed to the dramatic changes of the ^1J (C–C) values of angular methyl groups upon introduction of adjacent sp^2-hybridized carbon centers. Thus, ^1J $(C_{10}-C_{19})$ and $^1J(C_{13}-C_{18})$ in stigmasterol are reduced from 35.2 and 35.4 Hz to 32.9 and 32.7 Hz, respectively, in 1,4-androstadiene-3,11,17-trione. It is also worth noting that ^1J (C_8-C_9) is smaller by 2.2 Hz in 1,4-androstadiene-3,11,17-trione than in stigmasterol. The presence of the oxygen atom at C-3 in stigmasterol and the carbonyl group in ring D in 1,4-androstadiene-3,11,17-trione enhance the ^1J (C–C) values by about 3 Hz.

The ^{13}C NMR spectra of lithocholic acid, chenodesoxycholic acid, desocycholic acid and cholic acid have been analyzed and are discussed in the literature [570].

5.2.3 Cardenolides and Sapogenins

Comparison with the complete ^{13}C NMR signal assignments of steroids enabled the interpretation of the ^{13}C NMR spectra of structurally related cardenolides and sapogenins [596, 597]. Besides single-frequency off-resonance decoupling and low power noise decoupling, spectra of specifically deuterated compounds were used as additional aids for the signal identifications. The ^{13}C chemical shifts are collected in Table 5.12 and the

5.2 Steroids 359

Solv. Ref. C-	Digitoxi-genin CDCl₃/CD₃OD [596]	Gitoxi-genin CDCl₃/CD₃OD [596]	Digoxi-genin CDCl₃/CD₃OD [596]	Strophan-tidin CDCl₃/CD₃OD [596]	Deoxy-tigogenin CDCl₃ [597]	Tigogenin CDCl₃ [597]	Diosgenin CDCl₃ [597]	Hecogenin CDCl₃ [597]	Smila-genin CDCl₃ [597]	Sarsa-sapo-genin CDCl₃ [597]
1	30.0	30.0	30.0	24.8	38.7	37.0	37.2	36.5	29.9	29.9
2	28.0	28.0	27.9	27.4	22.2	31.5	31.6	31.2	27.8	27.8
3	66.8	66.8	66.6	67.2	26.8	71.2	71.5	70.7	67.0	67.0
4	33.5	33.5	33.3	38.1	29.0	38.2	42.2	37.8	33.6	33.6
5	35.9	36.4	36.4	75.3	47.1	44.9	140.8	44.6	36.6	36.5
6	27.1	27.0	26.9	37.0	29.0	28.6	121.3	28.3	26.5	26.6
7	21.6	21.4	21.9	18.1	32.4	32.3	32.0	31.4	26.5	26.6
8	41.9	41.8	41.3	42.2	35.2	35.2	31.4	34.4	35.3	35.2
9	35.8	35.8	32.6	40.2	54.8	54.4	50.1	55.5	40.3	40.3
10	35.8	35.8	35.5	55.8	36.3	35.6	36.6	36.0	35.3	35.3
11	21.7	21.9	30.0	22.8	20.7	21.1	20.9	37.8	20.9	20.9
12	40.4	41.2	74.8	40.2	40.2	40.1	39.8	213.0	39.9	39.9
13	50.3	50.4	56.4	50.1	40.6	40.6	40.2	55.0	40.7	40.6
14	85.6	85.2	85.8	85.3	56.5	56.3	56.5	55.8	56.5	56.4
15	33.0	42.6	33.0	32.2	31.8	31.8	31.8	31.5	31.8	31.7
16	27.3	72.8	27.9	27.5	80.8	80.7	80.7	79.1	80.9	80.9
17	51.5	58.8	46.1	51.4	62.3	62.2	62.1	53.5	62.4	62.1
18	16.1	16.9	9.4	16.2	16.5	16.5	16.3	16.0	16.4	16.5
19	23.9	23.9	23.8	195.7	12.3	12.4	19.4	12.0	23.8	23.9
20	177.1	171.8	177.1	177.2	41.6	41.6	41.6	42.2	41.6	42.1
21	74.5	76.7	74.6	74.8	14.5	14.5	14.5	13.2	14.4	14.3
22	117.4	119.6	117.0	117.8	109.0	109.0	109.1	109.0	109.1	109.5
23	176.3	175.3	176.3	176.6	31.4	31.4	31.4	31.2	31.4	27.1
24					28.9	28.9	28.8	28.8	28.8	25.8
25					30.3	30.3	30.3	30.2	30.3	26.0
26					66.7	66.7	66.7	66.8	66.8	65.0
27					17.1	17.1	17.1	17.1	17.1	16.1

structures and numberings of the carbon atoms are seen from the following formulae:

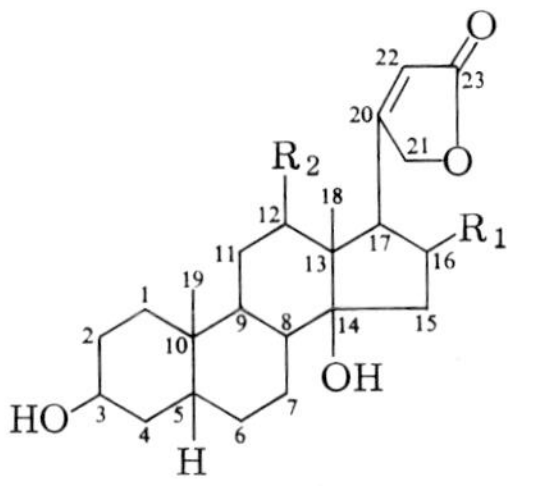

Digitoxigenin ($R_1 = R_2 = H$)
Gitoxigenin ($R_1 = OH, R_2 = H$)
Digoxigenin ($R_1 = H, R_2 = OH$)

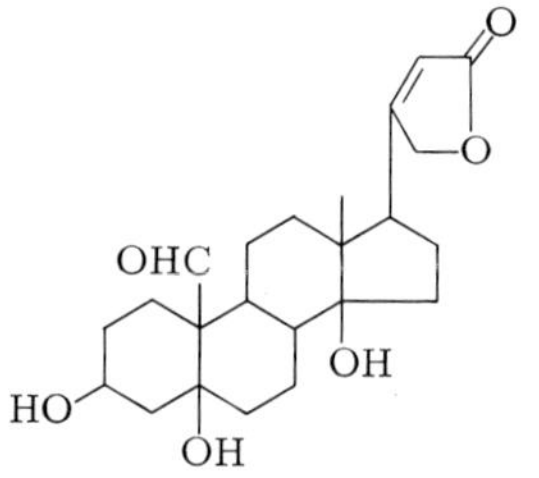

Strophantidin

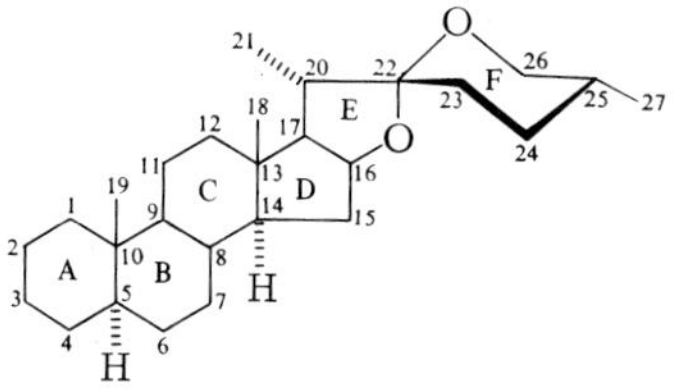

(25 R)-5 α-Spirostane

Deoxytigogenin ((25 R)-5 α-spirostane)
Tigogenin ((25 R)-5 α-spirostan-3 β-ol)
Diosgenin (Δ^5-(25 R)-5 α-spirosten-3 β-ol)
Hecogenin (3 β-hydroxy-(25 R)-5 α-spirostan-12-one)
Smilagenin ((25 R)-5 β)-spirostan-3 β-ol)
Sarsasapogenin ((25 S)-5 β-spirostan-3 β-ol)

5.3 Alkaloids

Numerous ^{13}C NMR investigations on alkaloids have been reported in the literature [598, 599]. In Table 5.13 the ^{13}C chemical shifts and structures of representative alkaloids of different types are collected: Pyrrolidine, piperidine and pyridine [600–602], tropane [600, 603–605], izidine [606–612], indole [600, 603, 613–633], isoquinoline [599, 630, 634–647], quinolinic [648–656], imidazole [657], yuzurimine alkaloids [658], alkaloids with exocyclic nitrogen [659, 660], diterpenoid [661–663], steroid [664–666] and peptide alkaloids [667–671]. The complete signal assignment for the alkaloids given in Table 5.13 was achieved using the correlations between ^{13}C NMR spectral parameters and structural properties and the ^{13}C chemical shift values of model compounds described in Chapters 3 and 4 of this monograph.

5.3 Alkaloids 361

Table 5.13. ^{13}C Chemical Shifts (δ_C in ppm) of Representative Alkaloids.

A Pyrrolidine, Piperidine, and Pyridine Alkaloids

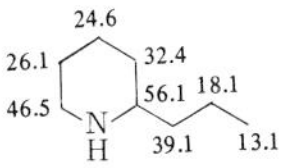

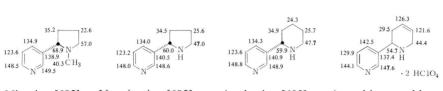

Coniine [600] 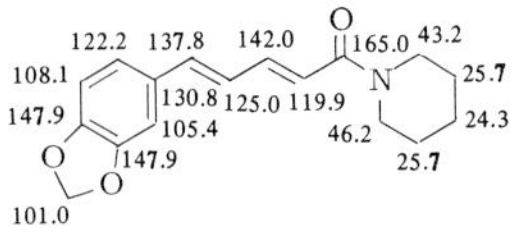   Piperine [601] 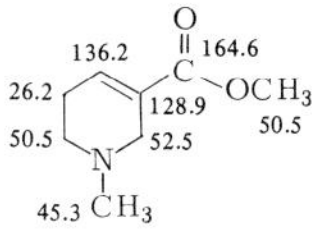   Arecoline [600]

Nicotine [602]   Nornicotine [602]   Anabasine [602]   Anatabine perchlorate [602]

B Tropane Alkaloids

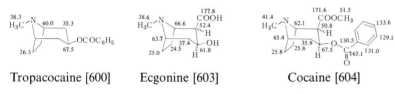

Tropacocaine [600]   Ecgonine [603]   Cocaine [604]

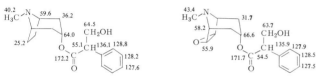

Atropine [605]   Scopolamine [600]

C Izidine Alkaloids

   Pyrrolizidines

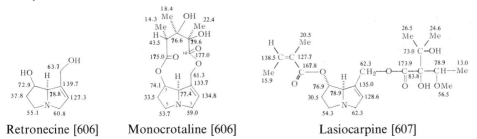

Retronecine [606]   Monocrotaline [606]   Lasiocarpine [607]

Table 5.13. (continued).

Quinolizidines

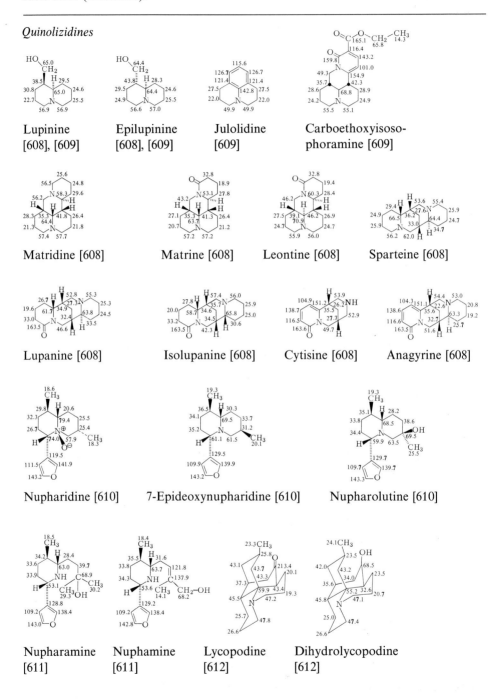

Table 5.13. (continued).

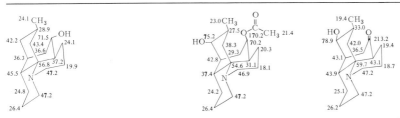

Epidihydrolycopodine [612]   α-Lofoline [612]   Clavolonine [612]

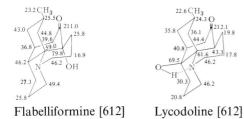

Flabelliformine [612]   Lycodoline [612]

D Indole Alkaloides
Substituted Indoles

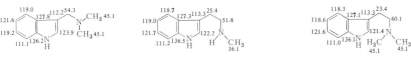

Gramine [600] N_b-Methyltryptamine [613]   N,N-Dimethyltryptamine [614, 615]

2,3-Dihydroindoles

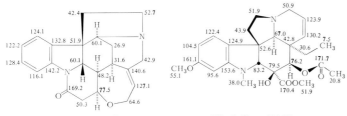

Strychnine [616, 617]   Vindoline [618]

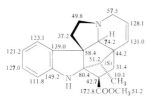

(19 S)-Vindolinine [619]   (19 R)-Vindolinine [620]

Table 5.13. (continued).

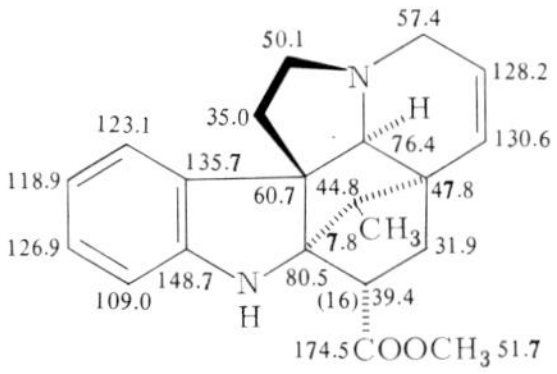

16-Epivindolinine [620]

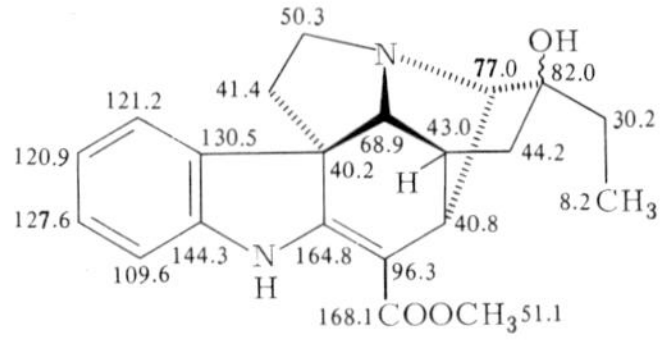

Pandine [621]

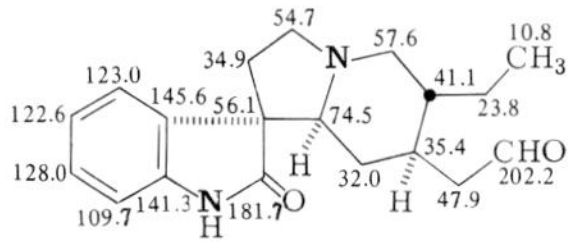

Rhyncophyllal [600]

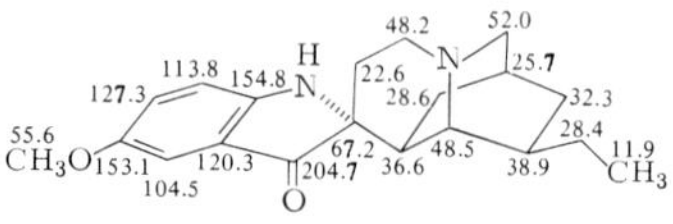

Iboluteine [622]

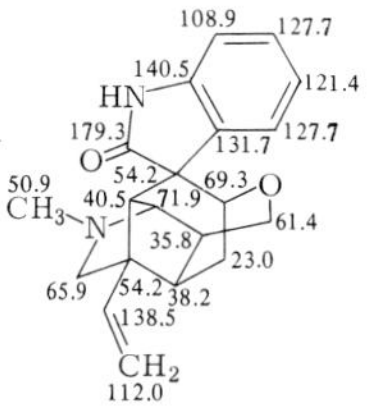

Gelsemine [623]

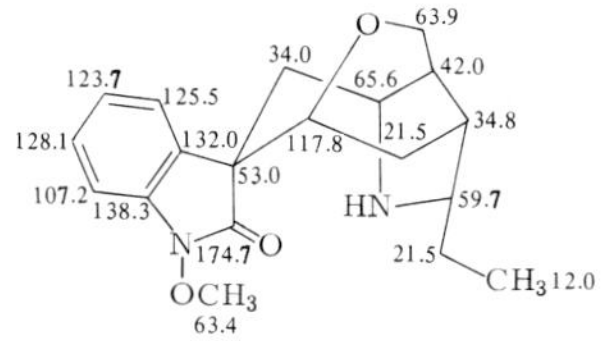

Gelsedine [624]

Tetrahydro-β-carbolins

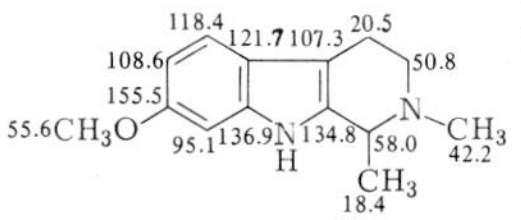

N_b-Methyltetra-hydroharmine [600]

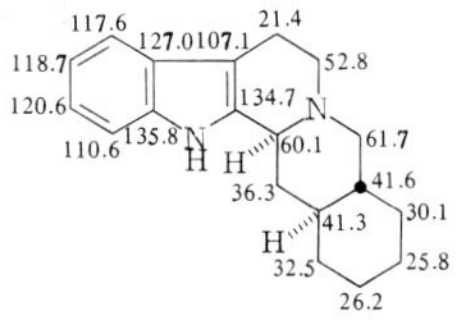

Yohimbane [614]

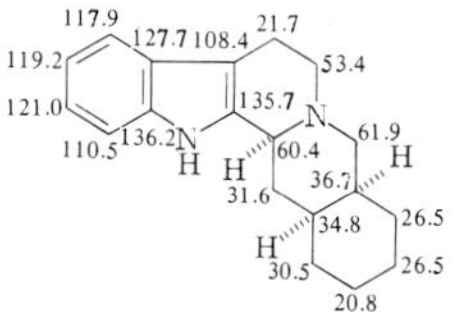

Alloyohimbane [614]

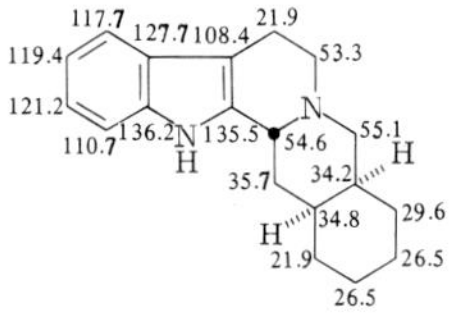

Epialloyohimbane [614]

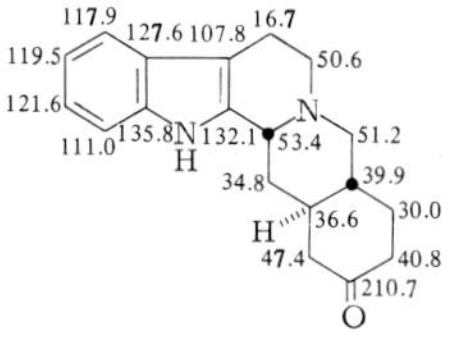

Pseudoyohimbone [614]

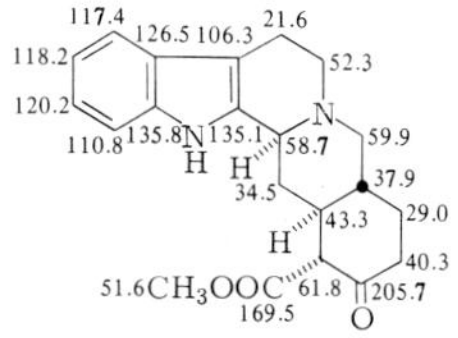

Yohimbinone [614]

5.3 *Alkaloids* 365

Table 5.13. (continued).

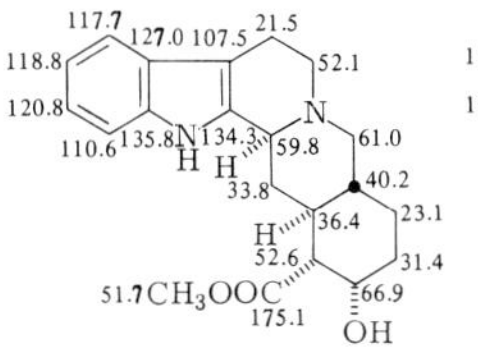

Yohimbine [614] 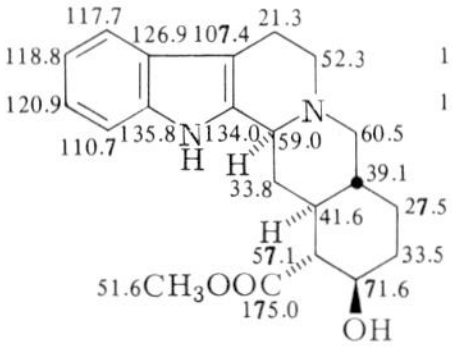   β-Yohimbine [614] 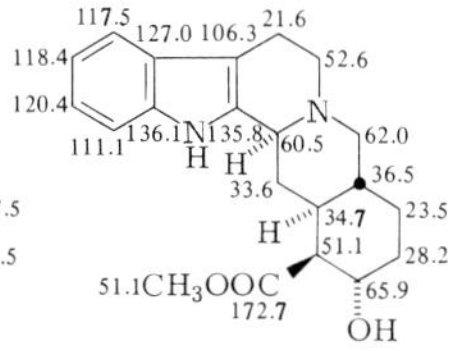   Corynanthine [614]

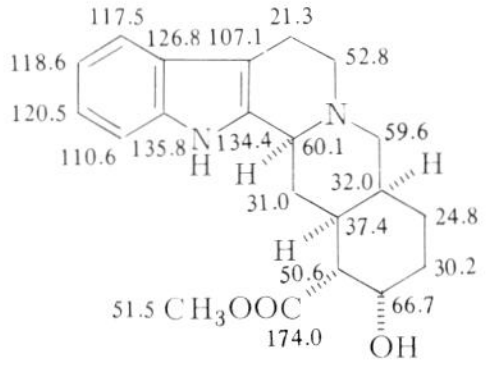

Alloyohimbine [614] 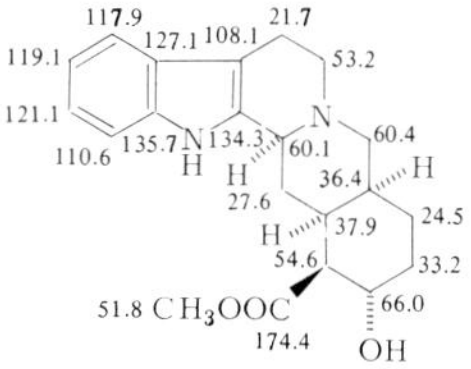   α-Yohimbine [614] 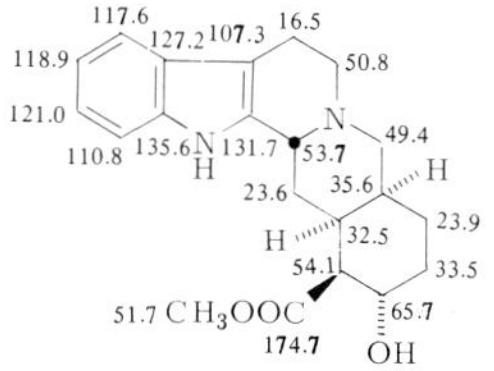   3-Epi-α-yohimbine [614]

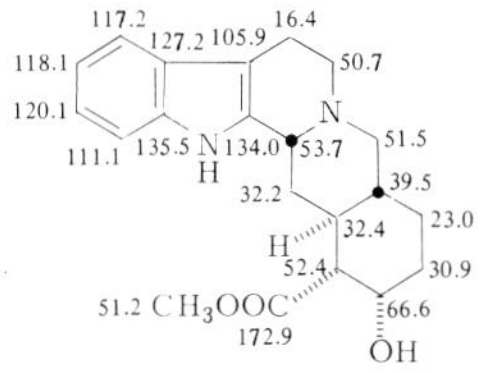

Pseudoyohimbine [614] 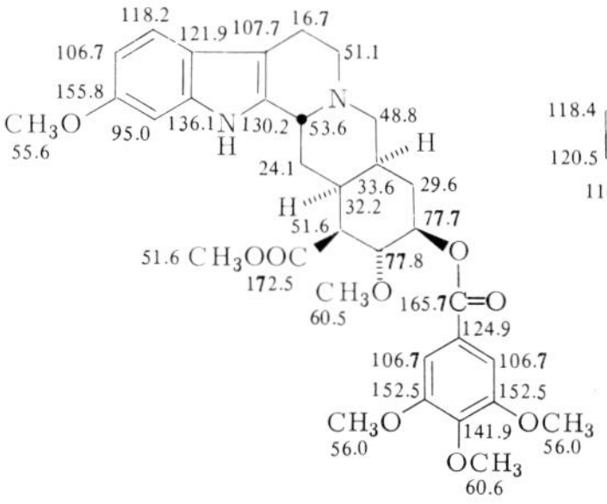   Reserpine [614] 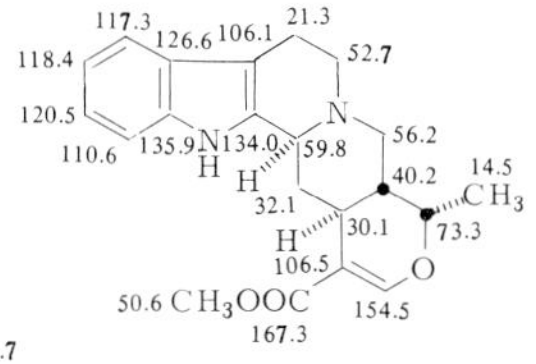   Ajmalicine [614]

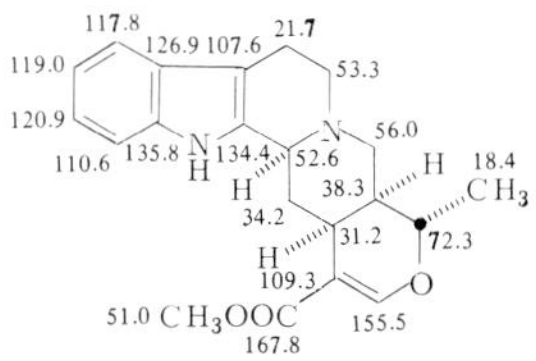

Tetrahydroalstonine [614] 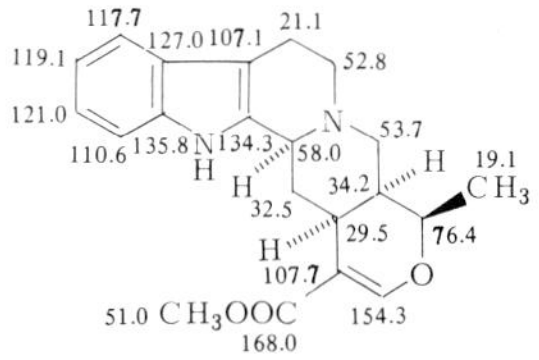   Rauniticine [614] 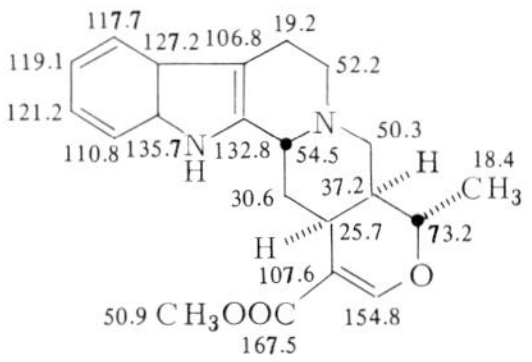   Akuammigine [614]

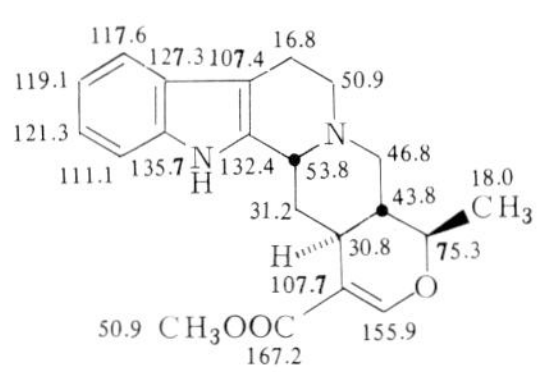

3-Iso-19-epiajma-
licine [614] 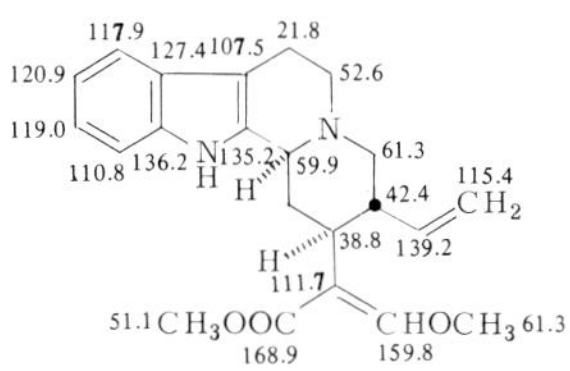   Corynantheine [600] 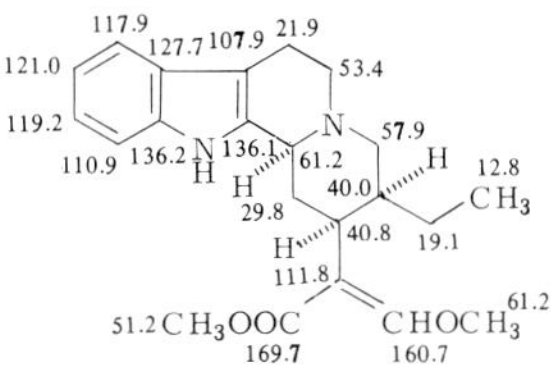   Corynantheidine [600]

Table 5.13. (continued).

Vincamine [625, 626]   16-Epivincamine [626]

Ibogaine [627]   Catharanthine [627]

Indolenine Alkaloids

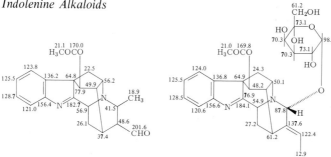

Perakine [628]   Raucaffricine [628]

Ergolines (Lysergic Acid Derivatives)

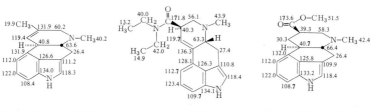

Agroclavine [629]   Lysergic acid diethyl-amide ("LSD") [603]   Methyl 9,10-di-hydrolysergate [629]

Table 5.13. (continued).

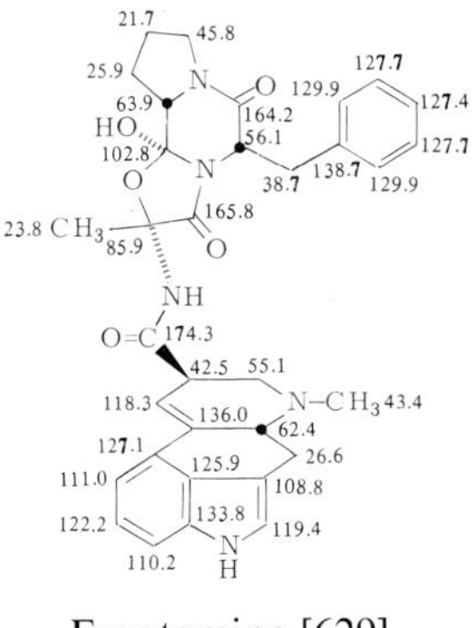

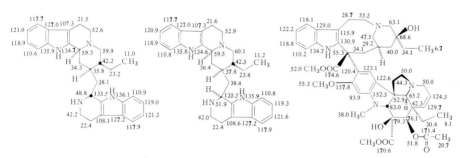

Ergotamine [629]

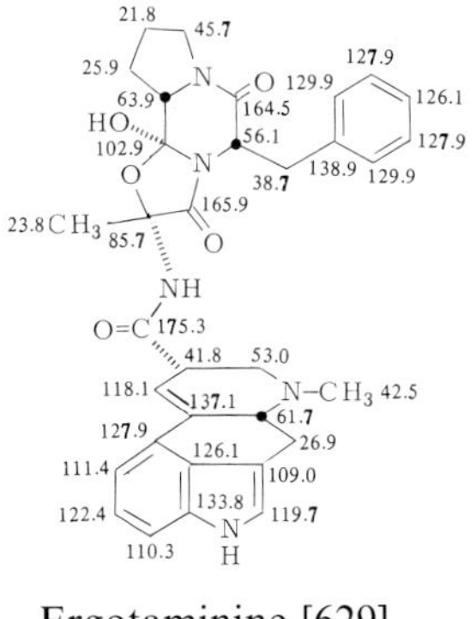

Ergotaminine [629]

Bisindole Alkaloids

Ochrolifuanine A [630]

Ochrolifuanine B [630]

Vincaleucoblastine [631]

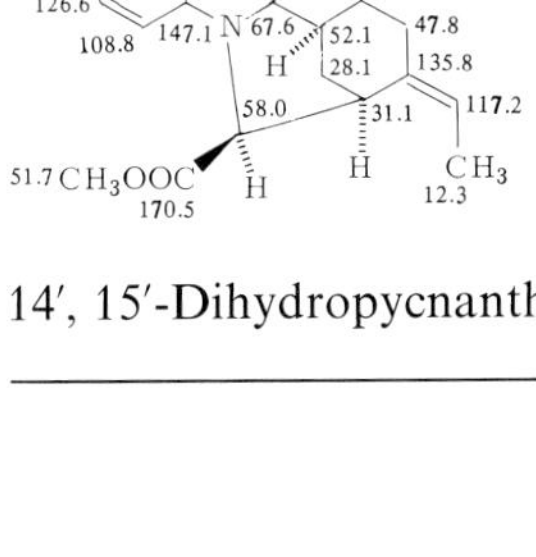

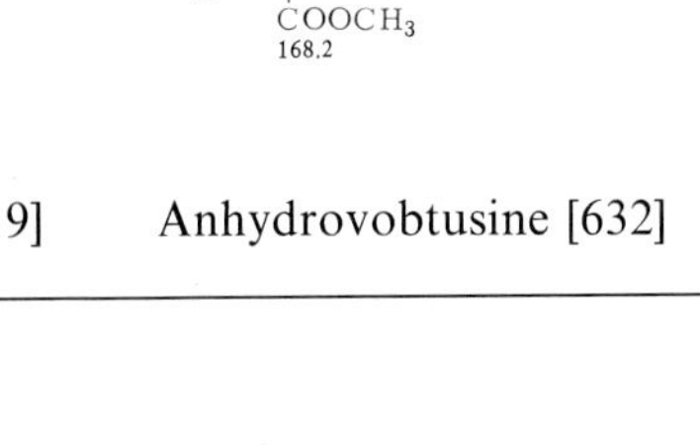

14′,15′-Dihydropycnanthine [619]

Anhydrovobtusine [632]

368   5 ^{13}C NMR Spectra of Natural Products

Table 5.13. (continued).

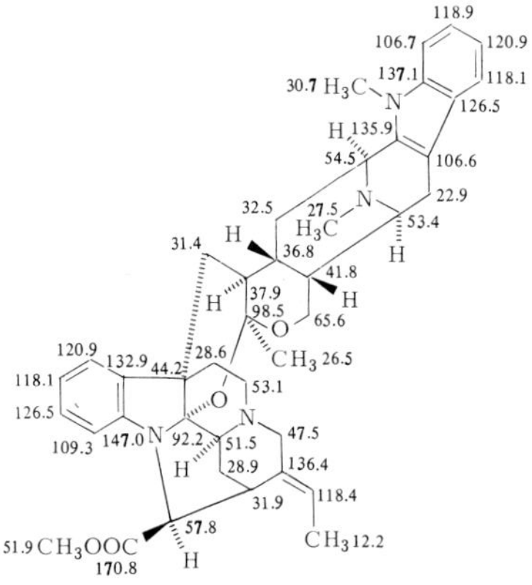

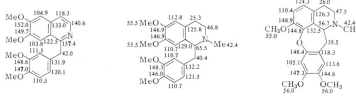

Villalstonine [633]

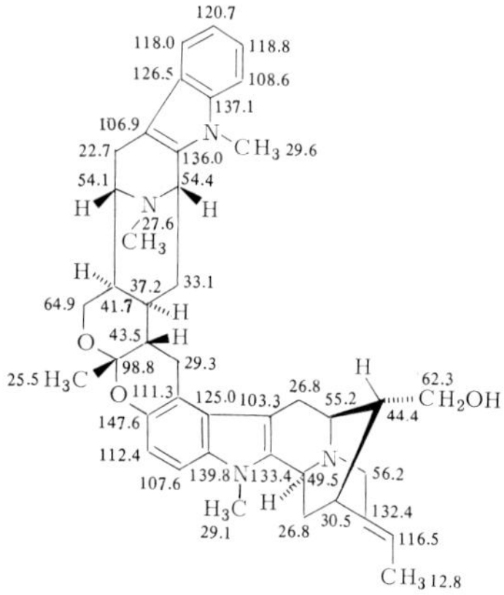

Macralstonidine [633]

E *Isoquinoline Alkaloids*

Benzylisoquinolines

Papaverine [634]   Laudanosine [635, 636]   Cularine [635]

Phthalidisoquinolines

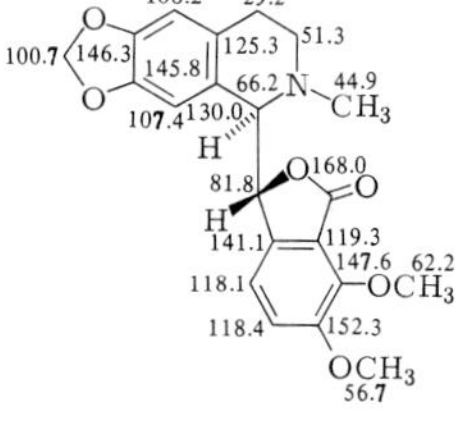

α-Hydrastine [637] 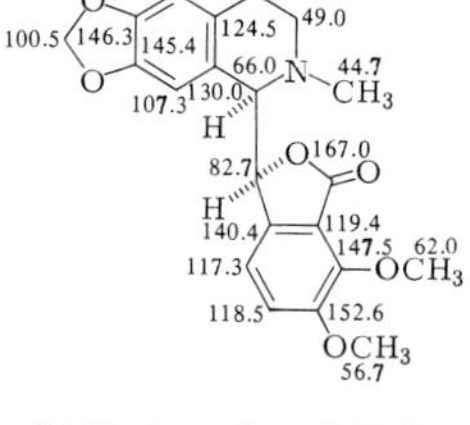   β-Hydrastine [637] 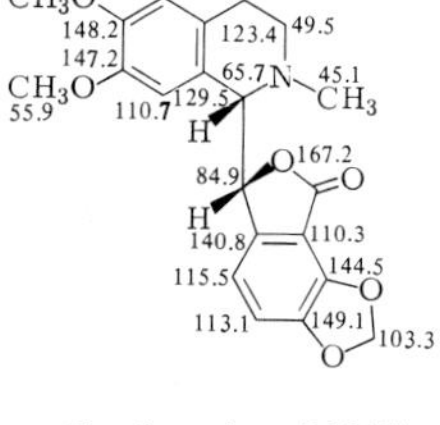   Corlumine [637]

Table 5.13. (continued).

Benzoquinolizidines

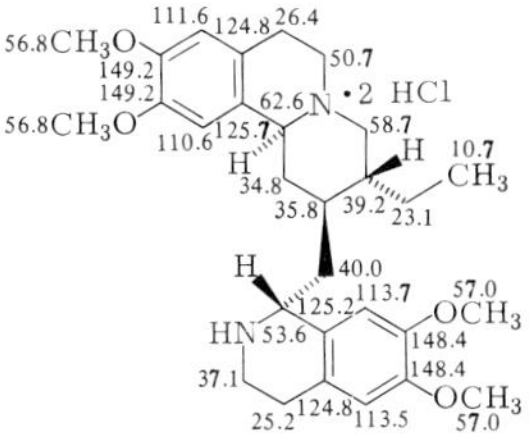

Emetine [630, 638] 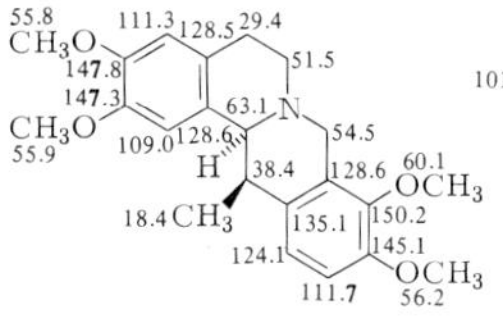   Corydaline [637, 639] 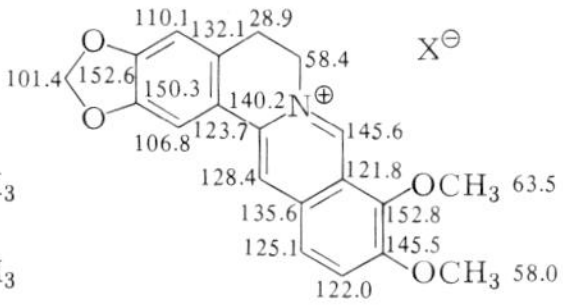   Berberine cation [599]

Protopine Alkaloids

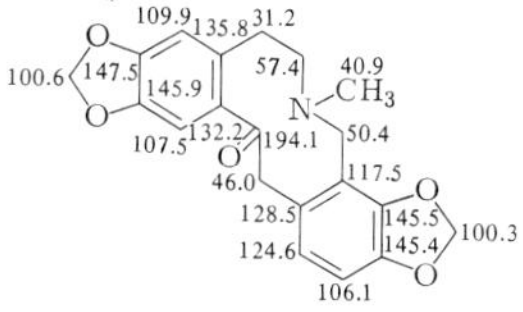

Protopine [640] 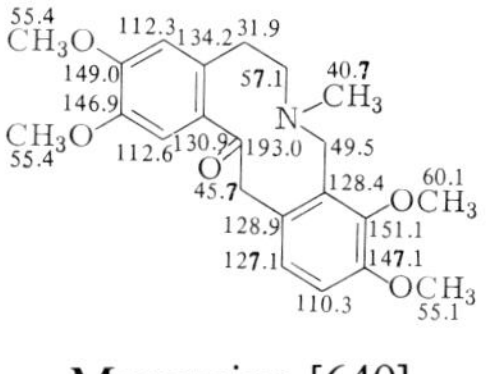   Muramine [640]

Aporphines

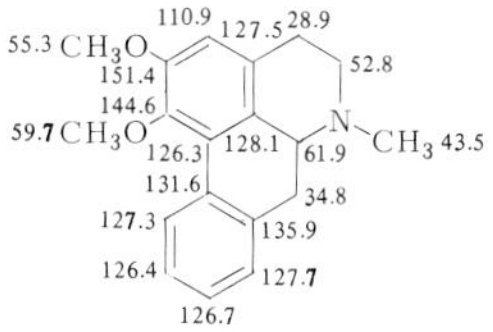

Nuciferine [635] 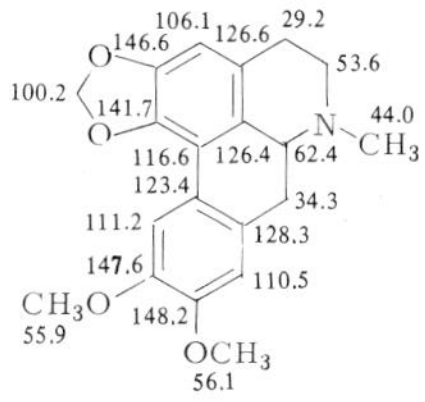   Dicentrine [641] 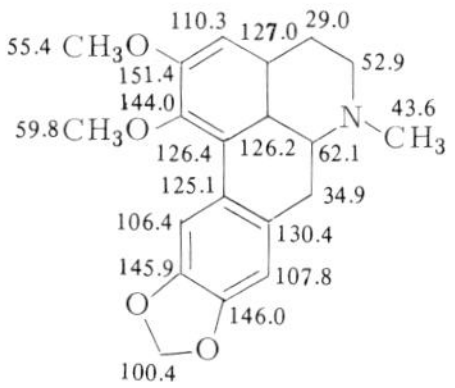   Nantenine [635]

Proaporphines

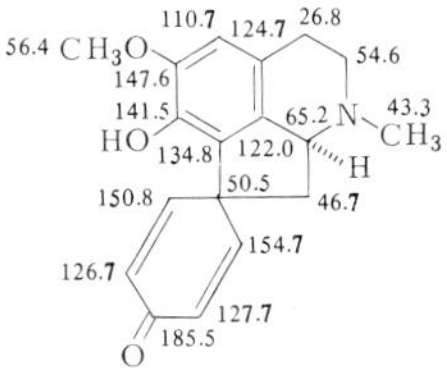

Glaziovine [636] 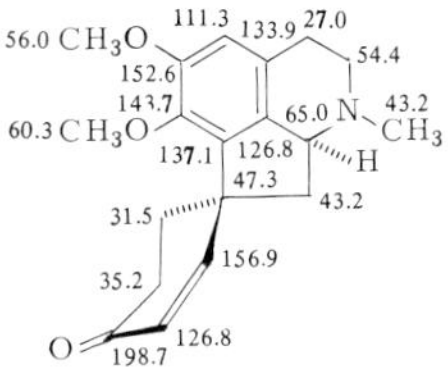   Amuronine [636]

Table 5.13. (continued).

Morphines

Morphine [642] Codeine [643] Thebaine [643]

Sinomenine [643] 6 α-Naltrexol [642] 6 β-Naltrexol [642]

Etorphine [642] Cancentrine [644]

Amarylidaceae Alkaloids

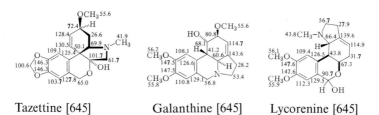

Tazettine [645] Galanthine [645] Lycorenine [645]

5.3 Steroids   371

Table 5.13. (continued).

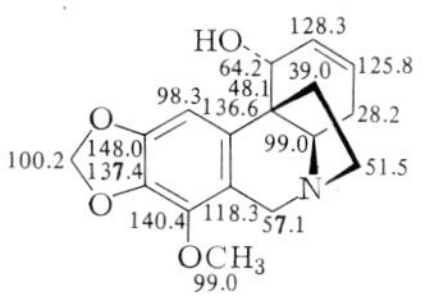

Buphanamine [645] 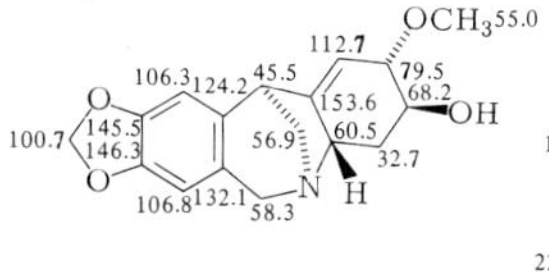   Montanine [645] 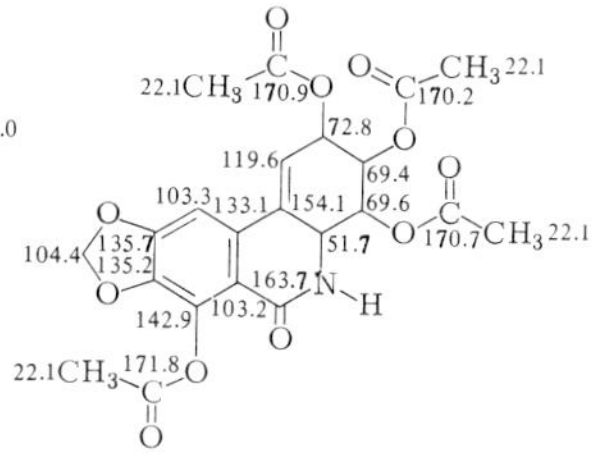   Narciclasine tetraacetate [646]

Bisisoquinolines

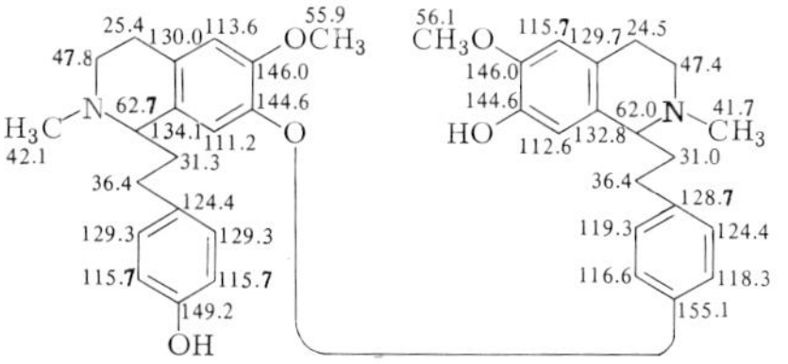

Yolantinine [647]

F *Quinolinic Alkaloids*

Perhydroquinolines, Furoquinolines, Quinolones

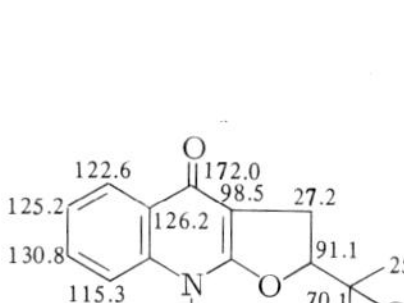

Pumiliotoxin C hydrochloride [648] 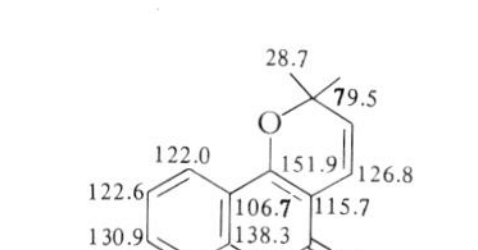   Skimmianine [649] 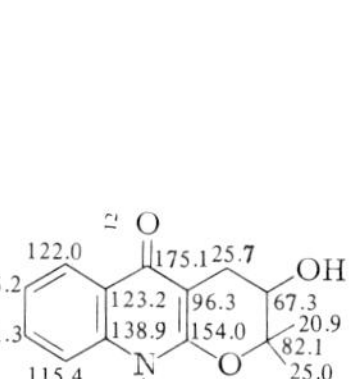   Araliopsine [650]

Isoplatydesmine [650]   Flindersine [651]   Ribalinine [650]

Table 5.13. (continued).

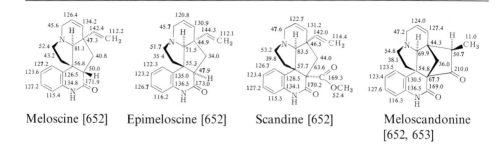

Meloscine [652]   Epimeloscine [652]   Scandine [652]   Meloscandonine [652, 653]

Quinine Derivatives

Quinine [654]   Dihydroquinine [654]   Epiquinine [654]

Cinchonidine [654]   Quinidine [654]   Dihydroquinidine [654]

Epiquinidine [654]   Dihydroepiquinidine [645]   (3 S)-3-Hydroxyquinidine [655, 656]

5.3 Alkaloids 373

Table 5.13. (continued).

G *Imidazole Alkaloids*

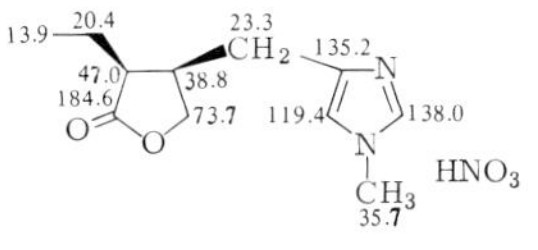

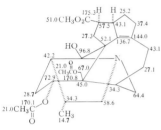

Pilocarpine nitrate [657] 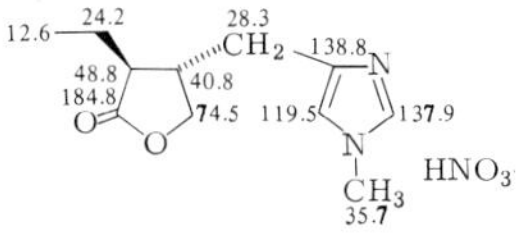   Isopilocarpine nitrate [657]

H *Yuzurimine Alkaloids*

Yuzurimine [658] 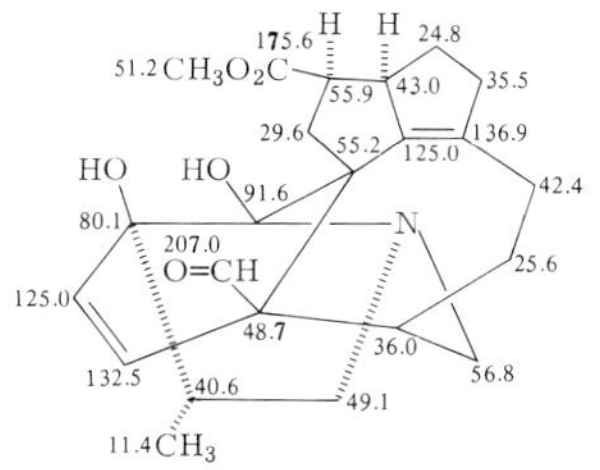   Yuzurimine-C [658]

I *Alkaloids with Exocyclic Nitrogen*

Colchicine Derivatives

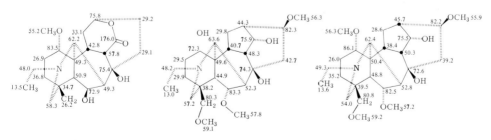

Colchicine [659]   Demecolcine [660]   Isodemecolcine [660]

J *Diterpenoid Alkaloids*

Heterasine [661]   Neoline [661]   Chasmanine [662]

Table 5.13. (continued).

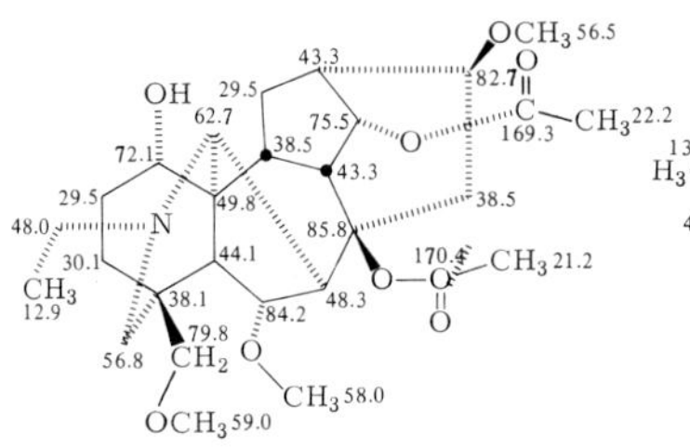

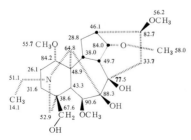

Delphisine [661]

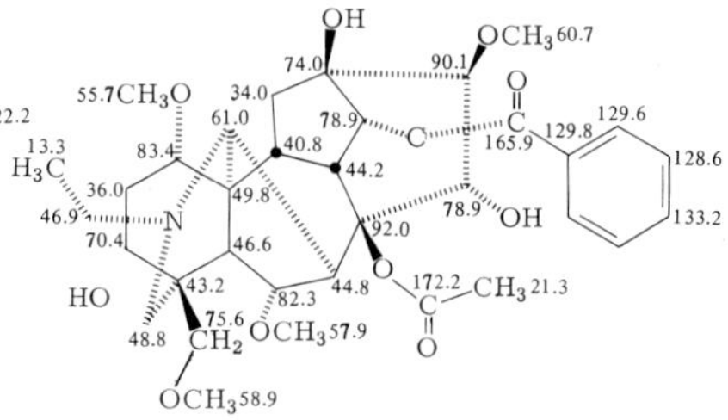

Aconitine [662]

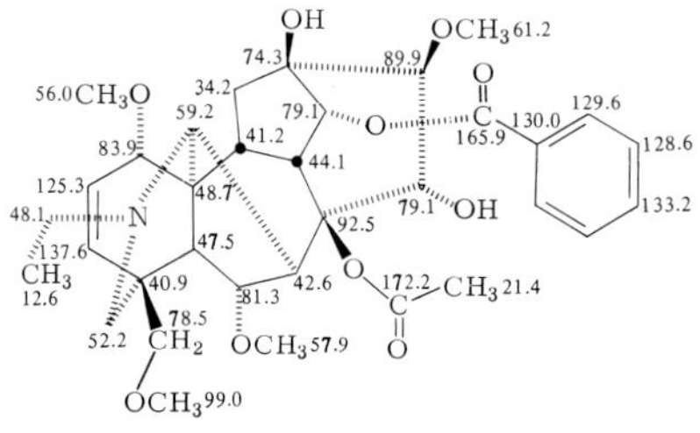

Anhydroaconitine [662]

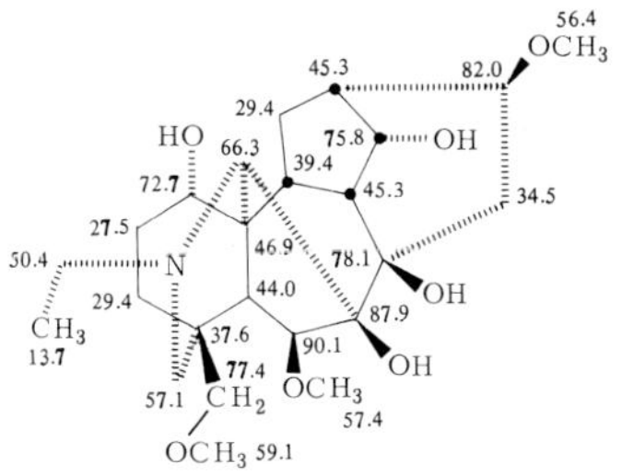

Delcosine [663]

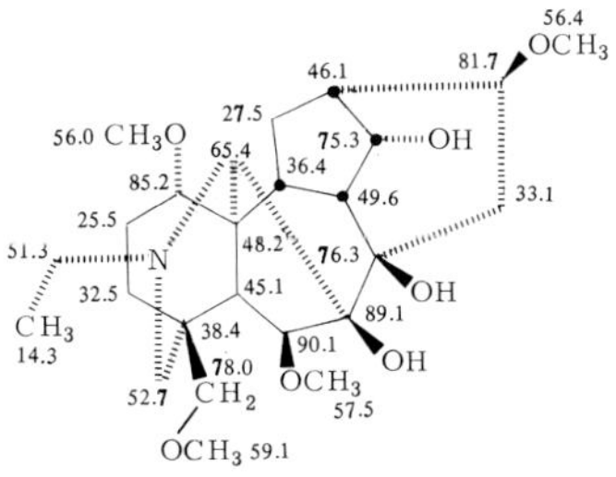

Browniine [663]

Lycoctonine [663]

Table 5.13. (continued).

K Steroid Alkaloids

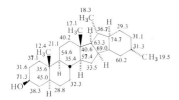

Demissidine [664]   Soladulcidine [664]

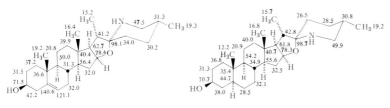

Solasodine [665]   Tomatidine [665]

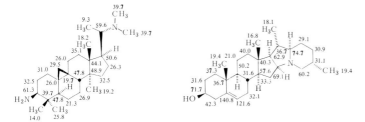

Cycloprotobuxine-F [666]   Solanidine [664]

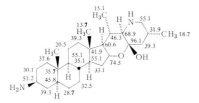

Solanocapsine [664]

376 5 ^{13}C NMR Spectra of Natural Products

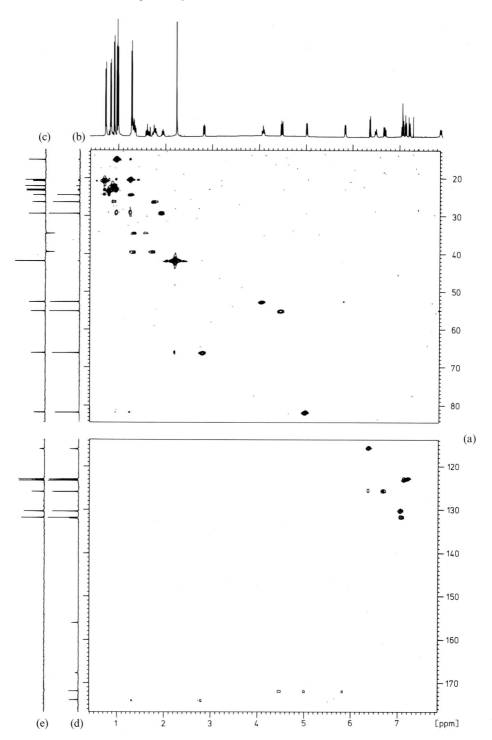

Using lanthanide shift reagents may be helpful for the signal assignment, as was demonstrated for the chemical shifts of piperine's aromatic and olefinic α, β, γ and δ carbons. These could be differentiated after investigations of the ^{13}C NMR spectra of the parent compound and its La(dpm)$_3$ and Eu(dpm)$_3$ complexes [601]. All Δ_{Eu} values with the exception of that for the carbonyl group agreed with the R and Θ relationship found for pseudocontact-shifts [672, 673]. Using copper acetyl acetonate as a paramagnetic broadening agent [674] the signals of C-23 and C-24 of tomatidine could be differentiated [665]. As was found for cyclohexylamines [674], the signals of carbons in positions α and β to the nitrogen broadened much faster than others [665].

Labeling by deuterium or carbon-13 always leads to unequivocal signal identification of specific carbon atoms [600, 602, 640]. After the administration of [5,6-^{14}C-^{13}C$_2$]nicotinic acid to *Nicotiana tabacum* and *N. glauca* labeled anabasine anatabine, nicotine and nornicotine could be isolated. The satellites at the resonances of the labeled natural products helped to complete the signal assignments of the ^{13}C NMR spectra of these alkaloids [602].

Protonation of the nitrogen atom in alkaloids causes significant shift changes for the signals of the neighboring carbon atoms [612, 616, 634]. In nitrogen aromatics, carbons β to nitrogen are shielded while the α and γ carbons are deshielded [94, 99, 100]. In general, N-protonation causes a shielding effect for methyl and methylene carbons of saturated nitrogen heterocycles; more substituted carbon atoms, however, may even be deshielded [612, 675].

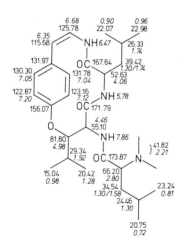

Fig. 5.3. Structure and complete ^{13}C (δ_H in ppm) and ^{1}H (δ_C in ppm) signal assignments of the peptide alkaloid franganine, based on data from Fig. 5.4 and Table 5.14.

Fig. 5.4. ^{13}C NMR spectra of franganine at 100.576/400.133 MHz (^{13}C/^{1}H), in deuteriochloroform (30 mg/0.4 mL) at 30 °C; (a) two-dimensional CH correlation of sp^3 (top) and sp^2 carbon nuclei (bottom) intense correlation signals due to one-bond CH couplings; weak correlations due to two- and three-bond CH couplings obtained in an additional experiment; (b) DEPT-subspectrum of sp^3 CH carbon nuclei; (c) DEPT-subspectrum of CH and CH$_3$ carbons (positive) and of CH$_2$ carbons (negative amplitude); (d, e) partial spectra of sp^2 carbon nuclei, proton broadband-decoupled.

N-Quaternization and the formation of N-oxides of alkaloids cause strong paramagnetic deshielding effects for the carbon atoms neighboring the nitrogen atom; however the carbons located at β-position usually become more shielded [600, 606, 634, 643].

An example using modern techniques of carbon-13 NMR (see Section 2.10) for structure elucidation of alkaloids is given for the peptide alkaloid franganine (Figs. 5.3 and 5.4, Table 5.14). Fig. 5.4 shows a two-dimensional carbon-proton shift correlation and the DEPT-subspectra for the determination of the CH_n multiplicities of franganine. The complete correlation of carbon-13 and proton shifts, identified via $^1J_{CH}$, $^2J_{CH}$ and $^3J_{CH}$, is seen from Table 5.14. Based on these experiments, complete ^1H and ^{13}C signal identifications for franganine are achieved (Fig. 5.3) and demonstrate well the enormous utility of two-dimensional NMR techniques for the structure elucidation of natural products.

Table 5.14. Carbon-13 and Correlating Proton Shifts (in ppm) of Franganine in Deuteriochloroform at 30 °C and 100.576/400.133 MHz ($^{13}C/^1H$).

δ_C	Kind of carbon	Correlating proton(s) via J_{CH} δ_H	Correlating proton(s) via $^{2,3}J_{CH}$ δ_H
173.87	CO	—	1.30 (CH); 2.80 (CH)
171.79	CO	—	4.46 (CH); 4.98 (CH); 5.78 (NH)
167.64	CO	—	
156.07	C	—	
131.97	C	—	
131.78	CH	7.04[a]	
130.30	CH	7.05[b]	
125.78	CH	6.68[c]–6.47 (NH)[c]	6.35 (CH)
123.16	CH	7.12[a]	
122.87	CH	7.20[b]	
155.88	CH	6.35[c]	
81.80	CH	4.98[d]	1.28 (CH_3)
66.20	CH	2.80[e]	2.21 ($NCCH_3)_2$)
55.10	CH	4.46[d]–7.86 (NH)[d]	
52.63	CH	4.06[f]–5.78 (NH)[f]	
41.82	$N(CH_3)_2$	2.21	
39.42	CH_2	1.30[f] AB 1.74[f]	
34.54	CH_2	1.30[e] AB 1.58[e]	
29.34	CH	1.92[g]	0.98 (CH_3); 1.28 (CH_3)
26.33	CH	1.75[f]	0.90 (CH_3); 0.96 (CH_3)
24.46	CH	1.30[e]	0.72 (CH_3); 0.81 (CH_3)
23.24	CH_3	0.81[e]	0.72 (CH_3)
22.98	CH_3	0.96[f]	0.90 (CH_3)
22.07	CH_3	0.90[f]	0.96 (CH_3)
20.75	CH_3	0.72[e]	0.81 (CH_3)
20.42	CH_3	1.28[g]	0.98 (CH_3)
15.04	CH_3	0.98[g]	1.28 (CH_3)

[a–g] similarly labeled proton shifts correlate with each other in the COSY experiment.

5.4 Carbohydrates

^{13}C NMR has proven to be one of the most efficient spectroscopic methods for configurational and conformational investigations in carbohydrate chemistry. The signal assignments of mono-, di- and polysaccharides, inositols and polyols are carried out on the following basis:

(1) Primary (*e.g.* C-6 of rhamnose), secondary (*e.g.* C-6 of glucose), tertiary (*e.g.* CHOH of a pyranose ring carbon atom) and quaternary carbons (*e.g.* of branched-chain sugars) can be distinguished by proton off-resonance decoupling [89, 676–678] or polarization transfer.

(2) Most of the signals are assigned by comparison of the chemical shifts of analogous series of sugars and their derivatives such as glycosides and nucleosides.

(3) The ^{13}C resonances shift downfield with increasing alkyl substitution and increasing number of electron withdrawing substituents A [88, 89, 676–684].

$$\delta_{CCH_3} < \delta_{CCH_2C} < \delta_{CCH_2A} < \delta_{CCHA_2}.$$

(4) Carbons attached to methoxy groups resonate at lower field than corresponding carbon atoms with free hydroxy groups [88, 89, 676–684].

(5) Ring carbon atoms with axial hydroxy groups generally [402] absorb at higher field than corresponding carbons with equatorial hydroxy groups [88, 89, 676–684].

(6) If an equatorial hydroxy group on a saturated six-membered ring is changed to axial orientation, the γ-carbons with axially oriented hydrogen atoms on the same side of the ring shift to higher field [88, 89, 676, 680].

(7) Recording ^{13}C NMR spectra before and after the mutarotational equilibration has been shown to be a valuable method for assigning signals of anomers (α or β) [676, 677, 682–684, 685].

(8) The signals of adjacent carbons bearing hydroxy groups in *cis* configuration to each other can be identified by a downfield shift observed on addition of boric acid as a complex shift reagent [686].

(9) Substitution of a β proton by a deuterium atom causes an isotope effect of 0 to -0.1 ppm. Disappearance of the α-carbon atom resonance and the β effect shift are helpful for the unequivocal signal assignments of carbohydrates [687].

(10) Using the differential isotope shift (DIS) technique by applying a dual coaxial NMR cell differential shift positions are received simultaneously for the ^{13}C resonances of carbohydrates dissolved in H_2O and D_2O. Deuterium substitution of OH causes the largest shifts to higher field for carbons bearing hydroxy groups (β shifts, 0.11 to 0.14 ppm). The γ shifts are in the range of 0.03 to 0.06 ppm, additive and thus permit one to calculate the isotope shift parameters [688].

(11) Because of scalar carbon-carbon coupling, specifically ^{13}C-labeled carbohydrates allow one to identify unequivocally the labeled and also more remote carbon resonances [689].

5.4.1 Monomeric Aldoses and Ketoses

The conformations and ^{13}C chemical shifts of the most common aldo- and ketopyranoses [89, 132, 689–694] and aldo- and ketofuranoses [132, 682, 690, 693, 695] are surveyed in Tables 5.15 and 5.16; the chemical shifts of selected monosaccharide derivatives are given in Table 5.17 [88, 89, 677, 680–682, 694, 696–725]. The signals were assigned according to rules (1) to (11).

Upon 3-O-methylation of D-galactose the resonance of C-4 shifts about 4.5 ppm upfield. This effect is a general rule: Methylation of pyranose hydroxy groups causes upfield shifts of the β-carbons with axial hydroxy groups of about 4.5 ppm [705].

A mutarotational equilibration is demonstrated in Fig. 5.5 by the time dependent ^{13}C NMR spectrum of a solution of D-glucose. Initially, the solution contains almost pure α-D-glucose, but the signals at lower field corresponding to the β-anomer soon appear. At equilibirium, the relative signal intensities indicate that the β-anomer is favored (60 % β, 40 % α; Fig. 5.5) [685].

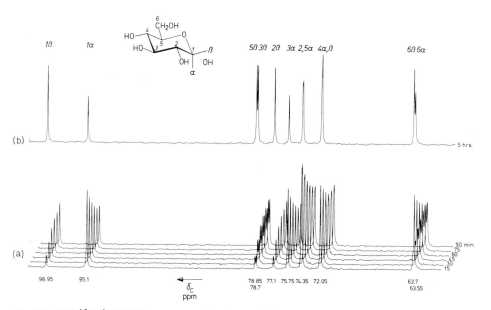

Fig. 5.5. PFT ^{13}C{^{1}H} NMR spectrum of D-glucose, 20 MHz, 1 mol/L in D$_2$O, temperature 30 °C, 90 °C pulses; accumulation of 200 pulse interferograms; (a) series of spectra recorded 15, 30, 45, 60, 75 and 90 min. after sample preparation; (b) spectrum of the same sample, recorded 12 hrs. after sample preparation.

Table 5.15. Conformations and ^{13}C Chemical Shifts (δ_C in ppm) of Selected Aldo- and Ketopyranoses.

Compound	Conformation (β-Anomer)	α-Pyranose						β-Pyranose						Ref.
		C-1	C-2	C-3	C-4	C-5	C-6	C-1	C-2	C-3	C-4	C-5	C-6	
D-Arabinose	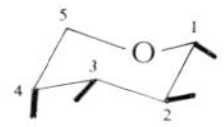	97.4	69.7	73.6	73.1	67.5	—	93.3	69.8	69.9	69.3	63.8	—	[690]
D-Lyxose		94.8	71.2	71.7	68.8	64.4	—	94.9	71.1	73.8	67.9	65.3	—	[89]
D-Ribose		96.3	72.8	72.0	70.0	65.7	—	96.6	73.8	71.7	70.0	65.7	—	[132]
D-Xylose		92.3	71.6	73.0	69.5	61.0	—	96.7	74.1	75.9	69.3	65.2	—	[690]
D-Allose	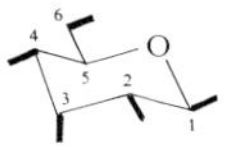	93.1	67.6^a	72.2	66.7	67.5^a	61.4	93.7	71.8	71.7	67.4	74.1	61.8	[687, 691]
D-Altrose		92.9	70.8	69.7^a	64.8	69.8^a	60.3	90.9	70.3^a	70.0^a	63.9	73.5	61.3	[692]
D-Galactose		93.8	70.0	70.8	70.9	72.0	62.8	98.0	73.6	74.4	70.4	76.6	62.6	[689]
D-Glucose		93.6	73.2	74.5	71.4	73.0	62.3	97.4	75.9	77.5	71.3	77.4	62.5	[689]
D-Gulose	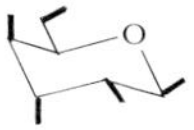	95.0	66.9	71.6	73.0	68.6	63.3	96.0	71.3^a	71.6^a	73.4	76.0	63.3	[693]
D-Idose		94.9	71.2	74.3	73.3	69.3	60.0	94.2	71.7^a	74.3	71.2^a	76.3	62.7	[693]

Table 5.15. (continued).

Compound	Conformation (β-Anomer)	α-Pyranose						β-Pyranose						Ref.
		C-1	C-2	C-3	C-4	C-5	C-6	C-1	C-2	C-3	C-4	C-5	C-6	
D-Mannose		95.5	72.3	71.9	68.5	73.9	62.6	95.2	72.8	74.8	68.3	77.6	62.6	[689]
D-Talose		94.0	70.5	65.0	69.4	70.9	61.4	93.5	71.3	68.2[a]	68.4[a]	75.2	61.1	[693]
D-Fucose		93.8	69.8	71.0	73.5	67.8	17.2	97.8	71.1	74.6	73.1	72.3	17.2	[689]
D-Rhamnose	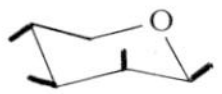	95.0	71.9	71.1	73.3	69.4	18.0	94.6	72.4	73.8	72.9	73.0	18.0	[690]
D-Fructose	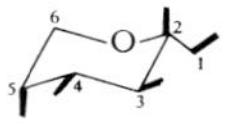	65.9	99.1	70.9	71.3	62.0	61.9	64.7	99.1	68.4	70.5	70.0	64.1	[694]
D-Psicose	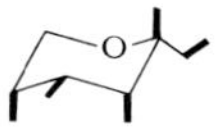	64.0	98.4	66.4	72.6	66.7	58.8	64.8	99.2	71.2	65.9	69.8	65.0	[694]
L-Sorbose	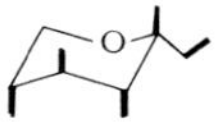	64.5	98.6	71.4	74.8	70.3	62.7	64.6	99.5	70.2[a]	73.6	71.1[a]	59.8	[694]
D-Tagatose	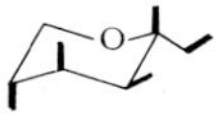	64.8	99.0	70.7	71.8	67.2	63.1	64.4	99.1	64.6	70.1[a]	70.7[a]	61.0	[694]

[a] assignments may be reversed.

Table 5.16. Conformations and ^{13}C Chemical Shifts (δ_C in ppm) of Selected Aldo- and Ketofuranoses.

Compound	Conformation (β-Anomer)	α-Furanose						β-Furanose						Ref.
		C-1	C-2	C-3	C-4	C-5	C-6	C-1	C-2	C-3	C-4	C-5	C-6	
D-Erythrose	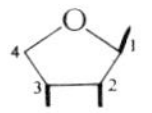	98.0	73.7	71.8	74.1	—	—	103.8	78.9	72.9	73.7	—	—	[682]
D-Arabinose	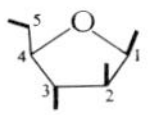	101.7	82.5	76.9	84.0	62.2	—	95.9	77.4	75.6	82.4	63.8	—	[690]
D-Lyxose		101.7	78.0	72.3	80.9	61.7	—	94.9	72.7	71.5	81.2	62.5	—	[682]
D-Ribose		99.0	73.7	72.8	85.8	64.1	—	103.7	78.0	73.2	85.2	65.0	—	[132]
D-Allose	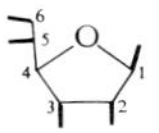	96.3	72.2	69.9	83.8	71.4	62.9	100.9	75.9	71.4	82.7	73.0	63.0	[682]
D-Altrose		100.8	80.8	75.3	82.6	71.1	62.0	94.4	76.0	74.5	80.5	72.1	62.0	[693]
D-Galactose		96.6	77.9	75.9	82.4	73.3	64.2	102.6	82.9	77.4	83.7	72.3	64.4	[682]
D-Gulose		99.9	73.9	72.3	81.9	71.9	64.8	102.7	79.5	a	81.7	72.6[a]	64.7	[693]
D-Idose		103.6	83.0[a]	76.4	84.4[a]	72.4	64.1	97.3	77.7	76.6	79.4	71.6	64.0	[693]

Table 5.16. (continued).

Compound	Conformation (β-Anomer)	α-Furanose						β-Furanose						Ref.
		C-1	C-2	C-3	C-4	C-5	C-6	C-1	C-2	C-3	C-4	C-5	C-6	
D-Mannose		102.4	79.0	71.0	80.1	70.7	64.6	96.9	74.4	71.2[a]	80.2	71.0[a]	64.5	[682]
D-Talose	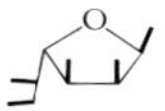	100.2	74.9	70.3	81.4	71.5	62.5	95.8	70.9	70.3	81.9	70.5	62.8	[693]
D-Fucose		95.7	77.7	76.0	85.7	68.5	20.1	101.5	82.5	77.5	87.2	70.1	20.4	[693]
D-Ribulose	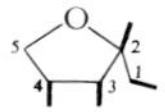	63.3	102.9	70.9	70.5	71.8	—	61.8	106.0	75.8	70.9	71.1	—	[693]
D-Fructose	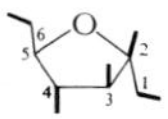	63.7	105.5	82.9	77.0	82.2	61.9	63.6	102.6	76.4	75.4	81.6	63.2	[695]
D-Psicose		64.2	104.0	71.2	71.2	83.6	62.2	63.3	106.4	75.5	71.8	83.6	63.7	[695]
L-Sorbose		64.2	102.5	77.0	76.3	78.6	61.6	63.6	106.3	82.3	74.2	81.1	62.1	[695]
D-Tagatose	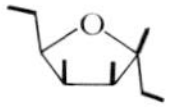	63.3	105.7	77.6	71.9	80.0		63.5	103.3	71.7	71.8	80.9	61.9	[695]

[a] Assignments may be reversed.

Table 5.17. ^{13}C Chemical Shifts (δ_C in ppm) of Selected Monosaccharide Derivatives.

Compound	C-1	C-2	C-3	C-4	C-5	C-6	OCH$_3$ etc.	Ref.
Methyl α-D-allopyranoside	100.0	68.3	72.1	68.0	67.3	61.7	56.3	[696]
Methyl β-D-allopyranoside	101.9	72.2	71.4	68.0	74.8	62.2	58.0	[696]
1,6-Anhydro-β-D-allopyranose	101.5	70.2	63.5	70.1	76.8	65.4		[697, 698]
Methyl-4,6-O-benzylidene-α-D-allopyranoside	100.00	67.70	68.70	77.30	56.80	68.70	101.40 (C-7); 55.70 (OCH$_3$); C$_{arom.}$: 136.80 (C$_{subst.}$); 128.70 (p-); 127.80, 125.80 (o-, m-)	[699]
Allal	146.2	101.3	62.5	67.0	74.8	61.3		[700]
Methyl α-D-altropyranoside	101.1	70.0	70.0	64.8	70.0	61.3	55.4	[88]
Methyl β-D-altropyranoside	100.4	70.7	70.2	65.6	75.6	61.7	57.7	[696]
Methyl 4,6-O-benzylidene-α-D-altropyranoside	101.50	69.30	68.60	75.80	57.60	68.60	101.50 (C-7); 54.80 (OCH$_3$); C$_{arom.}$: 136.80 (C$_{subst.}$); 128.70 (p-); 127.80, 125.80 (o-, m-)	[699]
Methyl 4,6-O-benzylidene-β-D-altropyranoside	98.90	70.40	68.20	76.00	62.40	68.20	101.40 (C-7); 56.30 (OCH$_3$); C$_{arom.}$: 136.80 (C$_{subst.}$); 128.70 (p-); 127.80, 125.80 (o-, m-)	[699]
Methyl 4,6-O-benzylidene-3-amino-3-deoxy-α-D-altropyranoside	101.80	69.40	51.30	75.80	57.60	68.70	101.50 (C-7); 54.80 (OCH$_3$); C$_{arom.}$: 136.80 (C$_{subst.}$); 128.70 (p-); 127.80, 125.80 (o-, m-)	[699]
Methyl 3-O-acetyl-4,6-O-benzylidene-2-deoxy-2-C-(1,3-dithiane-2-yl)-α-D-altropyranoside	102.20	46.10	74.80	69.70	69.50	59.00	55.80 (OCH$_3$); 99.90 (C-7); 21.40 170.50 (−O−C−); $\overset{\parallel}{O}$ (CH$_3$); 44.30 (C-2'); 27.80 (C-4'); 25.30 (C-5'); 28.00 (C-6'); C$_{arom.}$: 137.60 (C$_{subst.}$); 129.20 (p-); 128.20 (o-); 126.10 (m-)	[699] [701]

Table 5.17. (continued).

Compound	C-1	C-2	C-3	C-4	C-5	C-6	OCH$_3$ etc.	Ref.
Methyl 2-O-acetyl-4,6-O-benzyl-idene-3-deoxy-3-C-(1,3-dithiane-2-yl)-α-D-altropyranoside	103.10	77.20	45.00	71.50	69.70	59.60	55.70 (OCH$_3$); 99.50 (C-7); 169.90 (−O−C$\overset{\overset{\displaystyle O}{\|}}{−}$); 21.10 (CH$_3$); 44.20 (C-2′); 30.20 (C-4′); 25.70 (C-5′); 30.60 (C-6′) C$_{arom.}$: 138.00 (C$_{subst.}$); 128.60 (o-); 127.00 (m); 129.30 (p-) 55.50 (OCH$_3$)	[701]
Methyl α-D-arabinopyranoside	105.1	71.8	73.4	69.4	67.3		58.1	[696]
Methyl β-D-arabinopyranoside	101.0	69.4	69.9	70.0	63.8		56.3	[696]
Methyl α-D-arabinofuranoside	109.2	81.8	77.5	84.9	62.4		56.0	[696]
Methyl β-D-arabinofuranoside	103.1	77.4	75.7	82.9	62.4		56.3	[696]
2-Deoxy-α-D-arabinohexopyranose	92.1	38.3	68.8	72.0	72.8	61.6		[702]
2-Deoxy-β-D-arabinohexopyranose	94.1	40.5	71.4	71.7	76.8	61.9		[702]
L-Ascorbic acid	174.2	118.9	156.6	77.3	60.0	63.4		[703, 704]
Methyl α-D-galactopyranoside	99.50	69.60	68.30	69.30	70.80	61.30	55.15 (OCH$_3$)	[705]
Methyl β-D-galactopyranoside	103.90	70.80	72.85	68.75	75.20	61.05	57.30 (OCH$_3$(1))	[705]
Methyl 3-O-methyl-β-D-galactopyranoside	103.90	69.80	82.00	64.20	75.10	61.20	56.20 (OCH$_3$(3)); 57.30 (OCH$_3$(1))	[705]
Methyl 2,6-O-dimethyl-α-D-galactopyranoside	96.80	77.45	68.60	69.50	73.15	71.85	55.05 (OCH$_3$(1)); 57.70, 58.50 (OCH$_3$(2,6))	[705]
Methyl 2,3,4,6-tetra-O-methyl-β-D-galactopyranoside	103.35	79.75	82.20	73.05	74.90	71.00	57.10 (OCH$_3$(1)); 57.10, 58.50, 60.20, 60.95 (OCH$_3$(2,3,4,6))	[705]
p-Nitrophenyl α-D-galactopyranoside	100.75	75.50	71.20	70.45	72.05	63.00	C$_{arom.}$: 119.55, 128.15, 144.15, 165.05	[680]
m-Nitrophenyl β-D-galactopyranoside	104.20	73.25	76.05	71.10	78.65	63.45	C$_{arom.}$: 113.90, 119.55, 126.20, 133.55, 173.05, 160.65	[680]
o-Nitrophenyl β-D-galactopyranoside	104.10	73.15	76.25	71.10	78.65	63.45	C$_{arom.}$: 119.95, 124.70, 127.50, 137.00, 152.55, 143.15	[680]

Table 5.17. (continued).

Compound	C-1	C-2	C-3	C-4	C-5	C-6	OCH$_3$ etc.	Ref.
Phenyl β-D-galactopyranoside	104.00	73.55	76.40	71.30	78.30	63.65	C$_{arom.}$: 119.30, 124.90, 132.50, 160.40	[680]
1,6-Anhydro-β-D-galactopyranose	101.3	71.9	70.8	64.9	74.9	64.1		[697, 698]
Methyl 3,4-anhydro-α-D-galactopyranoside	97.0	64.4	54.0	51.8	67.1	62.1	56.3	[696]
Galactal	144.8	103.4	64.8	65.5	77.9	62.1		[696]
Methyl α-D-glucopyranoside	99.60	71.90	73.60	70.10	71.60	61.20	55.30	[88]
Methyl β-D-glucopyranoside	103.40	73.40	75.20	70.00	75.20	61.40	57.50	[88]
3-O-Methyl-α-D-glucopyranose	92.10	71.10	82.70	69.10	71.30	60.70	59.90 (OCH$_3$)	[89]
3-O-Methyl-β-D-glucopyranose	96.00	73.70	85.20	68.90	75.70	60.70	59.60 (OCH$_3$)	[89]
α-D-Glucopyranose-pentaacetate	88.60	68.90	69.50	67.70	69.50	61.10	169.80, 169.40, 169.20, 168.80, 168.30 (C=O); 20.30, 19.80, 19.30 (CH$_3$)	[681]
β-D-Glucopyranose-pentaacetate	91.30	70.00	72.30	67.60	72.30	61.30	169.80, 169.30, 169.80, 168.70, 168.40 (C=O); 19.90, 19.80 (CH$_3$)	[681]
p-Nitrophenyl α-D-glucopyranoside	100.45	74.10	76.90	72.50	75.75	63.45	C$_{arom.}$: 119.65, 128.25, 144.35, 164.85	[680]
p-Nitrophenyl β-D-glucopyranoside	102.70	75.95	80.05	72.40	79.30	63.45	C$_{arom.}$: 119.45, 128.50, 144.55, 165.15	[680]
m-Nitrophenyl β-D-glucopyranoside	103.55	76.05	79.95	72.60	79.20	63.55	C$_{arom.}$: 113.90, 119.55, 126.00, 133.45, 151.35, 160.55	[680]
o-Nitrophenyl β-D-glucopyranoside	103.25	75.95	80.05	72.40	79.50	63.45	C$_{arom.}$: 119.85, 124.70, 127.50, 136.90, 143.05, 152.40	[680]
Phenyl β-D-glucopyranoside	103.05	75.80	79.50	72.40	79.30	63.55	C$_{arom.}$: 119.00, 124.60, 132.15, 153.50	[680]
Methyl 4,6-O-benzylidene-α-D-glucopyranoside	99.80	72.20	70.40	80.60	61.90	68.40	54.80 (OCH$_3$); 101.40 (C-7); C$_{arom.}$: 136.80 (C$_{subst.}$); 128.70 (p-); 127.80, 125.80 (o-, m-)	[699]

Table 5.17. (continued).

Compound	C-1	C-2	C-3	C-4	C-5	C-6	OCH$_3$ etc.	Ref.
Methyl 4,6-O-benzylidene-3-amino-3-deoxy-α-D-glucopyranoside	98.40	72.10	52.40	81.00	62.20	68.70	55.00 (OCH$_3$); 101.50 (C-7); C$_{arom.}$: 136.80 ($_{subst.}$); 128.70 (p-); 127.80, 125.80 (o-, m-)	[699]
Methyl 4,6-O-benzylidene-β-D-glucopyranoside	103.80	73.90	72.60	80.00	65.70	67.90	56.70 (OCH$_3$); 101.30 (C-7); C$_{arom.}$: 136.80 (C$_{subst.}$); 128.70 (p-); 127.80, 125.80 (o-, m-)	[699]
1,6-Anhydro-β-D-glucopyranose	102.1	70.9	73.3	71.6	76.9	65.8		[698, 706, 707]
Methyl 3,6-anhydro-α-D-glucopyranoside	99.5	71.8	72.0	70.4	76.4	69.8	58.5	[706, 707]
Methyl 3,6-anhydro-β-D-glucopyranoside	101.4	72.5	72.8	71.8	75.3	70.2	56.5	[706, 707]
2-O-Methyl-α-D-glucopyranose	90.1	81.3	72.8	70.5	72.0	61.4	58.4	[702, 708]
2-O-Methyl-β-D-glucopyranose	96.5	84.4	76.6	70.5	76.1	61.5	60.9	[702, 708]
3-O-Methyl-α-D-glucopyranose	93.4	72.6	84.1	70.6	72.8	62.3	61.3	[708]
3-O-Methyl-β-D-glucopyranose	97.2	75.1	86.7	70.4	77.3	62.3	61.3	[708]
4-O-Methyl-α-D-glucopyranose	93.2	73.0	73.9	80.5	71.7	62.1	61.6	[708]
4-O-Methyl-β-D-glucopyranose	97.1	75.8	76.7	80.5	76.1	62.1	61.6	[708]
6-O-Methyl-α-D-glucopyranose	93.3	73.0	74.3	71.4	71.4	72.6	60.3	[708]
6-O-Methyl-β-D-glucopyranose	97.3	75.8	77.2	71.4	75.8	72.6	60.3	[708]
α-D-Glucopyranose 1-phosphate	96.3	72.9	74.3	70.9	74.1	74.1	61.9	[709]
β-D-Glucopyranose 1-phosphate	98.9	75.3	76.9	71.2	77.9	62.4		[709]
α-D-Glucopyranose 6-phosphate	93.0	72.2	73.3	69.9	71.3	64.8		[696]
β-D-Glucopyranose 6-phosphate	96.7	74.8	76.3	69.9	75.6	64.8		[696]
2-Amino-2-deoxy-α-D-glucopyranose (hydrochloride)	89.9	55.3	70.5	70.5	72.4	61.3		[702, 710]
2-Amino-2-deoxy-β-D-glucopyranose (hydrochloride)	93.5	57.8	72.8	70.5	76.9	61.3		[702, 710]

Table 5.17. (continued).

Compound	C-1	C-2	C-3	C-4	C-5	C-6	OCH$_3$ etc.	Ref.
6-Chloro-6-deoxy-α-D-glucopyranose	93.4	72.5	73.6	71.3	71.4	45.6		[711]
6-Chloro-6-deoxy-β-D-glucopyranose	97.1	75.2	76.5	71.2	75.6	45.1		[711]
2,6-Dibromo-2,6-dideoxy-α-D-glucopyranose	93.1	53.5	70.9	73.3	73.6	34.4		[712]
2,6-Dibromo-2,6-dideoxy-β-D-glucopyranose	96.6	56.3	74.9	73.0	76.7	33.6		[712]
Glucal	144.6	103.8	69.7	69.2	79.1	61.0		[700]
Methyl α-D-gulopyranoside	100.4	65.5	71.4	70.4	67.3	62.0	56.3	[713]
Methyl β-D-gulopyranoside	102.6	69.1	72.3	70.5	74.9	62.1	58.1	[714]
1,6-Anhydro-β-D-glopyranose	101.7	70.5	70.5	69.9	74.9	63.8		[697, 698]
Methyl β-D-fructopyranoside	61.8	101.4	69.3	70.5	70.0	64.7	49.3	[694]
Methyl α-D-fructofuranoside	58.7	109.1	81.0	78.2	84.0	62.1	49.1	[694]
Methyl β-D-fructofuranoside	60.0	104.7	77.7	75.9	82.1	63.6	49.8	[694]
β-D-Fructopyranose 1-phosphate	67.4	99.0	69.0	70.4	70.1	64.4		[715]
α-D-Fructofuranose 1-phosphate			83.0	77.0	83.0	62.6		[715]
β-D-Fructofuranose 1-phosphate	66.0		77.4	75.2	81.3	63.3		[715]
α-D-Fructofuranose 6-phosphate	63.8	105.3	82.6	76.9	81.4	64.5		[715]
β-D-Fructofuranose 6-phosphate	63.8	102.4	76.2	75.4	80.8	65.4		[715]
Methyl α-D-idopyranoside	101.5	70.9	71.8	70.3	70.8	60.2	55.8	[88]
Methyl α-D-lyxopyranoside	102.0	70.4	71.6	67.7	63.3		55.9	[716]
Methyl α-D-lyxofuranoside	109.2	77.0	72.2	81.4	61.5		56.9	[682]
Methyl β-D-lyxofuranoside	103.3	73.2	71.0	82.1	62.7		56.7	[682]
Methyl 2,3-anhydro-α-D-lyxopyranoside	96.4	51.1	56.8	61.4	59.9		56.4	[696]
Methyl 2,3-anhydro-α-D-lyxofuranoside	101.5	53.7	55.6	76.6	59.6	54.6		[717]

Table 5.17. (continued).

Compound	C-1	C-2	C-3	C-4	C-5	C-6	OCH₃, etc.	Ref.
Methyl 2,3-anhydro-β-D-lyxofuranoside	101.8	54.4	55.1	76.4	59.6	55.7		[717]
Methyl α-D-mannopyranoside	101.9	71.2	71.8	68.0	73.7	62.1	55.9	[718]
Methyl β-D-mannopyranoside	101.3	70.6	73.3	67.1	76.6	61.4	56.9	[718]
Methyl α-D-mannofuranoside	109.7	77.9	72.5	80.5	70.6	64.5	57.2	[682]
Methyl β-D-mannofuranoside	103.6	73.1	71.2	80.7	71.0	64.4	56.8	[682]
1,6-Anhydro-β-D-mannopyranose	101.9	66.6	70.9	72.2	76.4	65.3		[697, 698]
2-Amino-2-deoxy-α-D-mannopyranose (hydrochloride)	91.1	55.3	67.7	67.1	72.8	61.2		[719]
2-Amino-2-deoxy-β-D-mannopyranose (hydrochloride)	91.8	56.4	70.3	67.0	76.9	61.2		[719]
Methyl α-D-psicopyranoside	61.1	100.7	67.3	72.1	66.7	58.9	49.1	[694]
Methyl β-D-psicopyranoside	57.7	102.6	69.7	65.7	69.9	65.4	48.7	[694]
2-Deoxy-α-D-ribopyranose	91.65	35.15	64.65	67.50	62.85			[677]
2-Deoxy-β-D-ribopyranose	93.85	33.80	66.50	67.35	66.00			[677]
2-Deoxy-α-D-ribofuranose	91.65	41.15	70.95	85.30	61.55			[677]
		41.30	71.25	85.85	62.85			[677]
2-Deoxy-β-D-ribofuranose	98.05	41.15	70.95	85.30	61.55			[677]
		41.30	71.25	85.85	62.85			[677]
Methyl-β-D-ribopyranoside	103.85	70.40	69.85	72.65	65.55		58.30 (OCH₃)	[677]
Methyl-β-D-ribofuranoside	108.00	74.35	70.85	82.95	62.90		55.30 (OCH₃)	[677]
Methyl-4,6-O-benzylidene-3-deoxy-α-ribohexopyranoside	98.90	66.90	32.50	76.00	63.40	68.90	101.20 (C-7); 54.60 (OCH₃); C_arom.: 136.80 (C_subst.); 128.70 (p-); 127.80, 125.80 (o-, m-)	[699]
Methyl-3,4-O-isopropylidene-3-C-(1,3-dithiane-2-yl)-β-D-ribopyranoside	99.80	72.05	70.80	72.95	57.30	108.20	25.45 (C-7); 54.70 (C-2'); 29.80, 30.30 (C-4'); 24.80 (C-5'); 29.80, 30.30 (C-6'); 55.45 (OCH₃)	[720]
α-D-Ribofuranose 5-phosphate	97.5	71.9	71.3	83.6	65.8			[721]
β-D-Ribofuranose 5-phosphate	102.4	76.4	71.7	82.5	66.6			[721]

Table 5.17. (continued).

Compound	C-1	C-2	C-3	C-4	C-5	C-6	OCH$_3$, etc.	Ref.
Methyl 2,3-anhydro-β-D-ribo-pyranoside	96.3	52.7	52.4	62.3	61.9	56.6		[696]
Methyl 2,3-anhydro-α-D-ribo-furanoside	101.1	55.1	56.0	78.5	61.5	55.3		[717]
Methyl 2,3-anhydro-β-D-ribo-furanoside	101.7	54.8	55.4	78.6	61.7	54.8		[717]
Methyl α-D-sorbopyranoside	61.2	100.9	72.0	74.5	70.1	63.0	49.2	[694]
Methyl α-D-sorbofuranoside	60.7	104.2	80.0	76.5	78.8	61.6	49.9	[694]
Methyl β-D-sorbofuranoside	57.7	109.9	80.3	77.2	83.4	62.1	49.4	[694]
Methyl α-D-xylopyranoside	100.6	72.3	74.3	70.4	62.0		56.0	[718]
Methyl β-D-xylopyranoside	105.1	74.0	76.9	70.4	66.3		58.3	[718]
Methyl α-D-xyxlofuranoside	103.0	77.8	76.2	79.3	61.6		56.7	[682]
Methyl β-D-xylofuranoside	109.7	81.0	76.0	83.6	62.2		56.4	[682]
Ethyl 4-azido-2,3,4-trideoxy-α-D-threo-hex-2-enopyranoside	93.8	130.2	123.5	52.6	68.0	64.8	64.0 ($-$O$-$CH$_2-$), 15.1 (CH$_3$)	[722]
Ethyl 4-acet-amido-2,3,4,6-tetra-deoxy-α-D-threo-hexopyrano-side	96.8	23.8	23.4	46.9	62.6	17.5	64.9 ($-$O$-$CH$_2-$) 15.0 (CH$_3$) 170.0 (CO); 24.4 (CH$_3$CO)	[722]
Ethyl 4,6-diazido-2,3,4,6-tetra-deoxy-α-D-threo-hex-2-eno-pyranoside	93.8	131.1	121.8	52.7	69.4	51.8	64.1 ($-$O$-$CH$_2-$), 15.1 (CH$_3$)	[722, 723]
Ethyl 4,6-di-amino-2,3,4,6-tetra-deoxy-α-D-threo-hexopyra-noside	95.6	20.7	20.3	47.3	63.6	42.9	64.3 ($-$O$-$CH$_2-$), 14.3 (CH$_3$)	[722]
Ethyl 4,6-diazido-2,3,4,6-tetra-deoxy-α-D-erythrohex-2-eno-pyranoside	98.5	131.2	120.8	54.7	67.6	54.3	64.3 ($-$O$-$CH$_2-$), 14.9 (CH$_3$)	[722, 723]
Ethyl 4,6-diamino-2,3,4,6-tetra-deoxy-α-D-erythrohexo-pyranoside	96.6	20.8	20.4	45.5	61.4	43.2	64.6 ($-$O$-$CH$_2-$), 14.4 (CH$_3$)	[722]

Table 5.17. (continued).

Compound	C-1	C-2	C-3	C-4	C-5	C-6	OCH$_3$, etc.	Ref.
Methyl α-D-erythrofuranoside	103.6	72.8	69.9	73.6			56.7	[682]
Methyl β-D-erythrofuranoside	109.6	76.4	71.4	72.6			56.6	[682]
Methyl α-D-threofuranoside	109.4	80.5	76.4	73.7			55.5	[682]
Methyl β-D-threofuranoside	103.8	77.4	75.8	72.0			56.2	[682]
D-Ribulose (open chain)	66.8	212.9	76.2	73.0	63.2			[694]
L-Sorbose (open chain)	66.8	213.4	76.6	72.7	72.1	63.6		[694]
D,L-Glyceral-dehyde(hydrate)	89.9	74.0	62.0					[724]
D-Glyceraldehyde 3-phosphate (hydrate)	91.3	74.9	66.0					[721]
Dihydroxyacetone	211.9	64.8						[724]
Glycolaldehyde (hydrate)	89.9	64.5						[724]
D-Glycolaldehyde phosphate (hydrate)	90.7	68.2						[721]
Formaldehyde (hydrate)	83.3							[725]

5.4 Carbohydrates 393

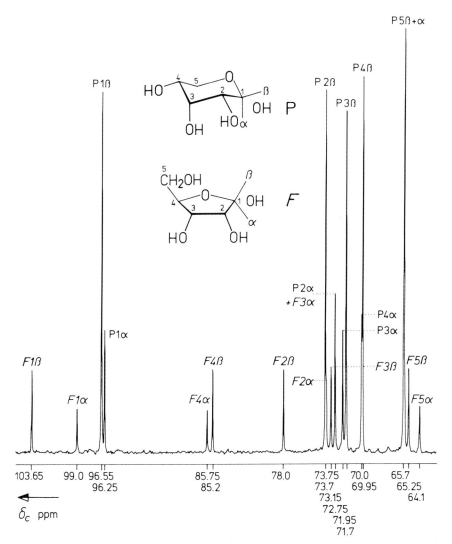

Fig. 5.6. PFT ^{13}C{^{1}H} NMR spectra of D-ribose, 22.63 MHz, 1 g/2 mL D$_2$O, temperature 30 °C, accumulation of 2000 pulse interferograms (6 K data points), 90 pulses, pulse interval 6 s, 2500 Hz, the numbers of the signals refer to the numbering of the C atoms. A quantitative evaluation of the spectrum gave 62% of β-ribopyranose (P β), 20.3% of α-ribopyranose (P α), 11.6% of β-ribofuranose (F β) and 6.1% of α-ribofuranose (F α) [132b].

Fig. 5.6 illustrates that D-ribose equilibrates in α- and β-pyranoses and furanoses when dissolved in water [132b]. High-field ^{13}C NMR also detected 0.6% α- and 0.3% β-D-mannofuranose in aqueous D-mannose solution [132c].

Hydroxylated furanose ring carbons resonate at lower field than the corresponding pyranose carbons, as is demonstrated by the PFT ^{13}C NMR spectrum of ribose in Fig. 5.6 [677].

An anomeric carbon with an axial hydroxy group usually resonates at higher field than the corresponding C-1 with an equatorial hydroxy group (rule (5)). This rule allows one to prove, experimentally, that arabinose exists in the $^1C_4$ and not in the $^4C_1$ conformation, in which most sugars occur in solution [88, 726]. However, rule (5) is not valid for the mannose, rhamnose and talose anomers. This was attributed to the presence of three closely spaced dipoles (Reeves effect) present in the β-anomeric configuration of these three sugars [683, 684] (Fig. 5.7).

Comparing the shifts of the methoxy resonances of anomeric pairs of methyl glycosides is a more reliable method for establishing the pyranose conformations. The methyl carbons of axial methoxy groups in methyl glycosides are found at 1.5 to 2 ppm higher field than those of equatorially oriented anomeric methoxy groups [726]. The ^{13}C resonances of a series of trans fused methyl 4,6-O-benzylidene-D-aldohexopyranosides (Table 5.17) were assigned by the general chemical shift rules discussed above [699]. The carbon of this series resonate from higher to lower field in the following order: $OCH_3 >$ C-5 $>$ C-6 $>$ C-2 $\approx$ C-3 $>$ C-4 $>$ C-1.

An almost direct relationship between ^{13}C chemical shifts of glucose and xylose carbons and their electron densities is found for C-1, C-3 and C-5 (Fig. 5.8); the shifts of C-2 and C-4, however, deviate substantially. Trans-annular interactions might be responsible for this "excess" shielding of C-2 and C-4 [88].

In reference [88] a relationship between the free energies of sugars and the sums of their chemical shifts is reported. For instance glucose and xylose have, compared to idose or allose, low free energy values, which correspond to low sums of chemical shifts and low values in total shielding for these molecules. Thus for the most stable sugars the lowest values in total shielding are calculated and *vice versa*.

Another discovery [88] merits mention here: Comparing ^{13}C and ^1H chemical shift differences of pairs of anomeric carbons or protons show that ^{13}C and ^1H shifts are affected conversely, *i.e.* a low field shift of ^{13}C is accompanied by a high field shift of the corresponding proton.

5.4.2 Di- and Polysaccharides

The ^{13}C NMR properties of oligo- and polysaccharides have been reviewed in several articles [691, 727–730]. Signal assignments of di- and polysaccharides were made following rules (1)–(11) discussed at the beginning of Section 5.4.

^{13}C chemical shifts and long range $^{13}C-^1H$ and $^{13}C-^{13}C$ couplings are suitable for conformational studies in di- and oligosaccharides [728]. Determination of $^3J_{^{13}C-^1H}$ couplings, however, yields more exact results for the determination of the conformation of the glycosidic linkage [731–734]. A relationship between the torsional angle and the size of coupling allows one to determine the glycosidic torsional angles [735–739].

$^{13}C-O-C-^1H$ couplings have been determined in methyl β-cellobioside after deuterium substitution of all protons on carbon atoms with a free hydroxy group. Assuming a Karplus type relationship between $^3J_{C-H}$ and the dihedral angle in ^{13}C-O-C-^1H groups, the torsional angles Φ and Ψ of methyl β-cellobioside were determined to be 25–30° [734].

Fig. 5.7. Configuration of α-, and β-C-1,2 of D-talose for the demonstration of the Reeves effect.

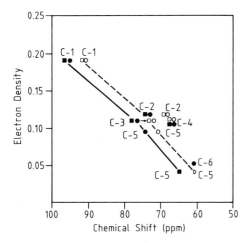

Fig. 5.8. Plot of ^{13}C chemical shift values of xylose and glucose carbons *versus* their calculated electron densities [88].
α-D-glucose ○,
α-D-xylose ◊,
β-D-glucose ●,
β-D-xylose ◆.

As discussed in Section 3.3.3.4 NT_1 value determination can be used for sugar sequence determination in oligosaccharides: With increasing distance from the aglycone strophantidine the corresponding NT_1 values of the carbon nuclei of the pyranose residues of k-strophantoside increase [172]. The concentration dependence of T_1 values can be used for the detection of intramolecular hydrogen bonds, as was demonstrated for saccharose [166].

In Table 5.18 the ^{13}C chemical shifts of selected disaccharides are given [708, 728]. Using smaller units of amylose as models led to the successful interpretation of its ^{13}C NMR spectrum at pD 14 [740, 741]; assigning C-2: 73.8, C-3: 75.4 and C-5: 72.6 ppm. No changes are observed for the chemical shifts of C-1 and C-4 of maltose and its oligomers; however, for amylose the signals shift 0.4 and 0.5 ppm, respectively, to higher field, an effect which is thought to be caused by a definite polysaccharide conformation [742]. Low solubilities and excessive linewidths have so far prevented the recording of interpretable ^{13}C NMR spectra of amylopectin and glycogen [708].

Table 5.18. Structures and ^{13}C Chemical Shifts (δ_C in ppm) of Selected Disaccharides.

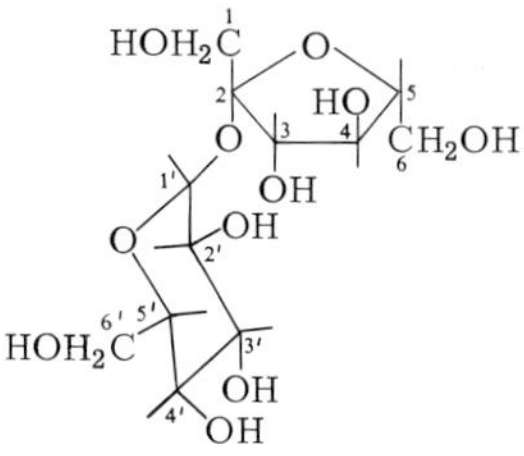

Sucrose (α-D-gluco-pyranosyl-β-D-fructofuranoside)

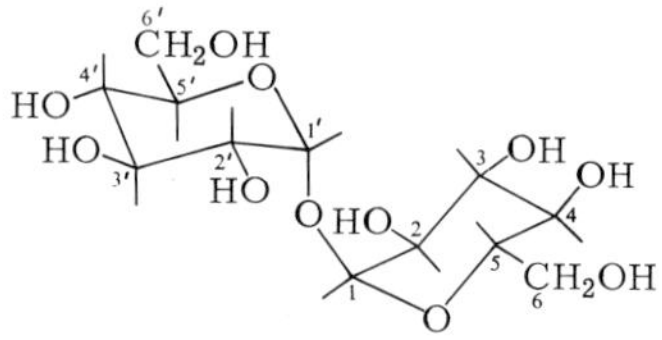

Trehalose (α-D-gluco-pyranosyl-α-D-gluco-pyranoside)

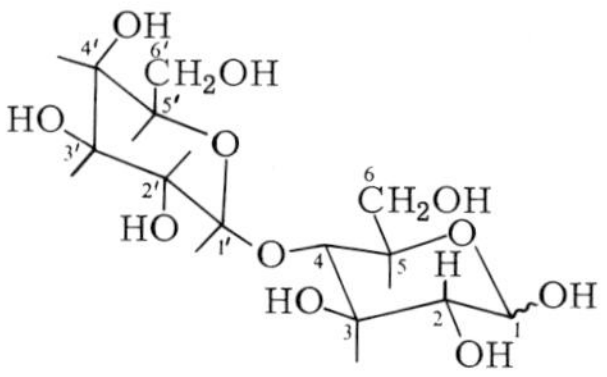

Lactose (4-O-β-D-galactopyranosyl-D-glucose)

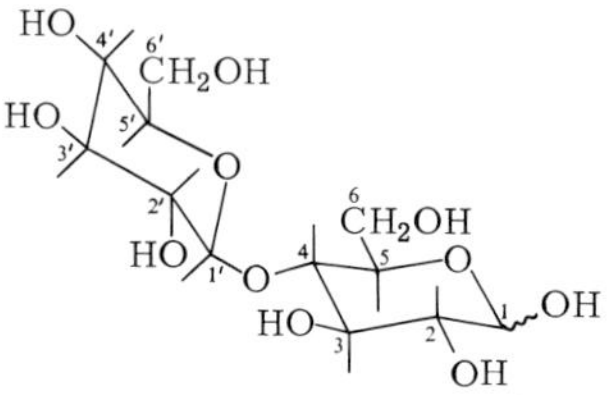

Cellobiose (4-O-β-D-glucopyranosyl-D-glucose)

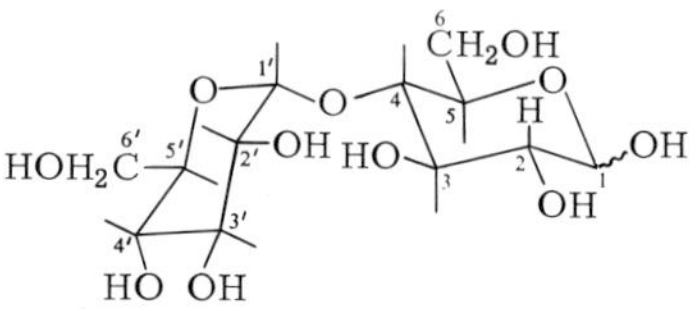

Maltose (4-O-α-D-gluco-pyranosyl-D-glucose)

Compound	C-1	C-2	C-3	C-4	C-5	C-6	Ref.
Sucrose	92.9	72.0	73.6	70.2	73.3	61.1	[728]
	63.3	104.4	77.4	75.0	82.2	63.4	
α,α-Trehalose	94.0	72.0	73.5	70.6	73.0	61.5	[728]
β,β-Trehalose	100.7	74.2	77.3	71.1	77.3	62.5	[708]
α,β-Trehalose	100.9	72.4	73.8	70.4	73.6	61.6	[728]
	103.6	74.1	76.4	70.4	77.0	62.0	
α-Lactose	103.6	72.0	73.5	69.5	76.2	62.0	[728]
	92.7	72.2	72.4	79.3	71.0	61.0	
β-Lactose	103.7	72.0	73.5	69.5	76.2	62.0	[728]
	96.6	74.8	75.3	79.2	75.6	61.1	
α-Cellobiose	103.3	74.1	76.5	70.4	76.8	61.6	[728]
	92.7	72.3	72.3	79.6	71.0	61.1	
β-Cellobiose	103.3	74.1	76.5	70.4	76.8	61.6	[728]
	96.6	74.9	75.3	79.5	75.6	61.1	
α-Maltose	101.1	73.2	74.3	70.8	74.0	62.0	[728]
	93.2	72.7	74.5	78.9	71.4	62.0	
β-Maltose	101.1	73.1	74.3	70.8	74.0	62.0	[728]
	97.2	75.4	77.5	78.6	76.0	62.2	

Comparison with the spectra of maltose, maltotriose, isomaltose, panose, amylose and methyl 6-O-methyl-α-D-glucopyranoside allows one to identify most of the signals in the ^{13}C NMR spectrum of dextran [740]. Single-unit side chains of branched-chain dextrans show larger T_1 values than the carbons in the main chains [743]. The T_1 values of C-3 and C-1 in the main chain of dextran B-742 fraction S ((1 → 6)-linked α-D-glucopyranosyl residues, substituted at 0–3 by α-D-glucopyranosyl groups) are 0.11 and 0.15 s, respectively; for C-1 of the side chains a T_1 value of 0.25 s is observed. The signals of the ^{13}C NMR spectrum of cellulose, dissolved in DMSO (80%)/4-methylmorpholine N-oxide (20%) and measured at 100 °C are assigned as follows: C-1: 102.5 ppm, C-4: 79.3 ppm and C-6: 60.5 ppm. The resonances at 73.1, 74.7 and 75.3 ppm, however, could not be individually identified as C-2, C-3 and C-5 [744].

The ^{13}C NMR spectrum of a β-D-(1 → 4)-linked D-mannopyranan, isolated from ivory nuts, is interpreted on the basis of spectral comparison with β-D-mannose: C-1: 101.7, C-2: 72.2, C-3: 73.8, C-4 and C-5: 78.7, and C-6: 62.1 ppm [727].

5.4.3 Polyols

Several ^{13}C NMR studies on unsubstituted polyols [678, 711, 745–747], O-methylated galactitols [705] and ribitols substituted with isoalloxazine moieties [147, 748] have been reported in the literature. The signals of the unsubstituted polyols were assigned mainly by comparison with the spectra of tetritols, pentitols and hexitols of different configuration [678]. Specifically methylated polyols, however, allow a more significant signal assignment [705]. 3-O-Methylation of galactitol causes a well-known downfield shift for the resonance of C-3 (about 9 ppm). The signals of the neighboring carbons C-2 and C-4 are shifted in the same direction, but only by 0.6 to 0.9 ppm. The lines of the γ-carbons are shifted upfield by 0.3 to 0.5 ppm relative to C-3, C-1 and C-5, and the resonance of C-6 is not influenced by the 3-O-methylation. The same rules can be used for the signal assignments of 3,4-di-O-methyl-D-galactitol. From the comparison of the galactitol spectra with those of methylated derivatives the signal assignments for the parent compound from higher to lower field are as follows: C-1,6, C-2,5 and C-3,4.

The ^{13}C chemical shifts of a series of polyols are listed in Table 5.19.

5.4.4 Aldonic Acids

So far only a table of the ^{13}C chemical shifts of aldonic acid salts and aldonolactones has been published in the literature (Table 5.20, [696]). The carbons of the carboxylate ion groups of all D-aldonic acid salts resonate at 180 ± 0.7 ppm. Upon γ-lactone formation an upfield shift for C-1 and a downfield shift for the γ-carbon is observed throughout.

Table 5.19. ^{13}C Chemical Shifts (δ_C in ppm) of Polyols.

Compound	–CH$_3$	–CH$_2$	CH$_2$OH	2–CHOH	3-CHOH	4-CHOH	5-CHOH	OCH$_3$	Ref.
Glycol			67.30						[678]
1,2-Propanediol	22.95		71.60	72.70					[678]
1,3-Butanediol	26.90	44.80	63.20		69.30				[678]
1,4-Butanediol		31.70	65.50						[678]
Glycerol			66.90	76.40					[678]
Erythritol			66.20	75.30	75.30				[678]
Pentaerythritol			64.30						[678]
Arabinitol			64.4 (C-1) 64.3 (C-5)	71.6	71.9	72.3	64.3		[696]
Ribitol			63.8 (C-1,5)	73.5	73.6	73.5			[696]
Xylitol			63.9 (C-1,5)	73.2	72.0	73.2			[696]
Allitol			63.7 (C-1,6)	73.5	73.7	73.7	73.5		[696]
Altritol			64.4 (C-1) 63.4 (C-6)	71.8	72.2	73.0	74.0		[696]
Galactitol			64.5 (C-1,6)	71.5	70.7	70.7	71.5		[696]
3-O-Methyl-D-galactitol			62.9 (C-1) 63.4 (C-6)	70.2	79.4	70.8	68.8	60.2	[705]
3,4-Di-O-methyl-D-galactitol			62.9 (C-1,6)	70.7	78.8	78.8	70.6	60.3	[705]

Table 5.19. (continued).

Compound	–CH₃	–CH₂	CH₂OH	2–CHOH	3-CHOH	4-CHOH	5-CHOH	OCH₃	Ref.
Glucitol			63.8 (C-1) 64.2 (C-6)	74.3	71.0	72.6	72.5		[696]
Iditol			64.1 (C-1,6)	73.1	72.5	72.5	73.1		[696]
Mannitol			64.6 (C-1,6)	72.2	70.7	70.7	72.2		[696]

Table 5.20. ^{13}C Chemical Shifts (δ_C in ppm) of D-Aldonic Acid Salts (pH 14) and Aldonolactones [696].

Compound	C-1	C-2	C-3	C-4	C-5	C-6
D-Allonic acid	179.5	74.7	74.6	73.6	72.7	63.4
D-Altronic acid	180.5	74.0	73.9	72.7	72.6	63.3
D-Galactonic acid	180.6	72.4	72.4	71.1	70.7	64.3
D-Gluconic acid	179.9	75.2	72.4	73.8	72.0	63.6
D-Gulonic acid	180.4	75.6	73.9	73.6	71.7	63.6
D-Idonic acid	179.5	73.5	73.2	72.5	71.9	63.9
D-Mannonic acid	180.3	75.4	72.2	72.2	71.6	64.0
D-Talonic acid	180.7	75.3	74.0	72.6	72.0	64.4
D-Arabinonic acid	180.5	73.4	72.6	72.3	64.1	
D-Lyxonic acid	180.0	75.1	72.9	72.3	64.0	
D-Ribonic acid	179.9	75.0	74.6	73.0	63.9	
D-Xylonic acid	180.4	74.2	74.0	73.6	63.6	
D-Erythronic acid	179.6	74.4	74.2	62.8		
D-Threonic acid	179.4	73.6	73.1	63.1		
D-Allono-1,4-lactone	178.7	70.9	69.5	86.9	69.0	62.8
D-Altrono-1,4-lactone	176.8	74.8	73.3	81.3	71.2	62.3
D-Galactono-1,4-lactone	176.7	74.5	73.7	80.9	69.8	62.9
D-Glucono-1,4-lactone	177.9	73.4	73.8	80.5	71.2	63.2
D-Gulono-1,4-lactone	178.8	71.7	71.0	82.2	70.4	62.4
D-Idono-1,4-lactone	176.5	74.0	71.9	79.9	68.8	62.9
D-Mannono-1,4-lactone	178.8	71.5	70.2	79.3	68.5	63.4
D-Talono-1,4-lactone	179.3	71.2	71.0	86.9	69.4	62.7
D-Arabinono-1,4-lactone	176.9	74.6	73.2	82.0	60.1	
D-Lyxono-1,4-lactone	179.0	71.3	70.3	82.4	60.6	
D-Ribono-1,4-lactone	179.3	70.3	69.8	87.5	61.4	
D-Xylono-1,4-lactone	177.9	73.9	72.9	81.2	59.9	
D-Erythrono-1,4-lactone	179.3	73.7	70.5	69.7		
D-Threono-1,4-lactone	178.0	74.0	73.1	70.4		

5.4.5 Inositols

The ^{13}C resonances of this class of compounds, which are summarized in Table 5.21, were assigned using the rules discussed in Section 5.4.

As the six carbons of *scyllo*-inositol are equivalent, only one signal is observed in the ^{13}C NMR spectrum of this compound [679]. In the spectra of *myo*-, *epi*- and 1,3-di-O-methyl-*myo*-inositol two pairs of carbons always resonate at different δ values for symmetry reasons; these signals can be easily identified by their double intensity [679].

The spectra of D-1-O-methyl-*myo*-inositol and L-2-O-methyl-*chiro*-inositol show that the resonances of β-carbons bearing axial hydroxy groups shift 4.5 ppm to higher field upon methylation of $C_\alpha - OH$ [679].

Table 5.21. ^{13}C Chemical Shifts (δ_C in ppm) of Inositols and O-Methylated Derivatives [679].

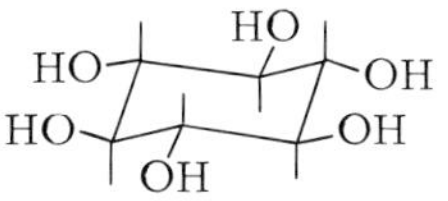

scyllo-Inositol

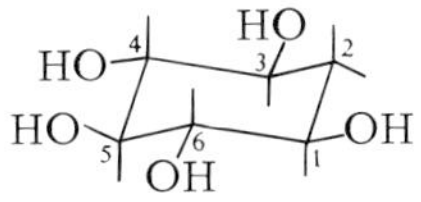

myo-Inositol

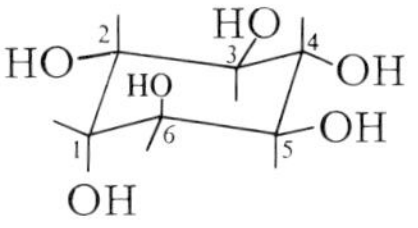

L-chiro-Inositol

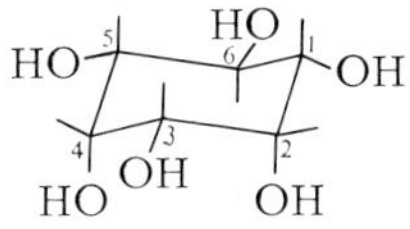

epi-Inositol

Compound	C-1	C-2	C-3	C-4	C-5	C-6	CH$_3$
scyllo-Inositol	73.70	73.70	73.70	73.70	73.70	73.70	
myo-Inositol	72.40	72.20	72.40	71.10	74.30	71.10	
L-chiro-Inositol	71.60	70.50	72.80	72.80	70.50	71.60	
epi-Inositol	71.70	74.50	70.10	74.50	71.70	66.80	
D-1-O-Methyl-myo-inositol	80.50	68.00	72.30	71.10	74.40	71.60	56.90
1,3-Di-O-methyl-myo-inositol	80.40	63.30	80.40	71.40	74.40	71.40	57.40
1,4-Di-O-methyl-myo-inositol	80.30	67.80	71.70	82.20	73.70	70.50	59.70 (4) 56.70 (1)
1,2-Di-O-methyl-myo-inositol	81.00	78.10	72.60	71.40	74.40	71.80	61.50 (2) 57.40 (1)
L-2-O-Methyl-chiro-inositol	67.20	80.10	71.90	72.80	70.40	71.30	56.80
D-3-O-Methyl-chiro-inositol	71.70 71.40	69.80	82.50	72.10	70.60	71.70 71.40	59.40

5.5 Nucleosides and Nucleotides

Owing to the great biological importance of this class of compounds, the ^{13}C NMR spectra of a considerable number of nucleosides and nucleotides and their aglycones have been reported in the literature [147, 428, 460, 676, 735, 748–760]. As demonstrated by the spectra of flavin adenine dinucleotide (Fig. 5.9), the ^{13}C NMR spectra of nucleosides and nucleotides have two groups of signals:

(1) The pyranose and furanose ring carbon resonances are observed between 40 and 100 ppm.

(2) The resonances of the heterocyclic carbon atoms are found between 90 and 170 ppm.

402   5 ^{13}C NMR Spectra of Natural Products

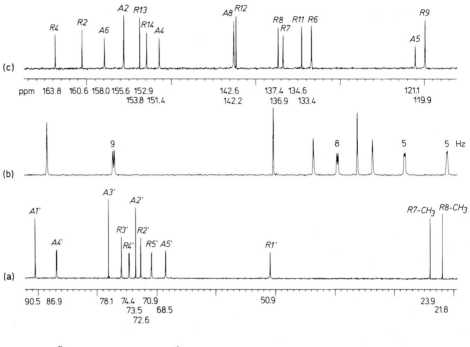

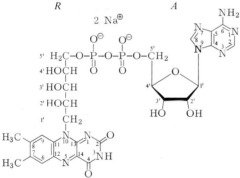

Fig. 5.9. ^{13}C{^{1}H} NMR spectrum of flavin adenine dinucleotide (FAD), disodium salt, 30 mg in 1 mL of deuterium oxide; 30 °C; 100.576/400.133 MHz (^{13}C/^{1}H); 5000 interferograms (32 K/15 000 Hz); (a) sp^3 carbon partial spectrum (20–92 ppm) with expanded output (b) to display carbon-phosphorus couplings ($^2J_{CP}$ for A5' and R5', $^3J_{CP}$ for A4' and R4' carbon nuclei); (c) partial spectrum of heterocyclic carbon nuclei (117–167 ppm). Signals are assigned according to ref. [147].

5.5.1 Assignment of the Purine Resonances

The assignments of the purine resonances [460, 749, 751, 758, 759, 761] of proton broadband-decoupled ^{13}C NMR spectra of nucleosides and nucleotides are made using the following aids [749, 761]: Correlation with the signals of the parent bases and analogous nucleosides or nucleotides; comparison of the ^{13}C NMR spectrum with that of a specifically analogue; correlations of the chemical shift values with the π-electron densities [749, 761] and, in addition, proton off-resonance decoupling, which affords the identification of quaternary carbon atoms.

5.5 Nucleosides and Nucleotides

Some of the common purine nucleosides are symbolized by the following formula:

Nebularine	($R^2 = R^6 = H$)
Adenine	($R^2 = H$; $R^6 = NH_2$)
Inosine	($R^2 = H$; $R^6 = OH$)
Guanosine	($R^2 = NH_2$; $R^6 = OH$)

Converting purine to the nucleoside nebularine causes considerable shifts of resonances in the spectrum of the parent base (cf. Table 5.22).

In contrast, substitution of 9-H of adenine by the ribofuranose moiety causes only very small shifts of the signals of the base. Proton off-resonance decoupling allows the differentiation between the signals of the quaternary carbon atoms C-4, C-5, and C-6 and the tertiary carbons C-2, and C-8. The signal at highest field is assigned to C-5 as this carbon atom is assumed to have the highest π-electron density. The resonance of C-8 is easily identified by comparison with the spectrum of 8-deuterio-adenosine. The second doublet signal in the off-resonance decoupled spectrum is thus attributed to C-2. The resonances due to C-4 and C-6 are assigned by comparison with the spectral data of analogous nucleosides.

C-8 of guanosine is the only tertiary carbon atom and may be easily detected and identified by off-resonance decoupling. Comparison of the spectral data of guanosine with those of 6-thioguanosine leads to the assignments of the resonances of C-6 and C-5: Both signals are shifted downfield by 10 to 20 ppm in the spectrum of the thio compound, measured in dimethyl sulfoxide (DMSO). However, oxidation of 6-thioguanosine by this solvent during the measurement cannot be excluded.

Comparison of the spectrum of inosine with that of guanosine leads to the assignments of the signals of C-2 and C-4: both resonances are shifted upfield in inosine by 5.4 ppm because of the lack of an electron-withdrawing amino group at C-2.

The signal of the quaternary carbon atom C-4 of inosine can be easily distinguished from that of C-2 by proton off-resonance decoupling. Comparison of the spectrum of inosine with the spectral data of 8-deuterioinosine and 6-thioinosine leads to the unequivocal signal assignment of the base residue of this nucleoside [749] (see Table 5.22).

^{13}C NMR spectroscopy is also suitable for determining the binding sites of metal ions in nucleosides and nucleotides: Thus, one concluded from ^{13}C NMR studies on guanosine in the presence of metal ions that hard metal ions bind to O-6 and soft metal ions to N-1 after its deprotonation [754].

Furthermore, ^{13}C NMR data allow one to determine the self-association of nucleotides, as was reported for disodium guanosine 5'-monophosphate (5'-GMP): With increasing concentration of 5'-GMP all purine carbon signals shift upfield, which is interpreted in terms of stack formation [753, 762].

Table 5.22. ^{13}C Chemical Shifts (δ_C in ppm) of Purines, Pyrimidines, Nucleosides and Nucleotides.

Compound	C-2	C-4	C-5	C-6	C-8	C-1'	C-2'	C-3'	C-4'	C-5'		Solvent	Ref.
Uracil	152.7	165.2	100.9	143.0								DMSO	[760]
1-Methyluracil	151.2	163.9	100.5	146.4							35.1 (CH$_3$)	DMSO	[760]
3-Methyluracil	151.7	163.4	99.6	140.3							26.4 (CH$_3$)	DMSO	[760]
1,3-Dimethyluracil	151.5	162.8	99.6	144.6							27.1 (CH$_3$, N-3) 36.3 (CH$_3$, N-1)	DMSO	[760]
2-Thiouracil	175.9	160.9	105.2	142.0								DMSO	[760]
1-Methyl-2-thio-uracil	176.4	160.4	105.9	146.6							42.0 (CH$_3$)	DMSO	[760]
3-Methyl-2-thio-uracil	176.7	160.4	103.8	140.5							32.9 (CH$_3$)	DMSO	[760]
4-Thiouracil	148.6	191.3	111.6	138.3								DMSO	[760]
2,4-Dithiouracil	172.7	187.6	117.0	136.4								DMSO	[760]
Uridine	152.40	164.70	103.05	142.20		89.10	71.10	74.85	85.90	62.20		DMSO	[760]
4-Thiouridine	149.10	191.05	113.65	136.90		89.60	70.55	74.90	86.05	61.70		DMSO	[749]
2,4-Dithiouridine	173.75	186.95	118.65	135.50		94.30	69.60	75.60	85.65	60.55		DMSO	[749]
5-Hydroxyuridine	150.60	161.70	121.05	133.55		88.55	71.30	74.15	85.85	62.35		DMSO	[749]
Uridine-5'-mono-phosphate	152.90	167.30	103.70	143.20		89.60	71.20	75.10	85.10	64.60		DMSO	[749]
2,5'-Anhydro-2,3'-isopropylidene-uridine	157.50	171.10	109.65	143.80		97.60	85.15	84.65	82.00	75.10		DMSO	[749]
Thymidine	151.55	164.85	110.50	137.30		85.05	40.40	71.65	88.40	62.35	13.20 (CH$_3$)	DMSO	[749]
Thymidine-5'-mo-nophosphate	151.50	166.20	111.40	137.40		84.80	38.50	71.20	85.90	64.00	11.70 (CH$_3$)	H$_2$O	[750]
4-Thiothymidine	148.70	191.50	118.80	134.15		85.80	39.90	71.15	87.80	62.12	17.85 (CH$_3$)	DMSO	[749]

Table 5.22. (continued).

Compound	C-2	C-4	C-5	C-6	C-8	C-1'	C-2'	C-3'	C-4'	C-5'		Solvent	Ref.
1-(2,3-Dideoxy-β-D-glycero)-pent-2-enofuranosyl-thymine	152.10	165.20	110.30	137.95		88.70	136.05	127.15	90.15	63.65	13.35	DMSO	[749]
Cytidine	156.90	166.65	95.65	142.80		90.10	70.55	75.10	85.25	61.85		DMSO	[749]
Deoxycytidine	157.00	166.95	95.85	142.45		86.55	40.60	71.75	88.55	62.60		DMSO	[749]
2-Thiocytidine	180.85	161.05	98.95	142.65		94.40	69.35	76.00	85.10	60.75		DMSO	[749]
6-Methylcytidine	157.30	166.15	96.50	155.35		92.90	71.10	71.85	86.15	63.25	20.85 (CH$_3$)	DMSO	[749]
Cytidine-5'-monophosphate	157.50	166.20	96.50	141.80		89.10	69.70	74.30	83.30	63.30		H$_2$O	[750]
Deoxycytidine-5'-monophosphate	157.40	166.00	96.50	141.80		85.60	39.30	70.90	85.90	63.60		H$_2$O	[750]
Purine	151.60	154.40	128.10	144.40	147.50							DMSO	[749] [603]
6-Chloropurine	151.35	147.80	129.05	154.20	146.10							DMSO-d$_6$	[603]
6-Bromopurine	151.50	146.65	131.85	152.90	146.00							DMSO-d$_6$	[603]
6-Iodopurine	151.80	136.50	120.30	149.95	145.10							DMSO-d$_6$	[603]
6-Methoxypurine	151.50	155.15	118.25	159.35	142.75						53.85 (OCH$_3$)	DMSO-d$_6$	[603]
6-Dimethylaminopurine	151.80	151.15	118.90	154.30	137.80						37.85 (CH$_3$)	DMSO-d$_6$	[603]
6-Methylmercaptopurine	151.60	150.20	129.35	158.60	143.15						11.35 (CH$_3$)	DMSO-d$_6$	[603]
Hypoxanthine	150.20	155.50	118.00	158.90	144.80							DMSO-d$_6$	[603]
Adenine	152.45	151.35	117.60	155.35	139.40							DMSO-d$_6$	[603]
6-Cyanopurine	152.25	133.45	127.85	154.95	149.30						114.45 (CN)	DMSO-d$_6$	[603]
2-Aminopurine	160.60	155.10	125.50	147.70	141.60							DMSO-d$_6$	[758]

Table 5.22. (continued).

Compound	C-2	C-4	C-5	C-6	C-8	C-1'	C-2'	C-3'	C-4'	C-5'		Solvent	Ref.
2-Fluoropurine	158.30	158.20	128.80	147.25	148.00							DMSO-d$_6$	[758]
2-Chloropurine	152.70	157.70	129.10	146.90	147.80							DMSO-d$_6$	[758]
2,6-Dichloropurine	151.00	156.20	128.50	148.10	147.40							DMSO-d$_6$	[758]
2-Methyl-6-amino-purine	160.70	151.80	115.80	154.90	138.60						25.30 (CH$_3$)	DMSO-d$_6$	[758]
2-Amino-6-methyl-purine	160.10	154.30	124.40	157.25	140.00						19.00 (CH$_3$)	DMSO-d$_6$	[758]
2-Methylthio-6-aminopurine	163.90	152.20	115.40	154.40	138.50						13.60 (CH$_3$)	DMSO-d$_6$	[758]
2-Amino-6-methyl-thiopurine	159.65	151.65	124.00	159.20	138.70						10.80 (CH$_3$)	DMSO-d$_6$	[758]
2-Fluoro-6-amino-purine	158.80	153.40	115.50	156.80	140.10							DMSO-d$_6$	[758]
2-Chloro-6-amino-purine	152.80	152.80	116.20	155.90	140.20							DMSO-d$_6$	[758]
6-Methylpurine	151.60	153.85	145.10	155.45	144.45						19.40 (CH$_3$)	DMSO-d$_6$	[603]
7-Methylpurine	152.00	159.90	125.80	140.70	149.80						31.60 (CH$_3$)	DMSO	[460]
9-Methylpurine	151.90	151.40	133.50	147.40	147.40						29.35 (CH$_3$)	DMSO	[460]
9-Methyladenine	152.50	149.90	118.70	156.00	141.50						29.40 (CH$_3$)	DMSO	[460]
7-Methylhypo-xanthine	144.40	157.00	115.50	154.60	144.40						33.30 (CH$_3$)	DMSO	[460]
7-Methylpurine-6-thione	144.70	152.70	125.90	170.40	148.30						34.70 (CH$_3$)	DMSO	[460]
Xanthine	151.30	148.80	106.70	155.50	140.40							DMSO-d$_6$	[759]
Uric acid	150.10	136.60	97.10	153.40	152.20							DMSO-d$_6$	[759]

5.5 Nucleosides and Nucleotides   407

Table 5.22. (continued).

Compound	C-2	C-4	C-5	C-6	C-8	C-1'	C-2'	C-3'	C-4'	C-5'	Solvent	Ref.
Nebularine	151.90	152.95	135.10	148.95	146.35	89.05	71.40	75.00	86.80	62.25	DMSO-d$_6$	[749]
6-Chloro-9-(β-D-ribofuranosyl)-purine	150.25	152.40	132.15	152.40	146.35	89.35	71.00	75.05	86.45	61.70	DMSO	[749]
Inosine	149.20	146.95	125.30	157.65	139.90	88.75	71.20	75.15	86.60	62.25	DMSO	[749]
Deoxyinosine	148.50	146.25	125.20	157.35	139.30	84.35	40.20	71.35	88.70	62.20	DMSO	[749]
6-Thioinosine	146.10	144.70	136.30	176.70	141.90	88.45	70.85	75.10	86.40	61.95	DMSO	[749]
Inosine-5'-monophosphate	146.70	148.20	123.10	158.30	139.80	87.60	70.60	74.80	84.20	63.50	H$_2$O	[750]
Adenosine	147.40	145.65	118.55	150.50	142.40	88.45	74.25	70.25	85.55	61.30	0.1 mol/L HCl	[677]
2'-Deoxyadenosine	147.50	145.20	118.25	150.10	142.30	84.80	39.15	70.90	87.50	61.40	0.1 mol/L HCl	[677]
9-β-D-Ribopyranosyladenine	148.40	144.90	118.80	150.20	143.15	81.15	68.60	66.05	70.80	64.95	0.1 mol/L HCl	[677]
9-α-D-Arabinopyranosyladenine	148.45	145.00	18.45	150.20	142.85	84.15	72.80	69.80 (69.50)	69.50 (69.80)	68.50	0.1 mol/L HCl	[676]
9-β-D-Arabinopyranosyladenine	147.70	144.70	117.95	150.10	143.30	80.40	69.80	65.60	69.40	63.35	0.1 mol/L HCl	[676]
9-α-D-Arabinofuranosyladenine	148.00	145.00	118.90	150.30	142.85	89.20	79.75	74.85	85.25	60.95	0.1 mol/L HCl	[676]
9-α-D-Xylopyranosyladenine	147.80	145.10	117.80	150.30	143.50	80.70	69.25 (69.60)	67.95 (67.55)	69.60 (69.25)	67.55 (67.95)	0.1 mol/L HCl	[676]
9-β-D-xylopyranosyladenine	148.35	144.90	118.70	150.20	143.05	84.25	71.65	76.40	68.85	68.10	0.1 mol/L HCl	[676]
9-β-D-Xylofuranosyladenine	149.20	147.60	118.35	153.10	141.45	89.65	75.20	80.15	82.95	60.00	0.1 mol/L HCl	[676]
Adenosine-5'-monophosphate	152.40	148.40	118.00	155.00	139.90	87.00	70.50	74.50	84.20	63.50	H$_2$O	[750]
Adenosine-5'-triphosphate	152.30	148.30	118.00	154.80	139.60	86.90	70.10	74.20	83.80	65.20	H$_2$O	[750]

Table 5.22. (continued).

Compound	C-2	C-4	C-5	C-6	C-8	C-1'	C-2'	C-3'	C-4'	C-5'	Solvent	Ref.
2'-Deoxyadenosine-5'-monophosphate	152.20	148.10	118.10	155.00	139.80	83.50	39.10	71.40	86.10	63.90	H$_2$O	[750]
6-N-Methylamino-9-(β-D-ribofuranosyl)-purine	153.60	149.30	120.90	156.25	140.80	89.40	71.95	74.90	87.10	62.95	DMSO	[749]
2-Amino-9-(β-D-ribofuranosyl)-purine	160.55	154.85	124.75	150.70	142.55	88.40	71.35	74.80	86.50	62.25	DMSO	[749]
2,6-Diamino-9-(β-D-ribofuranosyl)-purine	161.25	152.55	83.30	157.60	114.95	89.00	72.30	75.00	87.35	63.35	DMSO	[749]
Guanosine	154.55	152.30	117.50	157.75	136.85	87.25	71.50	74.85	86.35	62.15	DMSO	[749]
6-Thioguanosine	154.30	148.90	129.40	176.10	139.45	87.90	71.40	75.95	86.40	62.40	DMSO	[749]
Deoxyguanosine	154.70	152.00	117.65	158.15	136.65	83.85	40.40	71.80	88.65	62.40	DMSO	[749]
Guanosine-5'-monophosphate	153.60	151.00	115.70	158.30	137.30	87.00	70.60	74.30	84.10	63.90	H$_2$O	[750]
Deoxyguanosine-5'-monophosphate	153.80	150.90	115.70	158.50	137.10	83.20	39.10	71.30	86.10	64.00	H$_2$O	[750]
2-Chloroadenosine	154.20	151.30	119.30	157.55	141.20	88.90	71.45	74.80	86.75	62.50	DMSO	[749]
Xanthosine	158.90	152.35	117.25	164.45	137.45	89.85	71.50	74.80	87.25	62.15	DMSO	[749]
6-Thioxanthosine	158.40	149.00	127.25	182.95	137.45	89.90	71.65	75.05	87.00	62.15	DMSO	[749]

5.5.2 Assignment of the Pyrimidine Resonances

Some of the common pyrimidine nucleosides are symbolized by the following formula:

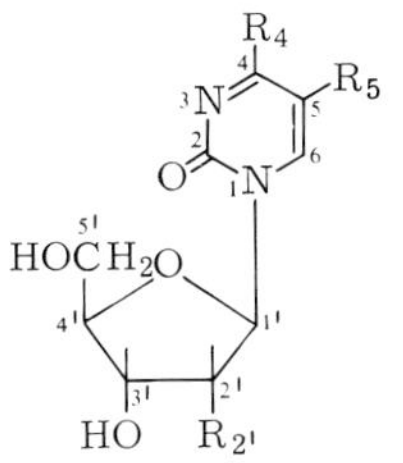

Uridine (R^4, R^2 = OH; R^5 = H)
Cytidine (R^4 = NH_2; R^5 = H; R^2 = OH)
Thymidine (R^4 = OH; R^5 = CH_3; R^2 = H)

Methods similar to those mentioned above have been used for the signal identification of the pyrimidine resonances [428, 735, 749, 757, 760, 765] (cf. Table 5.22).

Comparison of the spectrum of uridine with data of 4-thiouridine, 2,4-dithiouridine, thymidine, 4-thiothymidine, cytidine and 2-thiocytidine allows the assignment of all carbon resonances of these pyrimidine moieties. Further confirmation of these assignments is obtained from the proton off-resonance decoupled ^{13}C-spectra of these nucleosides, as has been demonstrated for thymidine: The pyrimidine CH_3 and CH resonances of the proton broadband-decoupled spectrum are split into a quartet and a doublet, respectively, in the proton off-resonance spectrum, and can thus be easily assigned.

5.5.3 Assignment of the Isoalloxazine Resonances

Due to the biological importance of the isoalloxazine moiety in enzyme cofactors, ^{13}C NMR studies of this heterocyclic compound have been reported [147, 748] (Fig. 5.9). The resonances of the methyl groups attached to R 7 and R 8 of the isoalloxazine moiety of riboflavin-5'-monophosphate [147] (cf. Fig. 5.9 for FAD) occur at highest field, having shifts corresponding to those of methyl carbons in methyl benzenes and showing quartets in the proton off-resonance decoupled spectra. By proton off-resonance decoupling, the signals of R 6 and R 9 are localized as doublets at higher field than any other isoalloxazine ring carbon. Since R 6 of this ring corresponds to C-8 of acridine (130 ppm) [456], R 6 is less shielded (133.4 ppm) than R 9 (119.9 ppm). The so-far unassigned signals are due to quaternary carbons, as they are singlets in the proton off-resonance decoupled ^{13}C spectrum. Of these, the carbonyl carbons R 2 and R 4 occur at lowest field. By comparison with the ^{13}C shifts of xanthosine [749], the signal of R 4 was found to be more deshielded than that of R 2 [748].

5.5.4 Assignment of the Sugar and Polyol Carbon Atoms

Most naturally occurring nucleosides and nucleotides contain a ribofuranose moiety. From detailed ^{13}C NMR studies of carbohydrates [88, 89, 696] it is known that primary CH_2OH groups resonate at higher field than the secondary ring CHOH groups, and thus

C-5' of the ribofuranose residue can easily be assigned. Another rule is found for the anomeric carbons C-1' of the ribofuranose residue which always absorb at lowest field [696]. C-4' can be assigned by noting the phosphorus-carbon long-range coupling in the proton broadband-decoupled ^{13}C NMR spectra of nucleotides [750] as demonstrated in Fig. 5.9.

The ribitol residue of the riboflavin moiety was assigned on the following basis (the same numbering of the C-atom as in Fig. 5.9 is used here):

As the ribitol residue is asymmetrical in riboflavin, all ribitol carbons show individual signals in contrast to the spectrum of the symmetrical parent compound [696]. The CH$_2$ carbons R 1' and R 5' are identified as triplets in the proton off-resonance decoupled ^{13}C spectrum. The signal of R 1' occurs at highest field of the ribitol resonances since R 1 is attached to the nitrogen of the isoalloxazine ring, and not, as R 5', to the stronger electron-withdrawing oxygen of the hydroxy group. Electron withdrawal at R 2' is weaker than at R 3' and R 4'. Therefore, the signal of R 2' is expected at higher field than the signals of R 3' and R 4'. R 4' can be differentiated from R 3' by phosphorus carbon long-range coupling in the proton broadband-decoupled ^{13}C NMR spectrum [147] (Fig. 5.9).

Several nucleosides containing other sugar residues, such as arabinopyranose, arabinofuranose, mannopyranose, talopyranose, xylopyranose or xylofuranose, have been investigated by ^{13}C NMR [676, 684]. The assignments of the signals of the carbohydrate moieties were mainly performed by comparison with the ^{13}C resonances of the parent sugars.

A further aid in assigning vicinal *cis* hydroxy groups in carbohydrate residues of nucleosides and nucleotides makes use of boric acid as a complex shift reagent at various pH values [686]. In the presence of boric acid and at pH > 7 all complexing $\geq$C–OH-carbons are shifted downfield by about 6 ppm, as is demonstrated for the nucleoside adenosine.

5.5.5 Correlations of ^{13}C Chemical Shifts with Other Physicochemical Parameters

From the plots of the π-charge densities of the carbon atoms of the purine and pyrimidine residues of nucleosides *versus* chemical shift values a gross correlation is obvious. Different calculations of the π-densities, however, yield different plots. The π-electron densities calculated by Hoffmann and Ladik [763] by the HMO method or those calculated by Veillard and Pullman [764] by the Pariser-Parr SCF method show the strongest deviation for C-4, most probably because amplified C–N bond polarization terms have to be used for the calculations.

Figure 5.10 shows the relation [749] between the ^{13}C chemical shifts and the π-electron density calculated by Fernández-Alonso [765] using the Pauling-Wheland method for the heterocyclic moieties of the purine and pyrimidine nucleosides. Only the resonances of the ^{13}C–H atoms obey the Spiesecke-Schneider relation [76] (160 ppm downfield shift

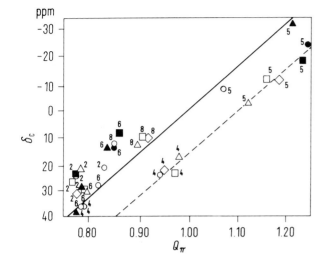

Fig. 5.10. Plot of ^{13}C chemical shifts *versus* π-charge densities [749], calculated by Fernández-Alonso [765]. (Adenosine ○, guanosine □, inosine △, xanthosine ◇, uridine ●, cytidine ▲, thymidine ■.)

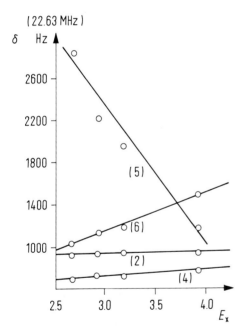

Fig. 5.11. ^{13}C Chemical shift values *versus* substituent electronegativity, E_x, for 5-fluoro-, 5-chloro-, 5-bromo- and 5-iodouracil and their different carbon atoms [756].

per π-electron). The points corresponding to quaternary carbon atoms are located close to a parallel (broken) line.

The different behavior of tertiary and quaternary carbon atoms seems to be due to either the complete neglect of overlap in these calculations or to polarization effects of the carbon-nitrogen bonds. Similar results are obtained for a series of 5-halouracils by plotting the ^{13}C NMR chemical shifs *versus* π-electron charge densities calculated by the extended Hückel theory [756]. Though for several nitrogen heterocycles a better correlation was found between ^{13}C chemical shift values and total $(\sigma + \pi)$ electron charge densities [465], the correlation did not improve in the case of the 5-halouracils. For the 5-halouracils, a good correlation exists between the ^{13}C chemical shifts and the substituent electronegativity, E_x [766], as can be seen from Fig. 5.11.

The largest substituent effect was observed for C-5, which is directly attached to the different substituents. Similar results were obtained in an investigation of 6-substituted purines [603].

5.5.6 Nucleic Acids

In the past most of the studies on structure determination of nucleic acids were performed by proton NMR [513, 767, 768]. However, recent ^{13}C NMR investigations have demonstrated that additional information can be obtained from the ^{13}C spectra [769].

The ^{13}C NMR signals in the spectrum of single-stranded homopolymeric polyuridylic acid (poly U) were assigned by spectral comparison with uridine, uridine-2'-phosphate, uridine-3'-phosphate and uridine-5'-phosphate (Table 5.23) [770]. Compared to uridine, the resonances of C-3' and C-5' of the polymer are shifted downfield by about 2.5 ppm due to phosphorylation. Stronger hydrogen bonds in the monomeric units are thought to be the reason for the upfield shifts of the carbonyl signals of the polymer compared to the monomers (Table 5.23) [771, 772]. As large $^3J_{^{31}P-^{13}C}$ couplings are expected for *trans* rotamers and small ones for *gauche* rotamers, uridine-5'-phosphate, uridine-2'-phosphate and uridine-3'-phosphate favor the 4'-*trans*, 1'-*trans* and 4'-*trans* rotamers, respectively. Correspondingly, it was concluded from the three-bond ^{13}C–^{31}P spin-spin coupling constants of polyuridylic acid that the 5'-phosphate and the 3'-phosphate have preferences for the 4'-*trans* and 2'-*trans* rotamers, respectively (Table 5.23, [770]). The transition of polyuridylic acid from random coil at room temperature to an ordered (helix) structure at lower temperature [773–776] can also be followed by ^{13}C NMR spectroscopy: The resonances are broadened due to intramolecular ^{1}H–^{13}C dipolar interactions and restricted motions in the ordered state, and the lines of C-5' and C-4' show large and those of C-6 and C-2 small upfield shifts [777]. In a temperature-dependent investigation of ^{13}C shifts and spin-lattice relaxation times of the homopolymeric polynucleotide polyadenylic acid the conclusion was drawn that at neutral pH and with decreasing temperature the polymer forms a more ordered structure [778, 779]. The ^{13}C NMR spectral comparison of poly(adenosine diphosphate ribose) with ribose, ribose 5-phosphate, adenosine 5'-monophosphate, adenosine diphosphate ribose and ribosyl adenosine 5',5"2-bis(phosphate) provides evidence for an α-(1" → 2')ribofuranosyl ribofuranoside moiety in the polymer [780]. In Table 5.24 the ^{13}C chemical shifts of randomly coiled poly(8-bromoadenylic acid) and polyadenylic acid are compared with the corresponding 5'-mononucleotides. From the C-3' upfield shift of poly(8-bromoadenylic acid) compared to that in polyadenylic acid a preferred C(3')*endo-syn* conformation is assumed for the brominated polymer [781].

Because of their relatively low molecular weight (70 to 90 nucleotide residues), transfer ribonucleic acids are of special interest for ^{13}C NMR investigations [769, 778, 782–784] of nucleic acids. Using a tube of 20 mm o.d., a sample of thermally denatured yeast

Table 5.23. ^{13}C Chemical Shifts (δ_C in ppm) and ^{13}P-^{13}C Coupling Constants (in Hz, in Parentheses) of Polyuridylic Acid and Related Monomers [770]; Solvent: D$_2$O, pD 7.8±0.6, Temperature: 37 °C; (d) = doublet(s), q = quartet).

C Atom	Uridine	Uridine-5'-phosphate	Uridine-3'-phosphate	Uridine-2'-phosphate	Polyuridylic acid
2	152.80	153.20	152.90	153.20	152.80
4	167.30	167.60	167.40	167.70	166.90
5	103.40	103.70	103.50	103.70	103.90
6	143.00	143.20	143.00	144.00	142.50
1'	90.50	89.60	90.30	90.30 ($^3J_{^{31}POC_2, ^{13}C_1}$ = 9 Hz (d))	89.60
2'	74.80	75.00	74.65 ($^3J_{^{31}POC_3, ^{13}C_2}$ = 2.5 Hz (d))	76.50 ($^2J_{^{31}PO^{13}C}$ = 4.5 Hz (d))	75.30 ($^3J_{^{31}POC_3, ^{13}C_2}$ = 5 Hz (d))
3'	70.55	71.10	73.10 ($^2J_{^{31}PO^{13}C}$ = 4.5 Hz (d))	71.00 ($^3J_{^{31}POC_2, ^{13}C_3}$ = 3 Hz (d))	73.95 ($^2J_{^{31}PO^{13}C}$ = 4.5 Hz (d))
4'	85.40	85.10 ($^3J_{^{31}POC_5, ^{13}C_4}$ = 8.5 Hz (d))	84.90 ($^3J_{^{31}POC_3, ^{13}C_4}$ = 6 Hz (d))	85.30	83.70 ($^3J_{^{31}POC_5, ^{13}C_4}$ = 7 Hz (q)); $^3J_{^{31}POC_3, ^{13}C_4}$ = 3 Hz (q))
5'	61.90	64.40 ($^2J_{^{31}PO^{13}C}$ = 4.5 Hz (d))	61.90	62.30	66.20 ($^2J_{^{31}PO^{13}C}$ = 4.5 Hz (d))

Table 5.24. ^{13}C Chemical Shifts (δ_C in ppm) of Randomly Coiled Poly(8-bromoadenylic Acid), Polyadenylic Acid and Corresponding 5'-Mononucleotides; Temperature 70 °C; Solvent: D$_2$O, pD 7.0–7.3 [781].

C Atom	8-Bromoadenylic Acid		Adenylic Acid	
	Polymer	Monomer	Polymer	Monomer
1'	92.40	92.30	90.40	90.10
2°	75.90	73.70	76.50	77.20
3'	73.80	72.90	76.20	73.40
4'	84.00	86.80	84.90	87.25
5°	68.40	66.90	67.70	66.50
2	155.40	155.90	155.30	155.80
4	152.40	153.40	151.10	152.00
5	121.60	122.20	121.20	121.60
6	156.50	157.30	157.80	158.50
8	130.00	130.95	141.90	143.10

transfer ribonucleic acid (tRNA) shows numerous resonances, which can be identified by comparison with the ^{13}C spectra of mononucleosides and mononucleotides. Also, the signals of C-4, -5 and -6 of dihydrouridine and C-5 of pseudouridine and/or ribosylthymine residues are observed. Primarily the C-4' resonance of the ribose moieties of denaturated tRNA (82 °C) is shifted appreciably ($\approx$ 1.5 ppm) upfield, compared to those in the mononucleotides. A further upfield shift of the C-4 resonance by about 1.5 ppm is observed in the spectrum of folded tRNA (recorded at 52 °C). Below 60 °C a dramatic line-broadening of the signals occurs, which is ascribed to tRNA aggregation. Up to 60 °C the ^{13}C spin-lattice relaxation times (T_1) of the ribose carbon atoms show no dependence on temperature. However, the T_1 values increase rapidly if recorded between 60 and 82 °C, indicating that the backbone of thermally denatured tRNA undergoes rapid segmental motions at elevated temperature.

5.6 Amino Acids

Correlations between the structures of amino acids and peptides and their ^{13}C chemical shifts are of general interest for bio-, enzyme- and peptide chemists, since these compounds occur in all cells of living organisms.

5.6.1 ^{13}C Chemical Shifts of Amino Acids

The results of several investigations of amino acids by ^{13}C NMR can be summarized as follows [97, 785–793] (Table 5.25):

(1) The ^{13}C signals of carboxy groups resonate between 168 and 183 ppm.

(2) The ^{13}C resonances of α-carbon atoms absorb in the range of 40 to 65 ppm.

(3) The chemical shift values of the β-carbons are spread across the range of 17 to 70 ppm.

(4) The signals of the γ- and δ-carbons are found between 17 and 50 ppm.

(5) The signals of the aromatic and heteroaromatic ring carbons resonate between 110 and 140 ppm.

For the synthesis of peptides, amino acid derivatives with protected amino or carboxy groups are used as starting materials. The application of ^{13}C NMR spectroscopy for the control of the synthesis of those protected amino acids has ben reported in the literature [791, 792, 794].

The carbon atoms of some important protecting groups resonate in the following ranges [791, 792, 794] (Table 5.25): The carbonyl carbon of the t-butyloxycarbonyl groups is found at 150 to 160, the quaternary carbon atom between 77 and 82 and the methyl group at 28 to 30 ppm. The aromatic carbons of benzyloxycarbonyl and benzyl groups

5.6 Amino Acids   415

Table 5.25. ^{13}C Chemical Shifts (δ_C in ppm) of Amino Acids and Derivatives. (For Abbreviations see [800]).

Compound	Amino acid moiety							Protecting groups							Solvent	Ref.
	COOH or COOR	$CONH_2$	C_α	C_β	C_γ	C_δ	$C_{arom.}$ or C_ϵ	C=O	Z or O-Bzl		t-Boc or OBu^t		Ac-, N-, COO–CH_3 OC_2H_5			
									CH_2	C_6H_5	C	CH_3				
H-Gly-OH	173.5		42.5												D_2O	[97]
H-Gly-NH_2·HCl		174.9	50.8												D_2O	[791]
CH_3-Gly-OH	170.0		50.2										78.45		D_2O	[791]
Z-Gly-OH	171.0		42.6					156.0	65.8	127.2 127.95 136.65					DMSO-d_6	[791]
H-Gly-OC_2H_5	167.65		39.8											14.25 61.80	DMSO-d_6	[794]
t-Boc-Gly-OH	172.1		41.9					156.25			78.05	27.95			DMSO-d_6	[791]
H-L-Ala-OH	176.8		51.6	17.3											D_2O	[97]
Z-DL-Ala-OH	174.0		49.55	17.55				155.15	65.5	127.2 127.75 136.55					DMSO-d_6	[791]
t-Boc-L-Ala-OH	175.1		48.95	17.3				155.9			78.1	28.05			DMSO-d_6	[791]
H-β-Ala-OH	183.4		35.7	38.45											D_2O	[791]
H-L-Val-OH	175.3		61.6	30.2	17.9										D_2O	[97]
H-L-Val-NH_2·HBr		168.55	57.25	29.65	17.8 18.6										D_2O	[791]
Ac-L-Val-OMe			59.3	31.9	21.0 20.5										D_2O	[791]
															DMSO	[801]
t-Boc-L-Val-OH	172.4		58.95	29.85	18.2 19.25			151.5			78.0	28.25			DMSO-d_6	[791]
H-L-Leu-OH	176.6		54.7	41.0	25.4	22.1 23.2									D_2O	[97]
Ac-L-Leu-OMe			51.9	43.0	26.0	25.0 23.0									DMSO	[801]
t-Boc-L-Leu-OH	173.55		52.0	40.5	24.5	21.3 22.9		154.7			77.6	28.25			DMSO-d_6	[791]

Table 5.25. (continued).

Compound	Amino acid moiety							Protecting groups							Solvent	Ref.
	COOH or COOR	$CONH_2$	C_α	C_β	C_γ	C_δ	$C_{arom.}$ or C_ϵ	C=O	Z or O-Bzl		t-Boc or OBu^t		Ac-, N-, COO—CH_3 OC_2H_5			
									CH_2	C_6H_5	C	CH_3				
H-L-Ile-OH	175.2		60.9	39.7	CH_3 15.9 CH_2 25.7	12.5									D_2O	[97]
t-Boc-L-Ile-OH	173.9		58.3	40.55	CH_3 15.95 CH_2 24.8	11.75		154.95			78.0	28.55			DMSO-d_6	[791]
H-L-Ser-OH	173.1		57.4	61.3											D_2O	[97]
H-L-Thr-OH	174.0		61.5	67.1	20.5										D_2O	[97]
t-Boc-L-Thr (Bzl)-OH	171.4		58.2	70.3	16.7			155.15	74.7	126.8 127.55 138.05	78.3	28.35			DMSO-d_6	[791]
H-L-Met-OH	175.3		55.3	31.0	30.1	15.2									D_2O	[97]
H-L-CySH-OH · HCl	171.95		57.45	27.4											D_2O	[791]
Ac-L-CySH-OH	179.4		57.2	27.3									23.75		DMSO-d_6	[791]
(H-DL-Homocy-S-OH)$_2$	173.8		54.2	35.3	31.8			178.6							DMSO-d_6	[791]
H-L-CySH-OMe · HCl	173.75		57.2	26.5									56.1		DMSO-d_6	[791]
H-L-CySH-OMe · HCl	174.2		57.35	25.95									56.9		D_2O	[791]
H-L-CySO$_3$H-OH	175.1		51.7	52.5											D_2O	[791]
(H-L-CyS-OH)$_2$ · 2HCl	175.6		54.7	39.0											D_2O/DCl	[791]
(H-L-CyS-OMe)$_2$ · 2HCl	174.0		53.15	42.85									56.25		DMSO-d_6	[791]
(H-L-CyS-OMe)$_2$ · 2HCl	174.55		54.55	38.45									57.1		D_2O	[791]
(t-Boc-L-CyS-OH)$_2$	178.2		55.8	42.75				160.6			81.75	30.35			DMSO-d_6	[791]
(Ac-L-CyS-OH)$_2$	178.85		54.35	41.1				179.3					23.65		D_2O	[791]
H-L-Orn-OH · HCl	179.4		56.6	29.2	24.5	41.1									D_2O	[791]
Ac-L-Orn-OMe			55.7	32.2	26.2	42.2									DMSO	[801]

Table 5.25. (continued).

Compound	Amino acid moiety							Protecting groups						Solvent	Ref.
	COOH or COOR	CONH$_2$	C$_\alpha$	C$_\beta$	C$_\gamma$	C$_\delta$	C$_{arom.}$ or C$_\varepsilon$	C=O	Z or O-Bzl		t-Boc or OBu^t		Ac-, N-, COO–CH$_3$ OC$_2$H$_5$		
									CH$_2$	C$_6$H$_5$	C	CH$_3$			
H-L-Arg-OH	175.2	157.5	55.1	28.5	24.9	41.5								D$_2$O	[97]
t-Boc-L-Arg(NO$_2$)·OH	173.0	158.55	53.3	28.35	25.45	40.25		154.85			78.0	28.35		DMSO-d$_6$	[791]
H-L-Lys-OH	175.4		55.3	27.2	22.4	30.7								D$_2$O	[97]
H-Asp(OH)-OH	α 170.35 171.55		49.20	34.85										DMSO-d$_6$	[794]
Z-Asp(OH)-OH	α 171.95 173.05		50.80	36.35				156.20	65.80	127.95 128.70 137.25				DMSO-d$_6$	[794]
Z-Asp-O	α 170.10 172.40		50.70	35.00				156.10	66.55	128.15 128.70 136.55				DMSO-d$_6$	[794]
Z-Asp(OH)-OC$_2$H$_5$	α 171.65 171.25		50.80	36.15				156.00	65.70	127.75 128.50 137.00			14.05 60.95	DMSO-d$_6$	[794]
Z-Asp(OBu^t)-OH	α 170.75 174.00		53.05	39.60				155.45	65.15	127.60 128.35 137.45	79.30	27.85		DMSO-d$_6$	[794]
H-Asp(OBu^t)-OH	α 170.35 170.35		50.35	35.85							82.05	27.95		DMSO-d$_6$	[794]
t-Boc-Asp(OBu^t)-OH	α 169.50 172.85		50.25	37.55				155.25			78.30 80.25	27.70 28.30		DMSO-d$_6$	[794]
t-Boc-Asp(OH)-OH	α 172.50 173.60		50.60	36.70				155.80			78.85	28.70		DMSO-d$_6$	[794]

Table 5.25. (continued).

Compound	Amino acid moiety								Protecting groups							Solvent	Ref.
	COOH or COOR	CONH$_2$	C$_\alpha$	C$_\beta$	C$_\gamma$	C$_\delta$	C$_{arom.}$ or C$_\epsilon$		C=O	Z or O-Bzl		t-Boc or OBu^t		Ac-, N-, COO—CH$_3$ OC$_2$H$_5$			
										CH$_2$	C$_6$H$_5$	C	CH$_3$				
t-Boc-L-Asp(OBzl)-OH	α 168.85 171.3		49.85	36.05					154.1	65.25	126.8 127.0 127.45 135.35	77.8	28.05			DMSO-d$_6$	[791]
H-L-Asn-OH	174.2	174.45	52.65	36.05												D$_2$O/DCl	[791]
H-L-Glu-OH	175.6 182.3		55.7	28.1	34.5											D$_2$O	[97]
t-Boc-L-Glu(OBzl)-OH	α 171.0 172.5		52.65	26.45	30.15				154.4	65.05	126.8 127.0 127.45 135.35	77.6	28.05			DMSO-d$_6$	[791]
t-Boc-L-Glu(OBu^t)-OH	α 170.35 172.5		52.45	26.45	31.35				154.5			77.6 79.2	28.15 27.7	(t-Boc) (O-Bu^t)		DMSO-d$_6$	[791]
H-L-Gln-OH	179.5	174.45	55.2	28.3	33.0				179.85 179.2							D$_2$O/DCl	[791]
Ac-L-Gln-OH	175.55	179.85 179.2	54.6	29.15	33.7									24.4		DMSO-d$_6$	[791]
t-BOC-L-Gln-OH	172.7	172.7	53.05	26.85	31.45				154.4			77.6	28.15			DMSO-d$_6$	[791]
H-Pyr-OH	180.2		58.35	25.35	29.65		181.2									H$_2$O, pH 6.3	[792]
H-L-Pro-OH	174.6		61.6	29.7	24.4	46.5										D$_2$O	[97]
t-Boc-L-Pro-OH (cis)	176.3		58.5	30.5	23.3	46.4			154.8			80.1	27.9			DMSO-d$_6$	[791]
t-Boc-L-Pro-OH (trans)	177.0		58.7	29.4	24.0	46.1			153.8 153.05								
Formyl-L-Pro-OH (cis)	175.90		59.40	29.10	22.20	44.00			175.90							H$_2$O	[793]
Formyl-L-Pro-OH (trans)	175.20		56.70	29.10	23.40	47.00			175.20								
Ac-L-Pro-OH (cis)	173.95		59.55	30.95	22.55	45.95			168.45					22.25		DMSO-d$_6$	[792]
Ac-L-Pro-OH (trans)	173.60		58.25	29.25	24.50	47.35			168.75					22.25		DMSO-d$_6$	[792]

5.6 Amino Acids   419

Compound	Amino acid moiety						Protecting groups							Solvent	Ref.
	COOH or COOR	CONH$_2$	C$_\alpha$	C$_\beta$	C$_\gamma$	C$_\delta$	C$_{arom.}$ or C$_\epsilon$	C=O	Z or O-Bzl		t-Boc or OBu^t		Ac-, N-, COO—CH$_3$ OC$_2$H$_5$		
									CH$_2$	C$_6$H$_5$	C	CH$_3$			
H-L-Phe-OH	175.0		57.3	37.5			129.5 130.7 131.7							D$_2$O	[97]
Ac-L-Phe-OMe			55.4	38.7										DMSO	[801]
t-Boc-L-Phe-OH	172.7		55.2	36.9			125.85 127.55 128.6 137.5	154.7			78.0	28.35		DMSO-d$_6$	[791]
H-L-Tyr-OH	175.0		57.3	37.5			117.5 130.5 156.3							D$_2$O	[97]
t-Boc-L-Tyr(OBzl)-OH	172.4		55.1	35.65			113.65 126.6 126.8 127.45 129.25 136.3	154.3	68.9	113.65 126.6 126.8 127.45 129.25 136.3	77.6	27.95		DMSO-d$_6$	[791]
H-L-Try-OH	174.7		56.0	28.2			114.25 120.5 121.8 124.5 127.6 128.9 138.6							D$_2$O/DCl	[791]
t-Boc-L-Try-OH	172.6		54.25	26.85			109.35 110.55 117.5 119.95 135.25	154.1			77.6	28.05		DMSO-d$_6$	[791]
H-L-His-OH	173.7		54.7	27.85			136.05 (C-5) 131.50 (C-2) 116.95 (C-3)							H$_2$O pH 7.2	[792]

Table 5.26. ^{13}C Chemical Shifts (δ_C in ppm) of Penicillin Derivatives [795].

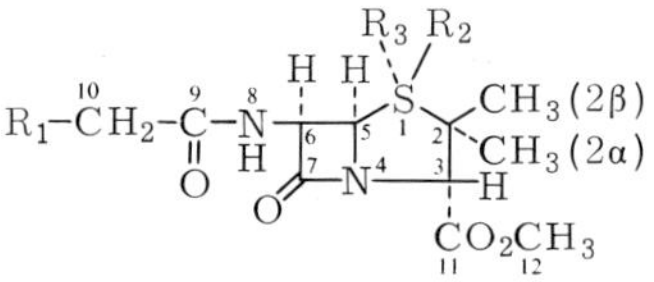

Penicillins

	$R^1 = H$ $R^2, R^3 = $ lone pair	$R^1 = H$ $R^2 = O$ $R^3 = $ lone pair	$R^1 = H$ $R^2 = $ lone pair $R^3 = O$	$R^1 = OC_6H_5$ $R^2 = $ lone pair $R^3 = $ lone pair	$R^1 = OC_6H_5$ $R^2 = O$ $R^3 = $ lone pair
C-2	63.6	74.1	68.1	64.0	74.9
C-3	69.7	65.1	63.9	70.0	65.8
C-5	67.1	75.5	78.7	67.4	76.0
C-6	58.3	55.5	57.0	58.4	54.8
2β-CH$_3$	30.0	18.7	23.4	31.0	19.0
2α-CH$_3$	26.1	17.6	15.3	26.6	18.0
C-12	51.7	52.2	52.2	52.1	52.8
C-10	20.8	22.1	21.6	65.6	67.0
C-7, C-9		168.1	167.3	167.6	167.3
C-9, C-7		169.2	169.5	167.6	167.7
C-11		173.8	170.2	172.6	173.8

absorb at 110 to 140, the CH_2-carbons at 60 to 75 and the carbonyl groups at lowest field (150 to 160 ppm). The signals of acetyl groups occur at 175 to 185 (CO) and at about 25 ppm (CH_3). The resonances of methoxy and methyleneoxy groups of methyl and ethyl ester are found at 55 to 59 and 59 to 63 ppm, respectively. Most of the shifts reported in the literature were measured using dimethyl sulfoxide, D_2O or DCl as solvents (Table 5.25).

Due to their relationship to amino acids, the ^{13}C NMR spectra of derivatives of penicillin [795] are discussed in this chapter. Proton off-resonance decoupling, comparison of the spectra of different analogues and applications of the equation derived by Ernst [32] were used as additional aids in signal assignment (Table 5.32).

Substitution of one of the lone pair electrons (R^2, R^3) by oxygen causes downfield shifts of the signals of C-2 and C-5 due to the expected decrease in electron density at these carbons. Similar downfield shifts are observed for the β-carbon atom upon converting cysteine to cystine and further to cysteic acid [790].

5.6.2 pH-Dependence of the ^{13}C Chemical Shift Values of Amino Acids

The first measurements of pH-dependences of ^{13}C spectra of amino acids were carried out by INDOR spectroscopy [96]. Pulse Fourier transform ^{13}C NMR measurements are less tedious than INDOR experiments and in a series of papers the pH-dependences of ^{13}C chemical shifts of amino acids were reported [84, 98].

The results may be summarized as follows: Deprotonation of the NH_3^+, SH or COOH functions usually causes downfield shifts of the signals of neighboring carbon atoms. Shifts to lower field may be larger for the signals of vicinal carbons than for α-carbons. Generally, the largest effects are observed on dissociation of the the NH_3^+ group. pK and pI values obtained by ^{13}C NMR measurements agree well with the results of electrometric or photometric studies. The pH-dependences of the chemical shifts of the nonequivalent methyl resonances of valine and penicillamine differ from each other. The low-field methyl resonance of penicillamine shifts downfield in two steps with increasing hydroxide concentration; the signal of the more shielded methyl carbon shifts to lower and finally to higher field with decreasing proton concentration.

The changes of chemical shifts upon deprotonation of the amino acids could be rationalized using shift calculations by CNDO/2 molecular orbital calculations [796]. Furthermore, the influence of pH on the coupling constants of amino acids was investigated, and populations of different rotamers were determined [797]. Also T_1 values of amino acids, such as glycine, D-,L-alanine and γ-aminobutyric acid, were measured as a function of pD and concentration, e.g. T_1 varying for the carboxy carbon in D,L-alanine from 19.0 sec at pD = 0.8 to 5.1 sec at pD = 6.2. The strong pD-dependence of T_1 of the carboxy carbon is interpreted in terms of intermolecular associations. The relaxation times of the residual carbon atoms, however, are relatively independent of changes in concentration and pD. For the relaxation mechanism of the carbonyl carbons spin rotation and for the other carbons of the amino acids dipole-dipole interactions are made responsible [788–798]. pH-dependent ^{13}C-chemical shift measurements and CNDO/2 MO calculations for phenylglycine and phenylalanine indicate that the electronic distribution of the phenyl group depends upon the distance of charged groups from the phenyl ring [799].

5.6.3 Prediction of Carbon Shifts and Their Correlation with Other Physicochemical Parameters

According to Grant's additivity rule [85], substitution of a hydrogen atom in alkanes by a methyl group causes downfield shifts of approximately 9 ppm for the α- and β-carbons and an upfield shift of approximately 2.5 ppm for the γ-carbons. If the hydrogen atom is replaced by an amino group the chemical shift parameters are 29.0 (C-α), 11.4 (C-β) and − 4.6 (C-γ); for COOH groups values of 21.5 (C-α), 2.0 (C-β) and − 2.5 (C-γ) are observed. On the basis of the additivity of these effects, the chemical shift values for the carbons of amino acids were predicted [96] and found to be in fairly good agreement with the observed ones.

A semi-empirical molecular orbital method for the correlation of charge distributions with ^{13}C shifts in amino acids was described [95]. Plotting of chemical shift parameters *versus* charge density changes of α-carbon atoms relative to the corresponding atoms in the parent hydrocarbons permits prediction of the chemical shifts of the α-carbons with an accuracy of about 10%. However, the slope (280 ppm per electron) in Fig. 5.12 is

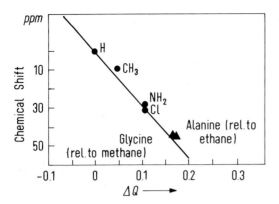

Fig. 5.12. Charge density change (ΔQ) of the α-carbons relative to the corresponding carbon atoms of the parent hydrocarbons *versus* chemical shifts [96].

larger than it should be according to the relationship of Spiesecke and Schneider (160 to 200 ppm per electron) [76]. Chemical shift parameters for β-carbons still cannot be explained theoretically, as most probably inductive, field, electron delocalization and steric effects are involved simultaneously.

According to the calculations mentioned above [95], a decrease of the positive charge distribution at the α-carbon and the α-hydrogen should be expected upon dissociation of the ammonium group. In fact, the dissociation leads to an upfield shift only for the α-hydrogen, and a downfield shift for the α-carbon. According to the calculations [95], the strongest charge density changes should occur at the α-carbon upon dissociation of the NH_3^+ group; experimental investigations show, however, that in many cases the chemical shift changes are larger for the *vicinal* carbons [84]. The model of Del Re, therefore, is not accurate enough to describe the dissociation processes in amino acids. Polarization of the C—H bond by induction through space could be a possible explanation for these dissociation effects [84].

5.7 Peptides

5.7.1 Oligopeptides

It is now well established that the ^{13}C chemical shifts of amino acid residues removed more than two units from the chain ends in unstructured polypeptides are not influenced, except by proline moieties, by the peptide sequence (Table 5.27, [96, 791, 792, 802–810]). Chemical shift changes of carbons of the same amino acid at different positions within a peptide or protein are therefore due to effects caused by secondary or tertiary structure.

A ^{13}C NMR study of the series glycine, glycyl-glycine, glycyl-glycyl-glycine demonstrates the dependence of the chemical shifts of the α-carbon atoms on the chain length (the Greek letters above glycine and the peptides are used for the characterization of the methylene resonances): In glycyl-glycine the signal of the α-methylene group appears

5.7 Peptides   423

Table 5.27. ^{13}C Chemical Shifts (δ_C in ppm) of Selected Oligopeptides [800, 813].

Compound	Formula with δ values	Solvent	Ref.
Z-Gly-DL-Ala-N$_2$H$_3$	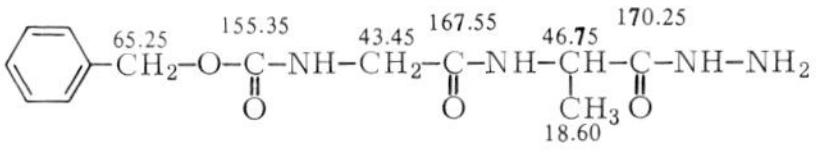 Ph–CH$_2$(65.25)–O–C(155.35)(=O)–NH–CH$_2$(43.45)–C(167.55)(=O)–NH–CH(46.75)(CH$_3$ 18.60)–C(170.25)–NH–NH$_2$	DMSO-d$_6$	[603]
Z-DL-Ala-Gly-N$_2$H$_3$	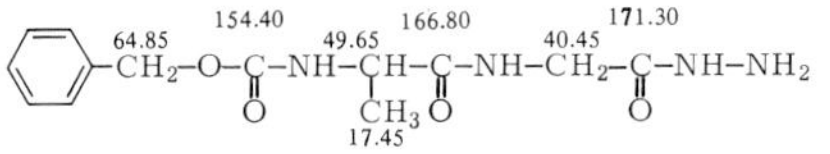 Ph–CH$_2$(64.85)–O–C(154.40)(=O)–NH–CH(49.65)(CH$_3$ 17.45)–C(166.80)(=O)–NH–CH$_2$(40.45)–C(171.30)–NH–NH$_2$	DMSO-d$_6$	[603]
Z-L-Asp(α-OEt)-Gly-OEt	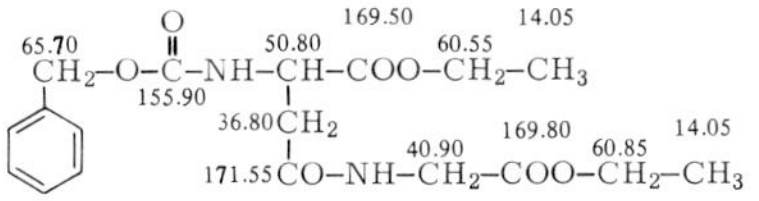 Ph–CH$_2$(65.70)–O–C(155.90)(=O)–NH–CH(50.80)–COO–CH$_2$(60.55)–CH$_3$(14.05); side: CH$_2$(36.80)–CO(171.55)–NH–CH$_2$(40.90)–COO–CH$_2$(60.85)–CH$_3$(14.05); C=O 169.50, 169.80	DMSO-d$_6$	[794]
Boc-L-Val^2-OH	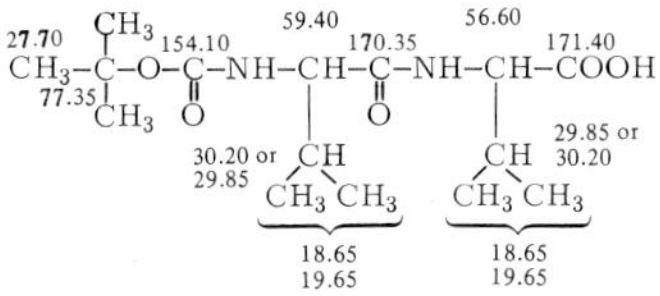 (CH$_3$)$_3$(27.70)C(77.35)–O–C(154.10)(=O)–NH–CH(59.40)–C(170.35)(=O)–NH–CH(56.60)–COOH(171.40); isopropyl CH 30.20 or 29.85, CH$_3$ 18.65, 19.65 (both Val)	DMSO-d$_6$	[791]
H-DL-Leu-DL-Leu-OH · HCl	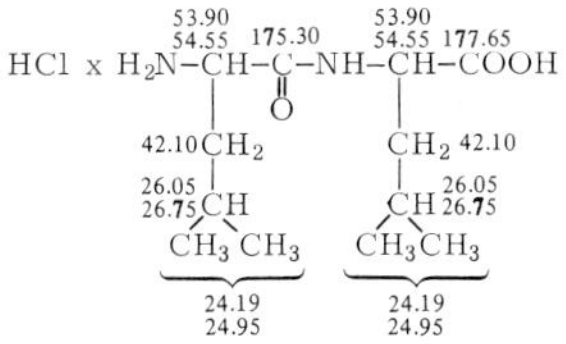 HCl × H$_2$N–CH(53.90/54.55)–C(175.30)(=O)–NH–CH(53.90/54.55)–COOH(177.65); CH$_2$ 42.10; CH 26.05/26.75; CH$_3$ 24.19, 24.95	D$_2$O	[791]
β-Ala-His (Carnosine)	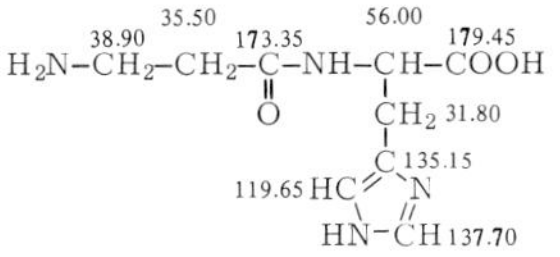 H$_2$N–CH$_2$(38.90)–CH$_2$(35.50)–C(173.35)(=O)–NH–CH(56.00)–COOH(179.45); CH$_2$ 31.80; imidazole: C 135.15, HC(119.65), HN–CH(137.70)	D$_2$O	[811]
Z-L-Cys(Bzl)-Gly-OEt	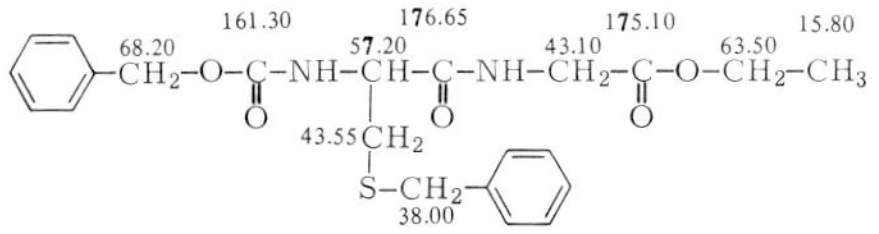 Ph–CH$_2$(68.20)–O–C(161.30)(=O)–NH–CH(57.20)–C(176.65)(=O)–NH–CH$_2$(43.10)–C(175.10)(=O)–O–CH$_2$(63.50)–CH$_3$(15.80); CH$_2$ 43.55; S–CH$_2$(38.00)–Ph	DMSO-d$_6$	[791]
Z-L-Asp(α-OH)-L-Cys(Bzl)-Gly-OH (Protected aspar-thione)	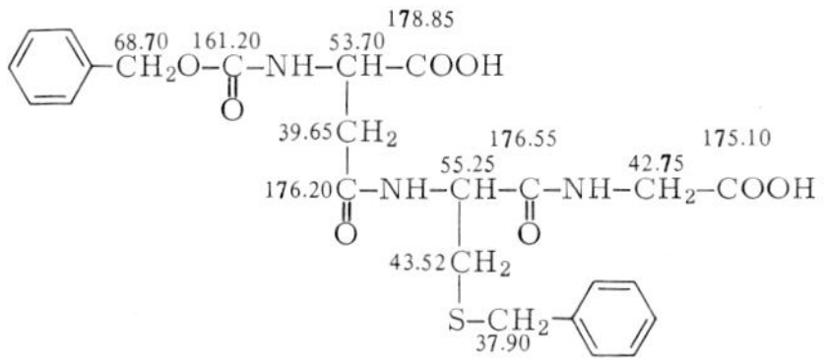 Ph–CH$_2$(68.70)–O–C(161.20)(=O)–NH–CH(53.70)–COOH(178.85); CH$_2$(39.65)–C(176.20)(=O)–NH–CH(55.25)–C(176.55)(=O)–NH–CH$_2$(42.75)–COOH(175.10); CH$_2$ 43.52, S–CH$_2$(37.90)–Ph	DMSO-d$_6$	[791]
L-Glu(α-OH)-L-Cys-Gly-OH (Glutathione)	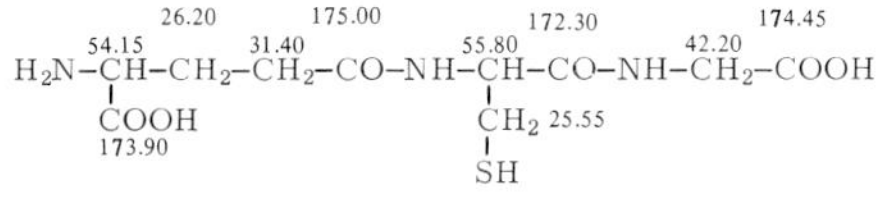 H$_2$N–CH(54.15)–CH$_2$(26.20)–CH$_2$(31.40)–CO(175.00)–NH–CH(55.80)–CO(172.30)–NH–CH$_2$(42.20)–COOH(174.45); COOH 173.90; CH$_2$ 25.55–SH	H$_2$O (pH 3.52)	[98]
L-Glu(α-OH)-L-Cys-Gly-OH (Glutathione oxidized)	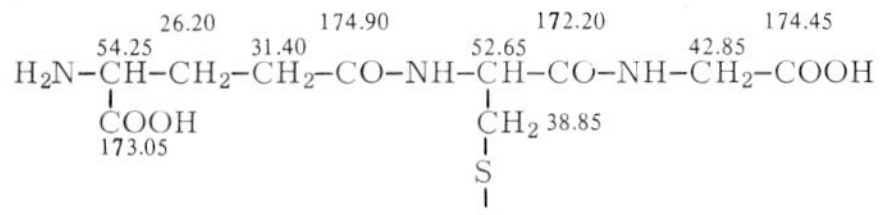 H$_2$N–CH(54.25)–CH$_2$(26.20)–CH$_2$(31.40)–CO(174.90)–NH–CH(52.65)–CO(172.20)–NH–CH$_2$(42.85)–COOH(174.45); COOH 173.05; CH$_2$ 38.85–S	H$_2$O (pH 3.52)	[98]

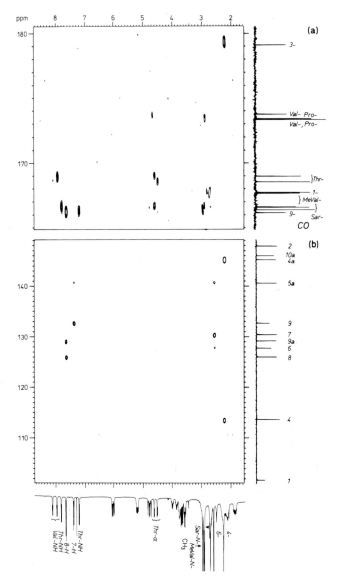

Fig. 5.13. State of the art carbon-13 signal assignment of actinomycin D [603] (40 mg in 0.4 mL of deuteriochloroform, 30 °C, 100.576 MHz for ^{13}C, 400.133 MHz for ^{1}H; two overnight experiments for one-bond and longer-range carbon-proton shift correlations).

(a) Carbonyl carbon nuclei with carbon-proton shift correlations *via* two- and three-bond couplings (example: 9-CO at 166.13 ppm is correlated with 8-H at 7.7 ppm).

(b) Heterocyclic carbon nuclei with carbon-proton shift correlations *via* two- and three-bond couplings (example: C-5a at 140.50 ppm is correlated with 6-CH$_3$ at 2.55 ppm and with 7-H at 7.36 ppm).

(c) sp^3 carbon nuclei of amino acid residues with carbon-proton shift correlations *via* one-bond couplings (example: C-α of sarcosine at 51.40–51.43 ppm is correlated with its attached *AB*-type methylene protons at 3.58 and 4.74 ppm).

5.7 Peptides

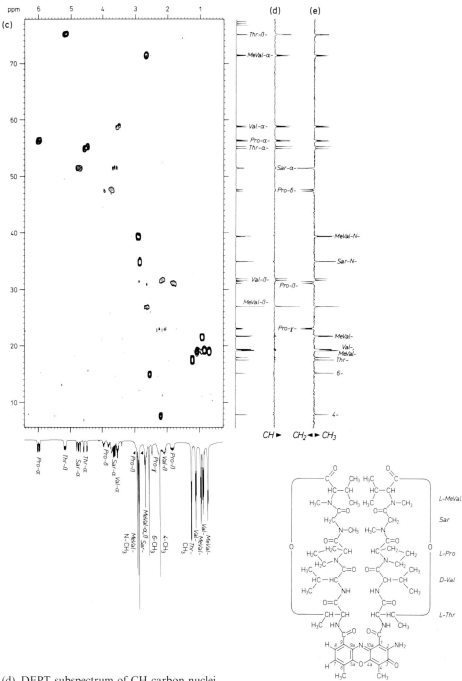

(d) DEPT subspectrum of CH carbon nuclei.
(e) DEPT spectrum of CH$_2$ carbon nuclei with negative amplitudes and CH$_3$ in addition to CH carbons (experiment (d)) with positive amplitudes.
Table 5.28 summarizes all signal assignments derived from experiments (a–e).

Table 5.28. Carbon-13 Shift Assignments of Actinomycin D, Derived from Fig. 5.13 and the Proton Shift Assignment (δ_C and δ_H in ppm) from ref. [814].

δ_C	CH_x[a]	δ_H	Assignment	δ_C	δ_H	COLOC with[b]	Assignment
7.79	CH_3	2.20	4-CH_3	101.90	–	–	C-1
15.05	CH_3	2.55	6-CH_3	113.54	–	4-CH_3	C-4
17.41	CH_3	1.24	Thr-CH_3	125.91	7.64	–	C-8
17.86	CH_3			127.65	–	6-CH_3	C-6
19.08	CH_3	0.74	MeVal-CH_3	129.11	–	8-H	C-9a
19.15	CH_3	1.10	Val-CH_3	130.33	7.36	6-CH_3	C-7
19.29	CH_3	0.86	Val-CH_3	132.60	–	7-H	C-9
19.34	CH_3			140.50	–	7-H, 6-CH_3	C-5a
21.60	CH_3	0.92	MeVal-CH_3	145.15	–	4-CH_3	C-4a
21.70	CH_3			145.92	–	–	C-10a
22.89	CH_2	2.10	Pro-γ	147.79	–	–	C-2
23.05	CH_2	2.22					
26.89	2 CH	2.65	MeVal-β				
31.01	CH_2	1.83		166.13	–	8-H, Thr-α, –NH	C-9-CO
31.32	CH_2	2.70	Pro-β	166.37	–	MeVal-α, –NCH_3	Sar-Co
		3.00		166.57	–	MeVal-α, –NCH_3	Sar-CO
31.56	CH	2.18	Val-β	166.59	–	MeVal-α,β, –CH_3	MeVal-CO
31.84	CH	2.25		167.64	–	MeVal-α,β	MeVal-CO
34.87	CH_3	2.87	Sar-N-CH_3	167.72	–		C-1-CO
34.92	CH_3			168.51	–	Thr-α, Val-NH	Thr-CO
39.24	CH_3	2.91	MeVal-N-CH_3	168.95	–	Thr-α, Val-NH	Thr-CO
39.35	CH_3			173.33	–	Sar-NCH_3	Pro-CO
47.36	CH_2	3.80	Pro-δ				Val-CO
47.64	CH_2	3.95		173.40	–	Sar-NCH_3	Pro-CO
51.40	CH_2	3.58	Sar-α	173.75	–	–	Val-CO
51.43	CH_2	4.74		179.12	–	4-CH_3	C-3
54.96	CH	4.63	Thr-α				
55.31	CH	4.47					
56.26	CH	5.99	Pro-α				
56.42	CH	5.96					
58.72	CH	3.54	Val-α				
58.88	CH						
71.32	CH	2.65	MeVal-α				
71.47	CH						
75.03	XH	5.16	Thr-β				
75.08	CH						

[a] From DEPT spectra (d) and (e); [b] correlation *via* two- and three-bond carbon-proton couplings.

1.5 ppm at lower field, that of the β-methylene group 1.5 ppm at higher field compared to the resonance of the α-carbon of glycine. In triglycine the β-methylene carbon resonates at about the same δ value as that of glycine; the α-methylene resonance appears at lower field and the γ-methylene signal at higher field than the line of the β-carbon [96].

Straightforward assignments of the ^{13}C signals of synthetic oligopeptides are therefore achieved by proton off-resonance decoupling and by comparison with the spectra of the constituent amino acids (see Section 5.6). The signals of the selected oligopeptides of Table 5.27 were identified using these assignment aids [98, 790–794, 811].

In future, the methods of two-dimensional NMR spectroscopy (Section 2.10) will be the ones used for the unequivocal signal identifications of longer chain peptides. Complete ^{13}C shift assignments are achieved from carbon-proton shift correlations *via* two- and three-bond couplings, as is demonstrated for the orange-red chromopeptide antibiotic actinomycin D (Fig. 5.13, Table 5.28, [603, 812]).

The oxidation state of cysteine in peptides such as glutathione may easily be recognized by the downfield shift of the signal of C_β in the cysteine residue upon oxidation from $-CH_2SH$ to $-CH_2-S-S-$ [89, 790]. The signal of the $-CH_2-S-S-CH_2-$ moiety at 41.6 ppm is also recognized in the spectra of oxytocin, vasopressin and insulin [790].

A large number of ^{13}C NMR studies on proline derivatives and proline peptides have appeared in the literature [815–830]. As the electron charge density of *cis*-proline carbons is different from that of *trans*-proline carbons, these isomers can be differentiated by ^{13}C NMR spectroscopy [826, 830]. On the basis of calculations Tonelli [831] predicted four conformations for the dipeptide Boc-Pro-Pro-OBzl, three of which could be detected by ^{13}C NMR spectroscopy [826, 830]. In proline-containing peptides the stereochemistry of the proline residue plays an important role for the conformation of these oligomers. The ^{13}C chemical shift data of *cis* and *trans* proline derivatives, collected in Table 5.29, are useful to determine the stereochemistry of the amino acid-proline bond, *e.g.* in cyclo-(Pro-Gly)$_3$, melanocyte-stimulating hormone release-inhibiting factor or thyrotropin-releasing hormone.

Table 5.29. ^{13}C Chemical Shifts (δ_C in ppm) of Proline in Derivatives and Peptides.

Compound	C-α	C-β	C-γ	C-δ	C=O	Other	Solvent	Ref.
Proline (*trans*)	62.2	29.8	24.7	47.1	175.3		D$_2$O	[793]
N-Formylproline (*cis*)	60.5	30.2	23.3	45.1	177.0	CHO 165.3	D$_2$O	[793]
N-Formylproline (*trans*)	57.8	30.2	24.5	48.1	176.3	CHO 164.2	D$_2$O	[793]
N-Acetylproline (*cis*)	61.8	31.7	23.4	47.5	177.2	CO 173.9 CH$_3$ 22.2	D$_2$O	[793]
N-Acetylproline (*trans*)	59.9	30.3	25.1	49.3	177.1	CO 173.6 CH$_3$ 22.2	D$_2$O	[793]
N-Acetylproline amide (*cis*)	62.1	32.5	23.3	47.8	178.1	CO 173.8 CH$_3$ 22.3	D$_2$O	[793]
N-Acetylproline amide (*trans*)	60.6	30.9	24.9	49.4	178.1	CO 173.8 CH$_3$ 22.1	D$_2$O	[793]
Glycylproline (*cis*)	62.5	32.4	23.2	48.1	179.1	Gly(CO) 166.4 Gly(α) 41.3	D$_2$O	[793]
Glycylproline (*trans*)	62.8	30.4	25.1	47.5	179.8	Gly(CO) 165.8 Gly(α) 41.5	D$_2$O	[793]
t-Butyloxycarbonyl-glycylproline (*cis*)	58.5	31.3	22.1	46.9	173.7	Gly(CO) 168.0 Gly(α) 42.9	CDCl$_3$	[818]

Table 5.29. (continued).

Compound	C-α	C-β	C-γ	C-δ	C=O	Other	Solvent	Ref.
t-Butyloxycarbonyl-glycylproline (trans)	59.2	28.7	24.5	46.1	174.2	BOC(CO) 156.5 C-quat. 79.9 CH$_3$ 28.2 Gly(CO) 168.4 Gly(α) 42.9	CDCl$_3$	[818]
t-Butyloxycarbonyl-glycylproline (cis)	58.3	31.4	22.1	46.9	173.7	BOC(CO) 156.1 C-quat. 79.9 CH$_3$ 28.2 Gly(CO) 167.7 Gly(α) 42.6	DMSO	[818]
t-Butyloxycarbonyl-glycylproline (trans)	58.9	29.0	24.7	45.8	173.7	BOC(CO) 156.0 C-quat. 78.2 CH$_3$ 28.5 Gly(CO) 167.7 Gly(α) 42.6	DMSO	[818]
t-Butyloxycarbonyl-glycylproline (cis)	61.7	31.5	22.1	47.2	179.5	BOC(CO) 156.0 C-quat. 78.2 CH$_3$ 28.5 Gly(CO) 169.8 Gly(α) 42.4	H$_2$O	[818]
t-Butyloxycarbonyl-glycylproline (trans)	61.7	29.3	24.2	46.5	178.7	BOC(CO) 157.9 C-quat. 81.2 CH$_3$ 27.5 Gly(CO) 169.1 Gly(α) 42.4	H$_2$O	[818]
t-Butyloxycarbonyl-alanylproline (trans)	58.9	28.2	24.7	47.0 47.7	174.2	BOC(CO) 157.9 C-quat. 81.2 CH$_3$ 27.5 Ala(CO) 172.6 Ala(α) 47.0 47.7 Ala(β) 17.7 BOC(CO) 155.3 CH$_3$ 28.2	CDCl$_3$	[818]

Table 5.29. (continued).

Compound	C-α	C-β	C-γ	C-δ	C=O	Other	Solvent	Ref.
t-Butyloxycarbonyl-alanylproline (trans)	58.7	28.8	24.8	46.5 47.7	173.6	Ala(CO) 171.4 Ala(α) 46.5 47.7 Ala(β) 17.0 BOC(CO) 155.3 C-quat. 78.2 CH$_3$ 28.4	DMSO	[818]
t-Butyloxycarbonyl-alanylproline (trans)	62.0	29.4	24.5	47.4 48.3	179.5	Ala(CO) 173.1 Ala(α) 47.4 48.3 Ala(β) 15.6 BOC(CO) 157.2 C-quat. 81.3 CH$_3$ 27.5	H$_2$O (pH 7)	[818]
t-Butyloxycarbonyl-proline	62.0	30.1	23.7	46.5	181.0	BOC(CO) 156.1 C-quat. 81.2 CH$_3$ 27.9	H$_2$O	[818]
Polyproline I	59.9	32.2	22.5	49.0	172.7		H$_2$O	[802]
Polyproline II	59.3	28.7	25.4	48.4	172.3		H$_2$O	[802]
Polyhydroxyproline	58.3	36.8	71.1	55.8	172.4		H$_2$O	[802]
Poly(Pro-Gly) (trans)	61.8	30.8	25.4	48.3	175.8	Gly(CO) 170.4 Gly(α) 43.0	H$_2$O	[802]
Poly(Pro-Gly) (cis)	61.8	33.0	23.2	48.3	175.8	Gly(CO) 170.8 Gly(α) 43.0	H$_2$O	[802]
Poly(Hyp-Gly) (trans)	60.3	38.3	70.9	55.6	175.2	Gly(CO) 170.4 Gly(α) 43.0	H$_2$O	[802]
Poly(Hyp-Gly) (cis)	60.3	40.7	69.1	55.6	175.2	Gly(CO) 170.8 Gly(α) 43.0	H$_2$O	[802]
L-Prolyl-L-proline diketopiperazine	60.6	28.1	23.8	45.5	166.8		DMSO	[828]
Cyclo-(Pro-Gly)$_3$-Na	59.7	29.3	24.8	46.4	173.3		DMSO	[793]
Melanocyte-stimulating hormone release-inhibiting factor (melanostatin)	61.1	31.6	26.5	47.6	179.0		D$_2$O (pH 4)	[802]
Thyrotropin-releasing hormone (thyroliberin)	61.6	33.1	25.7	49.1	177.81		D$_2$O (pH 4)	[832]

Several ^{13}C NMR studies on the hormone thyroliberin (TRH, Pyr-His-Pro-NH$_2$) have appeared in the literature [822, 832–838]. The total signal assignment of TRH was achieved by comparison with the spectra of the constituent amino acids and model peptides such as Pyr-His-OCH$_3$ or Pyr-His(NCH$_3$)$_2$. According to the ^{13}C NMR studies in aqueous solution the hormons exists in a *cis/trans* ratio of approximately 14:86; however, this ratio is solvent-dependent and, *e.g.*, in pyridine-d$_5$ no *cis*-isomer is detectable.

From the pH-dependences of the ^{13}C chemical shifts of the histidine carbons in TRH it can be concluded that the imidazole moiety exists as the N^τ–H tautomer in basic solution. Protonation of the imidazole ring has only a slight effect on the spin-lattice relaxation times of carbons in TRH and therefore there is almost no change in the relative mobilities of the different units in the hormone. Judging from the NT_1 values of the α carbons of the histidyl (0.39 sec., pH 9.9), prolyl (0.42 sec., pH 9.9) and pyroglutamyl (0.58 sec., pH 9.9) residues in TRH, the latter has the largest mobility. From the NT_1 values of the carbons of the pyrrolidine ring (βCH$_2$: 0.73 sec., γCH$_2$: 0.86 sec., δCH$_2$: 0.52 sec.) of the proline residue, it is concluded that the β and γ carbons have the greatest mobility.

Both ^{13}C chemical shifts and spin-lattice relaxation times of the carbons of melanostatin (melanocyte-stimulating hormone release-inhibiting factor, MIF, Pro-Leu-Gly-NH$_2$) and its dimethylamide derivative (Pro-Leu-Gly-N(CH$_3$)$_2$) have been reported in different communications [832, 833, 839, 840]. Corrected amino acid chemical shifts [97, 805] have been used to calculate the ^{13}C chemical signal positions of melanostatin (Table 5.30). Because the proline residue in melanostatin is located at the N-terminal position no *cis-trans* isomerism is observed. Table 5.30 shows that the T_1 values of the carbons of melanostatin are much shorter if determined in DMSO instead in D$_2$O. From the T_1 values it can be concluded that MIF is a conformationally flexible molecule in water and more rigid in DMSO solution; the CH$_2$ group of Gly is less hindered than are the α-carbons of Pro and Leu, the side chain of Leu undergoes segmental motion and its methyl groups show hindered rotation, and the γ carbon of proline flips into and out of the plane of the pyrrolidine ring.

The ^{13}C NMR data of two further hypothalamic hormones, luliberin [841–843] (luteinizing hormone-releasing hormone, LH–RH, Pyr-His-Trp-Ser-Tyr-Gly-Leu-Arg-Pro-Gly-NH$_2$) and somatostatin [844, 845] (Ala-Gly-Cys-Lys-Asn-Phe-Phe-Trp-Lys-Thr-Phe-Thr-Ser-Cys-OH) have been extensively interpreted in the literature. The identification of all ^{13}C resonances of LH–RH is based on spectral comparison with constituent peptides [832, 833, 841], monosubstituted analogues [842] and the calculated spectrum from the corrected shifts of the constituent amino acids [833]. According to the ^{13}C NMR studies the prolyl residue of LH–RH in aqueous solution predominantly exists in the *trans* conformation [837]. From T_1 measurements of LH–RH it follows that in solution the decapeptide amide molecule undergoes segmental motion along the backbone and the nonaromatic side chains, whereas the aromatic moieties are more restricted. In two recent communications heteronuclear 2D carbon-proton shift correlation, 2D *J*-resolved spectroscopy and homonuclear correlation spectroscopy (SECSY [846]) lead to almost complete signal assignments in the spectra of the somatostatin analogs des(Ala,Gly)-somatostatin [847] and cyclo[Phe-D-Trp-Lys(Z)-Thr-(Gly,Phe)-Pro]

Table 5.30. ^{13}C Chemical Shifts (δ_C in ppm) and Spin-Lattice Relaxation Times (T_1 in sec) of Pro-Leu-Gly-NH$_2$ and its Dimethyl Derivative [839].

C Atom		Pro-Leu-Gly-NH$_2$				Pro-Leu-Gly-N(CH$_3$)$_2$			
		δ (D$_2$O)	T_1 (D$_2$O)	δ (DMSO-d$_6$)	T_1 (DMSO-d$_6$)	δ (D$_2$O)	T_1 (D$_2$O)	δ (DMSO-d$_6$)	T_1 (DMSO-d$_6$)
Pro	α	61.07	0.86	61.69	0.26	61.12	0.65	61.65	0.82
	β	31.63	1.01	31.91	0.33	31.66	0.62	32.03	0.54
	γ	26.48	1.69	27.31	0.47	26.12	0.82	27.31	0.68
	δ	47.62	1.03	48.20	0.33	47.63	0.60	48.24	0.54
	C=O	178.99	10.0	175.87	5.3	178.33	11.0	175.39	4.3
Leu	α	53.63	0.74	52.17	0.29	53.38	0.52	51.87	0.45
	β	40.81	0.47			41.14	0.42		
	γ	25.53	1.02	25.89	0.72	25.55	0.61	25.88	0.73
	δ	23.77	0.85	24.59	0.60	23.32	0.66	24.64	0.67
	δ'	21.86	1.03	23.24	0.63	21.84	0.69	23.28	0.79
	C=O	176.52	10.8	173.69	3.3	175.97	11.2	173.50	4.3
Gly	α	43.25	0.66	43.39		42.03	0.48	43.26	0.36
	C=O	175.26	12.2	172.34	4.7	170.86	12.2	169.46	6.7
N–CH$_3$							1.5	36.59	≈ 4
N–CH$_3$							1.2	37.22	≈ 2

[848, 849]. The ^{13}C NMR spectra of numerous protected and unprotected peptides with the natural sequences of oxytocin and vasopressin have been recorded and interpreted with the aim of the total signal assignment of the ^{13}C resonances of the two neurohypophyseal hormones: Pro-Leu-Gly-NH$_2$ [821], benzyloxycarbonyl-Pro-Leu-Gly-OEt [850], benzyloxycarbonyl-Pro-Leu-Gly-NH$_2$ [850], Cys(benzyl)-Pro-Leu-Gly-NH$_2$ [850], benzyloxycarbonyl-Cys(benzyl)-Pro-Leu-Gly-NH$_2$ [850], Asn-Cys(benzyl)-Pro-Leu-Gly-NH$_2$ [850], benzyloxycarbonyl-Asn-Cys(benzyl)-Pro-Leu-Gly-NH$_2$ [821, 850], Gln-Asn-Cys(benzyl)-Pro-Leu-Gly-NH$_2$ [850], benzyloxycarbonyl-Gln-Asn-Cys(benzyl)-Pro-Leu-Gly-NH$_2$ [821, 850], Ile-Gln-Asn-Cys(benzyl)-Pro-Leu-Gly-NH$_2$ [850], benzyloxycarbonyl-Ile-Gln-Asn-Cys(benzyl)-Pro-Leu-Gly-NH$_2$ [821, 850], Tyr-Ile-Gln-Asn-Cys(benzyl)-Pro-Leu-Gly-NH$_2$ [850], benzyloxycarbonyl-Tyr(benzyl)-Ile-Gln-Asn-Cys(benzyl)-Pro-Leu-Gly-NH$_2$ [821, 850] or benzyloxycyrbonyl-Cys(benzyl)-Tyr-Ile-Gln-Asn-Cys(benzyl)-Pro-Leu-Gly-NH$_2$ [821, 850]. Moreover, the ^{13}C NMR spectrum of oxytocin [815, 821, 827, 850] was compared with spectra of lysine vasopressin, arginine vasopressin, arginine vasotocin, bacitracin, [4-glycine]-oxytocin or deamino-oxytocin. The deuterated analogues [α,β,β-^{2}H$_3$]-D,L-Cys(benzyl)-Tyr(benzyl)-Ile-Gln-Asn-Cys(benzyl)-Pro-Leu-Gly-NH$_2$ and Cys(benzyl)-Tyr(benzyl)-Ile-Gln-Asn-[α,β,β-^{2}H$_3$]-D,L-Cys(benzyl)-Pro-Leu-Gly-NH$_2$ enabled one to assign the signals at138.5 and 159.4 ppm to the α and β carbons of cysteine-1 and the resonances at 142.3 and

Table 5.31. ^{13}C Chemical Shifts (δ_C in ppm) of Oxytocin, Arginine Vasotocin, Arginine, and Lysine Vasopressin in D_2O [833].

C Atom		Oxytocin	Arginine vasotocin	Arginine vasopressin	Lysine vasopressin
Gly	α CH	43.31	43.22	43.20	43.27
	C=O	175.16	175.02	175.17	175.08
Leu	α CH	53.80			
	β CH$_2$	40.53			
	γ CH	25.62			
	δ CH$_3$	23.35			
	δ CH$_3$	21.91			
	C=O	176.31			
Lys	α CH				54.88
	β CH$_2$				27.36
	γ CH$_2$				23.21
	δ CH$_2$				31.24
	ε CH$_2$				40.46
	C=O				175.50
Arg	α CH		54.64	54.66	
	β CH$_2$		29.01	29.00	
	γ CH$_2$		25.83	25.56	
	δ CH$_2$		41.67	41.66	
	ε C		N.O.	N.O.	
	C=O		175.24	175.17	
Pro	α CH	61.85	61.87	61.80	61.80
	β CH$_2$	30.37	30.45	30.42	30.48
	γ CH$_2$	25.86	25.83	25.85	25.87
	δ CH$_2$	49.06	49.08	49.09	49.09
	C=O	175.42	175.43	175.36	175.43
Asn	α CH	51.62	51.64	51.34	51.38
	β CH$_2$	37.43	37.16	37.58	37.66
	γ C=O	175.72	175.67	175.64	175.77
	C=O	172.97	173.18	173.37	173.16
Gln	α CH	56.17	56.43	56.18	55.28
	β CH$_2$	27.07	26.97	26.89	26.97
	γ CH$_2$	32.29	32.19	32.19	32.26
	δ C=O	178.79	178.80	178.86	178.97
	C=O	174.20	174.19	174.19	174.09
Ile	α CH	60.66	61.32		
	β CH$_2$	36.84	37.02		
	γ CH$_2$	25.53	25.59		

Table 5.31. (continued).

C Atom		Oxytocin	Arginine vasotocin	Arginine vasopressin	Lysine vasopressin
Ile	γ CH$_3$	16.19	16.03		
	δ CH$_3$	11.92	11.77		
	C=O	174.27	174.42		
Phe	α CH			54.84	56.40
	β CH$_2$			37.09	37.03
	C$_1$ ar.			137.41	137.48
	C$_2$ ar.			130.13	130.11
	C$_3$ ar.			130.26	130.35
	C$_4$ ar.			128.48	128.48
	C=O			174.03	173.92
Tyr	α CH	56.33	56.43	56.40	56.01
	β CH$_2$	37.43	37.16	37.58	37.28
	C$_1$ ar.	129.27	129.02	128.97	129.14
	C$_2$ ar.	131.71	131.66	131.69	131.66
	C$_3$ ar.	116.77	116.85	116.75	116.73
	C$_4$ ar.	155.75	155.80	155.77	155.72
	C=O	173.93	173.83	173.37	173.78
Cys-6	α CH	52.84	52.41	52.55	52.60
	β CH$_2$	40.09	39.36	39.56	40.07
	C=O	170.91	170.98	170.88	170.86
Cys-1	α CH	54.54	53.31	53.48	54.46
	β CH$_2$	42.76	40.93	41.02	42.64
	C=O	174.23	168.86	168.68	172.96

N.O. not observed.

160.6 ppm to the α and β carbons of cysteine-6 [821]. The ^{13}C chemical shifts of oxytocin, arginine vasotocin, arginine and lysine vasopressin are collected in Table 5.31. Concerning the conformational properties of oxytocin and lysine vasopressin, the T_1 values [851, 852] (Fig. 5.14) give the following information:

1. The T_1 values of the α carbons of the ring portions are similar, indicating a similar rigidity for the ring carbons.

2. Increasing T_1 values in going from the α to the β to the γ carbons of glutamine indicate segmental motion in the side chains of Glu, which is stronger in oxytocin compared to lysine vasopressin.

3. Segmental motion of the side chain of lysine in lysine vasopressin is assumed for the same reasons as discussed in 2. for glutamine. According to the T_1 values a similar

A

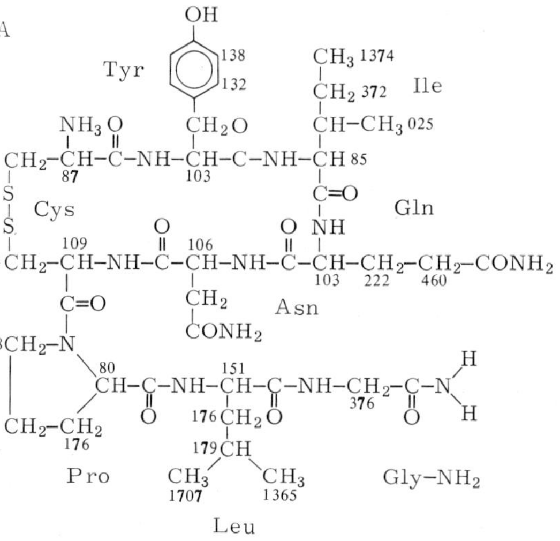

```
                OH
                 |
         Tyr    ⬡ 138
                ⬡ 132
    NH₃ O      CH₂O        CH₃ 1374
     |   ||     |           |
CH₂-CH-C-NH-CH-C-NH-CH      CH₂ 372    Ile
 |   87        103  |        |
 S                  CH-CH₃ 025
 |  Cys             |
 S                  CH 85
 |     109     O   106    O   NH
 |      |      ||   |     ||   |
84 CH₂-CH-NH-C-CH-NH-C-CH-CH₂-CH₂-CONH₂
        |           |     103  222   460
        C=O         CH₂        Gln
        |           |
    88 CH₂-N        CONH₂   Asn
        |
        |    80              151                        H
        |    CH-C-NH-CH-C-NH-CH₂-C-N
        |   /   ||    |   ||        376  ||   \
   CH₂-CH₂     O  176 CH₂  O             O    H
      176            |
                179 CH
                 /    \
          Pro   CH₃    CH₃              Gly-NH₂
               1707   1365
                  Leu
```

B

```
                OH
                 |
         Tyr    ⬡ 188         ⬡ 165  Phe
                ⬡ 160         ⬡ 165
    NH₃⁺ O    168 CH₂ O       CH₂ 168
     |   ||     |      ||      |
123 CH₂-CH-C-NH-CH-C-NH-CH 82
     |  126        82   |
     S                  C=O
     |  Cys             |
     S                  NH       Gln
     |     113    O   103    O   |
     |      |     ||    |    ||  |
80 CH₂-CH-NH-C-CH-NH-C-CH-CH₂-CH₂-CONH₂
        |           |     106  214   270
        C=O        168 CH₂
        |           |
    118 CH₂-N       CONH₂   Asn
        |
        |    130             97          468          H
        |    CH-C-NH-CH-C-NH-CH₂-C-N
        |   /   ||    |   ||              ||   \
   280 CH₂-CH₂ O  262 CH₂  O               O    H
      202            |
                380 CH₂
                    |
                680 CH₂                  Gly-NH₂
           Pro      |
                1244 CH₂
                    |
                    NH₃⁺
                  Lys
```

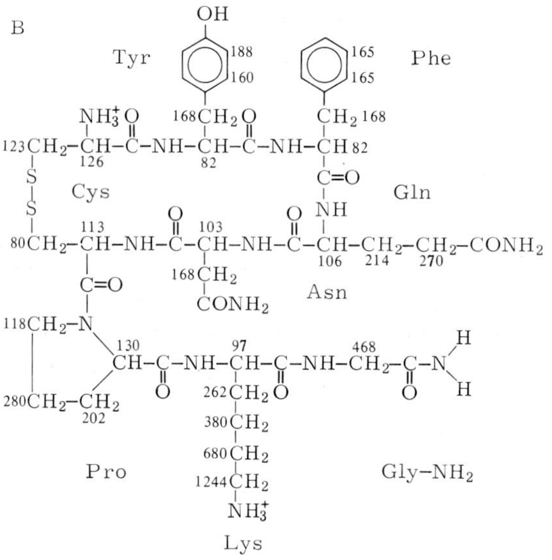

Fig. 5.14. Structures and NT_1 values (in ms), observed for oxytocin (A) and lysine vasopressin (B) (100 mg/mL, D_2O) [851].

mobility for the side chains of lysine in oligo- and polylysine [853, 854] and lysine vasopressine is suggested.

4. As the T_1 values of the secondary butyl group of isoleucine are similar to those in the free amino acid [855] the mobility of the side chain of Ile seems not to be influenced by other residues of the oxytocin molecule.

5. Increasing T_1 values of the α carbons of the acyclic terminal tripeptides with enlarging distance from the point of attachment indicate that the mobility increases in the direction of the C terminus.

6. The mobility of the acyclic terminal tripeptide in lysine vasopressin is larger compared to that in oxytocin, which is coincident with a model suggested from ^{1}H NMR studies [856].

In a similar way the ^{13}C NMR spectra and/or T_1 values were interpreted for numerous other peptide hormones or biologically active peptides and their derivatives, such as bradykinin (Arg-Pro-Pro-Gly-Phe-Ser-Pro-Phe-Arg) [857], tetragastrin (Trp-Met-Asp-Phe-NH$_2$) [858], eledoisin (Pyr-Pro-Ser-Lys-Asp-Ala-Phe-Ile-Gly-Leu-Met-NH$_2$) [837], angiotensin II (Asp-Arg-Val-Tyr-Val (or Ile)-His-Pro-Phe) [820, 859, 860], enkephalin (Tyr-Gly-Gly-Phe-Met (or Leu) [861–863] or insulin [864].

^{13}C NMR spectroscopy also gives valuable information on the conformation of cyclic peptides. The effect of aromatic ring currents on ^{13}C chemical shifts was studied by recording the ^{13}C NMR spectra of cyclo(Leu-Gly), cyclo(Leu-Leu), cyclo(Leu-Trp), cyclo(Trp-Trp), cyclo(Trp-Gly), cyclo(Phe-Gly), cyclo(Phe-Val), and cyclo(Tyr-Gly) [865]. There exists experimental proof that the ring current effect causes similar upfield shifts for ^{13}C and ^{1}H signals if the geometry, hybridization and charge of two molecular conformations are identical [866]. However, due to conformational changes in diketopiperazines the ^{13}C chemical shifts of cyclic dipeptides containing aromatic side chains do not always reflect the ring-current effects [865]. Only one signal for each carbon of the proline residue is found in the ^{13}C NMR spectrum (in dimethyl sulfoxide and chloroform) of cyclo(Pro)$_3$, indicating the presence of a C_3 symmetry or rapid interconversion between conformers [867]. From ^{13}C NMR studies it follows that in polar solvents (water, dimethyl sulfoxide) cyclo(Pro-Gly)$_3$ exists in an asymmetric conformation with one Gly-Pro *cis* peptide bond; however, in nonpolar solvents (dioxane, chloroform), the molecule adopts a symmetric C_3 conformation with *trans* peptide bonds only [868]. Specifically deuterated cyclo(Gly-Pro-Gly(d$_2$)$_2$) allowed one to determine that this cyclopeptide adopts a type II β turn instead of a type II′ β turn conformation in aqueous solution and that the glycine residue preceding the proline residue is intramolecularly hydrogen-bonded (Fig. 5.15); surprisingly it was found that the carbonyl carbons involved in hydrogen bonding resonate at highest field. The magnetically anisotropic peptide group in the center of the β turn is made responsible for this effect [825, 869]. From ^{1}H and ^{13}C NMR studies it was concluded that cyclo(Gly-Pro-Ser-D-Ala-Pro) adopts an all-*trans* conformation with a 1 ← 4 intramolecular hydrogen bond in a type II′ D-Ala-Pro β turn. Moreover, the existence of a 1 ← 3 hydrogen bond (γ turn) is assumed on the basis of a high-field Pro

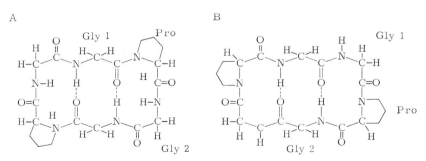

Fig. 5.15. Structures of type II β turn (A) and type II′ β turn (B) of cyclo(Gly-Pro-Gly)$_2$.

C^β resonance [870]. On a similar basis it was found that cyclo(Gly-Pro-Gly-D-Ala-Pro) preferentially adopts a conformation containing $1 \leftarrow 3$ and $1 \leftarrow 4$ intramolecular hydrogen bonds (β and γ turns) [871].

^{13}C NMR investigations of several cyclic and acyclic peptide antibiotics and toxins have been reported in the literature, e.g., gramicidin S-A (cyclo(Phe-Pro-Val-Orn-Leu)$_2$) [176, 801, 872–875], viomycin (cyclopeptide constituted of L-serine, α,β-L-diaminopropionic acid, β-L-lysine and viomycidine) [876], bacitracin A (Ile-Cys-Leu-D-Glu-Ile-Lys-D-Orn-Ile) [815], alamethicin (Ac-Aib-Pro-Val-Aib-Val-Aib-Aib-Ala-Aib-

Asp-His-D-Phe

D-Asn)

Aib-Gln-Aib-Leu-Aib-Gly-Leu-Aib-Pro-Val-Aib-Aib-Glu(Pheol)-Gln) [877–879], valinomycin cyclo(Lac-Val-D-HyIv-D-Val)$_3$ [305, 880–884] or antamanide cyclo(Val-Pro-Pro-Ala-Phe-Phe-Pro-Pro-Phe-Phe) [824]. For example, the signals of alamethicine were assigned by comparison with the carbon shifts of synthetic partial sequences, by selective proton decoupling and by T_1 measurement using the inversion-recovery technique, which permits the localization of some closely spaced signals of nonequivalent α carbons. Moreover, the methyl T_1 values increase with growing length of the amino acid side chain (Leu-C$_\delta$ > Val-C$_\gamma$ > Ala-C$_\beta$) [877–879], reflecting an increased degree of mobility. While the chemical shifts of the amino acid residues in the rigid α-helical part remain constant within the experimental error, those of carbons belonging to the more flexible portions of the alamethicine sequence are affected by conformational changes when the temperature is elevated up to 100 °C in dimethyl sulfoxide solution. To conclude, the α-helix is preserved even at 100 °C [877–879]. The complete signal assignment of gramicidin S-A was achieved using selective biosynthetic enrichment techniques [873]. The T_1 values of gramicidin S are comparable with those determined for lysine vasopressin and oxytocin and give information about the rigidity of the ring carbons and segmental motions of the amino acid side chains [176]. Upon complexation of valinomycin with K^+, Rb^+ and Cs^+ ions the carbonyl carbons are shifted 4–5 ppm downfield. The change in chemical shift is much smaller upon complexation with Na^+, which is interpreted as being due to the smaller ionic radius of the sodium ion [882].

5.7.2 Homopolymeric Polypeptides

^{13}C NMR spectroscopy was applied in extensive studies for the investigation of homopolymeric polypeptides [178, 179, 180, 823, 829, 854, 885–892]. The helix-coil transitions of poly-L-aspartic acid [893], poly-L-glutamic acid [887], poly(γ-benzyl-L-glutamate) [178, 180, 893, 894], poly-L-lysine [854], poly-L-methionine [179], poly(N-δ-carbobenzoxy-L-ornithine) [886], poly(N^5-3-hydroxypropyl)-L-glutamine) [889, 890, 895] or copoly(L-glutamic acid, L-lysine, L-alanine) [893] were studied with special interest. To determine the influence of helix formation the pH-dependences of the ^{13}C resonances of L-glutamic acid were compared with those of poly-L-glutamic acid, in which the polymer adopts a helical conformation at pH values < 5.5. Upon helix formation a pronounced downfield shift of the peptide carbonyl (− 2.25 ppm) and the α carbon (− 2.31 ppm) of the polymer are observed. As these results are in contrast to the titration behavior of glutamic acid the shift changes of the polymer are attributed to the change in secondary structure [887]. In CDCl$_3$ solution poly(N-δ-carbobenzoxy-L-ornithine) (PCBO) adopts a purely helical conformation. Upon addition of trifluoroacetic acid (TFA) a helix-coil transition is observed when the TFA volume fraction is ~ 0.1. Due to association among helices the ^{13}C NMR spectrum of PCBO in pure CDCl$_3$ shows extremely broad signals; at high TFA concentration narrowing of the signals and an upfield shift of the α and peptide carbonyl carbons are observed [886]. Similarly, converting helical poly(γ-benzyl-L-glutamate) to a random coil peptide causes a 3 ppm upfield shift for the signal of the α-carbon atom and a 2.7 ppm shift to higher field for the signal of the amide carbon [178]. The helix-coil transition of poly(N^5-(3-hydroxypropyl)-L-glutamine) (PHPLG) was followed by ^{13}C NMR spectroscopy in methanol-water mixtures; the polymer adopts a helical conformation at high methanol concentration. Double signals were observed for the α, β and peptide carbonyl carbons during the helix-coil transition. The CO main-chain signal, observed in methanol-rich mixtures, moves upfield with decreasing secondary structure. The side chain CO resonance is not influenced by the helix-coil transition. In the non-helical state only one resonance is observed for the α carbons of the glutamine residue; upon helix formation a further α carbon signal arises from atoms in the helical part of the polymer [889, 890].

Spin-lattice relaxation times and ^{13}C chemical shifts were used to study conformational changes of poly-L-lysine, which undergoes a coil-helix transition in a pH range from 9 to 11. In order to adopt a stable helical structure, a minimum number of residues for the formation of hydrogen bonds between the C=O and NH backbone groups is necessary; therefore for the polypeptide dodecalysine no helix formation was observed. Comparison of the pH-dependences of the ^{13}C chemical shifts of the carbons of poly-L-lysine and (L-Lys)$_{12}$ shows very similar values for both compounds; therefore downfield shifts of the α, β and peptide carbonyl carbons can only be correlated with caution with helix formation and are mainly due to deprotonation effects. On the other hand, a sharp decrease of the T_1 values of the carbonyl and some of the side chain carbons is indicative for helix formation [854].

5.7.3 Proteins

Although the ^{13}C NMR spectra and spin-lattice relaxation times of numerous proteins and their derivatives such as, *e.g.*, ribonuclease [41, 896–901], cytochrome c [902], myoglobins [788, 903–906], hemoglobins [903, 907–913], lysozyme [896, 914–917], α-lytic protease [918, 919], chymotrypsins [920–922], tryptophan synthetase [923, 924], ferredoxin [925], carbonic anhydrase [926, 927], gelatin [928] or collagen [929, 930] have been reported in the literature, resonance broadening and a larger variation of the signal positions compared to amino acids and smaller peptides often allow only tentative signal assignments (Table 5.32) [778, 802, 804, 931]. As was extensively shown for the case of ribonuclease, the ^{13}C NMR spectra of native proteins differ a great deal from those of denaturated ones. The ^{13}C NMR spectrum of native ribonuclease A shows very broad signals and only few resonances can be identified [41, 897]. Denaturating ribonuclease by

Table 5.32. Tentative Assignments (δ_C in ppm) of the ^{13}C Chemical Shifts of Proteins [802].

C Atom		Ribo-nuclease	Myo-globin	Lyso-zyme	α-Lact-albumin	α-Chymo-trypsin
Ala	$C\alpha$	43.7		43.3	41.0	43.5
	$C\alpha$	50.9		53.8–50.3	53.7–47.9	51.0–50.3
	C_β	17.2	17.5	16.5	16.9	17.4
Ser	C_α	56.4–53.9		56.1–56.0	55.5	56.1
	C_β	61.9–60.1		61.5	62.9–60.4	61.8
Cys	C_α	56.4–53.9		53.8–50.3	53.7–47.9	51.0–50.3
	C_β	37.0		36.5–30.0	36.3	36.7–30.8
Phe	C_α	56.4–53.9		56.1–56.0	55.5	56.1
	C_β	37.0		36.5–30.0	37.3	36.7–30.8
	C_γ (ar.)	137.2	137.2	136.7	136.0	136.4
	C_ε (ar.)	129.4	129.7	129.3	129.3	129.5
	C_δ (ar.)			128.9	128.9	129.1
	C_ζ (ar.)	128.0	128.0	127.3	127.2	127.4
His	C_α	56.4–53.9		53.8–50.03	54.7–47.9	54.7–53.1
	C_β	27.1	27.3	26.7	27.1–26.7	27.0
	C_γ	129.4	131.4			
	C_δ	118.2	121.4			
	C_ε	134.5	138.2		135.0	134.0
Asp	C_α	56.4–53.9		53.8–50.3	53.7–47.9	51.0–50.3
	C_β	37.0		36.5–30.0	36.3	36.7–30.8
	C_γ	177.2	178.2			
Asn	C_α	56.4–53.9		53.8–50.3	53.7–47.9	51.0–50.3
	C_β	37.0		36.5–30.0	36.3	36.7–30.8
	C_γ	177.2	177.9			
Val	C_α	61.9–60.1		60.0–59.1	59.0	60.0
	C_β	32.0–30.9	31.2	36.5–30.0	31.2–29.7	36.7–30.8
	$C_{\gamma 1}$	19.6	19.7	18.7	18.7	19.1
	$C_{\gamma 2}$	19.6		18.0	18.1	18.6
Thr	C_α	61.9–60.1		60.0–59.1	59.0	60.0
	C_β	67.9		67.0	67.4	67.4
	C_γ	19.6	19.7	19.2	19.0	19.6
Met	C_α	56.4–53.9		53.8–50.3	53.3–47.9	54.7–53.1
	C_β	32.0–30.9		36.5–30.0	31.2–29.7	36.7–30.8
	C_γ	32.0–30.9		36.5–30.0	31.2–29.7	36.7–30.8
	C_ε	15.3		14.7	14.8	13.0

Table 5.32. (continued).

C-Atom	Ribo-nuclease	Myo-globin	Lyso-zyme	α-Lact-albumin	α-Chymo-trypsin
Glu C_α	56.4–53.9		53.8–50.3	53.7–47.9	54.7–53.1
C_β	28.7	29.0	28.3	27.9	27.0
C_γ	32.0–30.9	34.5	36.5–30.0	33.7	36.7–30.8
C_δ	178.2	181.2			
Gln C_α	56.4–53.9		53.8–50.3	53.7–47.9	54.7–53.1
C_β	27.1		26.7	27.1–26.7	27.0
C_γ	32.0–30.9		36.5–30.0	31.2–29.7	36.7–30.8
C_δ	177.2				
Leu C_α	56.4–53.9		53.8–50.3	53.7–47.9	54.7–53.1
C_β	40.3	40.3	39.9	39.7	40.1
C_γ	25.3	25.5	27.7	24.5	25.0
C_{δ_1}	22.9	23.3	22.6	22.5	22.8
C_{δ_2}	21.7	21.9	21.4	20.9	21.4
Ile C_α	61.9–60.1		60.0–59.1	59.0	60.0
C_β	37.0		36.5–30.0	36.3	36.7–30.8
C_{γ_1}		25.5	24.7	24.5	25.0
C_{γ_2}	15.8	15.7	15.2	14.8	15.3
C_{δ_1}	11.2		10.8	10.7	11.1
Arg C_α	56.4–53.9		53.8–50.3	53.7–47.9	54.7–53.1
C_β	28.7	29.0	28.3	27.9	27.0
C_γ	25.3		24.7	24.5	25.0
C_δ	41.1	41.4	41.1	41.0	
C_ζ	157.7	157.7	157.3	156.8	
Lys C_α	56.4–53.9		53.8–50.3	53.7–47.9	54.7–53.1
C_β	32.0–30.9	31.2	36.5–30.0	31.2–29.7	36.7–30.8
C_γ	22.9	23.3	22.6	22.5	22.8
C_δ	27.1	27.3	26.7	27.1–26.7	27.0
C_ε	40.3	40.3	39.9	39.7	40.1
Pro C_α	61.9–60.1		60.9	62.9–60.4	61.8
C_β	32.0–30.9		36.5–30.0	31.2–29.7	36.7–30.8
C_γ	25.3		24.7	24.5	25.0
C_δ			56.1–56.0	55.5	56.1
Tyr C_α	56.4–53.9		36.5–30.0	36.3	36.7–30.8
C_β	37.0		128.2	127.9	127.4
C_γ (ar.)	128.8				
C_δ (ar.)	131.2		130.6	130.5	130.8
C_ε (ar.)	116.4		116.0	115.6	115.9
C_ζ	155.5	155.7	154.9	154.6	
CO	178.7–170.7	175.8–172.7	175.6	181.1–171.8	177.7–170.4

Table 5.33. Selected Carbon-13 Spin-Lattice Relaxation Times (in ms) and Rotational Correlation Times (in ns) of Ribonuclease A (H_2O [41]).

Type of carbon	Native protein		Denatured protein	
	T_1	τ_R	T_1	τ_R
Carbonyls	416		539	
α Carbons	42	30	120	0.40
β Carbons, Thr	~40	~30	99	0.48
Rigid side chains	~30			
ε Carbons, Lys	330	0.070	306	0.076

acid [41], heat, dithiothreitol or performic acid [901] causes narrowing of the lines due to increased motional flexibility and allows the assignment of many resonances by spectral comparison with amino acids and smaller peptides [41, 897, 898]. The importance of carbon-13 spin-lattice relaxation time (T_1) measurements was first demonstrated for ribonuclease (Table 5.33).

The data in Table 5.33 permit one to draw the following conclusions:

1. Some of the side chains, e.g. the β carbon moiety of Thr, have relatively low segmental mobility.

2. The large T_1 values of the ε carbons of lysine indicate a large degree of internal motion for the side chain of Lys.

3. The correlation times of the α carbons in the native protein compared to those of the denatured biomolecule are much larger, indicating an increased segmental motion in the backbone of denatured ribonuclease.

4. The T_1 values of the ε carbons of Lys are similar in native and denatured ribonuclease; therefore, the segmental motions of the side chains of lysine are similar in both forms of the protein.

Using 20 mm tubes the sensitivity of natural abundance PFT ^{13}C NMR can be increased drastically, as was demonstrated first for hen egg-white lysozyme. For example, the ^{13}C NMR spectrum recorded under these conditions shows 22 signals for 28 non-protonated aromatic carbons, however, with a broad background arising from 59 protonated aromatic carbons [916].

Though not based on T_1 measurements, the differences in the ^{13}C NMR spectra of variously liganded human and rabbit hemoglobins were interpreted in terms of different flexibilities of the two hemoglobin species [907]. Binding of enriched carbon monoxide to various heme globins (e.g., sperm whale myoglobin, adult human heme globin, fetal human hemoglobin, mouse hemoglobin, rabbit hemoglobin) was investigated by several groups [908–910, 912, 913]. All carboxyhemoglobins studied showed two distinct resonances of equal intensity around 206 ppm, indicating that carbon monoxide is bound either to the α or β subunits. The low-field ^{13}CO resonance of rabbit carboxyhemoglobin decreases much faster upon exposure to oxygen, and therefore, based on kinetic studies with stopped-flow techniques [932], this signal could be assigned to the carbon monoxide group bound to the α subunit [908].

5.8 Porphyrins

Numerous ^{13}C NMR spectroscopic studies on synthetic and natural free base prophyrins and metalloporphyrins have been reported in the literature [933–956]. The signal assignments in Table 5.34 were made mainly by correlation of the ^{13}C NMR spectra of different analogues and by proton off-resonance decoupling. The quaternary carbons of the pyrrole rings resonate within a range of 147.5 to 135.5 ppm. Although C-2, C-4 and the *meso* carbons of deuterioporphyrin dimethyl ester are in *β*-positions relative to the nitrogens, only the former carbons absorb in the olefinic range. This efficient shielding of the *meso* carbons can be explained in terms of cyclic conjugation in the 16-membered porphyrin ring [933].

In Table 5.35 ^{13}C chemical shifts of chlorophyll a and some derivatives are collected. A complete assignment was achieved by combined application of several assignment techniques: Chemical shift comparison of related compounds, long-range selective ^{1}H decoupling with low-power irradiation using known values of the proton chemical shifts of chlorin derivatives, fully coupled spectra, the "center-line INDOR" technique [937], a modification of the usual heteronuclear INDOR technique and chemical shift changes caused by solvent effects. The values of Table 5.35 demonstrate that Mg coordination causes a markedly larger change for the carbons (C-2a, C-2b) of the vinyl group than for those of the other side chains, indicating that the metal coordination also perturbs the π^* levels. This suggestion is supported by MO calculations [957, 958], which show a substantial vinyl group contribution to the π system of ethyl pheophorbide a and ethyl chlorophyllide a. Furthermore, the change from the 10 R to the 10 S configuration induces alterations, not only in rings IV and V but also, though much weaker, in the whole macrocycle, as is indicated by the changes in chemical shifts of Table 5.35 [956].

5.9 Coumarins

The ^{13}C chemical shifts of the parent compound were assigned by comparison with the spectrum of cinnamic acid (see Section 4.11.2) and from the knowledge that for benzenoid methine carbons introduction of an acyloxy function into the benzene ring causes small shifts of the *meta* carbons, and shifts the signals of the *para* carbons to higher field, while the *ortho* carbons become even more shielded (Fig. 5.16) [635].

The signals of a large number of coumarin derivatives [418] were assigned using the single-frequency off-resonance decoupling, J-modulated spin echo and INEPT techniques.

Table 5.34. ^{13}C Chemical Shifts (δ_C in ppm) of Porphyrines (Solvent: $CHCl_3$) [933].

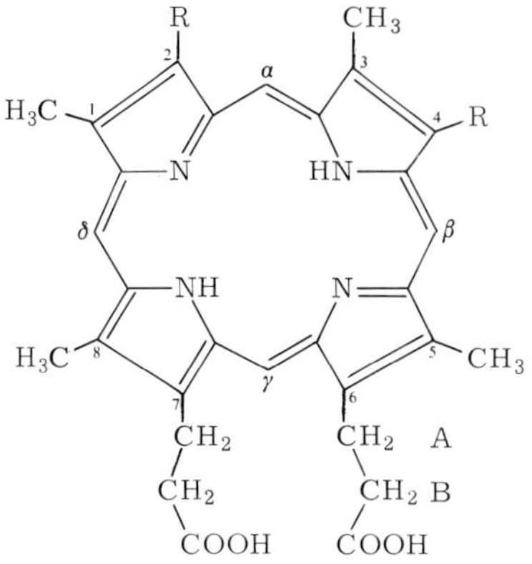

Deuterioporphyrin: R = ^2H
Mesoporphyrin: R = $^{2',4'}CH_2-CH_3$
Protoporphyrin: R = $^{2',4'}CH=CH_2$

C Atom	Deuterio-porphyrin dimethyl ester	Deuterio-porphyrin diethyl ester	2,4-Di-acetyl-deuterio-porphyrin dimethyl ester	2,4-Di-propionyl-deuterio-porphyrin dimethyl ester	Meso-porphyrin dimethyl ester	Proto-porphyrin diethyl ester
1,3-CH_3	13.60	13.60	13.60	13.80	17.80	12.50
C-2,4	128.60	128.80			19.90	130.40
C-2',4'			197.80			
5,8-CH_3	11.60	11.50	11.50	11.30	11.60	11.80
C-6A, 7A	21.90	21.90	21.50	21.30	22.10	22.00
C-6B, 7B	37.40	37.00	36.60	38.00 36.20	37.50	37.40
meso Carbons	100.20 99.40 97.00 96.00	100.00 99.20 96.90 95.90	101.90 99.70 96.80 95.30	101.70 99.80 96.80 95.10	96.30	96.90
CO (ester)	173.50 to 173.20	173.50 to 173.20	173.50 to 173.20	173.50 to 173.20	173.50 to 173.20	173.50 to 173.20
C_{quat} (pyrrole)	147.50 to 135.50	147.50 to 135.50	147.50 to 135.50	147.50 to 135.50	147.50 to 135.50	147.50 to 135.50
CH_3 (methyl ester)	51.80 to 51.60		51.80 to 51.60	51.80 to 51.60	51.80 to 51.60	
CH_3 (ethyl ester)		14.40				14.40
CH_2 (ethyl ester)		60.80				60.80
CH_3 (acetyl)			33.10 to 32.80			
CH_2 (propionyl)				38.00 36.20		
CH_3 (propionyl)				9.10		
CH_3 (ethyl)					11.60	
CH_2 (vinyl)						120.10

5.8 Porphyrins   443

Table 5.35. Structures and ^{13}C Chemical Shifts (δ_C in ppm) of Chlorophyll a (I) and a' (II), Pheophytin a (III) and a' (IV), Pyrochlorophyll a (V) and Pyropheophytin a (VI) (Solvent: Tetrahydrofuran-d$_8$) [956].

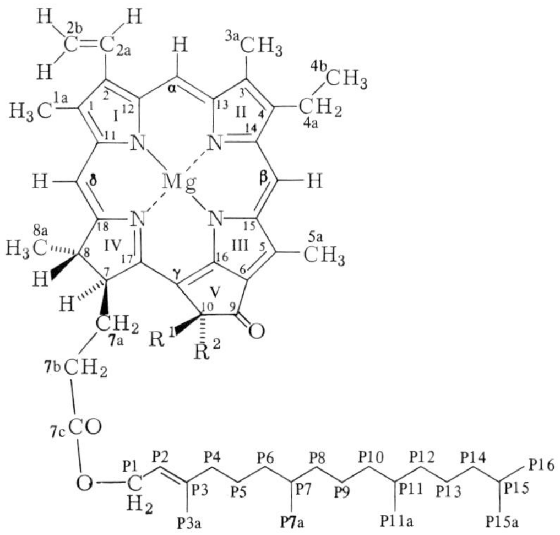

I: (10 R) Chlorophyll a; $R^1 = H$, $R^2 = CO_2CH_3$
II: (10 S) Chlorophyll a = Chlorophyll a'; $R^1 = CO_2CH_3$, $R^2 = H$
III: (10 R) Pheophytin a; $R^1 = H$, $R^2 = CO_2CH_3$; formula without Mg and with NH groups in rings I and III
IV: (10 S) Pheophytin a = Pheophytin a'; $R^1 = CO_2CH_3$, $R^2 = H$; formula without Mg and with NH groups in rings I and III
V: Pyrochlorophyll a; $R^1 = R^2 = H$
VI: Pyropheophytin a; $R^1 = R^2 = H$; formula without Mg and with NH groups in rings I and III

C Atom	Chlorophyll a	Chlorophyll a'	Pheophytin a	Pheophytin a'	Pyrochlorophyll a	Pyropheophytin a
1	135.5	135.5	132.1	132.2	135.2	131.7
2	139.0	139.1	136.4	136.3	138.8	136.1
3	134.0	134.0	136.0	136.1	133.7	135.7
4	144.1	144.1	145.2	145.2	143.8	145.0
5	134.2	134.2	128.9	128.8	133.3	128.0
6	131.9	131.5	130.1	130.0	133.5	131.4
7	51.6	52.3	52.1	52.8	52.1	52.6
8	50.0	50.9	50.8	51.6	50.1	50.7
9	189.3	189.5	189.2	189.3	195.2	194.7
10	66.2	66.2	66.4	66.4	49.2	48.4
11	154.0	154.2	142.2	142.4	153.5	141.6
12	148.0	147.9	136.5	136.6	148.1	136.3
13	151.4	151.6	155.6	155.7	151.0	155.1

Table 5.35. (continued).

C Atom	Chloro-phyll a	Chloro-phyll a'	Pheo-phytin a	Pheo-phytin a'	Pyro-chloro-phyll a	Pyro-pheo-phytin a
14	146.1	146.1	151.4	151.3	145.9	151.2
15	147.7	147.8	138.5	138.3	147.4	138.4
16	161.4	162.1	149.9	150.5	161.3	149.0
17	155.8	155.4	162.1	161.9	155.1	161.4
18	167.4	168.0	172.8	173.1	167.0	171.9
α	100.0	100.0	97.5	97.5	99.7	97.3
β	107.1	107.3	104.4	104.4	106.8	103.9
γ	106.2	106.7	106.8	107.3	106.4	107.2
δ	92.8	92.9	93.9	93.9	92.6	93.7
1a	12.6	12.6	11.9	11.9	12.6	12.0
2a	131.5	131.5	129.7	129.7	131.7	130.0
2b	118.9	118.9	122.1	122.1	118.6	121.9
3a	11.2	11.2	10.8	10.7	11.2	11.0
4a	20.0	20.0	19.5	19.5	20.2	19.7
4b	18.0	18.0	17.4	17.4	18.1	17.6
5a	12.6	12.6	11.7	11.6	12.6	11.5
7a	30.9	30.9	31.7	31.2	31.0	31.6
7b	30.1	30.1	30.2	29.8	30.7	30.0
7c	172.1	172.1	172.9	172.9	172.9	172.9
8a	23.9	23.9	23.3	23.5	23.8	23.3
10a	171.0	172.1	169.9	171.1		
10b	52.0	52.0	52.5	52.4		
P1	61.3	61.1	61.4	61.4	61.3	61.4
P2	119.4	119.4	119.5	119.5	119.5	119.5
P3	142.2	142.3	142.2	142.3	142.4	142.5
P3a	16.2	16.2	16.2	16.2	16.2	16.3
P4	40.4	40.4	40.5	40.5	40.5	40.5
P5	25.8	25.8	25.8	25.8	25.9	25.8
P6	37.4	37.4	37.4	37.4	37.4	37.4
P7	33.4	33.4	33.5	33.5	33.5	33.5
P7a	20.0	20.0	19.9	19.9	20.0	20.0
P8	38.0	38.0	38.1	38.1	38.2	38.1
P9	25.2	25.2	25.2	25.2	25.2	25.2
P10	38.0	38.0	38.1	38.1	38.2	38.1
P11	33.6	33.6	33.6	33.6	33.6	33.6
P11a	20.0	20.0	19.9	19.9	20.0	20.0
P12	38.0	38.0	38.1	38.1	38.2	38.1
P13	25.6	25.6	25.6	25.6	25.6	25.6
P14	40.1	40.1	40.2	40.2	40.2	40.2
P15	28.7	28.7	28.8	28.8	28.8	28.8
P15a	22.9	22.9	22.9	22.9	22.9	22.9
P16	23.0	23.0	23.0	23.0	23.0	23.0

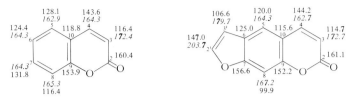

Fig. 5.16. Structures, ^{13}C chemical shifts (in ppm) and one-bond carbon-hydrogen coupling constants (in Hz, *italic types*) of coumarin (left side) and the furanocoumarin psoralen (right side).

Recording the ^{13}C NMR spectra of coumarins in different solvents, such as deuteriochloroform, dimethyl sulfoxide [959, 960] or methanol [961], shows that the shifts change only by about ± 1 ppm; however, protonation of the carbonyl group in 96% sulfuric acid causes dramatic solvent shifts which are, compared to those determined in chloroform, for C-2: 13.2, C-3: − 5.3, C-4: 16.5, C-5: 4.3, C-6: 7.1, C-7: 8.0, C-8: 3.4, C-9: 0.6 and C-10: 3.7 ppm [962].

Using titanium tetrachloride as complexing reagent causes strong downfield shifts for the signals of C=O (4.9) and C-4 (4.4) and much less deshielding for C-10 (0.2), C-5 (1.0), C-6 (1.5), C-7 (1.9) and C-8 (0.5). The resonances of C-3 (− 0.8) and C-9 (− 1.0 ppm) are shifted to higher field. An enhanced contribution of a dipolar mesomeric form is made responsible for this effect [963, 964].

Several approaches have been made to calculate ^{13}C chemical shifts of coumarins by MO methods. Good correlations were found between the ^{13}C chemical shift values of coumarin (also protonated) and the π charge densities calculated by the CNDO/2 method [962], and of coumarins with π charge densities calculated by the Hückel MO method (which, however, fails for methoxylated coumarins) [965]. Chemical shifts of mono- and dimethoxycoumarins have been correlated with parameters determined by refined INDO MO calculations, in which π bond orders, atom-atom polarizabilities, excitation energies and electron-nucleus distances were taken into consideration [966]. In 3-substituted 4-hydroxy and 4-hydroxy-7-methoxycoumarins chemical shifts were found to be related to Swain and Lupton's parameters $\mathscr{F}$ and $\mathscr{R}$ [388], according to equation 5.4 (SE = Substitution Effect):

$$SE = a\mathscr{F} + b\mathscr{R} + cQ + d \qquad (5.4) \, [388, 967, 975]$$

A good correlation was found between the HMO atom-atom polarizabilities π_{ij} and substitution effects (SE) for mono- and dimethoxycoumarins [965]:

$$SE = K\pi_{ij} \quad (K_{OCH_3} = 80.13) \qquad (5.5)$$

The α and β substitution effects correspond to those found for 1- and 2-substituted naphthalenes [392], however, so far unexplainable exceptions were found for 7-methyl-, 4-methoxy-, 6-cyano- and 7-nitrocoumarin [418].

Besides the signal identification methods mentioned at the beginning of this chapter, one-bond and three-bond carbon-hydrogen coupling constants are often useful parameters for the signal assignments in the ^{13}C NMR spectra of coumarins [959, 962, 965,

446 5 ^{13}C NMR Spectra of Natural Products

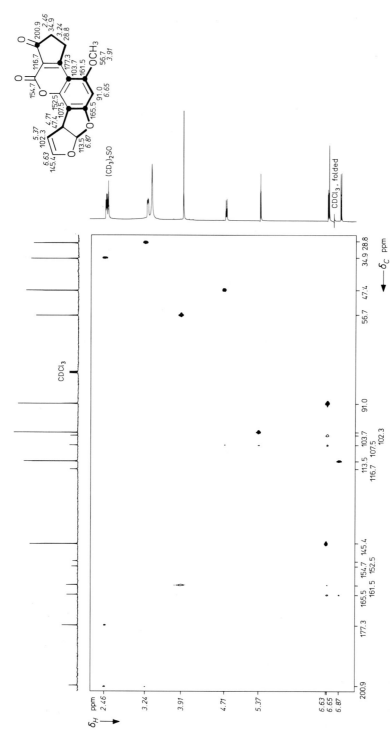

Fig. 5.17. Carbon-13 signal assignment of aflatoxin B_1 [603] by two-dimensional carbon-proton shift correlation (30 mg in 0.4 mL of hexadeuteriodimethyl sulfoxide, 30 °C, 100.576 MHz for ^{13}C, 400.133 MHz for ^{1}H; full and strong contours: correlations via one-bond couplings; empty and weaker contours: correlations via two- and three-bond couplings). Boldface printed substructures in the formula can be directly derived from this figure; the carbon nucleus at 91.0 ppm, for example, is correlated with the proton at 6.65 ppm via one-bond coupling; this proton is additionally correlated with the adjacent carbon nuclei at 165.5, 161.5, 107.5, and 103.7 ppm as indicated by correlation signals via two- and three-bond couplings.

Table 5.37. Structures and ^{13}C Chemical Shifts (δ_C in ppm) of Natural Coumarin Derivatives [635].

Osthol

Marmesin

Prantschimgin

Xanthotoxol

Isooxypeucedanin

Imperatorin

Peuarenarin

Peuarenin

Xanthalin

Esculin

Colladin

Colladonin

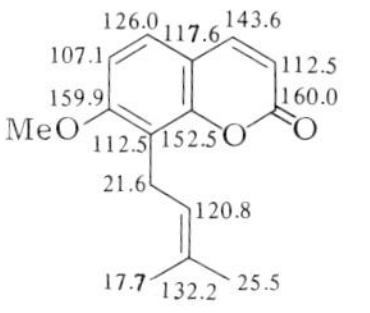

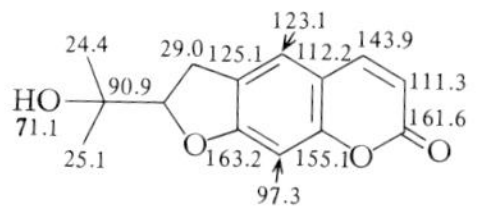

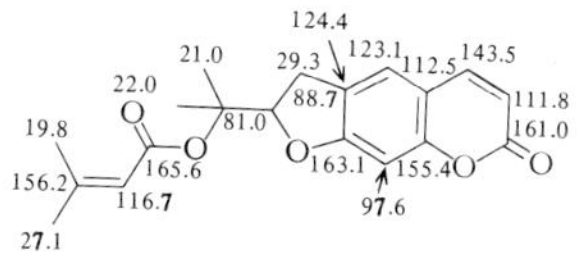

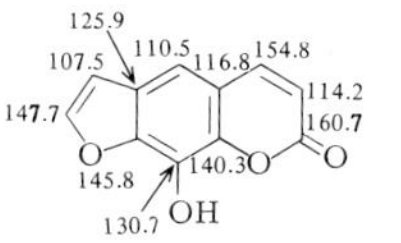

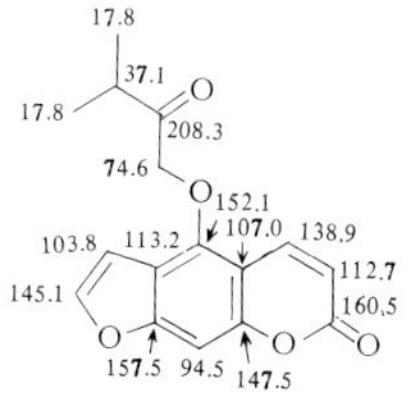

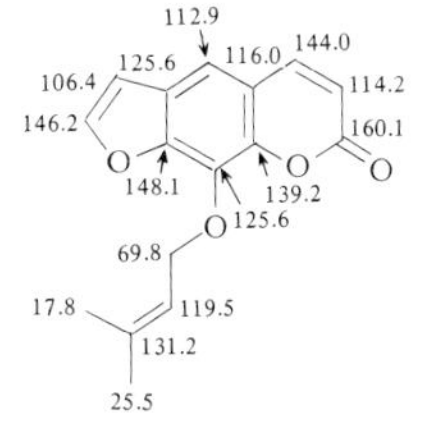

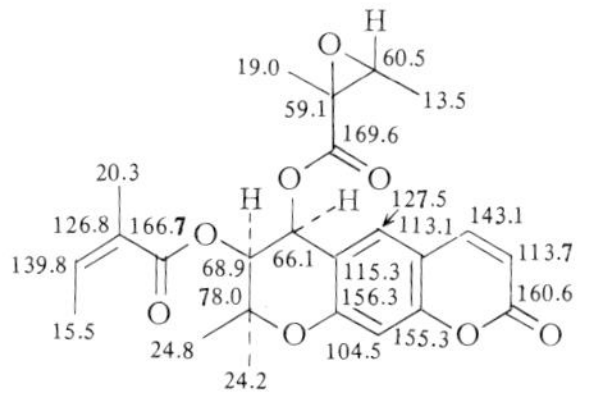

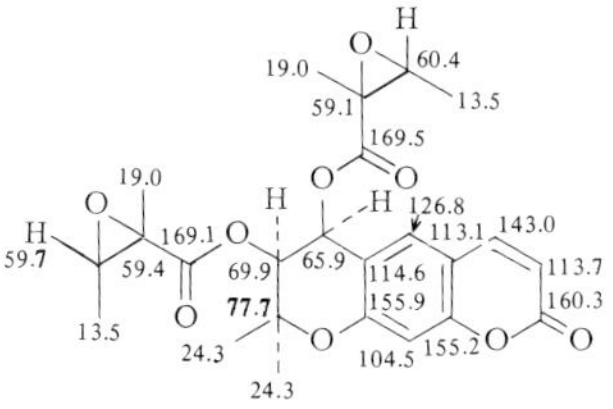

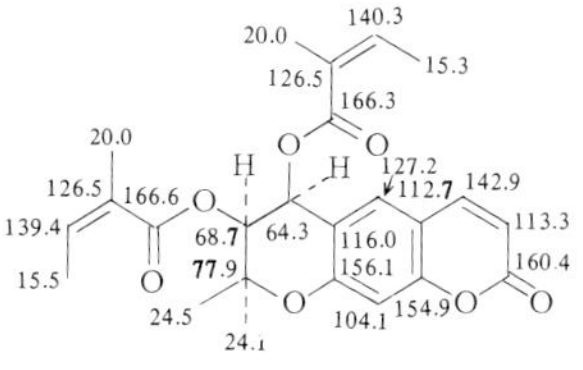

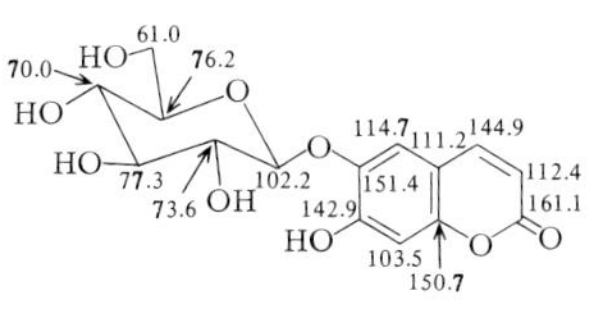

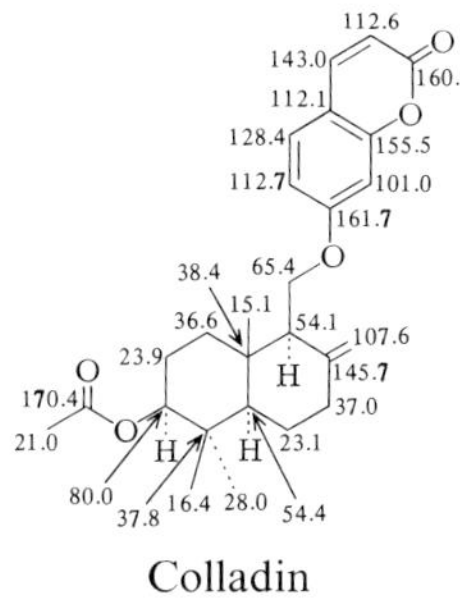

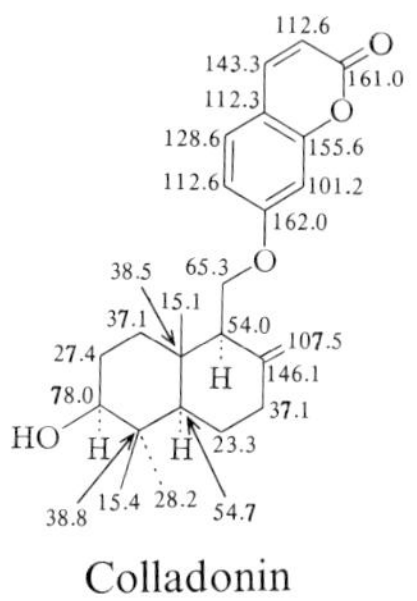

Table 5.36. ^{13}C Chemical Shifts (δ_C in ppm) of Monosubstituted Coumarin Derivatives.

Compound	C-2	C-3	C-4	C-5	C-6	C-7	C-8	C-9	C-10	Solvent	Ref.
Coumarin	160.4	116.4	143.6	128.1	124.4	131.8	116.4	153.9	118.8	CDCl$_3$	[635, 659–962, 972–974]
1-Thiocoumarin	185.4	126.0	143.7	130.0	124.2	131.6	126.5	137.7	126.2	CDCl$_3$	[421]
3-Methylcoumarin	162.1	125.7	139.2	127.0	124.3	130.4	116.3	153.2	119.6	CDCl$_3$	[970, 972]
3-Carboxycoumarin	157.2	118.2	148.5	130.3	124.9	134.4	116.2	154.6	118.1	DMSO-d$_6$	[973]
3-Hydroxycoumarin	158.5	141.8	115.0	126.3	124.5	127.5	115.6	149.2	120.7	DMSO	[971]
3-Chlorocoumarin	156.5	120.8	140.7	127.9	124.9	131.8	116.1	152.2	118.8	CDCl$_3$/DMSO-d$_6$	[418]
3-Bromocoumarin	156.3	111.0	144.0	126.9	124.5	131.6	116.0	152.6	118.8	CDCl$_3$	[469]
4-Methylcoumarin	160.5	115.1	152.3	124.6	124.2	131.7	116.9	153.5	120.0	CDCl$_3$	[970, 972]
4-Hydroxycoumarin	162.9	91.3	160.0	123.3	123.5	132.1	116.1	153.7	116.1	CDCl$_3$	[962, 977, 978]
4-Methoxycoumarin	162.5	90.0	166.2	122.8	123.7	132.2	116.5	153.1	115.5	CDCl$_3$	[971, 974, 979]
5-Methylcoumarin	160.6	115.9	140.4	136.3	125.7	131.6	114.7	154.6	117.7	CDCl$_3$	[980]
6-Methylcoumarin	160.9	116.5	143.4	127.8	134.1	132.8	116.5	152.2	118.6	CDCl$_3$	[961, 970]
6-Carboxycoumarin	161.2	117.3	144.3	130.7	127.8	133.6	117.3	157.0	119.0	CDCl$_3$ + CH$_3$OH (2:1)	[961]
6-Cyanocoumarin	160.2	118.7	143.2	133.3	108.9	135.4	118.7	156.9	120.9	CDCl$_3$ + CH$_3$OH (2:1)	[961]
6-Hydroxycoumarin	160.1	116.1	143.8	112.4	153.7	119.7	116.9	146.8	119.1	DMSO	[961, 971]
5-Methoxycoumarin	160.1	117.1	142.6	109.8	155.9	119.0	117.7	148.4	119.1	CCl$_4$ + CDCl$_3$ (3:2)	[961, 965]
6-Aminocoumarin	162.8	116.4	144.6	112.2	144.4	120.5	117.5	147.2	119.7	CDCl$_3$ + CH$_3$OH (2:1)	[961]

Table 5.36. (continued).

Compound	C-2	C-3	C-4	C-5	C-6	C-7	C-8	C-9	C-10	Solvent	Ref.
6-Nitrocoumarin	160.0	118.8	143.4	124.4	144.6	127.0	118.4	157.9	119.5	CDCl$_3$ + CH$_3$OH (2:1)	[961]
6-Chlorocoumarin	159.6	117.6	141.9	126.8	129.4	131.4	118.0	152.1	119.6	CDCl$_3$	[961, 962]
6-Bromocoumarin	59.4	117.5	141.8	129.9	116.7	134.2	118.3	152.7	120.2	CDCl$_3$	[969]
7-Methylcoumarin	160.9	115.4	143.4	127.6	125.6	143.1	116.9	154.2	116.5	CDCl$_3$	[961, 970]
7-Carboxycoumarin	161.6	118.0	143.8	128.6	126.0	134.6	118.5	154.0	122.6	CDCl$_3$ + CH$_3$OH (2:1)	[961]
7-Aminocoumarin	163.6	109.6	145.2	129.5	112.6	152.5	100.5	156.6	110.3	CDCl$_3$ + CH$_3$OH (2:1)	[961]
7-Nitrocoumarin	160.1	120.1	142.7	129.5	119.5	148.2	112.7	154.1	124.1	CDCl$_3$ + CH$_3$OH (2:1)	[961]
7-Chlorocoumarin	161.0	116.5	143.8	129.3	125.5	138.1	117.3	154.5	117.9	CDCl$_3$ + CH$_3$OH (2:1)	[961]
7-Bromocoumarin	159.4	116.6	142.4	128.6	127.6	125.5	119.8	154.1	117.6	CDCl$_3$	[969]
8-Methylcoumarin	160.9	116.3	143.8	125.6	124.0	133.2	126.3	152.4	118.6	CDCl$_3$	[970]
8-Hydroxycoumarin	160.0	116.1	144.5	118.4	124.4	118.4	144.7	142.4	119.7	DMSO	[971]
8-Methoxycoumarin	159.4	116.9	143.0	119.0	123.9	113.6	147.2	144.8	119.3	CCl$_4$ + CDCl$_3$ (3:2)	[965]

968–976]. With the exception of C-3 ($J = 172.4$ Hz), the $^1J(CH)$ values of coumarin are all in the range of 163–165.5 Hz; so the closely spaced signals of C-3 and C-8 (≈ 116.4 ppm) can be differentiated.

^{13}C-enriched coumarin derivatives can be obtained by biosynthesis, as was demonstrated for aflatoxin B$_1$ [981–985]. For these specifically ^{13}C-labeled compounds enhanced signal intensities facilitate the resonance assignments. Complete signal assignments may also be achieved by two-dimensional carbon-proton shift correlations, as demonstrated in Figure 5.17. The ^{13}C chemical shifts of a series of further aflatoxins and related sterigmatocystins, such as aflatoxin B$_2$, B$_{2a}$, B$_3$, D$_1$, G$_1$, G$_2$, G$_{2a}$ and O-methyl-, dihydro-, and O-methyldihydrosterigmatocystin are reported in ref. [984]. The structures and ^{13}C chemical shifts of selected natural coumarin derivatives are given in Table 5.37.

5.10. Flavonoids

Flavonoids are compounds widely found in all green plants, new representatives of which are still being isolated today. The ^{13}C NMR data collected in Tables 5.38 and 5.39 of chalcones, flavones, flavonols, isoflavones, dihydroflavones, dihydroflavonols, flavans and homoisoflavanones (aglycones) provide material for the structure elucidation of naturally occurring flavonoids. Further information on the ^{13}C NMR parameters of flavonoids is to be found in refs. [422, 986–997]. The signal assignments of flavonoids could be performed straightforwardly by comparison of the proton broadband-decoupled and off-resonance decoupled spectra of methyl benzoate, hydroxylated acetophenones, cin-

Table 5.38 a. ^{13}C Chemical Shifts (δ_C in ppm) of the Sugar Carbons in Flavone Glycosides. G Refers to Glucose, R to Rhamnose; Solvent: DMSO-d$_6$) [990].

Compound	C-1 R+G	C-2 G	C-3 G	C-4 G	C-5 G	C-6 G	C-2 R	C-3 R	C-4 R	C-5 R	C-6 R
Rutin (Quercetin 3-O-rutinoside)	101.5 100.7	74.2	76.8	70.4	76.1	67.1	70.4	70.8	72.2	68.2	17.5
Quercitrin (Quercetin 3-O-rhamnoside)	101.9						70.4	70.6	71.5	70.1	17.3
Hesperidin (Hesperetin 7-O-rutinoside)	100.7 99.8	73.3	76.6	69.9	75.8	66.4	70.6	71.0	72.4	68.6	18.2
Pseudobaptisin	100.5 100.7	73.2	76.7	70.2	75.9	66.7	70.4	71.0	72.3	68.4	17.7
Naringin (Naringenin 7-O-neohesperidoside)	100.4 98.0	76.7	77.2	70.1	77.0	60.8	70.4	70.7	72.1	68.2	17.8

namic acids, 4-methoxystyrene, coumarins and related flavonoids, also taking empirical substituent additivity parameters into consideration. In J-modulated spin-echoe experiments quaternary and secondary carbons cause negative, and tertiary and primary positive singlets, which facilitates the assignment of the resonances of flavonoids [996]. Table 5.38 a shows the ^{13}C chemical shifts of the sugar moieties of some flavone glycosides of Table 5.38 b; the assignments were performed according to the guidelines given in Section 5.4.

From the data given in Table 5.38 b the following guidelines can be set up, which are helpful for the interpretation of the ^{13}C NMR spectra of new flavonoids:

1. In 5,7-dihydroxyflavonoids carbons 6 and 8 are much more shielded compared to benzene carbons.

2. Based on three-bond carbon-hydrogen coupling constants, specific selective proton decoupling experiments and investigations of specifically deuterated compounds, the resonances of C-6 and C-8 in 5,7-dihydroxyflavonoids appear in the range of 90 to 100 ppm and C-8 is always more shielded compared to C-6. The chemical shift differences found are small for flavanones ($\sim$ 1 ppm) and larger for flavones and flavonols ($\sim$ 5 ppm).

3. Methylation of phenolic hydroxy groups (except C_5-OH) generally shifts the phenolic α carbon resonance downfield ($\sim$ 0.1–3 ppm) and that of the *ortho* carbon signal upfield ($\sim$ 1–4 ppm).

4. The ^{13}C chemical shifts of the carbonyl resonances of 5,7-dihydroxyflavonoids are characteristic of the different types of flavonoid compounds: For flavones and isoflavones the C-4 resonance generally appears around 181 ppm, for dihydroflavones around 196 ppm and for flavonols around 176 ppm.

5. Glycosylation of a flavonoid hydroxy group shifts the phenolic carbon resonance 1–2 ppm upfield and the *ortho*-related carbon resonance up to almost 10 ppm (*e.g.* quercetin) downfield.

5.11 Elucidation of Biosynthetic Pathways

The application of ^{13}C NMR spectroscopy has become one of the most informative standard methods for the elucidation of biosynthetic pathways. This is demonstrated by numerous examples reported in the literature [998–1003] and by some selected examples summarized in this section. By adding ^{13}C-enriched precursors to culture media, specifically labeled biosynthetic materials can be isolated. ^{13}C-enriched positions are easy to identify as they cause larger signal intensities in the ^{13}C NMR spectrum. In most cases an enrichment of 0.5% ^{13}C above natural abundance is sufficient to detect the labeled site.

Table 5.38b. Structures and ^{13}C Chemical Shifts (δ_C in ppm) of Selected Chalcones, Flavones, Flavonols, Isoflavones, Dihydroflavones, Dihydroflavonols and Flavans (Aglycones). Spectra were Recorded in DMSO-d_6. Except those of Flavone (CDCl$_3$) and 2'-Hydroxy-, and 2,2'-Dihydroxychalcone (D$_2$O/DMSO-d_6 (2:7)) [989, 990].

Chalcones

Compound	Subst.
Isoliquiritigenin	2',4,4'(OH)$_3$
2,2'-Dihydroxychalcone	2,2'(OH)$_2$
2'-Hydroxychalcone	2'(OH)

Compound	C-β	C-α	C=O	C-2'	C-3'	C-4'	C-5'	C-6'	C-1'	C-1	C-2	C-3	C-4	C-5	C-6	–OCH$_3$
Isoliquiritigenin	143.8	117.8	191.4	164.6	102.6	165.4	107.9	132.3	113.3	125.8	130.6	115.8	159.9	115.8	130.6	
2,2'-Dihydroxychalcone	141.8	117.3	195.0	162.7	118.7	137.3	120.4	131.2	122.2	121.0	158.2	120.4	133.7	121.0	130.1	
2'-Hydroxychalcone	145.5	118.3	194.7	161.7	118.6	137.2	120.2	131.0	122.2	136.8	126.9	129.3	131.5	129.3	126.9	

Flavones and Flavonols

Compound	Subst.
Chrysin	5,7(OH)$_2$
Apigenin	4',5,7(OH)$_3$
Acacetin	5,7(OH)$_2$; 4'(OCH$_3$)
Thevetiaflavone	4,7(OH)$_2$; 5(OCH$_3$)
Genkwanin	4,5(OH)$_2$; 7(OCH$_3$)
Luteolin	3',4',5,7(OH)$_4$
Chrysoeriol	4',5,7(OH)$_3$; 3'(OCH$_3$)
Diosmetin	3',5,7(OH)$_3$; 4'(OCH$_3$)

Compound	Subst.
Kaempferol	3,4',5,7(OH)$_4$
Quercetin	3,3',4',5,7(OH)$_5$
Azaleatin	3,3',4,7(OH)$_4$; 5(OCH$_3$)
Rhamnetin	3,3',4,5(OH)$_4$; 7(OCH$_3$)
Ombuin	3,3',5(OH)$_3$; 4',7(OCH$_3$)$_2$
Quercitrin	3',4',5,7(OH)$_4$; 3-O-rhamnosyl
Rutin	3',4',5,7(OH)$_4$; 3-O-rutinosyl

Table 5.38b. (continued).

Compound	C-2	C-3	C-4	C-5	C-6	C-7	C-8	C-9	C-10	C-1'	C-2'	C-3'	C-4'	C-5'	C-6'	–OCH$_3$
Flavone	163.2	107.6	178.4	125.7	125.2	133.7	118.1	156.3	124.0	131.8	126.3	129.0	131.6	129.0	126.3	
Chrysin	163.0	105.0	181.6	157.3	98.9	164.3	94.0	161.5	103.9	130.6	126.0	128.0	131.7	128.0	126.0	
Apigenin	164.1	102.8	181.8	157.3	98.8	163.7	94.0	161.5	103.7	121.3	128.4	116.0	161.1	116.0	128.4	
Acacetin	163.9	103.9	182.3	157.9	99.4	164.8	94.3	162.2	104.4	123.5	128.4	114.8	162.8	114.8	128.4	55.5
Thevetiaflavone	160.4	107.1	175.7	159.0	96.4	162.4	95.2	159.9	106.0	121.5	127.7	115.8	160.6	115.8	127.7	55.8
Genkwanin	164.6	103.4	182.3	157.7	98.2	165.6	92.9	161.8	105.0	121.6	128.8	166.3	161.8	116.3	128.8	56.0
Luteolin	164.5	103.3	182.2	157.9	99.2	164.7	94.2	162.1	104.2	119.3	113.8	146.2	150.1	116.4	122.1	
Chrysoeriol	163.7	103.8	181.8	157.4	98.8	164.2	94.0	161.6	103.3	120.4	110.2	150.8	148.0	115.8	121.7	56.0
Diosmetin	163.6	104.0	181.8	157.5	99.0	164.4	94.0	161.7	103.7	118.7	113.1	146.9	151.2	112.1	123.3	55.8
Kaempferol	146.8	135.6	175.9	156.2	98.2	163.9	93.5	160.7	103.1	121.7	129.5	115.4	159.2	115.4	129.5	
Quercetin	146.9	135.8	175.9	156.2	98.3	164.0	93.5	160.8	103.1	122.1	115.2	145.1	147.7	115.7	120.1	
Azaleatin	142.0	137.1	171.1	158.1	96.0	162.6	94.8	160.6	105.2	122.4	114.6	145.1	147.1	115.7	119.3	56.0
Rhamnetin	147.3	136.0	175.9	156.0	97.4	164.9	91.8	160.4	103.7	121.9	115.2	145.0	147.8	115.6	120.1	55.9
Ombuin	146.7	136.4	176.0	156.0	97.4	164.9	91.8	160.4	104.0	123.4	114.8	146.2	149.4	111.7	119.8	55.6 55.9
Quercitrin	156.4	134.4	177.7	157.0	98.6	164.0	93.5	161.2	104.2	130.3	115.4	145.1	148.3	115.8	121.0	
Rutin	156.4	133.6	177.4	156.6	98.8	164.0	93.6	161.2	105.2	121.6	115.3	144.6	148.3	116.5	121.6	

Table 5.38b. (continued).

Isoflavones

Compound	Subst.
Genistein	4′,5,7(OH)₃
Pseudobaptisin	3′,4′(−O−CH₂−O); 7-O-rutinosyl

Compound	C-2	C-3	C-4	C-5	C-6	C-7	C-8	C-9	C-10	C-1′	C-2′	C-3′	C-4′	C-5′	C-6′	−OCH₃
Genistein	153.6	121.4	180.2	157.6	98.6	164.3	93.7	157.6	104.6	122.4	130.0	115.2	162.1	115.2	130.0	
Pseudobaptisin	153.7	123.4	174.5	126.9	115.6	161.4	103.9	156.9	118.7	125.6	109.3	147.0	107.9	122.2		

Dihydroflavones

Compound	Subst.
Pinocembrin	5,7(OH)₂
Naringenin	4′,5,7(OH)₃
Isosakuranetin	5,7(OH)₂; 4′(OCH₃)
Eriodictyol	3′,4′,5,7(OH)₄

Compound	Subst.
Homoeriodictyol	4′,5,7(OH)₃; 3′(OCH₃)
Hesperetin	3′,5,7(OH)₃; 4′(OCH₃)
Naringin	4′,5(OH)₂; 7-O-neohesperidosyl
Hesperidin	3′,5(OH)₂; 4′(OCH₃); 7-O-rutinosyl

Compound	C-2	C-3	C-4	C-5	C-6	C-7	C-8	C-9	C-10	C-1′	C-2′	C-3′	C-4′	C-5′	C-6′	−OCH₃
Pinocembrin	87.4	42.2	195.8	163.6	96.1	166.6	95.1	162.7	101.9	138.0	126.5	128.5	128.5	128.5	126.5	
Naringenin	78.4	42.0	196.2	163.6	95.9	166.7	95.0	162.9	101.8	128.9	128.2	115.2	157.8	115.2	128.2	
Isosakuranetin	77.8	41.9	196.1	163.5	95.8	166.6	94.9	162.7	101.7	130.5	128.6	113.5	159.4	113.5	128.6	55.0
Eriodictyol	78.3	42.2	196.2	163.4	95.7	166.6	94.8	162.8	101.7	129.4	114.2	145.1	145.6	115.3	117.8	
Homoeriodictyol	78.7	42.1	196.3	163.5	95.8	166.6	95.0	162.9	101.8	129.4	111.1	147.5	146.9	115.2	119.6	55.6
Hesperetin	78.5	42.1	196.2	163.8	96.2	166.9	95.4	163.0	102.1	131.4	114.3	146.7	148.1	112.1	118.0	55.9
Naringin	78.6	42.0	196.7	162.9	96.5	164.9	95.4	162.7	103.5	128.7	128.0	115.3	157.7	115.3	128.0	
Hesperidin	78.4	42.0	196.7	163.0	96.7	165.2	95.8	162.5	103.5	131.2	114.3	146.7	148.1	112.7	117.8	

Table 5.38b. (continued).

Dihydroflavonols

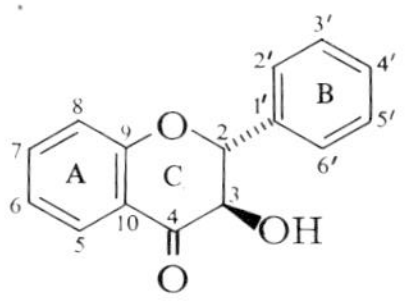

Compound	C-2	C-3	C-4	C-5	C-6	C-7	C-8	C-9	C-10	C-1'	C-2'	C-3'	C-4'	C-5'	C-6'	–OCH$_3$
Taxifolin	83.1	71.7	197.1	163.3	96.1	166.8	95.1	162.5	100.6	128.1	115.3	144.9	145.7	115.3	119.2	
Tetramethyltaxifolin	83.6	72.5	189.7	163.7	93.8	165.7	93.1	161.7	103.7	129.9	112.5	148.9	149.5	112.3	120.6	55.9 55.6

Subst.: Taxifolin 3',4',5,7(OH)$_4$; Tetramethyltaxifolin 3',4',5,7(OCH$_3$)$_4$

Flavans

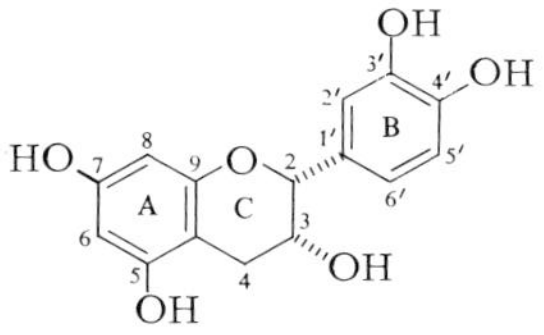

(−)-Epicatechin

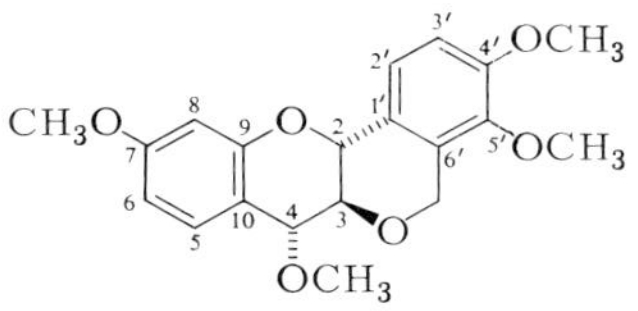

Tetramethylpeltagynol

Compound	C-2	C-3	C-4	C-5	C-6	C-7	C-8	C-9	C-10	C-1'	C-2'	C-3'	C-4'	C-5'	C-6'	–OCH$_3$
(−)-Epicatechin	78.1	65.1	28.0	156.0	95.6	156.3	94.5	155.7	98.8	130.7	118.1	144.4	144.5	115.0	118.1	
Tetramethylpeltagynol	77.5	71.6	68.0	129.0	110.1	159.6	100.7	154.3	118.5	127.3	108.1	148.2	149.1	108.1	124.6	56.1 55.9 55.2 67.0 (CH$_2$)

Table 5.39. Structures and ^{13}C Chemical Shifts (δ_C in ppm) of Representative Homo-isoflavanones (Solvent: CD_3OD) [997].

Compound	R^1	R^2	R^3	R^4
1	OCH_3	OH	H	H
2	OCH_3	OCH_3	H	H
3	OCH_3	OCH_3	OH	H
4	H	OH	H	H
5	H	OCH_3	H	H
6	H	OH	OCH_3	H
7	H	OCH_3	OH	H
8	H	OCH_3	OCH_3	H
9	H	OCH_3	$OCOCH_3$	H
10	H	OH	H	OH and 4'OCH_3

C Atom	1	2	3	4	5	6	7	8	9	10
C-2	70.2	70.6	70.6	70.1	70.4	70.5	70.5	70.4	70.8	70.2
C-3	49.1	48.1	48.5	48.2	48.4	48.0	48.8	48.8	48.0	47.9
C-4	199.9	200.5	200.9	199.4	200.2	199.3	200.1	199.8	199.9	199.3
C-4a	102.7	103.7	104.8	101.8	103.0	102.8	103.2	103.3	103.1	102.8
C-5	156.7	156.2	149.2	165.8	165.6	161.0	158.2	161.3	161.3	165.8
C-6	130.4	131.4	135.1	97.1	95.8	97.2	93.5	93.8	93.8	97.1
C-7	161.1	162.3	151.2	168.2	169.4	161.6	158.1	162.5	160.9	168.2
C-8	95.9	92.6	131.4	95.9	94.6	130.1	127.6	130.3	121.3	95.8
C-8a	160.0	160.6	146.4	164.7	164.6	155.6	149.2	154.9	154.5	164.7
C-9	32.9	32.9	32.6	32.9	32.9	33.1	32.8	32.0	32.9	33.1
C-1'	130.0	130.1	130.0	130.2	130.2	129.8	130.2	129.7	130.0	132.4
C-2' (6')	131.0	131.2	131.1	131.2	131.2	131.1	131.1	131.0	131.2	117.0
C-3' (5')	116.3	116.5	116.4	116.4	116.5	116.4	116.4	116.4	116.5	147.9
C-4'	157.1	157.3	157.0	157.2	157.3	157.1	157.2	157.2	157.4	147.6
C-5'										113.0
C-6'										121.4
6-OCH_3	60.9	61.1	61.4							
7-OCH_3		56.7	61.6		56.2		56.7	56.7	57.0	
8-OCH_3						1.5		61.5		
4'-OCH_3										56.8
8-$OCOCH_3$									170.7	
8-$OCOCH_3$									20.1	

5.11.1 Radicinin

The ^{13}C NMR spectrum of radicinin, isolated from cultures of *Stenphylium radicinum* containing sodium [2-^{13}C]-acetate, showed strong signals for the carbons C-3, -5, -9, -11, -12 and -14 only. *Stenphylium radicinum* cultures grown in media containing sodium [1-^{13}C]-acetate yielded a radicinin giving intense resonances for C-2, -4, -6, -8, -10 and -13 only [1004]. These ^{13}C NMR experiments prove that only acetate is needed for the biosynthesis of radicinin, as is demonstrated by the following scheme:

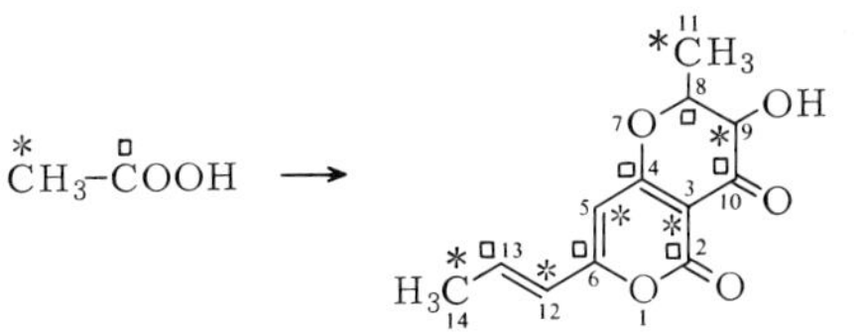

$$\overset{*}{C}H_3-\overset{\square}{C}OOH \longrightarrow \quad \text{Radicinin}$$

5.11.2 Asperlin

In a similar manner the pentaacetyl origin of the antibiotic asperlin could be proved experimentally. The ^{13}C NMR spectrum of the antibiotic lactone isolated from *Aspergillus nidulans* grown in a culture medium supplemented with sodium[2-^{13}C]-acetate shows resonances of increased intensity for the carbons 2, 4, 6, 8 and 10 [1005]. The origin of the different carbon atoms is demonstrated in the following scheme:

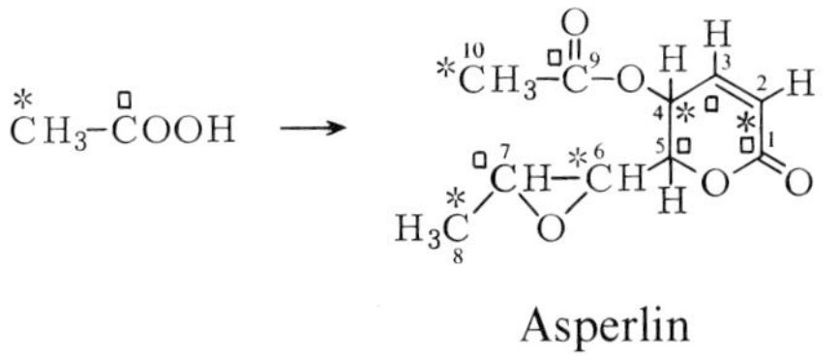

$$\overset{*}{C}H_3-\overset{\square}{C}OOH \longrightarrow \quad \text{Asperlin}$$

5.11.3 Cycloheximide

The antibiotic cycloheximide, isolated from *Strepomyces griseus*, is used as an antifungal agent and for inhibition of protein synthesis. The biosynthesis of cycloheximide *via* the polyacetate pathway is easily demonstrated by culturing *S. griseus* in the presence of [1-^{13}C]- and [2-^{13}C]-acetate and, after the isolation of the antibiotic, by comparing the natural abundance ^{13}C NMR spectrum with those of labeled cycloheximide [1006]. Table 5.40 demonstrates that six carbons (C-4, C-6, C-8, C-10, C-12 and C-14) are incorporated from the carbonyl carbon and six from the methyl carbon (C-3, C-5, C-7, C-9, C-11 and C-13) of acetate.

The labeling experiments show that [1-^{13}C]- as well as [2-^{13}C]-acetate cause an enrichment of C-2, indicating a degradation of the acetates to CO_2 and a reincorporation through carboxylation of acetate to malonate by the *S. griseus* culture. This suggestion could be confirmed by a sample of cycloheximide isolated from an *S. griseus* culture to which Na_2 $^{13}CO_3$ had been added: The ^{13}C NMR spectrum of the antibiotic showed an enrichment only for C-2. The ^{13}C NMR spectrum of cycloheximide isolayted from *S. griseus* cultures in the presence of [1, 2-^{13}C]-acetate clearly shows that 12 of the 15 carbon resonances appear as triplets. The methyl carbons (C-15, C-16) and one of the imide carbonyl signals (at 172.0 ppm), however, are singlets, indicating that these three carbons are not derived from acetate. From this ^{13}C NMR spectrum of the ^{13}C-enriched antibiotic the carbon-carbon coupling constants of cycloheximide can be determined, which are helpful for the ^{13}C chemical shift assignment of cycloheximide (Table 5.40) [1006].

Table 5.40. Biosynthetic Pathway, Structure, ^{13}C Chemical Shifts (δ_C in ppm) and ^{13}C–^{13}C Coupling Constants (J_{CC} in Hz) of Cycloheximide (Solvent: $CDCl_3$) [1006].

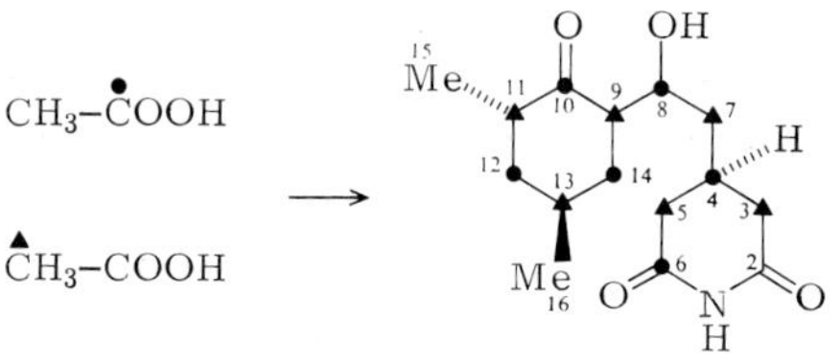

C Atom	δ_C	Coupled carbons	J_{CC}
C-2	172.0		
C-3	38.5	C(3)–C(4)	33.4
C-4	27.4	C(3)–C(4)	33.5
C-5	37.2	C(5)–C(6)	44.3
C-6	171.8	C(5)–C(6)	44.2
C-7	38.0	C(7)–C(8)	39.4
C-8	66.5	C(7)–C(8)	38.8
C-9	50.2	C(9)–C(10)	37.4
C-10	216.7	C(9)–C(10)	37.5
C-11	40.4	C(11)–C(12)	29.6
C-12	42.6	C(11)–C(12)	29.5
C-13	26.7	C(13)–C(14)	33.4
C-14	33.1	C(13)–C(14)	33.5
C-15	14.2		
C-16	18.4		

5.11.4 Averufin, Versicolorin A and their Relation to Aflatoxin B_1

The biosynthetic pathway of aflatoxin B_1, seen from Table 5.41 [983, 1007–1009], was elaborated by comparison of the natural abundance ^{13}C NMR spectra of averufin and versicolorin A with samples isolated from cultures of a mutant of *Aspergillus parasiticus* treated with [1-^{13}C]-, [2-^{13}C]-, and [1, 2-^{13}C]-acetate. With these spectra full assignment of all signals was achieved, and the carbon-carbon coupling constants could be determined (Table 5.41) [1008, 1009]. The signal assignment of the ^{13}C NMR spectrum of aflatoxin B_1 is given in Chapter 5.9. The ^{13}C NMR spectrum of averufin, biosynthesized from [1-^{13}C]-acetate, shows enhanced signals for C-1', C-3', C-5', C-1, C-3, C-6, C-8, C-9, C-11 and C-14; in the spectrum of [2-^{13}C] acetate-derived averufin the residual ten signals have increased intensity. On the basis of these studies a mechanism for the biotransformation of averufin to aflatoxin B_1 is suggested: Ring opening of averufin, followed sequentially by dehydration, epoxidation, rearrangement of the epoxide to a benzylic aldehyde and removal of the terminal acetyl group by a Baeyer-Villiger oxidation leads to versiconal acetate and then to versicolorin A (Table 5.41) [1008].

5.11.5 Virescenosides

The theory of terpene biosynthesis was confirmed by ^{13}C NMR measurements on derivatives of the diterpenic compounds virescenoside A and B isolated from cultures of the mushroom *Oospora virescens* (*Link*) *Wallr.* inoculated with [1-^{13}C]-acetate and [2-^{13}C]-acetate [1010].

Measurements on these ^{13}C-enriched compounds and the application of common assignment aids [73] led to the total signal identification of the following diterpenes; the ^{13}C shifts are outlined in Table 5.42.

5.11.6 Methyl Palmitoleate

The ^{13}C NMR spectrum of specifically ^{13}C-labeled methyl palmitoleate isolated from *Saccharomyces cerevisiae* was a great help in obtaining the complete ^{13}C signal assignment of this lipid [1011]. The shifts are listed in Table 5.43.

5 ^{13}C NMR Spectra of Natural Products

Table 5.41. Biosynthetic Pathway, Structures, ^{13}C Chemical Shifts (δ_C in ppm) and Carbon-Carbon Coupling Constants (J_{CC} in Hz) of Averufin and Versicolorin A [1008, 1009].

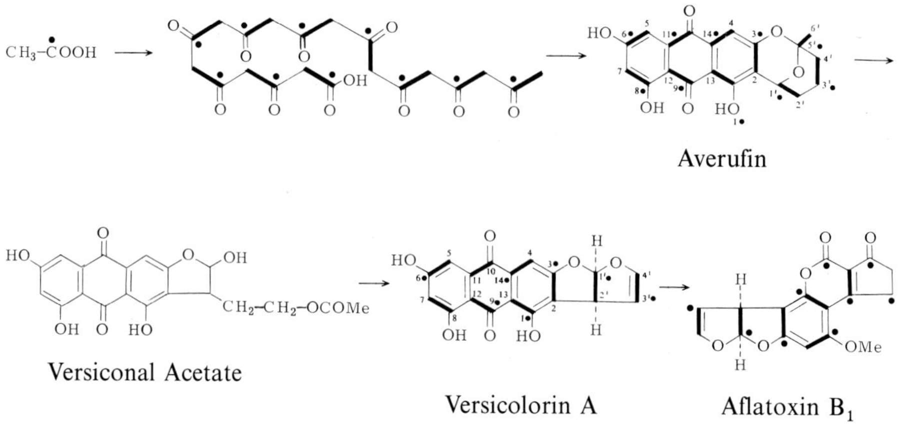

Averufin

Versiconal Acetate

Versicolorin A

Aflatoxin B$_1$

C Atom	Averufin		Versicolorin A	
	δ_C	J_{CC}	δ_C	J_{CC}
1	158.1	64.1	158.2	62.7
2	115.8	65.3	120.4	62.6
3	159.9	65.2	163.4	61.5
4	107.6	65.1	101.6	63.9
5	109.1	62.1	108.8	62.4
6	164.9	63.3	165.0	62.7
7	108.0	70.0	107.8	69.8
8	164.1	70.1	164.0	69.6
9	188.7	57.7	188.7	58.5
10	180.7	53.6	180.4	54.6
11	134.6	53.8	134.6	53.8
12	108.6	58.3	108.4	58.8
13	108.2	64.2	111.3	61.0
14	132.9	65.0	135.1	63.1
1'	66.3	34.8	112.8	33.2
2'	27.1	34.6	47.2	33.0
3'	15.6	31.6	101.2	75.0
4'	35.5	31.8	145.3	74.9
5'	100.9	48.8		
6'	27.5	49.0		

Table 5.42. ^{13}C Chemical Shifts (δ_C in ppm) of Diterpenes (Solvent: CHCl$_3$) [1010].

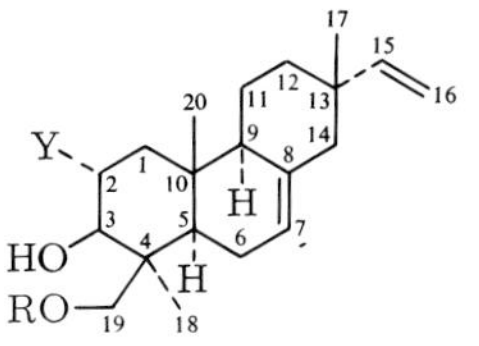

Aglycone of virescenoside A: R = H; Y = OH
Aglycone of virescenoside B: R = Y = H

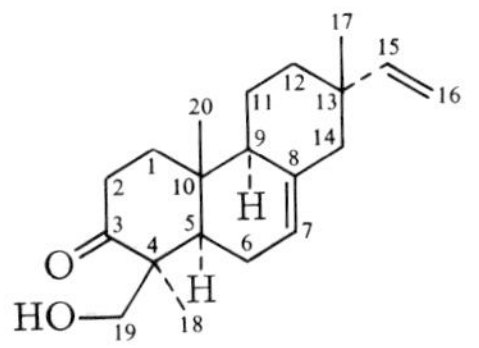

Aglycone of virescenoside C

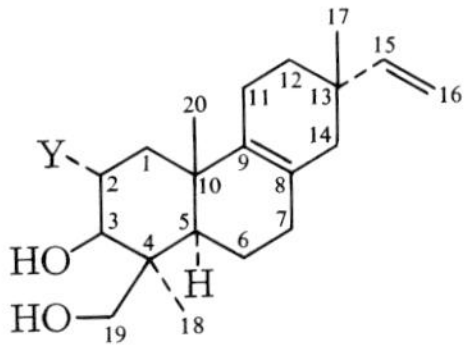

Isovirescenol A: Y = OH
Isovirescenol B: Y = H

C Atom	Aglycone of Viresce- noside A	Aglycone of Viresce- noside B	Aglycone of Viresce- noside C	Isovires- cenol A	Isovires- cenol B
C-1	43.3	38.0	36.8	42.6	35.4
C-2	69.1	28.0	35.5	69.6	29.2
C-3	85.8	81.4	217.6	85.4	81.0
C-4	43.3	42.1	52.9	43.2	43.0
C-5	51.8	51.4	53.3	51.9	52.1
C-6	23.6	23.2	24.2	21.6	21.6
C-7	121.8	121.7	121.7	32.8	33.1
C-8	135.9	136.0	136.2	125.0	125.1
C-9	52.5	52.1	50.8	134.9	136.7
C-10	37.4	35.2	35.5	38.4	37.4
C-11	21.2	20.6	21.1	19.6	19.1
C-12	36.7	36.3	36.5	35.0	34.9
C-13	37.0	37.0	37.2	35.3	35.1
C-14	46.4	46.1	46.3	42.0	42.0
C-15	150.6	150.6	150.3	146.1	146.2
C-16	110.2	109.7	109.9	111.3	111.1
C-17	22.2	21.7	21.9	28.1	28.3
C-18	24.0	23.0	22.5	23.3	22.9
C-19	66.0	64.8	66.4	65.5	64.7
C-20	17.8	16.4	16.0	21.6	20.6

Table 5.43. ^{13}C Chemical Shifts (δ_C in ppm) of Methyl Palmitoleate (Solvent: CDCl$_3$) [1011].

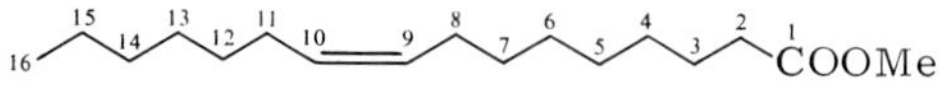

Methyl palmitoleate

C Atom	δ_C	C Atom	δ_C	C Atom	δ_C
1	173.90	7	28.80	13	29.40
2	33.70	8	29.40	14	31.50
3	24.70	9	129.60	15	22.40
4	26.90	10	129.60	16	13.70
5	28.80	11	29.40	OMe	51.10
6	28.80	12	28.80		

5.11.7 Sepedonin

The biosynthetic incorporation of [1-^{13}C]-acetate, [2-^{13}C]-acetate and [^{13}C]-formate into the molecule of the fungal tropolone sepedonin was investigated by ^{13}C NMR spectroscopy [1012]. These measurements demonstrated that carbon atom 8 is the only one originating from the formate precursor, as is seen in the scheme in Table 5.44.

Moreover, these investigations led to a total signal assignment of this natural tropolone. The shifts are listed in Table 5.44.

Table 5.44. ^{13}C Chemical Shifts (δ_C in ppm) of Sepedonin (Solvent: Pyridine) [1012].

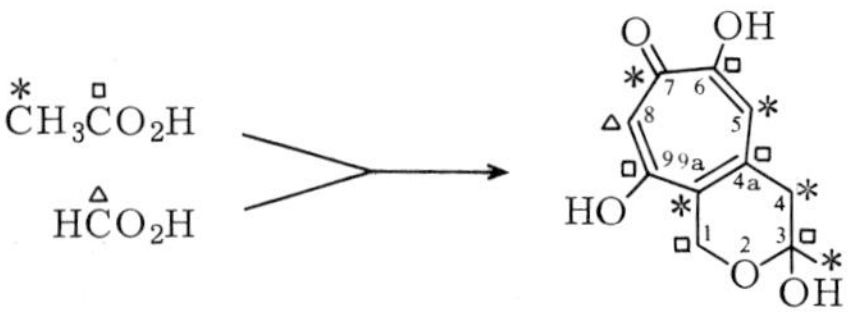

C Atom	δ_C	C Atom	δ_C	C Atom	δ_C
C-1	60.6	C-6	161.6	C-9	161.6
C-3	93.6		166.0		166.0
C-4	44.0	C-7	174.3	C-9a	128.4
C-4a	140.6	C-8	113.5	CH$_3$	29.0
C-5	115.6				

5.11.8 Antibiotic X-537 A

The ^{13}C NMR spectrum of the antibiotic X-537 A isolated from fermentations containing [1-^{13}C]-butyrate had three signals of very strong intensity caused by the carbons 13, 17 and 21. These ^{13}C NMR experiments demonstrated that all ethyl residues stem from butyrate units [1013]. The ^{13}C chemical shift values of the antibiotic X-537 A are listed in Table 5.45.

Sidechain signals were found at: 34.5, 34.0, 33.5, 31.2, 19.7, 16.2, 15.8, 15.5, 13.6, 13.3, 12.7, 12.1, 9.4 and 7.0 ppm.

Table 5.45. ^{13}C Chemical Shifts (δ_C in ppm) of the Antibiotic X-537 A (Solvent: CHCl$_3$) [1013].

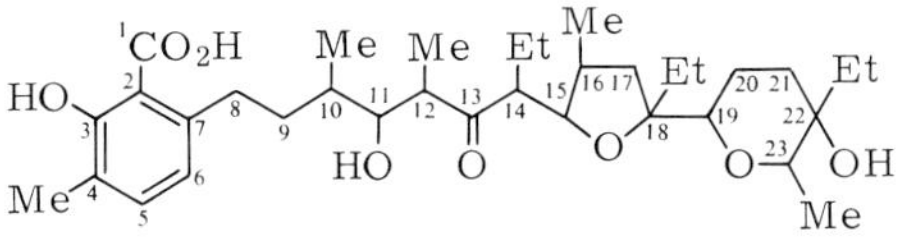

Antibiotic X-537 A

C Atom	δ_C	C Atom	δ_C	C Atom	δ_C
C-1	177.0	C-8	49.0	C-18	87.7
C-2	118.7	C-9	37.8	C-19	68.7
C-3	166.0	C-11	70.7	C-21	38.5
C-4	123.2	C-12	56.0	C-22	71.3
C-5	131.3	C-13	218.4	C-23	77.3
C-6	119.8	C-15	83.4		
C-7	143.3	C-17	29.5		

5.11.9 Cephalosporin

Part of the biogenesis of cephalosporins was confirmed by isolating ^{13}C-labeled compounds [1014]. Adding [2-^{13}C]-acetate to cultures of *Cephalosporium acremonium* yielded an antibiotic labeled at C-11, -12, -13, -14 and -19 (marked with asterisks in the formula); [1-^{13}C]-acetate, on the other hand, yielded a compound labeled at positions 10, 15 and 18 (symbolized by □ in the formula).

The specifically ^{13}C-enriched compounds were useful for the signal identification of the ^{13}C NMR spectrum of this microbial product. Spectral comparison of cephalosporin C with α-aminoadipic acid-N-ethylamide, cephalexin, 3-methyl-7(2-phenoxyacetamido)-3-cepham and 7-amino cephalosporanic acid led to the total signal identification of this antibiotic. The shifts are listed in Table 5.46.

Table 5.46. ^{13}C Chemical Shifts (δ_C in ppm) of Cephalosporin and Derivatives [1014].

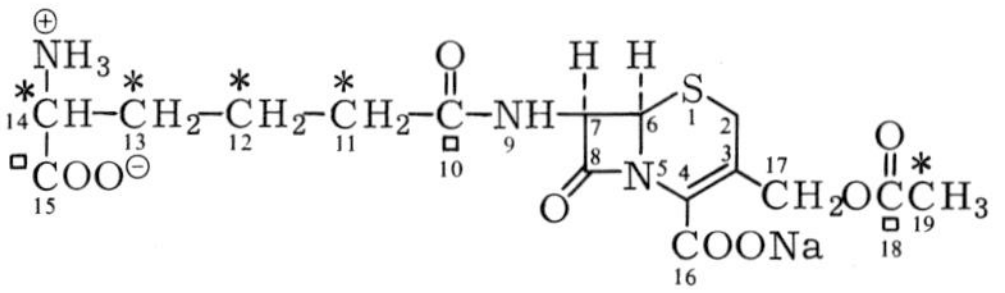

Cephalosporin C

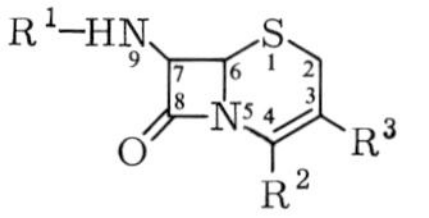

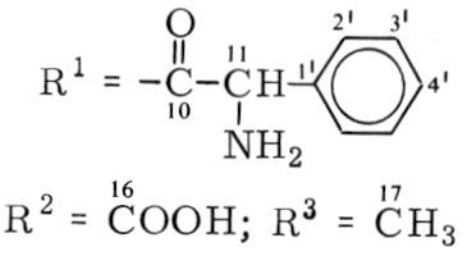

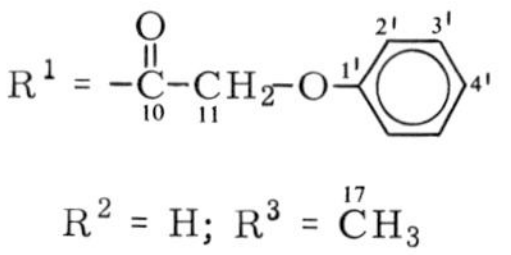

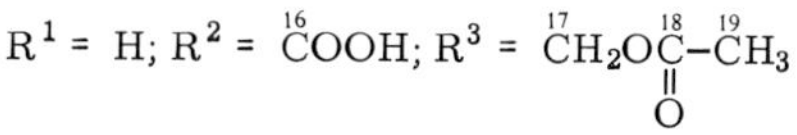

Cephalexin   3-Methyl-7(2-phenoxy-   7-Amino cephalosporanic acid
                acetamido)-3-cepham

Solvent C Atom	Cephalo- sporin C H$_2$O	D-1α-Amino- adipic acid- N-ethylamide H$_2$O	Cephalexin H$_2$O	3-Methyl- 7-(2-phenoxy- acetamido)- 3-cepham CHCl$_3$	7-Amino- cephalo- sporanic acid 3% NaHCO$_3$
2	24.70		28.00	27.80	25.50
3	134.00		142.60	116.70	132.10
4	118.00		124.10	122.70	116.70
6	57.30		57.10	56.40	62.50
7	59.30		58.60	58.20	58.80
8	168.20		167.40	163.60	163.50
10	180.50	181.50	179.60	169.60	
11	34.40	29.50	58.60	67.10	
12	19.90	17.40			
13	29.40	24.70			
14	54.60	56.70			
15	178.40	177.90			
16	171.90		173.00		168.50
17	64.70		17.70	20.60	64.50
18	177.90				174.90
19	19.30				20.50
1'		34.80 \} N−C$_2$H$_5$	128.90	158.00	
2'		10.60	131.20	115.10	
3'			128.90	130.20	
4'			130.40	118.60	

5.11.10 Prodigiosin

In a similar manner, the biosynthetic incorporation of acetate into prodigiosin was investigated by isolating the pyrrole derivative from cultures of the bacterium *Serratia marcescens* growing on media containing sodium [1-^{13}C]- or [2-^{13}C]-acetate [1015]. The ^{13}C NMR spectra of the different ^{13}C-labeled prodigiosins clearly show that carbons B 3, C 3, C 5, 2' and 4' originate from the carboxy part, in contrast to the carbons B 4, C 4, 1', 3' and 5', which originate from the methyl part of the acetate precursor (Table 5.47):

Table 5.47. ^{13}C Chemical Shifts (δ_C in ppm) of Prodigiosin Hydrochloride (Solvent: CHCl$_3$) [1015].

C Atom	δ_C	C Atom	δ_C	C Atom	δ_C
A-2	122.20	B-5	147.75	C-5	128.25
A-3	117.25	Methoxyl	58.80	C-2-methyl	12.30
A-4	111.75	Bridge	115.75	1'	25.30
A-5	126.55	head (1")		2'	29.70
B-2	146.30	C-2	120.80	3'	31.50
B-3	165.80	C-3	125.15	4'	22.50
B-4	93.05	C-4	128.25	5'	14.00

5.11.11 Myxovirescin A$_1$

Feeding *Myxococcus virescens* cultures with different ^{13}C-enriched precursors and recording the ^{13}C NMR spectra of differently labeled myxovirescins led to the following conclusions (Table 5.48) [1016]:

1. [1,2-^{13}C]-acetate: Thirteen acetate units are incorporated, eleven of which form a polyketide chain from C-1 to C-22, while two are incorporated in the section C-26 to C-36.

2. [Methyl-^{13}C] methionine: Four carbons show increased intensity: C-29, C-30, C-34 and C-37, indicating that all methyl groups except that of C-32 originate from methionine.

3. [1-^{13}C] glycine: The resonance enhancement of C-23 only demonstrates that only one unit of glycine is incorporated in the myxovirescin molecule.

Table 5.48. Structure, Schematic Representation of the Biosynthetic Incorporation of [1-^{13}C]-Acetate (●), [2-^{13}C]-Acetate (■), [Methyl-^{13}C]methionine (▲), and [1-^{13}C]Glycine (△) ([2-^{13}C]-Acetate Incorporation at C-31, C-32, C-33 (□) See Text) and ^{13}C Chemical Shifts (δ_C in ppm; Solvent: CDCl$_3$) of Myxovirescin A$_1$ [1016].

C Atom	δ_C	C Atom	δ_C
1	176.01	19	30.42
2	37.18	20	73.24
3	40.95	21	71.65
4	30.42	22	36.02
5	36.44	23	69.05
6	26.49	24	45.39
7	23.76	26	171.14
8	42.55	27	73.62
9	212.50	29	17.52
10	43.13	30	19.76
11	22.08	31	28.42
12	34.69	32	11.87
13	45.27	33	71.08
14	139.64	34	58.38
15	125.91	35	34.01
16	130.05	36	18.18
17	134.60	37	13.73
18	30.20		

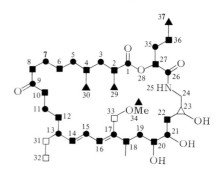

4. [2-^{13}C]-acetate: In addition to expected resonance enhancements, increased signal intensities were observed corresponding to C-2 of intact acetate for C-31, C-32 (3% enrichment, each) and C-33 (9% enrichment). This result indicates that C-33 originates from C-2 of a cleaved acetate and C-31 and C-32 arise from C-2 of acetate by a so far unknown biosynthetic pathway.

5.12 Appendix

The following table includes carbon-13 shifts (in ppm) and supplemental references of selected natural products with physiological activity. Some representatives have been portrayed in figures. Examples are colchicine (Fig. 2.19), biotin (Fig. 2.52), actinomycin D (Fig. 5.13) and aflatoxin B_1 (Fig. 5.17).

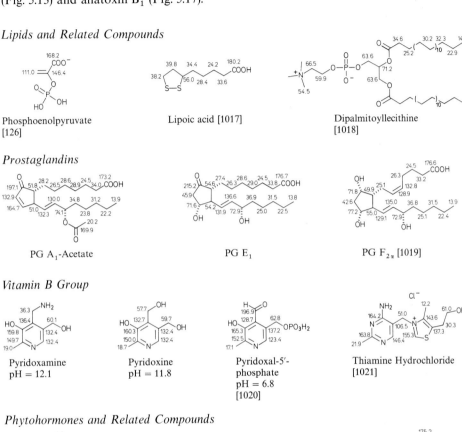

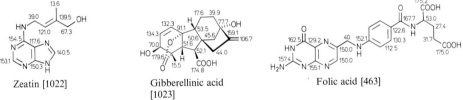

Fungal Metabolites and Antibiotics

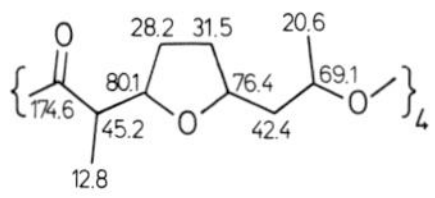

Nonactine [1025]

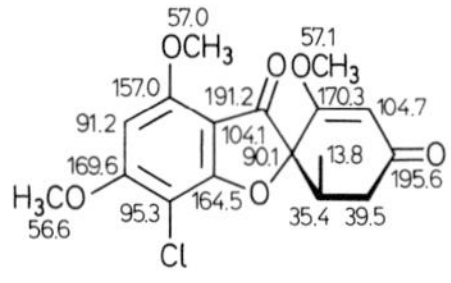

Griseofulvine

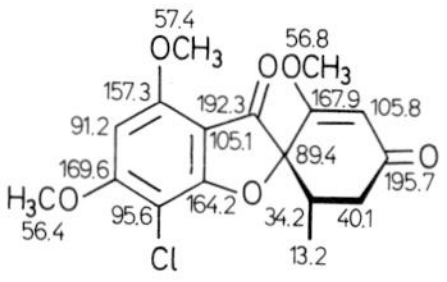

epi-Griseofulvine [1026]

Fungal Metabolites and Antibiotics

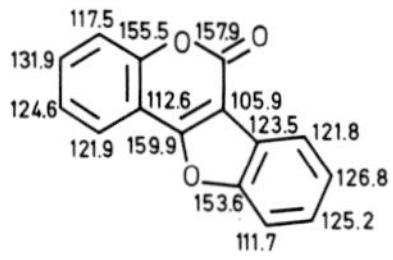

Coumestane

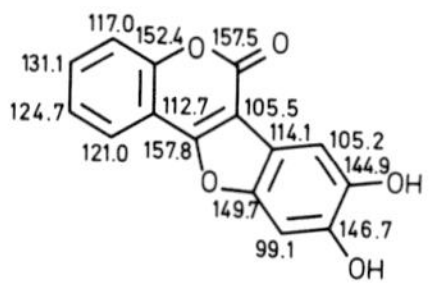

4′,5′-Dihydroxycoumestane

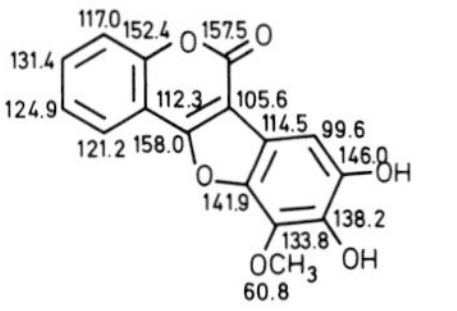

4′,5′-Dihydroxy-3′-methoxy-coumestane [1027]

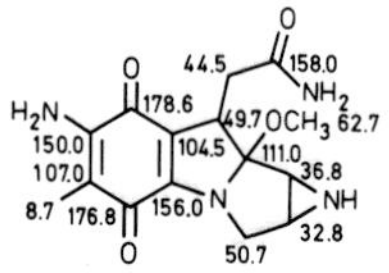

Mitomycin C [1028]

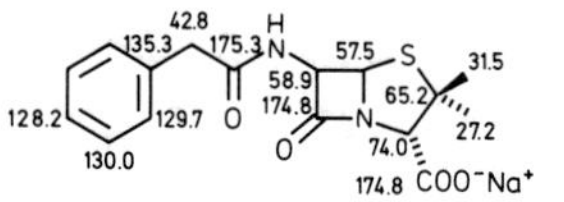

Penicillin G, sodium salt [126, 1029]

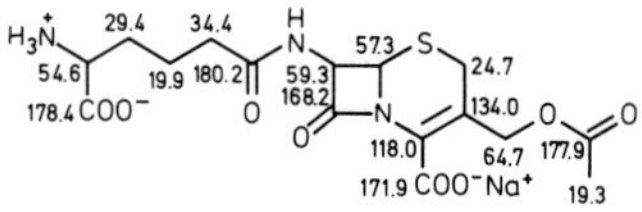

Cephalosporin C [1029]

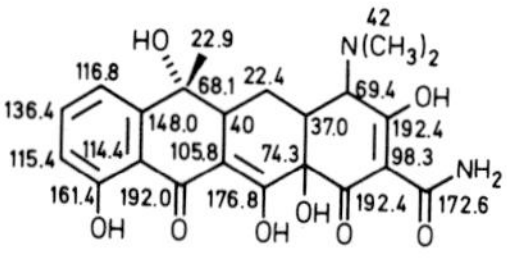

Tetracycline

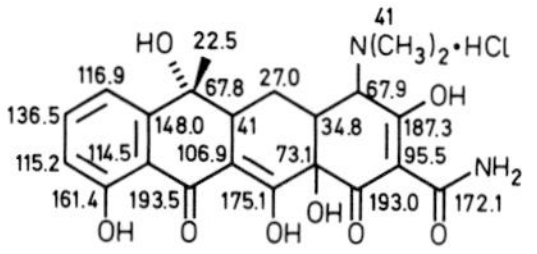

Tetracycline Hydrochloride

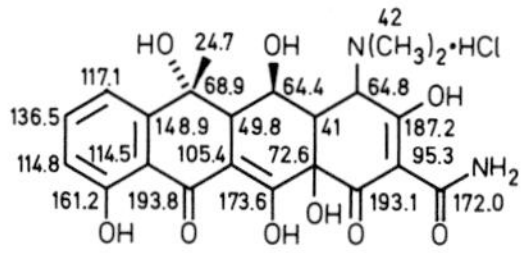

Oxytetracycline Hydrochloride [1030]

Macrolide Antibiotics

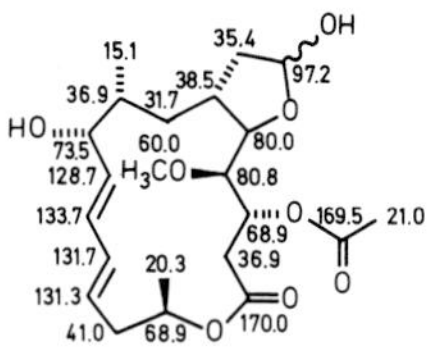

Leuconolide-A$_3$-1,18-hemiacetal

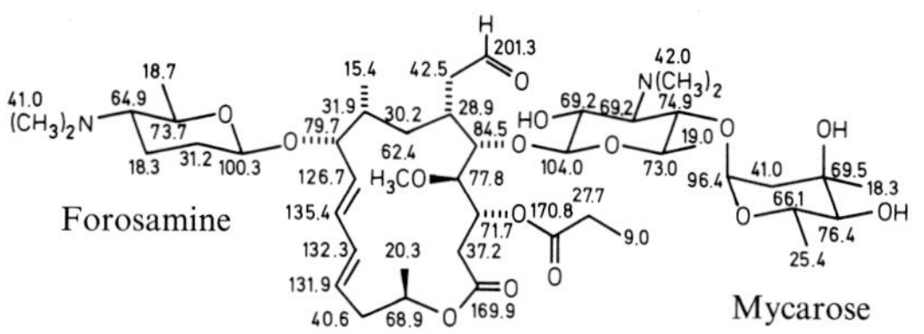

Forosamine

Spiramycin-III [1031]

Mycarose

Rifamycins

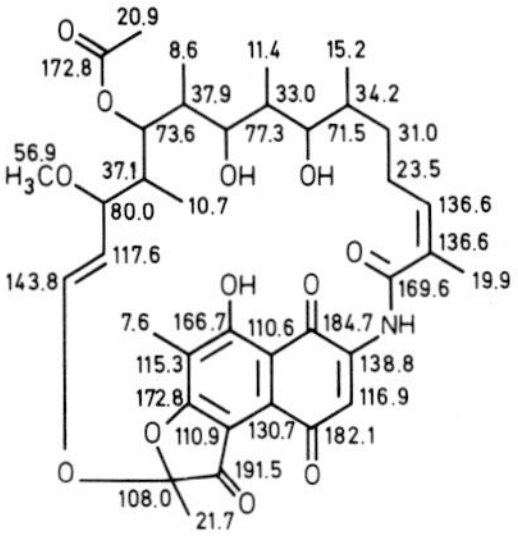

18,19-Dihydrorifamycin S

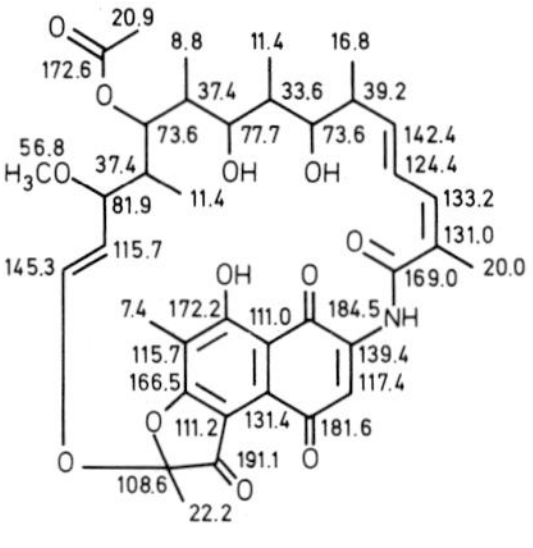

Rifamycin S [1032]

References

[1] J.A. Pople, W.G. Schneider, and H.J. Bernstein: High Resolution Nuclear Magnetic Resonance. Mc Graw-Hill, New York 1959.
[2] H. Suhr: Anwendungen der Kernmagnetischen Resonanz in der Organischen Chemie. Springer, Berlin, Heidelberg, New York 1965.
[3] E.D. Becker: High Resolution NMR. Academic Press, New York 1969.
[4] F.A. Bovey: Nuclear Magnetic Resonance Spectroscopy. Academic Press, New York 1969.
[5] J.W. Emsley, J. Feeney, and L.H. Sutcliffe: High Resolution Nuclear Magnetic Resonance Spectroscopy, Vols. 1 and 2. Pergamon Press, Oxford 1965, p. 166.
[6] F. Bloch: Phys. Rev. *70*, 460 (1946); F. Bloch, W.W. Hansen, and M. Packard: Phys. Rev. *70*, 474 (1946).
[7] T.C. Farrar and E.D. Becker: Pulse and Fourier Transform NMR. Introduction to Theory and Methods. Academic Press, New York 1971.
[8] W. Finkelnburg: Einführung in die Atomphysik, 7th and 8th Ed., Springer, Berlin, Göttingen, Heidelberg, New York 1962, p. 251.
[9] W.E. Lamb: Phys. Rev. *60*, 917 (1941).
[10] N.F. Ramsey and E.M. Purcell: Phys. Rev. *85*, 143 (1952); N.F. Ramsey: Phys. Rev. *91*, 303 (1953).
[11] P. Diehl, H. Kellerhals, and E. Lustig: NMR, Basic Principles and Progress, Vol. 6, Computer Assistance in the Analysis of High-Resolution NMR-Spectra. Springer, Berlin, Heidelberg, New York 1972.
[12] E.G. Hoffmann, W. Stempfle, G. Schroth, B. Weinmann, E. Ziegler, and J. Brandt: Angew. Chem. 84, 400 (1972); Int. Ed. Engl. *11*, 375 (1972).
[13] N. Boden: Pulsed NMR, in F.C. Nachod and J.J. Zuckerman, (eds.), Determination of Organic Structures by Physical Methods, Vol. 4. Academic Press, New York, London 1971, Chap. 2.
[14] H. Margenau and G.M. Murphy: The Mathematics of Physics and Chemistry. Van Nostrand, Princeton, N.J. 1943, Chap. 8.
[15] I.J. Lowe and R.E. Norberg: Phys. Rev. *107*, 46 (1957).
[16] A. Abragam: The Principles of Nuclear Magnetism. Clarendon Press, Oxford 1951, pp. 32, 114.
[17] R.R. Ernst and W.A. Anderson: Rev. Sci. Instr. *37*, 93 (1966); R.R. Ernst: Chimia *26*, 53 (1972).
[18] R.B. Blackman and J.W. Tuckey: The Measurement of Power Spectra. Dover, New York 1958; S. Goldman: Information Theory. Prentice Hall, Englewood Cliffs, N.J. 1953.
[19] J.W. Cooley and J.W. Tukey: Math. Comput. *19*, 296 (1965).
[20] R. Klahn and R.R. Shively: Electronics 124 (1968).
[21a] R.R. Ernst: Adv. Magn. Res. *2*, 1 (1966).

[21b] R.R. Ernst: Adv. Magn. Res. *2*, 59 (1966).
[22] J.S. Waugh: J. Mol. Spectrosc. *35*, 298 (1970).
[23] R.L. Streever and H.Y. Carr: Phys. Rev. *121*, 20 (1961).
[24] E.D. Becker, J.A. Ferretti, and T.C. Farrar: J. Am. Chem. Soc. *91*, 7784 (1969).
[25] A. Allerhand and D.W. Cochran: J. Am. Chem. Soc. *92*, 4482 (1970).
[26] P. Fellgett: Thesis. Cambridge 1951.
[27] P.L. Richards, in D.H. Martin, (ed.), Spectroscopic Techniques. North Holland Publishers, Amsterdam 1967.
[28] W.A. Anderson: Phys. Rev. *102*, 151 (1956).
[29] W. McFarlane: Nuclear Magnetic Double Resonance, in F.C. Nachod and J.J. Zuckerman, (eds.), Determination of Organic Structures by Physical Methods, Vol. 4. Academic Press, New York, London 1971, Chap. 3.
[30] W. v. Philipsborn: Angew. Chem. *83*, 470 (1971) and Int. Ed. Engl. *10*, 472 (1971).
[31] W.A. Anderson and F.A. Nelson: J. Chem. Phys. *39*, 183 (1963).
[32] R.R. Ernst: J. Chem. Phys. *45*, 3845 (1966); J. Mol. Phys. *16*, 241 (1969).
[33] K.F. Kuhlmann and D.M. Grant: J. Am. Chem. Soc. *90*, 7355 (1968).
[34] J.H. Noggle and R.E. Schirmer: The Nuclear Overhauser Effect. Chemical Applications. Academic Press, New York, London 1971.
[35] R. Freeman, J. Chem. Phys. *53*, 457 (1970).
[36] O.A. Gansow and W. Schittenhelm: J. Am. Chem. Soc. *93*, 4294 (1971).
[37] G.N. La Mar: J. Am. Chem. Soc. *93*, 1040 (1971); R. Freeman, K.G.R. Pachler, and G.N. La Mar: J. Chem. Phys. *55*, 4586 (1971).
[38] S. Barza and N. Engstrom: J. Am. Chem. Soc. *94*, 1762 (1972).
[39] R.L. Vold, J.S. Waugh, M.P. Klein, and D.E. Phelps: J. Chem. Phys. *48*, 3831 (1968).
[40] R. Freeman and R.C. Jones: J. Chem Phys. *52*, 465 (1970); R. Freeman and H.D.W. Hill: J. Chem. Phys. *53*, 4103 (1971).
[41] W. Bremser, H.D.W. Hill, and R. Freeman: Meßtechnik *78*, 14 (1971).
[42] A. Allerhand, D. Doddrell, V. Gluschko, D.W. Cochran, E. Wenkert, P.J. Lawson, and F.R.N. Gurd: J. Am. Chem. Soc. *93*, 544 (1971).
[43a] R. Freeman and H.D.W. Hill: J. Chem. Phys. *54*, 3367 (1971).
[43b] J.L. Markley, W.J. Horsley, and M.P. Klein: J. Chem. Phys. *55*, 3604 (1971).
[43c] G.C. McDonald and J.S. Leigh: J. Magn. Res. *9*, 358 (1973).
[44] E.L. Hahn: Phys. Rev. *80*, 580 (1950).
[45] H.Y. Carr and E.M. Purcell: Phys. Rev. *94*, 630 (1954).
[46] S. Meiboom and D. Gill: Rev. Sci. Instr. *29*, 688 (1958).
[47] R. Freeman and H.D.W. Hill: J. Chem. Phys. *55*, 1985 (1971).
[48] P.C. Jurs: Anal. Chem. *43*, 364 (1971); P.C. Jurs, B.R. Kowalski and T.L. Isenhour: Anal. Chem. *41*, 21 (1969); L.E. Wangen, N.M. Frew, and T.L. Isenhour: Anal. Chem. *43*, 845 (1971); T.L. Isenhour and P.C. Jurs: Anal. Chem. *43*, 20 A (1971). See also ref. [73n].
[49] J.I. Kaplan: J. Chem. Phys. *27*, 1426 (1957).
[50a] C. LeCocq and J.Y. Lallemand: J. Chem. Soc. Chem. Commun. *1981*, 150.
[50b] D.L. Rabenstein and T.T. Nakashima: Anal. Chem. *51*, 1465 A (1979); D.W. Brown, T.T. Nakashima, and D.L. Rabenstein, J. Magn. Res. *45*, 302 (1981).
[50c] J. Wesener, P. Schmitt, and H. Günther: J. Am. Chem. Soc. *106*, 10 (1984).
[50d] J.N. Schoolery: APT for ^{13}C NMR Spectrum Interpretation. Research and Application Notes, Varian NMR Applications Laboratory, Palo Alto, California, USA 1981.
[51] K.G.R. Pachler and P.L. Wessels: J. Magn. Res. *12*, 337 (1973).
[52] S. Sorensen, R.S. Hansen, and H.J. Jacobsen: J. Magn. Res. *14*, 243 (1974).
[53] A.A. Chalmers, K.G.R. Pachler, and P.L. Wessels: Org. Magn. Res. *6*, 445 (1974).
[54a] A.A. Maudsley and R.R. Ernst: Chem. Phys. Lett. *50*, 368 (1977).
[54b] G.A. Morris and R. Freeman, J. Am. Chem. Soc. *101*, 760 (1979).

[55]   D.M. Doddrell, D.T. Pegg, and M.R. Bendall, J. Magn. Res. *48*, 323 (1982); J. Chem. Phys. *77*, 2745 (1982).
[56]   M.R. Bendall, D. Doddrell, D.T. Pegg, and W.E. Hull: "DEPT", Brochure with experimental details, Bruker Analytische Meßtechnik GmbH, Karlsruhe 1982.
[57]   J. Wei, L. Chang, J. Wang, W.S. Chen, E. Friedrichs, H. Puff, and E. Breitmaier: Planta medica *1984*, 47.
[58]   A. Bax, R. Freeman, and S.P. Kempsell: J. Am. Chem. Soc. *102*, 4851 (1980).
[59]   O.W. Sorensen, R. Freeman, T. Frenkiel, T.H. Mareci, and R. Schuck: J. Magn. Res. *46*, 180 (1982).
[60]   L. Müller, A. Kumar, and R.R. Ernst: J. Chem. Phys. *63*, 5940 (1975); first application.
[61a] R. Freeman, S.P. Kempsell, and M.H. Levitt: J. Magn. Res. *34*, 663 (1979).
[61b] G. Bodenhausen, R. Freeman, and D.L. Turner: J. Chem. Phys. *65*, 839 (1976).
[62]   M. Ikura and K. Hikichi: Org. Magn. Res. *20*, 266 (1982).
[63a] A.A. Maudsley, L. Müller, and R.R. Ernst: J. Magn. Res. *28*, 463 (1977).
[63b] G. Bodenhausen and R. Freeman: J. Magn. Res. *28*, 471 (1977); J. Am. Chem. Soc. *100*, 320 (1978).
[63c] R. Freeman and G.A. Morris: J. Chem. Soc. Chem. Commun. *1978*, 684.
[64]   R.E. Hurd: "Relaxation Times", Vol. 2, No. 3, Brochure, Nicolet Instruments Inc., Madison Wisc., USA 1981.
[65a] H.J. Reich, M. Jautelat, M.T. Messe, F.J. Weigert, and J.D. Roberts: J. Am. Chem. Soc. *91*, 7445 (1969).
[65b] F.W. Wehrli, in T. Axenrod and G.A. Webb , (eds.): Nuclear Magnetic Resonance Spectroscopy of Nuclei other than Protons. J. Wiley & Sons, New York, London, Sydney, Toronto 1974, p. 175.
[66]   A. Bax: J. Magn. Res. *53*, 517 (1983).
[67a] W.E. Hull: Two-Dimensional NMR. This brochure of Bruker Analytische Meßtechnik GmbH, Karlsruhe 1982, includes experimental details, microprograms, examples and further literature references.
[67b] H. Kessler, C. Griesinger, J. Zarbock, and H. Loosli: J. Magn. Res. *57*, 331 (1984).
[68a] R. Benn and H. Günther: Angew. Chem. *95*, 381 (1983) with 160 literature references.
[68b] H. Kessler and D. Ziessow: Nachr. Chem. Tech. Lab. *30*, 488 (1982) with 58 literature references.
[68c] A. Bax: Two-Dimensional NMR in Liquids. Delft University Press – D. Reidel Publishing Company, Dordrecht, Boston, London 1982.
[68d] G. A. Morris: Magn. Res. Chem. *24*, 371 (1986).
[69a] A. Bax and R. Freeman: J. Magn. Res. *42*, 164 (1981); *44*, 542 (1981).
[69b] J. Jeener developed the basic idea of the COSY experiment during a lecture presented for Ampere International Summer School, Basco Polje 1971.
[69c] W. Aue, E. Bartholdi, and R.R. Ernst: J. Chem. Phys. *64*, 2229 (1976); mathematical fundamentals.
[70]   P.H. Bolton: J. Magn. Res. *48*, 336 (1982).
[71a] G. Bodenhausen: Prog. Nucl. Magn. Res. Spectrosc. *14*, 113 (1980).
[71b] A. Bax, R. Freeman, and T.A. Frenkiel: J. Am. Chem. Soc. *103*, 2102 (1981).
[72a] T.H. Mareci and R. Freeman: J. Magn. Res. *48*, 158 (1982).
[72b] A. Bax, R. Freeman, T.A. Frenkiel, and M.H. Levitt: J. Magn. Res. *43*, 478 (1981).
[73]   Reviews, Monographs, Data Collections:
[73a] J.B. Stothers: Quat. Rev. Chem. Soc. *19*, 144 (1965).
[73b] P.S. Pregosin and E.W. Randall: ^{13}C Nuclear Magnetic Resonance, in F.C. Nachod and J.J. Zuckerman, (eds.), Determination of Organic Structures by Physical Methods, Vol. 4. Academic Press, New York, London 1971, Chapt. 6.
[73c] H. Günther: Chem. Unserer Zeit *8*, 45 (1974).

[73d] E. Breitmaier and G. Bauer: Pharm. Unserer Zeit 5, 113 (1976).
[73e] J.B. Stothers: Carbon-13 NMR Spectroscopy. Academic Press, New York, London 1972.
[73f] G.C. Levy, R.L. Lichter, and G.L. Nelson: Carbon-13 Nuclear Magnetic Resonance Spectroscopy, 2nd. Ed. Wiley Interscience, New York, London, Sydney, Toronto 1980.
[73g] J.T. Clerc, E. Pretsch, and S. Sternhell: ^{13}C-Kernresonanzspektroskopie. Akademische Verlagsgesellschaft, Frankfurt am Main 1973 .
[73h] F.W. Wehrli and T. Wirthlin: Interpretation of Carbon-13 NMR Spectra. Heyden & Son Ltd., London, New York, Rheine 1976.
[73i] E. Breitmaier and G. Bauer: ^{13}C-NMR-Spektroskopie - eine Arbeitsanleitung mit Übungen. Georg Thieme Verlag, Stuttgart 1977.
[73j] H.O. Kalinowski, S. Berger, and S. Braun: ^{13}C-NMR- Spektroskopie. Georg Thieme Verlag, Stuttgart 1984.
[73k] E. Breitmaier, G. Haas, and W. Voelter: Atlas of Carbon-13 NMR Data, Vols. 1 and 2. Heyden & Son Ltd., London, New York 1975 and 1978.
[73l] V. Formacek, L. Desnoyer, H.P. Kellerhals, and J.T. Clerc: ^{13}C Data Bank, Vol. 1. Bruker Analytische Meßtechnik GmbH, Karlsruhe 1976.
[73m] L.F. Johnson and W.C. Jankowski: Carbon-13 NMR Spectra (Collection). J. Wiley & Sons, New York 1972.
[73n] W. Bremser, B. Franke, and H. Wagner: Chemical Shift Ranges in Carbon-13 NMR Spectroscopy. Verlag Chemie GmbH, Weinheim, Deerfield Beach, Basel 1982; W. Bremser, L. Ernst, B. Franke, R. Gerhards, and A. Hardt: Carbon-13 NMR Spectral Data, A "Living" COM-Microfiche Collection of Reference Material, Verlag Chemie GmbH, Weinheim, Deerfield Beach, Basel 1978-1985. This data base contains the data of more than 50000 carbon-13 NMR spectra. Searches can be performed *via* Fachinformationszentrum Energie, Physik, Mathematik GmbH, D-7514 Eggenstein-Leopoldshafen 2, FRG.
[74] D. Ziessow and M. Carrol: Ber. Bunsenges. Phys. Chem. 76, 61 (1972).
[75] M. Karplus and J.A. Pople: J. Chem. Phys. 38, 2803 (1963).
[76] H. Spiesecke and W.G. Schneider: Tetrahedron Lett. 468 (1961).
[77a] G.E. Maciel and D.A. Beatty: J. Phys. Chem. 69, 3920 (1965).
[77b] F.W. Wehrli, J.W. de Haan, A.I. Keulemans, D. Exner, and W. Simon: Helv. Chim. Acta 52, 103 (1969).
[78] R.B. Bates, S. Brenner, C.M. Cole, E.W. Davidson, G.D. Forsythe, D.A. Combs, and A.S. Roth: J. Am. Chem. Soc. 95, 926 (1973).
[79] G.A. Olah and A.M. White: J. Am. Chem. Soc. 91, 5804 (1969); G.A. Olah: Angew. Chem. 85, 183 (1973); Int. Ed. Engl. 12, 173 (1973).
[80a] C.G. Kreiter and V. Formacek: Angew. Chem. 84, 155 (1972); Int. Ed. Engl. 11, 141 (1972).
[80b] H.L. Retcofsky, E.N. Frankel, and H.S. Gutowsky: J. Am. Chem. Soc. 88, 271 (1966).
[81] J. Firl, W. Runge, and W. Hartmann: Angew. Chem. 86, 274 (1974); Int. Ed. Engl. 13, 270 (1974).
[82] D.M. Grant and V.B. Cheney, J. Am. Chem. Soc. 89, 5315 (1967).
[83] R.H. Levin and J.D. Roberts: Tetrahedron Lett. 135 (1973).
[84a] J.G. Batchelor, J.H. Prestgard, R.J. Cushley, and S.R. Lipsky: J. Am. Chem. Soc. 95, 6358 (1973).
[84b] J.G. Batchelor, J. Feeney, and G.C.K. Roberts: J. Magn. Res. 20, 19 (1975).
[84c] J.G. Batchelor: J. Am. Chem. Soc. 97, 3410 (1975).
[84d] L. Flohe, E. Breitmaier, W.A. Günzler, W. Voelter, and G. Jung: Hoppe-Seyler's Z. Physiolog. Chem. 353, 1159 (1972).
[85] E.G. Paul and D.M. Grant: J. Am. Chem. Soc. 85, 1701 (1963); D.M. Grant and E.G. Paul: J. Am. Chem. Soc. 86, 2984 (1964).
[86] W.M. Litchman and D.M. Grant: J. Am. Chem. Soc. 90, 1400 (1968).
[87] D.K. Dalling and D.M. Grant: J. Am. Chem. Soc. 89, 6612 (1967).

[88] A.S. Perlin, B. Casu, and H.J. Koch: Can. J. Chem. *48*, 2599 (1970).
[89] D.E. Dorman and J.D. Roberts: J. Am. Chem. Soc. *92*, 1355 (1970).
[90] R.A. Friedel and H.L. Retcovsky: J. Am. Chem. Soc. *85*, 1300 (1963).
[91 a] H. Spiesecke and W.G. Schneider: J. Chem. Phys. *35*, 722 (1961).
[91 b] J.K. Becconsall and P. Hampson: J. Mol. Phys. *10*, 21 (1965).
[92] R.L. Lichter and J.D. Roberts: J. Chem. Phys. *74*, 912 (1970).
[93] G.L. Nelson, G.C. Levy, and J.D. Cargioli: J. Am. Chem. Soc. *94*, 3089 (1972).
[94] E. Breitmaier and K.H. Spohn: Tetrahedron *29*, 1145 (1973).
[95] G. Del Re, B. Pullman, and T. Yonezawa: Biochem. Biophys. Acta *75*, 153 (1963).
[96] W.J. Horsley and H. Sternlicht: J. Am. Chem. Soc. *90*, 3738 (1968).
[97] W.J. Horsley, H. Sternlicht, and J.S. Cohen: J. Am. Chem. Soc. *92*, 680 (1970).
[98] G. Jung, E. Breitmaier, and W. Voelter: Eur. J. Biochem. *24*, 438 (1972).
[99] R.J. Pugmire and D.M. Grant: J. Am. Chem. Soc. *90*, 697 (1968).
[100] R.J. Pugmire and D.M. Grant: J. Am. Chem. Soc. *90*, 4232 (1968).
[101] J.S. Cohen, R.L. Shrager, M. McNeel, and A.N. Schechter: Nature *228*, 642 (1970).
[102] E.G. Finer: Macromolecules and Solids, in R.K. Harris, (ed.), Nuclear Magnetic Resonance, Vol. 1. The Chemical Society London 1972, p. 281.
[103] R. v. Ammon and R.D. Fischer: Angew. Chem. *84*, 737 (1972); Int. Ed. Engl. *11*, 675 (1972) and 168 references cited therein.
[104] B. Birdsall, J. Feeney, J.A. Glasel, R.J.P. Williams, and A.V. Xavier: J. Chem. Soc. Chem. Commun. *1971*, 1473.
[105] O.A. Gansow, M.R. Willcott, and R.E. Lenkinski: J. Am. Chem. Soc. *93*, 4295 (1971).
[106] K. Beyer: Thesis. Tübingen 1973.
[107] F.A.L. Anet and R. Anet: Configuration and Conformation by NMR, in F.C. Nachod and J.J. Zuckerman, (eds.): Determination of Organic Structures by Physical Methods, Vol. 3. Academic Press, New York, London 1971, chapt. 7.
[108] H. Kessler: Angew. Chem. *82*, 237 (1970); Int. Ed. Engl. *9*, 219 (1970) and 158 references cited therein.
[109] O.A. Gansow, J. Killough, and A.R. Burke: J. Am. Chem. Soc. *93*, 4297 (1971).
[110] F.A.L. Anet and J.J. Wagner: J. Am. Chem. Soc. *93*, 5266 (1971).
[111] F.A.L. Anet, C.H. Bradley, and G.W. Buchanan: J. Am. Chem. Soc. *93*, 258 (1971).
[112] H.J. Schneider, R. Price, and T. Keller: Angew. Chem. *83*, 759(1971); Int. Ed. Engl. *10*, 730 (1971).
[113] L.F. Johnson: Private communication cited in ref. [107].
[114] O.A. Gansow, A.R. Burke, and W.D. Vernon: J. Am. Chem. Soc. *94*, 2552 (1972).
[115] G.E. Maciel, J.W. McIver, Jr., N.S. Ostlund, and J.A. Pople: J. Am. Chem. Soc. *92*, 1 and 11 (1970) and references cited therein.
[116] D.M. Grant and W.M. Litchman: J. Am. Chem. Soc. *87*, 3994 (1965).
[117] N. Cyr and T.J. Cyr: J. Chem. Phys. *47*, 3082 (1967).
[118] E.R. Malinowski: J. Am. Chem. Soc. *83*, 4479 (1961).
[119] A.W. Douglass: J. Chem. Phys. *40*, 2413 (1964).
[120] G.S. Handler and H. Anderson: Tetrahedron *2*, 345 (1958); ref. [206].
[121] G.L. Gloss: Proc. Chem. Soc. (London) *1962*, 152.
[122] E. Lippert and H. Prigge: Ber. Bunsenges. Phys. Chem. *67*, 415 (1963).
[123] T. Yonezawa and I. Morishima: J. Mol. Spectrosc. *27*, 210 (1968); V.S. Gil, and A.C.P. Alves: J. Mol. Phys. *16*, 527 (1969).
[124] H. Yoder, R.H. Tuck, and R.F. Hess: J. Am. Chem. Soc. *91*, 539 (1969).
[125] E.F. Mooney and P.H. Winson, in E.F. Mooney, (ed.): Annual Review of NMR Spectroscopy, Vol. 2. Academic Press, New York, London 1969, p. 153.
[126] This book.
[127] G.A. Olah and A.M. White: J. Am. Chem. Soc. *89*, 7072 (1967).

[128] L. Ernst, V. Wray, V.A. Chertkow, and N.M. Sergeyew: J. Magn. Res. *25*, 123 (1977).
[129] J.L. Marshall: Carbon-Carbon and Carbon-Proton NMR Couplings: Applications to Organic Stereochemistry and Conformational Analysis; Verlag Chemie International, Deerfield Beach, Florida, USA 1983.
[130] F.J. Weigert and J.D. Roberts: J. Am. Chem. Soc. *90*, 3543 (1968); *91*, 4940 (1969).
[131] R. Wasylishen and T. Schäfer: Can. J. Chem. *50*, 3686 (1972); *51*, 961 (1973).
[132a] E. Breitmaier: Chimia *28*, 120 (1974).
[132b] E. Breitmaier and U. Hollstein: Org. Magn. Res. *8*, 573 (1976).
[132c] D.J. Wilbur, C. Williams, and A. Allerhand: J. Am. Chem. Soc. *99*, 5450 (1977).
[133a] F.J. Weigert and J.D. Roberts: J. Am. Chem. Soc. *94*, 6021 (1972).
[133b] M.A. Ihrig and J.L. Marshall: J. Am. Chem. Soc. *94*, 1756 (1972).
[134] A.S. Perlin and B. Casu: Tetrahedron Lett. *1969*, 2921.
[135] G.C. Levy and R.L. Lichter: Nitrogen-15 Nuclear Magnetic Resonance Spectroscopy. Wiley Interscience, New York, Chichester, Brisbane, Toronto 1979, and references cited therein.
[136] A. Römer: Org. Magn. Res. *21*, 130 (1983).
[137] G. Binsch, J.B. Lambert, B.W. Roberts, and J.D. Roberts: J. Am. Chem. Soc. *86*, 5564 (1964).
[138] E. Bullock, D.G. Tuck, and E. Woodhouse: J. Chem. Phys. *38*, 2318 (1963).
[139] R.L. Lichter: ^{15}N Nuclear Magnetic Resonance, in F.C. Nachod and J.J. Zuckerman, (eds.): Determination of Organic Structures by Physical Methods, Vol. 4. Academic Press, New York, London 1971, chapt. 4.
[140] W.M. McFarlane: Mol Phys. *10*, 603 (1966).
[141] D.M. Jerina, D.R. Boyd, L. Paolillo, and E.D. Becker: Tetrahedron Lett. *1970*, 1484.
[142] P.R. Wells: NMR Spectra of the Heavier Elements, in F.C. Nachod and J.J. Zuckerman, (eds.): Determination of Organic Structures by Physical Methods, Vol. 4. Academic Press, New York, London 1971, chapt. 5.
[143] F.R. Jerome and K.L. Servis: J. Am. Chem. Soc. *94*, 5896 (1972).
[144] G.A. Gray: J. Am. Chem. Soc. *93*, 2132 (1971).
[145] G.A. Gray and S.E. Cremer: J. Chem. Soc. Chem. Commun. *1972*, 367; Tetrahedron Lett. *1971*, 3061.
[146] P.D. Lapper, H.H. Mantsch, and J.P.C. Smith: J. Am. Chem. Soc. *94*, 6243 (1972).
[147] E. Breitmaier and W. Voelter: Eur. J. Biochem. *31*, 234 (1972).
[148] G.C. Levy: Acc. Chem. Res. *6*, 161 (1973).
[149] T.D. Alger and D.M. Grant: J. Phys. Chem. *75*, 2538 (1971).
[150] G.C. Levy, J.D. Cargioli, and F.A.L. Anet: J. Am. Chem. Soc. *95*, 1527 (1973).
[151] G.C. Levy: J. Chem. Soc. Chem. Commun. *1972*, 47.
[152] J. Grandjean, P. Laszlo, and R. Price: Mol. Phys. *25*, 695 (1973); dipolar and spin-rotational contributions to carbon-13 spin-lattice relaxation of camphor.
[153] R. Freeman, K.G.R. Pachler, and G.N. La Mar: J. Chem. Phys. *55*, 4586 (1971).
[154] C.F. Brewer, H. Sternlicht, D.M. Marcus, and A.P. Grollmann: Biochemistry *12*, 4448 (1973).
[155] J.R. Lyerla, Jr. and D.M. Grant: Int. Rev. Sci. Phys. Chem. Ser. *1*, 4, 155 (1972).
[156] D. Doddrell, V. Gluschko, and A. Allerhand: J. Chem. Phys. *56*, 3683 (1972).
[157] W.T. Huntress, Jr.: J. Chem. Phys. *48*, 3524 (1968).
[158] D.E. Woessner: J. Phys. Chem. *36*, 1 (1962); *42*, 1855 (1965).
[159] K.F. Kuhlmann, D.M. Grant, and R.K. Harris: J. Chem. Phys. *52*, 3439 (1970).
[160] S. Berger, F.R. Kreissl, and J.D. Roberts: J. Am. Chem. Soc. *96*, 4348 (1974).
[161] D.E. Woessner: J. Chem. Phys. *37*, 647 (1962).
[162] K.H. Spohn: Thesis. Tübingen 1974.

[163] D.M. Grant, R.J. Pugmire, E.P. Black, and K.A. Christensen: J. Am. Chem. Soc. 95, 8465 (1973).
[164] S. Berger, F.R. Kreissl, D.M. Grant, and J.D. Roberts: J. Am. Chem. Soc. 97, 1805 (1975).
[165] R.S. Becker, S. Berger, D.K. Dalling, D.M. Grant, and R.J. Pugmire: J. Am. Chem. Soc. 96, 7008 (1974).
[166] A. Allerhand, D. Doddrell, and R. Komoroski: J. Chem. Phys. 55, 189 (1971).
[167] G.C. Levy and R.L. Lichter: J. Am. Chem. Soc. 94, 4897 (1972).
[168] G.A. Gray and S.E. Cremer: J. Magn. Res. 12, 5 (1973).
[169] M. Imanari, M. Ohuchi, and K. Ishizu: J. Magn. Res. 14, 374 (1974).
[170] N.J.M. Birdsall, A.G. Lee, Y.K. Levine, J.C. Metcalfe, P. Partington, and G.C.K. Roberts: J. Chem. Soc. Chem. Commun. 1973, 757.
[171] C. Chachaty, Z. Wolkowski, F. Piriou, and G. Lukacs: J. Chem. Soc. Chem. Commun. 1973, 951.
[172] A. Neszmelyi, K. Tori, and G. Lukacs: J. Chem. Soc. Chem. Commun. 1977, 613.
[173] J.C. Metcalfe, N.J.M. Birdsall, J. Feeney, Y.K. Levine, and P. Partington: Nature 233, 199 (1971).
[174] Y.K. Levine, N.J.M. Birdsall, A.G. Lee, and J.C. Metcalfe: Biochemistry 11, 1416 (1972).
[175] A.G. Lee, N.J.M. Birdsall, and J.C. Metcalfe: Chem. Brit. 9, 116 (1973).
[176] A. Allerhand and R.A. Komoroski: J. Am. Chem. Soc. 95, 8228 (1973).
[177] A. Allerhand, D. Doddrell, V. Gluschko, D.W. Cochran, E. Wenkert, P.J. Lawson, and F.R.N. Gurd: J. Am. Chem. Soc. 93, 544 (1971).
[178] L. Paolillo, T. Tancredi, P.A. Temussi, E. Trivellone, E.M. Bradbury, and C. Crane-Robinson: J. Chem. Soc. Chem. Commun. 1972, 335.
[179] S. Tadorko, S. Fujiwara, and Y. Ichihara: Chem. Lett. 1973, 849.
[180] A. Allerhand and E. Oldfield: Biochemistry 12, 3428 (1973).
[181] J. Schäfer and D.F.S. Natusch: Macromolecules 5, 416 (1972).
[182] J. Schäfer: Macromolecules 5, 427 (1972).
[183] T.D. Alger, D.M. Grant, and J.R. Lyerla: J. Phys. Chem. 75, 2539 (1971).
[184] D. Doddrell and A. Allerhand: J. Am. Chem. Soc. 93, 1558 (1971).
[185] T.C. Farrar, S.J. Druck, R.R. Shoup, and E.D. Becker: J. Am. Chem. Soc. 94, 699 (1972).
[186] R.A. Goodman, E. Oldfield, and A. Allerhand: J. Am. Chem. Soc. 95, 7553 (1973).
[187] H. Saito, H.H. Mantsch, and J.P.C. Smith: J. Am. Chem. Soc. 95, 8453 (1973).
[188] I.M. Armitage, H. Huber, H. Pearson, and J.D. Roberts: Proc. Nat. Acad. Sci. USA 71, 2096 (1974).
[189] D.D. Ginanini, I.M. Armitage, H. Pearson, D.M. Grant, and J.D. Roberts: J. Am. Chem. Soc. 97, 3416 (1975).
[190] J.R. Lyerla, Jr., D.M. Grant, and R.B. Bertrand: J. Phys. Chem. 75, 3967 (1971).
[191] K.T. Gillen, M. Schwartz, and J.H. Noggle: Mol. Phys. 20, 899 (1971).
[192] H.W. Spiess, D. Schweitzer, U. Haeberlein, and K.H. Hauser: J. Magn. Res. 5, 101 (1971).
[193] A. Olivson and E. Lippma: Chem. Phys. Lett. 11, 241 (1971).
[194] J.R. Lyerla, Jr., D.M. Grant, and R.K. Harris: J. Phys. Chem. 75, 585 (1971).
[195] H.W. Spiess, D. Schweitzer, and U. Haeberlein: J. Magn. Res. 9, 444 (1973).
[196] L.J. Burnett and S.B. Roeder: J. Chem. Phys. 60, 2420 (1974).
[197] E. Goldammer, H.D. Luedemann, and A. Mueller: J. Chem. Phys. 60, 4590 (1974).
[198] T.K. Leipert, J.H. Noggle, and K.T. Gillen: J. Magn. Res. 13, 158 (1974).
[199] E. Goldammer, H.D. Luedemann, and O. Roeder: Chem. Phys. Lett. 26, 387 (1974).
[200] S. Berger, F.R. Kreissl, D.M. Grant, and J.D. Roberts: J. Am. Chem. Soc. 97, 1805 (1975).
[201] L.P. Lindeman and J.A. Adams: Anal. Chem. 43, 1245 (1971).
[202] J. Mason: J. Chem. Soc. A, 1971, 1038.
[203] J. Mason: J. Chem. Soc. Perkin Trans. 2, 1976, 1671.
[204] B.V. Cheney and D.M. Grant: J. Am. Chem. Soc. 89, 5319 (1967).

[205] W. Ritter, W.E. Hull, and H.J. Cantow: Tetrahedron Lett. *1978*, 3093.
[206] J.J. Burke and P.C. Lauterbur: J. Am. Chem. Soc. *86*, 1870 (1964).
[207] G. Schill, C. Zürcher, and H. Fritz: Chem. Ber. *111*, 2901 (1978).
[208] J.P. Monti, R. Faure, and E.J. Vincent: Org. Magn. Res. *8*, 611 (1976).
[209] E.L. Eliel and K.M. Pietrusiewicz: Org. Magn. Res. *13*, 193 (1980).
[210] M. Christl, H.J. Reich, and J.D. Roberts: J. Am. Chem. Soc. *93*, 3463 (1971).
[211] G. Mann, E. Kleinpeter, and H. Werner: Org. Magn. Res. *11*, 561 (1978).
[212] M. Christl and J.D. Roberts: J. Org. Chem. *37*, 3443 (1972).
[213] K. Wüthrich, S. Meiboom, and L.C. Snyder: J. Chem. Phys. *52*, 230 (1970).
[214] M. Christl and R. Herbert: Org. Magn. Res. *12*, 150 (1979).
[215] J.B. Stothers, C.T. Tan, and K.C. Teo: Can. J. Chem. *51*, 2893 (1973); *54*, 1211 (1976).
[216] R. Bicker, H. Kessler, and G. Zimmermann: Chem. Ber. *111*, 3200 (1978).
[217] J.B. Stothers and C.T. Tan: Can. J. Chem. *54*, 917 (1976).
[218] A. De Meijere, O. Schallner, C. Weitemeyer, and W. Spielmann: Chem. Ber. *112*, 908 (1979).
[219] J.A. Peter, J.M. Van der Toorn, and H. van Bekkum: Tetrahedron *33*, 349 (1977).
[220] M. Christl: Chem. Ber. *108*, 2781 (1975).
[221] P. Metzger, E. Casadevall, M.J. Pouet: Org. Magn. Res. *19*, 229 (1982).
[222] R. Kutschan: Tetrahedron *33*, 1833 (1977).
[223] K.B. Wiberg and F.W. Wacker: J. Am. Chem. Soc. *104*, 5239 (1982).
[224] P.G. Gassman and G.S. Proehl: J. Am. Chem. Soc. *102*, 6863 (1980).
[225] Z. Majerski, K. Mlinaric-Majerski, and Z. Meic: Tetrahedron Lett. *1980*, 4117.
[226] R.E. Pincock and F.N. Fung: Tetrahedron Lett. *1980*, 19.
[227] M. Christl: Chem. Ber. *108*, 2781 (1975); M. Christl and R. Herbert: Chem. Ber. *112*, 2022 (1979).
[228] G. Maier, S. Pfriem, U. Schäfer, and R. Matusch: Angew. Chem. *90*, 552 (1978).
[229] J.B. Grutzner, M. Jautelat, J.B. Dence, R.A. Smith, and J.D. Roberts: J. Am. Chem. Soc. *92*, 7107 (1970).
[230] D.K. Dalling, D.M. Grant, and E.G. Paul: J. Am. Chem. Soc. *95*, 3718 (1973).
[231] D.K. Dalling and D.M. Grant: J. Am. Chem. Soc. *96*, 1827 (1974).
[232] W. Hörbold, R. Keck, and R. Radeglia: J. Prakt. Chem. *317*, 1054 (1975).
[233] D.E. Dorman, M. Jautelat, and J.D. Roberts: J. Org. Chem. *36*, 2757 (1971).
[234] J.W. Haan and L.J.M. Van de Ven: Org. Magn. Res. *5*, 147 (1973).
[235] P.A. Couperus, A.D.H. Clague, and J.P.C.M. von Dongen: Org. Magn. Res. *8*, 426 (1976).
[236] S.H. Grover and J.B. Stothers: Can. J. Chem. *53*, 589 (1975).
[237] H. Brouwer and J.B. Stothers: Can. J. Chem. *50*, 1361 (1972).
[238] K.H. Albert and H. Dürr: Org. Magn. Res. *12*, 687 (1979).
[239] R.W. Hofmann and H. Kurz: Chem. Ber. *108*, 119 (1975).
[240] R. Wehner and H. Günther: Chem. Ber. *107*, 3152 (1974).
[241] R. Hollenstein, W. v. Philipsborn, R. Vögeli, and M. Neuenschwander: Helv. Chim. Acta *56*, 847 (1973); R. Hollenstein, A. Mooser, M. Neuenschwander, and W. v. Philipsborn: Angew. Chem. *86*, 595 (1973).
[242] H. Dürr and H. Gleiter: Angew. Chem. *90*, 591 (1978); H. Dürr, K.H. Albert, and M. Kausch: Tetrahedron *35*, 1285 (1979).
[243] R. Bicker, H. Kessler, A. Steigel, and W.D. Stohrer: Chem. Ber. *108*, 2708 (1975).
[244] A.K. Cheng, F.A.L. Anet, J. Nioduski, and J. Weinwald: J. Am. Chem. Soc. *96*, 2887 (1974).
[245a] J.F.M. Oth, K. Müllen, J.M. Gilles, and G. Schröder: Helv. Chim. Acta *57*, 1415 (1974).
[245b] H. Günther and J. Ulmen: Tetrahedron *30*, 3781 (1974).
[246a] S. Rang, T. Pehk, E. Lippma, and O. Eisen: Eesti NSV Tead. Akad. Toim. Keem. Geol. *16*, 346 (1967); *17*, 294 (1968).
[246b] N. Müller and D.E. Pritchard: J. Chem. Phys. *31*, 768, 1471 (1959).
[247] W. Hörbold, R. Radeglia, and D. Klose: J. Prakt. Chem. *318*, 519 (1976).

[248] M.T.W. Hearn and J.L. Turner: J. Chem. Soc. Perkin Trans. 2, *1976*, 1027.
[249] M.T.W. Hearn: J. Magn. Res. *22*, 521 (1976).
[250] H. Petersen and H. Meier: Chem. Ber. *113*, 2383, 2398 (1980).
[251] J.K. Crandall and S.A. Sojka: J. Am. Chem. Soc. *94*, 5084 (1972).
[252] W. Runge and J. Firl: Ber. Bunsenges. Phys. Chem. *79*, 907, 913 (1975).
[253] A. Marker, D. Doddrell, and N.V. Riggs: J. Chem. Soc. Chem. Commun. *1972*, 724.
[254] A.D. Buckingham: Can. J. Chem. *38*, 300 (1960).
[255] G.R. Somayajulu: J. Magn. Res. *33*, 559 (1979).
[256] G. Miyajima and K. Takahashi: J. Phys. Chem. *75*, 331 (1971).
[257] P.C. Lauterbur: Ann. Rep. New York Acad. Sci. *70*, 841 (1958).
[258] G.B. Savitsky, P.D. Ellis, K. Namikawa, and G.E. Maciel: J. Chem. Phys. *49*, 2395 (1968).
[259] G.B. Savitsky and K. Namikawa: J. Phys. Chem. *67*, 2754 (1963).
[260] D.D. Traficante and G.E. Maciel: J. Phys. Chem. *69*, 1348 (1965).
[261] G.E. Maciel: J. Phys. Chem. *69*, 1947 (1965).
[262] C.J. Carman, A.R. Tarpley, and J.H. Goldstein: J. Am. Chem. Soc. *93*, 2864 (1971).
[263] Y. Senda, J. Ishiyama, S. Imaizumi: Tetrahedron *31*, 1601 (1975).
[264] O.A. Subbotin and N.M. Sergeyew: J. Am. Chem. Soc. *97*, 1080 (1975).
[265] H.J. Schneider and V. Hoppen: J. Org. Chem. *43*, 3866 (1978).
[266] H.J. Schneider, W. Gschwendtner, D. Heiske, V. Hoppen, and F. Thomas: Tetrahedron *33*, 1769 (1977).
[267] V. Wray: J. Chem. Soc. Perkin Trans. 2, *1976*, 1598; J. Am. Chem. Soc. *103*, 2503 (1981).
[268] J.D. Roberts, F.J. Weigert, J.I. Kroschwitz, and H.J. Reich: J. Am. Chem. Soc. *92*, 1338 (1970).
[269] A. Ejchardt: Org. Magn. Res. *9*, 351 (1977).
[270] M.T.W. Hearn: Org. Magn. Res. *9*, 141 (1977).
[271] C. Konno and H. Hikino: Tetrahedron *32*, 325 (1976).
[272] M. Begtrup: J. Chem. Soc. Perkin Trans. 2, *1980*, 544.
[273a] S.H. Grover and J.B. Stothers: Can. J. Chem. *52*, 870 (1974).
[273b] L.M. Brown, R.E. Klink, and J.B. Stothers: Org. Magn. Res. *12*, 561 (1979), hydroxylated *cis*-decalins.
[274] J.B. Stothers and C.T. Tan: Can. J. Chem. *55*, 841 (1977).
[275] H. Masada and Y. Murotandi: Bull. Chem. Soc. Jpn. *52*, 1213 (1979).
[276] D.E. Dorman, D. Bauer, and J.D. Roberts: J. Org. Chem. *40*, 3729 (1975).
[277] A. Barabas, A.A. Botar, A.A. Gocan, N. Popovici, and F. Hodosan: Tetrahedron *34*, 2191 (1978).
[278] A.C. Rojas and J.K. Crandall: J. Org. Chem. *40*, 2225 (1975).
[279] E. Taskinen: Tetrahedron *34*, 425, 433 (1978).
[280] M. Herberhold, G.O. Wiedersatz, and C.G. Kreiter: Z. Naturforsch. *31* B, 35 (1976).
[281] J.B. Stothers and P.C. Lauterbur: Can. J. Chem. *42*, 1563 (1964).
[282] H. Brouwer and J.B. Stothers: Can. J. Chem. *50*, 601 (1972).
[283] L. Kozerski, K.K. Kaminska-Trela, and L. Kania: Org. Magn. Res. *12*, 365 (1979).
[284] G.E. Hawkes, K. Herwig, and J.D. Roberts: J. Org. Chem. *39*, 1017 (1974).
[285] L.M. Jackman and D.P. Kelly: J. Chem. Soc. B, *1970*, 102.
[286] M. Oha, J. Hinton, and A. Fry: J. Org. Chem. *44*, 3545 (1979).
[287] M. Yalpani, B. Modarai, and E. Koshdel: Org. Magn. Res. *12*, 254 (1979).
[288] F.J. Weigert and J.D. Roberts: J. Am. Chem. Soc. *92*, 1347 (1970).
[289] G.B. Savitzky, K. Namikawa, and G. Zweifel: J. Phys. Chem. *69*, 3105 (1965).
[290] M.J. Loots, L.R. Weingarten, and R.H. Levin: J. Am. Chem. Soc. *98*, 4571 (1976).
[291] D.H. Marr and J.B. Stothers: Can. J. Chem. *43*, 596 (1965); *47*, 3601 (1969).
[292] R. Bicker: Thesis. Frankfurt am Main 1977.
[293] J.H. Billman, S.A. Sojka, and R.P. Taylor: J. Chem. Soc. Perkin Trans. 2, *1972*, 2034.

[294] K.S. Dhami and J.B. Stothers: Can. J. Chem. *43*, 479, 498 (1965).
[295] S. Berger and A. Rieker: Tetrahedron *28*, 3123 (1972); Chem. Ber. *109*, 3252 (1976).
[296] R. Hollenstein and W. v. Philipsborn: Helv. Chim. Acta *56*, 320 (1973).
[297] R. Radeglia and S. Dähne: Z. Chem. *13*, 474 (1973).
[298] G. Höfle: Tetrahedron *33*, 1963 (1977).
[299] Y. Berger, M. Berger-Daguee, and A. Castonguay: Org. Magn. Res. *15*, 244, 303 (1981).
[300] M.T.W. Hearn and K.T. Potts: J. Chem. Soc. Perkin Trans. *2, 1974*, 1918.
[301] E. v. Dehmlow, R. Zeisberg, and S.S. v. Dehmlow: Org. Magn. Res. *7*, 418 (1975).
[302] H.O. Kalinowski, L.H. Franz, and G. Maier: Org. Magn. Res. *17*, 6 (1981).
[303] J.F. Bagli and M.S. Jacques: Can. J. Chem. *56*, 578 (1978).
[304] W. Städeli, R. Hollenstein, and W. v. Philipsborn: Helv. Chim. Acta *60*, 948 (1977).
[305] R. Hagen and J.D. Roberts: J. Am. Chem. Soc. *91*, 4504 (1969).
[306] E. Lippma, T. Pehk, K. Anderson, and C. Rappe: Org. Magn. Res. *2*, 109 (1970).
[307] D.L. Rabenstein and T.L. Sayer: J. Magn. Res. *24*, 27 (1976).
[308] A.B. Terentev, V.I. Dostovalova, and R.K.H. Freidlina: Org. Magn. Res. *9*, 301 (1977); *21*, 11 (1983).
[309] J. Batchelor, R.J. Cushley, and J.H. Prestegard: J. Org. Chem. *39*, 1698 (1974).
[310] P.A. Couperus, A.D.H. Clague, and J.P.C.M. von Dongen: Org. Magn. Res. *11*, 590 (1980).
[311] H.J. Schneider, W. Freitag, and E. Weigand: Chem. Ber. *111*, 2656 (1978).
[312] D.E. Dorman, D. Bauer, and J.D. Roberts: J. Org. Chem. *40*, 3729 (1975).
[313] D.E. Dorman and F.A. Bovey: J. Org. Chem. *38*, 1719 (1973).
[314] D.E. Torchia, J.D. Lyerla, and C.M. Deber: J. Am. Chem. Soc. *96*, 5009 (1974).
[315] H.O. Kalinowski, W. Lubosch, and D. Seebach: Chem. Ber. *110*, 3733 (1977).
[316] K.L. Williamson, M. Ul Hasan, and D.L. Clutter: J. Magn. Res. *30*, 367 (1978).
[317] W.J. Eliot and J. Fried: J. Org. Chem. *43*, 2708 (1978).
[318] D.E. James and J.K. Stille: J. Org. Chem. *41*, 1504 (1976).
[319] S.W. Pelletier, Z. Djarmati, and C. Pape: Tetrahedron *32*, 995 (1976).
[320] K.S. Dhami and J.B. Stothers: Can. J. Chem. *45*, 233 (1967).
[321] V. Galasso, G. Pellitzer, H. Le Bail, and G.C. Pappalardo: Org. Magn. Res. *8*, 457 (1976).
[322] G. Barbarella, P. Dembech, A. Garbesi, and A. Fava: Org. Magn. Res. *8*, 108 (1976).
[323] G. Dauphin and A. Cuer: Org. Magn. Res. *12*, 557 (1979).
[324] F. Freeman and C.N. Angeletakis: J. Org. Chem. *47*, 4194 (1982); Org. Magn. Res. *21*, 86 (1983).
[325] F. Freeman, C.N. Angeletakis, and T.J. Maricich: Org. Magn. Res. *17*, 53 (1981).
[326] J.B. Lambert and R.J. Nienhuis: J. Am. Chem. Soc. *102*, 6659 (1980).
[327] G.H. Schmidt: J. Org. Chem. *43*, 3767 (1978).
[328] A. Hoppmann, P. Weyerstahl, and W. Zummack: Liebigs Ann. Chem. *1977*, 1547.
[329] K. Pihlaja, M. Eskonmaa, R. Keskinen, A. Nikkilä, and T. Nurmi: Org. Magn. Res. *17*, 246 (1981).
[330] H.O. Kalinowski and H. Kessler: Angew. Chem. *86*, 43 (1974).
[331] R. Radeglia and S. Scheithauer: Z. Chem. *14*, 20 (1974).
[332] H.O. Kalinowski and H. Kessler: Org. Magn. Res. *6*, 305 (1974).
[333] H. Fritz, P. Hug, H. Sauter, T. Winkler, S.O. Lawesson, B.S. Pedersen, and S. Scheibye: Org. Magn. Res. *16*, 36 (1981).
[334] C.G. Andrieu, D. Debruyne, and D. Paquer: Org. Magn. Res. *11*, 528 (1978).
[335] W. Gombler: Z. Naturforsch. *36 B*, 1561 (1981).
[336] J. Dabrowski and L. Kozerski: Org. Magn. Res. *6*, 499 (1974).
[337] H. Eggert and C. Djerassi: J. Am. Chem. Soc. *95*, 3710 (1973).
[338] J.E. Sarneski, H.L. Surprenant, F.K. Molen, and C.N. Reilley: Anal. Chem. *47*, 2116 (1975).
[339] D.J. Hart and W.T. Ford: J. Org. Chem. *39*, 363 (1974).
[340] J.G. Batchelor: J. Chem. Soc. Perkin Trans. *2*, 1585 (1976).

[341] R. Murari and W.J. Baumann: J. Am. Chem. Soc. *103*, 1238 (1981).
[342] M.G. Ahmed, P.W. Hickmott, and R.D. Soelistyowati: J. Chem. Soc. Perkin Trans. 2, 372 (1978); 838 (1976).
[343] R. Radeglia, E. Gey, T. Steiger, S. Kulpe, R. Lück, M. Rutenberg, M. Stierl, and S. Dähne: J. Prakt. Chem. *316*, 766 (1974).
[344] J.E. Arrowsmith, M.J. Cook, and D.J. Hardstone: Org. Magn. Res. *11*, 160 (1978).
[345] R.R. Fraser, J. Banville, F. Akiyama, and N. Chuaqui-Offermanns: Can. J. Chem. *59*, 705 (1981).
[346] C.A. Bunnell and P.L. Fuchs: J. Org. Chem. *42*, 2614 (1977).
[347] B. Unterhalt: Arch. Pharm. *311*, 366 (1978).
[348] L.M. Jackman and T. Jeu: J. Am. Chem. Soc. *97*, 2811 (1975).
[349] H.O. Kalinowski and H. Kessler: Org. Magn. Res. *7*, 128 (1975).
[350] G.A. Olah and D.J. Donovan: J. Org. Chem. *43*, 860 (1978).
[351] C. Rabiller, J.P. Renou, and G.J. Martin: J. Chem. Soc. Perkin Trans. 2, 536 (1977).
[352] T. Yonemoto: J. Magn. Res. *12*, 93 (1973).
[353] J. Morishima, A. Mizuno, and T. Yonezawa: J. Chem. Soc. Chem. Commun. *1970*, 1321.
[354] R.W. Stephany, M.J.A. De Bie, and W. Drenth: Org. Magn. Res. *6*, 45 (1974).
[355] J. Firl, W. Runge, W. Hartmann, and H. Utical: Chem Lett. *1975*, 51.
[356] C. Collier and G.A. Webb: Org. Magn. Res. *12*, 659 (1979).
[357] T.A. Albright and W.J. Freeman: Org. Magn. Res. *9*, 75 (1977).
[358] R.O. Duthaler, H.G. Förster, and J.D. Roberts: J. Am. Chem. Soc. *100*, 4947 (1978).
[359] F.A.L. Anet and I. Yavari: Org. Magn. Res. *8*, 327 (1976).
[360] I. Ruppert: Angew. Chem. *89*, 336 (1977).
[361] P.S. Pregosin and E.W. Randall: J. Chem. Soc. Chem. Commun. *1971*, 399.
[362] A.R. Farminer and G.A. Webb: Tetrahedron *31*, 1521 (1975).
[363] A. Ejchardt: Org. Magn. Res. *10*, 263 (1977).
[364] B.E. Mann: J. Chem. Soc. Perkin Trans. 2, 30 (1972).
[365] T.A. Albright, M.D. Gordon, W.J. Freeman, and E.E. Schweizer: J. Am. Chem. Soc. *98*, 6249 (1976); Org. Magn. Res. *8*, 489 (1976).
[366] G.A. Gray: J. Am. Chem. Soc. *95*, 5092, 7736 (1973).
[367] S. Aime, R.K. Harris, E.M. McVicker, and M. Field: J. Chem. Soc. Dalton Trans. 2114 (1976).
[368] L. Ernst: Org. Magn. Res. *9*, 35 (1977).
[369] G.A. Gray: J. Am. Chem. Soc. *93*, 2132 (1971).
[370] A.J. Ashe, R.R. Sharp, and J.W. Tolan: J. Am. Chem. Soc. *98*, 5451 (1976).
[371] H. Schmidbauer, W. Richter, W. Wolf, and F.H. Köhler: Chem. Ber. *108*, 2649 (1975).
[372] R.M. Lequan, M.J. Pouet, and M.P. Simonnin: Org. Magn. Res. *7*, 392 (1975).
[373] H. Schmidbauer, W. Buchner, and D. Scheutzow: Chem. Ber. *106*, 1251 (1973); H. Schmidbauer, A. Schier, B. Milewski-Mahrla, and U. Schubert: Chem. Ber. *115* 722 (1982).
[374] M. Haemers, R. Ottinger, D. Zimmermann, and J. Reisse: Tetrahedron Lett. *1973*, 224.
[375] D.G. Gorenstein: Progr. Nucl. Magn. Res. *16*, 1 (1983).
[376] G.W. Buchanan and J.H. Brown: Can. J. Chem. *55*, 604 (1977); G.W. Buchanan and C. Benezra: Can. J. Chem. *54*, 231 (1976).
[377] G.A. Gray, S.E. Cremer, and K. Marsi: J. Am. Chem. Soc. *98*, 2109 (1976).
[378] G.A. Gray and S.E. Cremer: J. Org. Chem. *37*, 3458, 3470 (1972).
[379a] V. Galasso: J. Magn. Res. *34*, 119 (1979).
[379b] K. Dimroth: Acc. Chem. Res. *15*, 199 (1979).
[380] L.D. Quin, M.J. Gallagher, G.T. Cunkle, and D.B. Chesnut: J. Am. Chem. Soc. *102*, 3136 (1980).
[381] T.D. Alger, D.M. Grant, and E.G. Paul: J. Am. Chem. Soc. *88*, 5397 (1966).
[382] D.J. Sardella: J. Am. Chem. Soc. *98*, 2100 (1976).

[383] D.E. Ewing: Org. Magn. Res. *12*, 499 (1979) and references cited therein; this work compiles the substituent effects on carbon-13 shifts of 709 monosubstituted benzenes.
[384] R.H. Martin, J. Moriau, and N. Dehay: Tetrahedron *30*, 179 (1974).
[385] M.J. Shapiro: J. Org. Chem. *41*, 3197 (1976).
[386] H. Spiesecke and W.G. Schneider: J. Chem. Phys. *35*, 731 (1961).
[387] J.E. Bloor and D.L. Breen: J. Phys. Chem. *72*, 716 (1968).
[388] C.G. Swain and E. Lupton: J. Am. Chem. Soc. *90*, 432 (1968).
[389] W.G. Buchanan, G. Montaudo, and G. Finocchiaro: Can. J. Chem. *52*, 767 (1974).
[390] D. Leibfritz: Chem. Ber. *108*, 3014 (1975).
[391] A. Solladie and G. Solladie: Org. Magn. Res. *10*, 235 (1977).
[392] L. Ernst: J. Magn. Res. *20*, 544 (1975); *22*, 279 (1976); Z. Naturforsch *30 B*, 788, 794 (1975); Chem. Ber. *108*, 2030 (1978)
[393] J. Seita, J. Sandström, and T. Drakenberg: Org. Magn. Res. *11*, 239 (1978).
[394] P.E. Hansen: Org. Magn. Res. *12*, 109 (1979) and references cited therein.
[395] I. Schuster: J. Org. Chem. *46*, 5110 (1981).
[396a] H. Günther, G. Jikeli, H. Schmickler, and J. Prestien: Angew. Chem. *85*, 826 (1973); Int. Ed. Engl. *12*, 762 (1973).
[396b] W. Adcock, B.D. Gupta, T.C. Khor, D. Doddrell, D. Jordan, and W. Kitching: J. Am. Chem. Soc. *96*, 1595 (1974).
[397] A.P. Jones, P.J. Garrat, and K.P.C. Volhardt: Angew. Chem. *85*, 260 (1973).
[398] T. Sato, K. Torizuka, R. Komaki, and H. Atote: J. Chem. Soc. Perkin Trans. *2*, 561 (1980).
[399] G.W. Buchanan and R.S. Ozubko: Can. J. Chem. *52*, 2493 (1974).
[400] A.J. Jones, T.D. Alger, D.M. Grant, and W.M. Litchman: J. Am. Chem. Soc. *92*, 2386 (1970).
[401a] E. Vogel, U.H. Brinker, K. Nachtkamp, J. Wassen, and K. Müllen: Angew. Chem. *85*, 760 (1973).
[401b] H. Günther, H. Schmickler, U.H. Brinker, K. Nachtkamp, J.Wassen, and E. Vogel: Angew. Chem. *85*, 762 (1973).
[402] F.J. Weigert and J.D. Roberts: J. Am. Chem. Soc. *93*, 2361(1971).
[403] M. Christl and W. Buchner: Org. Magn. Res. *11*, 461 (1978).
[404a] H.A. Staab, H. Brettschneider, and H. Brunner: Chem. Ber. *103*, 1101 (1970).
[404b] H.A. Staab, K.S. Rao, and H. Brunner: Chem. Ber. *104*, 2634 (1971).
[405a] H.A. Staab and M. Haenel: Chem. Ber. *103*, 1095 (1970).
[405b] A.T. Balaban and D. Farcasiu: J. Am. Chem. Soc. *89*, 1958 (1967).
[406] W.D. Crow and M.N. Paddon-Row: J. Am. Chem. Soc. *94*, 4746 (1972).
[407] J. Prestien and H. Günther: Angew. Chem. *86*, 278 (1974); Int. Ed. Engl. *13*, 276 (1974).
[408] E.L. Eliel and K.M. Pietrusiewicz: Topics in Carbon-13 NMR Spectroscopy *3*, 172 (1979) and references cited therein.
[409] D.R. Paulson, F.Y.N. Tang, G.F. Moran, A.S. Murray, B.P. Pelka, and E.M. Vasquez: J. Org. Chem. *40*, 184 (1975).
[410] E.L. Eliel, W.F. Bailey, L.D. Kopp, R.L. Willer, D.M. Grant, R. Bertrand, K.A. Christensen, D.K. Dalling, M.W. Duch, E. Wenkert, F.M. Schell, and D.W. Cochran: J. Am. Chem. Soc. *102*, 3698 (1980).
[411] R.R. Fraser and B.T. Grindley: Can. J. Chem. *53*, 2465 (1975).
[412] E. Block, A.A. Bazzi, J.B. Lambert, and S.M. Wharry: J. Org. Chem. *45*, 4807 (1980).
[413] E.L. Eliel, V.S. Rao, and F.G. Riddell: J. Am. Chem. Soc. *98*, 3583 (1976).
[414] F.W. Vierhapper and E.L. Eliel: J. Org. Chem. *42*, 51 (1977).
[415] F.W. Vierhapper and R.L. Willer: J. Org. Chem. *42*, 4024 (1977).
[416] P. Deslongchamps, R. Chenevert, R.J. Taillefer, C. Moreau, and J.K. Saunders: Can. J. Chem. *53*, 1601 (1974).
[417] T. Imagawa, A. Haneda, and M. Kawanisi: Org. Magn. Res. *13*, 224 (1980).

[418]   H. Duddeck and M. Kaiser: Org. Magn. Res. *20*, 55 (1982).
[419]   F.T. Oakes and J.F. Sebastian: J. Org. Chem. *45*, 4959 (1980).
[420a] H. Günther and G. Jikeli: Chem. Ber. *106*, 1863 (1973).
[420b] S. Berger and A. Rieker: Org. Magn. Res. *6*, 78 (1974).
[421a] M.S. Chauhan and I.W.J. Still: Can. J. Chem. *53*, 2880 (1975).
[421b] I.W.J. Still, N. Plavac, D.M. McKinnon, and M.S. Chauhan: Can. J. Chem. *54*, 280 (1976).
[422]   C.A. Kingsbury and J.H. Looker: J. Org. Chem. *40*, 1120 (1975).
[423a] A. Pelter, R.S. Ward and R.J. Bass: J. Chem. Soc. Perkin Trans. *1*, 666 (1978).
[423b] H.C. Jha, F. Zilliken, and E. Breitmaier: Can. J. Chem. *58*, 1211 (1980).
[424]   A.W. Frahm and R.K. Chaudhuri: Tetrahedron *35*, 2035 (1979); *36*, 3273 (1980).
[425a] J.P. Marchal, J. Brondeau, and D. Canet: Org. Magn. Res. *19*, 1 (1982).
[425b] G. Fronza, R. Mondelli, E.W. Randall, and G.P. Gardini: J. Chem. Soc. Perkin Trans. *2*, 1746 (1977).
[426]   P.G. Gassmann, D.P. Gilbert, and T.Y. Luh: J. Org. Chem. *42*, 1340 (1977).
[427]   K.A. Kovar, D. Linden, and E. Breitmaier: Arch. Pharm. *314*, 186 (1981).
[428]   P.D. Ellis, R.B. Dunlap, A.L. Pollard, K. Seidman, and A.D. Cardin: J. Am. Chem. Soc. *95*, 4398 (1973); ref. [439].
[429]   A.J. Marsaioli, E.A. Ruveda, F.D.A.M. Reis: Phytochemistry *17*, 1655 (1978).
[430]   D. Bergental, I. Mester, Z.S. Roszsa, and J. Reisch: Phytochemistry *18*, 161 (1979).
[431]   K. Isomura, H. Taniguchi, M. Mishima, M. Fujio, and Y. Tsuno: Org. Magn. Res. *9*, 559 (1977).
[432]   W. Grahn: Tetrahedron *32*, 1931 (1976).
[433a] A. Kiewiet, J. De Wit, and W.D. Weringa: Org. Magn. Res. *6*, 461 (1974).
[433b] M. Maegi, E. Lippma, E. Lukevics, and N.P. Ercak: Org. Magn. Res. *9*, 297 (1977).
[434a] K. Takahashi, T. Sone, and K. Fujieda: J. Phys. Chem. *74*, 2765 (1970).
[434b] G. Musumarra and F. Ballisteri: Org. Magn. Res. *14*, 384 (1980).
[435a] R. Abraham, R.D. Lapper, K.M. Smith, and J.F. Unsworth: J. Chem. Soc. Perkin Trans *2*, 1004 (1974).
[435b] M.V. Sigalow, B.A. Trofimow, and A.I. Mikhaleva: Tetrahedron *37*, 3051 (1981).
[436a] H. Hiemstra, H.A. Houwing, O. v. Possel, and A.M. van Leusen: Can. J. Chem. *57*, 3168 (1977).
[436b] D.R. Chrisope, R.A. Keel, A.L. Baumstark, and D.W. Boykin: J. Heterocycl Chem. *18*, 785 (1981).
[437a] R. Garnier, R. Faure, A. Babadjamian, and E.E. Vincent: Bull. Soc. Chim. France 1972, *3*, 1040.
[437b] R. Faure, J.R. Llinas, E.J. Vincent, and M. Rajzmann: Can. J. Chem. *53*, 1677 (1975).
[438]   E. Gonzales, R. Faure, E.J. Vincent, M. Espada, and J. Elguero: Org Magn. Res. *12*, 587 (1979).
[439]   R. Faure, E.J. Vincent, G. Assef, J. Kister, and J. Metzger: Org. Magn. Res. *9*, 688 (1977).
[440]   J.H. Looker, N.A. Khatri, R.B. Patterson, and C.A. Kingsbury: J. Heterocycl. Chem. *15*, 1383 (1978).
[441]   M. Begtrup, R.M. Claramunt, and J. Elguero: J. Chem. Soc. Perkin Trans. *2*, 99 (1978).
[442]   R.M. Butler and T.M. McEvoy: J. Chem. Soc. Perkin Trans. *2*, 1087 (1978).
[443]   N. Platzer, J.J. Basselier, and P. Demerseman: Bull. Soc. Chim. France *1974*, 905.
[444a] P.D. Clark, D.F. Ewing, and R.M. Scrowston: Org. Magn. Res. *8*, 252 (1976).
[444b] P. Geneste, J.L. Olivé, S.N. Ung, M.E.A. El Faghi, H. Beierbeck, and J.K. Saunders: J. Org. Chem. *44*, 2887 (1979).
[445a] R.G. Parker and J.D. Roberts: J. Org. Chem. *35*, 996 (1970).
[445b] K.H. Park, G.A. Gray, and G.D. Davis: J. Am. Chem. Soc. *100*, 7475 (1978).
[446]   L. Stefaniak: Org. Magn. Res. *11*, 385 (1978).
[447]   R. Faure, J. Elguero, E.J. Vincent, and R. Lazaro: Org. Magn. Res. *11*, 617 (1978).

[448] P. Bouchet, A. Fruchier, G. Joucheray, and J. Elguero: Org. Magn. Res. 9, 716 (1977).
[449a] L.J. Mathias and C.G. Overberger: J. Org. Chem. 43, 3526 (1978).
[449b] V.A. Lopyrev, L.I. Larina, T.I. Vakul'Skaya, M.F. Larin, O.B. Netedowa, E.F. Shibanowa, and M.G. Voronkow: Org. Magn. Res. 15, 219 (1981).
[450a] S. Florea, W. Kimpenhaus, and V. Farcasan: Org. Magn. Res. 9, 133 (1977).
[450b] J. Giraud and C. Marzin: Org. Magn. Res. 12, 647 (1979).
[451] A.R. Katritzky, R.T.C. Brownlee, G. Musumara: Heterocycles 12, 775 (1979); Tetrahedron 36, 1643 (1980).
[452a] E. Pretsch, J.T. Clerc, J. Seibl, and W. Simon: Strukturaufklärung organischer Verbindungen. Springer, Berlin, Heidelberg, New York 1976 (substituent increments).
[452b] A. Marker, A.J. Cauty, and R.T.C. Brownlee: Aust. J. Chem. 31, 1255 (1978).
[453] H.P. Fritz, F.H. Köhler, and B. Lippert: Chem. Ber. 106, 2918 (1973).
[454] G.W.H. Cheeseman, C.J. Turne, and J.D. Brown: Org. Magn. Res. 12, 213 (1979).
[455] M. Matsuo, S. Matsumoto, T. Kurikawa, Y. Akita, T. Watanabe, and A. Okta: Org. Magn. Res. 13, 172 (1980).
[456] R.J. Pugmire, M.J. Robins, D.M. Grant, and R.K. Robins: J. Am. Chem. Soc. 91, 6381 (1969).
[457a] L.R. Isbrandt, R.K. Jensen, and L. Petrakis: J. Magn. Res. 12, 143 (1973).
[457b] V. Galasso, K.J. Irgolic, and G.C. Pappalardo: J. Organomet. Chem. 181, 329 (1979).
[458a] R.J. Pugmire, M.J. Robins, D.M. Grant, and R.K. Robins: J. Am. Chem. Soc. 93, 1887 (1971).
[458b] A.J. Jones, P. Hanisch, M.L. Heffernan, and G.M. Irvine: Aust. J. Chem. 33, 499 (1980).
[459] U. Hollstein and E. Breitmaier: J. Org. Chem. 41, 2104 (1976); ref. [136].
[460] M.T. Chenon, R.J. Pugmire, D.M. Grant, R.P. Panzica, and L.B. Townsend: J. Am. Chem. Soc. 97, 4627 (1975).
[461] L. Goodrich-Dunell and D.J. Hodgson: Org. Magn. Res. 10, 1 (1977).
[462a] M. Hirota, H. Masuda, Y. Hamada, and J. Takeuchi: Bull. Chem. Soc. Jpn. 47, 2083 (1974).
[462b] H.C. van der Plas, A. Feldhuizen, M. Wozniak, and P. Smith: J. Org. Chem. 43, 1673 (1978).
[463a] U. Ewers, H. Günther, and L. Jaenicke: Chem. Ber. 106, 3951 (1973).
[463b] J. Geerts, A. Nagel, and H.C. van der Plas: Org. Magn. Res. 8, 607 (1976).
[464] S. Braun, J. Kinkeldei, and L. Walther: Tetrahedron 36, 1353 (1980).
[465a] W. Adam, A. Grimison, and G. Rodriguez: Tetrahedron 23, 2513 (1967).
[465b] T. Tokihuro and G. Fraenkel: J. Am. Chem. Soc. 91, 5005 (1969).
[466a] P.D. Ellis, and R. Ditchfield: Topics Carbon-13 NMR Spectrosc. 2, 433 (1976).
[466b] J. Jokisaari, J. Kuonanoja, and A.M. Häkkinen: Z. Naturforsch. 33 A, 7 (1978).
[467a] H. Günther, A. Gronenborn, U. Ewers, and H. Seel, in B. Pullmann, (ed.): NMR Spectroscopy in Molecular Biology. D. Reidel Publ. Corp., Dordrecht 1978.
[467b] H. Günther and A. Gronenborn: Heterocycles 11, 337 (1978).
[467c] H. Seel and H. Günther: J. Am. Chem. Soc. 102, 7051 (1980).
[468a] Y. Takeuchi and N. Dennis: Org. Magn. Res. 7, 244 (1975).
[468b] F. Cavagna and H. Pietsch: Org. Magn. Res. 11, 204 (1978).
[469a] R.L. Lichter and R. Wasylishen: J. Am. Chem. Soc. 97, 1808 (1975).
[469b] V. Galasso: Org. Magn. Res. 12, 318 (1979).
[470] B.E. Mann: Adv. Organomet. Chem. 12, 135 (1974).
[471] L.J. Todd and J.R. Wilkinson: J. Organomet. Chem. 77, 1 (1974).
[472] M.H. Chisholm and S. Godleski: Prog. Inorg. Chem. 20, 299 (1976).
[473] O.A. Gansow and W.D. Vernon: Topics Carbon-13 NMR Spectrosc. 2, 269 (1976).
[474] B.E. Mann and B.F. Taylor: ^{13}C NMR Data for Organometallic Compounds. Academic Press, London, New York, Toronto, Sydney, San Francisco 1981.
[475] P.W. Jolly and R. Mynott: Adv. Organomet. Chem. 19, 257 (1981).

[476] F.H. Köhler: J. Organomet. Chem. *64*, 27 (1974); *91*, 57 (1975); Z. Naturforsch. *29 B*, 708 (1974).
[477] N.F. Ramsey: Phys. Rev. *91*, 303 (1953).
[478] D. Seebach, R. Hässig, and J. Gabriel: Helv. Chim. Acta *66*, 308 (1983).
[479] D. Leibfritz, B.O. Wagner, and J.D. Roberts: Liebigs. Ann. Chem. *763*, 173 (1972).
[480] K. Takahashi, Y. Kondo, and R. Asami: J. Chem. Soc. Perkin Trans. *2*, 577 (1978).
[481a] B. Wrackmeyer: Prog. Nucl. Magn. Res. Spectrosc. *12*, 227 (1979).
[481b] H. Nöth and B. Wrackmeyer: Chem. Ber. *114*, 1150 (1981).
[482a] L. Ernst: J. Organomet. Chem. *82*, 319 (1974).
[482b] S. Uemura, H. Miyoshi, and M. Okano: J. Organomet. Chem. *165*, 9 (1979)
[483a] E.A. Williams and J.D. Cargioli: Ann. Rep. NMR Spectrosc. *9*, 221 (1979).
[483b] B. Wrackmeyer and W. Biffar: Z. Naturforsch. *34 B*, 1270 (1979).
[484a] B. Wrackmeyer: J. Organomet. Chem. *145,* 183 (1978); *166*, 353 (1979); J. Magn. Res. *42*, 287 (1981).
[484b] T.N. Mitchell: Org. Magn. Res. *8*, 34 (1976).
[484c] W. Kitching: Org. Magn. Res. *20*, 123 (1982).
[485a] D. De Vos: J. Organomet. Chem. *104*, 193 (1979).
[485b] R.H. Cox: J. Magn. Res. *33*, 61 (1979).
[485c] D.C. Van Beelen, E.E.J. Van Kampen, and J. Wolters: J. Organomet. Chem. *187*, 43 (1980).
[486] H. Kuivila, J.L. Considine, R.H. Sharma, and R.J. Mynott: J. Organomet. Chem. *111*, 179 (1976).
[487a] D. Doddrell, I. Burfitt, W. Kitching, M. Bullpit, C.H. Lee, R.J. Mynott, J.L. Considine, H.G. Kuivila, and G. Sarma: J. Am. Chem. Soc. *96*, 1640 (1974).
[487b] W. Kitching: Org. Magn. Res. *20*, 123 (1982).
[488] O.A. Gansow, A.R. Burke, and W.D. Vernon: J. Am. Chem. Soc. *98*, 5817 (1976).
[489a] E.G. Hoffmann, P.W. Jolly, A. Küsters, R. Mynott, and G. Wilke: Z. Naturforsch. *31 B*, 1712 (1976).
[489b] M. Green, J.A.K. Howard, J.L. Spencer, and F.G.A. Stone: J. Chem. Soc. Dalton Trans. *1977*, 271.
[490] B. Henc, P.W. Jolly, R. Salz, G. Wilke, R. Benn, E.G. Hoffmann, R. Mynott, G. Schroth, K. Seevogel, J.C. Sekutowski, C. Krüger: J. Organomet. Chem. *191*, 425 (1980).
[491a] N.K. Wilson, R.D. Zehr, and P.D. Ellis: J. Magn. Res. *21*, 437 (1976).
[491b] P.F. Barron, D. Doddrell, and W. Kitching: J. Organomet. Chem. *139*, 361 (1977).
[491c] D.K. Breitinger, K. Geibel, W. Kress, and R. Sendelbeck: J. Organomet. Chem. *191*, 7 (1980).
[492] G.A. Olah et al.: J. Am. Chem. Soc. *96*, 3548 (1974); J. Org. Chem. *46*, 3751 (1981) and more than 240 other publications of this group about carbocations.
[493] G.A. Olah and G. Liang: J. Am. Chem. Soc. *95*, 194 (1973).
[494] G.A. Olah and G. Liang: J. Am. Chem. Soc. *94*, 6434 (1972).
[495a] K. Takeuchi: Tetrahedron *36*, 2939 (1980).
[495b] A.O.K. Nieminen, H. Ruotsalainen, and P.O.I. Virtanen: Org. Magn. Res. *19*, 118 (1982).
[496a] G.A. Olah, J.M. Denis, and P.W. Westerman: J. Org. Chem. *39*, 1206 (1974).
[496b] J.W. Larsen and P.A. Bonis: J. Am. Chem. Soc. *97*, 4418 (1975).
[497a] G.A. Olah: Acc. Chem. Res. *9*, 41 (1976).
[497b] G.A. Olah and D.J. Donovan: J. Am. Chem. Soc. *99*, 5026 (1977).
[497c] A.A. Cheremesin and P.V. Schastner: Org. Magn. Res. *14*, 327 (1980).
[498] G.A. Olah, J.S. Staral, G. Asencio, G. Liang, D.A. Forsyth, and G.D. Maleescu: J. Am. Chem. Soc. *100*, 6299 (1980).
[499] G.A. Olah, G.K.S. Prakash, M. Arvanaghi, and F.A.L. Anet: J. Am. Chem. Soc. *104*, 7105 (1982).
[500] D. Lenoir: Nachr. Chem. Tech. Lab. *31*, 889 (1983) and references cited therein.

[501] D.P. Kelly, G.R. Underwood, and P.F. Barron: J. Am. Chem. Soc. *98*, 3106 (1976).
[502] H.H. Vogt and R. Gompper: Chem. Ber. *114*, 2884 (1981).
[503] M. Schlosser and M. Stähle: Angew. Chem. *92*, 497 (1980).
[504] S. Bywater and D.J. Worsfold: J. Organomet. Chem. *159*, 229 (1978).
[505] T.W. Ford and M. Newcomb: J. Am. Chem. Soc. *96*, 309 (1974).
[506] J.P.C.M. Van Dongen, H.W.D. Dijkman, and M.J.A. De Bie: Recl. Trav. Chim. Pays Bas *93*, 29 (1974).
[507] D.H. O'Brien, C.R. Russell, and A.J. Hart: J. Am. Chem. Soc. *97*, 4410 (1975); *98*, 7427 (1976).
[508] G.A. Olah, G. Asenzio, H. Mayr, and P.v.R. Schleyer: J. Am. Chem. Soc. *100*, 4347 (1978).
[509a] G.A. Taylor and P.E. Rakita: Org. Magn. Res. *6*, 644 (1974).
[509b] U. Edlund: Org. Magn. Res. *9*, 593 (1977).
[510] A.G. Abatjoglon, E.L. Eliel, and L.F. Kuyper: J. Am. Chem. Soc. *99*, 8262 (1977).
[511] K.A. Kovar and E. Breitmaier: Chem. Ber. *111*, 1646 (1978).
[512] R.B. Bates, S. Brenner, C.M. Cole, E.W. Davidson, G.D. Forsythe, D.A. McCombs, and A.S. Roth: J. Am. Chem. Soc. *95*, 926 (1973).
[513] F.A. Bovey: High Resolution NMR of Macromolecules. Academic Press, New York, London 1972.
[514] J. Schäfer: Topics Carbon-13 NMR Spectrosc. *1*, 149 (1974).
[515] A.R. Katritzky and D.E. Weiss: Chem. Brit. *12*, 45 (1976).
[516] J.C. Randall: Polymer Sequence Determination. Academic Press, New York, San Francisco, London 1977.
[517a] F.A. Bovey, F.C. Schilling, T.K. Kwei, and H.L. Frisch: Macromolecules *10*, 559 (1977).
[517b] F.A. Bovey, F.C. Schilling, and W.H. Starnes: Polym. Prep. *20*, 160 (1979).
[518a] M. Mehring: High Resolution NMR Spectroscopy in Solids. Springer, Berlin, Heidelberg, New York 1976.
[518b] U. Haeberlen: Adv. Magn. Res. Suppl. *1*, 1976.
[519] J. Schäfer and E.O. Stejskal: Topics Carbon-13 NMR Spectrosc. *3*, 283 (1979).
[520] R.D. Miller and C.S. Yannoni: J. Am. Chem. Soc. *102*, 7396 (1980).
[521] A.E. Tonelli and F.C. Schilling: Acc. Chem. Res. *14*, 233 (1981) and references cited therein.
[522a] U.W. Sutter and P.J. Flory: Macromolecules *8*, 765 (1975).
[522b] A.E. Tonelli: Macromolecules *12*, 83 (1979).
[522c] A.E. Tonelli and F.C. Schilling: Macromolecules *13*, 270 (1980).
[523a] J. Schäfer: Macromolecules *4*, 105 (1971).
[523b] Y. Inoue and A. Nishioka: Polym. J. *3*, 149 (1972).
[524] J. Schäfer: Macromolecules *4*, 107 (1971).
[525] L.F. Johnson, F. Heatley, and F.A. Bovey: Macromolecules *3*, 175 (1970).
[526a] J. Schäfer: Macromolecules *2*, 533 (1969).
[526b] J. Schäfer and R. Yaris: J. Chem. Phys. *51*, 4469 (1969).
[527a] K. Matsuzaki, T. Kanai, T. Kawamura, S. Matsumoto, and T. Hryu: J. Polym. Sci. *11*, 961 (1973).
[527b] E. Klesper, A. Johnson, W. Gronski, and F.W. Wehrli: Makromol. Chem. *176*, 1071 (1975).
[528] J.C. Randall: Polym. Prep. *20*, 235 (1979).
[529] M.W. Duch and D.M. Grant: Macromolecules *3*, 165 (1970).
[530] F. Conti, A. Segre, P. Pini, and L. Porri: Polymer *15*, 5 (1974).
[531] A. Allerhand and R.K. Hailstone: J. Chem. Phys. *56*, 3718 (1972).
[532] D.E. Axelson, L. Mandelkern, and G.C. Levy: Macromolecules *10*, 557 (1977).
[533] V.W. Sutter and P. Neuenschwander: Macromolecules *14*, 528 (1981).
[534a] J.R. Lyerla, T.T. Horokawa, and D.E. Johnson: J. Am. Chem. Soc. *99*, 2463 (1977).
[534b] K. Hatada, T. Kitayama, Y. Ihamoto, K. Ohta, Y. Umemura, and H. Yuhi: Makromol. Chem. *179*, 485 (1978).

[535] F. Bohlmann, R. Zeisberg, and E. Klein: Org. Magn. Res. 7, 426 (1975).
[536] M. Jautelat, J.B. Grutzner, and J.D. Roberts: Proc. Nat. Acad. Sci. 65, 288 (1970).
[537] N. Chatterjee: J. Magn. Res. 33, 241 (1979).
[538] N.H. Werstiuk, R. Taillefer, and R.A. Bell: Can. J. Chem. 50, 2146 (1972).
[539] D.G. Morris and A.M. Murray: J. Chem. Soc. Perkin Trans. 2, 539 (1975).
[540] D.A. Forsyth, S. Mahmoud, and B.C. Giessen: Org. Magn. Res. 19, 89 (1982).
[541] Bruker ^{13}C Data Bank, Vol. 1. Bruker Analytische Meßtechnik GmbH, Karlsruhe 1976.
[542] E.J. Brunke and F.J. Hammerschmidt: Tetrahedron Lett. *1980*, 2405.
[543] P. Brun, J. Casanova, and J. Hatem: Org. Magn. Res. 12, 537 (1979).
[544] C.M. Holden and D. Whittaker: Org. Magn. Res. 7, 125 (1975).
[545] W. Offermann and A. Mannschreck: Tetrahedron Lett. *1981*, 3227.
[546] E. Lippmaa, T. Pehk, J. Paasivirta, N. Belikova, and A. Platé: Org. Magn. Res. 2, 581 (1970).
[547] M.C. Hall, M. Kinns, and E.J. Wells: Org. Magn. Res. 21, 108 (1983).
[548] W. Offermann: Org. Magn. Res. 20, 203 (1982).
[549] E. Wenkert, A.O. Clouse, D.W. Cochran, and D. Doddrell: J. Chem. Soc. Chem. Commun. *1969*, 1433.
[550] E. Wenkert and B.L. Buckwalter: J. Am. Chem. Soc 94, 4367 (1972).
[551] J.R. Hanson, G. Savona, and M. Siverns: J. Chem. Soc. Perkin Trans. 1, 2001 (1974).
[552] J.R. Hanson, M. Siverns, F. Piozzi, and G. Savona: J. Chem. Soc. Perkin Trans. 1, 114 (1976).
[553] I. Wahlberg, S.O. Almqvist, T. Nishida, and C.R. Enzell: Acta Chem. Scand. B 29, 1047 (1975).
[554] B.L. Buckwalter, I.R. Burfitt, A.A. Nagel, E. Wenkert, and F. Näf: Helv. Chim. Acta 58, 1567 (1975).
[555] R. Rowan and B.D. Sykes: J. Am. Chem. Soc. 96, 7000 (1974).
[556] Y. Inoue, A. Takahaski, Y. Tokito, and R. Chujo: Org. Magn. Res. 6, 487 (1974).
[557] G. Englert: Helv. Chim. Acta 58, 2367 (1975).
[558] B.V. Cheney: J. Am. Chem. Soc. 90, 5386 (1968).
[559] M.E. Van Dommelen, L.J.M. Van de Ven, H.M. Buck, and J.W. de Haan: Rec. J. Royal Netherlands Chem. Soc. 96, 295 (1977).
[560] W. Bremser and J. Paust: Org. Magn. Res. 6, 433 (1974).
[561] G.P. Moss: Pure Appl. Chem. 47, 97 (1976).
[562] P. Granger, B. Maudinas, R. Herber, and J. Villoutreix: J. Magn. Res. 10, 43 (1973).
[563] H. Cadosch, U. Vögeli, P. Rüedi, and C.H. Eugster: Helv. Chim. Acta 61, 783 (1978).
[564] J.W. Blunt and J.B. Stothers: Org. Magn. Res. 9, 439 (1977).
[565] N.S. Bhacca, D.D. Giannini, W.S. Jankowski, and M.E. Wolff: J. Am. Chem. Soc. 95, 8421 (1973).
[566] H. Eggert and C. Djerassi: J. Org. Chem. 38, 3788 (1973).
[567] R.J. Cushley and J.D. Filipenko: Org. Magn. Res. 8, 308 (1976).
[568] W.B. Smith, D.L. Deavenport, J.A. Swanzy, and G.A. Pate: J. Magn. Res. 12, 15 (1973).
[569] J.W.A. Simon, H. Beierbeck, and J.K. Saunders: Can. J. Chem. 51, 3874 (1973).
[570] D. Leibfritz and J.D. Roberts: J. Am. Chem. Soc. 95, 4996 (1973).
[571] D.J. Chadwick and D.H. Williams: J. Chem. Soc. Perkin Trans. 2, 1903 (1974).
[572] W.B. Smith and D. Deavenport: J. Magn. Res. 6, 256 (1972).
[573] R.J. Abraham and J.R. Monasterios: J. Chem. Soc. Perkin Trans. 2, 662 (1974).
[574] G. Engelhardt, G. Schneider, I. Weisz-Vincze, and A. Vass: J. Prakt. Chem. 316, 391 (1974).
[575] H. Eggert, C.L. Van Antwerp, N.S. Bhacca, and C. Djerassi: J. Org. Chem. 41, 71 (1976).
[576] F.W. Wehrli: Adv. Mol Relaxation Processes 6, 139 (1974).
[577] R.H. Solomon, K. Nakanishi, W.E. Fallon, and Y. Shimizu: Chem. Pharm. Bull. 22, 1671 (1974).

[578] Q. Khuong-Huu, G. Lukacs, A. Pancrazi, and R. Goutarel: Tetrahedron Lett. *1972*, 3579.
[579] G. Lukacs, X. Lusinchi, E.W. Hagaman, B.L. Buckwalter, F.M. Schell, and E. Wenkert: C. R. Acad. Sci. Ser. *C 274*, 1458 (1972).
[580] J.R. Hanson and M. Siverns: J. Chem. Soc. Perkin Trans. *1*, 1956 (1975).
[581] Q. Khuong-Huu, A. Pancrazi, and I. Kabore: Tetrahedron *30*, 2579 (1974).
[582] F.F. Hill, J. Schindler, R.D. Schmid, R. Wagner, and W. Voelter: Eur. J. Microbiol. Biotechnol. *15*, 25 (1982).
[583] T.A. Wittstruck and K.I.H. Williams: J. Org. Chem. *38*, 1542 (1973).
[584] G.W. Buchanan, D.A. Ross, and J.B. Stothers: J. Am. Chem. Soc. *88*, 4301 (1966).
[585] G.W. Buchanan and J.B. Stothers: Can. J. Chem. *47*, 3605 (1969).
[586] C.L. Van Antwerp, H. Eggert, G.D. Meakins, J.O. Miners, and C. Djerassi: J. Org. Chem. *42*, 789 (1977).
[587] G. Lukacs, F. Khuong-Huu, C.R. Bennett, B.L. Buckwalter and E. Wenkert: Tetrahedron Lett. *1972*, 3515.
[588a] G. Lukacs and C.R. Bennett: Bull. Soc. Chim. France *1972*, 3996.
[588b] E. Berman, N. Friedman, Y. Mazur, and M. Sheves: J. Am. Chem. Soc. *100*, 5626 (1978).
[589] B. Balogh, D.M. Wilson, and A.L. Burlingame: Nature *233*, 261 (1971).
[590] A. Neszmélyi and G. Lukacs: J. Am. Chem. Soc. *104*, 5342 (1982).
[591] A.C. Pinto, M.L.A. Goncalves, R.B. Filho, A. Neszmélyi, and G. Lukacs: J. Chem. Soc. Chem. Commun. *1982*, 293.
[592] A.C. Pinto, S.K. Do Prado, R.B. Filho, W.E. Hull, A. Neszmélyi, and G. Lukacs: Tetrahedron Lett. *1982*, 5267.
[593] G. Lukacs and A. Neszmélyi: J. Chem. Soc. Chem. Commun. *1981*, 1275.
[594] R. Jacquesy, C. Narbonne, W.E. Hull, A. Neszmélyi, and G. Lukacs: J. Chem. Soc. Chem. Commun. *1982*, 409.
[595] A. Neszmélyi, W.E. Hull, G. Lukacs, R. Wagner, and W. Voelter: Z. Naturforsch., in press.
[596] K. Tori, H. Ishii, Z.W. Wolkowski, C. Chachaty, M. Sangaré, F. Piriou, and G. Lukacs: Tetrahedron Lett. *1973*, 1077.
[597] H. Eggert and C. Djerassi: Tetrahedron Lett. *1975*, 3635.
[598] T.A. Crabb: Nuclear Magnetic Resonance of Alkaloids in G.A. Webb, (ed.): Annual Reports on NMR Spectroscopy, Vol. 13. Academic Press, New York 1982, p. 60.
[599] M. Shamma and D.M. Hindenlang: Carbon-13 NMR Shift Assignments of Amines and Alkaloids. Plenum Press, New York, London 1979.
[600] E. Wenkert, J.S. Bindra, C.J. Chang, D.W. Cochran, and F.M. Schell: Acc. Chem. Res. *7*, 46 (1974).
[601] E. Wenkert, D.W. Cochran, E.W. Hagaman, R.B. Lewis, and F.M. Schell: J. Am. Chem. Soc. *93*, 6271 (1971).
[602] E. Leete: Bioorg. Chem. *6*, 273 (1977).
[603] E. Breitmaier, this reference.
[604] V.I. Sternberg, N.K. Narain, and S.P. Singh: J. Heterocycl. Chem. *14*, 225 (1977).
[605] P. Hanisch, A.J. Jones, A.F. Casey, and J.E. Coates: J. Chem. Soc. Perkin Trans. *2*, 1202 (1977).
[606] E.J. Barreiro, A. de Lima Pereira, L. Nelson, L.F. Gomes, and A.J.R. de Silva: J. Chem. Res. (S) *1980*, 330.
[607] N.V. Mody, R.S. Sawhney, and S.W. Pelletier: J. Natural Products *42*, 417 (1979).
[608] F. Bohlmann and R. Zeisberg: Chem. Ber. *108*, 1043 (1975).
[609] E. Wenkert, B. Chauncy, K.G. Dave, A.R. Jeffcoat, F.M. Schell, and H.P. Schenk: J. Am. Chem. Soc. *95*, 8427 (1973).
[610] R.T. Lalonde, T.N. Donvito, and A.I.M. Tsai: Chem. Pharm. Bull. *53*, 174 (1975).
[611] Y. Itatani, S. Yasuda, M. Hanaoka, and Y. Arata: Pharm. Bull. Tokyo *24*, 2521 (1976).
[612] T.N. Nakashima, P.P. Singer, L.M. Browne, and W.A. Ayer: Can. J. Chem. *53*, 1936 (1975).

[61]   C. Poupat, A. Ahond, and T. Sévenet: Phytochem. *15*, 2019 (1976).
[614] E. Wenkert, C.J. Chang, H.P.S. Chawla, D.W. Cochran, E.W. Hagaman J.C. King, and K. Orito: J. Am. Chem. *98*, 3645 (1976).
[615] L. Töke, K. Honty, L. Szabó, B. Blaskó, and C. Szántay: J. Org. Chem. *38*, 2496 (1973).
[616] P.R. Srinivasan and R.L. Lichter: Org. Magn. Res. *8*, 198 (1976).
[617] R. Verpoorte, P.J. Hylands, and N.G. Bisset: Org. Magn. Res. *9*, 567 (1977).
[618] E. Wenkert, D.W. Cochran, E.W. Hagaman, F.M. Schell, N. Neuss, A.S. Katner, P. Potier, C. Can, M. Plat, M. Koch, H. Mehri, J. Poisson, N. Kunesch, and Y. Rolland: J. Am. Chem. Soc. *95*, 4990 (1973).
[619] P. Rasoanaivo and G. Lukacs: J. Org. Chem. *41*, 376 (1976).
[620] A. Ahond, M.M. Janot, N. Langlois, G. Lukacs, P. Potier, P. Rasoanaivo, M. Sangaré, N. Neuss, M. Plat, J. Le Men, E.W. Hagaman, and E. Wenkert: J. Am. Chem. Soc. *96*, 633 (1974).
[621] J. Le Men, M.J. Hoizey, G. Lukacs, L. Le Men-Olivier, and J. Lévy: Tetrahedron Lett. *1974*, 3119.
[622] E. Wenkert and H.E. Gottlieb: Heterocycles *7*, 753 (1977).
[623] E. Wenkert, C.J. Chang, A.O. Clouse, and D.W. Cochran: J. Chem. Soc. Chem. Commun. *1970*, 961.
[624] E. Wenkert, C.J. Chang, D.W. Cochran, and R. Pellicciari: Experientia *28*, 377 (1972).
[625] N. Neuss, H.E. Boaz, J.L. Occolowitz, E. Wenkert, F.M. Schell, P. Potier, C. Kan, M.M. Plat, and M. Plat: Helv. Chim. Acta *56*, 2660 (1973).
[626] E. Bombardelli, A. Bonati, B. Gabetta, E.M. Martinelli, G. Mustich, and B. Danieli: Tetrahedron *30*, 4141 (1974).
[627] E. Wenkert, D.W. Cochran, H.E. Gottlieb, E.W. Hagaman, R.B. Filho, F.J. de A. Matos, and M.I.L.M. Madruga: Helv. Chim. Acta *59*, 2437 (1976).
[628] M.A. Khan, H. Horn, and W. Voelter: Z. Naturforsch. *37 B*, 494 (1982).
[629] N.J. Bach, H.E. Boaz, E.C. Kornfeld, C.J. Chang, H.G. Floss, E.W. Hagaman, and E. Wenkert: J. Org. Chem. *39*, 1272 (1974).
[630] M.C. Koch, M.M. Plat, N. Préaux, H.E. Gottlieb, E.W. Hagaman, F.M. Schell, and E. Wenkert: J. Org. Chem. *40*, 2836 (1975).
[631] E. Wenkert, E.W. Hagaman, B. Lal, G.E. Gutowski, A.S. Katner, J.C. Miller, and N. Neuss: Helv. Chim. Acta *58*, 1560 (1975).
[632] Y. Rolland, N. Kunesch, J. Poisson, E.W. Hagaman, F.M. Schell, and E. Wenkert: J. Org. Chem. *41*, 3270 (1976).
[633] B.C. Das, J.P. Cosson, G. Lukacs, and P. Potier: Tetrahedron Lett. *1974*, 4299.
[634] S. Ferreira Fonseca, J. de Paiva Campello, L.E.S. Barata and E.A. Rúveda: Phytochemistry, in press; see also ref. [429].
[635] E. Wenkert, B.L. Buckwalter, I.R. Burfitt, M.J. Gasic, H.E. Gottlieb, E.W. Hagaman, F.M. Schell, and P.M. Wovkulich, in G.C. Levy, (ed.): Topics in Carbon-13 NMR Spectroscopy, Vol. *2*. Wiley Interscience, New York 1976.
[636] G.S. Ricca and C. Casagrande: Org. Magn. Res. *9*, 8 (1977).
[637] D.W. Hughes, H.L. Holland, and D.B. MacLean: Can. J. Chem. *54*, 2252 (1976).
[638] A. Buzas, R. Cavier, F. Cossais, J.P. Finet, J.P. Jaquet, L. Lavielle, and N. Platzer: Helv. Chim. Acta *60*, 2122 (1977).
[639] R.H.F. Manske, R. Rodrigo, H.L. Holland, D.W. Hughes, D.B. MacLean, and J.K. Saunders: Can. J. Chem. *56*, 383 (1978).
[640] T.T. Nakashima and G.E. Maciel: Org. Magn. Res. *5*, 9 (1973).
[641] M. Shamma and J.L. Moniot: Isoquinoline Alkaloids Research 1972–1977. Plenum Press, New York 1978.
[642] F.I. Carrol, C.G. Moreland, G.A. Brine, and J.A. Kepler: J. Org. Chem. *41*, 996 (1976).
[643] Y. Terui, K. Tori, S. Maeda and Y.K. Sawa: Tetrahedron Lett. *1975*, 2853.

[644] H.L. Holland, D.W. Highes, D.B. MacLean, and R.G.A. Rodrigo: Can. J. Chem. *56*, 2467 (1978).
[645] W.O. Crain, Jr., W.C. Wildman, and J.D. Roberts: J. Am. Chem. Soc. *93*, 990 (1971).
[646] L. Zetta, G. Gatti, and G. Fuganti: Tetrahedron Lett. *1971*, 4447.
[647] A.M. Ismailow, M.K. Yosynov, and X.A. Aslanov: Khim. Prir. Soedin *13*, 422 (1977).
[648] L.E. Overman and P.J. Jessup: J. Am. Chem. Soc. *100*, 5179 (1978).
[649] A. Ahond, F. Picot, P. Potier, and T. Sévenet: Phytochem. *17*, 166 (1978).
[650] N.M.D. Brown, M.F. Grundon, D.M. Harrison, and S.A. Surgenor: Tetrahedron *36*, 3579 (1980).
[651] F.R. Sternitz and I.A. Sharifi: Phytochem. *16*, 2003 (1977).
[652] M. Daudon, M.H. Mehri, M.M. Plat, E.W. Hagaman, F.M. Schell, and E. Wenkert: J. Org. Chem. *40*, 2838 (1975).
[653] M. Daudon, M.H. Mehri, M.M. Plat, E.W. Hagaman, and E. Wenkert: J. Org. Chem. *41*, 3275 (1976).
[654] C.G. Moreland, A. Philip, and F.I. Carroll: J. Org. Chem. *39*, 2413 (1974).
[655] F.I. Carroll, D. Smith, M.E. Wall, and C.G. Moreland: J. Med. Chem. *17*, 985 (1974).
[656] F.I. Carrol, A. Philip, and M.C. Coleman: Tetrahedron Lett. *1976*, 1757.
[657] H.Y. Aboul-Enein and R.F. Borne: Chem. Biomed. Environ. Instrum. *10*, 231 (1980).
[658] S. Yamamura, H. Irikawa, Y. Okumura, and Y. Hirata: Bull. Chem. Soc. Japan *48*, 2120 (1975).
[659] S.P. Singh, S.S. Parmar, V.I. Sternberg, and S.A. Farnum: Spectrosc. Lett. *10*, 1001 (1971).
[660] C.D. Hufford, H.C. Capraro, and A. Brossi: Helv. Chim. Acta *63*, 50 (1980).
[661] S.W. Pelletier, N.V. Mody, A.J. Jones, and M.H. Benn: Tetrahedron Lett. *1976*, 3025.
[662] S.W. Pelletier and Z. Djarmati: J. Am. Chem. Soc. *98*, 2626 (1976).
[663] S.W. Pelletier, N.V. Mody, R.S. Sawhney, and J. Bhattacharyya: Heterocycles *7*, 327 (1977).
[664] R. Radeglia, G. Adam, and H. Ripperger: Tetrahedron Lett. *1977*, 903.
[665] R.J. Weston, H.E. Gottlieb, E.W. Hagaman, and E. Wenkert: Aust. J. Chem. *30*, 917 (1977).
[666] M. Sangaré, F. Khuong-Huu, D. Herlem, A. Milliet, B. Septe, G. Berenger, and G. Lukacs: Tetrahedron Lett. *1975*, 1791.
[667] R. Herzog, A. Morel, J. Biermann, and W. Voelter: Hoppe-Seyler's Z. Physiol. Chem. *365*, 1351 (1984).
[668] R. Herzog, A. Morel, J. Biermann, and W. Voelter: Chemiker-Ztg. *108*, 406 (1984).
[669] W. Voelter and A. Morel: Liebigs Ann. Chem., in press.
[670] P. Hennig, A. Morel, and W. Voelter: Z. Naturforsch. *41 b*, 1180 (1986).
[671] W. Voelter, P. Hennig, and A. Morel: Z. Naturforsch., in press.
[672] H.M. McConnell and R.E. Robertson: J. Chem. Phys. *29*, 1361 (1958).
[673] G.N. La Mar: J. Chem. Phys. *43*, 1085 (1965).
[674] D. Doddrell, I. Burfitt, and N.V. Riggs: Aust. J. Chem. *28*, 369 (1975).
[675] I. Morishima, K. Yoshikawa, K. Okada, T. Yonezawa, and K. Goto: J. Am. Chem. Soc. *95*, 165 (1973).
[676] E. Breitmaier and W. Voelter: Tetrahedron *29*, 227 (1973).
[677] E. Breitmaier, G. Jung, and W. Voelter: Chimia *26*, 136 (1972).
[678] W. Voelter, E. Breitmaier, G. Jung, T. Keller, and D. Hiss: Angew. Chem. *82*, 812 (1970); Int. Ed. Engl. *9*, 803 (1970).
[679] D.E. Dorman, S.J. Angyal, and J.D. Roberts: J. Am. Chem. Soc. *92*, 1351 (1970).
[680] E. Breitmaier, W. Voelter, G. Jung, and C. Tänzer: Chem. Ber. *104*, 1147 (1971).
[681] D.E. Dorman and J.D. Roberts: J. Am. Chem. Soc. *93*, 4463 (1971).
[682] R.G.S. Ritchie, N. Cyr, B. Korsch, H.J. Koch, and A.S. Perlin: Can. J. Chem. *53*, 1424 (1975).
[683] W. Voelter, V. Bilik, and E. Breitmaier: Coll. Czech. Chem. Commun. *38*, 2054 (1973).
[684] W. Voelter and E. Breitmaier: Org. Magn. Res. *5*, 311 (1973).

[685] W. Voelter, E. Breitmaier, and G. Jung: Angew. Chem. *83*, 1011 (1971); Int. Ed. Engl. *10*, 935 (1971).
[686] W. Voelter, C. Bürvenich, and E. Breitmaier: Angew. Chem. *84*, 589 (1972); Int. Ed. Engl. *11*, 539 (1972).
[687] P.A.J. Gorin: Can. J. Chem. *52*, 458 (1974).
[688] P.E. Peffer, K.M. Valentine, and F.W. Parrish: J. Am. Chem. Soc. *101*, 1265 (1979).
[689] T.E. Walker, R.E. London, T.W. Whaley, R. Barker and N.A. Matwiyoff: J. Am. Chem. Soc. *98*, 5807 (1976).
[690] P.A. Gorin, M. Mazurek, and H.J. Koch: Can. J. Chem. *48*, 2596 (1970).
[691] K. Bock and H. Thogersen: Ann. Rep. NMR Spectrosc. *13*, 1 (1982).
[692] A.S. Perlin and B. Casu: Can. J. Chem. *53*, 1212 (1975).
[693] G.J. Wolff: Thesis. Bonn 1979.
[694] S.J. Angyal and G.S. Bethell: Aust. J. Chem. *29*, 1249 (1976).
[695] S.J. Angyal, G.S. Bethell, D.E. Cowley, and V.A. Pickles: Aust. J. Chem. *25*, 1239 (1976).
[696] K. Bock and C. Pedersen: Adv. Carbohydr. Chem. Biochem. *41*, 27 (1983).
[697] H. Paulsen, V. Sinnwell, and W. Grewe: Carbohydr. Res. *49*, 27 (1976).
[698] R.G.S. Ritchie, N. Cyr, and A.S. Perlin: Can. J. Chem. *54*, 2301 (1976).
[699] E. Conway, R.D. Guthrie, S.D. Gero, G. Lukacs, A.M. Sepulchre, E.W. Hagaman, and E. Wenkert: Tetrahedron Lett. *1972*, 4879.
[700] A.I.R. Burfitt, R.D. Guthrie, and R.W. Irvine: Aust. J. Chem. *30*, 1037 (1977).
[701] A.M. Sepulchre, G. Lukacs, G. Vass, and S.D. Gero: Angew. Chem. *84*, 111 (1972); Int. Ed. Engl. *11*, 148 (1972).
[702] K. Bock and C. Pedersen: J. Chem. Soc. Perkin Trans. *2*, 293 (1974).
[703] S. Berger: Tetrahedron *33*, 1587 (1977).
[704] T. Ogawa, J. Uzawa, and M. Matsui: Carbohydr. Res. *59*, C32(1977).
[705] E. Breitmaier, W. Voelter, E.B. Rathbone, and E.M. Stephen: Tetrahedron *29*, 3845 (1973).
[706] P.A.J. Gorin and M. Mazurek: Carbohydr. Res. *27*, 325 (1973).
[707] P.A.J. Gorin and M. Mazurek: Can. J. Chem. *51*, 3277 (1973).
[708] R. Usui, N. Yamaoka, K. Matsuda, K. Tuzimuru, H. Sugiyama, and S. Seto: J. Chem. Soc. Perkin Trans. *1*, 2475 (1973).
[709] J.V. O'Conner, H.A. Nunez, and R. Barker: Biochemistry *18*, 500 (1979).
[710] T.E. Walker and R. Barker: Carbohydr. Res. *64*, 266 (1978).
[711] R. Colson, K.N. Slessor, H.J. Jennings, and I.C.P. Smith: Can. J. Chem. *53*, 1030 (1975).
[712] K. Bock, I. Lundt, and C. Pedersen: Carbohydr. Res. *90*, 7 (1981).
[713] H. Naganawa, Y. Muraoka, T. Takita, and H. Umezawa: J. Antibot. Ser. *A*, *30*, 388 (1977).
[714] S. Jacobsen and O. Mols: Acta Chem. Scand. Ser. *B*, *35*, 163 (1981).
[715] T.A.W. Koerner, Jr., R.J. Voll, L.W. Cary, and E.S. Younathan: Biochemistry *19*, 2795 (1980).
[716] K. Bock and C. Pedersen: Acta Chem. Scand. Ser. *B*, *29*, 258 (1975).
[717] K.S. Kim, D.M. Vyas, and W.A. Szarek: Carbohydr. Res. *72*, 25 (1979).
[718] P.A.J. Gorin and M. Mazurek: Can. J. Chem. *53*, 1212 (1975).
[719] T. Yadomae, N. Ohno, and T. Miyazaki: Carbohydr. Res. *75*, 191 (1979)
[720] G. Lukacs, A.M. Sepulchre, A. Gateau-Olesker, G. Vass, S.D. Gero, R.D. Guthrie, W. Voelter, and E. Breitmaier: Tetrahedron Lett. *1972*, 5163.
[721] A.S. Serianni, J. Pierce, and R. Barker: Biochemistry *18*, 1192 (1979).
[722] A. Malik, N. Afza, and W. Voelter: J. Chem. Soc. Perkin Trans. *1*, 2103 (1983).
[723] W. Fuchs and W. Voelter: Liebigs Ann. Chem. *1982*, 1920.
[724] G. Harsch, M. Harsch, H. Bauer, and W. Voelter: J. Chem. Soc. Pak. *1*, 95 (1979).
[725] A. Serianni, E.L. Clark, and R. Barker: Carbohydr. Res. *72*, 79 (1979).
[726] W. Voelter, E. Breitmaier, R. Price, and G. Jung: Chimia *25*, 168 (1971).
[727] P.A.J. Gorin: Adv. Carbohydr. Chem. Biochem. *38*, 13 (1980).

[728] B. Coxon: Dev. Food Carbohydr. *2*, 351 (1980).
[729] R. Barker, H.A. Nunez, P. Rosevear, and A.S. Serianni: Methods Enzymol. *83*, 58 (1982).
[730] F.W. Wehrli and T. Nishida: Fortschr. Chem. Org. Naturst. *36*, 1 (1979).
[731] R.U. Lemieux, K. Bock, L.T.J. Delbaere, S. Koto, and V.S. Rao: Can. J. Chem. *58*, 631 (1980).
[732] H. Thogersen, R.U. Lemieux, K. Bock, and B. Meyer: Can. J. Chem. *60*, 44 (1982).
[733] D.Y. Gagnaire, F.R. Taravel, and M.R. Vignon: Nouv. J. Chim. *1*, 423 (1977).
[734] G.K. Hamer, F. Balza, N. Cyr, and A.S. Perlin: Can. J. Chem. *56*, 3109 (1978).
[735] R.U. Lemieux, T.L. Nagabhushan, and B. Paul: Can. J. Chem. *50*, 773 (1972).
[736] R.G.S. Ritchie, N. Cyr, and A.S. Perlin: Can. J. Chem. *54*, 2301 (1976).
[737] N. Cyr and A.S. Perlin: Can. J. Chem. *57*, 2504 (1979).
[738] P.E. Hansen: Progr. NMR Spectrosc. *14*, 175 (1981).
[739] K. Bock and C. Pedersen: Acta Chem. Scand. Ser. *B*, *31*, 354 (1977).
[740] P. Colson, H.J. Jennings, and I.C.P. Smith: J. Am. Chem. Soc. *96*, 8081 (1974).
[741] H.J. Jennings and I.C.P. Smith: J. Am. Chem. Soc. *95*, 606 (1973).
[742] H. Friebolin, N. Frank, G. Keilich, and E. Siefert: Makromol. Chem. *177*, 845 (1976).
[743] F.R. Seymour, R.D. Knapp, and S.H. Bishop: Carbohydr. Res. *72*, 229 (1979).
[744] D. Gagnaire and M. Vincendon: Bull. Soc. Chim. Fr. *1977*, 479.
[745] A.P.G. Kieboom, A. Sinnema, J.M. van der Toorn, and H. van Bekkum: J. Roy. Neth. Chem. Soc. *96*, 35 (1977).
[746] G.W. Schnarr, D.M. Vyas, and W.A. Szarek: J. Chem. Soc. Perkin Trans. *1*, 496 (1979).
[747] S.J. Angyal and R. Le Fur: Carbohydr. Res. *84*, 201 (1980).
[748] W. Voelter, G. Jung, E. Breitmaier, and R. Price: Hoppe-Seyler's Z. Physiol. Chem. *352*, 1034 (1971).
[749] A.J. Jones, D.M. Grant, M.W. Winkley, and R.K. Robins: J. Am. Chem. Soc. *92*, 4079 (1970).
[750] D.E. Dorman and J.D. Roberts: Proc. Nat. Acad. Sci. *65*, 19 (1970).
[751] S. Uesugi and M. Ikehara: J. Am. Chem. Soc. *99*, 3250 (1077).
[752] J.A. Elvidge, J.R. Jones, and S. Salih: J. Chem. Soc. Perkin II, *1979*, 1590.
[753] S.B. Petersen, J.D. Led, E.R. Johnston, and D.M. Grant: J. Am. Chem. Soc. *104*, 5007 (1982).
[754] L.G. Marzilli, B. de Castro, and C. Solorzano: J. Am. Chem. Soc. *104*, 461 (1982).
[755] S.J. Karlik, G.A. Elgavish, and G.L. Eichhorn: J. Am. Chem. Soc. *105*, 602 (1983).
[756] A.R. Tarpley, Jr. and J.H. Goldstein: J. Am. Chem. Soc. *93*, 3573 (1971).
[757] R.U. Lemieux: Ann. Rep. N. Y. Acad. Sci. *222*, 915 (1973).
[758] M.C. Thorpe, W.C. Coburn, Jr. and J.A. Montgomery: J. Magn. Res. *15*, 98 (1974).
[759] B. Coxon, A.J. Fatiadi, L.T. Sniegoski, H.S. Hertz and R. Schaffer: J. Org. Chem. *42*, 3132 (1977).
[760] I.W.J. Still, N. Plavac, D.M. McKinnon, and M.S. Chauhan: Can. J. Chem. *56*, 725 (1978).
[761] R.J. Pugmire, D.M. Grant, R.K. Robins, and G.W. Rhodes: J. Am. Chem. Soc. *87*, 2225 (1965).
[762] J.C.P. Smith, H.H. Mantsch, R.D. Lapper, R. Deslauriers, and T. Schleich: Jerusalem Symp. Quantum Chem. Biochem. *1973*, 381.
[763] T.A. Hoffmann and J. Ladik: Adv. Chem. Phys. *7*, 94 (1969).
[764] A. Veillard and B. Pullmann: J. Theor. Biol. *4*, 37 (1963).
[765] J.I. Fernández-Alonso: Adv. Chem. Phys. *7*, 1 (1964).
[766] B.P. Dailey and J.N. Schoolery: J. Am. Chem. Soc. *77*, 3977 (1955).
[767] F.A. Bovey: J. Pol. Sci. Macromol. Res. *9*, 1 (1974).
[768] T. Yamane in G.L. Cantoni and D.R. Davies, (eds.): Procedures in Nucleic Acid Research, Vol. 2, p. 262. Harper and Row, New York 1971.

[769] G. Govil and R.V. Hosur: Conformation of Biological Molecules in P. Diehl, E. Fluck, and R. Kosfeld, (eds.): NMR, Basic Principles and Progress, Vol. 20. Springer, Berlin, Heidelberg, New York 1982.
[770] H.H. Mantsch and J.C.P. Smith: Biochem. Biophys. Res. Commun. 46, 808 (1972).
[771] G.E. Maciel and G.C. Ruben: J. Am. Chem. Soc. 85, 3903 (1963).
[772] G.E. Maciel and J.J. Natterstad: J. Chem. Phys. 42, 2752 (1965).
[773] J.C. Thrierr and M. Leng: Biochem. Biophys. Acta 182, 575 (1969).
[774] M. Dourlent, J.C. Thrierr, F. Brun, and M. Leng: Biochim. Biophys. Res. Commun. 41, 1590 (1970).
[775] J.C. Thrierr, M. Dourlent, and M. Leng: J. Mol. Biol. 58, 815 (1971).
[776] D.L. Inners and G. Felsenfeld: J. Bol. Biol. 50, 373 (1970).
[777] G. Govil and J.C.P. Smith: Biopolymers 12, 2589 (1973).
[778] R.A. Komoroski, I.R. Peat, and G.C. Levy, in G.C. Levy, (ed.), Topics in Carbon-13 NMR Spectroscopy, Vol. 2, p. 179. Wiley Interscience, New York 1976.
[779] R.A. Komoroski: Thesis. Indiana University, Bloomington, Indiana 1973.
[780] M. Miwa, H. Saitó, H. Sakura, N. Saikawa, F. Watanabe, T. Matsushima, and T. Sugimura: Nucl. Acids Res. 4, 3997 (1977).
[781] G. Govil, C.L. Fisk, F.B. Howard, and H.T. Miles: Biopolymers 20, 573 (1981).
[782] R.A. Komoroski and A. Allerhand: Proc. Nat. Acad. Sci. 69, 1804 (1972).
[783] R.A. Komoroski and A. Allerhand: Biochemistry 13, 369 (1974).
[784] W.D. Hamill, Jr., D.M. Grant, W.J. Horton, R. Lundquist, and S. Dickman: J. Am. Chem. Soc. 98, 1276 (1976).
[785] E. Breitmaier, G. Jung, and W. Voelter: Angew. Chem. 83, 659 (1971); Int. Ed. Engl. 10, 673 (1971).
[786] W. Horsley, H. Sternlicht, and J.S. Cohen: Biochem. Biophys. Res. Commun. 37, 47 (1969).
[787] H. Sternlicht, G.L. Kenyon, E.L. Packer, and J. Sinclair: J. Am. Chem. Soc. 93, 199 (1971).
[788] F.R.N. Gurd, P.J. Lawson, D.W. Cochran, and E. Wenkert: J. Biol. Chem. 246, 3725 (1971).
[789] M. Christl and J.D. Roberts: J. Am. Chem. Soc. 94, 4565 (1972).
[790] G. Jung, E. Breitmaier, W. Voelter, T. Keller, and C. Tänzer: Angew. Chem. 82, 882 (1970); Int. Ed. Engl. 9, 894 (1970).
[791] W. Voelter, G. Jung, E. Breitmaier, and E. Bayer: Z. Naturforsch. 26 B, 213 (1971).
[792] W. Voelter, S. Fuchs, R.H. Seuffer, and K. Zech: Monatsh. Chem. 105, 1110 (1974).
[793] D.E. Dorman and F.A. Bovey: J. Org. Chem. 38, 2379 (1973).
[794] W. Voelter, K. Zech, W. Grimminger, E. Breitmaier, and G. Jung: Chem. Ber. 105, 3650 (1972).
[795] R.A. Archer, R.D.G. Cooper, P.V. Demarco, and L.R.F. Johnson: J. Chem. Soc. Chem. Commun. 1970, 1291.
[796] A.R. Quirt, J.R. Lyerla, Jr., I.R. Peat, J.S. Cohen, W.F. Reynolds, and M.H. Freedman: J. Am. Chem. Soc. 96, 570 (1974).
[797] S. Tran-Dinh, S. Fermandjian, E. Sala, R. Mermet-Bouvier, M. Cohen, and P. Fromageot: J. Am. Chem. Soc. 96, 1484 (1974).
[798] R. Freeman, H.D.W. Hill, and R. Kaptein: J. Magn. Reson. 7, 82 (1972).
[799] W.F. Reynolds, I.R. Peat, M.H. Freedman, and J.R. Lyerla, Jr.: Can. J. Chem. 51, 1857 (1973).
[800] Abbreviations according to IUPAC, Eur. J. Biochem. 1, 375 (1967).
[801] W.A. Gibbons, J.A. Sogn, A. Stern, L.C. Craig, and L.F. Johnson: Nature 227, 840 (1970).
[802] S.N. Rosenthal and J.H. Fendler: Adv. Phys. Org. Chem. 13, 279 (1976).
[803] A. Allerhand: Acc. Chem. Res. 11, 469 (1978).
[804] O.W. Howarth and D.M.J. Lilley: Progr. NMR Spectrosc. 12, 1 (1978).

[805] E. Schwertner, S. Herrling, E. Friderichs, S.M. Kim, and L. Flohé in W. Voelter and G. Weitzel, (eds.): Structure and Activity of Natural Peptides, p. 397. Walter de Gruyter, Berlin, New York 1981.
[806] P. Keim, R.A. Vigna, R.C. Marshall, and F.R.N. Gurd: J. Biol. Chem. *248*, 6104 (1973).
[807] P. Keim, R.A. Vigna, J.S. Morrow, R.C. Marshall, and F.R.N. Gurd: J. Biol. Chem. *248*, 7811 (1973).
[808] P. Keim, R.A. Vigna, A.M. Nigen, J.S. Morrow, and F.R.N. Gurd: J. Biol. Chem. *249*, 4149 (1974).
[809] C. Grathwohl and K. Wüthrich: J. Magn. Res. *18*, 191 (1975).
[810] C. Grathwohl and K. Wüthrich: J. Magn. Res. *13*, 217 (1974).
[811] G. Jung, E. Breitmaier, and W. Voelter: Hoppe-Seyler's Z. Physiol. Chem. *352*, 16 (1971).
[812] U. Hollstein, E. Breitmaier, and G. Jung: J. Am. Chem. Soc. *96*, 8036 (1974).
[813] Abbreviations according to IUPAC, Eur. J. Biochem. *17*, 193 (1970).
[814] A.B. Mauger and W.A. Thomas: Org. Magn. Res. *17*, 186 (1981).
[815] J.R. Lyerla, Jr. and M.H. Freedman: J. Biol. Chem. *247*, 8183 (1972).
[816] Yu.A. Ovchinnikow, V.T. Ivanov, V.F. Bystrov, and A.I. Miroshnikov, in J. Meienhofer, (ed.): Chemistry and Biology of Peptides. Ann Arbor Science Publ., Ann Arbor Michigan 1972.
[817] W.A. Thomas and M.K. Williams: J.Chem.Soc. Chem. Commun. *1972*, 994.
[818] W. Voelter and O. Oster: Chemiker Ztg. *96*, 586 (1972).
[819] K. Wüthrich, A. Tun-Kyi, and R. Schwyzer: FEBS Lett. *25*, 104 (1972).
[820] S. Zimmer, W. Haar, W. Maurer, and H. Rüterjans: Eur. J. Biochem. *29*, 80 (1972).
[821] A.I.R. Brewster, V.J. Hruby, A.F. Spatola, and F.A. Bovey: Biochemistrry *12*, 1643 (1973).
[822] R. Deslauriers, C. Garrigou-Lagrange, A.M. Bellocq, and J.C.P. Smith: FEBS Lett. *31*, 59 (1973).
[823] D.E. Dorman, D.A. Torchia, and F.A. Bovey: Macromolecules *6*, 80 (1973).
[824] D.J. Patel: Biochemistry *12*, 667 (1973).
[825] L.G. Pease, C.M. Deber, and E.R. Blout: J. Am. Chem. Soc. *95*, 258 (1973).
[826] W. Voelter, O. Oster, and E. Breitmaier: Z. Naturforsch. *28 B*, 370 (1973).
[827] R. Walter, K.U.M. Prasad, R. Deslauriers, and I.C.P. Smith: Proc. Nat. Acad. Sci. USA *70*, 2086 (1973).
[828] D.J. Patel: Biochemistry *13*, 1476 (1974).
[829] D.A. Torchia and J.R. Lyerla, Jr.: Biopolymers *13*, 97 (1974).
[830] O. Oster, E. Breitmaier, and W. Voelter in T. Axenrod and G.A. Webb, (eds.): NMR of Elements other than Hydrogen. Wiley Interscience, New York 1974.
[831] A.E. Tonelli: J. Am. Chem. Soc. *92*, 6187 (1970).
[832] R. Deslauriers, R. Walter, and I.C.P. Smith: Biochem. Biophys. Res. Commun. *53*, 244 (1973).
[833] I.C.P. Smith, R. Deslauriers, H. Saito, R. Walter, C. Garrigou-Lagrange, W.H. McGregor, and D. Sarantakis: Ann. N. Y. Acad. Sci. *222*, 597 (1973).
[834] W. Voelter, O. Oster, and K. Zech: Angew. Chem. *86*, 46 (1974); Int. Ed. Engl. *13*, 131 (1974).
[835] R. Deslauriers, W.H. McGregor, D. Sarantakis, and I.C.P. Smith: Biochemistry *13*, 3443 (1974).
[836] W. Voelter, C. Bürvenich, M. Eichner, C. Engelfried, S. Fuchs, H. Horn, H. Kalbacher, W. Klingler, E. Pietrzik, K. v. Puttkamer, K. Rager, K. Zech, and D. Gupta, in D. Gupta and W. Voelter, (eds.): Hypothalamic Hormones – Structure, Synthesis and Biological Activity. Verlag Chemie, Weinheim 1975.
[837] R. Deslauriers and I.C.P. Smith, in G.C. Levy, (ed.): Topics in Carbon-13 NMR Spectroscopy. John Wiley & Sons, New York, London, Sydney, Toronto 1976.
[838] W. Voelter: Fortschr. Chem Org. Naturst. *34*, 439 (1977).

[839] R. Deslauriers, R. Walter, and I.C.P. Smith: FEBS Lett. *37*, 27 (1973).
[840] T. Higashijima, M. Tasumi, T. Miyazawa, and M. Miyoshi: Eur. J. Biochem. *89*, 543 (1978).
[841] P.L. Wessels, J. Feeney, H. Gregory, and J.J. Gormley: J. Chem. Soc. Perkin Trans. 2, 1973, 1691.
[842] R. Deslauriers, G.C. Levy, W.H. McGregor, D. Sarantakis, and I.C.P. Smith: Biochemistry *14*, 4335 (1975).
[843] R. Deslauriers and R.L. Somorjai: J. Am. Chem. Soc. *98*, 1931 (1976).
[844] K. Hallenga, G. van Binst, A. Scarso, A. Michel, M. Knappenberg, C. Dremier, J. Brison, and J. Dirkx: FEBS Lett. *119*, 47 (1980).
[845] M. Knappenberg, A. Michel, A. Scarso, J. Brison, J. Zanen, K. Hallenga, P. Deschrijver, and G. van Binst: Biochem.Biophys.Acta *700*, 229 (1982).
[846] K. Nagayama, A. Kumar, K. Wüthrich, and R.R. Ernst: J. Magn. Res. *40*, 321 (1980).
[847] Y. Kobayashi, Y. Kyogoku, J. Emura, and S. Sakakibara: Biopolymers *20*, 2021 (1981).
[848] H. Kessler and V. Eiermann: Tetrahedron Lett. *23*, 4689 (1982).
[849] H. Kessler, M. Bernd, H. Kogler, J. Zarbock, O.W. Sorensen, G. Bodenhausen, and R.R. Ernst: J. Am. Chem. Soc. *105*, 6944 (1983).
[850] R. Deslauriers, R. Walter, and I.C.P. Smith: Biochem. Biophys. Res. Commun. *48*, 854 (1972).
[851] R. Deslauriers, I.C.P. Smith, and R. Walter: J. Am. Chem. Soc. *96*, 2289 (1974).
[852] R. Walter, I.C.P. Smith, and R. Deslauriers: Biochem. Biophys. Res. Commun. *58*, 216 (1974).
[853] H. Saitò and I.C.P. Smith: Arch. Biochem. Biophys. *163*, 699 (1974).
[854] H. Saitó and I.C.P. Smith: Arch. Biochem. Biophys. *158*, 154 (1973).
[855] R. Deslauriers, R. Walter, and I.C.P. Smith: FEBS Lett. *37*, 27 (1973).
[856] R. Walter, in M. Margoulies and F.C. Greenwood, (eds.): Structure- Activity Relationships of Protein and Polypeptide Hormones, Vol. 1, p. 181. Excerpta Medica Found., Amsterdam 1971.
[857] R.E. London, J.M. Stewart, J.R. Cann, and N.A. Matwiyoff: Biochemistry *17*, 2270 (1978).
[858] H.E. Bleich, J.D. Cutnell, and J.A. Glasel: Biochemistry *15*, 2455 (1976).
[859] R. Deslauriers, A.C.M. Paiva, K. Schaumburg, and I.C.P. Smith: Biochemistry *14*, 878 (1975).
[860] R. Delauriers, R.A. Komorowski, G.C. Levy, A.C.M. Paiva, and I.C.P. Smith: FEBS Lett. *62*, 50 (1976).
[861] E.R. Stimson, Y.C. Meinwald, and H.A. Scheraga: Biochemistry *18*, 1661 (1979).
[862] M.A. Khaled, M.M. Long, W.D. Thompson, R.J. Bradley, G.B. Brown and D.W. Urry: Biochem. Biophys. Res. Commun. *76*, 224 (1977).
[863] S. Combrisson, B.P. Roques, and R. Oberlin: Tetrahedron Lett. *38*, 3455 (1976).
[864] D.J. Sauners and R.E. Offord: FEBS Lett. *26*, 286 (1972).
[865] R. Deslauriers, Z. Grzonka, K. Schaumburg, T. Shiba, and R. Walter: J. Am. Chem. Soc. *97*, 5093 (1975).
[866] R. du Vernet and V. Boeckelheide: Proc. Nat. Acad. Sci. U.S. *71*, 2961 (1974).
[867] R. Deslauriers, M. Rothe, and I.C.P. Smith in J.Meienhofer and R. Walter, Eds.: Peptides: Chemistry, Structure and Biology, Proc. IVth American Peptide Symposium, Ann Arbor Science Publ., Ann Arbor, Mich. 1975.
[868] V. Madison, A. Atreyi, C.M. Deber, and E.R. Blout: J. Am. Chem. Soc. *96*, 6725 (1974).
[869] C. Grathwohl, R. Schwyzer, A. Tun-Kyi, and K. Wüthrich: FEBS Lett. *29*, 271 (1973).
[870] L.G. Pease, D.H. Niu, and G. Zimmermann: J. Am. Chem. Soc. *101*, 184 (1979).
[871] L.G. Pease and C. Watson: J. Am. Chem. Soc. *100*, 1279 (1978).
[872] L.F. Johnson: Anal. Chem. *43*, 28A (1971).
[873] J.A. Sogn, L.C. Craig, and W.A. Gibbons: J. Am. Chem. Soc. *96*, 3306 (1974).
[874] D.W. Urry: Res. Dev. *25*, 18 (1974).

[875] E.T. Fossel, W.R. Veatch, Y.A. Ovchinnikov, and E.R. Blout: Biochemistry *13*, 5246 (1974).
[876] P. Viglino, C. Franconi, A. Lai, E. Broscio, and F. Conti: Org. Magn. Res. *4*, 237 (1972).
[877] G. Jung, N. Dubischar, and D. Leibfritz: Eur. J. Biochem. *54*, 395 (1975).
[878] G. Irmscher and G. Jung: Eur. J. Biochem. *80*, 165 (1977).
[879] G. Jung, W.A. König, D. Leibfritz, T. Ooka, K. Janko, and G. Boheim: Biochem. Biophys. Acta *433*, 164 (1976).
[880] V.F. Bystrow, V.T. Ivanov, S.A. Koz'min, I.I. Mikhaleva, K.K. Khalilulina, Y.A. Ovchinnikow, E.I. Fedin, and P.V. Petrovskii: FEBS. Lett. *21*, 34 (1972).
[881] D.J. Patel: Biochemistry *12*, 496 (1973).
[882] E. Grell, T. Funck, and H. Sauter: Eur. J. Biochem. *34*, 415 (1973).
[883] M. Ohnishi, M.C. Fedarko, J.D. Baldeschwieler, and L.F. Johnson: Biochem. Biophys. Res. Commun. *46*, 312 (1972).
[884] V.F. Bystrov, S.L. Potnova, T.A. Balshova, S.A. Koz'min, Y.D. Gravilov, and V.A. Afanas'ev: Pure Appl. Chem. *36*, 19 (1973).
[885] D.A. Torchia and F.A. Bovey: Macromolecules *4*, 246 (1971).
[886] G. Boccalon, A.S. Verdini, and G. Giacometti: J. Am. Chem. Soc. *94*, 3639 (1972).
[887] J.R. Lyerla, B.H. Barber, and M.H. Freedman: Can. J. Biochem. *51*, 460 (1973).
[888] J. Schaefer: Macromolecules *6*, 882 (1973).
[889] R. Di Blasi and A.S. Verdini: Biopolymers *13*, 765 (1974).
[890] R. Di Blasi and A.S. Verdini: Biopolymers *13*, 2209 (1974).
[891] Y. Suzuki, Y. Inoue, and R. Chûjó: Rep. Pol. Phys. Jpn. *17*, 607 (1974).
[892] R. Boni, R. Di Blasi, and A.S. Verdini: Macromolecules *8*, 140 (1975).
[893] E.M. Bradbury, P.D. Cary, C. Crane-Robinson, and P.G. Hartmann: Pure Appl. Chem. *36*, 5 (1973).
[894] Y. Suzuki, Y. Inoue, and R. Chûjó: Biopolymers *14*, 1223 (1975).
[895] B. Perly, C. Chachaty, and A. Tsutsumi: J. Am. Chem. Soc. *102*, 1521 (1980).
[896] M.H. Freedman, J.R. Lyerla, I.M. Chaiken, and J.S. Cohen: Eur. J. Biochem. *32*, 215 (1973).
[897] A. Allerhand, D.W. Cochran, and D. Doddrell: Proc. Nat. Acad. Sci. USA *67*, 1093 (1970).
[898] V. Gluschko, P.J. Lawson, and F.R.N. Gurd: J. Biol. Chem. *247*, 2176 (1972).
[899] I.M. Chaiken, M.H. Freedman, J.R. Lyerla, and J.S. Cohen: J. Biol. Chem. *248*, 884 (1973).
[900] I.M. Chaiken: J. Biol. Chem. *249*, 1247 (1974).
[901] O.W. Howarth and L.Y. Lian: J. Chem. Soc. Chem. Commun. *1981*, 258.
[902] E. Oldfield and A. Allerhand: Proc. Nat. Acad. Sci. USA *70*, 3531 (1973).
[903] F. Conti and M. Paci: FEBS Lett. *17*, 149 (1971).
[904] A.M. Nigen, P. Keim, R.C. Marshall, V. Gluschko, P.J. Lawson, and F.R.N. Gurd: J. Biol. Chem. *248*, 3616 (1973).
[905] A.M. Nigen, P. Keim, R.C. Marshall, J.S. Morrow, R.A. Vigna, and F.R.N. Gurd: J. Biol. Chem. *248*, 3724 (1973).
[906] A.M. Nigen, P. Keim, R.C. Marshall, J.S. Morrow, and F.R.N. Gurd: J. Biol. Chem. *247*, 4100 (1972).
[907] R.B. Moon and J.H. Richard: Proc. Nat. Acad. Sci. USA *69*, 2193 (1972).
[908] R.B. Moon and J.H. Richards: J. Am. Chem. Soc. *94*, 5093 (1972).
[909] N.A. Matwiyoff, P.J. Vergamini, T.E. Needham, C.T. Gregg, J.A. Volpe, and W.S. Caughey: J. Am. Chem. Soc. *95*, 4429 (1973).
[910] P.J. Vergamini, N.A. Matwiyoff, R.C. Wohl, and T. Bradley: Biochem. Biophys. Res. Commun. *55*, 453 (1973).
[911] J.S. Morrow, P. Keim, R.B. Visscher, R.C. Marshall, and F.R.N. Gurd: Proc. Nat. Acad. Sci. USA *70*, 1414 (1973).
[912] E. Antonini, M. Brunori, F. Conti, and G. Geraci: FEBS Lett. *34*, 69 (1973).

[913] J.S. Morrow, R.J. Wittebort, and F.R.N. Gurd: Biochem. Biophys. Res. Commun. *60*, 1058 (1974).
[914] P.C. Lauterbur: Appl. Spectrosc. *24*, 460 (1970).
[915] J.C.W. Chien and J.F. Brandts: Nature, New Biology *230*, 209 (1971).
[916] A. Allerhand, R.F. Childers, and E. Oldfield: Biochemistry *12*, 1335 (1973).
[917] J.H. Bradbury and R.S. Norton: Biochem. Biophys. Acta *328*, 10 (1973).
[918] M.W. Hunkapiller, S.H. Smallcombe, D.R. Whitaker, and J.H. Richards: Biochemistry *12*, 4732 (1973).
[919] M.W. Hunkapiller, S.H. Smallcombe, D.R. Whitaker, and J.H. Richards: J. Biol. Chem. *248*, 8306 (1973).
[920] M.E. Ando, J.T. Gerig, and E.F. Weigand: J. Am. Chem. Soc. *104*, 3172 (1982).
[921] M.S. Matta, M.E. Landis, T.B. Patrick, P.A. Henderson, M.W. Russo, and R.L. Thomas: J. Am. Chem. Soc. *102*, 7151 (1980).
[922] P. Rodgers and G.C.K. Roberts: FEBS Lett. *36*, 330 (1973).
[923] D.T. Browne, G.L. Kenyon, E.L. Packer, D.M. Wilson, and H. Sternlicht: Biochem. Biophys. Res. Commun. *50*, 42 (1973).
[924] D.T. Browne, G.L. Kenyon, E.L. Packer, H. Sternlicht, and D.M. Wilson: J. Am. Chem. Soc. *95*, 1316 (1973).
[925] E.L. Packer, H. Sternlicht, and J.C. Rabinowitz: Proc. Nat. Acad. Sci. USA *69*, 3278 (1972).
[926] J.Feeney, A.S.V.Burgen and E.Grell: Eur. J. Biochem. *34*, 107 (1973).
[927] S.H. Koenig, R.D. Brown, T.E. Needham, and N.A. Matwiyoff: Biochem. Biophys. Res. Commun. *53*, 624 (1973).
[928] J.C.W. Chien and W.B. Wise: Biochemistry *12*, 3418 (1973).
[929] D.A. Torchia and K.A. Piez: J. Mol. Biol. *76*, 419 (1973).
[930] E. Hunt and H.R. Morris: Biochem. J. *135*, 833 (1973).
[931] E. Oldfield, R.S. Norton, and A. Allerhand: J. Biol. Chem. *250*, 6381 (1975).
[932] J.S. Olson, M.E. Andersen, and Q.H. Gibson: J. Biol. Chem. *246*, 5919 (1971).
[933] D. Doddrell and W.S. Caughey: J. Am. Chem. Soc. *94*, 2510 (1972).
[934] C.E. Strouse, V.H. Kollman, and N.A. Matwiyoff: Biochem. Biophys. Res. Commun. *46*, 328 (1972).
[935] T.E. Needham, N.A. Matwiyoff, T.E. Walker, and H.P.C. Hogenkamp: J. Am. Chem. Soc. *95*, 5019 (1973).
[936] N.A. Matwiyoff and B.F. Burnham: Ann. N.Y. Acad. Sci. *206*, 365 (1973).
[937] S.G. Boxer, G.L. Closs, and J.J. Katz: J. Am. Chem. Soc. *96*, 7058 (1974).
[938] H. Scheer and J.J. Katz in K.M. Smith, Ed.: Porphyrins and Metalloporphyrins, p. 482. Elsevier, Amsterdam 1975.
[939] S.S. Eaton and G.R. Eaton: Inorg. Chem. *15*, 134 (1976).
[940] R.J. Abraham, H. Pearson, and K.M. Smith: J. Am. Chem. Soc. *98*, 1604 (1976).
[941] H. Goff: J. Am. Chem. Soc. *99*, 7723 (1977).
[942] R.J. Abraham, F. Eivàzi, H. Pearson, and K.M. Smith: Tetrahedron *33*, 2277 (1977).
[943] H. Goff: J. Chem. Soc. Chem. Commun. *18*, 777 (1978).
[944] T.R. Janson and J.J. Katz, in D. Dolphin, (ed.): The Porphyrins, Vol. IV, p. 1. Academic Press, New York 1978.
[945] G.N. LaMar, D.B. Viscio, K.M. Smith, W.S. Caughey, and M.L. Smith: J. Am. Chem. Soc. *100*, 8085 (1978).
[946] H. Goff and L.O. Morgan: Bioinorg. Chem. *9*, 61 (1978).
[947] H. Falk, K. Grubmayr, and E. Haslinger: Monatsh. Chem. *110*, 1429 (1979).
[948] R.J. Abraham, S.C.M. Fell, H. Pearson, and K.M. Smith: Tetrahedron *35*, 1759 (1979).
[949] V. Wray, U. Jurgens, and H. Brockmann, Jr.: Tetrahedron Lett. *1979*, 2275.
[950] H. Goff: J. Am. Chem. Soc. *103*, 3714 (1981).
[951] S. Lötjönen and P.H. Hynninen: Org. Magn. Res. *16*, 304 (1981).

[952] K.M. Smith, R.J. Abraham, and H. Pearson: Tetrahedron 38, 2441 (1982).
[953] R.J. Abraham, J. Plant, and G.R. Bedford: Org. Magn. Res. 19, 204 (1982).
[954] A. Shirazi, E. Leum, and H.M. Goff: Inorg. Chem. 22, 360 (1983).
[955] T. Perkins, J.D. Satterlee, and J.H. Richards: J. Am. Chem. Soc. 105, 1350 (1983).
[956] S. Lötjönen and P.H. Hynninen: Org. Magn. Res. 21, 757 (1983).
[957] D. Spangler, G.M. Maggiora, L.L. Shipman, and R.E. Christoffersen: J. Am. Chem. Soc. 99, 7478 (1977).
[958] D. Spangler, G.M. Maggiora, L.L. Shipman, and R.E. Christoffersen: J. Am. Chem. Soc. 99, 7470 (1977).
[959] M.H.A. Elgamal, N.H. Elewa, E.A.M. Elkhrisy, and H. Duddeck: Phytochemistry 18, 139 (1979).
[960] H. Duddeck, M.H.A. Elgamal, F.K. Abd Elhady, and N.M.M. Shalaby: Org. Magn. Res. 14, 256 (1980).
[961] H.E. Gottlieb, R.A. de Lima, and F. delle Monache: J. Chem. Soc. Perkin Trans. 2, 435 (1979).
[962] S.A. Sojka: J. Org. Chem. 40, 1175 (1975).
[963] A.K. Bose, P.R. Srinivasan, and G. Trainor: J. Am. Chem. Soc. 96, 3670 (1974).
[964] A.K. Bose, H. Fujiwara, V.S. Kamat, G.K. Trivedi, and S.C. Bhattacharya: Tetrahedron 35, 13 (1979).
[965] H. Günther, J. Prestien, and P. Joseph-Nathan: Org. Magn. Res. 7, 339 (1975).
[966] G.I. Grigor and G.A. Webb: Org. Magn. Res. 9, 477 (1977).
[967] T. Schaefer, H. Hruska, and H.M. Hutton: Can. J. Chem. 45, 3143 (1967).
[968] N.J. Cussans and T.N. Huckerby: Tetrahedron Lett. 1975, 2445.
[969] N.J. Cussans and T.N. Huckerby: Tetrahedron 31, 2587 (1975).
[970] N.J. Cussans and T.N. Huckerby: Tetrahedron 31, 2591 (1975).
[971] N.J. Cussans and T.N. Huckerby: Tetrahedron 31, 2719 (1975).
[972] L. Ernst: J. Magn. Res. 21, 241 (1976).
[973] C.J. Chang, H.G. Floss, and W. Steck: J. Org. Chem. 42, 1337 (1977).
[974] K.K. Chan, D.D. Giannini, A.H. Cain, J.D. Roberts, W. Porter, and W.F. Trager: Tetrahedron 33, 899 (1977).
[975] A. Rabaron, J.R. Didry, B.S. Kirkiacharian, and M.M. Plat: Org. Magn. Res. 12, 284 (1979).
[976] A. Patra and A.K. Mitra: Org. Magn. Res. 17, 222 (1981).
[977] B.S. Kirkiacharian, A. Rabaron, and M.M. Plat: C. R. Acad. Sci., Ser. C, 284, 697 (1977).
[978] O. Convert, C. Deville, and J.J. Godfroid: Org. Magn. Res. 10, 220 (1977).
[979] P. Joseph-Nathan, J. Hidalgo, and D. Abramo-Bruno: Phytochemistry 17, 583 (1978).
[980] A.G. Osborne: Tetrahedron 37, 2021 (1981).
[981] D.P.H. Hsieh, J.N. Seiber, C.A. Reece, D.L. Fitzell, S.L. Yang, J.I. Dalezios, G.N. LaMar, D.L. Budd, and E. Motell: Tetrahedron 31, 661 (1975).
[982] P.S. Steyn, R. Vleggar, P.L. Wessels, and D.B. Scott: J. Chem. Soc. Chem. Commun. 1975, 193.
[983] K.G.R. Pachler, P.S. Steyn, R. Vleggar, P.L. Wessels, and D.B. Scott: J. Chem. Soc. Perkin Trans. 1, 1976, 1182.
[984] R.H. Cox and R.J. Cole: J. Org. Chem. 42, 112 (1977).
[985] T.J. Simpson, A.E. de Jesus, P.S. Steyn, and R. Vleggar: J. Chem. Soc. Chem. Commun. 1982, 632.
[986] P. Joseph-Nathan, J. Mares, M.C. Hernández, and J.N. Shoolery: J. Magn. Res. 16, 447 (1974).
[987] F.W. Wehrli: J. Chem. Soc. Chem. Commun. 1975, 663.
[988] B. Ternai and K.R. Markham: Tetrahedron 32, 565 (1976).
[989] H. Wagner, V.M. Chari, and J. Sonnenbichler: Tetrahedron 1976, 1799.

[990] K.R. Markham and B. Ternai: Tetrahedron *32*, 2607 (1976).
[991] A. Pelter, R.S. Ward, and T.I. Gray: J. Chem. Soc. Perkin Trans. *1, 1976*, 2475.
[992] L. Crombie, G.W. Kilbee, and D.A. Whiting: J. Chem. Soc. Perkin Trans *1, 1975*, 1497.
[993] V.M. Chari, M. Jordan, H. Wagner, and P.W. Thies: Phytochemistry *16*, 1110 (1977).
[994] V.M. Chari, M. Ilyas, A. Neszmélyi, F. Chen, L. Chen, and Y. Lin: Phytochemistry *16*, 1273 (1977).
[995] V.M. Chari, H. Wagner, and A. Neszmélyi, in L. Farkas, M. Gábor, and F. Kállay, (eds.): Flavonoids and Bioflavonoids, p. 49. Elsevier, Amsterdam, Oxford, New York 1977.
[996] W. Voelter, W. Stock, N. Afza, M. Przybylski, K.P. Voges, and G. Jung, in W. Voelter and G. Jung, (eds.): O-(β-Hydroxyethyl)-rutoside, p. 1. Springer, Berlin, Heidelberg, New York 1983.
[997] M. Adinolfi, R. Lanzetta, G. Laonigro, M. Parrilli, and E. Breitmaier: Magn. Res. Chem., in press.
[998] J.B. Grutzner: Lloydia *35*, 375 (1972).
[999] H.G. Floss: Lloydia *35*, 399 (1972).
[1000] G. Lukacs: Bull. Soc. Chim. Fr. *1972*, 351.
[1001] U. Séquin and A.I. Scott: Science *186*, 101 (1974).
[1002] A.I. Scott: Science *184*, 760 (1974).
[1003] A.G. McInnes, J.A. Walter, J.L.C. Wright, and L.C. Vining in, G.C.Levy, (ed.): Topics in Carbon-13 NMR Spectroscopy, Vol.2, p. 123. Wiley Interscience, New York 1976.
[1004] M. Tanabe, H. Seto, and L. Johnson: J. Am. Chem. Soc. *92*, 2157 (1970).
[1005] M. Tanabe, T. Hamasaki, D. Thomas, and L. Johnson: J. Am. Chem. Soc. *93*, 273 (1971).
[1006] P.W. Jeffs and D. McWilliams: J. Am. Chem. Soc. *103*, 6185 (1981).
[1007] R. Singh and D.P.H. Hsieh: Arch. Biochem. Biophys. *178*, 285 (1977).
[1008] C.P. Gorst-Allman, K.G.R. Pachler, P.S. Steyn, P.L Wessels, and D.B. Scott: J. Chem. Soc. Perkin Trans. *2, 1977*, 2181.
[1009] C.P. Gorts-Allman, P.S. Steyn, P.L. Wessels, and D.B. Scott: J. Chem. Soc. Perkin Trans. 1, *1978*, 961.
[1010] J. Polansky, Z. Baskevitch, N. Cagnoli-Bellavita, P. Ceccherelli, B.L. Buckwalter, and E. Wenkert: J. Am. Chem. Soc. *94*, 4369 (1972).
[1011] A.L. Burlingame, B. Balogh, J. Welch, S. Lewis, and D. Wilson: J. Chem. Soc. Chem. Commun. *1972*, 318.
[1012] A.G. McInnes, D.G. Smith, L.C. Vining, and L. Johnson: J. Chem. Soc. Chem. Commun. *1971*, 325.
[1013] J.W. Westley, D.L. Pruess, and R.G. Pitcher: J. Chem. Soc. Chem. Commun. *1972*, 161.
[1014] N. Neuss, C.N. Nash, P.A. Lembke, and J.B. Grutzner: J. Am. Chem. Soc. *93*, 2337 (1971).
[1015] R.J. Cushley, D.R. Anderson, S.R. Lipsky, R.J. Sykes, and H.H. Wasserman: J. Am. Chem. Soc. *93*, 6284 (1971).
[1016] W. Trowitzsch, K. Gerth, V. Wray, and G. Höfle: J. Chem. Soc. Chem. Commun. *1983*, 1174.
[1017] J.V.P. Paukstelis, E.F. Byrne, T.P. O'Connor, and T.E. Roche: J. Org. Chem. *42*, 3941 (1977).
[1018a] N.J.M. Birdsall, J. Feeney, Y.K. Levine, and J.C. Metcalfe: J. Chem. Soc. Perkin Trans. 1, 1441 (1972).
Glycolipids:
[1018b] P.L. Harris and E.R. Thornton: J. Am. Chem. Soc. *100*, 6738 (1978).
[1018c] L.O. Sillerud, J.H. Prestégard, R.K. Yu, D.E. Schäfer, and W.H. Königsberg: Biochemsitry *17*, 2619 (1978).
[1018d] T.A.W. Koerner, Jr., L.W. Cary, S.C. Li, Y.T. Li: J. Biol. Chem. *254*, 2326 (1979).
[1018e] L.O. Sillerud, R.K. Yu, and D.E. Schäfer: Biochemistry *21*, 1260 (1981).
[1018f] F. Sarmientos, G. Schwarzmann, and K. Sandhoff: Eur. J. Biochem. *146*, 59 (1985).

[1019a] G. Lukacs, F. Piriou, S.D. Gero, D.A. v. Dorp, E.W. Hagaman, and E. Wenkert: Tetrahedron Lett. *1973*, 515.
[1019b] G.I. Cooper and J. Fried: Proc. Nat. Acad. Sci. *70*, 1579 (1977).
[1019c] D.M. Rackhan, S.E. Cowdry, N.J.A. Gutteridge, and D.J. Osborne: Org. Magn. Res. *9*, 160 (1977).
[1020] T.H. Witherup and E.W. Abott: J. Org. Chem. *40*, 2229 (1975).
[1021] R.E. Echols and G.C. Levy: J. Org. Chem. *39*, 1321 (1974).
[1022] C.C. Duke and J.K. McLeod: Austr. J. Chem. *31*, 2219 (1978).
[1023] I. Yamaguchi, N. Takahashi, and K. Fujita: J. Chem. Soc. Perkin Trans. *1*, 992 (1975).
[1024a] E. Wenkert, D.W. Cochran, F.M. Schell, R.A. Archer, and K. Matsumoto: Experientia *28*, 250 (1972).
[1024b] R.A. Archer, D.W. Johnson, E.W. Hagaman, L.N. Moreno, and E. Wenkert: J. Org. Chem. *42*, 490 (1977).
[1025] E. Pretsch, M. Vasak, and W. Simon: Helv. Chim. Acta *55*, 1098 (1972).
[1026] S.G. Levine, R.E. Hicks, H.E. Gottlieb, and E. Wenkert: J. Org. Chem. *40*, 2540 (1975).
[1027] E.M. Halbach: Thesis. Bonn 1986.
[1028] J.W. Lown and A. Begleiter: Can. J. Chem. *52*, 2331 (1974).
[1029] D.E. Dorman, R.P. Srinivasan, and R.L. Lichter: J. Org. Chem. *43*, 2013 (1978).
[1030] G.L. Asleson and C.W. Frank: J. Am. Chem. Soc. *97*, 6246 (1975).
[1031a] J.G. Nourse and J.D. Roberts: J. Am. Chem. Soc. *97*, 4584 (1975).
[1031b] M.S. Puar and R. Brambilla: J. Chem. Soc. Perkin Trans. 1, 1847 (1980).
[1031c] S. Omura, A. Nakagawa, A. Nesmélyi, S.D. Gero, A.M. Sepulchre, F. Piriou, and G. Lukacs: J. Am. Chem. Soc. *97*, 4001 (1975).
[1032a] A. Fuhrer: Helv. Chim. Acta *56*, 2377 (1973).
[1032b] E. Martinelli, R.J. White, G.G. Gallo, and P.J. Beynon: Tetrahedron *29*, 3441 (1973).

Subject index

This index emphasizes chemical classes in *italic type*. Page numbers refer to carbon shifts if not otherwise specified. Individual compounds are included if their spectra are reproduced or if they are important parent skeletons.

Absorption-mode spectra 13f.
– example 16, 35
Absorption signal 5
Accumulation
– of spectra 21
– of FID signals 41
Acetals 220
Acetates 228 (table)
Acetic amide, N,N-dimethyltrichloro-
– temp. dep. spectra 130
Acetoacetate, ethyl-
– spectrum 232
Acetone
– spectrum 16
Acetophenones 221 f. (table)
Acetylacetone
– spectra 52
Acetylation shift
– of alcohols 207, 230
– of steroids 337, 340
Acetylenes 196
– see Alkynes
Acids
– see carboxylic and sulfonic acids
Aconitine alkaloids 374
Acquisition time 30
Actinomycin D
– 2D CH shift correlations 426 f.
Acyl cations 304
Adamantane
– spin-lattice relaxation times 61 (fig.)
Additivity relationships
– alkanes 184 f.
Aflatoxin B$_1$
– biosynthesis 459
– 2D CH shift correlation 446
Alcohols 206 ff.
– see alkanols and cycloalkanols
– spin-lattice relaxation times 179
Aldehydes–
– CH couplings–

– – one-bond 136 f.
– – three-bond 144
– – two-bond 141, 143
– shifts
– – aliphatic 114, 217 (table)
– – aromatic 221 (table), 260 (table)
– substituent increments 220
Aldofuranoses 383 f. (table)
Aldohexoses 380, 381 f. (table)
Aldonic acids 400 (table)
Aldonolactones 400 (table)
Aldopentoses 380 f. (table)
Aldopyranoses 381 f. (table)
Alkaliorganic compounds 295
Alkaloids 360 ff. (table)
– aconitines 374
– amarylidaceae 370
– aporphines 369
– bisindoles 367
– bisisoquinolines 371
– colchicines 373
– dihydroindoles 363
– diterpenes 373 f.
– ergolines 366 f.
– indoles 363 f.
– isoquinolines 368 ff.
– izidines 361 ff.
– lysergic acid derivatives 366 f.
– morphines 370
– peptides 377 f.
– piperidines 360 f.
– proaporphines 369
– pyridines 360 f.
– pyrrolidines 360 f.
– pyrrolizidines 361
– quinines 372
– quinolines 371
– quinolizidines 362 f.
– steroids 375
– tetrahydro-β-carbolins 364 f.
– tropanes 361

Alkaloids
- with exocyclic nitrogen 373
- yuzurimines 373

Alkanals 216ff.
- see aldehydes

Alkane derivatives
- shifts
- - substituent increments 315 (table)

Alkanes
- CC couplings 149
- CH couplings
- - one-bond 135f.
- - three-bond 143
- - two-bond 141
- shifts 111, 183 (table)
- - additivity relationship 184f.
- - increment system 184
- - steric effect 185f.

Alkanols
- acetylation shifts 207, 230
- protonation shifts 207
- shift increments 207
- shifts 208 (table)

Alkene-π-complexes 300f.

Alkenes
- CC couplings 150, 154, (E/Z)
- CH couplings
- - one-bond 135ff.
- - three-bond 144
- - two-bond 141
- shifts 111, 116, 192f. (table)
- - influence of configuration 192f.
- - substituent increments 194, 318 (table)

Alkenyl halides
- CH couplings 199 (table)
- shifts 199 (table)

Alkylation
- effect on spin-lattice relaxation 168
- shift of alcohols 213

Alkyl halides
- CF couplings 162 (table)
- CH couplings
- - one-bond 136
- shifts 111, 199 (table)
- solvent shifts 201 (table)

Alkynes
- CC couplings 150
- CH couplings
- - one-bond 135f.
- - three-bond 144
- - two-bond 141
- shift increments 197
- shifts 111, 197 (table)

Alkynols 208

Alkynylethers 215
- shifts 214 (table)

Alkynyl halides 199 (table)
Allenes 198 (table)
Allyl ions 302, 306
Alternately pulsed $^{13}C\{^1H\}$ NMR 50

Amides
- CN couplings 156
- shifts 229 (table)

Amide resonance 231

Amidines 241

Amines
- CN couplings 156
- pH (protonation) shifts 121, 236
- shift increments 236, 238 (table)
- shifts
- - aliphatic 237 (table)
- - aromatic 257 (table)
- - heterocyclic 273f. (tables)

Amino acids
- CN couplings 159
- pH shifts 121, 420f.
- protecting groups 414
- shifts 415ff. (table)
- - pH dependence 420

Ammonium salts 228, 237, 260

Androstanes
- CC couplings 338
- shifts 341 (table)

5β-Androstan-3,17-dione
- 2D CH correlation 95

Anilines
- CN couplings 156, 158
- shifts 113, 257 (table)

Angular momentum 1

Anhydrides 229 (table)

Anhydropanaxadiol
- spectra 84

Anisotropic rotation, effect on T_1 169f.

Annulenes
- bridged 266

Annulenones 224f.

Anthracenes 265f.

Antibiotics 457f., 463, 468

Aporphine alkaloids 369

APT (attached proton test) 75, 78

Aromatic compounds 254ff.
- CC couplings 268
- CF couplings 269
- CH couplings 145f., 266
- shifts 256ff. (table), 260f. (table), 263 (table)
- - substituent increments 264, 319f. (table)

Arrhenius equation 130, 181

Axial/equatorial configuration
- coupling differences 205
- shift differences 115, 203, 209ff. (table), 212, 316 (table), 379, 393

Subject index 501

Azacumulenes 244 f.
Azacycloalkanes
– CH couplings 288
– shifts 273 (table)
Azadecalins 276
Azapolycycles 277
Azines
– CH couplings 289
– shifts 283 (table)
Aziridines (Aziranes)
– CH couplings 138, 288
– shifts 273
Azo compounds 257
Azoles
– CH couplings 139, 146, 288
– shifts 282 (table)
Azoxy compounds
– CN couplings 156

Benzaldehydes
– CH couplings 143, 146, 267
– shifts 221 (table), 260
Benzene derivatives
– CC couplings 150 f. (table), 268 (table)
– – one-bond 150
– – three-bond 154
– – two-bond 151
– CF couplings 269 (table)
– CH couplings 139, 145 (table), 266
– – one-bond 139, 145
– – three-bond 145 f.
– – two-bond 142, 145
– shift prediction (examples) 321
– shifts 113, 256 ff. (tables)
– – alkyl-, alkenyl-, alkynyl- 256
– – disubstituted 260 f.
– – fused 265 f.
– – mixed disubstituted 256 f.
– – polyalkyl- 260
– – polysubstituted 261
– – substituent increments 319 f. (table)
– spin-lattice relaxation times 170 (table)
Benzenonium ions 305
Benzocycloalkenes
– CH couplings 265 (table)
– shifts 265 (table)
Benzoic acid esters 231 (table)
Benzoic acids 257, 260 f.
Benzyl derivatives 256
Benzyl ions 303, 307
Bicycloalkanones 219
Bicyclo[2,2,1]heptanes
– see norbornanes
Biopolymers 393, 423
Biosynthetic pathways 451 ff.

Biotin
– J-resolved 2D spectra 90 f.
Biphenyl, 4,4'-dimethyl-
– temp. dep. spin-lattice relaxation 182 (fig.)
Biphenyls
– shifts 256
– spin-lattice relaxation times 174, 182
Bipyridine, 2,2'-
– inversion-recovery spectra 292
Bipyrrole, 2,2'-
– spectra 291
Bisindole alkaloids 367
Bisisoquinoline alkaloids 371
Bloch equations 8 ff.
Boltzmann distribution of spins 5, 7, 22, 78
Boron compounds
– see organoboron compounds
Bulk susceptibilities 17
Butadiene, 1,3-, 3-(isopinocampheoxy)-2-methyl-
– DEPT, 2D INADEQUATE 105
– 2D CH correlation cover page
Butadienes
– see dienes

Calibration of NMR spectra 16
D-Camphor
– J-modulated spectra 77
– isotropic shifts 125 (fig.)
Camphor derivatives
– CH couplings 329 (table)
– shifts 331 (table)
Carbanions 112
– CN couplings 308
– shifts 112, 305 f. (table)
Carbenium ions
– CH couplings, averaging 304
– shifts 113, 303 (table)
Carbocations 113, 303 f.
Carbodiimides 245
β-Carbolin alkaloids 364 f.
Carbohydrates 379 ff.
– aldonic acids 400 (table)
– aldonolactones 400 (table)
– disaccharides 393 f. (table)
– glycosides 383 f. (table)
– inositols 401 (table)
– monosaccharides 380, 381 ff. (table)
– – CC couplings 148
– – CH couplings 27, 142
– polyols 398 (table)
– polysaccharides 393
Carbon chemical shifts 107 ff.
– comparison with proton shifts 108, 118
– conversion 108
– correlation with UV absorption 110

Carbon chemical shifts
- influence of
- - anisotropic fields   116, 255
- - carbon hybridization   111
- - charge density   111, 254, 281, 265, 411 (fig.), 422 (fig.)
- - configuration   115f., 203, 209ff., 229, 272, 274
- - conformation   116, 131, 274ff.
- - conjugation   114
- - crowded substituents   112
- - diastereotopic effects   205
- - dilution   120
- - electric fields   116, 121
- - electron deficiency   113
- - electron density   111, 254, 281, 395 (fig.)
- - electron donation   113, 205, 215, 222, 258, 281
- - electronegativity   111f., 411 (fig.)
- - electronic excitation energy   110, 233
- - electron releasing   113, 205, 215, 222, 258, 281
- - electron withdrawing   113f., 216, 224, 258, 281
- - heavy atoms   117, 233, 300
- - homoconjugation   194, 219
- - hybridization   111
- - hydrogen bonding   117f., 222
- - π-inductive effect   255
- - intramolecular fields   116, 121
- - isotopic substitution   117, 379
- - N-oxidation   286
- - pH (protonation)   121f., 236, 286, 420f.
- - resonance (mesomeric effects)   113, 255, 281
- - ring current   117
- - solvents   118, 120, 201
- - steric interactions (γ effects)   115f., 203, 213, 218, 220f., 255, 259, 307
- - temperature   131ff.
- - unshared electron pairs   112, 244
- prediction   314ff.
- references   17, 108
- survey   110ff., 119
Carbon coupling constants   133ff.
- calculation   134
- carbon-aluminum   293
- carbon-boron   161, 293, 297
- carbon-cadmium   161, 293
- carbon-carbon   147ff., 268
- - influence of
- - - bond angle   141
- - - bond order   153
- - - carbon hybridization   150f.
- - - configuration   152ff.
- - - electronegativity   152

- - longer-range   152ff.
- - measurement by INADEQUATE   84f., 350
- - one-bond   147ff.
- - three-bond   153ff.
- - - comparison with CH and HH couplings   153
- - two-bond   152
- carbon-deuterium   109 (table), 147
- carbon-fluorine   161f.
- - longer-range   162 (table), 206 (table), 269 (table), 340 (table)
- - - through-space mechanism   270
- - one-bond   162 (table), 206 (table), 269 (table), 340 (table)
- carbon-germanium   293, 298
- carbon-lead   293, 299 (table)
- carbon-lithium   293, 295
- carbon-mercury   161, 293, 301
- carbon-metal
- - prediction   294
- carbon-nitrogen   155ff.
- - as assignment aid   158
- - longer-range   157f.
- - lone pair influences   156ff.
- - one-bond   155f., 159 (table)
- carbon-phosphorus   161, 250ff.
- - longer range   251 (table), 410
- - - structural influences   252f. (table)
- - one-bond   251 (table)
- carbon-platinum   293
- carbon-proton   134ff., 266
- - influence of
- - - bond angle   141
- - - carbon hybridization   134f., 135 (fig.)
- - - carbon s character   135 (fig.)
- - - chelated hydrogen   147
- - - configuration   138, 142ff., 266
- - - distance   146
- - - electronegativity   134f., 139, 142, 146
- - - hydrogen bonding   118
- - - ring strain   138
- - - solvent   118 (table), 140
- - - substituent crowding   135, 136 (table), 138
- - longer-range   134ff.
- - - comparison with HH couplings   140 (table)
- - three-bond   143ff. (table)
- - two-bond   141f. (table)
- - one-bond   134ff. (tables)
- carbon-selenium   161
- carbon-silicon   161, 293, 299 (table)
- carbon-tellurium   161
- carbon-thallium   161, 293, 297 (table)
- carbon-tin   161, 298f. (tables)

- carbon-tungsten 293
- Carbon-13(12) labelling 270f., 457ff.
- Carbon *s* character
- influence on
- – CC couplings 151 (fig.)
- – CH couplings 135 (fig.)
- – CP couplings 252
- Carbon-13 satellites 18, 79, 85
- Carbon tetrachloride
- spin-lattice relaxation time 57 (fig.)
- spin-spin relaxation time 65 (fig.)
- *Carbonyl compounds* 215ff.
- aldehydes 216ff., 221, 260
- carboxylic acids and derivatives 226ff., 256, 260
- CH couplings
- – one-bond 137f.
- – three-bond 144
- – two-bond 141, 143
- ketones 216ff.
- phenones 116, 221f. (table)
- quinones 222ff.
- shifts
- – correlation with UV absorption 110, 219
- – in comparison with thiocarbonyl compounds 233
- *Carbonyls*
- see metal carbonyls
- *Carboxylic acids*
- CC couplings
- – one-bond 149, 152
- – three-bond 153f.
- – two-bond 152
- CH couplings
- – one-bond 137f.
- – two-bound 141
- pH shifts 121
- shift increments 227, 230
- shifts 225f. (table)
- – aliphatic 225f. (table)
- – aromatic 257, 260f. (table)
- – influence of hydrogen bonding 227
- solvent shifts 230
- α,β-unsaturated
- – substituent increments 230
- – shifts 228 (table)
- *Carboxylic acid derivatives*
- shifts 228f. (table)
- amides 229 (table), 231
- anhydrides 229 (table), 231
- esters 228 (table), 231
- halides 229 (table), 232
- *Cardenolides* 359 (table)
- *Carotenoids* 335 (table)
- Carr-Purcell-Meiboom-Gill spin echo (CPMGSE) experiments 63f.

CAT method 21
CC COSY 148
Cephalosporins
- biosynthesis 463
- shifts 464 (table), 468
Chalcones 452
π-Charge density
- vs. ^{13}C shift 111 (fig.), 254, 281
Chemical shift(s) 15f., 107ff.
- see carbon-13 chemical shifts
Chemical shift anisotropy relaxation 163
Chemical shift ranges 119 (fig.)
Chiral shift reagents 124
Chloroacetylation shifts
- of alcohols 230
Chloroform
- solvent shifts 118 (table), 120
- SPT spectra 79
Chlorophyll derivatives
- coordination shifts 441
- shifts 443f. (table)
Cholanes 340, 357 (table)
Cholestanes 340
- CC couplings 350
- shifts 351f. (table)
Cholestan-3-one, 5α-
- spectra 349
Chromones 279
cis-trans isomerism of
- alkenes 116, 192f. (table)
- cycloalkanes 115, 187 (table)
- cycloalkanols 210 (table)
- nitrosamines 246
- oximes 137, 241
- polymers 311f.
Coalescence temperature 129
Colchicine alkaloids 373
Colchicine
- spectra 45
COLOC 96
Complexation shifts
- boric acid 410
- titanium tetrachloride 445
Computer simulation of lineshapes 130 (fig.)
Configuration
- influence on
- – CF couplings 205f.
- – CH couplings 137
- – CP couplings 252
- – shifts 115f., 187, 192, 210, 203f., 209ff., 215, 225f., 229, 241, 246, 274
Conformation
- from spin-lattice relaxation times 172
- from temp. dep. spectra 131f., 186, 276
Conjugation
- influence on shifts 114

Subject index

Continuous wave (CW) NMR   22
– sensitivity relative to PFT   41
Contour plot   88
Conversion of ^{13}C shifts   108
Cope systems
– shifts   195 (table)
– – temp. dependence   196
Correlation
– see shift correlation
Correlation function   167 (fig.)
Correlation time, effective   166
– vs. T_1, T_2 167 (fig.)
COSY   96f.
– with delay   98
Coumarins
– CH couplings   288, 445
– shifts   278, 445, 448ff. (table)
Coumestanes   468
Coupling constants   18f., 133
– see carbon-13 coupling constants
– signs   19, 80
Coupling, residual   50
Cumulenes   198 (table)
Cyanides   243
– see nitriles
Cyanines
– CH couplings   240
– shifts   239
Cycloalkanes
– CC couplings   149, 151 (table), 153, 154
– – one-bond   149
– – three-bond   153f.
– – two-bond   151
– CF couplings   162 (table), 206
– CH couplings   139 (table)
– – one-bond   139 (table), 186 (table)
– – two-bond   140f.
– shifts   115 (*cis/trans*), 186f., (tables)
– shift increments   188
Cycloalkanols
– configurational isomerism   209 (fig.), 210 (table)
– shift increments   212 (table)
– shifts   210f. (table)
Cycloalkanones
– shifts   218 (table)
– – correlation with UV absorption   219
Cycloalkenes
– CH couplings   139
– shifts   117, 195 (table)
Cycloalkenones   219 (table)
Cycloalkenyl cations   302
Cycloalkyl halides
– CF couplings   206 (table)
– shift increments   203

– shifts   204 (tables)
Cycloalkynes   197f.
Cyclodienes   195 (table)
Cyclohexane, *cis*-1,2-dimethyl-
– temp. dep. spectra   132
Cyclohexanes
– CF couplings   206 (table)
– CH couplings   139, 186
– shifts   187 (table), 210 (table), 236
– – substituent increments   188, 316 (table)
Cyclohexanol, *cis*- and *trans*-4-*t*-butyl-
– spectrum   209
Cyclohexanols
– shift increments   212
– shifts   210 (table)
Cyclononane
– temp. dep. spectra   131
Cyclopentanes
– shift increments   188
– shifts   187 (table)
Cyclophanes   265
Cyclopolyene-π-complexes   294 (table)
Cyclopolyenes
– shifts   195 (table)
– – homoconjugative effect   194
Cyclopolyenones
– see annulenones
Cyclopolyenyl anions   306
Cyclopropanes
– CC couplings   149
– CH couplings   138f., 154
– shifts   187

Decalins
– shift increments   191
– shifts   190, 209, 211 (table)
Decoupler pulse   81
– calibration   82
Decoupling (spin decoupling)   43ff.
– basic concept   43f.
– broadband   45
– channel   71
– gated   50
– – for *J*-resolved 2D NMR   90
– inverse gated   50
– low-power noise   45
– noise   45
– off-resonance   47ff.
– pulsed   50f.
– selective   53f., 94
DEFT   39f.
DEPT   80f.
Degree of alkylation
– influence on
– – shifts   112

Subject index 505

– – spin-lattice relaxation times   168
Degree of Substitution
– influence on
– – CH coupling constants   135 f.
– – shift   112
– – spin-lattice relaxation time   168
Deshielding   17
Detection period   75, 87
Deuterated solvents
– shifts and couplings   109 (table)
Deuterium isotopic effects   117, 379
Deuterium labeling   337, 379
Diamines
– aliphatic   237
– aromatic   260
Diamagnetic shielding term   110
Diastereotopic shifts
– of alcohols   206
Diazoalkanes
– CN couplings   156
– shifts   114, 244
Dicarboxylic acid anhydrides   229 (table)
Dicarboxylic acids
– aliphatic   226 (table)
– aromatic   260
o-Dichlorobenzene
– T_1, T_2 67 (fig.)
Dienes
– shifts   114, 193 (table)
– – influence of
– – – configuration   194
– – – donor substituents   239
Dienyl anions   306 (table)
Digital filtering   36
Digital resolution   36 f.
Digitization   30
Dihalobenzenes   260
Dihydroflavones   454 (table)
Dilution shifts   120, 285
Diones (diketones)
– keto-enol tautomerism   220
– shifts   218, 257
Dipeptides   423 (table), 427 (table)
Diphenylether
– spin-lattice relaxation times   58 (fig.)
Diphosphines   249
Dipolar relaxation   46, 164, 166 f.
Dipole-dipole relaxation mechanism   46, 164, 166 f.
Disaccharides   393, 396 (table)
Dispersion spectrum   14
– example   35
Dissociation effects
– on CN couplings   156 ff.
– on shifts   121 f.
Disulfides   234

Diterpene alkaloids   373 (table)
Diterpenes
– biosynthesis   459, 461
– shifts   332 (table), 461 (table)
Dithianes, 1,3-
– shifts   233, 275 (table)
Dodecahedrane   189
Double quantum frequency   102
Double quantum transfer   86, 102
Double resonance techniques in
  ^{13}C NMR   43 ff.
Dwell time   30

Ecgonine
– 2D shift correlations   98 (fig.)
Editing of spectra   82 f.
Electron deficiency
– effect on shifts   113
Electronegativity
– effect on
– – CH coupling   134 f., 142, 144, 146
– – CP coupling   252, 254
– – ^{13}C shifts   111
Electron density vs. ^{13}C shifts   111, 395 (fig.)
Electron withdrawing
– effect on ^{13}C shifts   113
Enamines   239, 280
Enaminoaldehydes   240
Enantiomers
– signal separation   124
Endo/exo configuration
– coupling differences   205 f., 253, 301
– shift differences   203 f., 212, 236, 317 (table)
Energy of activation   130
Enolethers
– aliphatic   214 (table)
– heterocyclic
– – CH couplings   289
– – shifts   276, 278
Enones   114, 218 (table), 219
Enthalpy of activation   130
Equatorial/axial configuration
– shift difference   115, 203, 209 ff. (table), 212, 316 (table), 379, 393
Equilibrium magnetization   7, 22
Ergoline alkaloids   366 f.
Esterification shift
– of carboxylic acids   230
– of steroids   337
Esters   228 ff.
– see carboxylic acid derivatives
Estranes   338, 341 f. (table)
Ethane, 1,2-dimethoxy-
– spectra   49
Ethers
– aliphatic

Ethers
– – shift increments 213
– – shifts 214 (table)
– – steric influences 213
– aromatic 257, 261, 263
– heterocyclic
– – CH couplings 288
– – shifts 273, 277
Ethylbenzene
– spectrum 32
Evolution
– period 75, 87
– time 87
Exo/endo configuration
– see *endo/exo* configuration
Exponential multiplication 36 f.
External lock 72
External reference 72, 108
Eyring equation 128

Fellgett principle 42
Fermi contact mechanism
– of coupling 18 f., 133 f., 299
Fermi contact shift 124
Field-frequency lock 70
Filtering of frequencies 31
Flavin adenine dinucleotide
– spectra 402
Flavones (Flavonoids)
– CH couplings 288
– shifts 279, 452 f. (table)
Flavone glycosides 450
Fluoroalkanes
– CF couplings 162, 206
Folding back 31
Formic acid, NOE 46
Formyl compounds (aldehydes)
– CH couplings 137 f., 141, 143 f.
– shifts 114
Fourier series 29, 41
Fourier transform 29
Fourier transformation 33
Franganine
– 2D CH correlation 376
Free enthalpy of activation 130
Free induction decay (FID) 24 f.
– examples 26 f., 37
– driven equilibrium FID 40
Fulvenes 195 f.
Functional group shifts 118 f. (fig.)
Fungal metabolites 457 f., 467
Furanosides 383 f. (table)
Furans
– CH couplings 146
– shifts 282
Fused aromatic compounds
– CC couplings 154 f.
– CF couplings 270
– CH couplings 266
– shifts 266 (table)
Fused heterocycles
– shifts 284 (table)
– – nitrogen increments 324

D-**G**alactose
– spectrum 38
Gated decoupling 50
Gated spin-echo 75
Gaussian multiplication 36
Gibbs-Helmholtz equation 130
D-Glucose
– 1-^{13}C enriched
– – spectra 27
– mutarotational equilibration
– – spectra 380
– uniformly enriched
– – CC COSY 148
Glycylalanine
– isotropic shifts 126 (fig.)
Glykolipids 497
Glycosidation shifts 379, 385 f. (table), 452
Glycoside sequencing
– by spin-lattice relaxation times 175
Glycosides
– spin-lattice relaxation times 175
– shifts 385 f. (table), 450 (table)
Glycylalanine
– isotropic shifts
– – spectra 126
Grignard reagents 258, 296 (table)
– Schlenk equilibrium 295 f.
Guanidines 241 f.
Guanidinium ions 303
Guttapercha
– spectrum 312
Gyromagnetic ratio 2
– influence on line-broadening 308

Half-maximum intensity width 4
Halides
– see alkyl halides, haloalkanes
– – and carboxylic acid halides
Hammet σ constants *vs.* ^{13}C shifts 259 (fig.)
Haloalkanes
– CF couplings 162 (table)
– CH couplings
– – one-bond 136
– – three-bond 144
– shifts 111 f., 199 (table)
– solvent shifts 201 (table)
Haloalkenes
– CH couplings 199 (table)

Subject index   507

– shifts   199 (table)
Haloalkynes
– CH couplings   199 (table)
– shifts   199 (table)
Halobenzenes
– CF couplings   269
– CH couplings   145
– shifts   256
Halomethanes
– CH couplings   199 (table)
– shift prediction   200
– shifts   199 (table), 200 (fig.)
Halouracils   411
Heisenberg uncertainty relation   5
Hemiacetals   232
Henderson-Hasselbach equation   122
Heterocycloalkanes
– (*Heteroalicyclic compounds*)   272ff.
– CH couplings   139, 288
– shifts   273 (table)
Heterocycloalkenes
– CH couplings   288f. (table)
– shifts   278ff. (table)
Heteroaromatic compounds
– CC couplings   150
– CF couplings   291 (table)
– CH couplings   139, 142, 145f.
– – one-bond   139 (table), 145
– – three-bond   145, 146
– – two-bond   145, 146
– CN couplings   156ff.
– five-membered, benzo-fused   282
– five-membered, dibenzo-fused   283
– five-membered, monocyclic
– – CH couplings   288ff. (table)
– – shifts   282 (table)
– fused heterocycles
– – CH couplings   289f.
– – shifts   284 (table)
– six-membered, benzo-fused   283
– six-membered, dibenzo-fused   284
– six-membered, monocyclic
– – CF couplings   291 (table)
– – CH couplings   288ff. (table)
– – shifts   284 (table)
– shifts   282ff. (table)
– – nitrogen increments   324
– – substituent increments   322f.
Heteropolycycloalkanes   275, 277 (table)
Heteronuclear couplings   155ff., 160ff.
Heterospirocyclic compounds   277 (table)
Hexadeuteriodimethylsulfoxide
– spectrum   26
Hexamethylphosphoramide
– spectrum   37
Hindered rotation   130, 172f.

– of amides   229 (table), 231
Homonuclear couplings   147ff.
Homoisoflavanones   456 (table)
Hormones
– hypothalamic   430
– peptides   430, 432 (table)
– steroids   341ff.
– phyto-   467
Hybridization
– effects on
– – ^{13}C coupling constants   134f., 150f., 155f., 252
– – shifts   111
Hydrazones   240
Hydroaromatic compounds   265 (table)
Hydrocarbons   183ff.
– see alkanes, alkenes, alkynes, benzenes
Hydrogen bonding, intramolecular
– effect on
– – shift   117
– – CH coupling   147
– – spin-lattice relaxation time   178f.
Hyperconjugation
– influence on CH coupling   290

Imaginary spectra   14, 36
Imidazoles
– alkaloids   373
– CH couplings   288
– shifts   282
Imines   240ff.
– aldimines   242 (table)
– cyclic   280
– oximes   241 (table)
– ketimines   242 (table)
Imonium salts   242
INADEQUATE
– one-dimensional   84f.
– two-dimensional   102f.
– symmetrized   102, 105
Increment systems   313ff.
– see shift and substituent increments
Indoles   282
– alkaloids   363ff.
Indolizines   283
INEPT   80
Inositols   401 (table)
Intermolecular interaction
– effect on
– – spin-lattice relaxation time   176, 178f.
Internal reference   17, 72
Intramolecular mobility
– effect on
– – shifts   127ff.
– – spin-lattice relaxation time   166ff., 172f.
Inverse gated decoupling   50, 52

Inversion of rings   131f.
Inversion recovery techniques   55ff.
Ionic species   302ff.
Isoalloxazines   402, 409
Isocyanates   245
Isocyanides
– see isonitriles
Isoflavones
– CH couplings   288
– shifts   279, 454
Isonitriles
– CN couplings   159, 243 (table)
– shifts   243 (table)
3-(Isopinocampheoxy)-2-methyl-1,3-butadiene
– DEPT, 2D-INADEQUATE   105
Isoquinolines
– alkaloids   368ff.
– nitrogen increments   324
– shifts   283
Isothiocyanates
– CN couplings   159
– shifts   245
Isotope effect on ^{13}C shifts   117
Isotropic shifts   123ff.
Izidine alkaloids   361f.

J-modulation   75f., 88
– for multiplicity determination   76f.
– in two-dimensional NMR   89
J-modulated spin-echo (JMSE)   75ff.
J-resolved 2D NMR   89f.

Karplus-Conroy relations   143, 153, 205, 253, 395
Ketals   220
Ketenes   114
Ketenimines   244
Keto-enol tautomerism   52 (fig.), 220, 232 (fig.)
β-Ketoesters   232
Ketones
– shifts
– – aliphatic   217f. (table)
– – aromatic   221f. (table), 257
– – correlation with UV absorption   219
– – homoconjugative effects   219
– – steric effects   218
Ketofuranoses   383f.
Ketopyranoses   381f.
Ketoses   380f.

Lactams   279
Lactones   278
D-Lactose
– 2D CH correlation   95
Lamb formula   110

Lanthanoid shift reagents   124f.
Larmor equation   4, 15
Larmor frequencies of nuclei   69 (fig.)
Larmor frequency   4
Larmor precession   4
Lecithins
– shifts   467
– spin-lattice relaxation times   176
Line broadening
– due to fast relaxation   167
Line shape simulation   130 (fig.)
Line width   4, 29
– temperature dependence   129 (fig.)
Lipids
– shifts   467
– spin-lattice relaxation times   176
Lithiumorganic compounds
– see organolithium compounds
Lock, lock signals   70
Longitudinal magnetization   7, 23
Longitudinal relaxation time   8
Long-range couplings
– CC   152ff.
– CF   206, 270
– CH   140ff., 266f., 289f.
– CN   157ff.
Lorentzian line shape   14
Low power noise decoupling   45
Lysergic acid alkaloids   366

Macrolides   468
Magic acid as NMR solvent   302
Magnetic moment   2
Magnetic shielding constant   15, 110
Magnetic susceptibility corrections   17
Magnetization
– equilibrium   7, 22
– longitudinal   7, 23
– partially relaxed   56f.
– steady state   60
– transverse   12, 23
Magnetization vector   7, 23
Magnetogyric ratio (gyromagnetic ratio)   2
Magnitude spectrum   14, 36
– example   35
Magnets for NMR   69
Measurement of ^{13}C NMR parameters   21ff.
Mechanisms
– of coupling   18f.
– of spin-lattice relaxation   163f.
Mechanistic studies   270f.
Medium shifts   120f.
Meisenheimer anions   307
Menthane derivatives   328 (table)
Menthol, (–)-   328
– DEPT   83

Subject index

- 2D-INADEQUATE 103
- spectra (NOE) 53
Mercaptanes
- see thiols
Mercuryorganic compounds
- see organomercury compounds
Mesomeric effects on shifts 113
Metal carbonyls 294, 300
Metallocenes 294
Metalorganic compounds
- see organometal compounds
Methane derivatives
- CH couplings 136
- shifts 112, 183, 200 (fig.)
Methanol
- spectrum 33
Methanol, tetradeuterio-
- spectrum 37
6-Methoxy-1-tetralone
- spectra 54
O-Methylation shifts
- of carbohydrates 380
Methyl carbon shieldings 185
Methylesterification
- shifts 230
Methylglycosides 383 f. (table)
Methyl rotation
- influence on spin-lattice relaxation 172
Methyl substituent effects
- on cycloalkanes 188
- on decalins 190
Modulation
- see also *J*-modulation 75 f., 88
- of frequencies 45
- pulse interferograms 26
Molecular mobility
- and spin-lattice relaxation 172 ff.
Molecular size
- and spin-lattice relaxation 168
Monosaccharides 381 f. (table)
- derivatives 383 (table)
Monoterpenes 327 ff. (tables)
Morphine alkaloids 370
Motional narrowing 167
Multiple resonance 43 ff.
Multiplet analysis
- by DEPT 80 ff.
- - *J*-modulated spin-echo 75 ff.
- - off-resonance decoupling 47 ff.
Multiplet line intensities 20
- after SPT 80
Multiplicity 17 ff., 20
- analysis 75 f., 80 f., 82 f.
- selection 82 f.
Multipulse experiments
- one-dimensional 50 ff., 55 ff., 73 ff.

- two-dimensional 87 ff.
Mutarotation
- example, spectra 380

Naphtalene, 1,8-diiodo-
- spectra 268
Naphthalenes
- CC couplings 154 ff.
- CF couplings 270
- CH couplings 266
- shifts 263 (table)
- substituent increments 264
Naphthyridines 284, 325
Natural rubber
- spectrum 312
Natural products 327 ff.
Nickel, bis-(1,1-dimethylallyl)-isomers
- spectrum 301
Nicotine 361
- spectra 54, 107
Nitriles
- CN couplings 159 (table)
- shifts 113, 243 (table), 311
Nitrites (esters) 246
Nitroalkanes
- CN couplings 156
- shifts 247 (table)
Nitroalkenes 247
Nitrobenzenes 257, 261
Nitrosamines
- shifts 246 (table)
- - influence of
- - - configuration 246, 274
- - - conformation 274
NMDR 43 ff.
NMR 4
- detection 22, 24 ff., 41
- instrumentation 67 ff.
- in the rotating frame 10 f.
- signals 29
- spectrometer (schematic diagram) 68
- theory 1 ff.
- two-dimensional 87 ff.
NOE 46 f.
- enhancement 46 f., 50 f.
- measurement 51 f.
- suppression 50 f.
Noise decoupling 45
Noise modulation 45
Nonbenzenoid aromatic ions
- shifts 111, 254, 302, 306
- - correlation with
- - - π-charge density 111 (fig.)
Nonbonded (through-space) interactions
- in carbon-fluorine coupling 161

Norbornanes
– CF couplings 206 (table)
– CHg couplings 301
– CSn couplings 299
– CTl couplings 297
– shifts 190 (table), 204 (tables), 211 (table), 219, 236
– – substituent increments 317 (table)
Norbornanols 211 (table)
Norbornanones 219, 331 (table)
Norbornenes 195, 219
Norbornyl cation 305
Nuclear induction 13, 24
Nuclear Overhauser effect (NOE) 46, 164 f.
– NOE enhancement 46 f.
– measurement 51 f.
– quenching of 47
– relation to dipolar relaxation 164 f.
Nuclear precession 2, 23
Nuclear properties, Tab. 2
Nuclear shielding 15
Nuclear spin 1
Nucleobases 404 ff. (table)
Nucleic acids 412
– homopolymeric
– – CP couplings 413 (table)
– – helix random-coil tansition 412
– – shifts 413 (table)
Nucleosides 401, 404 ff. (table)
Nucleotides 401, 404 ff. (table)
Nyquist equation 30

Off-resonance decoupling 47
– frequency dependent 48
Olefins 192 ff.
– see alkenes
Oligopeptides 423 (table), 427 (table)
Oligosaccharides
– CH couplings 393
– shifts 396 (table)
One-bond coupling 19, 134 ff.
– CC 147 ff. (table)
– CD 147
– CF 162 (table), 206
– CH 19, 134 ff. (tables), 288 f. (table)
Organoboron compounds 297
Organoelement compounds
– group V 249 (table)
Organolead compounds
– couplings 293, 299 (table)
– shifts 299 (table)
Organolithium compounds
– CLi couplings 295
– shifts 258, 278, 295
Organometallic compounds 293 ff.
– carbon metal couplings 161, 293 f.

– group I 295
– group II 295
– group III 296
– group IV 297
– methyl-metal compounds
– – carbon-metal couplings 293 (table)
– – shifts 293 (table)
Organomercury compounds
– couplings 301
– shifts 293
Organonitrogen compounds 236 ff.
Organophosphorus compounds
– CP couplings 161, 251 (table), 413
– – influence of structural features 252 f. (table)
– shifts 248 (table)
Organosilicon compounds
– CSi couplings 293, 299 (table)
– shifts 293
Organosulfur compounds
– shift increments 233
– shifts 234 (table)
Organothallium compounds
– couplings 297
Organotin compounds
– couplings 299 (table)
– shifts 298 (table)
Organotransition metal compounds 293 (table)
Orthoesters 220
Overhauser effect
– see nuclear Overhauser effect
Oxacycloalkanes
– CH couplings 288
– shifts 273 (table)
Oxacycloalkenes
– CH couplings 288
– shifts 278 (table)
Oxapolycycles 277 (table)
Oximes
– CN couplings 157
– shifts 241 (table)
– – influence of configuration 157, 160, 241
– spin-lattice relaxation times 173
Oxiranes
– CH couplings 138, 288
– shifts 273
Oxonium salts 215
Oxygen
– influence on spin-lattice relaxation 165 f.

Paramagnetic additives 47
Paramagnetic compounds
– as shift reagents 123 ff.
– influence on spin-lattice relaxation times 47
Paramagnetic relaxation 165 f.
Paramagnetic shielding term 110

Partially relaxed Fourier transform spectra 56
Penicillins 420 (table), 468
Peptides
 – alkaloids 367, 377 f.
 – hormones 432 f. (table)
 – proline- 427 f. (table)
 – shifts 423 f. (table), 432 f. (table)
 – spin-lattice relaxation times 431
Phase correction 33 ff.
Phase cycling 86
Phase memory time 8
Phase of NMR signals 29
Phase selection 30
pH (protonation) shifts 121 f., 236, 286
Phenanthrenes 265 f.
Phenazine, 1-methoxycarbonyl-
 – spectra 158
Phenazines
 – CN couplings 158
 – shifts 284
Phenols 257, 260 f. (table)
Phenones
 – shifts 116, 221 f. (table)
 – – influence of conformation 116, 259, 262
Phosphabenzenes
 – couplings 252, 254
 – shifts 249
Phosphates 248, 251
Phosphetanes
 – CP couplings 252 f.
 – spin-lattice relaxation times 174
Phosphines (*phosphanes*) 248, 251
Phosphinoxides 248, 251
Phosphites 248, 251
Phosphonates 248, 251
Phosphonium salts 248, 251
Phosphonium ylides 248 f., 251
Phosphoramide, hexamethyl-
 – spectrum 37
Phosphorus compounds
 – CP coupling constants 160 (fig.), 250 ff. (tables)
 – shifts 248 f. (table)
Phthalic acid derivatives 231 f.
Phytohormones 467
Piperidine (INADEQUATE) 86
Piperidines
 – alkaloids 302
 – CC couplings 86
 – shifts 274 (table)
 – – influence of
 – – – configuration 272, 274
 – – – protonation 274
Piperidinopentadienal
 – 2D shift correlation 101
pK Determination 122

Polarization transfer 78, 92, 94
 – non-selective (DEPT, INEPT) 80 f.
 – selective (SPT) 79
Polybutadiene
 – partial spectra 312
Polycycloalkanes
 – shift increments 191
 – shifts 189, 190 (table)
Polycycloalkenes 195 (table)
Polycyclic (*fused*) *nitrogen heterocycles* 284
Polyenones 224
 – see annulenones
Polyisoprenes
 – *cis-trans*-isomerism 177
 – spectra 312
Polymers
 – *cis-trans* isomerism 178, 311 f.
 – segmental mobility 177, 313
 – spin-lattice relaxation times 177 f., 313
 – tacticity (stereosequence) 309 f.
Polyols 398 (table), 409
Polypeptides
 – helix random-coil transitions 436
 – shifts 423 f. (table), 427 f. (table), 432 f. (table)
 – spin-lattice relaxation times 431, 437
Polypropylene
 – spectra 310
Polysaccharides 393
Population transfer 78 ff.
 – see polarization transfer
Porphyrins 442 (table)
Power spectra 14
Pregnanes 338
 – CF couplings 340 (table)
 – shifts 341 (table)
Preparation period 75
Proaporphine alkaloids 369
Progressive saturation 60 ff.
Propane 183
 – three-bond coupling 143 (fig.)
Propane, 1,1,3,3-tetraethoxy-
 – spectra 49
Propynol
 – spin-lattice relaxation times 62 (fig.)
Prostaglandins
 – shifts 467
 – spin-lattice relaxation times 179
Proteins
 – helix-random coil transitions 177
 – segmental mobility (T_1) 177, 440
 – shifts 438 f. (table)
Protonation shifts 121 f., 236, 286
Proton decoupling 43 ff.
 – gated 50
 – inverse gated 50 f.

Proton decoupling
- low-power noise 45
- noise (broad-band) 44 f.
- off-resonance (single frequency off-resonance) 47 ff.
- pulsed 50 ff.
- – for determination of NOE enhancements 51 f.
- selective 53 f., 94
Pseudo contact shift 124
Pteridines 284
Pulse angle 12, 22 ff.
- adjustement 32 f.
Pulsed NMR 22 ff.
Pulse Fourier Transform NMR 28 ff.
Pulse frequency 31
Pulse interferograms 25
- examples 26 f., 37
Pulse interval 30
Pulse sequences
- Carr-Purcell-Meiboom-Gill 63 f.
- COSY (Jeener) 97
- DEFT 39 f.
- DEPT 80 f.
- Inversion recovery 55 f.
- INADEQUATE 85
- Progressive saturation 60 f.
- Saturation recovery 59 f.
- Spin-echo 73
- Two-dimensional 90, 93, 97
Pulse width 22 ff.
- adjustment 32, 33 (fig.)
Purines
- CF couplings 291
- CH couplings 290
- shifts 284, 402, 404 f. (table)
Pyranosides 385 ff. (table)
Pyrazole, 3-methyl-5-oxo-1-phenyl-
- spectra, tautomerism 287
Pyrazoles
- CH couplings 288
- shifts 282
Pyrenes
- CC couplings 155
- CF couplings 270
- shifts 266
Pyridines
- alkaloids 361
- CC couplings 150
- CF couplings 291
- CH couplings 145
- – one-bond 139, 145, 288
- – three-bond 145, 290
- – two-bond 145, 289
- CN couplings 156 ff.
- dilution shifts 285 (fig.)

- protonation shifts 286
- shifts 283, 286
- – substituent increments 323 (table)
Pyrimidine, 2,4,6-trichloro-
- spectra 51
Pyrimidines
- CH couplings 145, 288 ff.
- shifts 282, 404 ff. (table), 409
Pyrones, 4-
- CH couplings 288
- shifts 278
Pyrroles
- CH couplings 146, 288 ff.
- CN couplings 157
- shifts 282
Pyrrolidine alkaloids 361
Pyrrolizidine alkaloids 361
Pyrylium salts 283

Quadrature detection (QD)
- digital quadrature detection) 32, 42
Quadrupolar coupling constants
- from spin-lattice relaxation times 180
Quadrupolar nuclei
- line-broadening 295
Quantum number 1, 2
Quinine alkaloids 372
Quinoline
- pH *vs.* shifts 122 (fig.)
- pK determination 123 (fig.)
- solvent shifts 120 (fig.)
- spectra 42
Quinolines
- alkaloids 371
- CH couplings 289
- CN couplings 157 f.
- shifts 283
- – nitrogen increments 324
Quinolizidine alkaloids 362 f.
Quinones 223 (table)

Rate constants
- from shift difference 129 (fig.)
Reaction mechanisms
Real spectra 33 f.
Reeves effect 393
Referencing of ^{13}C shifts 17, 108
Relaxation 5 f.
- in the rotating frame 12
- see also spin-lattice and spin-spin relaxation
Relaxation delay 39, 75
Relaxation mechanisms 163 ff.
Relaxation times 5 f.
- measurement 55 ff.
RELAY 100
Residual couplings 48 f.

Resolution enhancement 36
Resolution in PFT NMR 36 f.
Resonance effects on shifts 113
Restricted rotation 127
– influence on
– – spin-lattice relaxation times 173
Rf field 4
Ribose, D-
– spectrum 394
Rifamycins 468
Ring inversion 127, 131 f.
Ring size
– influence on
– – CH coupling 138 f.
– – shift 118
Rotating frame of reference 9 ff., 24
Rotations

Saccharose (Sucrose)
– spin-lattice relaxation time 182 (table)
Sample preparation 71 f.
Sapogenins 359 (table)
Satellite signals 18, 79
– due to
– – carbon-carbon coupling 85
– – carbon-metal coupling 298
Saturation 6
Saturation-recovery method 59 f.
Scalar (coupling) relaxation 163
s-Character of carbon
– influence on coupling constants 135 (fig.), 151 (fig.)
SEFT 75
Segmental mobility
– effect on
– – spin-lattice relaxation times 174 f.
Selective proton decoupling 53 f.
Sensitivity (signal:noise)
– enhancement in ^{13}C NMR
– – by polarization transfer 79 ff.
– – for protonated carbons 46
– – for slowly relaxing carbons 47
SFORD 47 ff.
Shielding constant 15, 110
Shielding of nuclei 15, 17, 110
Shift correlation, two-dimensional
– CC (INADEQUATE) 102 ff.
– CH (CH COSY)
– – *via* one-bond coupling 92 ff.
– – *via* two- and three-bond coupling 96
– HH (COSY) 96 ff.
– relayed coherence transfer 100 f.
– survey 106
Shift increments 111, 313 ff.
– alkane drivatives 315 (table)
– alkanes 184 f.

– alkanols 207
– alkenes 194
– alkynes 197
– amines 236
– benzenes 264, 319 f. (table)
– carbonyl compounds 220
– carboxylic acids 227, 230
– cycloalkanes 188
– cycloalkanols 212
– cycloalkyl halides 203
– cyclohexanes 316 (table)
– decalins 191
– naphthalenes 264
– norbornanes 317 (table)
– organosulfur compounds 233
– pyridines 281, 323 f. (table)
– survey 313 ff.
Shift reagents 124
Shift references in ^{13}C NMR 17, 108
Shifts
– see carbon chemical shifts
Signal enhancement
– by NOE 46 f., 50 f.
– by polarization transfer 78 ff.
Signal intensity 4
– influence of
– – spin-lattice relaxation time 39
– responding to polarization transfer 82
Signal to noise ratio 21
– see also sensitivity
Siliconorganic compounds
– see organosilicon compounds
Single frequency proton decoupling (off-resonance decoupling) 47 ff.
Solvents, deuterated
– shifts and couplings 109 (table)
Solvent shifts 120 (fig.), 200, 201, 230
– *vs.* dielectricity constant 200
Solvent effects
– on carbon-proton coupling 118
sp carbon
– coupling constants 135, 136, 138, 150 f.
– shifts 111, 119
sp^2 carbon
– coupling constants 135, 137 ff., 145, 150 f.
– shifts 111, 119
sp^3 carbon
– coupling constants 135, 136, 149, 150
– shifts 111, 119
Spectral width 30
– to line width ratio 108
Spin 1
Spin alignments 3
Spin-decoupling 43 ff.
– see decoupling, double resonance

Spin-echo 63f., 73f.
– *J*-modulated 75
Spin-echo techniques 41, 63f., 75f.
Spin-lattice relaxation 5, 8
– influence of molecular size 168
– influence on signal:noise 39, 50
– mechanisms 163ff.
Spin-lattice relaxation time (T_1) 6
– as assignment aid 168
– correlation with
– – mobility and constitution 166ff.
– – spin-spin relaxation time 167
– definition 8
– influence of
– – anisotropic rotation 169f.
– – association 178
– – configuration 173
– – conformation 171f.
– – degree of alkylation 168
– – dissolved oxygen 166
– – methyl rotation 172
– – molecular motion 166ff.
– – molecular size 168
– – number of attached hydrogens 168
– – paramagnetic compounds 165
– – segmental mobility 174f., 313
– – solvent 181
– – steric interactions 173
– – temperature 181
– – viscosity 181
– influence on
– – signal intensity 39
– methods of measurement 55ff.
– relation to
– – spin-spin relaxation time 167 (fig.)
Spin-locking FT experiments 66f.
Spin population 3, 78
Spin-quantum number 1, 2
Spin-rotation relaxation 163f.
Spin-spin coupling 17
– mechanism 18f., 133f.
Spin-spin relaxation 6
Spin-spin-relaxation time (T_2) 6
– definition 8
– influence on line-width 6
– measurement 63ff.
– relation to
– – spin-lattice relaxation time 167 (fig.)
Spirocyclic compounds 277, 467
SPT 79
Stabilization of magnetic fields 70
Stacked plots 88
Steric effects
– on shifts 115
– on spin-lattice relaxation 173
Steroids 337ff.

– alkaloids 375 (table)
– androstanes 338, 341f. (table)
– cardenolides 359 (table)
– cholanes 340, 357 (table)
– cholestanes 340, 351f. (table)
– estranes 338, 348 (table)
– pregnanes 441f. (table)
– – fluorinated
– – – CF couplings 340 (table)
– spin-lattice relaxation times 172
Structure elucidation
– ^{13}C NMR strategy 104ff.
Subspectra (DEPT) 82f.
Substituent effects (increments) 118
– applications 313ff.
– of alkoxy groups 213
– – alkyl groups 184f., 188, 194
– – amino groups 236
– – carbonyl groups 220
– – carboxy groups 227
– – halogens 111, 200, 203
– – hydroxy groups 207, 212
– – sulfur containing groups 233
– on alkane carbons 111, 184, 233, 227, 315 (table)
– – alkene carbons 193f., 227, 230, 318 (table)
– – alkyne carbons 197
– – benzene carbons 258f., 264, 319f. (table)
– – cyclohexane carbons 188, 203, 212, 316 (table)
– – decalin carbons 191, 212
– – naphthalene carbons 264
– – norbornane carbons 203, 212, 317 (table)
– – pyridine carbons 281, 323f. (table)
– – steroid carbons 339
– survey 313ff.
Sulfides (thioethers)
– aliphatic 234
– aromatic 257
Sulfinic acids 234
Sulfones 234, 275
Sulfonic acids and derivatives
– aliphatic 234
– aromatic 257
Sulfonium ions 234
Sulfoxide, hexadeuteriodimethyl-
– spectrum 26
Sulfoxides 234, 275
Sugars
– see carbohydrates
Superconducting solenoids 70

Tacticity 202, 308f.
Tautomerism
– carbodiimides 245

Subject index   515

- keto-enol   52, 220, 232
- pyrazolone (spectra)   287
- pyridinees   290
Temperature dependence
- of shifts   129 (fig.), 131 f., 300
Temperature effects on
- spin-lattice relaxation times   181 f.
Terpenes   327 ff.
- bicyclic   328 f. (table)
- carotenoids   335 (table)
- cyclic   328 (table)
- di-   332 (table)
- ionones   334 (table)
- menthane derivatives   328 (table)
- mono-   328 f. (tables)
- open-chain   327 (table)
- retinals   334 (table)
- spin-lattice relaxation times   172 f., 179
- tetra-   335 (table)
Tetracyclins   468
Tetradeuteriomethanol
- spectrum   37
1,1,3,3-Tetraethoxypropane
- spectra   49
Tetrahedrane, tetra-*t*-butyl-   189
Tetrahydrocannabinol derivatives   467
Tetralone, 1-, 6-methoxy-
- spectrum   37
Tetraterpenes   335 (table)
Thermodynamic data of interconversion   129 f.
Thiacycloalkanes
- CH couplings   288
- shifts   273 f. (tables)
Thiadecalins   276
Thiapolycycles   277
Thiazoles
- CH couplings   288
- shifts   282
Thiirans
- CH couplings   138, 288
- shifts   273
Thioacetals   233
Thioaldehydes   235
Thiocarbonyl compounds   233, 235 (table)
Thiocyanates   245
Thioethers
- aliphatic   234
- aromatic   257
- heterocyclic   273
Thioketals   233
Thioketones   233, 235 (table)
Thiols (mercaptans)
- aliphatic   234
- aromatic   257
Thiophenes
- CH couplings   146, 288 ff.

- shifts   281 f.
Thioureas   235
Titration curves   122
Transition metal compounds
- see organotransition metal compounds
Transverse magnetization   12, 23
- components   28
Transverse relaxation function   24
Transverse relaxation time   12
2,4,6-Trichloropyrimidine
- spectra   51
Tripeptides   423 (table)
Tropane alkaloids   361
Tropolone derivatives   224, 462
Two-dimensional NMR   87 ff.
- basic concept   87
- columns   89
- *J*-resolved   89
- rows   88
- shift correlations
- - CC   102 ff.
- - CH   92 ff.
- - HH   96 ff.
- - long-range CH   96
- - relayed coherence transfer   100

U-mode   14
Uracils
- CH couplings   288
- shifts   279, 404 f. (table), 409, 411
Ureas
- CN couplings   156
- shifts   279

Vanilline
- spectra   267
Variable temperature studies   130 f.
V-mode   14
Vinyl halides
- CH couplings   199 (table)
- shifts   199 (table)
Virescenosides   459
Viscosity effects on T_1 181
Vitamins   334, 351, 467

Width of NMR lines
- influence of
- - T_2 6
- - rate constants and temperature   128 ff.
Woessner's equations   169 f.

Ylides   251, 253
Y stabilization
- of carbanions   302
- of carbenium ions   308
Yuzurimine alkaloids   373